ADVANCES AND TRENDS IN OPTIMIZATION WITH ENGINEERING APPLICATIONS

MOS-SIAM Series on Optimization

This series is published jointly by the Mathematical Optimization Society and the Society for Industrial and Applied Mathematics. It includes research monographs, books on applications, textbooks at all levels, and tutorials. Besides being of high scientific quality, books in the series must advance the understanding and practice of optimization. They must also be written clearly and at an appropriate level for the intended audience.

Editor-in-Chief

Katya Scheinberg
Lehigh University

Editorial Board

Santanu S. Dey, *Georgia Institute of Technology*
Maryam Fazel, *University of Washington*
Andrea Lodi, *University of Bologna*
Arkadi Nemirovski, *Georgia Institute of Technology*
Stefan Ulbrich, *Technische Universität Darmstadt*
Luis Nunes Vicente, *University of Coimbra*
David Williamson, *Cornell University*
Stephen J. Wright, *University of Wisconsin*

Series Volumes

Terlaky, Tamás, Anjos, Miguel F., and Ahmed, Shabbir, editors, *Advances and Trends in Optimization with Engineering Applications*

Todd, Michael J., *Minimum-Volume Ellipsoids: Theory and Algorithms*

Bienstock, Daniel, *Electrical Transmission System Cascades and Vulnerability: An Operations Research Viewpoint*

Koch, Thorsten, Hiller, Benjamin, Pfetsch, Marc E., and Schewe, Lars, editors, *Evaluating Gas Network Capacities*

Corberán, Ángel, and Laporte, Gilbert, *Arc Routing: Problems, Methods, and Applications*

Toth, Paolo, and Vigo, Daniele, *Vehicle Routing: Problems, Methods, and Applications, Second Edition*

Beck, Amir, *Introduction to Nonlinear Optimization: Theory, Algorithms, and Applications with MATLAB*

Attouch, Hedy, Buttazzo, Giuseppe, and Michaille, Gérard, *Variational Analysis in Sobolev and BV Spaces: Applications to PDEs and Optimization, Second Edition*

Shapiro, Alexander, Dentcheva, Darinka, and Ruszczyński, Andrzej, *Lectures on Stochastic Programming: Modeling and Theory, Second Edition*

Locatelli, Marco and Schoen, Fabio, *Global Optimization: Theory, Algorithms, and Applications*

De Loera, Jesús A., Hemmecke, Raymond, and Köppe, Matthias, *Algebraic and Geometric Ideas in the Theory of Discrete Optimization*

Blekherman, Grigoriy, Parrilo, Pablo A., and Thomas, Rekha R., editors, *Semidefinite Optimization and Convex Algebraic Geometry*

Delfour, M. C., *Introduction to Optimization and Semidifferential Calculus*

Ulbrich, Michael, *Semismooth Newton Methods for Variational Inequalities and Constrained Optimization Problems in Function Spaces*

Biegler, Lorenz T., *Nonlinear Programming: Concepts, Algorithms, and Applications to Chemical Processes*

Shapiro, Alexander, Dentcheva, Darinka, and Ruszczyński, Andrzej, *Lectures on Stochastic Programming: Modeling and Theory*

Conn, Andrew R., Scheinberg, Katya, and Vicente, Luis N., *Introduction to Derivative-Free Optimization*

Ferris, Michael C., Mangasarian, Olvi L., and Wright, Stephen J., *Linear Programming with MATLAB*

Attouch, Hedy, Buttazzo, Giuseppe, and Michaille, Gérard, *Variational Analysis in Sobolev and BV Spaces: Applications to PDEs and Optimization*

Wallace, Stein W. and Ziemba, William T., editors, *Applications of Stochastic Programming*

Grötschel, Martin, editor, *The Sharpest Cut: The Impact of Manfred Padberg and His Work*

Renegar, James, *A Mathematical View of Interior-Point Methods in Convex Optimization*

Ben-Tal, Aharon and Nemirovski, Arkadi, *Lectures on Modern Convex Optimization: Analysis, Algorithms, and Engineering Applications*

Conn, Andrew R., Gould, Nicholas I. M., and Toint, Phillippe L., *Trust-Region Methods*

ADVANCES AND TRENDS IN OPTIMIZATION WITH ENGINEERING APPLICATIONS

Edited by

Tamás Terlaky
Lehigh University
Bethlehem, Pennsylvania

Miguel F. Anjos
Polytechnique Montréal
Montréal, Quebec
Canada

Shabbir Ahmed
Georgia Institute of Technology
Atlanta, Georgia

Society for Industrial and Applied Mathematics
Philadelphia

Mathematical Optimization Society
Philadelphia

Publisher	David Marshall
Acquisitions Editor	Paula Callaghan
Developmental Editor	Gina Rinelli Harris
Managing Editor	Kelly Thomas
Production Editor	Lisa Briggeman
Copy Editor	Julia Cochrane
Production Manager	Donna Witzleben
Production Coordinator	Cally Shrader
Compositor	Lumina Datamatics, Inc.
Graphic Designer	Lois Sellers

Library of Congress Cataloging-in-Publication Data
Names: Terlaky, Tamás, editor. | Anjos, Miguel F., editor. | Ahmed, Shabbir, 1969- editor.
Title: Advances and trends in optimization with engineering applications / Tamás Terlaky, Lehigh University, Bethlehem, Pennsylvania, Miguel F. Anjos, Polytechnique Montréal, Montréal, Quebec, Canada, Shabbir Ahmed, Georgia Institute of Technology, Atlanta, Georgia, editors.
Description: Philadelphia : Society for Industrial and Applied Mathematics ; Philadelphia : Mathematical Optimization Society, [2017] | Series: MOS-SIAM series on optimization ; 24 | Includes bibliographical references and index.
Identifiers: LCCN 2016052905 (print) | LCCN 2016053308 (ebook) | ISBN 9781611974676 (print) | ISBN 9781611974683 (e-book)
Subjects: LCSH: Mathematical optimization.
Classification: LCC QA402.5 .A36 2017 (print) | LCC QA402.5 (ebook) | DDC 519.6–dc23
LC record available at https://lccn.loc.gov/2016052905

 is a registered trademark.

 is a registered trademark.

List of Contributors

Shabbir Ahmed
Georgia Institute of
Technology, GA, U.S.A.

Edoardo Amaldi
Politecnico di Milano, Italy

Cristina H. Amon
University of Toronto, ON,
Canada

Miguel F. Anjos
Polytechnique Montreal, QC,
Canada

Egon Balas
Carnegie Mellon University,
PA, U.S.A.

Lorenz T. Biegler
Carnegie Mellon University,
PA, U.S.A.

Aldo Bischi
Politecnico di Milano, Italy

Sergiy Butenko
Texas A&M University, TX,
U.S.A.

Sonia Cafieri
École Nationale de l'Aviation
Civile, France

Melih Çelik
Middle East Technical
University, Turkey

Timothy C. Y. Chan
University of Toronto, ON,
Canada

Brian Y. Chen
Lehigh University, PA,
U.S.A.

Ruobing Chen
Bosch Research and
Technology Center, CA,
U.S.A.

Ana Luísa Custódio
Universidade Nova de Lisboa,
Portugal

Özlem Ergun
Northeastern University,
MA, U.S.A.

Matteo Fischetti
University of Padova, Italy

Christodoulos A. Floudas†
Princeton University, NJ,
U.S.A. (†Deceased August
2016)

Albin Fredriksson
RaySearch Laboratories,
Sweden

Vikas Goel
ExxonMobil, TX, U.S.A.

Ignacio E. Grossmann
Carnegie Mellon University,
PA, U.S.A.

Didier Henrion
Université de Toulouse,
France

C. T. Kelley
North Carolina State
University, NC, U.S.A.

Pınar Keskinocak
Georgia Institute of
Technology, GA, U.S.A.

George A. Khoury
Princeton University, NJ,
U.S.A.

Mustafa R. Kılınç
Carnegie Mellon University,
PA, U.S.A.

Michal Kočvara
University of Birmingham,
U.K.

Vitaliy Krasko
Colorado School of Mines,
CO, U.S.A.

Jim Y. J. Kuo
California State University,
Los Angeles, CA, U.S.A.

Leonardo Lamorgese
SINTEF, Norway

Eva K. Lee
Georgia Institute of
Technology, GA, U.S.A.

Miguel Lejeune
George Washington
University, DC, U.S.A.

Jinchao Li
University of California, Los
Angeles, CA, U.S.A.

Zukui Li
University of Alberta, AB,
Canada

Carlo Mannino
SINTEF, Norway

Philip Allen Mar
University of Toronto, ON,
Canada

Emanuele Martelli
Politecnico di Milano, Italy

Joaquim R. R. A. Martins
University of Michigan, MI,
U.S.A.

Helmut Mausser
IBM, ON, Canada

Yanfeng Ouyang
University of Illinois at
Urbana-Champaign, IL,
U.S.A.

Panos M. Pardalos
University of Florida, FL,
U.S.A.

Edouard Pauwels
Université Toulouse 3,
France

Andy Philpott
University of Auckland, New
Zealand

Mauro Piacentini
University of Rome, Italy

Jose M. Pinto
Praxair Inc., CT, U.S.A.

Sreekanth Rajagopalan
Carnegie Mellon University,
PA, U.S.A.

Steffen Rebennack
Colorado School of Mines,
CO, U.S.A.

Oleksandr Romanko
IBM, ON, Canada

David A. Romero
University of Toronto, ON,
Canada

Nikolaos V. Sahinidis
Carnegie Mellon University,
PA, U.S.A.

Katya Scheinberg
Lehigh University, PA,
U.S.A.

James Smadbeck
Princeton University, NJ,
U.S.A.

Lawrence V. Snyder
Lehigh University, PA,
U.S.A.

Mallory Soldner
UPS, GA, U.S.A.

Mathias Stolpe
Technical University of
Denmark, Denmark

X. Andy Sun
Georgia Institute of
Technology, GA, U.S.A.

Yifan Sun
University of California, Los
Angeles, CA, U.S.A.

Julie Swann
Georgia Institute of
Technology, GA, U.S.A.

Christopher L. E. Swartz
McMaster University, ON,
Canada

Leonardo Taccari
Politecnico di Milano, Italy

Tamás Terlaky
Lehigh University, PA,
U.S.A.

Francisco Trespalacios
Carnegie Mellon University,
PA, U.S.A.

Lieven Vandenberghe
University of California, Los
Angeles, CA, U.S.A.

Luis Nunes Vicente
University of Coimbra,
Portugal

Andreas Wächter
Northwestern University, IL,
U.S.A.

John M. Wassick
The Dow Chemical
Company, MI, U.S.A.

Stefan M. Wild
Argonne National
Laboratory, IL, U.S.A.

Owen Q. Wu
Indiana University, IN,
U.S.A.

Victor M. Zavala
University of
Wisconsin-Madison, WI,
U.S.A.

Peter Y. Zhang
University of Toronto, ON,
Canada

Hui Zhao
Pennsylvania State
University, PA, U.S.A.

Contents

List of Figures

List of Tables

List of Algorithms

Foreword

Few areas of scholarly activity have had broader impact over the second half of the 20th century than the field of mathematical optimization. In 2000, the editors of SIAM selected their top 10 algorithms of the 20th century, and George Dantzig's simplex algorithm was on the list. Although the problem of minimizing or maximizing a differentiable function is as old as calculus itself, the golden age of optimization in which we now find ourselves has been enabled by developments in three main areas: computing capability, data, and methods. It is not coincidental that the development of the digital computer and its enormous impact on engineering, businesses, government, and decision making in general also began in the mid-20th century. The first general purpose all-electronic computer was built in 1946 at the University of Pennsylvania by a team led by John Presper Eckert Jr. and John William Mauchly. Since then, the power of computers and their price performance have consistently improved exponentially over time; this phenomenon is informally referred to as Moore's law. (Moore's original 1965 paper noted that the density of transistors on silicon was doubling approximately every two years and conjectured that this would continue for the next decade. A consequence of this phenomenon was a similar increase in computing power of processor chips. Today the term *Moore's law* is often used broadly to refer to any aspect of computer performance or price performance that grows exponentially over time.)

Computers were initially developed to perform high-speed computation. It was quickly discovered that they also could collect, manage, and analyze data coming from a variety of sources. Punched cards and paper tape gave way to magnetic tape followed by disk storage and then solid-state storage devices. Each new technology increased storage capacity as well as the speed with which the data could be managed or manipulated. The combination of exponentially increasing computing power and massive amounts of data, human generated and machine generated, from both repositories and streaming sources, is creating a broad variety of new opportunities.

The implementation of optimization algorithms on today's computer systems can be viewed as an activity that is valuable but of less importance than the development of new algorithms. Indeed, in 1969, Martin Beale said, "I would much rather work with today's algorithms on yesterday's computers than with yesterday's algorithms on today's computers." However, in addition to computers providing optimization platforms, their architectures have significantly influenced the development of optimization algorithms. The product form of the inverse for linear optimization, developed in the 1960s, was a perfect approach for magnetic tape systems. Higher-order updates emerged as direct-access storage became common.

Over the last two decades, the exponential growth of computer power has been achieved through parallel processing. Part of the attraction of interior-point methods, which led to their rapid development and acceptance, was that they worked well on these computer systems. These methods perform a relatively small number of large matrix operations, which enabled easier exploitation of the computers of that time than variants of the simplex algorithm, which performed many matrix operations on smaller matrices. Operations on large matrices map more easily onto large parallel systems than long sequences of operations on smaller matrices. Today's supercomputers make use of massive parallelism (tens or hundreds of thousands of processors) to obtain the desired performance. We are already beginning to see how this is affecting the design of algorithms for linear and integer optimization.

Part of the attraction of mathematical optimization is that it has deep roots in a variety of fields of applied mathematics, including real analysis, geometry, and linear and affine algebra. It also draws on probability and statistics as its domain broadens to include nondeterministic situations. Moreover, these advanced approaches really do result in better methods, which provide measurably better and more robust solutions to a broad variety of problems. In some cases, the mathematical foundations provide a duality theory that provides a bound guaranteeing solution quality even if optimality cannot be certified. This ability to establish the quality of a solution to the problem being attacked is very valuable when it is possible.

Engineering itself is changing from a collection of specialized solution approaches for specific types of problems to an approach for dealing with a broad range of applications, many of which do not fit neatly into traditional taxonomies. Their solutions often require integrating a variety of different methods. For example, a system designed to minimize the cost of running a commercial steel production facility required integrating 12 different algorithms for 12 different subproblems.

Financial engineering, managing the complexity of the world of financial instruments and alternatives to an organization, has become a standard part of the operations of companies and individuals. The challenge of measuring and managing risk in uncertain times is rapidly becoming a major area of activity.

The field of healthcare is being transformed by applications of optimization. Answers to traditional questions of how to provide services in an efficient, cost-effective way are being recognized as key to containing rapidly growing costs. Other questions now being considered include predicting the effectiveness of alternative treatments to a problem and thereby a selection of approaches that improve outcomes while reducing undesired side effects. Developments in this area are being enabled by the incorporation of genetic data in patient records. This can be a key factor in selecting and applying treatments.

There is a requirement to share information between healthcare researchers and clinicians. This requires integrating massive amounts of data coming from many different sources. IBM's Watson project is already exhibiting the potential for automated ingestion of this data and then using it in a coherent fashion to aid with diagnosis and selection of treatment protocols.

Structural engineering is supported by optimization models. Traditionally, a goal has been to create structures with a desired strength while minimizing the weight or cost of materials. Formulating such problems as optimization problems has led to innovative designs with improved price/performance over traditional methods. In addition, building structures that can survive disruptive events such as earthquakes is becoming required by new building codes.

Energy optimization is an area of increasing importance. Electrical power is provided by a variety of power sources, some traditional, such as coal, hydroelectric, or nuclear, and some green, such as wind or solar. Demand is difficult to predict, and energy is difficult to store. In addition, the increasing use of hybrid and electric automobiles is shifting demand from hydrocarbon-based fuels that go into a vehicle's tank to the electrical power grid, which needs to support recharging.

Urban planning creates a complex set of challenges. Many of the networks required to support growing city populations are expensive and difficult to upgrade. Think of an urban metro system or water supply. As populations shift to new areas, demand can change, often significantly. Cities are already applying optimization to maximize the efficiency of road networks by controlling traffic lights. Some cities already analyze historical traffic data to predict congestion and gridlock.

The challenges presented by engineering applications are significant and real. The systems being optimized are often complex and require many different subsystems to be linked. The objective functions themselves are sometimes ill defined. How do you measure the effectiveness of an urban transit system? How do you measure the impact that a new generation of a dual-source CT scan machine will have on an individual's health or the state of health of the population of a city? How do you balance the trade-off between guaranteeing a result that is optimal with generating a result with the best expected value? And, in the latter case, what distribution of possible outcomes should be used to compute that expected value?

Constraints come in many forms. Some are hard (cannot be violated) and some are soft (can be violated but not by too much). Some are linear and some are nonlinear. Some variables must be integer and others may be fractional. How can historical data regarding the problem being modeled be incorporated into the model?

Finally, if a reasonably accurate mathematical formulation of a problem can be constructed, can it be solved quickly enough to provide an answer in a timely fashion?

Advances and Trends in Optimization with Engineering Applications serves several important functions: First, it provides a broad, up-to-date summary of the state of the art in the areas of optimization most important to engineering practice. Second, it presents detailed descriptions of applications of these methods to a very broad range of real problems. Third, it reports on computational results of the methods as applied to these problems. It is remarkable in that it successfully deals with all three issues.

This book provides a solid foundation for engineers and mathematical optimizers alike who want to understand the importance of optimization methods to engineering, the current methods themselves, and their capabilities. I also believe that it will provide a springboard for continued progress over the next decade.

William R. Pulleyblank, NAE
Department of Mathematical Sciences
United States Military Academy, West Point

Preface

Optimization is an area of critical importance in engineering and applied sciences. When designing products, materials, factories, production processes, manufacturing or service systems, and financial products, engineers strive for the best possible solutions, the most economical use of limited resources, and the greatest efficiency. As system complexity increases, these goals mandate the use of state-of-the-art optimization methodology and computational tools.

The theory and computational methodology of optimization have seen revolutionary improvements in the past three decades. Novel algorithmic concepts and high-performance modeling methodologies have been developed. Moreover, the exponential growth in computational power along with the availability of multicore computing with virtually unlimited memory and storage capacity has fundamentally changed what engineers can do to optimize their designs.

This is a two-way process: engineers benefit from developments in optimization methodology, and challenging new classes of optimization problems arise from engineering applications.

This handbook reviews the major areas of optimization and provides a sampling of their engineering applications. It is organized into 10 parts, encompassing both well-established areas and emerging trends that are significant for engineering. There are four chapters in each part, where the first chapter provides an overview of the state of the art, and the subsequent three present illustrative applications.

Due to its broad coverage, we anticipate that the readership of this handbook will span large portions of the engineering, optimization, and operations research communities, from doctoral students to experienced researchers. In particular, this handbook will be of interest to two main audiences:

- For engineers, it provides a handy reference to a huge variety of state-of-the-art optimization techniques essential to the various fields of engineering.
- For optimization researchers, it provides numerous examples of successful engineering applications as well as a broad range of research areas to further explore.

Many different contributors participated in this project. We are especially grateful to the coordinators who worked with us to shape the content of each part: E. Balas, M. Fischetti, P. Pardalos, S. Butenko, A. Waechter, N. Sahinidis, T. Chan, L. Snyder, and K. Scheinberg.

We also thank all those who prepared chapters. We were deeply saddened to learn that one of the contributors, Christodoulos (Chris) Floudas, died just as the handbook was being completed. Professor Floudas was a leading expert in the application of mathematical optimization to complex systems and was consistently at the forefront of developments in both chemical engineering and global optimization. He was the author of more than 300 scientific publications; among his top honors were his elected memberships to the U.S. National Academy of Engineering, the Academy of Athens, and the U.S. National Academy of Inventors. He will be sorely missed.

All the authors benefited tremendously from the advice of the anonymous reviewers who read one or more chapters and provided critical and constructive comments. Their generous support of this project has had a considerable impact on the outcome, and we gratefully acknowledge their time and effort.

Tamás Terlaky, Fellow of INFORMS, Fellow of The Fields Institute
George N. and Soteria Kledaras '87 Endowed Chair Professor, and Department Chair
Department of Industrial and Systems Engineering
Lehigh University, U.S.A.

Miguel F. Anjos, Fellow of the Canadian Academy of Engineering
NSERC/Hydro-Quebec/Schneider Electric Senior Industrial Research Chair
Inria International Chair
Department of Mathematics and Industrial Engineering
Polytechnique Montréal, Canada

Shabbir Ahmed
College of Engineering Dean's Professor
H. Milton Stewart Faculty Fellow
School of Industrial & Systems Engineering
Georgia Institute of Technology, U.S.A.

Bethlehem, Montreal, and Atlanta
February 2017

Part I

Linear and Quadratic Optimization

Chapter 1

Algorithms for Linear and Quadratic Optimization

Tamás Terlaky

1.1 ▪ Introduction

Linear optimization (LO) and linearly constrained convex quadratic optimization (QO) problems as the bedrock of more general optimization problems, play a fundamental role in modeling and solving optimization problems in arguably all areas of engineering. We begin this book with a brief introduction to LO and QO. We briefly review the two most powerful algorithmic paradigms, the families of pivot methods and interior-point methods (IPMs), and give a review of existing software tools.

The subsequent three chapters demonstrate the power of LO and QO in modeling and solving engineering optimization problems in three engineering disciplines:

- mechanical and aerospace engineering, in particular structural and truss topology optimization, which covers optimization models and numerical methods for optimally designing such structures as bridges, buildings, cars, and airplanes;
- electrical engineering, optimal designs in signal and image processing, electricity network and VLSI design, and control and communication systems; and
- chemical engineering, where optimization is extensively used both to optimize system designs and also in the control and optimal operation of existing systems.

The general LO problem [549, 1591] is considered in the standard primal form

$$\min\left\{c^T x : Ax = b, \ x \geq 0\right\}, \tag{PLO}$$

together with its standard dual

$$\max\left\{b^T y : A^T y \leq c\right\}, \tag{DLO}$$

where $A : m \times n$, $b \in \mathbb{R}^m$, and $c \in \mathbb{R}^n$ represent the problem data, while $x \in \mathbb{R}^n$ and $y \in LR^m$ are the unknowns of the problem. It is well known from the *strong duality theorem* [549, 1591] that if both the primal and dual problems have feasible

3

solutions, then both have optimal solutions, x^* and y^*, respectively. By the strong duality theorem, for optimal solutions we have $c^T x^* = b^T y^*$, which is equivalent to

$$(x^*)^T \left(c - A^T y^*\right) = 0.$$

Since both $x^* \geq 0$ and $s^* := c - A^T y^* \geq 0$, the above optimality conditions are equivalent to

$$x_i^* \left(c - A^T y^*\right)_i = x_i^* s_i^* = 0 \qquad \text{for all} \ \ i = 1, \ldots, n,$$

which are referred to as the *complementarity conditions*.

For LO it is also well known [822, 1591] that if both the primal and dual problems have feasible solutions, then they not only have optimal solutions but always have *strictly complementary optimal solutions* x^* and (y^*, s^*) satisfying $x^* + s^* > 0$. This property does not hold for more general optimization problems, not even for QO problems.

The general QO problem [230, 579, 889, 1815] is considered in the standard primal form

$$\min \left\{ c^T x + \frac{1}{2} x^T Q x \ : \ Ax = b, \ x \geq 0 \right\}, \tag{PQO}$$

together with its standard dual

$$\max \left\{ b^T y - \frac{1}{2} x^T Q x \ : \ A^T y - Qx \leq c \right\}, \tag{DQO}$$

where A, b, c, x, y are given as in the case of LO and Q is an $n \times n$ symmetric positive semidefinite (PSD) matrix. It is well known from the strong duality theorem [579, 1815] that if both the primal and dual problems have feasible solutions, then both have optimal solutions, x^* and (x^*, y^*), respectively. By the strong duality theorem, for optimal solutions we have $c^T x^* + \frac{1}{2} x^T Q x = b^T y^* - \frac{1}{2} x^T Q x$, which is equivalent to

$$(x^*)^T (c + Qx - A^T y^*) = 0.$$

Since both $x^* \geq 0$ and $s^* := c + Qx - A^T y^* \geq 0$, the above optimality conditions are equivalent to

$$x_i^* (c + Qx - A^T y^*)_i = x_i^* s_i^* = 0 \qquad \text{for all} \ \ i = 1, \ldots, n,$$

which are referred to as the complementarity conditions.

Strictly complementary optimal solutions do not exist in general for QO. However, if both the primal and dual problems have feasible solutions, then they have so-called *maximally complementary optimal solutions* x^* and (x^*, y^*, s^*) that are maximal w.r.t. the number of positive coordinates of $x^* + s^*$, i.e., $\text{supp}(x_i^* + s^*)$ is maximal.

1.2 ■ From Duality Theory to Algorithmic Concepts

The brief introduction presented in Section 1.1 reveals that both LO and QO problems share the extremely important strong duality property. These problem classes are unique in the sense that only the existence of primal and dual feasible solutions is needed to have optimal solutions for both the primal and dual problems with equal objective values. In other words, we have zero duality gap at optimality while the optimal objective value is attained by all optimal solutions. This strong property is essential when designing algorithms for LO and QO.

1.2.1 ▪ Algorithmic Concepts for LO

To illustrate the foundations of designing algorithms, the various elements of optimality conditions for the case of LO are presented in Figure 1.1.

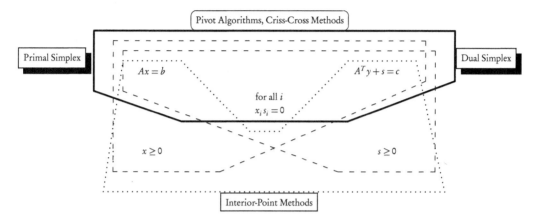

Figure 1.1. *Algorithms keep some of the optimality criteria valid while working toward the others. Reprinted with permission from Elsevier* [975].

LO problems are solved by iterative algorithms. The algorithm designs are based on the optimality conditions. The initial solution is required to satisfy some of the conditions, those conditions are preserved through all the iterates, and the algorithms strive to finally satisfy all the optimality conditions. Here we discuss briefly the two most popular, and in computational practice most efficient, algorithm classes.

Pivot algorithms utilize basis solutions of the system of equations $Ax = b$. Whenever a basis A_B, an $m \times m$ nonsingular matrix from A, is selected, this choice immediately allows us to satisfy the primal and dual equality constraints and the complementarity conditions. By partitioning matrix A and vectors c, x, s according to the basis coordinates, indicated by subscript $\{.\}_B$, and to nonbasis coordinates, indicated by subscript $\{.\}_N$, we have

- $Ax = b$, where $x = (x_B^T, x_N^T)^T$ with $x_B = A_B^{-1}b$ and $x_N = 0$, and
- $A^T y + s = c$, where $c = (c_B^T, c_N^T)^T$, $s = (s_B^T, s_N^T)^T$
  with $s_B = 0$, $y = (A_B^T)^{-1}c_B$, and $s_N = c_N - A_N^T y$.

Clearly, for every basis, solution $x_i s_i = 0$ holds for all coordinates i, i.e., complementarity holds. A *pivot* means that the algorithm moves from a basis solution (and the corresponding primal-dual solutions) to a neighboring basis solution, where the two bases are neighbors if all but one column vector of A_B are identical.

Pivot algorithms utilize various *pivot rules* that specify how the algorithm moves from one basis to another. Criss-cross type algorithms [1760] may start from any basis, usually define a unique pivot, and finally reach an optimal basis solution in a finite number of steps. Primal (dual) simplex methods [549, 1762] require a primal (dual) feasible basis, i.e., $x_B \geq 0$ ($s_N \geq 0$), to start with. Such a feasible basis solution can be obtained by solving a so-called first-phase problem. Primal (dual) simplex method pivot rules are designed to preserve primal (dual) feasibility and so ensure monotonic improvement of the primal (dual) objective function. In general, especially when *the LO problems are degenerate*, i.e., if either x_B or s_N or both have zero coordinates, simplex rules have considerable freedom to select the next basis that can be exploited to increase efficiency in practice. However, simplex methods might cycle [174, 947], i.e., return to the same basis repeatedly, unless anticycling rules are used. As will be discussed in the following sections, simplex algorithms allow highly efficient implementations and so are broadly used to solve engineering optimization problems in practice.

IPMs for LO are based on different principles than pivot algorithms. The initial solution, and all iterates of the most popular primal-dual IPMs, strictly satisfy the inequality conditions, i.e., $x > 0$ and $s > 0$. *Feasible IPMs* [1591], for all iterates, also satisfy the primal and dual equality constraints $Ax = b$ and $A^T y + s = c$, while the iterates of *infeasible IPMs* [1905, 1926] satisfy only the strict inequality conditions $x > 0$ and $s > 0$. Having such a solution, the complementarity gap $x^T s$ (and infeasibility of the equality constraints for infeasible IPMs) is systematically reduced to close to zero. This is done by applying Newton steps for the feasibility and the perturbed complementarity conditions:

$$\begin{array}{rcll} Ax & = & b, & x > 0, \\ A^T y + s & = & c, & s > 0, \\ x_i s_i & = & \mu & \forall i, \end{array} \qquad \text{(central-path)}$$

where $\mu > 0$ is the so-called central path parameter. During this process, μ is gradually reduced to zero, and so $x_i s_i \to 0$. Let us assume that matrix A has full row rank and that a feasible solution pair with $x > 0$ and $s > 0$ exists. The latter condition is referred to as the *interior-point condition (IPC)*. If the IPC holds, then for all $\mu > 0$, system (central-path) has a unique solution. The set of solutions $\{x(\mu) : \mu > 0\}$ and $\{(y(\mu), s(\mu)) : \mu > 0\}$ define [1591] the primal and dual central paths, respectively.

The Newton system for system (central-path) for a given $x > 0$, $s > 0$, y, and $\mu > 0$ is defined as

$$\begin{array}{rcll} A\Delta x & = & b - Ax, \\ A^T \Delta y + \Delta s & = & c - A^T y - s, & \text{(Newton system)} \\ x_i \Delta s_i + \Delta x_i s_i & = & \mu - x_i s_i & \forall i. \end{array}$$

Note that for feasible IPMs, $b - Ax = 0$ and $c - A^T y - s = 0$. The above Newton system, for all $\mu > 0$, $x > 0$, $s > 0$, has a unique solution, and the new iterate is calculated as $x := x + \alpha\Delta x$, $y := y + \alpha\Delta y$, and $s := s + \alpha\Delta s$ for some appropriately chosen step length $\alpha > 0$. After one or more Newton steps, the parameter μ is reduced. Modern polynomial-time IPMs carefully choose the reduction of μ and the step length α. They maintain $x > 0$ and $s > 0$ for each step and finally measure and carefully control the magnitude of infeasibility w.r.t. the equations (central-path). The reader can find numerous variants and their theoretical complexity proofs in the previously cited books and papers.

The most effective IPM variants in computational practice are based on Mehrotra's [1312, 1591] predictor-corrector algorithm, where at each step of the IPM, equation system (central path) is solved twice with the same left-hand side matrix. For the predictor step, $\mu = 0$ is chosen. The predictor direction $(\Delta^a x, \Delta^a y, \Delta^a s)$ is calculated; however, the predictor step is not executed. The predictor direction is used to calculate the anticipated best μ value for the corrector direction. Because system (central-path) is solved twice with the same coefficient matrix with different right-hand sides, one factorization of the coefficient matrix is sufficient to calculate solutions with different right-hand sides, while through the predictor direction the algorithm learns about the local properties of the central path. For information about LO software packages, please consult Section 1.6.

1.2.2 ▪ Algorithmic Concepts for QO

For QO the design of algorithms is based on similar concepts, though the design and actual performance of pivot algorithms for QO face more challenges.

Pivot algorithms utilize basis solutions of the primal-dual system of equations

$$
\begin{aligned}
-Ax &= -b, & x \geq 0, \\
A^T y - Qx + s &= -c, & s \geq 0.
\end{aligned}
$$

Analogous to the LO case, a basis A_B of A is chosen that allows us to construct a basis and a primal-dual solution for this larger system for which complementarity holds. Then, pivot rules are constructed so that complementarity, i.e., $x_i s_i = 0 \ \forall i$, is preserved throughout the algorithm. Such algorithms are called *complementarity pivot algorithms* [517, 518]. To prove the correctness and finiteness of complementarity pivot algorithms is more challenging than the proofs for the analogous simplex methods for LO. The performance of such algorithms when solving large-scale QO problems is weaker than the performance of their counterparts for LO. Consequently, the leading software packages for QO do not include, or do not choose implementations of, complementarity pivot algorithms as the default solver for QO, with the exception being if mixed-integer QO problems are solved.

IPMs for QO are analogous to IPMs for LO [579, 1905, 1926]. Their design, structure, and computational properties are analogous. Initialization with an interior-point solution pair follows the same route. The structure and organization of both for primal-dual infeasible and feasible algorithms are the same.

The perturbed complementarity (central path) conditions for QO are as follows:

$$
\begin{aligned}
Ax &= b, & x &> 0, \\
A^T y - Qx + s &= c, & s &> 0, \\
x_i s_i &= \mu & \forall i,
\end{aligned}
\qquad \text{(central-path-QO)}
$$

where $\mu > 0$ is the so-called central path parameter, which is gradually reduced to zero. Let us assume that matrix A has full row rank and that a feasible solution pair with $x > 0$ and $s > 0$ exists. Then for all $\mu > 0$, system (central-path-QO) has a unique solution. The set of solutions $\{x(\mu) : \mu > 0\}$ and $\{(y(\mu), x(\mu), s(\mu)) : \mu > 0\}$ defines the primal and dual central paths, respectively. The Newton system for the system (central-path) with a given $x > 0$, y, $s > 0$, and $\mu > 0$ is given as

$$
\begin{aligned}
A\Delta x &= b - Ax, \\
A^T \Delta y - Q\Delta x + \Delta s &= c - A^T y - s, \\
x_i \Delta s_i + \Delta x_i s_i &= \mu - x_i s_i & \forall i.
\end{aligned}
\qquad \text{(Newton system-QO)}
$$

All the elements of IPMs are the same as those for LO. Due to the presence of matrix Q in the dual equation system, solving system (Newton system-QO) requires more computation, and IPMs for QO usually need somewhat more iterations than when solving LO problems. It should be mentioned that Mehrotra's predictor-corrector methodology is generalized and implemented for solving QO too.

The discussions in this subsection show that for QO problems, software developers made the right choice in making IPM-based solvers the default in their packages. For more information on software, see Section 1.6.

1.3 ▪ Complexity and Solutions Obtained

The paper [975] presents a comprehensive evaluation of the advantages and disadvantages of pivot and interior point algorithms. Here we address only two issues: complexity and solutions obtained.

1.3.1 ▪ Complexity

Computational complexity bounds for algorithms give bounds on how the algorithms may behave in the worst case and may also indicate how the respective algorithms may behave when solving problems in practice. There is no significant difference between the complexity results for LO and QO, so in the following we present a unified discussion that is valid for both LO and QO.

1.3.1.1 ▪ Pivot Algorithms

As was mentioned earlier, when an LO problem is degenerate, pivot algorithms may cycle; thus they might not even solve LO problems. The cycling phenomenon is not observed when solving LO problems in computational practice. Nevertheless, several anticycling pivot rules were designed to ensure the finiteness of pivot algorithms [1762].

The theoretical worst-case complexity of both primal and dual simplex methods, even with anticycling rules, is exponential. For references, see, e.g., [975, 1591]. The Klee–Minty cube [1071] and some of its variants provide an example that pivot algorithms may require 2^n pivots to solve an LO problem in dimension n. We need to stress that such exponential behavior of primal or dual simplex methods in high-quality professional software is never observed. In practice, the number of pivots needed to solve LO problems tends to be linear in the problem dimension, which is close to the expected number of pivots that Borgwardt [318] proves under appropriate probabilistic assumptions.

1.3.1.2 ▪ IPMs

For both LO and QO, both feasible and infeasible IPMs provide polynomial-time algorithms [230, 579, 1591, 1926]. The theoretical complexity results for QO are usually a constant factor weaker than for LO. Furthermore, note that feasible IPMs have a factor $\sqrt{n}$ better complexity than infeasible IPMs.

Infeasible IPMs can be initiated by any vector y and any positive vectors x and s. On the other hand, feasible IPMs require both primal and dual solutions to be strictly feasible. Such solutions might not exist in general, so additional methods are needed to reformulate the corresponding problems so that the resulting problems satisfy the IPC. The elegant models of homogeneous self-dual embedding [1591, 1926, 1928] provide the best solution to date for the problem of initialization. Any positive vectors x and s can be used in deriving a self-dual model. The solution of the self-dual model provides strictly complementary solutions for the original primal-dual LO problems and maximally complementary solutions[1] for QO problems. The price for this elegant solution is a constant factor in the worst-case complexity bound. The embedding model approach allows efficient implementation too [72, 75, 77, 78, 1591, 1919].

---

[1]An optimal solution pair is maximally complementary if $x_i s_i = 0 \; \forall \, i$ and the number of coordinates for which $x_i + s_i > 0$ is maximal among all optimal solutions. Obviously, strictly complementary solutions are maximally complementary.

It is known that neither the theoretical complexity nor the performance of IPMs in practice is influenced by the degeneracy of the problem. However, redundant constraints may severely impact IPMs. In a series of papers [591–593], the authors provide redundant variants of the Klee–Minty cube [1071] that force the central path to trace the exponentially long pivot path of simplex algorithms. These examples demonstrate that the worst-case complexity bound for IPMs is tight.

1.4 ▪ Basis Solution versus Interior Solution

1.4.1 ▪ Pivot Algorithms Provide an Optimal Basis Solution

Optimal basis solutions have several desired properties: basis solutions usually have a low number of positive coordinates, basis solutions represent extremal points (vertices) of the feasible sets, and basis solutions are needed to generate cutting planes when solving mixed-integer LO and QO problems. If a strictly complementary optimal solution pair is needed, the Balinski–Tucker [156] procedure is available to generate strictly complementary solutions. The Balinski–Tucker procedure repeatedly calls the simplex method; thus it is not a polynomial-time algorithm.

1.4.2 ▪ IPMs Provide an Interior Optimal Solution

More precisely, the limit point of the central path, as μ goes to zero, is a maximally complementary solution pair in the relative interior of the set of optimal solutions [1591, 1926].

In practice, the iteration sequence of IPMs provides an ϵ-optimal solution, i.e., for which $x^T s \leq \epsilon$. For practical purposes $\epsilon \leq 10^{-8}$ is sufficient. Then, within computer precision, the given solution is optimal. To get an exact optimal solution, theoretically, ϵ needs to be smaller in fact so small that an exact optimal solution can be found by a strongly polynomial[2] rounding procedure [1591, 1926]. The result of the rounding procedure is an *exact, strictly complementary primal-dual optimal solution pair* for LO problems and an *exact, maximally complementary primal-dual optimal solution pair* for QO problems. If an optimal basis solution is needed, a strongly polynomial optimal basis identification procedure [1591, 1926] provides an optimal basis. Thus, the total complexity of the solution process—interior-point algorithm, rounding procedure, optimal basis identification—is polynomial both for LO [1591, 1926] and for QO [230]. An efficient implementation strategy of the rounding and basis identification procedures is presented in [76].

1.5 ▪ Computational Methods of, and Software for, LO and QO

The implementation of simplex methods has a long history with extensive literature. For computational methods of the simplex method, the reader may consult the papers [914, 1425] and the comprehensive books of Maros [1283] and Pan [1439]. IPMs not only allowed the development of a rich theory, but they also have been implemented with great success for solving LO, QO, and more general conic (see Part III) and nonlinear optimization (NLO) (see Part V) problems. It is now common sense that

---

[2] *Strongly polynomial* means that the complexity of the algorithm depends only on the dimension, independent of problem data. The strongly polynomial rounding procedure of IPMs requires only a few matrix-vector multiplications and the solution of a linear system of equations.

for large-scale sparse structured LO and QO problems, IPMs are the method of choice. All major commercial optimization software systems contain implementations of feasible and infeasible IPMs utilizing Mehrotra's predictor-corrector method. Computational and implementation issues of IPMs for LO and QO are thoroughly discussed in [72, 825, 1309, 1312, 1919]. The books [1591, 1905, 1926] also devote a chapter to that subject.

Modern optimization software packages integrate powerful solvers for both LO and QO. Leading vendors incorporate solver engines based on both primal and dual simplex methods for LO, as well as feasible and infeasible IPM solvers for LO and QO. The solvers allow the user to choose which algorithm to use, and the latest versions may utilize multicore architecture to run algorithm variants on different processors in a competing mode. The IPM solvers also offer the possibility of producing optimal basis solutions. Bixby and Saltzman in their paper [288] describe a numerically stable crossover procedure from interior to simplex methods, while Andersen and Ye [76] present an intriguing implementation of the rounding and optimal basis identification procedures with the possibility of crossover to simplex algorithms if needed.

For implementations of IPMs for NLO problems (see Part V), the reader may consult [77, 376, 1826, 1858, 1859]. For discussions of the implementation strategies and documentation of available software for conic optimization (see Part III), the papers [73, 75, 316, 1735, 1774] are a rich source of information.

1.6 ▪ Available Software

The development of computational methods for simplex algorithms spans over six decades. Intensive research on the theory and computational methods of modern IPMs spans over three decades. Consequently, the optimization community has developed a good understanding of both the theory and the practice of simplex algorithms and IPMs. Several sophisticated implementations exist of simplex algorithms and IPMs for both LO and QO. Today, all leading commercial software packages include both simplex and IPM solvers for LO. Each of the listed commercial software packages features both primal and dual simplex methods, IPMs, and crossover from IPMs to simplex methods. To solve QO problems, in each of these packages, only IPMs are implemented. The solvers are supported by most modeling languages [748], both general and domain specific, and are available for most platforms and operating systems.

These high-performance solvers are capable of solving LO and QO problems on a PC in seconds or minutes and can efficiently solve problems that were hardly solvable on a supercomputer a few decades ago.

Below we list the most broadly used commercial software packages:

CPLEX: CPLEX Optimization, Inc.: http://www-01.ibm.com/software/commerce/optimization/cplex-optimizer/index.html

GuRoBi: GuRoRi Optimization: http://www.gurobi.com

XPRESS: FICO XPRESS Optimization Suite: http://www.fico.com/en/products/fico-xpress-optimization-suite

MOSEK: MOSEK ApS: https://mosek.com/

A few free academic solvers are listed below. All the solvers in the first group use simplex methods:

SoPLex: Zuse Institute Berlin (ZIB): http://soplex.zib.de/

GLPK: Free Software Foundation: `https://www.gnu.org/software/glpk/`
CLP: COIN-OR: `https://projects.coin-or.org/Clp`

The following list presents IPM-based solvers:

IPOPT: COIN-OR: `https://projects.coin-or.org/Ipopt`
PcX: S. Wright: `http://pages.cs.wisc.edu/~swright/PCx/`
LIPSOL: Y. Zhang: `http://www.caam.rice.edu/~yzhang/`
BPMPD: Cs. Mészáros: `http://www.sztaki.hu/~meszaros/bpmpd/`
LOQO: R. Vanderbei: `http://www.princeton.edu/~rvdb/loqo/LOQO.html`
SeDuMi: download from Lehigh U.: `http://sedumi.ie.lehigh.edu`

More information about commercial or research LO codes can be found at

- `http://neos-guide.org/content/lp-faq`
- `http://plato.asu.edu/sub/nlores.html#LP-problem`
- `http://plato.asu.edu/sub/nlores.html#QP-problem`

1.7 ▪ Extensions

The impact of the rapid development of modern IPMs goes far beyond LO and QO. Some of the previously mentioned books [579, 989, 1396, 1761, 1926] discuss extensions of IPMs for classes of nonlinear problems (see Part V) and to the recently developed area of conic optimization (see Part III). Much research has been devoted to developing the theory and computational methodology of IPMs for NLO problems, in particular for conic linear optimization (CLO). The best-known classes of conic problems are second-order and semidefinite optimization (SDO), and both of these problem classes have numerous interesting applications. Applications reach far beyond such traditional areas as combinatorial optimization [50] and control [211, 1498]. The impact is profound on various areas of engineering, including structural [211, 568] and electrical engineering [332, 1821]. For surveys on algorithmic and complexity issues, the reader may consult [567, 569–571, 1396, 1397, 1499, 1508, 1734, 1898, 1926] and the subsequent chapters of this handbook.

1.8 ▪ Summary

The three decades of IPMs, in Margaret Wright's words [1904], the "Interior Point Revolution," have lasting consequences and have brought the theory and practice of LO and QO to a new level. LO and QO problems with sizes in the hundreds of thousands and even millions are solved routinely today on personal computers. Solving LO and QO problems of such size was beyond imagination three decades ago. The improvement in the efficiency of optimization software just in the 1990s was measured on the order of 10^6 or more [284]. Computer hardware and the advances in optimization theory contributed equally to the dramatic advances of our capability to solve very large scale engineering optimization problems efficiently and reliably. Not only duality theory and the theory of algorithms, but practical usage-oriented theories, such as parametric and sensitivity analysis [889, 1091, 1591], were rejuvenated. The theory, computational practice, and engineering applications of optimization are rich areas of research and high-impact applications, as the following chapters demonstrate.

Thirty years after the publication of Karmarkar's path-breaking paper [1038], the theory and computational practice of LO, QO, and their extensions, as enabling technologies, has matured to a level that has ignited transformative changes in engineering applications of optimization.

Chapter 2

Truss Topology Design by Linear Optimization

Mathias Stolpe

2.1 ▪ Introduction

Structural optimization is a multidisciplinary area that covers theory, mathematical models, and numerical methods for optimal design of (parts of) load-carrying structures such as bridges and cars. Structural *topology* optimization is a subfield that is intended to be used in the conceptual design phase to propose innovative designs; see, e.g., [219]. The main strength of topology optimization is its capability of determining both the geometry and the topology of the structure. A classical subject in optimal structural design is the optimization of grid-like continua, pioneered in [1320]. This field has since developed into the well-established layout theory for frame systems; see, e.g., [1516]. A closely related area, and the main topic of this chapter, is numerical topology optimization of truss structures.[3] This topic first appeared in the early 1960s in, e.g., [619] and [930]. Structural topology optimization has since evolved into a mature multidisciplinary research field, and the developed models and methods are today extensively used in industrial applications; see, e.g., [219]. Lately, structural topology optimization has been introduced in the field of architecture for structural design of highrise buildings [185].

The early works on truss topology optimization used a so-called ground structure, and this approach is still the predominant method in the field. In the ground structure approach for truss topology optimization, a number of nodes with fixed positions are distributed over a given design domain. The nodes are then connected by a normally large number of potential bars, creating what is often referred to as a structural universe. Associated with each bar is a continuous design variable. The design variable represents the cross-sectional area or the volume of the bar. By allowing the design variables to attain the value zero, the topology, and to a limited extent also the geometry, of the truss can be determined by solving a structural sizing problem. Two examples of ground structures and the corresponding optimal designs are illustrated

---

[3]A truss is a structure that consists of slender bars connected at joints. The joints are assumed to be frictionless and they are therefore not able to carry any moments.

(a) *A coarse ground structure with* 21 × 6 *nodes and* 425 *potential bars. Only neighboring nodes are connected by a potential bar.*

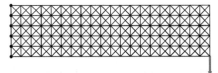

(b) *A denser ground structure with* 21 × 6 *nodes and* 1,125 *potential bars. Nodes that are no more than two rows or columns apart are connected.*

(c) *Optimal design corresponding to the ground structure in Figure* 2.1(a).

(d) *Optimal design corresponding to the ground structure in Figure* 2.1(b).

Figure 2.1. *Two different ground structures with boundary conditions and external load for a two-dimensional cantilever beam design problem. The optimal designs were obtained by solving the problem of minimizing the structural volume while satisfying stress constraints in all bars.*

in Figure 2.1. These particular examples clearly illustrate that the choice of ground structure can be very important for the complexity of the optimal design.

This chapter begins with a short introduction to the required mechanical assumptions and the structural analysis equations and an introduction to the notation used. The chapter then presents a brief review of some of the classical single-load truss topology optimization problems. The entry point is the problem of finding the truss with minimum volume while satisfying stress constraints in all bars, which was stated in, e.g., [619]. Several of the formulations are linear optimization (LO) problems. Some of the known theoretical results about these problems are stated. In particular, based on LO theory, it is shown that there exist optimal designs with a limited number of bars. In engineering terms, there are optimal designs that are statically determinate. Furthermore, it is shown that the optimal designs are fully stressed, i.e., the stress in each present bar is at either the lower or the upper allowed bound. The minimum volume problem is also closely related to the important and extensively studied minimum compliance problem, i.e., the maximum stiffness problem [16]. Under certain assumptions, it is possible to show that the two problems are equivalent in the sense that an optimal solution to one problem is, up to scaling, an optimal design of the other problem.

The applicability of the problem formulations is illustrated by solving several large-scale two-dimensional truss topology optimization problems using commercial software for LO. The chapter concludes with a short survey of the limitations of LO in truss topology optimization and outlines some challenging and open research areas.

2.1.1 ▪ Structural Analysis of Trusses

Truss structures are assemblies of straight slender bars connected at frictionless nodes (joints). The bars are assumed to remain straight in deformation. Truss structures in this chapter are assumed to be subject to a single static external load that is applied only at some of the nodes. These assumptions imply that the bars only carry axial loads that are constant throughout the length and over the cross-section of the individual bars [507]. For trusses, the behavior of the members is, due to the structural assumptions, independent of the choice of geometry of the cross-sections. It is assumed that all bars are made of linear elastic materials.

Let $a_j \geq 0$ and $l_j > 0$ denote the cross-sectional area and the undeformed length of the jth bar, respectively. Associated with each node in a planar (space) truss are two (three) components of the displacement vector $u \in \mathbb{R}^d$. The total number of degrees of freedom, d, is for a planar (space) truss given by two (three) times the total number of nodes minus the number of degrees of freedom, which have prescribed fixed displacements (boundary conditions). One common boundary condition is that nodes are fixed to a solid wall. The member force in the jth bar is denoted by $q_j \in \mathbb{R}$. The bar is in compression if $q_j < 0$ and in tension if $q_j > 0$. Associated with each bar is a vector $r_j \in \mathbb{R}^d$, which is determined solely by the position of the end nodes of the bar, containing the direction cosines [507]. Let the compatibility matrix $R = (r_1 \; r_2 \; \cdots \; r_n) \in \mathbb{R}^{d \times n}$. Similarly, collect the bar forces in the vector $q = (q_1 \; q_2 \; \cdots \; q_n)^T \in \mathbb{R}^n$ and the bar areas in the vector $a = (a_1 \; a_2 \; \cdots \; a_n)^T \in \mathbb{R}^n$. With these, static equilibrium of the bar forces at the nodes is expressed as

$$\sum_{j=1}^{n} q_j r_j = Rq = f. \tag{2.1}$$

The linearized elongation of the jth bar is given by $r_j^T u$, and the relative elongation, i.e., the strain, is $\epsilon_j = (r_j^T u)/l_j$. For linear elastic materials, the bar stress $\sigma_j \in \mathbb{R}$ is determined by Hooke's law, $\sigma_j = E_j \epsilon_j$, where E_j is a material constant (Young's modulus[4]) for the jth bar. Strains are unitless, displacements are measured in meters, and stresses are measured in Pa (N/m^2), although normally MPa (N/mm^2) is used. The bar forces are determined by the kinematic compatibility equations

$$q_j = a_j \sigma_j = a_j \frac{E_j}{l_j} r_j^T u \; \forall \; j. \tag{2.2}$$

Combining force equilibrium (2.1) and kinematic compatibility (2.2) gives the elastic equilibrium equations

$$K(a)u = f, \tag{2.3}$$

where the global stiffness matrix $K(a) : \mathbb{R}^n \to \mathbb{R}^{d \times d}$ of the truss is linear in the design variables, i.e.,

---

[4]The elastic modulus is approximately $200 \cdot 10^9$ Pa for steel and $70 \cdot 10^9$ Pa for aluminum alloys.

$$K(a) = \sum_{j=1}^{n} a_j K_j = \sum_{j=1}^{n} a_j \frac{E_j}{l_j} r_j r_j^T. \tag{2.4}$$

The stiffness matrix is thus given as the sum of dyadic matrices and is symmetric and positive semidefinite for all $a \geq 0$.

2.2 ▪ Problem Formulations

Traditionally, single-load truss topology design problems are formulated as LO problems in terms of bar forces and only considering *plastic* analysis, i.e., a formulation excluding the kinematic compatibility equations (2.2); see, e.g., [619] and [930].

In this chapter, an alternative approach, which was presented in [1724], is followed. The truss topology design problem is first formulated as a nonlinear optimization (NLO) problem including the *elastic* equilibrium equations (2.3). The problem is written in terms of the design variables $a \in \mathbb{R}^n$ and the state variables $u \in \mathbb{R}^d$ and $q \in \mathbb{R}^n$. The complicating constraints causing the nonlinearity are then relaxed. In the process, both the displacement variables and the elastic moduli of the bars are removed from the formulation. The resulting LO problem is then reformulated into standard form. Based on the solution of this problem, it is possible to construct a feasible, and thus optimal, point to the original nonlinear problem. The dual of the linear problem is also stated and interpreted in mechanical terms.

2.2.1 ▪ Minimum Volume Problems in Areas and Member Forces

A relevant problem in structural optimization in general, and in truss topology design in particular, is to minimize the total volume of the structure while satisfying elastic equilibrium and certain constraints on the local stresses in the individual bars. Let $\overline{\sigma}_j > 0$ and $\underline{\sigma}_j > 0$ denote the given stress limits[5] in tension and compression, respectively. One of several possible formulations of this problem is cast using the design variables $a \in \mathbb{R}^n$ together with the state variables $q \in \mathbb{R}^n$ and $u \in \mathbb{R}^d$:

$$\begin{aligned}
\min_{a \in \mathbb{R}^n, q \in \mathbb{R}^n, u \in \mathbb{R}^d} \quad & \sum_{j=1}^{n} a_j l_j \\
\text{s.t.} \quad & Rq = f, \\
& q_j = a_j (E_j/l_j) r_j^T u, \quad j = 1, \dots, n, \\
& -a_j \underline{\sigma}_j \leq q_j \leq a_j \overline{\sigma}_j, \quad j = 1, \dots, n, \\
& a \geq 0.
\end{aligned} \tag{2.5}$$

In (2.5) the stress constraints are modeled using the member force variables. These constraints imply that the stress in the jth bar, σ_j, satisfies $-\underline{\sigma}_j \leq \sigma_j \leq \overline{\sigma}_j$ whenever $a_j > 0$, and the bar force $q_j = 0$ when $a_j = 0$. The stress constraints are thus effectively removed when the corresponding design variable is zero, i.e., when the bar is not present in the structure. Note that (2.5) is an NLO problem due to the bilinear terms in the kinematic compatibility equations.

---

[5]For classical engineering materials such as steel and aluminum alloys, the stress limits can, to a large extent, be chosen since they depend on the particular alloy composition and the heat and mechanical treatment of the metal [102]. This is in sharp contrast to the elastic moduli, which are essentially constant for these metals. For different steel types, the yield strength can vary between 260 and 1300 MPa, and for aluminum alloys the yield strength can vary between 100 and 600 MPa; see, e.g., [102].

The traditional formulation in truss topology design is reached by removing the kinematic compatibility equations from (2.5). The resulting problem is the LO problem

$$
\min_{a\in\mathbb{R}^n, q\in\mathbb{R}^n} \quad \sum_{j=1}^{n} a_j l_j
$$

$$
\text{s.t.} \quad Rq = f,
$$

$$
-a_j \underline{\sigma}_j \le q_j \le a_j \overline{\sigma}_j, \quad j = 1,\ldots,n, \tag{2.6}
$$

$$
a \ge 0.
$$

The change of variables $q = q^+ - q^-$ with $q^+ \ge 0$ and $q^- \ge 0$ and the observation that due to optimality $a_j = q_j^+/\overline{\sigma}_j + q_j^-/\underline{\sigma}_j$ imply that problem (2.6) can be written in standard form:

$$
\min_{q^+\in\mathbb{R}^n, q^-\in\mathbb{R}^n} \quad \sum_{j=1}^{n} l_j \left(\frac{q_j^+}{\overline{\sigma}_j} + \frac{q_j^-}{\underline{\sigma}_j} \right)
$$

$$
\text{s.t.} \quad R(q^+ - q^-) = f, \tag{2.7}
$$

$$
q^+ \ge 0, \quad q^- \ge 0.
$$

The variable q_j^+ represents the bar force in tension, while q_j^- represents the bar force in compression.

Throughout the following, natural assumptions on the problem data are made. Similar assumptions can be found in, e.g., [14] and [1724].

(A1) The external load vector, $f \in \mathbb{R}^d$, satisfies $f \ne 0$, and f is independent of a, u, and q.

(A2) The matrix $R = \begin{pmatrix} r_1 & r_2 & \cdots & r_n \end{pmatrix} \in \mathbb{R}^{d\times n}$ has full row rank, and $n \ge d$.

(A3) The stress limits $\underline{\sigma}_j > 0$ and $\overline{\sigma}_j > 0$ for all $j = 1,\ldots,n$.

The first assumption excludes trivial design situations. Assumption (A2) is equivalent to the common assumption that the stiffness matrix $K(a)$ is positive definite for all $a > 0$. The second assumption excludes mechanisms and rigid-body motions. For truss topology optimization problems based on the ground structure approach, the number of design variables, n, is normally much larger than the number of degrees of freedom, d. This is also the case for all problem instances solved in this chapter (cf. below). The third assumption is included to simplify the presentation. An extension of the problem formulations and the theoretical results with $\underline{\sigma}_j = 0$ or $\overline{\sigma}_j = 0$ can be found in [14].

Since the compatibility matrix R due to assumption (A2) has d linearly independent rows, there is at least one solution to $R(q^+ - q^-) = f$. The feasible set of the linear problem (2.7) is thus nonempty. Furthermore, the objective function in (2.7) is, because of assumption (A3), bounded from below by zero. These two observations guarantee the existence of an optimal solution to (2.7); see, e.g., [1239].

Lemma 2.1. *There exists an optimal solution to* (2.7).

The equivalence between the LO problems (2.6) and (2.7) outlined in the lemma below is stated and proved in, e.g., [1724].

Lemma 2.2. *If (q^+, q^-) is an optimal solution to (2.7), then (a,q), with $a_j = q_j^+/\overline{\sigma}_j + q_j^-/\underline{\sigma}_j$ for all j and $q = q^+ - q^-$, is an optimal solution to (2.6).*

If an LO problem has an optimal solution, it also has an optimal basic solution; see, e.g., [1239]. An immediate consequence of Lemma 2.1 is thus that (2.7) has an optimal basic solution. A remarkable result is that, based on an optimal basic solution to the linear problem (2.7), and thus to the linear problem (2.6), it is possible to construct a feasible point to the nonlinear problem (2.5). Since (2.6) is a relaxation of (2.5), it follows that this point is also optimal to (2.5).

Let $B \in \mathbb{R}^{d \times d}$ denote a basic matrix consisting of linearly independent columns from the matrix $(R \quad -R)$ corresponding to basic variables in an optimal basic solution (q^+, q^-) to the linear problem (2.7). Construct the vector $b \in \mathbb{R}^{2n}$ in the following manner. Set $b_j = l_j \overline{\sigma}_j/E_j$ and $b_{n+j} = l_j \underline{\sigma}_j/E_j$ for $j = 1, \ldots, n$. Let $r \in \mathbb{R}^d$ denote the vector consisting of elements in b corresponding to the basic variables. Finally, a suitable displacement vector $u \in \mathbb{R}^d$ is given by the unique solution to the linear system

$$B^T u = r. \tag{2.8}$$

This implies that $r_j^T u = l_j \overline{\sigma}_j/E_j$ for all j such that q_j^+ is basic and $r_j^T u = l_j \underline{\sigma}_j/E_j$ for all j such that q_j^- is basic. The coupling between (2.7) and (2.5) is described in the following lemma from [1724].

Lemma 2.3. *Let (q^+, q^-) be an optimal basic solution to the linear problem (2.7). Furthermore, let u be the corresponding solution to (2.8), $q = q^+ - q^-$, and $a_j = q_j^+/\overline{\sigma}_j + q_j^-/\underline{\sigma}_j$ for $j = 1, \ldots, n$. Then (a, q, u) is an optimal solution to the nonlinear problem (2.5).*

Proof. The proof follows the corresponding proof in [1724]. It is enough to show that the point (a, q, u) satisfies the relaxed constraints $q_j = a_j(E_j/l_j)r_j^T u$ for all j. Consider a j for which $q_j^+ > 0$. For such bars, $q_j = q_j^+ = a_j \overline{\sigma}_j$ and $r_j^T u = l_j \overline{\sigma}_j/E_j$. Thus, $q_j = a_j(E_l/l_j)r_j^T u$. Similar arguments show that the relaxed constraints are satisfied if $q_j^- > 0$. Finally, consider bars for which $q_j^+ = q_j^- = 0$. For such bars, $q_j = 0$ and $a_j = 0$. Again, it follows that $q_j = a_j(E_j/l_j)r_j^T u$. □

An immediate consequence of the proof is that at most d force variables are nonzero in an optimal basic solution to (2.7). The structure described by the corresponding design variables therefore has at most d bars present. The structure is *statically determinate*.

Corollary 2.4. *There exists an optimal design to (2.5) with at most d bars present, i.e., a statically determinate structure.*

Remark 2.1. For a basic feasible solution to (2.7), at most one of q_j^+ and q_j^- is nonzero. From Lemma 2.3 it follows that the optimal design variables are determined such that the bar stress is at either the lower or the upper stress limit. The optimal design is thus fully stressed.

2.2.2 ▪ Formulations in Displacement-Like Variables

The dual of (2.7) is

$$
\max_{\lambda \in \mathbb{R}^d} \quad f^T \lambda
$$
$$
\text{s.t.} \quad -\frac{l_j}{\underline{\sigma}_j} \leq r_j^T \lambda \leq \frac{l_j}{\overline{\sigma}_j}, \quad j = 1, \ldots, n. \tag{2.9}
$$

Note that the dual (2.9) has a nonempty feasible set since $\lambda = 0$ is feasible. In fact, due to the strong duality theorem and the fundamental theorem of LO (see, e.g., [1239]), there exists an optimal basic solution to (2.9).

Under certain additional assumptions, the dual (2.9) can be interpreted in mechanical terms. Assume that the elastic modulus is identical for all bars, i.e., $E_j = E$ for all j. Furthermore, assume that the stress limits are identical in compression and tension, i.e., $\overline{\sigma}_j = \underline{\sigma}_j = \overline{\sigma}$ for all j. Then, after a slight change, the variables in (2.9) can be viewed as (scaled) displacements and the objective as the minimization of the static compliance with lower and upper bounds on the bar strains.

The optimality conditions for (2.7) include

$$
r_j^T \lambda = l_j / \overline{\sigma} \quad \forall\, j : q_j^+ > 0,
$$
$$
r_j^T \lambda = -l_j / \overline{\sigma} \quad \forall\, j : q_j^- > 0.
$$

Let (q^+, q^-) be a basic optimal solution to (2.7) with Lagrange multipliers λ, and let the design variables be computed by $a_j = (q_j^+ + q_j^-)/\overline{\sigma}$ for all j. Then a suitable displacement vector is given by

$$
u = \frac{\overline{\sigma}^2}{E} \lambda.
$$

It follows, with this choice of displacement vector, that

$$
q_j^+ = a_j \overline{\sigma} = a_j \frac{\overline{\sigma}^2}{l_j} r_j^T \lambda = a_j \frac{E}{l_j} r_j^T u.
$$

Similarly,

$$
q_j^- = a_j \overline{\sigma} = -a_j \frac{\overline{\sigma}^2}{l_j} r_j^T \lambda = -a_j \frac{E}{l_j} r_j^T u.
$$

It follows that

$$
q_j = q_j^+ - q_j^- = a_j \frac{E}{l_j} r_j^T u,
$$

and hence (a, q, u) satisfies the kinematic compatibility equations.

2.3 ▪ On the Coupling to the Minimum Compliance Problem

Another relevant and well-studied problem in truss topology design is to find the minimum compliance truss for a given volume of material, i.e., to find the globally stiffest truss for the given load; see, e.g., [16] and [219]. This problem can be formulated as

$$\min_{a\in\mathbb{R}^n, u\in\mathbb{R}^d} \quad \frac{1}{2} f^T u$$
$$\text{s.t.} \quad K(a)u = f,$$
$$\sum_{j=1}^{n} a_j l_j = V,$$
$$a \geq 0,$$

(2.10)

where $V > 0$ is a given volume bound. For single-load truss topology optimization problems, there is a close relationship between the minimum compliance problem (2.10) and the minimum volume problem (2.5). In this section, this relationship is outlined. The precise details can be found in, e.g., [16].

The nonlinear minimum compliance problem (2.10) can be reformulated using the principle of minimum potential energy. In the following reformulation process, care should of course be taken when replacing inf by min and sup by max. Throughout this section we almost consistently use min and max and refer to [216] for the complete details. The potential energy, Π, of a structure subject to an external nodal force vector f, under the assumption of linear elastic material behavior, is given by

$$\Pi(a, u) = \frac{1}{2} u^T K(a)u - f^T u.$$

For a fixed design a with $a \geq 0$,

$$\inf_{u\in\mathbb{R}^d} \Pi(t, u) = \begin{cases} -\infty & \text{if } f \notin R(K(a)), \\ -\frac{1}{2} f^T u^* = \text{constant} & \forall \, u^* \text{ with } K(a)u^* = f. \end{cases}$$

The infimum is attained if and only if $f \in R(K(a))$. For details and a formal proof, see, e.g., [15]. The reformulation process presented herein closely follows the developments in [219, Chapter 4]. Problem (2.10) is thus equivalent to

$$\max_{a\geq 0, a^T l = V} \min_{u\in\mathbb{R}^d} \left\{ \frac{1}{2} u^T K(a)u - f^T u \right\}.$$

(2.11)

Problem (2.11) is concave-convex, and the constraint set in a is both convex and compact. The min and max operations can thus be interchanged, resulting in

$$\min_{u\in\mathbb{R}^d} \max_{a\geq 0, a^T l = V} \left\{ \frac{1}{2} u^T K(a)u - f^T u \right\}.$$

(2.12)

The inner maximization problem in (2.12) is a linear problem in the design variables. The inner problem can be solved by noting that

$$\max_{a\geq 0, a^T l = V} \left\{ \frac{1}{2} u^T K(a)u - f^T u \right\} \leq -f^T u + \frac{V}{2} \max_{1\leq j\leq n} \left\{ \frac{1}{l_j} u^T K_j u \right\}.$$

The inequality is satisfied with equality if all available material is distributed to the bar with largest $u^T K_j u / l_j$. Problem (2.10) can thus be reduced to the convex unconstrained but nonsmooth problem

$$\min_{u\in\mathbb{R}^d} \max_{1\leq j\leq n} \left\{ \frac{V}{2l_j} u^T K_j u - f^T u \right\}.$$

(2.13)

After introducing the additional variable τ, problem (2.13) can be reformulated as the convex quadratically constrained problem

$$\min_{u \in \mathbb{R}^d, \tau \in \mathbb{R}} \quad -f^T u + \tau$$
$$\text{s.t.} \quad \frac{V}{2l_j} u^T K_j u \leq \tau, \quad j = 1, \ldots, n. \tag{2.14}$$

For single-load truss topology design problems, the scaled element stiffness matrices K_j are dyadic, and the convex quadratic constraints in (2.14) can be written as

$$\frac{V}{2l_j} u^T K_j u = \left(\sqrt{\frac{V E_j}{2}} \frac{r_j^T u}{l_j} \right)^2 \leq \tau \quad \forall\, j = 1, \ldots, n.$$

Up to a scaling, problem (2.14) is thus equivalent to the LO problem

$$\min_{u \in \mathbb{R}^d} \quad -f^T u$$
$$\text{s.t.} \quad -\frac{l_j}{k_j} \leq r_j^T u \leq \frac{l_j}{k_j}, \quad j = 1, \ldots, n, \tag{2.15}$$

where

$$k_j = \sqrt{\frac{V E_j}{2}} \quad \forall\, j = 1, \ldots, n.$$

Note that contrary to the minimum volume problem (2.5), the minimum compliance problem (2.10) does not treat bars in compression or tension differently. An optimal structure obtained by solving (2.10) possesses the same stress limits in tension and compression.

The correspondence between the linear problem (2.15) and the minimum compliance problem (2.10) is described in the lemma below. The lemma slightly modifies, due to differences in notation and choice of design variables, the result, which can be found together with a formal proof in [16].

Lemma 2.5. *Let u^* be an optimal solution to (2.15) with dual variables (q^+, q^-). Then (a, u) with*

$$a_j = \frac{V}{V_C k_j} (q_j^+ + q_j^-) \quad \forall\, j = 1, \ldots, n$$

and

$$u = \frac{V_C}{2} u^*$$

with

$$V_C = \sum_{j=1}^{n} \frac{l_j}{k_j} (q_j^+ + q_j^-)$$

solve the minimum compliance problem (2.10).

If the stress limits are chosen as $\underline{\sigma}_j = \overline{\sigma}_j = \sqrt{V E_j/2}$ for all j, then the dual problem (2.9) and the problem (2.15) are identical. The above lemma thus establishes an equivalence between the minimum volume problem (2.5) and the minimum compliance problem (2.10) in the situation that the stress bounds in tension and compression are identical and appropriately chosen based on the problem data in (2.10).

2.4 ▪ Numerical Experiments

The LO problem formulation (2.7) of the minimum volume problem (2.5) is solved
for three classical benchmark problems from the literature. The construction of the
ground structure, the computation of the problem data, and the visualization are done
in MATLAB R2013a.

The primal simplex method and the sifting method implemented in IBM ILOG
CPLEX [974] version 12.1.1 are used for the numerical experiments. The sifting
method exploits the characteristics of the problems. For a truss topology design problem in the form (2.7), the number of columns is generally much larger than the number of rows. Furthermore, most of the primal variables are at their lower bound in an
optimal solution. All results are obtained on an AMD Opteron 6168 processor with
12 cores and running at 1.9 GHz. IBM ILOG CPLEX is allowed to use only one thread,
i.e., it runs in sequential mode. All other parameters in IBM ILOG CPLEX are set to
default values.

For simplicity, and to obtain well-scaled problems, the problem data have been
scaled such that the loads all have unit norm, $E_j = 1$, $\underline{\sigma}_j = -1$, and $\overline{\sigma}_j = 1$ for all j. In
all examples, the nodes in the ground structure are evenly distributed over the design
domain. All nodes are connected by a potential bar. The number of bars in the ground
structure is thus $N(N-1)/2$, where N is the number of nodes.

A commercial LO solver is chosen because of the favorable characteristics of the
linear problem (2.7). The number of degrees of freedom, i.e., the number of equality constraints, is modest even for large-scale examples. Most of the development of
efficient special purpose methods for truss topology optimization have targeted nonlinear formulations of the problems, see, e.g., [990]. Recently, there has been an effort
to solve truss topology design problems as LO problems with extremely many bars
in the ground structure with special purpose methods. In [800], problems with more
than 100,000,000 potential bars are solved by a special purpose column generation
method.

The first example is a classical Michell beam with (scaled and unitless) dimensions
3×2; see Figure 2.2(a). The left-hand side is entirely fixed to a rigid wall, and the vertical
unit load is applied on the middle node on the right-hand side. Similar problems are
solved in, e.g., [16], [990], and [800]. The optimal design has a close resemblance to
the designs obtained by exact layout theory; see, e.g., [218]. The second example is
a cantilever beam subjected to a vertical tip load applied at the lower right node and
with the left-hand side fixed to a rigid wall; see Figure 2.3(a). The unitless dimensions
of the design domain are 4×1. Cantilever design problems with different dimensions
are reported in, e.g., [16] and [217]. The third example also has a rectangular design
domain with dimensions 2×1, as shown in Figure 2.4(a). A vertical unit load is applied
at the middle node on the lower side. The bottom left node is fixed entirely, while the
bottom right node is fixed in the vertical direction, i.e., with a roller support. The
optimal design to this problem closely resembles half a wheel. Similar solutions can
be found in, e.g., [217] and [800].

For each of the design situations, two or more problem instances with an increasing number of nodes and potential bars in the structure are solved. The sizes of the
problem instances are listed in Table 2.1.

The computational results with the primal simplex method are listed in Table 2.2.
The optimal designs for the largest problem instances in Table 2.1 are depicted in
Figures 2.2(b), 2.3(b), and 2.4(b). The cross-sections are drawn to be circular in all fig-
ures. Note that only bars with area > 0.01 are plotted.

Table 2.1. *Problem characteristics for the single-load truss topology design problems.*

Problem	Number of nodes	Number of bars	Number of dof	nnz(R)
Cantilever	41×11	101,475	880	374,000
Cantilever	81×21	1,445,850	3,360	5,544,000
Michell	31×21	211,575	1,260	787,500
Michell	61×41	3,126,250	4,920	12,054,000
Wheel	21×11	26,565	459	98,690
Wheel	41×21	370,230	1,719	1,426,780
Wheel	61×31	1,786,995	3,779	6,972,270

(a) *Design domain with dimensions* 3×2*, boundary conditions, and external load.*

(b) *Optimal design for the Michell problem for a ground structure with* 61×41 *nodes and* $3,126,250$ *potential bars. Note that bars with small areas are not plotted.*

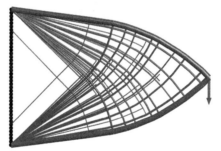

Figure 2.2. *Optimal design of a Michell beam.*

The numerical results presented in Tables 2.2 and 2.3 show that the sifting method is superior to the primal simplex method for all but one of the problem instances in Table 2.1. The sifting method could also solve some larger problem instances than reported in the tables for which the primal simplex method reported failure due to memory limitations. The solved problem instances are indeed large scale and are generally (much) larger than necessary in a practical design situation. The quest to solve larger problems is, for truss topology design, because the solutions can give insight into the optimal design for exact layout theory.

2.5 ▪ Limitations and Outlook

Truss topology optimization by LO can successfully be used for the fast optimal conceptual design of frame-like structures. The nonlinear minimum volume problem (2.5) can be generalized in several ways while maintaining the possibility of reformulating and solving it as an LO problem. Generalizations of the theoretical result for objective functions that model cost rather than volume and mass can be found in [14].

(a) *Design domain with dimensions 4 × 1, boundary conditions, and external load.*

(b) *Optimal design for the cantilever beam problem for a ground structure with 81 × 21 nodes and 1,445,850 potential bars.*

Figure 2.3. *Optimal design of a cantilever beam.*

(a) *Design domain with dimensions 2 × 1, boundary conditions, and external load.*

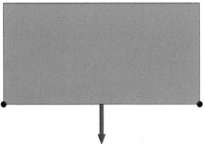

(b) *Optimal design for the wheel problem for a ground structure with 61 × 31 nodes and 1,786,995 potential bars.*

Figure 2.4. *Optimal design of (half) a wheel.*

Table 2.2. *Numerical results obtained by the primal simplex method in IBM ILOG CPLEX for the single-load truss topology design problems listed in Table 2.1.*

Problem	Number of nodes	Itn.	CPU (s)	Objective
Cantilever	41 × 11	71,823	10.6	22.5593
Cantilever	81 × 21	719,298	496.6	22.4726
Michell	31 × 21	199,292	48.9	9.0244
Michell	61 × 41	1,917,377	2335.6	9.0066
Wheel	21 × 11	38,906	3.9	3.1708
Wheel	41 × 21	700,623	230.2	3.1565
Wheel	61 × 31	3,097,263	2755.5	3.1499

Table 2.3. *Numerical results obtained by the sifting method in IBM ILOG CPLEX for the single-load truss topology design problems listed in Table 2.1.*

Problem	Number of nodes	Itn.	CPU (s)	Objective
Cantilever	41×11	19	9.3	22.5593
Cantilever	81×21	24	84.3	22.4726
Michell	31×21	20	19.1	9.0244
Michell	61×41	40	2841.6	9.0066
Wheel	21×11	11	1.3	3.1708
Wheel	41×21	13	16.9	3.1565
Wheel	61×31	15	122.2	3.1499

In the same article, the model is extended to allow the bars to have different stress limits and elastic moduli in tension and compression.

Truss topology design by LO unfortunately suffers from severe modeling limitations. In most practical design situations, more than one load must be accounted for. In the case of multiple loads, the optimal truss structure is generally neither statically determinate nor fully stressed. It is no longer possible to reformulate the optimal design problem as an LO problem. Even the seemingly innocent introduction of upper bounds on the design variables will in general render optimal structures that are neither statically determinate nor fully stressed. These limitations can be overcome by considering second-order cone or semidefinite optimization (SDO) formulations of truss topology design problems, as shown in, e.g., [207, 208] and [18]. Second-order cone formulations of truss topology design problems are the topic of Chapter 11 in this book.

Many of the structural requirements that arise in real-life design situations, such as bounds on structural eigenfrequencies, local and global buckling constraints, and contact in structural analysis, result in optimal design problems that cannot be reformulated as convex problems. Structural optimization problems are in general nonconvex and often also nondifferentiable. Generally, optimal structural design problems do not satisfy standard constraint qualifications. The latter is particularly true for stress constraints, which should be modeled as vanishing constraints; see, e.g., [1596, 1597, 1743], and [17]. Since optimal structural design problems are often very large scale, there are still many theoretical and numerical challenges in the field.

Chapter 3

Linear and Quadratic Optimization in Electrical Engineering

Jinchao Li, Yifan Sun, and Lieven Vandenberghe

3.1 ▪ Introduction

Linear and quadratic optimization (LO and QO) methods are widely used in all areas of electrical engineering. They are important as computational tools for optimal design in signal and image processing, circuit design, control, and communications and have been investigated for this purpose since the 1960s. This interest was motivated by the flexibility of constrained optimization formulations of design problems and the possibility of efficiently solving a range of problems for which no closed-form solutions exist. An early example is the use of LO and QO in digital filter design (see Section 3.2). As computing power has increased, a second type of application has emerged, in addition to optimal design of electrical engineering systems: the use of optimization as a data-processing algorithm. An example is image deblurring by regularized or penalized least-squares minimization (see Section 3.4.2). This evolution toward embedded optimization has reached a point where real-time embedded optimization in signal processing applications has become feasible [1298]. Recently, the importance of LO and QO in electrical engineering has increased dramatically with the development of ℓ_1-norm techniques in statistical signal processing, image processing, learning, and control.

Given the breadth of the field, it is impossible to give a comprehensive survey of electrical engineering applications of LO and QO. In this chapter, we therefore limit ourselves to a few representative applications. Perhaps the most important omission is control, which, however, will be covered elsewhere in this volume. Controller design methodologies based on linear and convex optimization have been developed in [330, 539], and online QO is essential to model-predictive control.

This chapter is restricted to LO and convex QO. Although most of the applications we discuss naturally lead to conic linear optimization (CLO; see Chapter 9), these extensions are only briefly mentioned.

27

3.2 ▪ Filter Design

A finite impulse response (FIR) filter of order n is a linear system described by a convolution operation

$$y(t) = \sum_{k=0}^{n-1} h_k u(t-k) \tag{3.1}$$

that transforms an input signal, $u : \mathbf{Z} \to \mathbb{R}$, into an output signal, $y : \mathbf{Z} \to \mathbb{R}$. The weights h_k are the filter coefficients. The frequency response of the filter is a periodic complex-valued function $H : \mathbb{R} \to \mathbf{C}$ defined as

$$H(\omega) = \sum_{k=0}^{n-1} h_k e^{-jk\omega}, \tag{3.2}$$

where $j = \sqrt{-1}$. The filter design problem is to select the filter coefficients or weights h_k according to an optimality criterion and subject to constraints on the frequency response (3.2) or the time-domain response (3.1). LO approaches to the design of FIR filters were first proposed around 1970 [392, 927, 1537, 1538, 1790]. Linear inequality constraints naturally arise when limiting the deviation between $H(\omega)$ and a desired response $\hat{H}(\omega)$ on a discretized frequency interval by imposing bounds on $|\Re H(\omega) - \Re \hat{H}(\omega)|$ and $|\Im H(\omega) - \Im \hat{H}(\omega)|$. Convex quadratic cost functions appear as the squared deviation $\int_{\omega \in \Omega} |H(\omega) - \hat{H}(\omega)|^2 \, d\omega$ between H and a desired frequency response $\hat{H}$ over a frequency band Ω, or the energy over a band (if we take $\hat{H}(\omega) = 0$), leading to QO formulations [21, 22]. On the other hand, upper bounds on the complex magnitude $|H(\omega)|$ cannot be expressed as linear inequalities in the filter coefficients, and lower bounds on $|H(\omega)|$, which are also very common, are in general not convex. We discuss two methods to circumvent these important restrictions: imposing symmetries on the filter coefficients and making a nonlinear change of variables.

3.2.1 ▪ Linear Phase Filter Design

If the coefficients h_k are symmetric or antisymmetric about their midpoint, the frequency response can be written as the product of a complex exponential and a real trigonometric polynomial. For example, if the filter order is odd ($n = 2N + 1$) and the filter coefficients are symmetric about their midpoint ($h(k) = h(2N - k)$ for $k = 0, \ldots, N-1$), we have

$$H(\omega) = e^{jN\omega} G(\omega), \qquad G(\omega) = h_n + 2 \sum_{k=1}^{N} h_{N-k} \cos k\omega.$$

Upper bounds on $|H(\omega)|$ now reduce to upper bounds on the absolute value of $|G(\omega)|$ and can be expressed as linear inequalities. As a simple example, a low-pass filter with passband $[0, \omega_p]$ and stopband $[\omega_s, \pi]$ can be designed by solving a semi-infinite QO problem

$$\begin{aligned}
\min \quad & \int_{\omega_s}^{\pi} G(\omega)^2 d\omega \\
\text{s.t.} \quad & 1/\delta_p \le G(\omega) \le \delta_p, \quad \omega \in [0, \omega_p], \\
& -\delta_s \le G(\omega) \le \delta_s, \quad \omega \in [\omega_s, \pi],
\end{aligned}$$

with variables h_k, where $\delta_p > 1$ is given [21]. The cost function is the stopband energy. The constraints include a limit on the peak deviation from one in the passband (using a logarithmic scale) and a limit on the peak value in the stopband [21]. (Note that without loss of generality we can assume that $G(\omega)$ is positive on $[0, \omega_p]$.) This semi-infinite problem can be approximated by a finite QO problem by discretizing the frequency axis. This basic problem can be extended in several ways, for example, by adding time-domain constraints, as noted in [1538].

3.2.2 ▪ Magnitude Filter Design

Magnitude or spectral mask constraints have the form $L(\omega) \leq |H(\omega)| \leq U(\omega)$, where U and L are frequency-dependent (for example, piecewise-constant) upper and lower bounds. Magnitude constraints are not convex in the filter coefficients but can be re-formulated as linear inequalities by making the nonlinear change of variables

$$r_k = \sum_{i=0}^{n-1-k} h_i h_{k+i}, \quad k = 0, \ldots, n-1.$$

The coefficients r_k are called the autocorrelation coefficients associated with the filter coefficients. It can be verified that

$$|H(\omega)|^2 = R(\omega) = r_0 + 2\sum_{k=1}^{n-1} r_k \cos k\omega.$$

Moreover, if $R(\omega) \geq 0$ for all ω, then the coefficients r_k are autocorrelation coefficients, and corresponding filter coefficients h_k can be computed by spectral factorization; see [52, 561, 636, 1030, 1910] for details. Therefore, if we change variables to the autocorrelation coefficients and discretize the frequency axis, the magnitude constraints turn into sets of linear inequalities.

This idea was applied to the design of infinite impulse response (IIR) or recursive filter design by Rabiner *et al.* in [1539]. The frequency response of an IIR filter takes the form

$$H(\omega) = \frac{N(\omega)}{D(\omega)} = \frac{\sum_{k=0}^{m} a_k e^{-jk\omega}}{\sum_{k=0}^{n} b_k e^{-jk\omega}}.$$

Magnitude constraints $L(\omega) \leq |H(\omega)| \leq U(\omega)$ are equivalent to $L(\omega)^2 Q(\omega) \leq P(\omega) \leq U(\omega)^2 Q(\omega)$, where $P(\omega) = |N(\omega)|^2$ and $Q(\omega) = |D(\omega)|^2$. These constraints are linear in the autocorrelation coefficients for the numerator and denominator polynomials. (Note that a normalization must be added to avoid a zero solution for P and Q. For example, one can fix the first autocorrelation coefficient of the denominator to be one.)

The spectral factorization method has also been used in the design of matched transmit and receive filters [1622, 1623], filter banks [636, 1099, 1360], and wavelet bases [1957].

3.2.3 ▪ Extensions

Early papers on filter design via LO often mention the flexibility of this approach as one of its principal advantages [1537, 1721]. This flexibility has been greatly enhanced by the more recent development of semidefinite and second-order cone optimization (SOCO; see Chapter 9), which can handle a wider class of constraints and objectives and avoid some of the approximations needed for LO formulation (such as, for example, the discretization of the frequency axis in constraints on nonnegative trigonometric polynomials); see [52, 561, 794, 1141, 1203, 1582, 1857, 1910] for examples. Surveys of CLO in signal processing are available in [560, 636, 1240].

3.3 ▪ Pattern Classification

Large-scale QO is used extensively in the training of support vector machine classifiers. The basic techniques have their roots in the study of linear and nonlinear pattern classification in the 1960s [520, 852, 1262, 1263, 1592] and became very prominent in machine learning research during the 1990s. In this section, we present some of the fundamental QO formulations. The reader is referred to the surveys in [359, 919, 1681, 1706, 1829] for a discussion of the statistical theory and specialized algorithms.

Let x_i, $i = 1, \dots, m$, be a set of feature vectors in $\mathbb{R}^n$ labeled with binary labels $y_i \in \{-1, 1\}$ and available as a training set for a binary classification algorithm. The simplest type of classifier evaluates a decision function of the form $f(x) = \text{sign}(w^T x + b)$ to assign a feature vector x to one of the two classes. This divides the feature space into two open halfspaces separated by the hyperplane $w^T x + b = 0$. The elements in the training set are classified correctly if $y_i(w^T x_i + b) > 0$ for $i = 1, \dots, m$. Note that the left-hand sides of these inequalities are homogeneous in the variables w, b, and therefore any feasible w, b can be scaled to satisfy $y_i(w^T x_i + b) \geq \alpha$, $i = 1, \dots, m$, where α is an arbitrary positive number. It follows that, without loss of generality, we can replace the strict inequalities with nonstrict inequalities $y_i(w^T x_i + b) \geq 1$. A linear classifier can therefore be computed by solving a set of linear inequalities in the parameters w, b.

If the classes are linearly separable, one can seek to increase the robustness of the classifier by maximizing the distance between the parallel hyperplanes $w^T x + b = \pm 1$. The distance is equal to $2/\|w\|_2$, where $\|w\|_2$ is the Euclidean norm, so the maximum margin-separating hyperplane of the two classes in the training set can be computed by solving the QO problem

$$
\begin{aligned}
\min \quad & (1/2) w^T w \\
\text{s.t.} \quad & y_i(w^T x_i + b) \geq 1, \quad i = 1, \dots, m.
\end{aligned}
$$

If the classes in the training set are not linearly separable, one can relax the strict separation requirement and replace it with a penalty function that penalizes the slack $1 - y_i(w^T x_i + b)$ in the violated inequalities. The most common choice is the piecewise-linear penalty used in the formulation

$$
\min \quad \frac{1}{2} w^T w + \gamma \sum_{i=1}^{m} \max\{0, 1 - y_i(w^T x_i + b)\},
$$

where γ is a positive parameter [221, 515]. This optimization problem is equivalent to the QO problem

$$
\begin{aligned}
\min \quad & (1/2) w^T w + \gamma \sum_{i=1}^{m} \xi_i \\
\text{s.t.} \quad & y_i(w^T x_i + b) \geq 1 - \xi_i, \quad i = 1, \dots, m, \\
& \xi_i \geq 0, \quad i = 1, \dots, m,
\end{aligned}
\tag{3.3}
$$

with variables $w \in \mathbb{R}^n$, $b \in \mathbb{R}$, $\xi \in \mathbb{R}^m$. It is usually solved via its dual, which is given by

$$\begin{array}{ll} \max & -(1/2)z^T Q z + y^T z \\ \text{s.t.} & 1^T z = 0, \\ & 0 \le \mathbf{diag}(y)z \le \gamma 1, \end{array} \tag{3.4}$$

with variable $z \in \mathbb{R}^m$, where Q is the $m \times m$ matrix with elements $Q_{ij} = x_i^T x_j$. The number of variables in the dual problem is equal to the size of the training set.

The dual formulation is particularly important when generalizing the QO approach to nonlinear classifiers, i.e., classifiers that divide the feature space into two regions separated by a nonlinear surface. In this extension, one replaces the decision function by a function $f(x) = \mathrm{sign}(w^T \phi(x) + b)$, where ϕ is a vector of nonlinear basis functions. The QO problems (3.3) and (3.4) still apply if one replaces the inner products $w^T x_i$ in (3.3) by $w^T \phi(x_i)$ and defines the Hessian matrix Q in (3.4) as $Q_{ij} = \phi(x_i)^T \phi(x_j)$. For many types of basis functions, inner products $\varkappa(x,y) = \phi(x)^T \phi(y)$ can be evaluated efficiently without explicitly forming $\phi(x)$ [326]. This is known as the kernel trick. When ϕ is very high dimensional (or infinite dimensional), the dual formulation (3.4), in combination with the kernel trick, is therefore more interesting for computational purposes than the primal.

3.4 ▪ 1-Norm Optimization

The last two decades have seen spectacular growth in optimization techniques that apply the ℓ_1-norm (sum of absolute values) in penalty functions and regularization terms to signal and image processing, statistics, identification, and control [162, 396, 446, 663, 920, 1599, 1769]. Many of these ℓ_1-norm optimization problems can be expressed as LO or QO problems. In most cases, the choice for the ℓ_1-norm over a sum-of-squares (SOS) penalty function is motivated by one of two important differences between the two functions. First, for large values, the absolute value function increases more slowly than the quadratic penalty. This feature is useful in robust estimation (Section 3.4.1). The second difference lies in the nondifferentiablity of the absolute value function at the origin. This property is exploited when the ℓ_1-norm is used in the cost function of an optimization problem to promote sparsity in the solution (Section 3.4.2). In some applications, only one of these two properties is desirable, and a combination of the quadratic and absolute value penalty functions is preferred. The best-known example is the Huber penalty function $f(x) = \sum_i \phi(x_i)$, where

$$\phi(u) = \begin{cases} u^2/2, & |u| \le c, \\ c|u| - c^2/2, & \text{otherwise,} \end{cases} \tag{3.5}$$

and c is a positive constant [967]. The Huber penalty behaves like the ℓ_1-norm for large values of x but is smooth and quadratic at the origin.

3.4.1 ▪ Robust Approximation

In a linear approximation or fitting problem, we approximate a vector b by a linear combination Ax of the columns of a matrix A. A general approach is to assign a penalty to the ith residual $(Ax - b)_i = a_i^T x - b_i$ and to minimize the sum of the penalties:

$$\min \quad \sum_{i=1}^m \phi(a_i^T x - b_i). \tag{3.6}$$

Different choices of the penalty ϕ can lead to very different distributions of the residuals $r_i = a_i^T x - b_i$ [332]. In particular, the ℓ_1-norm penalty ($\phi(u) = |u|$) tends to produce solutions that are less influenced by outliers than the least-squares solution, which uses $\phi(u) = u^2$. This is easily understood by comparing the two penalty functions for large u. Although the linear approximation problem (3.6) with ℓ_1-norm penalty has no closed-form solution, it is readily converted to the LO problem

$$\begin{aligned} \min \quad & \sum_{i=1}^{m} y_i \\ \text{s.t.} \quad & y_i \le a_i^T x - b_i \le y_i, \quad i = 1, \ldots, m, \end{aligned}$$

with variables x, y.

A penalty function that combines the quadratic dependence for small residuals with the linear growth for large values is the Huber function (3.5). The approximation problem (3.6) with the Huber penalty can be expressed as a QO problem in several ways. For example, noting that ϕ can be expressed as $\phi(u) = \inf_v(c|v| + (v-u)^2/2)$, one can show the equivalence with the QO problem

$$\begin{aligned} \min \quad & \sum_{i=1}^{m} \left(c y_i + \frac{1}{2} (v_i - a_i^T x + b_i)^2 \right) \\ \text{s.t.} \quad & -y \le v \le y, \end{aligned}$$

with variables v, y, and x.

A related technique is total variation regularization in image deblurring [905, 1599, 1856]. In total variation deblurring, we solve an optimization problem of the form

$$\min \quad \|A(X) - B\|_2^2 + \gamma \sum_{i,j=1}^{N} \|D(X)_{ij}\|, \qquad (3.7)$$

where the variable $X \in \mathbb{R}^{N \times N}$ represents the reconstructed image, $B \in \mathbb{R}^{N \times N}$ is an observed blurry and noisy image, A is a linear blurring operator, and $D(X)_{ij} = (X_{i+1,j} - X_{ij}, X_{i,j+1} - X_{ij})$ is a 2-vector of horizontal and vertical differences at pixel i, j (with appropriate boundary conditions to interpret $i, j = N + 1$). The norms in (3.7) are either Euclidean norms (in which case the penalty term is called the isotropic total variation) or ℓ_1-norms (anisotropic total variation). The anisotropic total variation deblurring problem can be cast as a QO problem. The role of the total variation penalty in (3.7) is to smooth the image by penalizing the derivatives $D(X)$. It grows more slowly for large values of $D(X)$ than the quadratic penalty $\sum_{ij} \|D(X)_{ij}\|_2^2$ and therefore tends to better preserve sharp edges in the reconstructed image.

3.4.2 ▪ Sparse Optimization

Sparse optimization problems arise naturally in many areas of signal processing and statistics. A fundamental example is the problem of finding a sparse solution of an underdetermined set of linear equations $Ax = b$. In basis pursuit, the equation $Ax = b$ represents the decomposition of a signal b as a linear combination of vectors in an overcomplete dictionary given by the columns of A [446]. In compressed sensing, the equations represent an underdetermined set of measurements of a sparse signal x [394, 397, 613]. Finding the sparsest solution x is very difficult in general, and a popular heuristic consists of solving the least ℓ_1-norm problem

$$\begin{aligned} \min \quad & \|x\|_1 \\ \text{s.t.} \quad & Ax = b. \end{aligned}$$

This problem is equivalent to an LO problem, and several efficient large-scale methods have been developed for its solution. As a variation, one can allow approximate solutions of the linear equations and minimize a weighted sum of the error $\|Ax - b\|_2$ in the linear equation and the ℓ_1-norm of x:

$$\min \quad \frac{1}{2}\|Ax - b\|_2^2 + \gamma\|x\|_1,$$

where $\gamma > 0$ is a scalar regularization parameter. This is equivalent to a QO problem.

The use of the ℓ_1-norm as a heuristic for sparse signal estimation has a long history in statistics and signal processing. Early papers in which this technique is mentioned include [479, 1625]. More recently, an extensive mathematical theory has been developed to explain when and why this approach works [394–397, 613–615, 1785]. It is impossible to give an adequate survey of this very rapidly developing area in a short chapter. Instead we refer the reader to several survey papers and books on this topic [162, 348, 398, 663, 664, 920, 1586, 1981].

A classic example of ℓ_1-norm optimization in statistics is the lasso method for feature selection in least-squares problems [1769]. The method is based on an ℓ_1-norm constrained least-squares problem of the form

$$\begin{aligned} \min \quad & \|Ax - b\|_2^2 \\ \text{s.t.} \quad & \|x\|_1 \leq \gamma. \end{aligned}$$

This problem is equivalent to a QO problem.

Regularization with ℓ_1-norm penalties on the derivatives of signals, with the purpose of obtaining piecewise-constant or piecewise-linear solutions, has been used for trend filtering in time series [1066] and segmentation of time-varying system models [1418].

The success of ℓ_1-norm techniques for sparse optimization has motivated the development of convex regularization techniques for other types of structures. An example is the group lasso, which uses the sum of ℓ_2-norms as a penalty to promote sparsity of groups of variables [1941]. These extensions cannot be handled by LO or QO, but they can still be solved efficiently by convex optimization.

3.4.3 ▪ Circuit Placement

The ℓ_1-norm arises naturally in wirelength minimization problems for integrated circuit placement. A simplified version of the problem is as follows. We consider a hypergraph with m vertices, representing modules or cells in an integrated circuit, and hyperlinks or nets representing groups of two or more connected cells. With each vertex we associate a two-dimensional coordinate vector (x_i, y_i). Some vertices have fixed positions; the positions of the other cells must be chosen to minimize a measure of the total size of the nets. For example, one can minimize the sum of the half-perimeters of the nets, defined as

$$\sum_{\text{nets } N_k} \left(\max_{i \in N_k} x_i - \min_{j \in N_k} x_j + \max_{i \in N_k} y_i - \min_{j \in N_k} y_j \right).$$

This minimization problem is equivalent to an LO problem. If each net has two elements (the hypergraph is a graph), the objective reduces to a sum of absolute values

$$\sum_{\text{edges } \{i,j\}} \left(|x_i - x_j| + |y_i - y_j| \right).$$

Methods based on this ℓ_1-norm objective are known as linear placement methods, [1694], as opposed to quadratic placement methods, which minimize a sum of squared Euclidean distances. A smooth approximation of the wirelength is often used to make the problem amenable to smooth unconstrained nonlinear optimization (NLO) algorithms [1018, 1179]. A difficult challenge in circuit placement is the removal of overlap between the cells. The nonoverlap constraints are highly nonconvex, and many different mechanisms exist to enforce them. In practice, the linear or quadratic placement problem is therefore only a subproblem in a much more complex algorithm; see [495, 1375] for surveys of circuit placement.

3.5 ▪ Information and Coding Theory

In this section we give a few interesting examples in which LO tools are used to derive bounds in information theory. Often, these are achieved using duality or relaxations.

3.5.1 ▪ Linear Optimization Bounds for Spherical Codes

A spherical (n, N, θ)-code is a set of N points in $\mathbb{R}^n$ that lie on a unit sphere and whose pairwise angles are at least equal to a given value θ. The maximum value of N, for given θ and n, determines attainable coding rates in certain communication channels and can be upper bounded using LO. The key reformulation follows from a condition by Schoenberg [1645] that states that a function $f : [-1, 1] \to \mathbb{R}$ satisfies $f(x_i^T x_j) \geq 0$ for all $x_i, x_j \in \mathbb{R}^n$ on the unit sphere if and only if it can be written as a nonnegative combination of *Gegenbauer polynomials* with parameter n:

$$f(t) = c_0 + c_1 G_1^n(t) + \cdots + c_k G_k^n(t), \quad c_i \geq 0.$$

Delsarte, Goethals, and Seidel [577] show that if $f(t)$ is written in this form, and additionally $f(t) \leq 0$ for all $-1 \leq t \leq \cos(\theta)$ and $f(0) > 0$, then the number of codewords is upper bounded as $N \leq f(1)/f(0)$. (See [672] for a simple proof and a discussion of the properties of these polynomials.) An upper bound of N is then obtained by solving the following semi-infinite LO problem:

$$
\begin{aligned}
\min \quad & f(1) \\
\text{s.t.} \quad & f(t) \leq 0 \quad \text{for } -1 \leq t \leq \cos\theta, \\
& c_1, \ldots, c_k \geq 0,
\end{aligned}
\tag{3.8}
$$

where $f(t) = 1 + \sum_k c_k G_k^n(t)$.

In practice, the semi-infinite LO problem (3.8) can be approximated and converted to a finite LO problem by discretizing the interval $[-1, \cos\theta]$. (It can also be reformulated exactly as a semidefinite optimization (SDO) problem.) Analytical bounds can be computed by making particular choices of basis polynomials [333]. This LO technique has also been applied to other problems; see [487, 577, 1367].

3.5.2 ▪ Information Theory Inequalities

Information-theoretic properties can often be written as simple linear equalities and inequalities; for example, the Shannon entropy $H(X)$ of a set S must satisfy the following properties:

- $H(\emptyset) = 0$,
- $A \subset B \Rightarrow H(A) \leq H(B)$, and
- $H(A) + H(B) \geq H(A \cap B) + H(A \cup B)$.

These and a few additional linear inequalities are sufficient to prove a wide range of important information-theoretic results [1935]. In [1934, 1935, 1962, 1963], this idea is used as the basis of a simple calculus for verifying unconstrained and constrained linear information identities and inequalities and to develop an automated *information-theoretic inequality prover* (ITIP) [978]. The information-theoretic properties are expressed as joint entropies of subsets of n random variables, which can be represented as points in $(2^n - 1)$-dimensional space. The prover solves the optimization problem

$$
\begin{aligned}
\min \quad & b^T v \\
\text{s.t.} \quad & Dv \geq 0,
\end{aligned} \tag{3.9}
$$

where v is a $(2^n - 1)$-dimensional variable; $Dv \geq 0$ enforces the information-theoretic axioms; and $b^T v$ represents a new proposed property, satisfied if $b^T v \geq 0$ for all feasible v. Since $v = 0$ is always feasible, the optimal value is upper bounded by zero. If the optimal value equals zero, then the new property is provably true. If it is less than zero, then its validity is not provable using this solver. From duality, it can be shown that this is true only if b is a nonnegative combination of the rows in D.

One of the main theoretical results to come out of this research is that no finite set of inequalities can fully characterize the space of valid information-theoretic statements. In [1962], the authors give information-theoretic inequalities that cannot be proved from the proposed axioms. Other "non-Shannon-type inequalities" were later found by [620, 1256, 1299]. In [1300], it is shown that a full characterization requires an infinite number of inequalities.

3.5.3 ▪ Decoding by LO

The maximum likelihood decoding problem of a binary linear code over a discrete memoryless channel can be written as the LO problem

$$
\begin{aligned}
\min \quad & c^T x \\
\text{s.t.} \quad & x \in \text{conv}(\mathscr{C}),
\end{aligned} \tag{3.10}
$$

where the vector variable x is the decoded codeword, $\text{conv}(\mathscr{C})$ is the convex hull of the codebook (set of valid codewords) $\mathscr{C}$, and $c = (c_1, \ldots, c_n)$ contains the negative log-likelihood ratio of a received signal $\hat{z}$ given a transmitted signal z. For example, in additive white Gaussian noise (AWGN) channels with noise variance σ^2, the coefficients are $c_i = (2/\sigma^2)\hat{z}_i$. As with all maximum likelihood decoding algorithms, solving this problem is intractable, since characterizing the set $\text{conv}(\mathscr{C})$ usually requires an exponential number of linear inequalities. However, approximate solutions can be computed by using tractable relaxations of the feasibility region, as in the *LP decoding* techniques of Feldman et al. [698–700]; see also [364]. In this approach, the relaxed polytope $\mathscr{Q}$ is characterized by a polynomial number of inequalities and has the key property that all the vertices of $\text{conv}(\mathscr{C})$ are vertices of $\mathscr{Q}$, and all vertices of $\mathscr{Q}$ that are not vertices of $\text{conv}(\mathscr{C})$ are nonintegral. The resulting LO decoder is a useful analysis tool for iterative decoding and provides useful bounds for maximum likelihood decoding.

3.6 ▪ Communication Networks

Many problems in communication networks can be described as network optimization problems, which are LO problems or include linear constraints. In particular, convex duality plays an important role in the study of communication network algorithms [455, 1052, 1228, 1229]. One example is the *network utility maximization* (NUM) problem defined as

$$
\begin{aligned}
\max \quad & \sum_k U_k(x_k) \\
\text{s.t.} \quad & Rx \le c, \\
& x \ge 0.
\end{aligned}
\tag{3.11}
$$

The variable x is a vector of flow rates through the network. The matrix R is the routing matrix, with $R_{ij} = 1$ if flow j passes through link i and $R_{ij} = 0$ otherwise, and c is a vector of link capacities. The cost function U_k is a concave utility function for flow k. If the utility functions are piecewise linear, this is equivalent to an LO problem. Important applications are found in optimal routing and congestion control [985, 1051]; see [1672] for a survey.

Chapter 4

Linear and Quadratic Optimization in Chemical Engineering

Christopher L. E. Swartz

4.1 ▪ Introduction

Optimization applications in chemical engineering fall into two main categories — design and operations. Detailed design calculations typically involve nonlinear models, whereas linear models have been widely used at certain levels of operations optimization. In this chapter, we will focus on two such application areas — production planning and process control. Production planning involves relatively long time horizons spanning weeks to months, and it also often involves a large spatial footprint, such as a collection of process units, a process plant, or an entire enterprise. As such, a coarse representation of the process behavior is generally adequate, with assumptions of linearity justified. Process control, on the other hand, involves short time scales of seconds to minutes. The majority of continuous chemical process operations operate around desired steady-state conditions, possibly with transitions between steady-state points. Such systems typically do not deviate significantly from the desired point under normal operating conditions, and as a result they are adequately described by linear dynamic models. The second class of applications considered here is model predictive control (MPC), which for systems described by linear dynamics can be formulated as a quadratic optimization (QO) problem that is repeatedly solved as time evolves. Finally, industrial implementations of MPC typically involve a higher-level, economics-driven linear optimization (LO) or QO layer that provides set-points to the model predictive controller. In what follows, we will describe these production planning and control problems more fully, present representative mathematical optimization formulations, and discuss a number of variations and extensions.

4.2 ▪ Production Planning

The focus of this application is on refinery production planning; it represents one of the earliest industrial applications of LO (see [291, 1744]), with LO-based methods remaining prevalent even today.

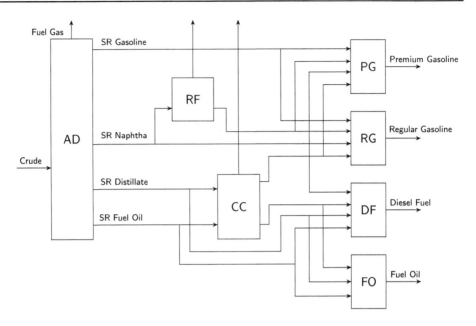

Figure 4.1. *Basic refinery configuration.*

4.2.1 ▪ Process Description and Basic LO Formulation

In oil refining, crude oil is separated into several fractions based primarily on the boiling point ranges of the constituent molecular components. Some of the streams are further processed via chemical reaction. The streams are then blended together to yield products with specified properties.

Figure 4.1, taken from [1483], illustrates the configuration of a basic refinery process. Crude oil is fed to an atmospheric distillation tower (AD), where it is separated into four liquid streams and fuel gas. The straight-run (SR) gasoline is blended to form premium or regular gasoline. Part of the SR naphtha is sent directly to blending, with the remainder fed to the reformer (RF), whose liquid product is blended into the gasoline products. The SR distillate and fuel oil streams are split between catalytic cracker (CC) feed and blending to diesel fuel and fuel oil. The CC liquid product streams are sent as feed streams to the blending operations.

Presented below is a basic formulation for a refinery LO planning model. It comprises material balance equations, including yield relationships; product quality specification constraints; supply, demand, and capacity constraints; and an objective function. The process is modeled as an interconnection of process blocks, with multiple input and output streams, and splitters, which split a single stream into multiple streams. We further categorize the process blocks into two different types: processing units (such as the AD tower, RF and CC) and blending units. Sets associated with the processing units, blending units, and splitters are defined respectively as *PR*, *BL*, and *SP*. Let K denote the set of all units and I the set of all streams. A volumetric basis is typically used for material flows in refinery planning models [870].

A material balance for the process and blending units gives

$$\sum_{i \in I_k} a_{ijk} F_i = F_j, \quad j \in O_k, \quad k \in PR \cup BL, \tag{4.1}$$

where F_i is the flow rate of stream i, k represents the process unit, I_k is the set of inlet streams to unit k, and O_k is the set of outlet streams from unit k. a_{ijk} is the fraction of inlet stream i that reports to outlet stream j in unit k (yield coefficient).

A node at which a stream splits into several streams (splitter) can be modeled as

$$F_j = \sum_{i \in O_k} F_i, \quad k \in SP, \quad j \in I_k, \tag{4.2}$$

Process streams are characterized by a set of properties. For final products, a number of properties (such as octane number, viscosity) have to be within specified limits. For the blending units, the stream properties are calculated as the average of the inlet stream properties, weighted by the corresponding flow rate:

$$p_{lj} = \frac{\sum_{i \in I_k} p_{li} a_{ijk} F_i}{\sum_{i \in I_k} a_{ijk} F_i}, \quad l \in L, \quad j \in O_k, \quad k \in BL, \tag{4.3}$$

where p_{lj} represents the value of property l of stream j and L is the set of stream properties. Note in the expression above that $a_{ijk} F_i$ is the amount of inlet stream i that reports to outlet stream j of unit k. Placing lower and upper bounds on the properties gives

$$p_{lj}^{\min} \le p_{lj} \le p_{lj}^{\max}, \quad j \in O_k, k \in BL. \tag{4.4}$$

Substituting (4.3) into (4.4) and multiplying by the denominator gives the linear constraints

$$p_{lj}^{\min} \sum_{i \in I_k} a_{ijk} F_i \le \sum_{i \in I_k} p_{li} a_{ijk} F_i, \tag{4.5}$$

$$\sum_{i \in I_k} p_{li} a_{ijk} F_i \le p_{lj}^{\max} \sum_{i \in I_k} a_{ijk} F_i. \tag{4.6}$$

We remark that the properties of the blending unit feed streams are assumed known. If they are calculated in a similar way to (4.3) in upstream units, this would introduce nonlinearity into the formulation — a manifestation of the so-called pooling problem [136]. To avoid this situation, average stream properties can be used based on operating or simulated data. A further point is that not all properties are amenable to the blending formula described in (4.3), which led to the development of linear blending indices [291]. A linear blending index is a transformation of a property such that the volume-averaged blending rule holds for the transformed property, from which the physical property of the mixture can then be computed. Examples of blending indices for pour point, smoke point, and viscosity can be found in [870].

Maximum and minimum bounds on the flow rates can be expressed as

$$F_i^{\min} \le F_i \le F_i^{\max}, \quad i \in I. \tag{4.7}$$

Capacity constraints can be placed on process units, based on the total feed, as follows:

$$\text{cpty}_k^{\min} \le \sum_{i \in I_k} F_i \le \text{cpty}_k^{\max}, \quad k \in PR \cup BL, \tag{4.8}$$

where $\text{cpty}_k^{\min}$ and $\text{cpty}_k^{\max}$ are the minimum and maximum capacities of unit k.

A typical objective is to maximize profit, giving the objective function

$$\phi = \sum_{i \in I_{prod}} c_i F_i - \sum_{i \in I_{fs}} c_i F_i - \sum_{k \in PR \cup BL} \left(c_k^{op} \sum_{i \in I_k} F_i \right), \qquad (4.9)$$

where c_i is the revenue or cost associated with stream i, c_k^{op} is the operating cost associated with unit k based on the amount of material processed, I_{prod} is the set of product streams, and I_{fs} is the set of feedstocks. A similar objective function is used in [42].

Equations (4.1)–(4.2) and (4.5)–(4.9) constitute an LO problem that can be solved to determine the optimal feed, intermediate and product stream rates that maximize profit subject to raw material availability, stream and process unit capacity, and product specification constraints. Similar formulations are given in [42, 1483, 1502].

4.2.2 ▪ Planning Model Extensions

Several extensions to the refinery model described above have been proposed, many of which are in industrial use. Process units often operate under different sets of conditions, or modes. One way to incorporate this is to index yields and streams by a mode, m. In this way, the amount of material processed via each mode can be determined. Dutta et al. [646] utilize this construct, with the resulting problem remaining an LO problem.

Other model enhancements often introduce nonlinearity into the formulation. Baker and Lasdon [136] identify four sources of nonlinearity in refining planning problems:

- *Pooling.* This refers to the construct of volume-averaged properties, which can induce nonlinearity in downstream unit calculations, as discussed earlier.
- *Nonlinear blending.* The blending relationship described in (4.3) may not hold for certain properties, and linear blending indices may not be available or adequate, requiring the use of nonlinear relationships. Such nonlinear behavior may be induced by the presence of other components, an example being lead, which significantly influences the relationship between the product octane number and that of the component streams.
- *Process yields.* It is often desirable to include process operating conditions as planning decision variables, which typically results in nonlinear yield relationships.
- *Cost functions.* The relationship between operating variables and cost is often nonlinear.

Successive LO has been widely adopted for handling nonlinearities in refinery planning formulations since its development in the 1960s [136, 291]. It essentially iterates between linearization of the nonlinear terms and solution of the linearized LO problem to generate an updated solution point. Baker and Lasdon [136] describe a variant in which a constraint violation penalty term is added to the objective function and adaptive bounds on the magnitudes of the nonlinear variable iterates are included.

Bodington and Baker [291] review the history of mathematical programming in the petroleum industry from the earliest applications in the 1940s and 1950s to the state of the art at the time of publication (1990). Recent contributions to refinery production planning include [42, 870, 1186].

Other key extensions to refinery planning problems are the inclusion of uncertainty and integration of planning and scheduling. A two-stage stochastic formulation for including the effects of uncertainty is presented in [1502], where the linearity of the formulation is retained. Inclusion of scheduling decisions in refinery planning, such as those related to the piping of material between the unit operations and blending tanks, typically includes binary variables to describe associated logic. A mixed-integer linear optimization (MILO) framework for planning and scheduling refinery operations is presented in [1490]. A recent review of scheduling and planning in refinery operations is given by Shah et al. [1668].

In summary, LO-based solution of refinery planning problems continues to be widely used, 60 years after the initial development and adoption of LO refinery planning formulations. This is also reflected in commercial refinery planning software, which is still largely LO based [42]. LO approaches have remained popular for several reasons. The 1970s saw model management becoming a key issue, with software and consulting firms starting to fill the breach [291]. The high maintenance and convergence difficulties with nonlinear optimization (NLO) models prompted companies to revert from in-house NLO models back to LO-based systems [42]. In addition to the advantages of speed and robustness, LO problems permit marginal values to be easily obtained, permitting useful sensitivity analysis [1010]. However, new developments in large-scale NLO solution techniques; continuously increasing computing power; demands for greater accuracy of planning models; and trends toward integration of planning, scheduling, and control, may lead to a shift from the domination of LO methods in this application area.

4.3 ▪ Model Predictive Control

MPC represents a class of control algorithms in which a dynamic model of the process is used directly in the calculation of the control law. Since its publication in the 1970s as two independently developed algorithms — model predictive heuristic control (or model algorithmic control, as it was later called) [1565] and dynamic matrix control (DMC) [534] — it has gained significant attention in both academia and industry.

In a survey on control practice in the Japanese process industry conducted approximately a decade after MPC was first published, MPC was reported to be the most widely used advanced control technique, with a satisfaction level equivalent to that of the basic and ubiquitous proportional-integral-derivative (PID) type control [1921]. Morari and Lee [1349] state that "there is probably not a single major oil company in the world where DMC (or a functionally equivalent product with a different trade name) is not employed in most new installations or revamps." A comprehensive review of industrial MPC algorithms and applications is given in [1528].

The operation of the algorithm is illustrated in Figure 4.2, which shows an input profile (u) and the corresponding response (y). Given the state of the system at time instant k, a dynamic model can be used to predict the output trajectory corresponding to a proposed future input profile. The MPC algorithm computes a future input trajectory as the solution of an optimization problem in which the deviation between the predicted and desired response, and the severity of the control action, suitably weighted, are minimized. Although a trajectory of future input moves is calculated, only those corresponding to the first time interval are implemented. At the end of the sample time, process measurements are compared to the predicted values, and the difference is used to adjust the model prediction in a new controller calculation, thus providing a feedback mechanism. This process is repeated at every sample time, with the prediction and control move horizons shifted, giving rise to a receding horizon implementation framework.

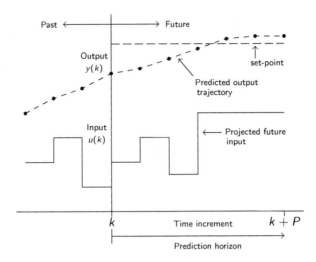

Figure 4.2. *Illustration of the MPC concept.*

Most MPC implementations utilize a linear dynamic model. The original DMC formulation was posed as a QO problem with only equality constraints relating the model inputs and outputs, which admits an analytical solution. A subsequent extension included inequality constraints on the inputs, input changes, and outputs [785].

Key advantages of MPC over prior approaches are its ability to directly account for interaction effects through a multivariable process model and its direct consideration of constraints in the control move calculation. Since the optimal economic operating point typically lies at the intersection of process constraints, the constraint-handling capability of MPC has been a key driver for its widespread adoption in the petrochemical process industry.

MPC has been the subject of hundreds of publications and several books. The purpose of this section is to provide an overview of MPC, highlighting in particular the optimization formulation, and also its position within the overall plant automation hierarchy.

4.3.1 ▪ Basic MPC Formulation

Different types of dynamic models have been used in MPC formulations. Here, we utilize a discrete-time state-space model of the form

$$\mathbf{x}(k+1) = A\mathbf{x}(k) + B\mathbf{u}(k), \tag{4.10}$$

$$\mathbf{y}(k) = C\mathbf{x}(k), \tag{4.11}$$

where $\mathbf{x} \in \mathbb{R}^{n_x}$ is a state vector, $\mathbf{u} \in \mathbb{R}^{n_u}$ is an input vector, and $\mathbf{y} \in \mathbb{R}^{n_y}$ is a vector of measured outputs. $A \in \mathbb{R}^{n_x \times n_x}$, $B \in \mathbb{R}^{n_x \times n_u}$, and $C \in \mathbb{R}^{n_y \times n_x}$ are linear(ized) discrete-time state-space matrices.

An optimization problem to be solved at each sample time, k, can be formulated as

$$\min_{\mathbf{u}(k),\dots,\mathbf{u}(k+M-1)} \phi = \sum_{i=1}^{P} \|\mathbf{y}(k+i) - \mathbf{y}_{\text{set}}(k+i)\|_Q + \sum_{i=0}^{M-1} \|\Delta\mathbf{u}(k+i)\|_R \tag{4.12}$$

s.t.

$$x(k+i) = Ax(k+i-1) + Bu(k+i-1), \quad i = 1, \ldots, M, \tag{4.13}$$

$$x(k+i) = Ax(k+i-1) + Bu(k+M-1), \quad i = M+1, \ldots, P, \tag{4.14}$$

$$y(k+i) = Cx(k+i) + d(k+i), \quad i = 1, \ldots, P, \tag{4.15}$$

$$\Delta u(k) = u(k) - u(k-1), \tag{4.16}$$

$$\Delta u(k+i) = u(k+i) - u(k+i-1), \quad i = 1, \ldots, M-1, \tag{4.17}$$

$$u_{\min} \le u(k+i) \le u_{\max}, \quad i = 0, \ldots, M-1, \tag{4.18}$$

$$\Delta u_{\min} \le \Delta u(k+i) \le \Delta u_{\max}, \quad i = 0, \ldots, M-1, \tag{4.19}$$

$$y_{\min} \le y(k+i) \le y_{\max}, \quad i = 1, \ldots, P. \tag{4.20}$$

The norms in (4.12) are defined as

$$\|x\|_Q = x^T Q x.$$

$u(k-1)$ is known, and $x(k)$ is assumed to be available, for example, by estimation using the process model applied to the previous time interval:

$$x(k) = Ax(k-1) + Bu(k-1). \tag{4.21}$$

$d(k+i)$ is a disturbance estimate that accounts for plant-model mismatch. In the original DMC formulation [534, 1249], it is taken as constant over the prediction horizon, P, and computed as the difference between the measured output at the current time step and the predicted output based on the model and prior information. Using the state-space model formulation, this becomes

$$d(k) = y_m(k) - Cx(k),$$
$$d(k+i) = d(k), \quad i = 1, \ldots, P.$$

y_{set} is the desired value of the output, or set-point. The optimization problem seeks to minimize the deviation of the predicted outputs from their set-points, without excessive control input movement, as reflected by the two terms in the objective function. Weighting matrices Q and R in the objective function are typically chosen to be diagonal; they can be used to weight specific output and input variables differently relative to each other. The magnitude of R relative to Q has a particularly strong influence on system stability and robustness. Large values of R place a heavier penalty on the input movement, resulting in a more sluggish response, which also has a stabilizing effect. These matrices are the key controller-tuning parameters. The other main user-specified parameters are the output prediction and control move horizons, P and M. The control move horizon is defined as that over which input changes are permitted and is typically much shorter than P.

We remark that more precise notation, such as $\hat{y}(k+i|k), \hat{x}(k+i|k)$, which signifies the predicted values of y and x at time step $k+i$ based on information available at time step k, is helpful for the application of more sophisticated state estimation schemes and to distinguish variables internal to the optimization from one controller execution to the next [135, 1249]. However, for clarity of exposition, the simpler notation above is used with no loss in rigor in the controller optimization problem formulation.

4.3.2 ▪ Variants and Extensions

A number of variants of the above basic MPC formulation have been proposed, many of which have been adopted in industrial implementations. Here, we consider two categories; for a more complete discussion on this topic, we refer the reader to the review article by Qin and Badgwell [1528].

4.3.2.1 ▪ Constraint Softening

Hard constraints on the outputs can result in an infeasible optimization problem, depending on the system dynamics and the disturbance estimate, and can also lead to closed-loop instability [1945]. Thus, soft constraints are typically used for the outputs, whereby constraint violations are penalized via an additional term in the objective function. In this scheme, constraints (4.20) in the MPC optimization problem would be replaced by

$$\mathbf{y}_{\min} - \epsilon_i \le \mathbf{y}(k+i) \le \mathbf{y}_{\max} + \epsilon_i, \quad i = 1,\dots,P, \tag{4.22}$$

with the following term added to the objective function:

$$\sum_{i=1}^{P} \|\epsilon_i\|_S, \tag{4.23}$$

where S is a diagonal weighting matrix and the ϵ_i are decision variables.

4.3.2.2 ▪ Stability Enhancement

The presence of inequality constraints in MPC prohibits direct use of linear stability analysis tools. Stability guarantees for the standard MPC formulation proved difficult but have been achieved with modified formulations. In particular, Keerthi and Gilbert [1046] proved that under certain assumptions, closed-loop stability can be guaranteed with the inclusion of the terminal constraint $\mathbf{x}(k+P) = \mathbf{0}$ (where the variables have been shifted such that the origin is the desired steady-state point). However, this result hinges on feasibility of the MPC optimization problem with the added terminal constraint, which is not known a priori for an arbitrarily specified prediction horizon, P. A subsequent approach is based on the attractive stability properties of infinite-horizon linear quadratic control. The infinite prediction horizon is replaced by a finite horizon with a penalty term, $\mathbf{x}(k+P)^T \overline{P} \mathbf{x}(k+P)$, with $\overline{P}$ determined as the solution of a matrix Lyapunov equation [1368, 1554]. A comprehensive treatment of these types of stabilizing MPC formulations can be found in [1303].

   While significant theoretical advances have been made in this area, these MPC formulations have not made inroads of any significance in industrial process control applications to date.

4.3.3 ▪ Position within Plant Automation Structure

Chemical and petrochemical process plants often employ real-time optimization (RTO) to determine economically optimal set-points that are provided to the regulatory controllers [550, 1282]. RTO systems are largely based on nonlinear steady-state models and are executed at a relatively low frequency (hours). There is typically an additional decision-making layer between the RTO and MPC systems that is executed at a higher frequency than RTO, and often at the same frequency as the underlying MPC systems. The function of this layer is to provide feasible set-points to the MPC layer, based on the set-point targets determined by the RTO system, but accounting for higher-frequency disturbances that may have entered the system between RTO executions.

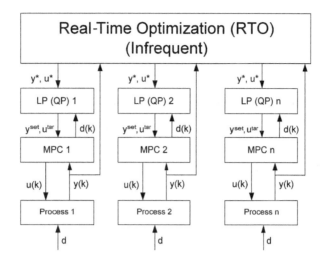

Figure 4.3. *Hierarchical decision-making structure in process operations automation. Reprinted with permission from Elsevier* [1401].

It typically utilizes a static version of the MPC model and is formulated as an LO or QO problem. This intermediate LO/QO layer is discussed in [1231, 1528, 1936] and referred to in [1231] as a coordination collar.

Figure 4.3, as given in [1936], illustrates the location of these decision-making systems within the overall process automation structure.

An LO formulation for the coordination collar takes the form

$$\min \quad c_y^T \bar{y} + c_u^T \bar{u} \tag{4.24}$$

$$\text{s.t.} \quad \bar{y} = G^{ss}\bar{u} + d, \tag{4.25}$$

$$u_{min} \leq \bar{u} \leq u_{max}, \tag{4.26}$$

$$y_{min} \leq \bar{y} \leq y_{max}, \tag{4.27}$$

where the overbar denotes the variable at its steady state and c_y and c_u are cost coefficients that could be determined from a linearization of the RTO problem at its solution. G^{ss} is the steady-state gain matrix, which can be determined from (4.10) and (4.11) at steady state:

$$\bar{x} = A\bar{x} + B\bar{u}, \tag{4.28}$$

$$\bar{y} = C\bar{x}. \tag{4.29}$$

Solving (4.28) for $\bar{x}$ and substituting into (4.29) gives

$$\bar{y} = C(I - A)^{-1}B\bar{u}, \tag{4.30}$$

from which we obtain $G^{ss} = C(I - A)^{-1}B$. The disturbance **d** is estimated as in MPC.

The optimal values of $\bar{y}$ and $\bar{u}$ are sent to the MPC system as set-points and input targets, respectively — the latter when there are fewer controlled outputs than manipulated inputs, resulting in excess degrees of freedom.

A QO formulation uses the same constraints as the LO formulation, but with a quadratic objective function. Minimization of the weighted squared deviation between the steady-state point and the RTO solution (y^*, u^*) gives

$$\min \quad (\bar{\mathbf{y}} - \mathbf{y}^*)^T \mathbf{C}_y (\bar{\mathbf{y}} - \mathbf{y}^*) + (\bar{\mathbf{u}} - \mathbf{u}^*)^T \mathbf{C}_u (\bar{\mathbf{u}} - \mathbf{u}^*). \tag{4.31}$$

Ying and Joseph [1936] include linear terms and suggest that the objective function coefficients be obtained from the RTO optimizer; this would be the case if a quadratic approximation of the RTO objective function were used. We remark that the output constraints can be softened, as discussed for MPC [1936].

Despite the wide use of the LO/QO layer in industrial applications, unsatisfactory performance has been reported, particularly with regard to high-frequency variation in the computed set-points (chattering) [1401]. It was shown that under certain conditions, noise could be amplified by the LO solution [1401].

Part II

Mixed-Integer Linear Optimization

Chapter 5

Integer (Linear) Optimization

Egon Balas and Matteo Fischetti

An amazing variety of activities and situations can be adequately modeled as linear optimization (LO) problems, also known as linear programming (LP) problems, or convex nonlinear optimization (NLO) problems. "Adequately" means that the degree of approximation to reality that such a representation involves is acceptable. However, as the world that we inhabit is neither linear nor convex, the most common obstacle to the acceptability of these models is their inability to represent nonconvexities or discontinuities of different sorts. This is where integer programming (IP) comes into play: it is a universal tool for modeling nonconvexities of various kinds.

To illustrate, suppose we have to put together an optimal production plan for a factory whose capacity limitations can be expressed through a system of linear constraints. This can be formulated as an LP problem, but when the latter gets solved, the resulting plan is likely to be unacceptable because it violates a number of other conditions, such as: threshold constraints (there is a minimum amount under which it is not worth producing an item); implications (if you are going to produce A, you need to also produce B); mutual exclusivity (you either do A or do B, but not both); and precedence constraints (action A must precede action B by a certain amount of time). It should be clear that conditions of this type are in no way exceptional. On the contrary, their presence is pervasive in most real-world situations.

An LO (NLO) problem whose variables are restricted to integer values is called a linear (nonlinear) integer (or discrete) optimization problem, or simply an *integer program* (IP) (linear unless otherwise stated). If only some of the variables are restricted to integer values, we have a *mixed-integer program* (MIP). Such a problem can be stated as $\min\{c^T x : Ax \geq b, x \geq 0, x_j$ integer for $j \in N_1 \subseteq N\}$, where A is a given $m \times n$ matrix, c and b are given vectors of conformable dimensions, $N := \{1,\ldots,n\}$, and x is a variable n-vector. The "pure" IP problem is the special case of MIP when $N_1 = N$.

IP or integer optimization as a field started in the mid-1950s. A number of excellent textbooks are available for its study [493, 1390, 1649, 1899].

5.1 ▪ Scope and Applicability

Applications of IP abound in all spheres of decision making. Some typical real-world problem areas where IP is particularly useful as a modeling tool are facility (plant, warehouse, hospital, fire station) location, scheduling (of personnel, production, other activities), routing (of trucks, tankers, aircraft), design of communication (road, pipeline, telephone, computer) networks; capital budgeting, project selection, and analysis of capital development alternatives. Various problems in science (physics: the Ising spin glass problem; genetics: the sequencing of DNA segments) and medicine (optimizing tumor radiation patterns) have been successfully modeled as IP problems. In engineering (electrical, chemical, civil, mechanical, biological, and medical), the sphere of applications is growing steadily; see below.

By far the most important special case of IP is the (pure or mixed) 0-1 *programming problem*, in which the integer-constrained variables are restricted to zero or one. This is so because a host of frequently occurring nonconvexities, such as the ones listed above, can be formulated via 0-1 variables. For instance, the threshold condition mentioned above (the amount, x, produced cannot be less than some quantity q), i.e., $x = 0$ or $x \geq q$, can be expressed through the inequalities $q\delta \leq x \leq Q\delta$, $\delta \in \{0, 1\}$, with the 0-1 variable δ enforcing the threshold: if $\delta = 0$, $x = 0$; if $\delta = 1$, $q \leq x \leq Q$, i.e., x will be between the threshold q and a large enough upper bound Q.

Next we present a few well-known pure and mixed-integer models.

The *fixed charge problem* asks for the minimization, subject to linear constraints, of a function of the form $\sum_i \gamma_i(x_i)$, with

$$\gamma_i(x_i) := \begin{cases} f_i + c_i x_i & \text{if } x_i > 0, \\ 0 & \text{if } x_i = 0. \end{cases}$$

As an example, consider the problem of installing a communications network whose arcs can have different capacities at different costs, with the objective of minimizing the total cost while satisfying certain overall capacity requirements. Whenever x_i is bounded by U_i and $f_i > 0$ for all i, such a problem can be restated as a (linear) MIP by setting for all i

$$\begin{aligned} \gamma_i(x_i) &= c_i x_i + f_i y_i, \\ x_i &\leq U_i y_i, \\ y_i &\in \{0, 1\}. \end{aligned}$$

Clearly, when $x_i > 0$, then y_i is forced to one, and when $x_i = 0$ the minimization of the objective function drives y_i to zero.

The *facility location problem* consists of choosing among m potential sites (and associated capacities) of facilities to serve n clients at a minimum total cost:

$$\begin{aligned} \min \quad & \sum_{i=1}^{m}\sum_{j=1}^{n} c_{ij}x_{ij} + \sum_{i=1}^{m} f_i y_i \\ \text{s.t.} \quad & \sum_{i=1}^{m} x_{ij} = d_j, \quad j = 1, \ldots, n, \\ & \sum_{j=1}^{n} x_{ij} \leq a_i y_i, \quad i = 1, \ldots, m, \\ & x_{ij} \geq 0, \quad i = 1, \ldots, m; \, j = 1, \ldots, n, \\ & y_i \in \{0, 1\}, \quad i = 1, \ldots, m. \end{aligned}$$

Here d_j is the demand of client j, a_i is the capacity of a potential facility to be located at site i, c_{ij} is the per-unit cost of shipments from facility i to client j, and f_i is the fixed cost of opening a facility of capacity a_i at location i. In any feasible solution, the indices i such that $y_i = 1$ designate the chosen locations for the facilities to be opened, while the continuous variables x_{ij} represent the quantity shipped from facility i to client j.

Variants of this problem include the *warehouse location problem* (which considers cheap bulk shipments from plants to warehouses and expensive packaged shipments to retailers) and various *emergency facility location* problems (where one chooses locations to minimize the maximum distance traveled by any user of a facility, rather than the sum of travel costs).

The *knapsack problem* is an IP problem with a single constraint:

$$\max\{c^T x : a^T x \leq b, x \geq 0 \text{ integer}\},$$

where c and a are positive n-vectors, while b is a positive scalar. It represents the choices one faces when trying to fill a container of capacity b with objects j of volume a_j and value c_j to maximize the value of the objects included. When the variables are restricted to zero or one, we have the 0-1 *knapsack problem*.

A variety of situations can be fruitfully modeled as *set covering problems*: given a set M and a family of weighted subsets $S_1, \ldots, S_n$ of M, find a minimum weight collection C of subsets whose union is M. If A is a 0-1 matrix whose rows correspond to the elements of M and whose columns are the incidence vectors of the subsets $S_1, \ldots, S_n$, and c is the n-vector of subset weights, the problem can be stated as

$$\begin{aligned} \min \quad & c^T x \\ \text{s.t.} \quad & Ax \geq 1, \\ & x \in \{0,1\}^n, \end{aligned}$$

where the right-hand side of the inequality is the m-vector of ones. This model and its close relative, the *set-partitioning problem* (in which $\geq$ is replaced by $=$) has been (and is still) widely used in airline, bus, and train crew scheduling (each row represents a leg of a trip that has to be covered; each column stands for a potential duty period of a crew). Another application is in medical diagnostics: each column represents a diagnostic test, and each row stands for a pair of diseases, with a one in column j if the pair's reactions to the tests are different, and a zero if they are the same, the goal being to select a minimum cost battery of tests guaranteed not to yield identical outcomes for any two diseases.

Unlike in the case of LO, integer optimization problems can typically be formulated in several different ways, and choosing a good formulation is of paramount importance. What should be the guiding criterion when comparing different formulations? In LO, the criterion is typically the number of variables and constraints; the fewer of each, the better. In integer optimization the situation is very different. Because of the central role of branch-and-bound methods in solving IPs (see below), whose efficiency depends on the strength of the bounds provided by the linear relaxation of the instance being solved, the leading criterion in choosing between different formulations is the tightness of the linear relaxation provided by each formulation. Thus, in choosing between two formulations of a minimization problem, one of which has twice as many variables and/or constraints as the other, but has an optimum over its linear relaxation higher than the other, the formulation with the higher LP optimum is usually preferable.

5.2 ▪ Combinatorial Optimization

A host of interesting combinatorial problems can be formulated as 0-1 programming problems defined on graphs, undirected or directed, vertex weighted or edge weighted. The joint study of these problems by mathematical programmers and computer scientists, starting from around 1960, has led to the development of the burgeoning field called *combinatorial optimization* [509]. Some typical problems of this field are edge matching (finding a maximum weight collection of pairwise nonadjacent edges) and edge covering (finding a minimum weight collection of edges that together cover every vertex), vertex packing (finding a maximum weight independent set, i.e., a collection of pairwise nonadjacent vertices) and vertex covering (finding a minimum weight collection of vertices that together cover every edge), maximum clique (finding a maximum cardinality complete subgraph) and minimum vertex coloring (partitioning the vertices into a minimum number of independent sets, i.e., coloring the vertices with a minimum number of colors such that all adjacent pairs differ in color), and the traveling salesman problem (TSP; finding a cycle of minimum total edge weight that meets every vertex).

We will briefly discuss two of the above problems, which in a sense span the universe of combinatorial optimization. At one end of the spectrum, the *matching problem* on a graph $G = (V, E)$ can be stated as

$$\max\{x(E) : x(\delta(v)) \leq 1, v \in V, x \geq 0 \text{ integer}\},$$

where $x(F) = \sum_{e \in F} x_e$ for $F \subset E$ and $\delta(v)$ is the set of edges incident with v. Here $x_e = 1$ if e is in the matching; $x_e = 0$ otherwise. The weighted version of the problem asks for maximizing $\sum_{e \in E} w_e x_e$, where w_e is the weight of edge e. This problem has the nice property that the integrality condition can be omitted if the above nonnegativity and degree constraints are supplemented with the inequalities

$$x(\gamma(S)) \leq \frac{|S| - 1}{2} \quad \text{for all } S \subseteq V, |S| \text{ odd},$$

where $\gamma(S)$ is the set of edges with both ends in S. In other words, the above "odd set inequalities," along with the nonnegativity and degree constraints, fully describe the convex hull of incidence vectors of matchings. The discovery of this remarkable phenomenon in the mid-1960s (due to Edmonds [653]) started a massive pursuit of facets of the convex hull of other combinatorial polyhedra and can be viewed as the inaugural step in the development of the field called *polyhedral combinatorics*. It can also be viewed as a precursor of the theory of NP-completeness in the early to mid-1970s [508, 1040], which brought fundamental clarity to the issue of computational complexity in that it classified problems into polynomially solvable ones (i.e., solvable in time polynomial in the problem size, like LP and the matching problem mentioned above) and those called NP-hard, for which no polynomially bounded procedure is known (and is never likely to be found, since if one of them were polynomially solvable, then all of them would be).

At the other end of the spectrum, one of the hardest and most thoroughly investigated combinatorial optimization problems is the TSP already mentioned, in which a salesman is looking for the cheapest tour of n cities, given the cost of travel between all pairs of cities. This problem, NP-hard according to the above classification, is the prototype model for situations dealing with the optimal sequencing of objects (e.g., items to be processed on a machine in the presence of sequence-dependent setup costs). The standard formulation on a complete directed graph with node set N and arc costs c_{ij} is

$$\min \quad \sum_{i \in N} \sum_{j \in N \setminus \{i\}} c_{ij} x_{ij}$$

$$\text{s.t.} \quad \sum_{j \in N \setminus \{i\}} x_{ij} = 1, \quad i \in N,$$

$$\sum_{i \in N \setminus \{j\}} x_{ij} = 1, \quad j \in N,$$

$$\sum_{i \in S} \sum_{j \in S \setminus \{i\}} x_{ij} \leq |S| - 1, \ S \subset N, 2 \leq |S| \leq n - 1,$$

$$x_{ij} \in \{0, 1\}, \quad i, j \in N, \quad i \neq j.$$

The first two sets of equations define an assignment problem whose solutions span unions of directed cycles. The third set, consisting of inequalities called subtour elimination constraints, exclude all cycles with fewer than $n = |N|$ arcs. The number of solutions — tours — is factorial in the number of nodes. The TSP has become a testbed for the development of, and experimentation with, various approaches to combinatorial optimization [91].

5.3 ▪ Solution Methods

Unlike LP problems, IP problems, including 0-1 programming and most combinatorial optimization problems, are NP-hard. The difficulty in solving IPs lies in the nonconvexity of the feasible set, which makes it impossible to establish global optimality from local conditions. The two principal approaches to solving IPs try to circumvent this difficulty in two different ways.

The first approach, which until the late 1980s was the standard way of solving IPs, is *enumerative* (branch-and-bound, implicit enumeration). It partitions the feasible set into successively smaller subsets tied together as nodes of a branch-and-bound tree, calculates bounds on the objective function value over each subset, and uses these bounds to discard certain subsets (nodes of the tree) from further consideration. The lower bounds (in a minimization problem) typically come from solving the linear relaxation corresponding to the given node (i.e., the LP obtained by removing the integrality condition), and the upper bounds come from integer solutions found at some of the nodes. The procedure ends when each subset either has produced a feasible solution or is shown to contain no better solution than the one in hand. The efficiency of the procedure depends crucially on the strength of the bounds. An early prototype of this approach is due to Land and Doig [1124]; another is [141].

The second approach, known as the *cutting-plane method*, is a convexification procedure: it approximates the convex hull of the set of feasible integer points by a sequence of inequalities that cut off (hence the term "cutting planes") part of the LP feasible set without removing any feasible integer point. The first finitely convergent procedure of this type, which uses modular arithmetic applied to the rows of the optimal simplex tableau to derive valid cutting planes for pure IPs, is due to Gomory [823], who later extended it to the mixed-integer case. Chvátal [474] has shown that the procedure can be viewed as one of *integer rounding*, in which positive multiples of $Ax \geq b$, $x \geq 0$ are added up and the coefficients of the resulting inequality are rounded up to the nearest integer. The resulting inequalities form the elementary closure of $Ax \geq b$,

$x \geq 0$. The procedure can then be applied to the elementary closure, and so on. The number of times the procedure needs to be iterated to obtain the convex hull of integer points within a polyhedron is called the Chvátal rank of the given polyhedron. No bound is known on the Chvátal rank of an arbitrary polyhedron. By contrast, the matching polyhedron has Chvátal rank one, since the so-called odd set inequalities can be obtained from the degree inequalities by integer rounding. The Gomory–Chvátal procedure has been extended to MIP and has been enhanced by the use of subadditive functions and group theory [823, 824].

A different approach, originating in the early 1970s, uses geometric concepts and convex analysis. Given a basic fractional solution $\bar{x}$ to the LP relaxation of a MIP, the inequalities that are tight at $\bar{x}$ form a cone, $C(\bar{x})$ (see Figure 5.1). If S is any convex set whose interior contains $\bar{x}$ but no feasible integer points, then the hyperplane through the intersection points of the n extreme rays of C with boundary S defines an inequality that cuts off $\bar{x}$ but no feasible integer point. Such an inequality is an intersection cut [142]. If the integer-constrained variables are basic in the optimal solution, with one of them, say x_k, fractional, and $S := \{x : \lfloor \bar{x}_k \rfloor \leq x_k \leq \lceil \bar{x}_k \rceil\}$ (where $\lfloor \ \rfloor$ and $\lceil \ \rceil$ mean rounding down and rounding up), then the intersection cut from the convex set S is the Gomory cut from the row of the simplex tableau associated with x_k.

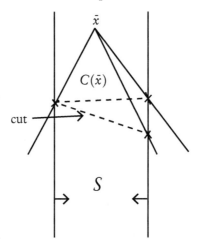

Figure 5.1. *The inequalities that are tight at $\bar{x}$ form a cone, $C(\bar{x})$.*

An intersection cut from a convex polyhedron like S can also be viewed as a cut from the disjunction that results from reversing the inequalities defining S, i.e., from the condition $x_k \leq \lfloor \bar{x}_k \rfloor$ or $x_k \geq \lceil \bar{x}_k \rceil$, which has to be satisfied by any feasible integer point. Viewed this way, the condition can be strengthened by amending each term of the disjunction with all inequalities valid for both, which gives rise to a condition of the form

$$\begin{pmatrix} Ax & \geq & b, x \geq 0 \\ x_k & \leq & \lfloor \bar{x}_k \rfloor \end{pmatrix} \vee \begin{pmatrix} Ax & \geq & b, x \geq 0 \\ x_k & \geq & \lceil \bar{x}_k \rceil \end{pmatrix}.$$

Such a disjunction is a union of two polyhedra, which has led to the study of disjunctive programming [143, 144], or LP with logical conditions (conjunctions, disjunctions, and implications involving inequalities). In this approach, which uses the tools of convex analysis, like polarity and projection, 0-1 programming (pure or mixed) is viewed as optimization over the (nonconvex) union of (convex) polyhedra, i.e., a set of the form

$\cup_{i \in Q} P_i$, where $P_i = \{x : A^i x \geq b^i\}$, $i \in Q$. There is a compact formulation of the convex hull $P_D := \operatorname{conv} \cup_{i \in Q} P_i$ in a higher-dimensional space whose projection onto the original space yields all the valid cutting planes. From a classical theorem of the alternatives known as the Farkas lemma, it follows that $\alpha^T x \geq \beta$ is a valid inequality for $\cup_{i \in Q} P_i$ if and only if $\alpha^T \geq u_i^T A^i$ and $\beta \leq u_i^T b^i$ for some $u_i \geq 0$, $i \in Q$. A central result of this approach is that an important class of disjunctive programs, called *facial*, which includes pure and mixed 0-1 programs, are *sequentially convexifiable*. For a 0-1 program (pure or mixed) with n 0-1 variables and an LP relaxation P_0, this means that one can impose the 0-1 condition on x_1 and generate the convex hull P_1 of $P_{00} \cup P_{01}$, where $P_{00} := \{x \in P_0 : x_1 = 0\}$, $P_{01} := \{x \in P_0 : x_1 = 1\}$. Then one can impose the 0-1 condition on x_2 and generate the convex hull P_2 of $P_{10} \cup P_{11}$, where $P_{10} := \{x \in P_1 : x_2 = 0\}$, $P_{11} := \{x \in P_1 : x_2 = 1\}$, etc. At the end of n steps, the convex hull P_n of $P_{n-1,0} \cup P_{n-1,1}$ turns out to be the convex hull of $\{x \in P_0 : x_j \in \{0,1\}, j = 1, \ldots, n\}$. This property does not hold for arbitrary IPs and is thus a main distinguishing feature of 0-1 programs. If one defines the disjunctive rank of a polyhedron as the number of times the above procedure has to be iterated to generate all of its facets, it follows that an arbitrary 0-1 programming polyhedron has disjunctive rank n.

5.4 ▪ Revolution in the State of the Art

IP was recognized early on as a miraculous tool that could be used to model almost any problem; however, the result of such an exercise was usually a model that provided insights but could not actually be solved. Of the two approaches described above, branch-and-bound was relatively straightforward to implement, and for about three decades it was the only one practically used to solve some real-world problems. However, the size of these problems — apart from some special structures — was limited to 30–50 variables. As to the cutting-plane approach, attempts at its computational implementation were unsuccessful. As one would generate a cut, add it to the constraints, and reoptimize the resulting LP and then repeat this cycle, numerical issues would arise and prevent the solution of even small problems. So for three decades cutting planes were the object of intense theoretical research, but the only practical tools for solving IPs were the branch-and-bound codes, commercial versions of which appeared by the 1970s. This situation persisted until the late 1980s; then, roughly in the 15-year period between 1990 and 2005, a true revolution occurred in the state of the art of integer optimization. A simple measure of this is the following. Over the years, a collection of IP instances was developed under the name of MIPLIB [1076]. Whereas earlier most MIPLIB instances remained unsolved, after the turn of the century most of them, including many new ones, got solved. Whereas earlier general IPs with more than 50 variables could not be solved, nowadays instances with thousands of variables are routinely solved (see below for some details).

How did this revolution occur? Several factors played a role. Among them were faster computers, more efficient LP codes, various heuristics, and more sophisticated branching rules, but the crucial factor was the embedding of cutting planes into a branch-and-bound context. Next we briefly outline how this came about.

The key impetus came from a revival of the disjunctive programming approach outlined above and its implementation into an efficient computational tool called *lift-and-project* [146]. The name conveys the idea of a higher-dimensional representation of the convex hull (lifting), which is then projected back to generate cutting planes. In the meantime, Lovász and Schrijver [1226] (see also Sherali and Adams [1686]) developed

a closely related procedure that derives higher-dimensional representations of a 0-1
programming polyhedron by multiplying the constraint set of P_0 by the inequalities
$x_j \geq 0$ and $1 - x_j \geq 0$, $j \in N$, and then linearizing the resulting quadratic forms
and projecting them back into the original space. As in the disjunctive programming
approach, n iterations of this procedure yield the convex hull of the 0-1 programming
polyhedron. However, the quadratic forms obtained during the procedure can also be
used to derive *positive semidefiniteness constraints* that are stronger than the inequalities
obtained by linearization.

Semidefiniteness constraints aside, a streamlined version of the Lovász–Schrijver
procedure, in which P_0 is multiplied at every iteration by just one pair of inequalities, $x_j \geq 0$, $1 - x_j \geq 0$, rather than by all pairs, was shown in [146] to be equivalent
to the disjunctive programming procedure for 0-1 polyhedra, in that the linearized
version of the quadratic constraints obtained by multiplication is exactly the same
as the higher-dimensional representation of the convex hull used in disjunctive programming [143, 144]. The paper [146] also showed how to use a cut-generating linear
program (CGLP) to obtain lift-and-project (or disjunctive) cuts that are deepest in a
well-defined sense. Most important, these cuts can be generated in a subspace, i.e., using only a subset of the variables, and then lifted to the full space. This has opened
the door to combining the enumerative and convexifying approaches into a *branch-
and-cut procedure*, which generates cutting planes as long as they "work" but branches
whenever the cut generating "stalls." This was done earlier for the special case of the
TSP [1435], but its extension to general MIPs was made possible by showing that cuts
generated at a node of the search tree can be lifted to be valid at any node. The outcome
was a robust procedure, considerably more efficient than either a branch-and-bound
or a cutting-plane algorithm by itself [147].

Experimentation with this procedure showed that the loss of accuracy and related
numerical problems that used to accompany the recursive cut-generating procedure of
reoptimizing the LP problem after the addition of every cut can be substantially miti-
gated by generating *rounds* of cuts, one from each row whose integer-constrained basic
variable is fractional, before reoptimizing. Finally, as a consequence of these experiments, it turned out that Gomory cuts, when embedded in a branch-and-bound procedure and generated in rounds, could also be computationally tractable [148]. Since
these are easier to generate, they were the first ones to make it into the commercial
codes in the 1990s. The stronger but more expensive lift-and-project cuts made it into
the commercial codes after a way was found [149] to solve the CGLP indirectly, without actually setting it up, by pivoting in the LP simplex tableau. Other cuts used include mixed-integer rounding, knapsack cover, flow cover, and clique cuts.

To give the reader an idea of the extent of the revolution in the computational
capability of IP solvers, it is worth citing a few numbers. According to the developers
of the MIP solver CPLEX [286], a comparison of the performance of CPLEX 1.2
(1991) and CPLEX 11.0 (2007) on 1,852 real-world instances of linear MIPs yielded the
following results. The speedup in computing time during those 16 years was 29,530-
fold. In other words, the 2007 version of CPLEX ran on average almost 30,000 times
faster than the 1991 version, which corresponds roughly to a doubling of the speed
every year. As a result, whereas in 1991 only 15% of the instances attempted could
be solved, in 2007 69% of the instances were solved. After 2007, the improvements
in computing power continued at a somewhat slower, but still steady, rate: the 2013
version (5.6) of the commercial code GUROBI is 21 times as fast as the 2009 version
(1.0) [285].

5.5 ▪ Heuristics

While solving integer optimization problems exactly was a hard nut to crack, and it took more than 30 years to reach a degree of success that made large-scale commercial applications possible, various heuristic procedures were developed over the years with the goal of finding more or less acceptable approximate solutions. By more or less acceptable we mean practical, differing from the optimum usually by a few percentage points, as opposed to the theoretical acceptability offered by a guaranteed bound that is a polynomial function of the optimum. Not surprisingly, the type of heuristics that enjoyed the earliest successes were those devoted to solving special structures. For instance, the notoriously hard TSP, a model for sequence optimization problems of various types, was quite successfully attacked by a type of local improvement heuristic based on interchanging some in-tour arcs with out-of-tour arcs (2-, 3-, k-interchange), which culminated in the *variable-depth interchange* heuristic of Lin and Kernighan [1198], whose latest version amended by Helsgaun [928] finds good approximate solutions to instances with hundreds of thousands of variables in a matter of seconds. Various combinations of primal and dual greedy heuristics, Lagrangian relaxation, and subgradient optimization solved large set covering and partitioning problems arising in airline and railway crew scheduling [145, 403]. A job-shop scheduling heuristic based on a disjunctive graph formulation and the idea of repeatedly identifying the bottleneck machine and tackling it locally, called the *shifting bottleneck procedure* [20], has become a basic staple of operations scheduling in various job-shop environments (see [1485, Chapter 5] or [1484, Chapter 7]).

By the 1990s, a number of metaheuristics had been proposed, like for instance *tabu search* [816], i.e., procedures based on some general heuristic principle, whose various incarnations are specific heuristics for certain problem classes. One such general principle is that of *neighborhood search*. Elements of such heuristics made their way into the leading mixed-integer solvers and were applied locally at various stages of the search procedure to speed up its convergence.

One of the procedures that proved to be successful in practice was the *feasibility pump* (FP), originally proposed for the 0-1 case [715] and then extended to general MIPs [238]. It is based on the observation that a feasible MIP solution is a point x of P that coincides with its rounding. Replacing "coincides with" with "is as close as possible to" leads to the following iterative scheme, to be described for the sake of simplicity, for pure 0-1 IPs. FP works with a pair of points (x^*, x'), with x^* in P and x' integer, that are iteratively updated with the aim of bringing them as close to each other as possible, i.e., minimizing

$$\Delta(x^*, x') := \sum_{j \in N} |x_j^* - x_j'| = \sum_{j \in N : x_j' = 0} x_j^* - \sum_{j \in N : x_j' = 1} (1 - x_j^*).$$

To be more specific, one starts with any x^* in P, e.g., an optimal LP solution, and initializes an integer x' as the rounding of x^*. At each FP iteration, called a pumping cycle, x' is fixed and one finds a point x^* in P that is as close to x' as possible by solving an auxiliary LP. If $\Delta(x^*, x') = 0$, x^* is integer and we are done; otherwise x' is replaced by the rounded x^* in an attempt to bring them closer, and the process is iterated. Several variants and extensions have been proposed and implemented [12, 295, 719].

A new generation of MIP heuristics emerged in the late 1990s. Their hallmark is the use of a "black-box" external MIP solver to explore a solution neighborhood defined by invalid linear constraints. The use of an exact MIP solver inside a MIP heuristic may appear naive at first glance, but it turns out to be effective in the cases where the added invalid constraints lead to a structural simplification of the MIP at hand and allow, e.g., for a more powerful instance of preprocessing and/or for extensive node pruning. The *local branching* (LB) scheme of [716] appears to be the first method to embed a MIP solver within a general MIP heuristic framework. Suppose a feasible *reference solution* $\tilde{x}$ of a 0-1 MIP is given, and one aims at finding an improved solution that is "not too far" from $\tilde{x}$. To this end, one can define the *k-OPT neighborhood* of $\tilde{x}$ as the set of the MIP solutions satisfying the invalid *LB constraint* $\Delta(x, \tilde{x}) \leq k$ for a small parameter k (typically, $k = 10$ or $k = 20$) and explore it by means of an external MIP solver, often heuristically, i.e., within a prefixed number of branch-and-bound nodes. The method is in the spirit of local search metaheuristics and in particular of *large neighborhood search* (see, e.g., [1680]), with the novelty that neighborhoods are obtained through invalid cuts to be added to the original MIP model. Diversification cuts can be defined in a similar way, thus leading to a flexible toolkit for the definition of metaheuristics for general MIPs.

The *relaxation-induced neighborhood search* (RINS) framework of [546] also uses the (heuristic) solution of a simplified MIP through an external MIP solver as a main ingredient, but it extends the idea by taking the solution of the LP relaxations into account. At specified nodes of the branch-and-bound tree, the current LP relaxation solution x^* and the incumbent $\tilde{x}$ are compared and all integer-constrained variables that agree in value are fixed and removed, and the solution of the resulting sub-MIP is attempted by invoking the MIP solver itself with a tight time/node limit.

5.6 ▪ The Impact of Contemporary MIP Technology

The revolution in the state of the art of MIP outlined above has brought this powerful modeling tool to bear on an increasing array of practical problems in industry, transportation, energy, finance, healthcare, and a host of other activities. The annual competition run by INFORMS for the prestigious Franz Edelman Award attracts companies from across the globe. The purpose of the award is to recognize and reward outstanding examples of real-world applications of operations research and management science techniques. Since the launching of the annual competition, cumulative benefits from the finalists' projects (of which there are six per year) have exceeded $210 billion. The most frequently used technique in these applications turns out to be MIP. According to Nemhauser [1388], 53% of all the finalists have used MIP techniques of one kind or another in their projects. Among the winners, the percentage of MIP users is even higher. All but two of the Edelman Prize awardees of the last decade use one form or another of MIP. To illustrate the use of these techniques, we briefly discuss two of the award-winning projects.

The 2008 winner was Netherlands Railways for the entry "The New Dutch Timetable: The OR Revolution" [1109]. By 2006, the volume of traffic on the Dutch passenger railway network had increased significantly; more and larger trains had been scheduled without changing the structure of the timetable, thus overloading the system and causing consumer nightmares. Operations researchers working with Netherlands Railways constructed an improved timetable. As a result, the percentage of trains arriving within three minutes of the scheduled time increased, commuter satisfaction improved, and the number of passengers grew. In 2007, this resulted in an additional annual profit of $60 million.

To construct a new timetable and its related resource schedules, a sequence of planning problems was defined and solved. As input, a railway line system was given. Then, the timetable was defined, including the detailed routings through the stations. Finally, rolling-stock and crew schedules were constructed. In each phase, MIP models were formulated and solved (often heuristically).

The model used for timetable generation describes the cyclic timetabling problem in terms of the periodic event-scheduling problem (PESP) constraints. This is a generic model for scheduling a set of periodic events, such as the event times in a cyclic timetable. All PESP constraints are expressed as differences of event times. For example, the running time of a train from one station to another is the difference between the arrival time and the departure time. Similarly, the headway time is the difference between the departure times of two consecutive trains on the same track. All time differences are computed modulo 60 to reflect the timetable cycle of one hour. The timetabling problem was solved through CADANS (a proprietary constraint programming software for finding a first feasible PESP solution) and STATIONS (a MIP-based solver to find detailed routings through the stations) systems.

The goal in scheduling rolling stock is to allocate an appropriate amount of the appropriate rolling-stock type to each train in the given timetable. A constraint on the number of train units available is imposed—during peak hours, most trains will simultaneously require more units. A further complexity is that demand varies substantially during the day and on a line. The goal is to find a balance between three conflicting objectives in rolling-stock scheduling: (1) service, (2) efficiency, and (3) robustness. In this context, service means offering as many passengers as possible a seat. Efficiency aims at minimizing the amount of rolling stock and the number of rolling-stock kilometers. Robustness is addressed by reducing the number of shunting movements and by having a line-based rolling-stock circulation. The resulting rolling-stock problem was solved through a proprietary ROSA system built on top of a commercial MIP solver (CPLEX).

Each train in the timetable requires a train driver and a number of conductors depending on the rolling-stock composition of the trains. Approximately 6,000 crew members (i.e., train drivers and conductors) operate from 29 crew bases throughout the Netherlands. Each crew member belongs to a specific crew base. A duty starts and ends in a crew base and describes the consecutive trips for a single crew member. For each day, a number of anonymous duties are first generated (crew scheduling phase) and then assigned to individual crew members on consecutive days (rostering phase).

The crew scheduling problem was solved by using TURNI, a commercial software based on the set covering (MIP) model. In such a model, there is a binary decision variable for each potential duty (one if the potential duty is selected and zero otherwise). The problem is then to select a subset of duties from a predetermined set of feasible ones such that it covers each trip by at least one duty, it satisfies all additional constraints at the crew-base level, and the total costs of the selected duties are as low as possible. Because the number of feasible potential duties is extremely large, a column generation procedure is used, where a pricing model generates the feasible potential duties on the fly whenever they are needed. A typical workday includes approximately 15,000 trips for drivers and 18,000 for conductors. The resulting number of duties in an optimal solution is approximately 1,000 for drivers and 1,300 for conductors. This leads to very difficult crew scheduling instances. Nevertheless, because of its highly sophisticated MIP algorithms, TURNI solves these cases in a few hours of computing time on a personal computer. Therefore, one can construct all crew schedules for all days of the week within just a few days.

In 2012, the Edelman prize was awarded to TNT Express for the entry "Supply Chain–Wide Optimization at TNT Express" [730]. TNT's Global Optimization (GO) program initiative led to the development of optimization solutions to assist the operating units of TNT Express and improve their package delivery in road and air networks, based mostly on MIP-driven software. Over a seven-year period, TNT Express used these optimization methods to save $207 million. In the framework of the GO program, three separate subprograms were developed and a portfolio of models, methodologies, and mostly MIP-based tools was designed.

Subprogram 1, named TNT Express Routing and Network Scheduling (TRANS), was concerned with the optimization of routes for the transportation of packages and vehicle tours. A transportation route defines the sequence of hubs, from the depot of origin to the destination depot, including scheduled times of arrival and departure at the hubs, that a package will visit. A tour describes the sequence of locations visited by a vehicle (and driver), including the times at which each location is visited. Because of the size of the networks that TNT Express operates, the problem was split into several subproblems, each supported by a specific module in TRANS: (i) a service capability analyzer determines the fastest feasible routes based on the prespecified movements in the network and is based on a multicommodity flow problem in a time-space network solved through fast heuristics; (ii) a routing module generates a set of routes (not only the fastest) and assigns the packages to the movements of these routes by using a branch-and-bound algorithm to generate routes that meet the service requirements; (iii) an optimal-path module determines the optimal paths for each package, given the current infrastructure of depots and hubs, and is based on a network-design MIP model.

Subprogram 2, named Tactical Planning in Pickup and Delivery (SHORTREC), was planned at the depot level and affected the first and last miles in the supply chain. Pickup and delivery account for more than 30% of operational costs, hence they were an important focus area of the GO program. At TNT Express, a round corresponds to a single vehicle starting at the depot, visiting customers in a certain sequence for collection or delivery of packages, and returning to the depot. Effectively organizing the whole process is challenging because millions of packages must be picked up and delivered each week. The optimization problem was to minimize the total pickup and delivery costs while meeting all service-level requirements. Constraints to be considered included vehicle capacity, service levels, driver regulations, and some softer restrictions to ensure repetitiveness in the rounds and workload balancing. Ad hoc optimization software was designed based on vehicle routing heuristics.

Subprogram 3 addressed the whole supply chain optimization (DELTA Supply Chain). Because the air network forms a crucial part of TNT's global service offering, a supply chain optimization project was started to reduce aircraft use and to preserve future growth capabilities without worsening service. The DELTA Supply Chain model enabled TNT to optimize the complete supply chain for a fixed depot and hub infrastructure under varying volumes and ways of working (e.g., cutoff times, road and air transport). The model was aimed at using road transport rather than air transport because the former generally results in lower costs and CO_2 emissions. For packages that can be shipped by road, the number of required movements was calculated based on the routings of the packages. For the packages that were unable to meet the service requirements via the road network, an air network was constructed by using a separate model to create a minimum cost air schedule between the airports in the network. The model starts by assigning depots to the airports in the air network according to some heuristic criteria. Based on these assignments, the model determines the packages to be transported from the airport to the air hub and vice versa. Next, a MIP is solved to determine the minimum cost air schedule. The model ensures that sufficient aircraft capacity is available to carry all the packages, and it balances the number of incoming and outgoing aircraft per aircraft type at each location. For airports, the MIP model includes the earliest permitted arrival or departure times, airport closing times, and the consideration that some airports do not permit multiple stops by TNT Express airplanes. With the road and air network complete, a binary IP model estimates the impact of the network movement arrival and departure times on the pickup and delivery cost. In a final step, the model calculates the total cost of the complete supply chain to support management decision making.

5.7 ▪ How to Use a MIP Code

Several powerful MIP codes are available, including the commercial solvers IBM ILOG CPLEX, Gurobi, and FICO Xpress, as well as the open-source solvers SCIP, COIN-OR CBC, and GLPK (just to mention some of them). Each of these solvers can be accessed through the Internet. Most solvers exploit multithreading to take advantage of all the available processing units (cores) of the CPU in use. In addition, a distributed version of the main MIP solvers is available; it distributes the computation over a set of independent computers in a cluster/network.

MIP codes can be used in command mode, meaning that a user can write his/her (integer) linear model into a text file according to suitable format (MPS or LP) and then read it in and solve by appropriate command. This approach is the easiest to apply as the solver is used as a black box, with limited control on its main parameters. In many practical cases, this is just what is needed to get a successful application.

A nonexhaustive list of the main MIP commands and parameters follows. First of all, one has the *read* command to load the MIP model to be solved from a text file, and the *write* command to write the optimal solution found onto the screen or a text file. The *run/optimize* command invokes the MIP solver on the loaded instance, and the control is returned to the command mode when some termination conditions are reached. The obvious termination condition is of course that the solver found a provably optimal solution. However, early termination can be enforced by changing some internal execution limits by using an appropriate *change parameter* command (to be executed before starting the optimization). The most common limits that a user can change include the overall computing time (say, in seconds), the number of branching nodes, the number of feasible solutions found (i.e., of incumbent updates), and the absolute/relative optimality gap (meaning that the execution terminates when the current-best integer solution is guaranteed to be sufficiently close to the true optimal one).

Modern MIP codes typically do not require an external fine-tuning of the parameters that determine the "aggressiveness" of the internal heuristics and of the cutting-plane generation procedures, in the sense that the default mode modifies these parameters automatically in an adaptive way. It should be noted, however, that the ultimate goal of the computation is assumed to be the proof of optimality of the best solution found. Hence, the default mode can be inadequate in the hard cases where the user would prefer to use the MIP solver to find good feasible solutions within short computing times. In this "heuristic mode," what matters is in fact the capability to quickly find and improve feasible solutions, rather than prove that they are (almost) optimal. To this end, one can consider increasing the aggressiveness of the internal heuristics by increasing the parameter controlling the *heuristic frequency* to invoke them every, say, 1 or 10 branching nodes. Moreover, some MIP solvers implement a final "clean-up"

postprocessing of the best solution, which is possibly improved by using ad hoc refining procedures. This final refining step is not applied by default, but it may be worth
trying with a short time limit. Deactivation of all cuts (through a suitable parameter
that controls the aggressiveness of each cutting-plane generation function) can sometimes be useful in the heuristic mode—note, however, that cuts can be beneficial for
heuristics too.

Recent MIP codes also exhibit a user's parameter to change the random seed internally used to generate random numbers to break ties, etc. As discussed in [718],
this parameter can randomly affect the computation and produce different branching
choices and hence very different (sometime better, sometimes worse) search paths. Although the random seed does not affect the optimality of the final solution, it can be
used to diversify the MIP solver behavior to hopefully produce better solutions within
shorter computing times. As a matter of fact, running the same MIP solver 10 times
(say) on the same input data with 10 different random seeds for (say) 5 minutes sometimes produces much better heuristic solutions than running the same MIP solver once
for 50 minutes. Hence, the random seed can be very useful in the heuristic mode, even
more so if runs with different seeds can be executed in parallel.

For parallel MIP codes, a main parameter is the type of parallel execution: *parallel
deterministic* (meaning that the run will be reproducible due to the presence of synchronization points that, however, can slow down computation), *parallel opportunistic*
(generally faster as no synchronization is required), and *distributed* (meaning that the
execution is distributed over a cluster of different computers).

A more advanced use of MIP technology consists of writing a program (in any high-
level programming language, such as C/C++, Java, Fortran, Python, MATLAB, etc.)
that internally generates the model and solves it by invoking appropriate functions
provided by the solver. In some cases, one can even be interested in customizing the
MIP solver by exploiting some problem-specific knowledge. To this end, modern MIP
solvers provide so-called callback functions to be invoked at the critical points of the
solution method. By default, the callbacks are not installed, meaning that they are
not invoked and the solver uses its default solution strategy. By installing his/her own
callbacks, an advanced user can therefore take control of and customize the solution
algorithm. We will next describe the most-used callback functions, which refer to a
generic MIP solver based on the branch-and-cut solution scheme.

The *informative* callback is invoked by the solver in several parts of the code to allow the user to retrieve (e.g., to take statistics or to print) information like the value of
the current-best solution (incumbent), computing time spent so far, number of branching nodes and cuts generated, etc.

The *lazy cut* callback is invoked whenever a feasible (integer) solution of the current
MIP model is going to update the incumbent: this is a sort of last checkpoint where a
user can discard a solution because it violates some conditions that are not explicitly
part of the model. This mechanism turns out to be very useful when the problem to be
solved involves an exceedingly large number of (complex) constraints, and one wants
not to include all of them explicitly in the initial model for the sake of easing its solution. If the solution passed to the lazy cut callback is not accepted for whatever reason,
its infeasibility must be certified by one or more violated cuts that are automatically
added on the fly to the current model. In this way, the MIP model is enriched during
the run, and only the "relevant constraints" for the instance at hand are discovered and
inserted explicitly in the model.

The *heuristic* callback allows a user to write a problem-specific heuristic, possibly
based on the current LP solution (which is made available on input to the callback

itself). The *user cut* callback is invoked at every node with a fractional LP solution and can generate one or more cuts to be added (on the fly) to the current model with the aim of excluding the current optimal LP solution. Additional callbacks are used to change the way the LP relaxation is solved at each node (*solve* callback), the choice of the branching variable (*branch* callback), or the choice of the next node to process (*node* callback), etc.

Chapter 6

Integer Optimization Techniques for Train Dispatching in Mass Transit and Main Line

*Leonardo Lamorgese, Carlo Mannino,
and Mauro Piacentini*

6.1 • Introduction

Trains moving in railway systems are often affected by delays or cancellations. This in turn may produce knock-on effects and propagate to other trains and other regions of the network. These undesired effects may be alleviated by suitably rerouting and rescheduling trains in real time. Train dispatching is thus a central task in managing railway systems because it allows recovery from undesirable deviations from the timetable and a better exploitation of railway resources. With few exceptions, dispatching is still almost entirely in the hands of human operators, despite the fact that it is a large and complex optimization problem that does not lend itself to manual solution. In this chapter, we describe how integer programming (IP) can be exploited to quickly find optimal solutions to large dispatching problems and describe real-life implementations of these ideas.

6.2 • Train Dispatching

Railway transportation represents a significant share of the overall transportation sector, accounting (in 2013) for 11% of freight and 9% of passenger transport (per kilometer) [679]. Perceived as the "greenest" transportation mode, railway is expected to play an increasing role in future communications. Indeed, railway operators over the world are renewing rolling stock, infrastructure, and signaling systems, as well as their organization. This in turn is demanding more research in managing and better exploiting the different components of railway systems. Optimization models have been studied and proposed for, e.g., personnel and crew scheduling, maintenance operations, train composition, timetable creation, and dispatching. IP, in particular, has often been exploited to model and solve the associated railway optimization problems. Such models are generally developed to provide middle- or long-term plans and are based on predictions of the number of passengers, personnel availability, train speed, etc. However, in daily operations, disturbances or disruptions may occur, generating deviations from the planned or expected behavior. As a consequence, actions must be taken to alleviate the effects of such deviations. A first, possible approach is to take uncertainty into consideration at the planning stage by utilizing *robust* or *stochastic* models [202]. Another, complementary, way is to carry out real-time replanning decisions when such deviations occur (for a very recent survey on rescheduling in railway operations, see [383]). *Train dispatching* may be viewed as a series of replanning actions carried out in real time when running trains deviate from the official timetable. In the following, we outline today's dispatching process in railway systems (see, e.g., [551]).

Railway systems can be classified as *mass transit* (typically urban systems such as subways) and *main line*. In either case, a railway is basically an intricate network of interconnected tracks. In the current *fixed block* signaling systems, tracks are segmented into *block sections*. Each block section can accommodate at most one train and is always preceded by a traffic light: incoming trains must stop at a red light. In railway networks we may identify regions (subnetworks) with specific features and roles. *Stations* are complex subnetworks where trains can perform various operations (*services*), such as embarking or alighting passengers and reversing direction. Sections in stations are classified as *stopping points* (where a train can actually stop, such as platforms) or *interlocking routes*, which are sequences of track segments connecting stopping points. The schematic representation of a small terminal station is given in Figure 6.1. In the scheme, stopping points are identified by a green label. In the figure, a possible train route is represented by the directed path going through stopping points En, P1, S2, P3, Ex. Each arc of the path corresponds to an interlocking route. So the train enters the station from stopping point En, runs the interlocking route from En to P1 (which is a platform) where it disembarks passengers, continues through the next interlocking route to stopping point S2, where it reverses direction, and then proceeds along the remaining part of the route to exit from exit point Ex.

Finally, we refer to tracks connecting distinct stations as *links*. Links are often *single track*, that is, trains must use the same physical connection in both directions. Other relevant subnetworks are *stops* (along links), where trains can embark and alight passengers; *junctions*, where a track splits into two, equipped with *switches* to allow trains to switch between alternative tracks; and so forth.

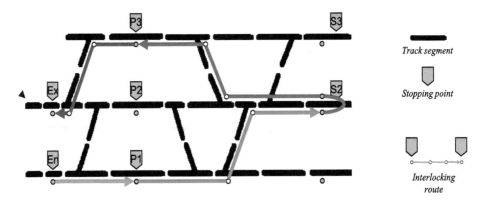

Figure 6.1. *A subway terminal station. Reproduced with permission. Copyright IN-FORMS,* http://www.informs.org [1266].

A train moving on the railway will run through a sequence of sections, called the *route*. The *official timetable* defines, for each train, when it should pass through specific sections of its route, as a platform in a station (i.e., arrival) or the exit point (departure). Timetables may also specify the entire route schedule. Generating good and robust timetables is a hard task, and an entire line of optimization research is devoted to it [384].

The task of controlling trains running through a railway network is carried out by experienced operators, *dispatchers*, with support from software tools and remote equipment called *train management systems* (TMSs). The railway network is typically subdivided into small control areas, each assigned to a dispatcher. Dispatchers are constantly updated about the status of the network, the position of trains, their speed, the status of signals and switches, etc. When trains are running according to the official timetable, dispatchers' role is limited to ensuring that signals and switches are set according to the plan; actually, in many modern route-setting systems, this task is performed by the TMS, while dispatchers only have supervising responsibilities. On the contrary, when one or more trains are running late (or some disruption occurs on the line), dispatchers are required to intervene and apply recovery or control actions. These include changing dwell times between trains, advising modified train speed, modifying train routes (e.g., by assigning a new platform in a station), changing train order at junctions, changing arrival or departure orders at stations, etc. To give an example, suppose train A is arriving at a station and, according to the timetable, will stop at platform 1. However, the platform is currently occupied by train B, which was supposed to leave the station before the arrival of A but is actually out of order and cannot release the platform. As a consequence, a dispatcher will establish a new platform for train A and possibly change the departure order of trains A and B. In a more complex setting, other trains may also be involved, forcing train A to wait at the signal before the station. Another typical situation occurs when stations 1 and 2 are connected by a single-track link, trains A and B run it in opposite directions with A running from 1 to 2, and A and B are supposed to meet in station 1. Now, assume B is running late. The dispatcher may decide to keep A waiting in station 1, or let B wait in station 2 until A reaches it. Or, if B is very late, the dispatcher may even decide to let B wait in a station preceding station 2 on its route. All such control actions are taken in the attempt to reduce the effects of deviations on the future. Based on a general assessment of the situation, on the relative (often simply perceived) importance of the trains involved, on a rough evaluation of the consequences of different recourse actions, on personal experience, or on rules of thumb and operational rules, the dispatcher is presented with the daunting task of making, in a few seconds, decisions that may impact the overall network for many hours to come. With some effort, it is possible to define the cost of a real-time train schedule as a function of its deviation from the official timetable. Such a cost function may take into account delays of trains in stations (weighted by train relevance), loss of connections between trains, headway between successive trains, etc. It follows that the decisions made by dispatchers may be interpreted as approximated solutions to a large optimization problem, which consists of finding a new feasible timetable of minimum cost. We refer to this problem as the *train-dispatching problem*.

Today's dispatching is supported in various ways by TMSs. Besides interfacing dispatcher decisions with the field, TMSs show the current status of the network in the dispatching area, the position of trains, and the status of signals and switches. They are also able to predict train movements and identify potential conflicts in the use of resources. A *conflict* is the simultaneous occupation by two trains of a block section. This must never occur, and, in general, it is prohibited by protocols and physical devices. However, when trains deviate from the official timetable, potential conflicts may arise. When identified, such conflicts are presented to dispatchers, who make and implement recovery decisions. Today's TMSs are in general unable to make such decisions and implement them autonomously. In Section 6.5 we describe a few exceptions in operation in Europe, based on the modeling and algorithmic ideas presented in Sections 6.3 and 6.4. Actually, according to two very recent studies [383, 1463], there are no other examples currently in operation.

6.3 ▪ Basic Mixed-Integer Linear Programming Models for the Train Dispatching Problem

In this section we limit ourselves to presenting the basic mixed-integer linear programming (MILP) models for the train-dispatching problem. For the sake of simplicity, we consider the case in which train routes are fixed; extensions to include the routing problem may be found in the literature [383]. To describe train movements, we associate, with every section of a train's route, a continuous variable that represents the time the train enters the section. The vector t of entry times, for all trains and all sections in their route, is called the *schedule*. Our target is to find a feasible schedule t^* that minimizes some cost function $c(t)$ (which in turn is the measure of the deviations from the official timetable earlier introduced). Next, we introduce the basic constraints that must be satisfied by any feasible schedule t. If l is the minimum running time of a train through a block section, t_u is the time the train enters the section, and t_v is the time it enters the next section, then $t_v - t_u \geq l$ holds. Other temporal constraints, such as release times and due dates, may be represented by similar simple precedence constraints. Now, when the routes of two distinct trains, say trains i and j, share a block section, one of the two trains must exit the section before the other one enters it. So, if t_z and t_u represent the time at which train i enters and exits the section, and t_v and t_w are the times at which train j enters and exits the section, then either $t_z \geq t_w$ or $t_v \geq t_u$ (see Figure 6.2). This is referred to as a *disjunctive precedence constraint* and is identified by two ordered pairs (u, v) and (w, z).

The number of such disjunctions may grow very large with the number of trains and the length of their routes in terms of block sections. Every feasible schedule t must

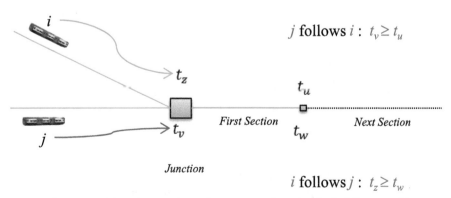

Figure 6.2. *A junction, two incoming trains, and associated scheduling variables.*

satisfy one of the two precedence constraints for every disjunction. So, we may write the dispatching problem as

$$\min \quad c(t)$$

s.t.

(i) $\quad t_v - t_u \geq l_{uv},$ $\qquad\qquad\qquad (u,v) \in A,$ $\qquad$ (6.1)

(ii) $\quad t_v - t_u \geq 0 \ \bigvee\ t_z - t_w \geq 0, \quad \{(u,v),(w,z)\} \in F,$

$t \in \mathbb{R}^V,$

where A, F, V are the set of indices of the simple precedence constraints, of the disjunctive precedence constraints, and of the variables, respectively. *Disjunctive programs* [143] such as the one above were introduced in the context of *job-shop scheduling* problems. Indeed, trains may be viewed as jobs performing certain *operations* (occupying a block section) on a set of machines (block sections). In particular, the dispatching problem belongs to the class of *blocking, no-wait job-shop scheduling*, which is known to be hard to solve in theory and in practice [1293].

Disjunctive programs may be represented and solved by means of mixed-integer programming (MIP). Two major approaches compete in the literature: *time-indexed formulations* and *big-M formulations*. In big-M formulations, we associate with each disjunctive constraint $\{(u,v),(w,z)\} \in F$ a binary variable x_{uvwz}, with $x_{uvwz} = 1$ if $t_v - t_u \geq 0$ holds and $x_{uvwz} = 0$ if $t_z - t_w \geq 0$ holds. To ensure this, we introduce a suitably large coefficient M (the notorious big M) and we replace each disjunctive constraint (6.1.ii) with two (conjunctive) constraints:

(iia) $\quad t_v - t_u \geq M(x_{uvwz} - 1), \quad \{(u,v),(w,z)\} \in F,$

(iib) $\quad t_z - t_w \geq -M x_{uvwz}, \qquad \{(u,v),(w,z)\} \in F.$ $\qquad$ (6.2)

It is easy to see that, when $x_{uvwz} = 1$, (6.2.iia) reduces to $t_v - t_u \geq 0$, while (6.2.iib), due to the large coefficient M, becomes redundant, as it is satisfied by any schedule t (satisfying all simple precedence constraints). Similarly, when $x_{uvwz} = 0$, (6.2.iia), becomes redundant. It is well known that this kind of formulation tends to provide weak lower bounds. Computing strong bounds is important because they provide quality assessments when a solution to a problem is at hand, and they are crucial in limiting the search space in enumerative solution methods.

Time-indexed formulations for scheduling problems were introduced precisely to cope with this issue. The idea is to discretize the time horizon H into a finite number of periods, i.e., $H = \{1, 2, \ldots\}$, and replace each original continuous variable t_v with a set of $|H|$ binary variables, $y_{v1}, y_{v2}, \ldots$, with the interpretation that $y_{vq} = 1$ implies $t_v = q$. Time-indexed formulations produce better bounds than big-M formulations (see [649]) but, in general, at the cost of increased computational burden. The trade-off typically depends on the ability to limit the number of time-indexed variables and to work with implicit representations of the overall formulation.

Both representations have been exploited in the literature on train dispatching. Time-indexed formulations are exploited, e.g., in [389, 1242], whereas big-M formulations are developed, e.g., in [97, 290, 552, 609, 1103, 1463, 1610, 1630, 1641]. However, in our experience, big-M formulations appear to be more suitable for train dispatching, as time-indexed formulations tend to grow too large and become intractable for instances of practical relevance, mostly due to the difficulty in fixing a small time horizon for the scheduling variables.

We remark that a fundamental requirement of a dispatching system is that solutions must be found and returned in real time or, more precisely, within the time limit set by the application. Typically, this implies that the time available for computation is limited to a few seconds. Now, a simple approach to solving a train-dispatching instance could consist of generating the corresponding big-M formulation and then invoking a state-of-the-art MILP problem solver to solve it. Unfortunately, for real-life instances of practical interest, this naive approach tends to fail in finding within the time limit not only the optimal solution but often even a feasible solution.

For this reason, almost all works presented in the literature resort to some type of heuristic, i.e., algorithms that search for feasible solutions but are unable to provide optimality certificates.[6] Instead, we decided to follow a different path and developed a decomposition approach that can return and certify optimal solutions to the dispatching problem.

6.4 ▪ The Decomposition Principle in Dispatching

Decomposition is a natural way to tackle complex control or optimization problems. In train dispatching, this is how normal operations are carried out in practice. Indeed, even though the railway network is a huge connected graph, each dispatcher controls only a small portion of it, possibly coordinating with other dispatchers who control adjacent areas. The size of each control area is carefully determined so that one dispatcher suffices to manage it. Regional lines, for example, are often controlled by one or two dispatchers. Large stations are typically subdivided into smaller substations, each controlled by a different dispatcher. Even if some communication with neighboring dispatchers is possible, a dispatcher has no authority on dispatching areas assigned to others and can only inherit decisions without any possibility of affecting them.

Another example of decomposition in practical dispatching is the way a dispatcher controls the assigned stretch of regional line. In a broad representation, such railway regions may be viewed as a sequence of stations connected by links. When potential conflicts are identified by the TMS, the dispatcher in charge decides where trains should meet or pass each other on the line (e.g., a station or a double-track link). In making this decision, a station is considered as an aggregated resource with a given capacity. Detailed decisions about the sequencing of trains in the station are made in a second phase, as trains approach the station.

Remarkably, this practical decomposition has a counterpart in MILP approaches to train dispatching. The original railway network is again decomposed into subnetworks, each with its own controlled stations and links. We associate with each subnetwork a corresponding dispatching problem and solve it, possibly to optimality. Then, the solutions to all of the subproblems must be recomposed into a unique solution for the overall original problem. Different implementations of the decomposition principle mainly differ in the way subproblems are generated, in the solution technique, in the way solutions of subproblems are recomposed, and in how the case in which subsolutions cannot be recomposed is handled.

One version of the decomposition paradigm is the so-called macroscopic/microscopic approach [319] recently introduced in the optimization literature for train timetabling and dispatching. At a microscopic level, all block sections are considered. At a macroscopic level, every subnetwork, such as a station or a dispatching area, is

---

[6]The literature on heuristics is very large, and we refer the reader to the comprehensive survey [383] for further details.

represented in each train route as one section. Indeed, at this macro level, for each train, only (tentative) arrival and departure times to and from each subnetwork are actually computed. In the standard approach, a good (possibly optimal) tentative macrosolution is found first. Then the macrosolution is turned into a global solution by extending it to the subregions. If this attempt fails, the initial macrosolution is modified and the scheme iterated. This approach is followed, e.g., in [608], where the macroproblem is modeled as an MILP problem, and where microproblems are solved by means of heuristics. In [609], both macro- and microproblems are modeled as MILP problems, but again combination and communication between levels are implemented by heuristics. Similar decomposition ideas are also exploited in [512], where the macroproblem is obtained by shrinking subnetworks corresponding to dispatching areas.

Next, we will show how the macro-micro approach, and in general the decomposition principle in dispatching, has a counterpart in a classical method in linear and integer optimization, namely Benders decomposition. We will actually refer to a generalization of the classical Benders decomposition to cope with integer subproblems, as presented in [486, 1822]. For simplicity's sake, we will consider the case of a single railway line and let the subnetworks in our decomposition be the stations of the line. However, the scheme may also apply to different hierarchies. If we let t be the scheduling vector and x be the binary variables corresponding to precedence decisions, then we may write the overall MILP problem as[7]

$$\min c(t)$$

s.t.

$$
\begin{array}{llll}
\text{(i)} & Ax_L + & Bt & \geq b, \\
\text{(ii)} & & Dt + Ex_S & \geq q, \\
\text{(iii)} & & t \text{ real}, \quad x_L, x_S \text{ binary.}
\end{array}
\qquad (6.3)
$$

The decision vector x has been written as $x = (x_L, x_S)$ to distinguish between decisions associated with stations (x_S) and decisions associated with tracks between stations (x_L). Correspondingly, we may identify the two blocks (i) and (ii) in program (6.3). In many relevant practical contexts, as in the real-life implementations of Section 6.5, the two blocks "communicate" with each other through a small subset of t variables. This happens when decisions made in the line regions and in the station regions do not "directly" affect each other but only have an effect through their consequences on the scheduling.[8] The idea in Benders decomposition is to solve a restricted problem (the *master*) where block (6.3.ii) is dropped. Let (x_L^*, t^*) be the optimal solution to the restricted problem. Then, if t^* can be extended to a feasible solution (t^*, x_S^*) to (6.3.ii) (*slave* problem), we are done as (x_L^*, t^*, x_S^*) is an optimal solution to (6.3). Otherwise (x_L^*, t^*) cannot be extended to a feasible solution for the whole problem, and we identify an inequality $q^T x_L + r^T t \leq k$ that is satisfied by all of the feasible solutions to (6.3) but violated by (x_L^*, t^*). This inequality is added to the master problem, and the process is iterated. In our experience with real-life dispatching instances, the number of iterations is quite small. A nice feature of the approach is that the slave problem further decomposes into a number of independent subproblems, one for each station.

---

[7]We assume here that the cost function $c(t)$ is linear.

[8]The situation is different when routing decisions for the line region may affect routing decisions in the stations. Consider for instance the case where a train can choose different routes in the line, involving different entry points in a given station, which in turn may force different routing decisions in the station.

We developed this approach in [1120–1122, 1265], enhancing the algorithm with delayed row and column generation [64]. This allowed us to solve to optimality a number of real-life instances from very different application contexts, as described in the next section.

6.5 ■ Real-Life Implementations

Systems based on the ideas sketched in this chapter (described in more detail in [1120–1122, 1265]) are already in operation on several lines in Europe. Also, new implementations are in progress or scheduled in the near future. In particular, we will briefly describe real-life implementations in Italy, Norway, and Latvia. Furthermore, we will mention an application for a large station in Italy and a mass transit system that was in operation in 2007–2009 in some terminal stations of the Milan underground.

All these TMSs are embedded in our optimization algorithms, either directly (Italy, Latvia) or in a collaborative framework (Norway). Although most of the input is acquired remotely, dispatchers may interact with the system by manually providing further information (e.g., train delays or cancellations, fixing dispatching decisions, network disruptions). The optimization algorithm will then return new dispatching decisions. Depending on local operative rules, such decisions may be (1) presented to dispatchers for validation or (2) forwarded directly to the field through remote equipment. The systems deployed in Italy and Norway fall in case (1), while both (1) and (2) apply to Latvia since freight trains are fully controlled by the system, whereas decisions regarding passenger trains require human validation. Remarkably, statistics show that even when the system does not control trains directly, dispatchers follow its suggestions in almost all cases. Another important question regards the "degree of freedom" of the algorithm. For instance, the Italian and Latvian railway operators have specific rules for resolving conflicts, which limit the potential impact of the algorithm because some dispatching solutions are forbidden.

6.5.1 ■ Regional Italian Lines

The first deployment on the main line of a TMS embedded in our dispatching algorithms was carried out in Italy in 2011, on a minor, single-track regional line: Trento–Bassano del Grappa line (23 stops or stations). Since then, the tool has been extended to other lines in southern, central, and northern Italy. Table 6.1 reports some data on these lines.

The algorithm identifies alternative solutions, which are ranked according to their cost and presented to dispatchers: in practice, the first solution in the ranking is the one often chosen. Since validation is left to the dispatchers, the tool is equipped with an exact procedure for detecting whether dispatching decisions lead to deadlock situations.

Table 6.1. *Regional lines in Italy.*

Regional Line	Stations/Stops	Network Complexity	Station Link
Trento–Bassano	23	Single Line	Single Track
Parma–San Zeno	17	Single Line	Single Track
Foligno	53	Interconnected Lines	Single and Double Tracks
Milano–Mortara	12	Single Line	Single Track
Sicilia	111	Interconnected Lines	Single Track

Table 6.2. *Average delay distribution of instances of the Foligno line over a week in January* 2013. *The time unit is minutes. Mean number of controlled trains is 86, and standard deviation is 27. Average number of trains running late is 9.*

Solution space	delay ≤ 5	delay ≤ 10	$10 <$ delay ≤ 15
Restricted	90%	91%	93%
Full	95%	98%	99%

Table 6.3. *Algorithmic information.*

Periods	Instances	Iterations	Time (s)	Conflicts	
				Potential	Solved
[04:08]	982	9	3.99	13777	154
[08:12]	941	9	4.58	11639	151
[12:16]	1127	9	3.85	7056	126
[16:20]	1433	5	4.02	3164	49
[20:24]	1232	5	4.57	3617	61

We have tried to quantify the benefit of relaxing current operative rules and allowing a larger solution space. In particular, we compared restricted and full solution space optimization on a test set of instances from the Foligno line over a week in January 2013. In this setting, an instance represents the status of network and trains at a given moment in time. In Table 6.2 we cluster trains into three groups according to their computed delay at final destination.

The results show an increase in the number of trains on time or with little delay (+5%), with virtually all trains (99%) arriving at the destination with at most 15 minutes' delay. On the other hand, restricted solution space solutions averaged 7% of trains arriving at the destination with more than 15 minutes' delay. In Table 6.3 we present some computational results and algorithmic information regarding a day of experiments in the above-mentioned week in January 2013. On this day, the average number of controlled trains on this line was 107, with the highest number throughout the day being 130 and the lowest 67, while the average number of trains simultaneously on the line was 10. Results are aggregated in five time ranges (four hours each).

The column "Periods" represents the time range (hh-hh), the column "Instances" is the number of distinct instances solved within that time range, the column "Iterations" expresses the *average* number of macroiterations (i.e. the master-slave) of the algorithm, the column "Time" shows the *average* computation time (in seconds), and finally the two subcolumns under "Conflicts" express the *average* number of potential conflicts and those solved by the algorithm, respectively. The information regarding how many conflicts are solved in practice with respect to their potential number is expressed to give a measure of the importance of using a delayed column generation approach. Assuming (roughly speaking) that each conflict is expressed in the model by an integer variable, the delayed approach saves us from adding around 98% of such integer variables on average, which in turn brings considerable computational benefits.

Although the current implementation represents an important step toward improved and automatic railway management, these figures show that further improvements may be obtained by applying new regulations and an effective, exact optimization algorithm. This was allowed for the first time in a recent implementation in Norway.

6.5.2 ▪ Norwegian Lines

A pioneering dispatching system that that exploits the decomposition approach described in the previous section has been tested on some regional lines in Norway (Trondheim–Dombås, Stavanger–Moi) and operated on one of these (Stavanger–Moi). The novelty of such a system lies in the possibility of exploring the full solution space, allowing a complete exploitation of the exact algorithm. The system was released in February 2014, displaying the real-time optimal train graph[9] on a screen in the dispatching center. Each time the network status changes (train delayed, deviation, etc.), a new solution is computed and the graph is updated accordingly. The dispatcher may then decide whether to accept the suggestion or to discard it. At the time of writing, the use of the system was on hold until the release of the tender for renovating Norway's entire signaling system, as the system is considered a competitor by the network operator.

We now give some information regarding the Stavanger–Moi line. The railway stretching from Moi to Stavanger is 123 km long and visits 16 stations. Every 12 hours, up to and, in some cases, over 100 trains run this line (on average). In particular, around 40% of this traffic is exclusive to the (entirely double-track) portion connecting Stavanger and Sandnes (stations that are more or less 15 km and 5 stops apart), while the remaining traffic also passes through this area. As a consequence, this portion of the network is particularly dense and presents a challenge for local dispatchers.

6.5.3 ▪ Freight Lines in Latvia

TMSs equipped with our optimization algorithms are in operation on an extended railway network in the east of Latvia. The overall network is composed of several lines: Daugavpils–Eglaine, Daugavpils–Krustpils, Rezekne–Krustpils, Zilupe–Krustpils, and Karzava–Rezekne. In total, there are 52 stations, with 10 communication points and 8 station gates. These lines are characterized by high traffic volumes and are mainly used for freight transportation. The TMS's dispatching decisions regarding freight trains are automatically forwarded to the field, without requiring a dispatcher's validation. Dispatchers accordingly will focus only on solving conflicts involving passenger trains, where, as described above, they will be presented with the best solutions identified by the algorithm.

6.5.4 ▪ A Large Station in Italy

TMSs also prove to be very useful for monitoring and controlling traffic in large stations. Monfalcone (in northern Italy) is being provided with one such automatic TMS, able to reroute and reschedule trains in an optimal way with respect to the timetable and the current network situation. The system is currently (February 2017) under commissioning. The system in Monfalcone is required to control three connecting "satellite" stations: Monfalcone Station (14 stopping points and 59 interlocking routes), Ronchi Nord Station (7 stopping points and 16 interlocking routes), Ronchi Sud Station (7 stopping points and 16 interlocking routes), and a communication point.

---

[9]A train graph is a standard graphical representation of a timetable.

6.5.5 ▪ Mass Transit

To the best of our knowledge, the first fully automatic dispatching tool operated in some terminal stations of the Milano Underground System from 2007 to 2009 [1266], managed by ATM (the major Italian municipal transport company). Our optimization algorithm was embedded into the TMS developed by Bombardier Transportation, a large multinational corporation of the transport sector. Prior to its activation, an extensive test campaign was carried out to compare the performance that of the system against that of dispatchers in charge at one of the terminal stations. Such direct comparisons are very rare in the literature and in practice and quite complicated to set up. The difficulty stems from the impossibility for the system and the dispatchers to have control over station and trains at the same time and compare them on exactly the same data input. To get around this issue, during the test campaign eight pairs of one-hour time slots with equivalent traffic patterns were identified by ATM engineers. For each of the eight pairs, one of the two time slots was assigned to the automatic system, whereas the other was assigned to dispatchers. Two objective functions were considered: the first measured deviations from the timetable; the second measured deviations in regularity, namely in the difference between actual and desired train headways. Final results showed that the system improved both objectives by an average of 8% over the dispatchers, despite the relatively small size of the terminal stations.

6.6 ▪ Concluding Remarks

The literature abounds with models and solution algorithms for real-time train traffic management—train dispatching and other related problems such as delay management—but presents very few applications. This shortcoming can be attributed in part to the operators' resistance to innovation, but also, in some measure, to the first attempts at using optimization methods in practice not delivering the bounty that had been "promised" to the industry (e.g., increased efficiency, lower costs). However, as shown in this chapter, the landscape is slowly shifting. Applications of optimization to traffic management are now starting to appear throughout Europe. Operators are increasingly aware of the potential of optimization-based traffic management, as proved in recent tenders, by explicitly requesting "intelligent" dispatching functionalities for new TMSs. In our tests and real-life applications, we show that much improvement over the current practice can be achieved by using optimization techniques, and mathematical programming in particular. The limitation with our exact approach is that some of our assumptions do not apply to every railway line. Therefore, work has to be done to extend this approach with a more general decomposition scheme.

Chapter 7

Applications of Mixed-Integer Linear Optimization in Chemical Engineering

Ignacio E. Grossmann, Vikas Goel, Jose M. Pinto, and John M. Wassick

7.1 • Introduction

Chemical engineering through the petroleum and chemical industry has been a very fertile area of application of mixed-integer linear programming (MILP). The major reason for this is that the potential for optimization of discrete and continuous decisions in these industries is very large. Ranging from design and investment decisions to the operation of process systems, the economic impact of optimizing the decisions in these problems can be very significant. This is mainly due to the large-scale nature of these industries. To provide an economic perspective, the 2013 revenues of companies that employ three of the co-authors were as follows: ExxonMobil, $452 billion; Dow Chemical, $57 billion; and Praxair, $11.2 billion.

Furthermore, many developments in mixed-integer programming (MIP) have been motivated by applications in process systems engineering (PSE), the subdiscipline in chemical engineering that is concerned with optimization of the design, operation, and control of process systems. Extensive applications of mixed-integer linear and nonlinear optimization (MILO and MINLO) methods have taken place in process synthesis [863]. Major examples include design of distillation sequences [379], synthesis of heat exchanger networks [453, 1930], design of integrated water networks [38], synthesis of steam and power plants [350], and general process flowsheets [1932]. Supply chain, planning, and scheduling of process systems have also become very active areas in the use of MILP models [861, 910], where for instance the State-Task Network model [1092] for batch scheduling and planning for capacity expansion of process networks [1616] are prime examples. Other areas of application include process control [194] and molecular design [10].

This chapter presents three major industrial applications of MILP in chemical engineering. The first application deals with the optimal production planning at Praxair for air separation plants under time-sensitive electricity prices; the second on a liquified natural gas (LNG) inventory routing problem at ExxonMobil, where ship schedules must be optimized; and the third the design of supply chains at Dow Chemical. For each application, we describe the specific problem, outline its mathematical formulation, and discuss its practical implementation and the benefits obtained. We conclude with remarks on some future barriers that must be overcome to make the application of MIP even more extensive and successful than it is currently.

7.2 ▪ Optimal Production Planning under Time-Sensitive Electricity Prices for Air Separation Plants

Power-intensive processes can lower operating expenses when adjusting production planning according to time-dependent electricity pricing schemes [1338]. This is especially significant in air separation plants since the raw material, air, is free, and the major expenses are involved in cryogenic separation, which requires large consumption of electricity by compressors. In the emerging area of demand-side management, the electricity consumption of industrial customers is influenced by the market price of electricity. Typical examples of time-sensitive electricity prices are time-of-use (TOU) rates and real-time pricing (RTP). While TOU rates are usually specified in terms of on-peak, mid-peak, and off-peak hours, real-time prices vary every hour and are quoted either on a day-ahead or an hourly basis. This allows industrial customers to perform production planning based on predefined hourly prices. In this section, we describe a discrete-time, deterministic MILP model that allows optimal production planning for air separation plants [1338] accounting for a deterministic forecast of RTP of electricity.

The problem faced by Praxair can be stated as follows. Given is a set of products $g \in G$ (e.g., liquid oxygen, liquid nitrogen, oxygen gas, nitrogen gas) that can be produced in different continuously operated plants $p \in P$. While liquid products can be stored on-site ($g \in Stor$), gas products must be directly delivered to customers ($g \in Nonstor$). The plants have to satisfy demand for the products, which can be specified on a weekly, daily, or hourly basis. The costs of production vary every hour, $h \in H$, and are related to electricity costs, which also undergo seasonal changes. It is assumed that a seasonal electricity price forecast for a typical week is available on an hourly basis and that the production decisions do not influence the electricity prices. The problem is to determine production levels, modes of operation (including start-up and shut-down of equipment), inventory levels, and product sales on an hourly basis, so that the given demand is met. The objective is to minimize the operating costs, mostly due to energy expenses.

The traditional way of modeling a process involves heat and mass balance, which requires the detailed description of the system's performance (e.g., thermodynamics, kinetics). The disadvantage of this approach is that the model can become prohibitively hard to solve due to its nonlinearities and its size for longer time horizons. An alternative approach is to model the plant in a reduced space, e.g., the product space. Hence, we shift the computational burden of evaluating the feasible region to off-line computations to obtain a surrogate model. In the absence of a detailed plant model, the surrogate model can also be built from historic plant data.

Equipment operation involves discrete as well as continuous operating decisions. A discrete operating decision refers to the state of equipment, e.g., "off," "production mode," or "ramp-up transition." A continuous operating decision refers to production levels, i.e., flow rates of material. To distinguish the discrete operating decisions of a plant, we formally introduce the concept of a mode, which is a set of operating points for which the same discrete operating decisions are active. The operating points vary in terms of the continuous variables in the product space. The feasible region of each mode is approximated by a convex region consisting of known operating points. In addition, energy consumption is approximated by a correlation for the entire feasible region of a mode. In any given time period, only one mode can be active; in other words the modes are disjoint. Off-line computations or plant measurements provide the data that describe the disjunctive set of modes for the feasible region. The data for each mode are represented as a collection of operating points (slates) that are the extreme points in terms of the products. These extreme points have to be determined a priori by using an appropriate tool, such as MATLAB [1296] or PORTA [468]. The convex combination of the extreme points determines the hourly production for each product.

The MILP model constraints can be classified into three sets [1338]. The first set deals with the previously mentioned surrogate description for the feasible region in the product space:

$$\sum_{i \in I} \lambda^h_{p,m,i} \, x_{p,m,i,g} = \overline{Pr}^h_{p,m,g} \qquad \forall \, p \in P, m \in M, g \in G, h \in H, \qquad (7.1\text{a})$$

$$\sum_{i \in I} \lambda^h_{p,m,i} = y^h_{p,m} \qquad \forall \, p \in P, m \in M, h \in H, \qquad (7.1\text{b})$$

$$0 \le \lambda^h_{p,m,i} \le 1 \qquad \forall \, p \in P, m \in M, i \in I, h \in H, \qquad (7.1\text{c})$$

$$Pr^h_{p,g} = \sum_{m \in M} \overline{Pr}^h_{p,m,g} \qquad \forall \, p \in P, g \in G, h \in H, \qquad (7.1\text{d})$$

$$\overline{Pr}^h_{p,m,g} \le \overline{M}_{p,m,g} \, y^h_{p,m} \qquad \forall \, p \in P, m \in M, g \in G, h \in H, \qquad (7.1\text{e})$$

$$\sum_{m \in M} y^h_{p,m} = 1 \qquad \forall \, p \in P, h \in H, \qquad (7.1\text{f})$$

where $x_{p,m,i,g}$ are the extreme points $i \in I$ of mode $m \in M$ of plant $p \in P$ in terms of the products $g \in G$, whose convex combination, with weight factors $\lambda^h_{p,m,i}$, ultimately determines the production $Pr^h_{p,g}$ at each hour $h \in H$. In the constraint set (7.1), $y^h_{p,m}$ determines whether plant $p \in P$ operates in mode $m \in M$ in hour $h \in H$, and parameter $\overline{Pr}^h_{p,m,g}$ denotes the production amount of product g in mode m, plant p, hour h.

The second set contains constraints in terms of 0-1 variables $z^h_{p,m',m}$ to indicate a transition from mode m to mode m' at plant p from hour h. Due to the hourly changing electricity prices, it might be economically desirable to run the plant at different time-dependent operating points. If the operating point is changed, a transition occurs. There are two major types of transitions: (1) between operating points of different modes and (2) between operating points that are both part of the same mode—rate of change constraints. The link between state $y^h_{p,m}$ and transition variables $z^h_{p,m',m}$ is given by

$$\sum_{m' \in M} z^h_{p,m',m} = y^h_{p,m} \qquad \forall \, p \in P, h \in H, m \in M, \qquad (7.2\text{a})$$

$$\sum_{m' \in M} z^h_{p,m,m'} = y^{h-1}_{p,m} \qquad \forall \, p \in P, h \in H, m \in M. \qquad (7.2\text{b})$$

We enforce a minimum stay in a mode for the set of allowed transitions AL:

$$y_{p,m'}^{h+\theta} \geq z_{p,m,m'}^{h} \quad \forall (p,m,m') \in AL,\, h \in H,\, \theta \in MinStay(m,m'). \tag{7.3}$$

We also enforce a maximum stay $K_{m,m'}^{max}$ in a mode:

$$1 - y_{p,m'}^{h+K_{m,m'}^{max}+1} \geq z_{p,m,m'}^{h} - \sum_{\theta=K_{m,m'}^{min}}^{K_{m,m'}^{max}} \sum_{m'' \neq m'} z_{p,m'',m'}^{h+\theta} \quad \forall (p,m,m') \in AL,\, h \in H. \tag{7.4}$$

Forbidden transitions are specified by

$$z_{p,m,m'}^{h} = 0 \quad \forall (p,m,m') \in DAL,\, h \in H, \tag{7.5}$$

while couplings between transitions are given in terms of $K_{m,m'}^{min}$, the minimum stay in mode m':

$$z_{p,m,m'}^{h-K_{m,m'}^{min}} - z_{p,m',m''}^{h} = 0 \quad \forall (p,m,m',m'') \in Trans,\, h \in H. \tag{7.6}$$

The constraint that limits the production in terms of the maximum rate $r_{p,m,g}$ is given by

$$\overline{Pr}_{p,m,g}^{h+1} - \overline{Pr}_{p,m,g}^{h} \leq r_{p,m,g} \quad \forall\, p \in P,\, m \in M,\, g \in G,\, h \in H. \tag{7.7}$$

Finally, the third and last set describes the mass balances that establish relationships on inventories $INV_{p,g}^{h}$ and sales $S_{p,g}^{h}$ that satisfy demands for liquid and gas products g. Equation (7.8a) represents the mass balance for the storable products. The inventory levels $INV_{p,g}^{h}$ are constrained in (7.8b) by the outage levels ($INV_{p,g}^{L}$, lower bound) and by the storage capacities ($INV_{p,g}^{U}$, upper bound). If a product g cannot be stored, the inventory level is simply $INV_{p,g}^{h} = 0\ \forall h$ and the mass balance becomes (7.8c). Constraint (7.8d) satisfies demand that is usually specified on an hourly basis. However, it is also possible to specify demand for the storable products on a daily or weekly basis ($d_{p,g}^{weekly}, d_{p,g}^{daily}$), e.g., if the exact sales points for the products are unknown, as in (7.8e) and (7.8f):

$$INV_{p,g}^{h} + Pr_{p,g}^{h} = INV_{p,g}^{h+1} + S_{p,g}^{h} \quad \forall\, p \in P,\, g \in Stor,\, h \in H, \tag{7.8a}$$

$$INV_{p,g}^{L} \leq INV_{p,g}^{h} \leq INV_{p,g}^{U} \quad \forall\, p \in P,\, g \in Stor,\, h \in H, \tag{7.8b}$$

$$Pr_{p,g}^{h} = S_{p,g}^{h} \quad \forall\, p \in P,\, g \in Nonstor,\, h \in H, \tag{7.8c}$$

$$S_{p,g}^{h} \geq d_{p,g}^{h,hourly} \quad \forall\, p \in P,\, g \in G,\, h \in H, \tag{7.8d}$$

$$\sum_{h} S_{p,g}^{h} \geq d_{p,g}^{weekly} \quad \forall\, p \in P,\, g \in Stor, \tag{7.8e}$$

$$\sum_{h \in day} S_{p,g}^{h} \geq d_{p,g}^{daily} \quad \forall\, day,\, p \in P,\, g \in Stor. \tag{7.8f}$$

The objective function minimizes the total cost that consists of production cost in terms of the power consumption and the production levels, inventory cost, and transition cost for every hour:

$$\min OBJ = \sum_{p,h} e_p^h \left(\sum_{m,g} \Phi_{p,m,g} \overline{Pr}_{p,m,g}^h \right) + \sum_{p,g} \delta_{p,g} \left(\sum_h INV_{p,g}^h \right)$$
$$+ \sum_{p,m,m'} \xi_{p,m,m'} \sum_h z_{p,m,m'}^h. \tag{7.9}$$

In (7.9), we denote by $\Phi_{p,m,g}$ the coefficients that typically have to be determined by a multivariate regression (in [power/volume]). The inventory cost coefficients are denoted by $\delta_{p,g}$ (in [\$/volume]), and the cost coefficients related to transition cost are described by $\xi_{p,m,m'}$ (in [\$]).

7.2.1 ▪ Case Study

Two different air separation plants are optimized—a plant with one liquefier (P1) and a plant with two liquefiers (P2)—that supply the liquid merchant market without on-site customers. The power consumption for cryogenic air separation is mostly due to the compressors that provide cooling for separation and liquefaction. Each operating mode can be related to a liquefier in the air separation plant. If the plant increases production, a transitional mode is active in which a specified amount of off-spec product is generated and vented. If the plant has a second liquefier installed and switches its operation from one to two liquefiers, the plant also undergoes a transitional mode for a few hours.

The computational performance of the MILP model was tested for five cases (A–E) that differ in the demand specified for the liquid products on a six-hour basis (in which a certain number of tractor trailers load products) for a time horizon of one week (168 hours). Due to confidentiality issues, we cannot disclose the actual LO2 and LN2 demands and production levels. Therefore, scales are omitted, but the total demand is reported as a function of the plants' production capacity in Table 7.1. The electricity price forecast of the day-ahead prices (see Figure 7.1) is given on an hourly basis.

To assess the economic impact of the MILP optimization model, solutions are compared for the noncyclic schedules with a heuristic that assumes constant operation throughout the week and finds a set-point that minimizes the energy consumption while satisfying the demand constraints. In Table 7.1, we can see that for both plants the potential savings increase with decreasing demand, which is reported as a percent of the total production capacity. For test cases A and B, the realized savings are mostly

Table 7.1. *Economic impact and computational performance for plants P1 and P2. Reprinted with permission from Springer [818].*

Case	Capacity	Savings	# constraints	# variables	# binary	CPU (s) CPLEX	CPU (s) XPRESS
P1A	82%	4.58%	22,508	16,129	1512	2	3
P1B	74%	12.02%				4	5
P2E	95%	3.76%				4	4
P2D	85%	4.90%	44,173	29,401	3528	4	4
P2C	72%	7.44%				7	10
P2B	51%	13.78%				135	47

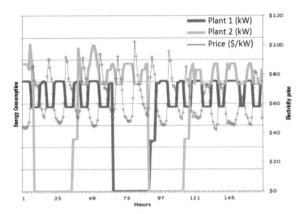

Figure 7.1. *Energy consumption and power pricing for air separation units. Reprinted with permission from Elsevier* [1338].

due to plant shutdowns, which are illustrated at the top of Figure 7.1 for the electricity consumption profiles of plants P1 and P2 for test case B. One can also see that the electricity consumption is reduced during hours of high electricity prices—due to higher operational flexibility, plant P2 can produce 5.2% more cheaply than plant P1.

The computational results can be found in Table 7.1. All test cases were solved with CPLEX 12.2.0.1 and XPRESS 21.01 (both with default settings; no parallel computing features were used) in GAMS 23.6.2 [342] on an Intel i7 (2.93 GHz) machine with 4 GB RAM, using a termination criterion of 0.1% optimality gap. CPLEX and XPRESS perform similarly, except for test cases A and B, with advantages on both solvers. It can be observed that with increasing demand the problems become easier to solve since operational flexibility decreases. This can be seen in terms of CPU time and also in terms of the tightness of the linear programming (LP) relaxation. For case P2E, with the highest demand, the initial gap is in fact 0%.

In conclusion, production planning based on time-sensitive pricing has significant potential for economic savings in cryogenic air separation, where electricity costs represent 40–50% of overall production costs and 90–95% of variable production costs. For an industrial gas company with an annual expenditure of $1.0–1.5 billion in electricity, the efficiencies reported in this work can lead to tens of millions of dollars in annual savings.

7.3 ▪ LNG Inventory Routing

To be transported as LNG, natural gas produced from a reservoir is first processed to remove impurities and then liquefied by cooling it to $-163\,°C$ at a liquefaction terminal (also referred to as a production terminal). The LNG is stored at dedicated storage facilities at the production terminal before being loaded on specially designed ships that transport it to regasification (regas) terminals where it is stored temporarily before being regasified (i.e., vaporized) and injected into the local natural gas pipeline grid.

Motivated by the need for a model that can assist in the design analysis of new LNG projects, we describe an LNG inventory routing problem for developing optimized ship schedules for a heterogeneous pool of ships that deliver LNG from a set of liquefaction terminals to a set of regas terminals. The problem accounts for constraints related to inventory storage, port operations, and contractual delivery obligations. We

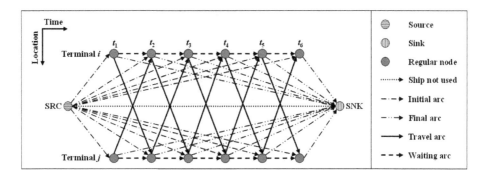

Figure 7.2. *Time-space network representation for LNG inventory routing model. Reprinted with permission from Elsevier* [1715].

describe an arc-flow formulation based on an MILP model and a solution strategy for addressing large-scale problems [818].

Formally, the problem considers a set of liquefaction and regas terminals with given storage capacities. Each production terminal has a prespecified production profile. Regas rates at regas terminals, on the other hand, can be adjusted within specified bounds on a daily basis. Profiles for the minimum and maximum regas rates over the planning horizon are specified for each regas terminal. In addition, total LNG demand over the planning horizon at each regas terminal is specified. Finally, each terminal has a limited number of berths to load and unload cargo. Each loading or unloading activity spans one time period and occupies a berth for that duration. We consider the case where each ship will fully load at a single production terminal and fully discharge at a single regas terminal.

The arc-flow formulation is based on a time-space network representation (Figure 7.2), where each node represents a terminal at a given time period. In addition to these location-based nodes, dummy source and sink nodes representing the initial and final locations of a ship are also represented in the network. Arcs in the network represent travel between terminals, waiting at a terminal, arrival of a ship at its initial location, and final departure of a ship to a sink node. Constraints related to voyage lengths between terminals, and logical relationships that constrain ships to travel from a liquefaction terminal to a regas terminal and vice versa are imposed on the structure of the network.

As explained in [818], the time-space network is defined individually for every ship. Binary variable x_a represents whether the ship associated with arc a will travel along that arc in the solution. The path followed by a ship in the network represents the overall schedule for the ship. The MIP model presented below can be viewed as an integer multicommodity network flow formulation with side constraints, where a ship is a commodity and a node represents a possible visit to a terminal at a particular time:

$$\min \sum_{(l,t)\in N|l\in L} w_{l,t}\, o_{l,t} + \sum_{(r,t)\in N|r\in R} w_{r,t}\, s_{r,t} + \sum_{r\in R} w_r^D \delta_r^F, \qquad (7.10a)$$

$$\text{s.t.} \sum_{a\in A_v\cap\delta_{\bar{n}}} x_a = \begin{cases} -1 : n = SNK,\, v\in V, \\ +1 : n = SRC,\, v\in V, \\ 0 : n\in \overline{N}\{SRC, SNK\},\, v\in V, \end{cases} \qquad (7.10b)$$

$$\sum_{v}\sum_{a\in(\cup_v^F)\cap\delta_n^+} x_a \le b_j \quad \forall j \in J, t \in T, n = (j,t) \in \overline{N}, \tag{7.10c}$$

$$I_{l,t} = I_{l,t-1} + p_{l,t} - \sum_{v}\sum_{a\in(A_v^T\cup A_v^F)\cap\delta_{n(l,t)}^+} c_v x_a - o_{l,t} \quad \forall l \in L, t \in T, \tag{7.10d}$$

$$I_{r,t} = I_{r,t-1} - d_{r,t} + \sum_{v}\sum_{a\in(A_v^T\cup A_v^F)\cap\delta_{n(r,t)}^+} c_{v,r} x_a + s_{r,t} \quad \forall r \in R, t \in T, \tag{7.10e}$$

$$0 \le I_{j,t} \le c_j \quad \forall j \in J, t \in T, \tag{7.10f}$$

$$d_r^L \le d_{r,t} \le d_r^U \quad \forall r \in R, t \in \{1,2,\ldots,|T|\}, \tag{7.10g}$$

$$\delta_r^D \ge D_r - \sum_{t}\sum_{v}\sum_{a\in(A_v^T\cup A_v^F)\cap\delta_{n(r,t)}^+} c_{v,r} x_a \quad \forall r \in R, \tag{7.10h}$$

$$o_{l,t}, s_{r,t}, \delta_r^D \ge 0 \quad \forall l \in L, r \in R, \tag{7.10i}$$

$$x_a \in \{0,1\} \quad \forall a \in A. \tag{7.10j}$$

Constraints (7.10b) impose network flow balance constraints for each ship v. Constraint (7.10c) limits the number of ships being loaded or unloaded at a terminal at a given time point, t, based on the number of berths, b_j, at that terminal, j. Constraints (7.10d)–(7.10f) impose inventory bounds at production and regas terminals, l, and r, respectively. Production rates $(p_{l,t})$ at each liquefaction terminal are assumed to be fixed, while regas rates $(d_{r,t})$ at each terminal are allowed to vary within bounds (constraint (7.10g)). At production terminals, the storage capacity restriction is modeled as a soft constraint to account for the fact that production may have to be curtailed if the inventory level approaches the storage capacity at that terminal. Similarly, on the regas side, the inventory balance constraint is modeled as a soft constraint to account for stockouts where gas may have to be purchased on the spot market if not enough inventory is available. Continuous variables $I_{j,t}$ represent the inventory level at terminal j at time t, and slack variables $o_{l,t}$ and $s_{r,t}$ represent lost production and stockouts at production and regas terminals, respectively. Finally, constraint (7.10h) includes a slack variable (δ_r^D) to account for violation of the contractual obligation of satisfying the overall demand at a regas terminal. The objective of the model is to develop a schedule that minimizes the weighted sum of lost production at production terminals, stockouts at regas terminals, and overall unmet demand at regas terminals.

Based on planning conventions in the LNG industry and typical loading and discharging durations, a year-long planning horizon with one-day time discretization generates realistic instances for the problem. Due to the large-scale and complex nature of the above problem, commercially available MILP problem solvers cannot solve industrial-scale instances of this problem within a reasonable amount of time. Reference [818] presents local neighborhood search heuristics with the goal of finding good solutions to the proposed model in a short amount of time. This approach can also be used to warm-start an exact solution method. The proposed algorithm is a three-step heuristic method, where the first step involves using a construction heuristic to build an initial feasible solution to the model. This construction heuristic builds a feasible solution by iteratively scheduling ship departures for the entire fleet on a daily basis. Ship departures on any given day are scheduled using an urgency-based greedy heuristic. Two improvement heuristics that employ different neighborhood structures are then used in sequence to improve this solution. The time-window improvement heuristic seeks to improve the solution obtained from the construction heuristic by allowing for departure dates for each voyage in the current solution to change within a small time window defined around the current departure dates. The two-ship improvement heuristic seeks to improve the existing solution by allowing a subset of the ships to change schedules over the entire horizon. The subproblems generated by both improvement heuristics can be solved as MILP problems using a commercial solver. The performance of the two-ship improvement heuristic strongly depends on the sequence in which the two-ship subproblems are solved. Reference [818] presents four approaches for sequencing the subproblems in the heuristic.

Table 7.2 reports the computational performance of the proposed approach compared to CPLEX 11.1 for 14 test instances. The table reports the dimensions for each problem, where $|L|$, $|R|$, and $|V|$ represent the number of production terminals, regas terminals, and ships, respectively. All the problems are defined over a one-year planning horizon. The number of continuous and binary variables and the number of constraints are also reported. All optimization models have been solved with CPLEX 11.1 on a Dell 7500 Windows PC with a 2.93 GHz dual-quad core processor and 24 GB RAM.

Table 7.2. *Computational performance on a set of 14 test problems. Reprinted with permission from Springer* [818].

	$|L|, |R|, |V|$	Cont vars	0-1 vars	Constr.	CPLEX Best feasible (BF)	CPLEX Time to FB	Heur-LP Best feasible (BF)	Heur-LP Time to BF	Speedup Heur-LP vs CPLEX
P1	(1,2,6)	2557	12812	6504	1172	17821	1187	171	4
P2	(1,3,8)	4019	16988	8635	294	1734	320	63	19
P3	(2,1,10)	2557	21029	9303	5650	195	5696	87	2
P4*	(3,1,13)	2922	27547	12196	0	2932	0	64	47
P5	(1,4,15)	5115	31449	14269	1701	4894	1970	311	1
P6	(4,1,4)	3652	8377	6406	151	35550	151	237	131
P7	(1,6,6)	5117	18335	10812	1750	26473	1650	1061	#N/A
P8	(1,1,14)	1827	28830	11205	116	7589	116	121	63
P9	(1,2,17)	2558	35142	14061	126	18145	126	200	91
P10	(1,4,27)	5115	95788	32744	32782	8837	29714	2960	#N/A
P11*	(1,5,14)	6228	34438	15625	4013	37941	0	150	#N/A
P12	(1,4,18)	3655	37752	16491	3574	46099	631	1297	#N/A
P13	(1,8,40)	6579	132995	47364	7988	19434	1453	19648	#N/A
P14	(1,10,69)	11691	232856	79425	331275	72003	54561	18501	#N/A

Table 7.2 compares CPLEX and the proposed heuristic in terms of the best feasible solutions and the time to get these solutions. Since CPLEX cannot prove optimality for most instances, CPU time limits of 5 hours, 10 hours, or 20 hours were applied, depending on the size of each instance. The table shows that for medium and large problems, the proposed heuristic generates significantly better solutions than CPLEX. For smaller problems, CPLEX can generate similar and, in some cases, better solutions than the proposed heuristic. The table also shows a speedup factor for the proposed heuristic, the ratio of CPU time taken by CPLEX to generate a solution at least as good as that generated by the proposed heuristic, relative to the CPU time taken by the heuristic. Problems for which CPLEX could not find a solution that is equivalent to or better than the heuristic are marked as "N/A." Based on this set of instances,

the proposed heuristic is 15.7 times as fast (geometric mean) as CPLEX in generating equivalent solutions.

Given the nature of the LNG market, most LNG is sold on long-term (~20-year) contracts, and new projects tend to secure dedicated shipping resources for the length of the contract. At the same time, infrastructure required for LNG projects is extremely capital intensive: liquefaction costs range between $200 and $700/ton of annual capacity (a typical LNG train processes 5 million tons/year); an LNG ship can cost ~$200 million.

7.4 ▪ Chemical Supply Network Optimization

The cost pressures in the chemical industry require cost reductions for manufacturers such as The Dow Chemical Company to remain competitive. Chemical supply chains are a fruitful area of cost reduction because they represent a significant portion of the total cost to serve customers, they constantly change, and their complexity often hides the lowest-cost option. In the application described below, we confine our attention to the movement and handling of manufactured materials from the production sites to customer locations in a multiechelon supply chain. The goal is to improve the design of an existing supply chain network made up of production sites and shipping terminals serving customer locations. An MILP model is described that is capable of optimizing a multiproduct supply chain network made up of production sites, an arbitrary number of echelons of distribution centers (DCs), and customer sites. The emphasis of the approach is on the redesign of existing supply chain networks [704].

The Dow Chemical Company invests significant time in developing supply chain network optimization models and maintains a large expertise center to ensure effective use of the technology. A typical project requires 300 hours of modeling time to fully develop and deliver the model to the business. Once implemented, these models can produce a value realization in excess of 10 times the invested dollars to the business. In addition to delivering significant financial benefits, network optimization models provide important insight into supply chain designs, such as

- assessment of current network to alternative/new optimized designs;
- understanding of the benefits and trade-offs between
  - production locations and costs,
  - warehousing/terminal locations and costs,
  - transportation modes and costs,
  - direct shipments from manufacturing plants versus routing products through DCs/warehouses/terminals;
- estimated network sustainability/carbon or energy footprint;
- confirmation that the network can meet service level requirements, where service level in a supply chain network is defined as the weighted-average distance from the last pick-up location;
- production and inventory planning over multitime periods in case of seasonal demand, planned shutdown, or limited capacity; and
- profit maximization, where certain customers may not be profitable to serve.

Network optimization has a well-established project methodology. Figure 7.3 shows the major stages for each project. Project scoping is very important for establishing the domain of the model and clarifying the business requirements. It may require revisiting the objective and goals a couple of times with businesses/clients to determine exactly what is needed. The data collection stage is supported by purpose-built databases that contain historical shipment information. Model development and alternative design analysis are carried out using commercial supply chain modeling software. Recommendations involve presenting the business with alternative courses of action that address the design trade-offs. One key approach adopted by Dow is to engage the businesses/clients early while finalizing the solution or recommendations. Thus, the solution or recommendations are agreed upon and fully implementable. This ensures the success of the overall project methodology.

Figure 7.3. *Project flow for a supply chain network optimization project.*

The supply chain networks considered by Dow are a collection of nodes comprising suppliers, customers, DCs, and production sources connected by transportation lanes for shipping material by truck, rail, vessel, or any other defined transport mode. Each shipping mode is distinguished by a different cost per mile traveled and may have a minimum shipping quantity. A supplier is a node in the network at a geographic location with a fixed capacity that produces a key raw material at a unit cost for production sources. A customer is defined as a node with a fixed demand over time of a single product at a unit price and resides at a geographic location. A customer can be given a status of either preferred or nonpreferred. A preferred customer's demand must be met, even if the cost of serving that customer is nonoptimal. The long-term value for a 100% service level for preferred customers is a business decision. The total supply to a nonpreferred customer is allowed to be less than or equal to the customer demand, including zero, which would indicate dropping the customer altogether. Customers with multiple product demands are broken down into separate entities with single product demands. DCs are storage locations that are geographically separate from the production plants and the customers. Bulk liquid terminals, rail-to-truck transloaders, and warehouses are common examples of DC. Each DC has a set of viable inbound transport modes (e.g., rail, truck, vessel), a set of viable outbound transport modes (e.g., rail, truck, vessel), a piecewise-linear cost structure with both fixed and variable components, and maximum volume limitations. The production sources (plants) supply product to the system to satisfy the demand of the customers. Each source has a maximum production capacity, but not all may produce every product. Each production plant ships product via various outbound transport modes, has maximum capacity to produce each product in its portfolio, and has a linear cost structure with both fixed and variable components. The optimization problem is to determine how to satisfy customer demand using the supply chain network such that either costs are minimized or the total value is maximized. The key continuous variables in the model are flows (shipments) in the network, production amounts of the plants, and inventory at the network nodes. Integer variables are used for facility fixed costs, enforcing lower limits on shipments, modeling fixed shipment size, and modeling tiered costs.

The optimization can be posed as a single time period where inventory is ignored or as a multitime period problem where inventory can be carried forward from one time period to the next. The size of the network can vary widely from one problem to the next. But a typical problem can contain 5–10 production plants, dozens of DCs, hundreds of customers, and over 100 products.

As stated earlier, the solution of the network optimization problem is carried out with commercial modeling software. Two popular products are LogicNet Plus from IBM and Supply Chain Guru from LlamaSoft. They contain powerful utilities for defining the network to be modeled, entering model parameters (including geocoding for facility location), optimizing various scenarios, and displaying and analyzing results. These products use a preconfigured MILP model to solve for the optimal network design [704]. The products rely on commercial solvers such as CPLEX and Xpress to solve the MILP problem. Table 7.3 lists the key equations and constraints found in network optimization models.

Table 7.3. *Important equations/constraints of a supply chain network optimization model.*

Customer mass balance	Shipments received are limited to demand.
DC mass balance	Replenishments equal shipments to customers.
Facility selection	Which facilities of a set of facilities are used in the network.
Product assignment	What products are produced/handled at each facility
Facility capacity limits	Product volume must respect lower and upper limits.
Network topology	Specifications on the combination of facilities in the network
Production time limits	Production volume is limited by time available for running the plant.
Total cost	Facility costs (production and DC) + transportation

A model as described in Table 7.3 can be formulated as follows. The model spans the following sets: K = facilities, DCs, or production sources; R = customers; P = products; M = modes of transport; $F^{plant}(K)$ = facilities that are production plants.

The continuous decision variables are $S_{k,r,p,m}$, which specifies the flow of product-p from facility k to customer r using transportation mode m, and $S_{k,k',p,m}$, which specifies the flow of product p from facility k to another facility k' using transportation mode m.

Customer mass balance specifies that total supply to a customer must be less than or equal to the customer's demand:

$$\sum_{k,p,m} S_{k,r,p,m} \leq D_r \quad \forall \, r \in R. \tag{7.11}$$

DC mass balance specifies that no inventory is accumulated, so the mass shipped to the DC must equal the mass leaving the DC. Mass into the DC can arrive from plants as well as from other DCs. Mass leaving the DC can go to customers, as well as to other facilities:

$$\sum_{k',m,p,k\neq k'} S_{k',k,p,m} = \sum_{r,m,p} S_{k,r,p,m} + \sum_{k',m,p,k\neq k'} S_{k,k',p,m} \quad \forall \, k \in K. \tag{7.12}$$

Facility selection is modeled using the big-M formulation, defining the binary variable U_k to be true if any product flows through a facility with the big-M constant set equal to the maximum demand of the product:

$$\sum_{k',m,p} S_{k,k',p,m} + \sum_{r,m,p} S_{k,r,p,m} \leq \sum_p D_p^{max} U_k \quad \forall \, k \in K, \, U_k \in \{0,1\}. \tag{7.13}$$

U_k is forced to be false if less than one unit of material is flowing through the facility:

$$\sum_{k',m,p} S_{k,k',p,m} + \sum_{r,m,p} S_{k,r,p,m} \geq U_k \quad \forall k \in K, \, U_k \in \{0,1\}. \tag{7.14}$$

In addition to facility use, it is useful to have another set of indicator variables, $U_{k,p}$, that define whether a facility is handling a certain product. $U_{k,p}$ is set false if less than one unit of product p flows through (DC) or from (production plant) facility k and true otherwise:

$$\sum_{k',m} S_{k,k',p,m} + \sum_{r,m} S_{k,r,p,m} \leq \sum_p D_p^{max} U_{k,p} \quad \forall k \in K, \, p \in P, \, U_{k,p} \in \{0,1\}, \tag{7.15}$$

$$\sum_{k',m} S_{k,k',p,m} + \sum_{r,m} S_{k,r,p,m} \geq U_{k,p} \quad \forall k \in K, \, p \in P, \, U_{k,p} \in \{0,1\}. \tag{7.16}$$

The binary variables defined in (7.13)–(7.16) can be used to enforce facility throughput constraints. The total throughput of all products moving through a facility is constrained to be between a minimum C_k^{min} and a maximum C_k^{max} limit:

$$C_k^{min} U_k \leq \sum_{r,m,p} S_{k,r,p,m} + \sum_{k',m,p} S_{k,k',p,m} \leq C_k^{max} U_k \quad \forall k \in K. \tag{7.17}$$

The amount of an individual product flowing through the facility can be similarly constrained:

$$C_{k,p}^{min} U_{k,p} \leq \sum_{r,m} S_{k,r,p,m} + \sum_{k',m} S_{k,k',p,m} \leq C_{k,p}^{max} U_{k,p} \quad \forall k \in K, \, p \in P. \tag{7.18}$$

The set of facilities used in the network can be constrained in general by

$$\Theta_i^{min} \leq \sum_{k \in G_i(K)} U_k \leq \Theta_i^{max} \quad \forall i \in I, \tag{7.19}$$

where index i is a constraint number to represent multiple constraints, with set $G_i(K)$ being the ith group of facilities whose total count is to be constrained, and Θ_i^{min} and Θ_i^{max} limits.

For example, consider a network with 12 DCs labeled T1 through T12. Equation (7.19) could represent a wide variety of constraints such as

- "use at least two of the terminals T1, T2, T3, T4":

$$2 \leq U_{T1} + U_{T2} + U_{T3} + U_{T4}; \tag{7.20}$$

- "use between two and four of terminals T1, T3, T6, T7":

$$2 \leq U_{T1} + U_{T3} + U_{T6} + U_{T7} \leq 4; \tag{7.21}$$

- "use any two of (T1, T2 and T3) AND either T8 or T12":

$$2 \leq U_{T1} + U_{T2} + U_{T3} \leq 2, \quad 1 \leq U_{T8} + U_{T12} \leq 1. \tag{7.22}$$

Finally, the total time used at a production facility cannot exceed the total time available for actual production. The actual production time is calculated using an average production rate $\dot{r}_{k,p}$ for product p at plant k:

$$\sum_{r,m,p} \frac{S_{k,r,p,m}}{\dot{r}_{k,p}} + \sum_{k',m,p} \frac{S_{k,k',p,m}}{\dot{r}_{k,p}} \leq T_k \quad \forall k \in F^{plant}(K). \tag{7.23}$$

The first summation covers all direct shipments from plants to customers, and the second summation covers replenishments to other facilities. The total time available for production is given by T_k.

Facility costs comprise fixed costs, $FC_{k,p}$, incurred if the facility is open, and variable costs, $VC_{k,p}$, dependent on the volume of product processed. As indicated, both costs can be product dependent. The facility costs are totaled across all facilities and products:

$$C_K = \sum_{k,p} FC_{k,p} U_{k,p} + VC_{k,p} \left(\sum_{r,m} S_{k,r,p,m} + \sum_{k',m} S_{k,k',p,m} \right). \tag{7.24}$$

The freight cost for each lane in the network connecting two facilities is calculated as the product of the quantity shipped between two facilities and the freight rate. The freight rates, $f_{k,k',m}$, are a function of the source facility, the destination facility, and the mode of transport. The total replenishment freight for the entire network, $C_{K \to K'}$, is the summation over all potential facility-to-facility moves:

$$C_{K \to K'} = \sum_{k,k',m,p} S_{k,k',m,p} f_{k,k',m}. \tag{7.25}$$

In a similar fashion, freight cost involving shipments to customers is calculated using facility-to-customer freight rates $f_{k,r,m}$:

$$C_{K \to R} = \sum_{k,r,m,p} S_{k,r,m,p} f_{k,r,m}. \tag{7.26}$$

The total network cost for the supply chain is the sum of the facility costs, replenishment freight cost, and customer freight cost:

$$J_{cost} = C_K + C_{K \to K'} + C_{K \to R}. \tag{7.27}$$

The optimization objective is to minimize the total network cost J_{cost}.

For Dow, supply chain network costs easily reach into the hundreds of millions of dollars for a typical business. Even modest reductions in supply chain costs represent significant savings. A recent project involved a performance product producer with the objective of optimizing supply chain and freight costs to support a "low cost to serve" business strategy. An important business objective was to determine the appropriate storage capacities to handle DC replenishments that would mitigate supply chain and plant reliability constraints. The project focused on the North America and Pacific area products. The study identified a short-term solution that produced a cost reduction greater than $500,000 over five years of future demand. The new network design improved the service level by 31%, and the carbon footprint was reduced by 1,120 mT.

7.5 ▪ Concluding Remarks

As shown with the three industrial applications at Praxair, ExxonMobil, and Dow Chemical, MILP is a very powerful tool that is being used extensively for making operational and design decisions. There is no question that the combination of very significant computational advances in MILP over the last decade [287], combined with the availability of modeling systems such as GAMS and AIMMS [1025], which have allowed the rapid development and deployment of these tools, have had an important impact in terms of improving the quality of the decisions and achieving large economic savings that have helped to make the chemical process industry become more competitive.

As for future needs and challenges, aside from the need to effectively solve ever larger MILP models, there is an increasing awareness of the need to account for uncertainty in these, mostly through robust optimization [245] and stochastic programming [1580, 1612]. However, current methods typically require one or two orders of magnitude greater computational effort, and therefore their practical applications in the industry have been very limited. The other major need that has been recognized is the need to model and solve MINLP problems [192, 1130]. Here again the level of complexity in these optimizations is greatly increased, although a few success stories have been reported with current MINLP methods [1579].

Chapter 8

Applications of Discrete Optimization in Medicine and Healthcare

Eva K. Lee

8.1 ▪ Introduction

Integer programming (IP) has long been a cornerstone for the advancement of business analytics in industrial, government, and military applications. In recent years, there has been an increasing use of IP in medicine, healthcare, public services, energy, and sustainability domains. In particular, many complex problems in medicine and healthcare can be formulated as IP instances, and many problems in the area of health logistics and operations are classic facility location, scheduling, and resource allocation problems with special characteristics. IP can also serve as a general tool for tackling classification problems, which are often encountered in biology, medical diagnosis, and early disease detection.

In Section 8.2, we briefly summarize some applications of IP to healthcare in recent years and cite a sample of references. Section 8.3 describes our own experience of two specific IP models derived for classification and optimal treatment design that have been used successfully in medicine. We conclude with some challenges in Section 8.4.

8.2 ▪ Overview of Integer Programming in Medicine and Healthcare

8.2.1 ▪ Diagnosis and Detection

Diagnosis and early detection of disease relate to classifying patients into various disease categories based on symptoms and/or noninvasive tests. In general, feature selection is first performed, followed by classification. Feature selection determines the features that can best discriminate populations, while pattern classification establishes the rule for classifying new observations.

Berretta et al. [236] formulate IP models to find a small subset of genes that relate well to the different types of cancer. Their classification solution maintains the similarities within groups while maximizing their dissimilarities. They apply the algorithms to microarray data related to Alzheimer's disease [1357].

Mixed-integer programming (MIP) has been used to determine the coefficients of linear discriminant functions. For two-group classification problems, IP models have been proposed to minimize the number of misclassifications, the expected cost of misclassification, the sum of error deviations, and the difference between the mean scores of two groups. The solution techniques for these problems include a heuristic algorithm based on linear programming (LP) relaxation, a decomposition method, and a divide and conquer algorithm. For multigroup classification problems, the approaches include MIP models involving a general single-function classification model and a general multiple-function classification model that minimize the total number of misclassifications.

In medicine and healthcare, various forms of MIP models have been proposed to solve two-group or multigroup classification problems [779, 1149, 1808], classification with imbalanced data [1480], classification with missing data [574], and robust classification [1660]. Applications include determining the presence and type of diseases including heart disease [1143, 1426], epigenetics in early cancer prediction [702], epilepsy [432], cancer [256, 1294], Alzheimer's disease [1155, 1357], deciding tumor types [538], predicting protein localization sites [960], identifying tumor shape and volume for treatment of sarcoma [1148], predicting cell disruption for drug delivery [1150], locating treatment units [516], and tracking nuclei for cell cycle analysis [1182].

8.2.2 ▪ Treatment Design

Optimization for radiotherapy treatment has been investigated by numerous researchers. Lee and Zaider [1158, 1159] formulate a model for brachytherapy, the implantation of radioactive seeds for cancer treatment via an MIP approach. The resulting instances are dense and difficult to solve via traditional approaches; this motivated the development of a new polyhedral theory related to hypergraphs [651, 1159]. External beam radiation therapy includes stereotactic radiation and radiosurgery, three-dimensional conformal planning, intensity-modulated radiation therapy (IMRT) [1145–1147, 1520, 1587], and intensity-modulated proton therapy (IMPT) [400].

Dose-volume constraints are critical for modeling radiation therapy treatment. IP is used by setting binary variables to enumerate the voxels that receive a dose lower or higher than a given threshold [184, 1127, 1145]. Such a representation often entails a large number of variables, and the solution can be approximated by relaxation [522].

Stereotactic radiation delivers a high radiation dose with great precision. Two problems are often considered: isocenter optimization and sector duration optimization. An isocenter is the focal point of radioactive sources. Isocenter optimization determines the number and locations of isocenters that optimize the treatment plan quality, where quality is gauged by metrics such as percentage of target volume receiving 100% of the prescribed dose [1145], dose inhomogeneity and conformity [1145, 1146], and treatment time [637]. Sector duration optimization decides the size of the collimator and radiation duration of each isocenter. The objective is commonly chosen as a weighted penalty function of the dose received beyond the allowable range [1232]. Different constraints are satisfied according to the clinician's preferences, including the lower and upper bounds of the dose to voxels and the radiation duration, approximate conformity measures, and number of isocenters [705, 1145]. Sampling is often performed to balance the computational effort with the realism of the models.

IMRT delivers radiation through modulation of beamlets by a multileaf collimator for each beam. The treatment is carefully planned by optimizing the beam direction and intensity of radiation (usually referred to as fluence map optimization). Constraints that are frequently added into the model include dose-volume histogram constraints [1127, 1145], normal tissue complication probability [1768], and equivalent uniform dose [1768]. In addition to fluence maps, other objectives have been studied, such as beam's eye view [1527], maximum angle separation [553], mean organ at risk data [634], and entropy and Fourier transforms [1712]. Multiobjective models have also been considered [948]. Many global optimization techniques have been applied, including heuristic algorithms such as simulated annealing [46] and genetic algorithms [682]. Ehrgott et al. [655] address the beam selection problem by IP models, in which the directions of radiation delivery are optimally selected. Wake et al. [1862] formulate an IP model to minimize the total treatment time using multileaf collimators. Lee et al. [1146, 1147] tackle IMRT beam orientation and intensity with dose volume via a single IP problem and solve it using a branch-and-cut approach.

8.2.3 ▪ Health Logistics and Operations

Healthcare logistics and operations can be formulated into classical models such as facility location, resource allocation, and scheduling. In healthcare center location problems, researchers consider practical constraints such as weather conditions [1364], seasonal changes [1383], distribution time [349], and geographical factors [912]. The objective may be maximizing coverage [1275, 1542], minimizing total costs including patient travel and treatment costs, and minimizing excess capacity utilization [516]. There may be multiple objectives [1733]. Case studies have been conducted in Ghana [946], Italy [349], Germany [1733], and the United States and Canada [1843]. Researchers have also investigated the best locations for ambulances and other emergency vehicles [93, 335].

Health logistics, including medicine inventory and allocation, have been investigated by many researchers. Pierskalla [1482] addresses inventory and allocation problems arising from blood bank supply chains. Van de Klundert et al. [1814] build an IP model combined with a dynamic programming DP technique to minimize the cost of transportation and inventory involved in the flow of sterile instruments. Rauner and Bajmoczy [1552] investigate the problem of allocating medical materials. They consider the costs of facility maintenance, technician training, and hospitality, as well as future costs. IP is proposed to maximize the survival and quality-of-life benefits. Oddoye et al. [1411] formulate a multiobjective IP model to minimize the allocation costs. They consider the number of death cases, incidence, prevalence of years of life lost, and loss of quality of life. In our own work [1144, 1153], MIP is used for resource allocation optimization within simulations where at each iteration the model simulates the operations logistics in the facility and at the same time optimizes the resource allocation for improved performance.

Scheduling problems arise for patients, providers, and processes. In patient scheduling, the waiting time and medical costs need to be minimized while the equipment utilization is optimized. Persson and Persson [1467] use IP to model the scheduling problem where patients may be transferred to other hospitals to decrease waiting times. Nurse scheduling involves assigning nurses with different skills to satisfy operational needs and personal preferences as much as possible [360, 566]. Operating room scheduling assigns operating rooms to minimize the idle time and overtime while maximizing the satisfaction of patients [289, 1433]. Solution strategies for scheduling include branch-and-cut [701], Lagrangian relaxation [163], heuristics [1720], genetic algorithms [1459], goal programming [1433], and tabu search [258].

8.2.4 ▪ Public Health

In public health, researchers are interested in methods to prevent the spread of diseases. Jacobson et al. [984] perform a pilot study using IP to show how to guide vaccine selection and procurement. They also build an IP model to evaluate the economic benefit of having a combination of vaccines available. In both models, binary variables are used to indicate the usage of each vaccine. Hall et al. [898] use an IP model to design schedules that minimize the weighted sum of the immunization cost and extra immunization cost for childhood immunization programs.

Resource allocation is often an important element in controlling infectious diseases. Parker [1452] design resource allocation models for controlling malaria. Stinnett and Paltiel [1722] develop an MIP framework to maximize the sum of health benefits subject to budget constraints. Brandeau [337] review operating room tools for allocating resources to control infectious diseases. Another issue involved in public health is the distribution of condoms to control the spread of sexually transmitted diseases. Lasry et al. [1129] use IP to develop a multilevel resource allocation model for HIV prevention. Lee et al. [1152, 1153] model resource allocation for large-scale medical countermeasure dispensing and disaster relief efforts via MIP within a real-time simulation framework.

8.3 ▪ Specific Applications

We now highlight two IP models that we developed for classification and personalized treatment planning, respectively. The models have some commonalities in that they are both denser than traditional IP instances, they are *NP-hard*, and they are computationally challenging to solve. Each requires advances in polyhedral theory and a specialized branch-and-cut implementation to solve practical instances.

8.3.1 ▪ Optimization-Based Classifier: Discriminant Analysis via MIP

Readmission is a key challenge in emergency department (ED) performance. In particular, avoidable readmissions (i.e., readmissions resulting from an adverse event that occurred during the initial admission or from inappropriate care coordination following discharge) through the ED have become a major burden on the U.S. health system; see [1326]. Recent research shows that nearly one in five patients is readmitted to the discharging hospital within 30 days of discharge; these readmissions accounted for $17.8 billion in Medicare spending in 2004 [1427].

Numerous studies have been conducted to identify frequently readmitted patients' characteristics and construct patient profiles to aid hospitals in predicting these patients. These studies have identified a number of demographic and clinical factors that are thought to significantly correlate with readmission. Other factors concerning hospital operations have also been investigated. Various statistical tools have been used to identify patient factors that are associated with readmission [54, 269, 918, 1068]. Westert et al. [1885] conduct an international study, including three U.S. states and three countries, to find patterns in the profiles of readmitted patients. The findings are divided into demographic and social factors, clinical factors [269, 1716], and hospital operations factors [215, 559, 1015, 1655, 1828, 1885]. A study of 26 readmission risk prediction models reviews 7,843 citations and concludes that none of the models analyzed could suitably predict future hospital readmissions [1036]. Allaudeen et al. [55] note that healthcare personnel could not accurately predict the readmission of patients discharged from their own hospitals; however, the conclusions from these studies may be premature, given that many of the analyses were performed via logistic regression on only subsets of data. We recently [1144, 1156] published a readmission study in which, for the first time, a predictive model can incorporate comprehensive factors related to demographics and socioeconomic status, clinical and hospital resources, operations and utilization, and patient complaints and risk factors for global prediction. We now briefly describe this optimization-based predictive model.

Suppose we have n entities (e.g., patients) from K groups (e.g., returning or nonreturning) with m features. Let $\mathbf{G} = \{1,\dots,K\}$ be the group index set, $\mathbf{O} = \{1,\dots,n\}$ be the entity index set, and $\mathbf{F} = \{1,\dots,m\}$ be the feature index set. Also, let $\mathbf{O}_k, k \in \mathbf{G}$, and $\mathbf{O}_k \subseteq \mathbf{O}$ be the entity set that belongs to group k. Moreover, let $\mathbf{F}_j, j \in \mathbf{F}$, be the domain of feature j, which could be the space of real, integer, or binary values. The ith entity, $i \in \mathbf{O}$, is represented as $(y_i, \mathbf{x}_i) = (y_i, x_{i1},\dots,x_{im}) \in \mathbf{G} \times \mathbf{F}_1 \times \cdots \times \mathbf{F}_m$, where y_i is the group to which entity i belongs and $(x_{i1},\dots,x_{im})$ is the feature vector of entity i. The classification model finds a function $\Psi : (\mathbf{F}_1 \times \cdots \times \mathbf{F}_m) \to \mathbf{G}$ to classify entities into groups based on a selected set of features.

Let π_k be the prior probability of group k and $f_k(\mathbf{x})$ be the conditional probability density function (pdf) for the entity $\mathbf{x} \in R^m$ of group $k, k \in \mathbf{G}$. Also let $\alpha_{hk} \in (0,1)$, $h,k \in \mathbf{G}, h \neq k$, be the upper bound for the misclassification percentage that group h entities are misclassified into group k. Discriminant analysis via MIP (DAMIP) seeks a partition $\{P_0, P_1,\dots,P_K\}$ of R^K, where $P_k, k \in \mathbf{G}$, is the region for group k and P_0 is the reserved judgment region with entities for which group assignments are reserved (for potential further exploration).

Let u_{ki} be a binary variable indicating whether or not entity i is classified into group k. Mathematically, DAMIP can be formulated as follows [779, 1146, 1149]:

$$\max \quad \sum_{i \in \mathbf{O}} u_{y_i i} \tag{8.1}$$

$$\text{s.t.} \quad L_{ki} = \pi_k f_k(\mathbf{x}_i) - \sum_{h \in \mathbf{G}, h \neq k} f_h(\mathbf{x}_i)\lambda_{hk} \quad \forall i \in \mathbf{O}, k \in \mathbf{G}, \tag{8.2}$$

$$u_{ki} = \begin{cases} 1 & \text{if } k = \text{argmax} \ \{0, L_{hi} : h \in \mathbf{G}\} \\ 0 & \text{otherwise} \end{cases} \quad \forall i \in \mathbf{O}, k \in \{0\}\cup\mathbf{G}, \tag{8.3}$$

$$\sum_{k \in \{0\}\cup\mathbf{G}} u_{ki} = 1 \quad \forall i \in \mathbf{O}, \tag{8.4}$$

$$\sum_{i:i\in\mathbf{O}_h} u_{ki} \leq \lfloor \alpha_{hk} n_h \rfloor \quad \forall h,k \in \mathbf{G}, h \neq k, \tag{8.5}$$

$$u_{ki} \in \{0,1\} \quad \forall i \in \mathbf{O}, k \in \{0\}\cup\mathbf{G},$$

$$L_{ki} \quad \text{unrestricted in sign} \quad \forall i \in \mathbf{O}, k \in \mathbf{G},$$

$$\lambda_{hk} \geq 0 \quad \forall h,k \in \mathbf{G}, h \neq k.$$

The objective function (8.1) maximizes the number of entities classified into the correct group. Constraints (8.2) and (8.3) govern the placement of an entity into each of the groups in $\mathbf{G}$ or the reserved judgment region. Thus, the variables L_{ki} and λ_{hk} provide the shape of the partition of the groups in the **G** space. Constraint (8.4) ensures that an entity is assigned to exactly one group. Constraint (8.5) allows the users to preset the desirable misclassification levels, which can be specified as overall errors for each group, pairwise errors, or overall errors for all groups together. With the reserved judgment in place, the mathematical system ensures that a solution that satisfies the preset errors always exists. Once a classification rule is obtained, a new entity's group status can be obtained by calculating the maximum value of L_{ki}.

In the ED readmission study, entities correspond to patients. The input attributes for each patient include comprehensive demographics, socioeconomic status, clinical information, hospital resources and utilization, and disease behavioral patterns. There are two statuses for patients: they come back to the hospital for visits (return group), or they do not come back (nonreturn group). The classification aims to uncover from the set of all attributes a set of discriminatory attributes that can classify each patient into the return or nonreturn group. We seek to identify the rule that offers the best predictive capability. The associated optimal decision variable values (L_{ki} and λ_{hk}) form the classification rule, which consists of the discriminatory attributes; examples include patient chief complaint, diagnosis, whether intravenous (IV) antibiotics were ordered, trainee and (or) resident involved, primary nurse, and time when the patient received an ED bed to time until first medical doctor arrived.

We give the results for two hospitals [1156]. There were 27,534 ED patients at hospital 1,996 (3.62%) of whom returned within 72 hours, and there were 39,327 at hospital 2,1523 (3.87%) of whom returned within 72 hours. All patients went home after the first ED visit. In our analysis, the training set is 15,000 and 20,000, respectively, and the blind prediction set consists of the rest of the patients. Table 8.1 contrasts the DAMIP results with other classification methods. Uniformly, other classification methods suffer from group imbalance, and the classifiers tend to place all entities into

Table 8.1. *Comparison of DAMIP results against other classification methods. Reprinted with permission from American Medical Informatics Association* [1156].

Hospital 1	Training set: 15,000			Blind prediction set: 12,534		
	10-fold cross-validation accuracy			Blind prediction accuracy		
Classification method	Overall	Nonreturn	Return	Overall	Nonreturn	Return
Linear discriminant analysis	96.3%	99.6%	5.5%	96.1%	99.6%	5.3%
Naïve Bayesian	51.6%	50.3%	87.0%	51.7%	50.2%	89.2%
Support vector machine	96.5%	100.0%	0.0%	96.2%	100.0%	0.0%
Logistic regression	96.5%	99.8%	5.9%	96.3%	99.8%	8.3%
Classification tree	96.6%	99.9%	4.4%	96.3%	100.0%	3.0%
Random forest	96.6%	100.0%	1.5%	96.3%	100.0%	1.9%
Nearest shrunken centroid	62.7%	62.9%	50.0%	48.7%	48.2%	64.7%
DAMIP	**83.1%**	**83.9%**	**70.1%**	**82.2%**	**83.1%**	**70.5%**
Hospital 2	Training Set: 20,000			Blind Prediction: 19,327		
	Overall	Nonreturn	Return	Overall	Nonreturn	Return
LDA	96.2%	100.0%	0.1%	96.0%	100.0%	0.3%
Naïve Bayesian	53.4%	52.2%	83.9%	54.4%	53.2%	84.2%
SVM	96.3%	100.0%	0.0%	96.0%	100.0%	0.0%
Logistic regression	96.3%	100.0%	0.0%	96.1%	99.9%	3.3%
Classification tree	96.2%	100.0%	0.0%	96.0%	100.0%	0.0%
Random forest	96.2%	100.0%	0.5%	96.1%	100.0%	0.5%
Nearest shrunken centroid	60.5%	60.6%	50.1%	45.8%	45.1%	61.2%
DAMIP	**80.1%**	**81.1%**	**70.1%**	**80.5%**	**81.5%**	**70.0%**

the nonreturn group. In particular, linear discriminant analysis, support vector machines, logistic regression, classification trees, and random forests placed almost all patients (>99%) into the nonreturn group by sacrificing the very small percentage of "return" patients. This table also shows the importance of reporting the classification accuracy for each group, in addition to the overall accuracy.

Mathematically, we have proved that DAMIP is *NP-hard* and that the resulting classification rule is *strongly universally consistent*, given that the Bayes optimal rule for classification is known; see [345]. The predictive model maximizes the number of correctly classified cases; therefore, it is robust and not skewed by errors committed by observation values or by outliers. Computationally, the classification rules from DAMIP appear to be insensitive to the specification of prior probabilities yet capable of reducing misclassification rates when the number of training entities from each group is different, and the DAMIP model generates stable classification rules regardless of the proportions of training entities from each group [344,779,1143,1154,1156,1533]. Furthermore, the misclassification rates using the DAMIP method are consistently lower than other classification approaches on both simulated data and real-world data. It has outperformed other classifiers in diverse biomedical applications, including epigenetics and methylation, vaccine immunogenicity, Alzheimer's disease, and the prediction of readmission and practice variance [702,1083,1156,1374,1533,1736].

8.3.2 ▪ Novel TCP-Driven PET-Image-Guided Treatment-Planning Model

Each year almost a million cancer patients in the United States receive some form of radiation therapy [67,105]. Radiation is delivered using either external beam technology or a procedure known as brachytherapy. Brachytherapy uses a radioactive substance sealed in needles, seeds, wires, or catheters, which are placed directly (permanently or temporarily) into or near the cancer. This allows a full tumoricidal effect to eradicate the tumor from within the cancer site, while ensuring that minimal radiation reaches healthy surrounding tissues.

We have recently developed a novel MIP model for high-dose-rate (HDR) brachytherapy treatment design [1157]. The model incorporates tumor control probability (TCP) into the planning objective, which depends upon a highly complex function that models the responses of cancer cells and normal cells to radiation. This is distinct from the dose-based planning that is commonly employed in current treatment design. Second, positron emission tomography (PET) information, which relates cancer cell proliferation and distribution, is incorporated within the objective and the constraints, facilitating targeted, escalated dose delivery to improve the overall clinical outcome of HDR treatment.

Specifically, we generalize the TCP based on a reliable biological model developed by [1946]. The formulas are derived using the birth and death processes. For brachytherapy, the TCP equation is

$$TCP(t) = \left[1 - \frac{S(t)e^{(b-d)t}}{1 + bS(t)e^{(b-d)t}\int_0^t \frac{dt'}{S(t')e^{(b-d)t'}}} \right]^n, \qquad (8.6)$$

where n is the initial number (at time $t = 0$) of tumor cells, $S(t)$ is the survival probability of tumor cells at time t, and b and d are the birth and death rates of these cells, respectively. In the TCP calculation, the time t in (8.6) is typically taken to be the duration of the treatment period or the expected remaining lifespan of the patient.

We employ a multiobjective MIP model for HDR brachytherapy treatment planning. The model incorporates the TCP as the objective function, in addition to the rapid dose fall-off function to ensure dose conformity to the tumor region.

Briefly, let x_j be a 0-1 indicator variable for recording placement or nonplacement of a seed in grid position j, and let t_j be the continuous variable for the dwell time of a seed in grid position j. The total radiation dose at voxel P is given by $\sum_j \dot{D}(\|P - X_j\|)t_j$, where X_j is a vector corresponding to the coordinates of grid point j, $\|\bullet\|$ denotes the Euclidean norm, and $\dot{D}(r)$ denotes the dose contribution per minute of a seed to a voxel that is r units away.

The target lower and upper bounds, L_P and U_P, for the radiation dose at voxel P are represented by the following dose constraints:

$$\sum_j \dot{D}(\|P - X_j\|)t_j \geq L_P,$$

$$\sum_j \dot{D}(\|P - X_j\|)t_j \leq U_P.$$

For each voxel P in each anatomical structure, a binary variable is used to capture whether or not the desired dose level is achieved.

The TCP-driven PET-image-guided dose-escalated multiobjective treatment model can be formulated as

$$\max \quad \sum_P (\eta_P v_P^L + \mu_P v_P^U),$$

$$\max \quad TCP(t)$$

$$\text{s.t.} \quad \sum_j \dot{D}(\|P - X_j\|)t_j \geq \text{PrDose} \cdot \lambda, \qquad P \in BTV, \qquad (8.7)$$

$$\sum_j \dot{D}(\|P - X_j\|)t_j + M_P(1 - v_P^L) \geq L_P, \qquad P \in PTV - BTV,$$

$$\qquad\qquad\qquad\qquad\qquad\qquad\qquad\qquad P \in OARs, \qquad (8.8)$$

$$\sum_j \dot{D}(\|P - X_j\|)t_j - N_P(1 - v_P^U) \leq U_P, \qquad P \in PTV - BTV,$$

$$\qquad\qquad\qquad\qquad\qquad\qquad\qquad\qquad P \in OARs, \qquad (8.9)$$

$$|BTV| + \sum_{p \in (PTV-BTV)} v_P^L \geq \alpha |PTV|, \qquad (8.10)$$

$$t_j \leq T_j x_j, \qquad (8.11)$$

$$\sum_j x_j \leq \text{MaxSeeds}, \qquad (8.12)$$

$$v_P^L, \quad v_P^U, \quad x_j \in \{0,1\}, t_j \geq 0,$$

Here, PTV represents the planning target volume and BTV represents the boost target volume, which pertains to the set of tumor voxels identified by the PET images. $OARs$ denote the organs at risk. PrDose represents the clinical prescribed dose to the tumor, and λ ($\lambda > 1$) represents the dose escalation factor. This factor is guided by clinicians as well as its effect on normal tissue complication. Constraint (8.7) ensures that the PET-identified tumor voxels receive escalated doses. In constraints (8.8) and (8.9), v_P^L and v_P^U are 0-1 variables. If a solution is found such that $v_P^L = 1$, then the lower bound for the dose level at point P is satisfied. Similarly, if $v_P^U = 1$, the upper bound at point P is satisfied (see constraint (8.9)).

The constants M_p and N_p are chosen appropriately for the *PTV* and for various *OARs*. For *PTV*, M_p corresponds to the underdose limit, whereas N_p corresponds to the overdose limit, and $L_p = \text{PrDose}$ corresponds to the prescription dose. M_p and N_p are strategically chosen so that the overall *PTV* dose remains relatively homogeneous (e.g., $(U_p + N_p)/(L_p - M_p) < 1.2$), as the clinicians desire. For the *OARs*, $U_p = 0$ and N_p represents the maximum dose tolerance that the organs can sustain, without inflicting severe and permanent harm. These values are determined from clinical findings and are part of the planning procedures and guidelines.

In constraint (8.10), α corresponds to the minimum percentage of tumor coverage required (e.g., $\alpha = 0.95$). Because all the PET-identified tumor voxels satisfy the prescribed dose bound (and beyond), we count the voxels in *PTV-BTV* and in *BTV* to ensure that overall they satisfy α percent of the tumor volume. Here, $|PTV|$ represents the total number of voxels used to represent the *PTV*. Constraint (8.10) thus corresponds to the coverage level that the clinician desires. In constraint (8.11), the duration t_j in grid position j is positive only when this position is selected. Its value is bounded by the maximum time limit T_j. The time is usually bounded by the length of the treatment session, which is usually between 20 and 30 minutes for HDR, depending on the tumor staging and prognosis condition. Constraint (8.12) limits the number of seeds used to *MaxSeeds*. The constant can be omitted; however, in some cases, clinicians know their desired numbers, which they tell the planner.

Note that *BTV* voxels are excluded in constraint (8.9) because there is no reason to place an upper bound on the dose to these tumor voxels. Constraint (8.7) ensures that no underdose for PET-identified voxels exists; thus, constraint (8.8) is unnecessary for these voxels.

The first objective is to find a treatment plan that satisfies as many bound constraints as possible; this helps to achieve rapid dose fall-off, ensuring conformity of the prescribed dose to the tumor. The parameters η_p and μ_p allow one to prioritize the importance of various anatomical structures. Using a weighted sum is important as it balances the volume of *PTV* versus the nearby *OARs*. The second objective function incorporates the TCP function, which depends on the time of the treatment, radioactive decay of the radioisotope, dose received, volume and density of tumor cells, and the biological radiosensitivity and radioresistancy of the normal and tumor cells.

We introduced an original branch-and-cut and local-search approach that couples new polyhedral cuts with matrix reduction and intelligent geometric heuristics. The result has been accurate solutions that are obtained rapidly [1157].

Table 8.2 highlights the TCP for three representative patients: *BTV* < 10% (small), 10–25% (medium), and > 25% (large) of *PTV*. In our study of 15 patients, the TCP of all escalated plans is over 70%. Specifically, when the BTV is boosted to over 40 Gy,[10] the resulting TCP is uniformly high (> 87%). We list the plans A–E according to improvements to the TCP. The best TCP (plan E) is achieved when we boost the PET-identified pockets to over 40 Gy while maintaining the *PTV* dose at 35 Gy [1157].

8.3.3 ▪ Solution Strategies

In both applications, we advance the computational strategies by using hypergraphic structures. Given a binary polytope $P = \{c^T x : Ax \leq b, x \in \{0, 1\}^n\}$ where A is a nonnegative $m \times n$ matrix and b is a nonnegative vector in R^m, we construct a conflict hypergraph for P where each variable in P corresponds to a node in the hypergraph;

---

[10]The gray (symbol: Gy) is a derived unit of ionizing radiation dose in the International System of Units (SI). It is defined as the absorption of one joule of radiation energy per kilogram of matter.

Table 8.2. *This table contrasts the TCPs in various plans. The boldface values represent the TCP associated with escalated plans.*

BTV/PTV ratio category	Small	Medium	Large
PTV (in cc)	82.5	131.5	89.7
PET-identified volume (*BTV* in cc)	5.2	25.5	26.0
Ratio: *BTV/PTV*	6.3%	19.39%	28.99%
Treatment-planning model	TCP		
A: Standard HDR plan (*PTV* dose = 35 Gy)	0.61	0.64	0.59
B: PET-guided escalated plan (*BTV* $\geq$ 37 Gy, *PTV* = 35 Gy)	0.87	0.94	0.97
C: PET-guided escalated plan (*BTV* $\geq$ 37 Gy, *PTV* = 33 Gy)	0.84	0.74	0.78
D: PET-guided escalated plan (*BTV* $\geq$ 40 Gy, *PTV* = 33 Gy)	0.98	0.95	0.96
E: PET-guided escalated plan (*BTV* $\geq$ 40 Gy, *PTV* = 35 Gy)	0.99	0.97	0.97

two or more nodes form an edge in the conflict hypergraph if and only if the associated variables cannot all be set to one (i.e., they form a dependent set). Conflict k-hypergraphs are introduced in Easton et al. [651] and generalized by Maheshwary [1254] and Lee and Maheshwary [1151]. A k-hypergraph is a hypergraph where the edges are subsets of k nodes. In [345] we establish linear systems and prove the conditions in which they form an edge in a conflict hypergraph for DAMIP. We use these results to generate polyhedral cutting planes and hypercliques and solve DAMIP instances within a branch-and-cut framework.

For the cancer instances, as reported in [1157], we tackle dose volume based and biological tumor control objectives simultaneously using a branch-and-cut and local-search approach. We caution that because TCP is highly nonlinear, it is difficult to convexify or linearize for actual branch-and-cut solution exploration. Specifically, we solve the MIP instance with the dose volume based objective via a branch-and-cut algorithm that couples new polyhedral cuts with matrix reduction and intelligent geometric heuristics algorithms. When we obtain an integer solution, or when a heuristic within the branch-and-cut setting returns a feasible solution, we perform a local search to examine the TCP values across the entire neighborhood. Given a seed configuration with dwell times, we calculate the associated TCP based on the resulting *PTV* and PET-pocket dose volume histograms. We then keep the best solution (i.e., the solution with the maximum TCP value) as the incumbent solution. The local search involves swapping and a hybrid genetic algorithm, where one can rapidly examine the neighborhood space to identify the best TCP-value solutions.

Such an approach guarantees the return of a feasible solution, while exploiting the best possible TCP values within the neighborhood feasible space. Recently, we have made further advances in directly addressing the TCP objective by forming mixed-integer nonlinear programming (MINLP) problems.

8.4 ▪ Open Challenges

Optimization has long been a cornerstone for the advancement of business analytics in industrial, government, and military applications. It is now becoming more common in advancing and transforming clinical and translational research and the healthcare delivery system. In particular, multisource data system modeling and big data analytics and technologies play an increasingly important role in modern healthcare enterprises. Many problems can be formulated as mathematical programming models with discrete decision variables and analyzed using sophisticated optimization theory and computational techniques. However, this human-centric system poses unique challenges with uncertainties in the description of problem parameters and in stochastic service times.

Further, biological knowledge, such as the TCP presented herein, can be complex and difficult to model. Challenges remain in the ability to develop realistic models that can capture the actual processes, and the ability to solve the instances and return solutions that are practical and of clinical relevance. The medical and healthcare domains offer a rich and stimulating environment for the theoretical and computational growth of the IP community. The ability of our methods to have an impact on healthcare and medicine and to be broadly adopted depends in large part on close collaboration with domain experts, including hospital administrators and frontline healthcare workers.

Part III

Conic Linear Optimization

Chapter 9

Conic Linear Optimization

Miguel F. Anjos

Conic linear optimization (CLO) refers to the problem of optimizing a linear function over the intersection of an affine space and a closed convex cone. These problems are thus a particular class of convex optimization problems. We focus particularly on the special case where the cone is chosen as the cone of positive semidefinite (PSD) matrices for which the resulting optimization problem is called a semidefinite optimization (SDO) problem. The class of SDO problems includes linear optimization (LO) problems as a special case, namely when all the matrices involved are diagonal. Another special case of SDO is second-order cone optimization (SOCO), which corresponds to optimizing over the second-order cone (SOC), also known as the Lorentz cone. As the case of LO is discussed in detail in Part I of this book, we focus here on the use of other cones. Although most research has focused on the PSD cone, the SOC is starting to receive more attention, due to its numerous practical applications, as shown for example in Chapter 12.

This chapter introduces CLO, and the subsequent three chapters present applications in control engineering, truss topology design, and financial engineering. The reader wishing to delve deeper into the area is referred to the books [87, 332, 1898] and their extensive bibliographies.

9.1 ▪ Fundamentals

9.1.1 ▪ Semidefinite Optimization

SDO is concerned with the optimization of a linear (or possibly convex quadratic) function of a matrix variable subject to linear constraints on the elements of the matrix, and the additional constraint that the matrix must be symmetric PSD, i.e., that all its eigenvalues are nonnegative.

SDO problems are an important class of optimization problems for several reasons. First, SDO problems are solvable in polynomial time. This means that any problem that can be expressed using SDO is also solvable in polynomial time. Second, SDO problems can be solved efficiently in practice. This can be done by using one of the software packages available, or alternatively by implementing a suitable algorithm. Third, SDO can be used to obtain tight approximations for hard problems in integer and global optimization. Fourth, SDO has a variety of applications, as exemplified in the companion chapters on systems and control (Chapter 10), structural engineering (Chapter 11), and finance (Chapter 12).

SDO has a number of similarities with LO. Like LO problems, SDO problems also come in pairs. One of the problems is referred to as the *primal* problem, and the second one is the *dual* problem. Either problem can be chosen as the primal, since the two problems are dual to each other. The most common standard formulation of SDO is as follows:

$$
\begin{array}{llll}
\text{(P)} & \inf & \langle C, X \rangle & \qquad \text{(D)} \quad \sup \quad b^T y \\
& \text{s.t.} & \langle A_i, X \rangle = b_i, i = 1, \dots, m, & \qquad \text{s.t.} \quad \sum_{i=1}^{m} y_i A_i + S = C, \qquad (9.1) \\
& & X \succeq 0, & \qquad \qquad \quad S \succeq 0,
\end{array}
$$

where (P) denotes the primal problem and (D) the dual problem; the variables X and S are in $\mathscr{S}^n$, the space of $n \times n$ real symmetric matrices; $X \succeq 0$ denotes that the matrix X is PSD; the data matrices $A_i \in \mathscr{S}^n$ and $C \in \mathscr{S}^n$ may be assumed to be symmetric without loss of generality; and $b \in \mathbb{R}^m$ and $y \in \mathbb{R}^m$ are column vectors. We use the inner product between two matrices in $\mathscr{S}^n$ defined as

$$
\langle R, S \rangle := \operatorname{trace}(RS) = \sum_{i=1}^{n} \sum_{j=1}^{n} R_{i,j} S_{i,j},
$$

where $\operatorname{trace} M$ denotes the trace of the square matrix M, which is the sum of the diagonal elements of M. It is normally assumed, without loss of generality, that the matrices A_i, $i = 1, \dots, m$, are linearly independent. When it is known from the context that the optimal values are attained, it is common to replace inf by min and sup by max. An example of an SDO problem for which the optimum is not attained is given in Example 9.6 below.

PSD matrices have numerous properties; see, e.g., [956, Section 7.7]. For a 2×2 symmetric matrix, the necessary and sufficient conditions for positive semidefiniteness are

$$
\begin{pmatrix} x_{11} & x_{12} \\ x_{12} & x_{22} \end{pmatrix} \succeq 0 \Longleftrightarrow x_{11} \geq 0, x_{22} \geq 0 \quad \text{and} \quad x_{11} x_{22} - x_{12}^2 \geq 0.
$$

This is a special case of Theorem 9.1 below. To state the theorem, we need the following definition.

Definition 9.1. *If $X \in \mathscr{S}^n$, then for every nonempty subset $I \subseteq \{1, 2, \dots, n\}$, the principal submatrix of X corresponding to I, denoted $X(I)$, is the square submatrix with rows and columns indexed by I. The determinant of $X(I)$ is called the* principal minor *of X corresponding to I.*

Theorem 9.1 [956]. *For $X \in \mathscr{S}^n$, X is PSD if and only if all the principal minors of X are nonnegative.*

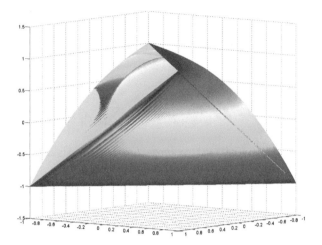

Figure 9.1. *Visualization of the feasible set of (9.2) for $n = 3$.*

Example 9.2. Our first example of an SDO problem is

$$
\begin{aligned}
\min \quad & \langle C, X \rangle \\
\text{s.t.} \quad & x_{ii} = 1, \quad i = 1, \ldots, n, \\
& X \succeq 0.
\end{aligned}
\tag{9.2}
$$

Problem (9.2) arises in SDO approaches to problems with binary variables; see, e.g., [1897].

The feasible set of (9.2) is the set of all symmetric $n \times n$ matrices that are PSD and have ones on the diagonal. Note that Theorem 9.1 implies that $x_{ij}^2 \leq 1$ holds for every off-diagonal element of X feasible for (9.2).

For $n = 3$, the feasible set of (9.2) has the form

$$
\left\{ \begin{pmatrix} x \\ y \\ z \end{pmatrix} \in \mathbb{R}^3 : \begin{pmatrix} 1 & x & y \\ x & 1 & z \\ y & z & 1 \end{pmatrix} \succeq 0 \right\}.
$$

Figure 9.1 visualizes this set in $\mathbb{R}^3$. The four "corners" in Figure 9.1 correspond to the matrices

$$
\begin{pmatrix} 1 & 1 & 1 \\ 1 & 1 & 1 \\ 1 & 1 & 1 \end{pmatrix}, \begin{pmatrix} 1 & 1 & -1 \\ 1 & 1 & -1 \\ -1 & -1 & 1 \end{pmatrix}, \begin{pmatrix} 1 & -1 & 1 \\ -1 & 1 & -1 \\ 1 & -1 & 1 \end{pmatrix},
$$

and

$$
\begin{pmatrix} 1 & -1 & -1 \\ -1 & 1 & 1 \\ -1 & 1 & 1 \end{pmatrix}. \quad\blacksquare
$$

The dual SDO problem in (9.1) can equivalently be written without using the dual variable S:

$$
\begin{aligned}
\sup \quad & b^T y \\
\text{s.t.} \quad & C - \sum_{i=1}^{m} y_i A_i \succeq 0.
\end{aligned}
$$

The inequality is interpreted according to the partial order induced on $\mathcal{S}^n$ by the PSD cone. This is called the Löwner partial order and is defined as

$$A \succeq B \quad \Leftrightarrow \quad A - B \succeq 0.$$

The constraint

$$C - \sum_{i=1}^{m} y_i A_i \succeq 0$$

is often called a *linear matrix inequality (LMI)*. LMIs can be used to express a wide array of problems in the areas of system and control theory, such as matrix norm inequalities and Lyapunov and convex quadratic matrix inequalities [331]. A simple application to eigenvalue optimization is shown in the next example.

Example 9.3. Suppose we wish to minimize the maximum eigenvalue of a symmetric matrix $A(y)$ whose entries depend linearly on the variable y:

$$A(y) = A_0 + y_1 A_1 + y_2 A_2 + \cdots + y_m A_m$$

for some $m \geq 1$. The problem of interest is thus

$$\min_{y \in \mathbb{R}} \lambda_{\max}(A(y)), \tag{9.3}$$

where $\lambda_{\max}(R)$ denotes the largest eigenvalue of the matrix R.

To formulate this problem using SDO, we use the observation that (9.3) is equivalent to

$$\begin{aligned} \min \quad & t \\ \text{s.t.} \quad & t \geq \lambda_{\max}(A(y)). \end{aligned}$$

Since $t \geq \lambda_{\max}(A(y))$ is equivalent to $\lambda_{\min}(tI - A(y)) \geq 0$, and hence to $tI - A(y) \succeq 0$, problem (9.3) can be equivalently expressed in the form

$$\begin{aligned} \max \quad & -t \\ \text{s.t.} \quad & tI - A_0 - y_1 A_1 - y_2 A_2 - \cdots - y_m A_m \succeq 0, \end{aligned}$$

which is an SDO problem in the dual form above.

By similar reasoning, the problem of minimizing the spectral norm of $A(y)$,

$$\min_{y \in \mathbb{R}} |\lambda_{\max}(A(y))|,$$

can be written as the following SDO problem:

$$\begin{aligned} \max \quad & -t \\ \text{s.t.} \quad & tI - A_0 - y_1 A_1 - y_2 A_2 - \cdots - y_m A_m \succeq 0 \\ & tI + A_0 + y_1 A_1 + y_2 A_2 + \cdots + y_m A_m \succeq 0. \end{aligned}$$ $\blacksquare$

9.1.2 ▪ Second-Order Cone Optimization

The $(n+1)$-dimensional SOC is defined as

$$\mathrm{SOC}^{n+1} = \{(x_0, x_1, \ldots, x_n) \in \mathbb{R}^{n+1} \mid x_0 \geq \sqrt{x_1^2 + \cdots + x_n^2}\}.$$

An equivalent expression can be given in the form of a rotated quadratic cone:

$$\text{SOC}^{n+1} = \{(x_0, x_1, \ldots, x_n) \in \mathbb{R}^{n+1} \mid 2x_0 x_1 \geq x_2^2 + \cdots + x_n^2, x_0 \geq 0, x_1 \geq 0\}.$$

Mathematically, the latter is a rotation of the former through an angle of 45° in the (x_0, x_1) plane, and for modeling purposes the rotated form is often more convenient.

A nonnegativity constraint $x_0 \geq 0$ is just an SOC constraint in a space of dimension one ($n = 0$); hence LO is a special case of SOCO. For dimension two, the SOC can be expressed as

$$\text{SOC}^2 = \{(x_0, x_1) \in \mathbb{R}^2 \mid x_0 \geq |x_1|\},$$

which is a rotated nonnegative orthant; hence SOCO in dimension two is also an LO problem. For dimensions three and greater, the SOC is not polyhedral, and hence in general, SOCO is not equivalent to LO. For instance, a 2×2 PSD constraint is equivalent to a three-dimensional SOC constraint [1065]:

$$\begin{pmatrix} x_{11} & x_{12} \\ x_{12} & x_{22} \end{pmatrix} \succeq 0 \Leftrightarrow x_{11} x_{22} \geq x_{12}^2 \text{ and } x_{11}, x_{22} \geq 0 \Leftrightarrow \begin{pmatrix} x_{11} + x_{22} \\ x_{11} - x_{22} \\ 2x_{12} \end{pmatrix} \in \text{SOC}^3.$$

SOCO is the problem of optimizing a linear (or possibly convex quadratic) function over the Cartesian product of SOCs, with additional linear constraints on the variables if desired. Note that, unlike for the PSD cone, the direct product of SOCs is not an SOC. The SOCO primal-dual form thus has the following form:

$$(\text{P}_{\text{SOC}}) \quad \inf \quad \sum_{j=1}^{\ell} c_j^T x_j \qquad\qquad (\text{D}_{\text{SOC}}) \quad \sup \quad b^T y$$

$$\text{s.t.} \quad \sum_{j=1}^{\ell} A_j x_j = b, \qquad\qquad\qquad \text{s.t.} \quad A_j^T y + s_j = c_j, j = 1, \ldots, \ell,$$

$$x_j \in \text{SOC}^{n_j+1}, j = 1, \ldots, \ell, \qquad\qquad s_j \in \text{SOC}^{n_j+1}, j = 1, \ldots, \ell,$$

$$(9.4)$$

where ℓ is the number of SOCs involved.

Membership in SOC^{n+1} is equivalent to the following PSD constraint with an arrow structure:

$$\begin{pmatrix} x_0 & & & & x_1 \\ & x_0 & & & x_2 \\ & & \ddots & & \vdots \\ & & & x_0 & x_n \\ x_1 & x_2 & \cdots & x_n & x_0 \end{pmatrix} \succeq 0, \qquad\qquad (9.5)$$

where all the unspecified entries are zero. In this sense, SOCO is a special case of SDO. However, solving SOCO problems by converting them to SDO problems by using (9.5) is much less efficient than making use of the specific SOC structure [75]. We refer the reader to [51] for an in-depth study of the SOC, of algorithms for SOCO, and of classes of optimization problems that can be formulated using SOCO.

9.1.3 • General Conic Optimization

The growing interest in using different cones justifies the definition of a general conic optimization problem.

Let V be a vector space endowed with the inner product $\langle \cdot, \cdot \rangle$. We say that $\mathcal{K} \subset V$ is a *cone* if $\mathcal{K}$ is closed under positive scalar multiplication:

$$\lambda X \in \mathcal{K} \text{ whenever } X \in \mathcal{K} \text{ and } \lambda > 0.$$

A cone $\mathcal{K}$ is said to be *pointed* if $\mathcal{K} \cap (-\mathcal{K}) = \{0\}$. Finally, a cone is said to be *proper* if it has nonempty interior and is closed, convex, and pointed.

The general conic optimization problem has the primal-dual form

$$(\text{P}_\mathcal{K}) \quad \inf \quad \langle C, X \rangle \qquad\qquad (\text{D}_\mathcal{K}) \quad \sup \quad b^T y$$
$$\text{s.t.} \quad \langle A_i, X \rangle = b_i, i = 1, \ldots, m, \qquad \text{s.t.} \quad \sum_{i=1}^{m} y_i A_i + S = C, \quad (9.6)$$
$$X \in \mathcal{K}, \qquad\qquad\qquad\qquad\qquad S \in \mathcal{K}^*,$$

where $\mathcal{K}$ is a proper cone and $\mathcal{K}^*$ is its dual cone.

For any convex set $\mathcal{K}$ in V, the *dual cone* $\mathcal{K}^*$ is defined as

$$\mathcal{K}^* := \{ y \in V : \langle x, y \rangle \geq 0 \quad \forall x \in \mathcal{K} \}.$$

It immediately follows that $\mathcal{K}^*$ is a closed convex cone regardless of the choice of $\mathcal{K}$ (even if $\mathcal{K}$ is not a cone). It also follows that the self-duality of the nonnegative orthant in LO is obvious, that of the SOC follows by the Cauchy–Schwarz inequality, and that of the PSD cone by Fejer's theorem; see, e.g., [956, Corollary 7.5.4].

Moreover, the PSD cone, the SOC, and the nonnegative orthant are members of the class of self-scaled cones [1397, 1398] that admit symmetric primal-dual algorithms. The self-scaled cones coincide with the class of symmetric (homogeneous self-dual) cones [1397, 1398]. In the remainder of this chapter, we shall use the term "symmetric cones" to refer to the three types of cones of interest here.

While symmetric cones are currently the most commonly used in applications, there is a growing interest in other convex cones. One such example is the *p-order* cone:

$$p\text{OC}^n = \left\{ (x_1, \ldots, x_n) \in \mathbb{R}^n \mid x_1 \geq \left(\sum_{i=2}^{n} |x_i|^p \right)^{\frac{1}{p}} \right\},$$

where $p \geq 1$. The *p*-order cone is a generalization of the SOC, as 2OC^n is precisely SOC^{n-1}. It was shown in [813] that it is possible to solve *p*-order cone optimization in polynomial time (up to a given accuracy). For $p > 1$, the dual cone of $p\text{OC}^n$ is $q\text{OC}^n$, where q satisfies $\frac{1}{p} + \frac{1}{q} = 1$. Hence, $p\text{OC}^n$ is self-dual when $p = 2$, i.e., for the SOC. The recent paper [65] shows that for a suitable choice of inner product, the *p*-order cone is self-dual under that inner product.

Two convex cones that are not self-dual, and hence not symmetric, are the completely positive cone,

$$\mathscr{C} := \left\{ X \in \mathscr{S}^n : X = \sum_{i=1}^{k} v_i v_i^T, v_i \geq 0 \right\},$$

and its dual cone, the copositive cone,

$$\mathscr{C}^* := \{ X \in \mathscr{S}^n : v^T X v \geq 0 \; \forall v \geq 0 \}.$$

A significant downside of the completely positive and copositive cones is that it is NP-hard to optimize over them; indeed, only checking membership in the copositive cone is co-NP-complete [1366], and it is also NP-hard for the completely positive cone [596]. Nevertheless, the corresponding area of copositive optimization is a rapidly expanding and fertile field of research, with a diversity of formulations using these cones being used in a variety of applications. We refer the reader to the recent surveys [299, 641] for the state of the art in this dynamic area of research.

9.1.4 ▪ Conic Duality and Constraint Qualification

For the general conic problem (9.6), we have (as in LO) a *weak duality* theorem ensuring that the dual (respectively primal) problem always provides a global upper (respectively lower) bound on the optimal value of the primal (respectively dual).

Theorem 9.4. *If $\tilde{X}$ is feasible for $(P_{\mathcal{K}})$ and $\tilde{y}, \tilde{S}$ for $(D_{\mathcal{K}})$, then $\langle C, \tilde{X}\rangle \leq b^T\tilde{y}$.*

However, because of the nonlinear conic constraint, strong duality, i.e., equality of the bounds at the optimum, does not automatically hold for conic optimization unless it is an LO problem. We illustrate this with the following two SDO examples from [1898, pp. 71–72]. Examples of SOCO problems where strong duality fails are given in [74].

Example 9.5 (positive duality gap). In LO, if both (P) and (D) are feasible, then there is no duality gap. This may fail for SDO problems. Consider the following primal-dual pair of SDO problems:

$$
\begin{aligned}
(\text{P}) \quad &\max \quad -ax_{11} \\
&\text{s.t.} \quad x_{11} + 2x_{23} = 1, \\
&\qquad\quad x_{22} = 0, \\
&\qquad\quad \begin{pmatrix} x_{11} & x_{12} & x_{13} \\ x_{12} & x_{22} & x_{23} \\ x_{13} & x_{23} & x_{33} \end{pmatrix} \succeq 0,
\end{aligned}
\qquad
\begin{aligned}
(\text{D}) \quad &\min \quad -y_2 \\
&\text{s.t.} \quad \begin{pmatrix} y_2 - a & 0 & 0 \\ 0 & y_1 & y_2 \\ 0 & y_2 & 0 \end{pmatrix} \preceq 0,
\end{aligned}
$$

where $a > 0$ is a constant. We claim that the optimal values of (P) and (D) are respectively $-a$ and 0, so that the duality gap is positive and equal to a.

To prove the claim, let us first examine the primal problem (P). The constraint $x_{22} = 0$ implies that $x_{23} = 0$ by applying Theorem 9.1 to the principal submatrix $\{2, 3\}$:

$$
\begin{pmatrix} 0 & x_{23} \\ x_{23} & x_{33} \end{pmatrix} \succeq 0 \Rightarrow -x_{23}^2 \geq 0.
$$

Hence, $x_{11} = 1$ for every feasible solution of the primal, and therefore (P) has optimal objective value $-a < 0$ provided that a feasible solution exists. But every feasible solution to (P) has the form

$$
\begin{pmatrix} 1 & x_{12} & x_{13} \\ x_{12} & 0 & 0 \\ x_{13} & 0 & x_{33} \end{pmatrix},
$$

so by setting, say, $x_{12} = 0$, $x_{13} = 0$, and $x_{33} = 0$, we obtain a PSD matrix and thus confirm the feasibility of problem (P).

The dual problem (D) has a constraint in the form of negative semidefiniteness; i.e., all the eigenvalues of the 3×3 matrix variable should be negative. This is equivalent to

$$
\begin{pmatrix} a - y_2 & 0 & 0 \\ 0 & -y_1 & -y_2 \\ 0 & -y_2 & 0 \end{pmatrix} \succeq 0.
$$

By Theorem 9.1, this implies that $-y_2^2 \geq 0$, so that $y_2 = 0$ for all feasible solutions of (D), and thus every feasible solution to (D) has the form

$$\begin{pmatrix} a & 0 & 0 \\ 0 & -y_1 & 0 \\ 0 & 0 & 0 \end{pmatrix}.$$

By setting y_1 to a negative value, say $y_1 = -1$, we confirm that (D) is feasible. Hence, the optimal objective value of (D) is zero. ∎

Example 9.6 (weak infeasibility). Even if there is no duality gap at optimality, the optimal value may not be attained for (P) or (D). Consider the primal-dual pair

(P) sup $2x_{12}$ (D) inf y_1

s.t. $\begin{pmatrix} 1 & x_{12} \\ x_{12} & 0 \end{pmatrix} \succeq 0,$ s.t. $\begin{pmatrix} y_1 & -1 \\ -1 & y_2 \end{pmatrix} \succeq 0.$

First, observe that by the same arguments as in Example 9.5, the optimal objective value of (P) is 0, and this value is attained because the matrix $\begin{pmatrix} 1 & 0 \\ 0 & 0 \end{pmatrix}$ is PSD.

For the dual problem (D), the semidefinite constraint is equivalent to $y_1 \geq 0$, $y_2 \geq 0$, and $y_1 y_2 \geq 1$. Therefore, while it is possible to set y_1 to a value arbitrarily close to zero, it is not possible to actually set $y_1 = 0$ because then it is impossible to satisfy the constraint $y_1 y_2 \geq 1$. Hence the infimum of y_1 equals zero, and thus there is no duality gap, but $y_1 = 0$ cannot be attained. ∎

We can avoid these difficulties by requiring that (P) and (D) satisfy a *constraint qualification* (CQ). This is a standard condition in nonlinear optimization (NLO) whose purpose is to ensure that strong duality holds: it is possible to achieve primal and dual feasibility with no duality gap. The most commonly used CQ is Slater's CQ.

Definition 9.2. *Slater's CQ holds if ($P_{\mathcal{K}}$) has a feasible solution in the interior of $\mathcal{K}$ and ($D_{\mathcal{K}}$) has a feasible solution in the interior of $\mathcal{K}^*$.*

We then have the following result.

Theorem 9.7. *Under Slater's CQ, both the primal problem and the dual problem have optimal solutions, and the duality gap is zero at optimality.*

For example, to verify that Slater's CQ holds for a given primal-dual SDO pair, it is necessary to exhibit for each of (P) and (D) a feasible matrix that is positive definite, i.e., a matrix with all of its eigenvalues strictly greater than zero, and that satisfies the constraints of the problem. Finding such matrices is often best done by taking into account the structure of the problem. For example, for the SDO problem (9.2), it is obvious that the $n \times n$ identity matrix is positive definite and feasible. Rewriting (9.2) in the form

$$\begin{aligned} \min \quad & \langle C, X \rangle \\ \text{s.t.} \quad & \langle e_i e_i^T, X \rangle = 1, \ i = 1, \dots, n, \\ & X \succeq 0, \end{aligned} \tag{9.7}$$

where e_i denotes a vector of length n with all zeros except for a one in the ith component, we deduce that its dual problem is

$$\min \quad \langle e, y \rangle$$
$$\text{s.t.} \quad \sum_{i=1}^{m} y_i e_i e_i^T + S = C, \tag{9.8}$$
$$S \succeq 0,$$

where e denotes the vector of all ones. To verify Slater's CQ for this problem, it suffices to choose $\bar{y}$ sufficiently large so that

$$\bar{y}_i > \sum_{j=1}^{n} C_{ij} \text{ for each } i = 1, \dots, n.$$

It then follows that $S = \text{Diag}(\bar{y}) - Q$ is positive definite (by diagonal dominance; see, e.g., [956, Theorem 6.1.10]) and hence feasible for (9.8).

When Slater's CQ fails, a number of undesirable phenomena can happen (except in the case of an LO problem). Some of these phenomena were illustrated in Examples 9.5 and 9.6, both of which fail to satisfy Slater's CQ. It is in principle possible to remedy the situation by applying a facial reduction algorithm; see, e.g., [454, 1458, 1863]. The use of facial reduction as a regularization technique for convex optimization was pioneered in [323–325], and recent practical applications include [362, 1105].

9.1.5 ▪ Optimality Conditions and Interior-Point Algorithms

The KKT optimality conditions can be the starting point for implementing primal-dual interior-point algorithms to solve conic optimization problems. This is discussed for LO problems in Section 1.2 of Chapter 1, so we focus here on the SDO and SOCO cases.

Let us start with SDO. From Theorem 9.4, we have that the duality gap equals

$$\langle C, X \rangle - b^T y = \left\langle \sum_{i=1}^{m} y_i A_i + S, X \right\rangle - \sum_{i=1}^{m} \langle A_i X \rangle y_i = \langle S, X \rangle.$$

Because both X and S are PSD, $\langle X, S \rangle = 0$ implies $XS = SX = 0$, and we obtain the following sufficient optimality conditions:

$$\langle A_i, X \rangle = b_i, \ i = 1, \dots, m, \text{ and } X \succeq 0 \quad \text{(primal feasibility)},$$
$$S + \sum_{i=1}^{m} y_i A_i = C, \text{ and } S \succeq 0 \quad \text{(dual feasibility)}, \tag{9.9}$$
$$\frac{SX + XS}{2} = 0 \quad \text{(complementarity)}.$$

If Slater's CQ holds, they are also necessary for optimality.

Because the optimality conditions (9.9) are hard to solve directly, we consider the perturbed system

$$\langle A_i, X \rangle = b_i, \ i = 1, \dots, m, \text{ and } X \succeq 0 \quad \text{(primal feasibility)},$$
$$S + \sum_{i=1}^{m} y_i A_i = C, \text{ and } S \succeq 0 \quad \text{(dual feasibility)},$$
$$\frac{SX + XS}{2} = \mu I \quad \text{(perturbed complementarity)}.$$

This system has a unique solution for given $\mu \geq 0$, and the set $\{(X(\mu), y(\mu), S(\mu))\}$ of these solutions is the *central path*. The fundamental idea in primal-dual interior-point methods is to start at a point on the central path corresponding to $\mu > 0$ and approximately follow the path to the optimal solution ($\mu = 0$) by iteratively decreasing μ and using Newton's method to compute the point corresponding to the new value of μ.

One of the key questions for primal-dual interior-point algorithms in SDO is the handling of the complementarity constraints. Because the product of two symmetric matrices is not symmetric in general, the complementarity constraints can be expressed in different ways, such as

$$\frac{SX + XS}{2} = 0,\ XS = 0,\ \text{or } SX = 0.$$

Depending on which form of the complementarity constraints is chosen, and on how the corresponding Newton step for the central path is defined, different search directions can be obtained for the primal-dual interior-point approach to SDO. Twenty different search directions and their theoretical properties are summarized in [1772].

For SOCO, given a primal solution $x \in \text{SOC}^{n+1}$ and a dual solution $s \in \text{SOC}^{n+1}$, the complementarity conditions have at first the same form as for LO, i.e., x and s are orthogonal:

$$x^T s = 0 \Leftrightarrow \sum_{i=0}^{n} x_i s_i = 0.$$

However, while for LO this implies $x_i s_i = 0$ for each i because all the variables are nonnegative, in SOCO the variables, except for the leading variables x_0 and s_0, can take on negative values. Hence, two cases are possible: either $x = 0$ and $z = 0$ as in LO, or x and z lie on opposite sides of the boundary of SOC^{n+1} and are thus nonzero and still orthogonal. Therefore, a more explicit statement of the complementarity constraints for SOCO is as follows.

Theorem 9.8 [51, Lemma 15]. *Suppose that $x \in \text{SOC}^{n+1}$ and $s \in \text{SOC}^{n+1}$. Then $x^T s = 0$ if and only if*

1. $x_0 s_0 + \sum_{i=1}^{n} x_i s_i = 0$ *and*

2. $x_0 \begin{pmatrix} s_1 \\ s_2 \\ \vdots \\ s_n \end{pmatrix} + s_0 \begin{pmatrix} x_1 \\ x_2 \\ \vdots \\ x_n \end{pmatrix} = 0.$

We do not present here specific interior-point algorithms for solving conic optimization problems. Extensive research and computational developments have led to the availability of excellent implementations of such algorithms that perform extremely well on even large-scale linear and SOC problems. Large-scale SDO problems, however, remain challenging for interior-point methods, and thus solvers based on alternative techniques have also been developed. Instead we present in Section 9.2 the state of the art with respect to computational solvers available for broad use. The reader may find descriptions of the algorithms implemented by each solver in the references cited in Section 9.2 and consult their bibliographies for references providing the theory behind the algorithms.

9.2 ▪ Software for Conic Optimization

We divide the solvers available into two groups: those that handle the three types of symmetric cones, and those that focus exclusively on solving SDO problems. Because nonnegativity and SOC constraints can be cast as PSD constraints, it is possible to handle these constraints using the SDO-specific solvers; however, they will be less efficient than the first two groups of solvers because they do not exploit the simpler structure of the other cones.

For reasons of brevity, we mention only some of the most commonly used software packages. Our intention is not to provide a catalog of all available tools but rather to provide the reader with an overview of the software available. Unless specified otherwise, all software mentioned here is available for free via the given URLs (at the time of publication).

A good point of entry for new users of conic optimization are the following MATLAB-based packages that can solve small- to medium-scale problems combining the three types of symmetric cones:

- SDPT3 [1773, 1796] is hosted at `http://www.math.nus.edu.sg/~mattohkc/sdpt3.html`. This software is very popular, fairly well tested, and actively maintained. The basic code is written in MATLAB, but key portions are written in C. SDPT3 actively exploits sparsity and block diagonal structure, as well as low-rank structure in the data. The current version 4.0 dates from 2009.

- SeDuMi [1735] is hosted and maintained at `http://sedumi.ie.lehigh.edu`. SeDuMi is very reliable and robust, and because the solutions are stored and updated in a product form, it can provide highly accurate solutions. Although the core code has not changed much since 2007, it remains a highly popular and efficient code.

Next we mention commercial interior-point solvers that can handle all symmetric cones. All have MATLAB interfaces and are available for academic use at little or no cost:

- CVX [838, 839] is available at `http://cvxr.com/cvx`. CVX was designed to support *disciplined convex programming*. This solver is growing in popularity and is actively developed and maintained. Beyond the solver, CVX provides an environment for modeling in MATLAB using convex optimization (see below).

- LOQO [1824,1825] is an NLO solver that can handle a large variety of problems, convex or nonconvex, constrained or unconstrained. Specific routines are provided to solve SOCO and SDO problems. We refer the reader to `http://www.princeton.edu/~rvdb/loqomenu.html` for more information.

- MOSEK [71] is a solver for general convex optimization problems, specifically based on optimization over symmetric cones. It also offers the capability to solve mixed-integer versions of some of these problems. We refer the reader to `https://mosek.com/products/mosek` for more information.

Other interior-point solvers that focus exclusively on SDO include the following:

- CSDP [314] is written in C and designed to be used as a callable subroutine. It also makes effective use of sparsity in the data. It is now a COIN-OR project: `https://projects.coin-or.org/Csdp`.

- DSDP [226] is an open-source implementation of a dual-scaling algorithm that exploits low-rank structure and sparsity in the data and requires relatively little memory [227]. A MATLAB interface is available. The latest version is hosted at http://www.mcs.anl.gov/hs/software/DSDP.
- SDPA [1922] is a family of solvers based on the algorithm described in [1923]: http://sdpa.sourceforge.net. Among the various codes in this family are a version that exploits sparsity in the data, another that exploits structure through a matrix completion technique, and three high-precision arithmetic versions.

All of these codes have parallel versions as well.

Two quantities are critical to determine the "size" of an SDO problem for computational purposes: the size of the matrix (n) and the number of linear constraints (m), both as defined in (9.1). While it is widely accepted that interior-point solvers are generally very efficient and robust for moderate values of m and n, for SDO problems with m large (say greater than 10,000) and n moderate (say up to 5,000), the limitations of interior-point approaches are severe due to the need to form and factor the so-called Schur complement matrix of size $m \times m$.

The limitations of interior-point methods motivated the development of SDO solvers that are based on other algorithmic frameworks. The software packages arising from this research include the following:

- PENOPT is a commercial family of solvers based on a generalized augmented Lagrangian method. Its PENNON solver [1080, 1081] is applicable to general NLO problems and is particularly aimed at large-scale problems with sparsity in the data. The PENSDP variant solves SDO problems, and other variants target bilinear matrix inequalities and free-material optimization. PENLAB is a free open-source MATLAB-based version of PENNON. We refer the reader to http://www.penopt.com for more information.
- SDPLR uses a first-order NLO algorithm that is based on the idea of low-rank factorization of the matrix variable in SDO as described in [357]. This approach transforms the SDO problem into a nonconvex problem but allows direct control of the rank of the solution and hence a significant reduction in the number of variables. The source code and documentation are available at http://sburer.github.io/files/SDPLR-1.03-beta.zip.
- ConicBundle is a library of C/C++ subroutines that implement a bundle method for minimizing the sum of convex functions that are given by first-order oracles or arise from the Lagrangian relaxation of conic optimization problems [925,926]. This library includes code specific to optimizing over the symmetric cones. The source code and documentation are available at https://www-user.tu-chemnitz.de/~helmberg/ConicBundle.
- SDPNAL is a recently developed SDO solver based on a semismooth Newton-CG augmented Lagrangian method [1969]. It is designed to solve large-scale SDO problems with the same input format as SDPT3, so users of the latter can easily use SDPNAL as well.

Many other codes have been developed for specific applications in areas such as combinatorial optimization and image processing. We do not cover them here for lack of space and instead refer the reader to the handbook [87].

We close this section with some notes of practical importance for using conic optimization. The first is that many of the solvers mentioned here can be accessed via the NEOS (Network-Enabled Optimization System) Server http://www.neos-server.org/neos/solvers/index.html. This is a free Internet-based service for solving most standard forms of optimization problems, including conic problems.

Second, many of the above solvers provide access not only via MATLAB, but also via Python, R, Julia, and other environments. Moreover, there are modeling environments that specifically support the use of conic optimization. Examples of such environments are as follows:

- CVX [838,839] supports modeling using a particular paradigm called disciplined convex programming [840] in which the problem to be solved is built using a base library of convex functions and sets using a small set of rules from convex analysis. While these conventions do not limit generality, they support an automated solution method based on the use of symmetric cones. SDPT3 and SeDuMi are included with the standard CVX distribution.
- YALMIP [1217] is a modeling language that supports a variety of optimization problems, including the three types of symmetric cones and their mixed-integer forms, as well as related classes of problems, such as sum-of-squares (SOS) problems (see Section 9.3 below) and robust optimization (see Chapter 25).

9.3 ▪ Polynomial Optimization

Until the mid-1990s, the standard techniques to handle models with nonlinearities were to either use a general NLO algorithm or devise a suitable LO approximation. While they have a strong theoretical basis and there exists reliable mature software to solve them, one main weakness of NLO algorithms is that the optimal solution returned by the algorithm is inescapably a local optimum determined by the initial point chosen (either by the user or by the software implementation) to initialize the algorithm. Hence, in the absence of favorable structural properties of such a convexity or a priori knowledge to guide the choice of the starting point, it is not possible for NLO approaches to guarantee convergence to the global optimum in general.

LO-based approaches can guarantee global optimality under certain conditions, and while there are well-known linearization techniques that have been highly successful in practice, their performance varies dramatically from one application to another.

The development of SDO provided a new means to convexify quadratic problems, and more generally polynomial optimization problems (POPs). A POP is a mathematical optimization problem whose objective and constraints are multivariate polynomials. Polynomial optimization generalizes several special cases that have been thoroughly studied in optimization, including mixed binary LO, convex and nonconvex quadratic optimization (QO), and linear complementarity problems.

The general approach to solve POPs is based on SOS certificates of nonnegativity for multivariate polynomials. Following the pioneering work of Lasserre [1133] and Parrilo [1454], this approach builds a hierarchy of relaxations leading to solving a sequence of SDO problems. Under mild conditions, these relaxations provide bounds that converge to the global optimal value of the original POP. One drawback of this approach is that the size of the SDO relaxations becomes exponential to ensure convergence to global optimality.

An SOS certificate is a sufficient condition for a given polynomial to be nonnegative. The idea is that a polynomial that is an SOS of (lower-degree) polynomials is always nonnegative. For example, the polynomial $2x^4+y^4-2x^3y-x^2y^2$ is a nonnegative polynomial for all values of x and y because

$$2x^4 + y^4 - 2x^3y - x^2y^2 = (x^2 - xy)^2 + (y^2 - x^2).$$

However, it is not true that all nonnegative polynomials can be expressed as an SOS. A famous counterexample is the Motzkin polynomial $x^4y^2 + x^2y^4 - 3x^2y^2 + 1$, which is nonnegative on $\mathbb{R}^2$ but cannot be expressed as an SOS [1564].

The connection between SOSs and SDO is that checking whether a given polynomial is an SOS is equivalent to checking the feasibility of a certain SDO problem. For our example above, to verify that our polynomial is nonnegative, we need to check that the equation

$$2x^4 + y^4 - 2x^3y - x^2y^2 = \begin{pmatrix} x^2 \\ xy \\ y^2 \end{pmatrix}^T M \begin{pmatrix} x^2 \\ xy \\ y^2 \end{pmatrix}$$

has a solution with M PSD. One possible choice is

$$M = \begin{pmatrix} 2 & -1 & -1 \\ -1 & 1 & 0 \\ -1 & 0 & 1 \end{pmatrix};$$

hence the polynomial is nonnegative for all values of x and y. We do not expand further on the rich and deep theory of polynomial optimization but rather refer the reader to the handbook [87] and the references therein.

From a practical point of view, it is important to know that software exists to automatically build and solve the hierarchy of semidefinite relaxations. GloptiPoly [934] and SOSTOOLS [1517] are two MATLAB-based software packages that provide this functionality, and they can be used in conjunction with several of the solvers presented in Section 9.2. This theory is finding its way into different areas of application, including finance, control theory, and signal processing. An application of this approach to nonlinear optimal control is given in Chapter 10.

Chapter 10

Conic Linear Optimization for Nonlinear Optimal Control

Didier Henrion and Edouard Pauwels

10.1 ▪ Introduction

In this chapter, we discuss infinite-dimensional conic linear optimization (CLO) for nonlinear optimal control problems. The primal linear problem consists of finding occupation measures supported on optimal relaxed controlled trajectories, whereas the dual linear problem consists of finding the largest lower bound on the value function of the optimal control problem. We develop various approximation results relating the original optimal control problem and its conic linear formulations. These results are relevant in the context of finite-dimensional semidefinite programming relaxations used to numerically approximate the solutions of the infinite-dimensional conic linear problems.

10.2 ▪ Motivation

In [1132, 1133], J.-B. Lasserre described a hierarchy of convex semidefinite programming (SDP) problems allowing us to compute bounds and find global solutions for finite-dimensional nonconvex polynomial optimization problems (POPs). Each step in the hierarchy consists of solving a primal moment SDP problem and a dual polynomial sum-of-squares (SOS) SDP problem corresponding to discretizations of infinite-dimensional conic linear problems, namely a primal linear programming (LP) problem on the cone of nonnegative measures and a dual LP problem on the cone of nonnegative continuous functions. The number of variables (number of moments in the primal SDP, or degree of the SOS certificates in the dual SDP) increases when progressing in the hierarchy, global optimality can be ensured by checking rank conditions on the moment matrices, and global optimizers can be extracted by numerical linear algebra. For more information on the moment-SOS hierarchy and its applications, see [1134].

This approach was then extended to polynomial optimal control in [1135]. Whereas the key idea in [1132, 1133] was to reformulate a (finite-dimensional) nonconvex polynomial optimization on a compact semialgebraic set into an LP problem in the (infinite-dimensional) space of probability measures supported on this set, the key idea in [1135], also developed in [777], was to reformulate an (infinite-dimensional) nonconvex polynomial optimal control problem with compact constraint set into an LP problem in the (infinite-dimensional) space of occupation measures supported on this set. Note that LP formulations of optimal control problems (on ODEs and PDEs) are classical and can be traced back to the work by L. C. Young, Filippov, and Warga and Gamkrelidze, amongst many others. For more details and a historical survey, see, e.g., [690, Part III]. We believe that what is innovative in [1135] is the observation that the infinite-dimensional linear formulations for optimal control problems can be solved numerically with a moment-SOS hierarchy of the same kind as those used in polynomial optimization.

The objective of this contribution is to revisit the approach of [1135] and to survey the use of occupation measures to linearize polynomial optimal control problems. This is an opportunity to describe duality in infinite-dimensional conic problems, as well as various approximation results on the value function of the optimal control problem. The primal LP problem consists of finding occupation measures supported on optimal relaxed controlled trajectories, whereas the dual LP problem consists of finding the largest lower bound on the value function of the optimal control problem. The value function is the solution (in a suitably defined weak sense) of a nonlinear PDE called the Hamilton–Jacobi–Bellman equation; see, e.g., [1782, Chapters 8 and 9] and [481, Chapters 19 and 24]. It is traditionally used to verify optimality and to explicitly compute of optimal control laws, but we do not describe these applications here.

To improve the readability of this chapter, we omit the technical proofs. They can be found in an extended version of this text on the webpage of the authors. We first describe the optimal control problem and concrete examples. The relaxed formulation of optimal control is introduced afterward, and important aspects of this formulation are explained in the context of Bolza's problem. All the approximation results are illustrated numerically on examples described in the following section.

10.3 ▪ Polynomial Optimal Control

We begin this section with a description of [1782, Example 4.2.2], which motivates the use of optimal control for engineering applications. Let x denote the position of a car along a straight line. Suppose that we control the acceleration of the car; in other words, we have the equation $\ddot{x} = u$, where u is a control variable. For a given time horizon T, we would like to balance the following two contradictory goals: the car should go as far as possible while minimizing the work spent on controlling the acceleration. This can be modeled by the following problem:

$$\inf_u \quad \int_0^T u(t)^2 dt - x(T) \tag{10.1}$$
$$\text{s.t.} \quad \ddot{x}(t) = u(t), \ x(0) = \dot{x}(0) = 0.$$

This is an optimization problem over an infinite-dimensional space (the set of trajectories) under differential equality constraints. It is an occurrence of a large class of control problems that we now describe.

10.3.1 ▪ General Description

We consider polynomial optimal control problems (POCPs) of the form

$$v^*(t_0, x_0) \; := \; \inf_u \; \int_{t_0}^{T} l(x(t), u(t))dt + l_T(x(T))$$

$$\text{s.t.} \quad \dot{x}(t) = f(x(t), u(t)), \; x(t_0) = x_0, \qquad (10.2)$$
$$x(t) \in X, \; t \in [t_0, T],$$
$$u(t) \in U, \; t \in [t_0, T],$$
$$x(T) \in X_T,$$

where the dot denotes the time derivative, $l \in \mathbb{R}[x, u]$ is a given Lagrangian (integral cost), $l_T \in \mathbb{R}[x]$ is a given terminal cost, $f \in \mathbb{R}[x, u]^n$ is a given dynamics (vector field), $X \subset \mathbb{R}^n$ is a given compact state constraint set, and $U \subset \mathbb{R}^m$ is a given compact control constraint set, and $X_T \subset X$ is a given compact terminal state constraint set. Also given are the terminal time $T \geq 0$, the initial time $t_0 \in [0, T]$, and the initial condition $x_0 \in X$. In POCP (10.2), the minimum is with respect to all control laws $u \in \mathscr{L}^\infty([t_0, T]; U)$, which are bounded functions of time t with values in U, and the resulting state trajectories $x \in \mathscr{L}^\infty([t_0, T]; X)$, which are bounded functions of time t with values in X.

Let $\mathscr{A} \subset [0, T] \times X$ denote the set of values (t_0, x_0) for which there is a controlled trajectory $(x, u) \in \mathscr{L}^\infty([t_0, T]; X \times U)$ starting at $x(t_0) = x_0$ and admissible for POCP (10.2). The function $(t_0, x_0) \mapsto v^*(t_0, x_0)$ defined in (10.2) is called the value function, and its domain is $\mathscr{A}$.

10.3.2 ▪ Examples

We describe more concrete examples of optimal control problems of the form of (10.2). They are taken from the literature and illustrate the relevance of the optimal control formulation for applications in engineering. These examples will be used in the numerical section to illustrate our theoretical results. Bolza's problem is an academic example that will be used to illustrate various aspects of the weak formulation of the control problem in later sections.

Example 10.1 (LQR). We introduce an instance of one of the most widespread classes of optimal control problems, the linear quadratic regulator (LQR). Suppose that x denotes the number of individuals of an unwanted animal species on an island. This number naturally grows proportionally to itself (reproduction), and we would like to limit this growth. Let u denote the number of hunters who will hunt the unwanted species; this number is a decision variable. These two effects are combined in the simple dynamical model $\dot{x} = x - u$. Given an initial population, x_0, and a time horizon, T, we would like to minimize both the growth of x and the number of hunters that should be involved:

$$\inf_u \; \int_0^1 (10x(t)^2 + u(t)^2)dt$$

$$\text{s.t.} \quad \dot{x}(t) = x(t) - u(t), \qquad (10.3)$$
$$x(t_0) = x_0.$$

Here the weight 10 could be arbitrary and balances the interest that we have in minimizing x versus the effort spent, u. This example will be used to illustrate the results of Theorem 10.13 in Section 10.6.2. Note that this example has the same structure as the one presented at the beginning of this section: the objective is a quadratic function and the dynamics is linear, hence the name of this class of problems. ∎

Example 10.2 (turnpike, [481, Section 2.22]). This example can be used to model the process of limiting the growth of a state. Suppose, for example, that x denotes the quantity of some weed in a field whose presence cannot be avoided but should still be as limited as possible. The natural growth of the weed implies a constant term (seeds brought by wind, birds) and a term that is proportional to x (reproduction). We have the possibility of putting in some effort u (manual work, pesticides) to destroy some of the weed, this effect being more efficient when more weeds are present. A simple dynamical model involving these three effects is $\dot{x} = 1 + x - ux$. Given a time horizon, T, we would like to minimize both the extension of the weed and the work spent in limiting its extension. We can add the constraint that the effort represented by u cannot exceed a certain threshold, three for example. This leads to the following optimal control problem:

$$\inf_u \quad \int_0^2 (x(t) + u(t))dt$$
$$\text{s.t.} \quad \dot{x}(t) = 1 + x(t) - x(t)u(t), \tag{10.4}$$
$$x(0) = x_0,$$
$$u(t) \in [0,3],$$

which has the same form as the general problem (10.2). This example will be used in Section 10.5.2 to illustrate the result of Theorem 10.12. ∎

Example 10.3 (Bolza). Bolza's problem is more academic in nature. It will be used to illustrate different abstract notions, such as occupation measures and Liouville's equation, on a well-identified simple optimal control problem:

$$\inf_u \quad \int_0^1 x(t)^4 + (1 - u(t)^2)^2 dt$$
$$\text{s.t.} \quad \dot{x}(t) = u(t), \tag{10.5}$$
$$x(0) = x(1) = 0. \qquad\qquad ∎$$

We now discuss the notion of infimum using Example 10.3. In problem (10.5), the cost is nonnegative, and therefore the infimum is also nonnegative. Consider the following sequence of control variables:

$$u_k : t \rightarrow \begin{cases} 1 & \text{if } t \in \left[\frac{2s}{2^k}, \frac{2s+1}{2^k}\right] \text{ for some } s \in \mathbb{N}, \\ -1 & \text{otherwise.} \end{cases}$$

It can be checked that the sequence u_k produces objective values in (10.5) that tend to zero as k goes to infinity. Thus the infimum in (10.5) is zero. However, there cannot exist a trajectory that achieves this value; indeed, this would lead to $x = 0$ and $|\dot{x}| = 1$ for almost all $t \in [0,1]$, which is contradictory. Hence we have the general infimum notation in (10.2). These types of phenomena motivate the introduction of a more general formulation of problem (10.2) over a larger space, which allows us to formally handle these types of inconveniences.

10.4 ▪ LP Problem Formulation

As explained in the introduction, to derive an LP formulation of POCP (10.2) we need to introduce measures on trajectories, the so-called occupation measures. The first step is to replace classical controls with probability measures, and for this we have to define additional notation.

Given a compact set $X \subset \mathbb{R}^n$, let $\mathscr{C}(X)$ denote the space of continuous functions supported on X, and let $\mathscr{C}_+(X)$ denote its nonnegative elements, the cone of nonnegative continuous functions on X. Let $\mathscr{M}_+(X) = \mathscr{C}_+(X)'$ denote its topological dual, the set of all nonnegative continuous linear functionals on $\mathscr{C}(X)$. By the Riesz representation theorem, these are nonnegative Borel-regular measures, or Borel measures, supported on X. The topology in $\mathscr{C}_+(X)$ is the strong topology of uniform convergence, whereas the topology in $\mathscr{M}_+(X)$ is the weak-star topology. The duality bracket

$$\langle v, \mu \rangle := \int_X v(x) \mu(dx)$$

denotes the integration of a function $v \in \mathscr{C}_+(X)$ against a measure $\mu \in \mathscr{M}_+(X)$. For background on weak-star topology, see, e.g., [1238, Section 5.10] or [169, Chapter IV]. For $x \in X$, the Dirac measure δ_x will denote the measure that represents functional evaluation at x for all $f \in \mathscr{C}(X)$, that is, $\langle \delta_x, f \rangle = f(x)$. Finally, let us denote by $\mathscr{P}(X)$ the set of probability measures supported on X consisting of Borel measures $\mu \in \mathscr{M}_+(X)$ such that $\langle 1, \mu \rangle = 1$.

10.4.1 ▪ Relaxed Controls

As illustrated by Example 10.3, the infimum in POCP (10.2) is not attained. To circumvent this issue, our next step is to assume that, at each time $t \in [t_0, T]$, the control is not a vector $u(t) \in U$ but a time-dependent probability measure $\omega(du \,|\, t) \in \mathscr{P}(U)$ that rules the distribution of the control in U. We use the notation $\omega_t := \omega(\cdot|t)$ to emphasize the dependence on time. This is called a relaxed control, or stochastic control, or Young measure in the functional analysis literature. POCP (10.2) is then relaxed to

$$v_R^*(t_0, x_0) \quad := \quad \min_\omega \quad \int_{t_0}^T \langle l(x(t), .), \omega_t \rangle dt + l_T(x(T))$$

$$\text{s.t.} \quad \dot{x}(t) = \langle f(x(t), .), \omega_t \rangle, \ x(t_0) = x_0, \tag{10.6}$$
$$x(t) \in X, \ t \in [t_0, T],$$
$$\omega_t \in \mathscr{P}(U), \ t \in [t_0, T],$$
$$x(T) \in X_T,$$

where the minimization is w.r.t. a relaxed control. Note that we replaced the infimum in POCP (10.2) with a minimum in relaxed POCP (10.6). Indeed, it can be proved that this minimum is always attained using (weak-star) compactness of the space of probability measures with compact support.

Since classical controls $u \in \mathscr{L}^\infty([t_0, T]; U)$ are a particular case of relaxed controls $\omega_t \in \mathscr{P}(U)$ corresponding to the choice $\omega_t = \delta_{u(t)}$ for a.e. $t \in [t_0, T]$, the minimum in relaxed POCP (10.6) is smaller than the infimum in classical POCP (10.2), i.e.,

$$v^*(t_0, x_0) \geq v_R^*(t_0, x_0).$$

Contrived optimal control problems (e.g., with overly stringent state constraints) can be cooked up such that the inequality is strict, i.e., $v^*(t_0, x_0) > v_R^*(t_0, x_0)$; see, e.g., the examples in [933, Appendix C]. We do not consider these examples to be practically relevant, so the following assumption will be made.

Assumption 10.4 (no relaxation gap). *For any relaxed controlled trajectory* (x, ω_t) *admissible for relaxed POCP (10.6), there is a sequence of controlled trajectories* $(x_k, u_k)_{k \in \mathbb{N}}$ *admissible for POCP (10.2), i.e., for each k,*

$$(x_k, u_k) \in \mathscr{L}^\infty([t_0, T], X \times U), \ \dot{x}_k = f(x_k, u_k) \ a.e.$$

such that

$$\lim_{k \to \infty} \int_{t_0}^T v(x_k(t), u_k(t)) dt = \int_{t_0}^T \langle v(x(t), \cdot), \omega_t \rangle dt$$

for every function $v \in \mathscr{C}(X \times U)$. *Then it holds that*

$$v^*(t_0, x_0) = v_R^*(t_0, x_0)$$

for every $(t_0, x_0) \in \mathscr{A}$.

Note that this assumption is satisfied under the classical controllability and/or convexity conditions used in the Filippov–Ważewski theorem with state constraints; see [757] and the discussions around Assumption I in [777] and Assumption 2 in [933]. However, let us point out that Assumption 10.4 does not imply that the infimum is attained in POCP (10.2). Conversely, if the infimum is attained, the values of POCP (10.2) and relaxed POCP (10.6) coincide, and Assumption 10.4 is satisfied.

Example 10.5. Consider Bolza's problem in Example 10.3. As we have seen in Section 10.3.2, the value $v^*(0, 0)$ is not attained for this problem. Set $\omega_t = \frac{1}{2}(\delta_{-1} + \delta_1)$ for $t \in [0, 1]$. In other words, for any continuous $f : \mathbb{R} \to \mathbb{R}$, it holds that $\langle f, \omega_t \rangle = \frac{f(-1) + f(1)}{2}$. It can be seen that $(0, \omega_t)$ is admissible for the relaxed problem (10.6) with value zero. This shows that the infimum in (10.6) is indeed attained and that $v^*(0, 0) = v_R^*(0, 0)$. This value of ω_t corresponds to the limiting measure (in the sense of Assumption 10.4) of the sequence u_k described in Section 10.3.2. Assumption 10.4 states that this approximation is always possible for problems (10.2) and (10.6) in their general form. ∎

10.4.2 ▪ Occupation Measure

Given initial data $(t_0, x_0) \in \mathscr{A}$, and given a relaxed control $\omega_t \in \mathscr{P}(U)$, the unique solution of the ODE

$$\dot{x}(t) = \langle f(x(t), \cdot), \omega_t \rangle, \ x(t_0) = x_0, \tag{10.7}$$

in relaxed POCP (10.6) is given by

$$x(t) = x_0 + \int_{t_0}^t \langle f(x(s), \cdot), \omega_s \rangle ds \tag{10.8}$$

for every $t \in [t_0, T]$. Let us then define

$$\mu(dt, dx, du) := dt \, \delta_{x(t)}(dx) \, \omega_t(du) \in \mathscr{M}_+([t_0, T] \times X \times U) \tag{10.9}$$

as the occupation measure concentrated uniformly in time on the state trajectory starting at x_0 at time t_0, for the given relaxed control ω_t. An analytic interpretation is that integration w.r.t. the occupation measure is equivalent to time integration along system trajectories, i.e.,

$$\int_{t_0}^{T} v(t, x(t)) dt = \int_{t_0}^{T} \int_X \int_U v(t, x) \mu(dt, dx, du) = \langle v, \mu \rangle,$$

given any test function $v \in \mathscr{C}([t_0, T] \times X)$.

Let us define the linear operator $\mathscr{L} : \mathscr{C}^1([t_0, T] \times X) \to \mathscr{C}([t_0, T] \times X \times U)$ by

$$v \mapsto \mathscr{L} v := \frac{\partial v}{\partial t} + \sum_{i=1}^{n} \frac{\partial v}{\partial x_i} f_i = \frac{\partial v}{\partial t} + \operatorname{grad} v \cdot f.$$

Given a continuously differentiable test function $v \in \mathscr{C}^1([t_0, T] \times X)$, notice that

$$
\begin{aligned}
v(T, x(T)) - v(t_0, x(t_0)) &= \int_{t_0}^{T} dv(t, x(t)) &&= \int_{t_0}^{T} \dot{v}(t, x(t)) dt \\
&= \int_{t_0}^{T} \mathscr{L} v(t, x(t)) dt &&= \langle \mathscr{L} v, \mu \rangle,
\end{aligned}
$$

which can be written more concisely as

$$\langle v, \mu_T \rangle - \langle v, \mu_0 \rangle = \langle \mathscr{L} v, \mu \rangle \tag{10.10}$$

upon defining respectively the initial and terminal occupation measures

$$\mu_0(dt, dx) := \delta_{t_0}(dt) \delta_{x(t_0)}(dx), \quad \mu_T(dt, dx) := \delta_T(dt) \delta_{x(T)}(dx). \tag{10.11}$$

Let us define the adjoint linear operator $\mathscr{L}' : \mathscr{C}([t_0, T] \times X \times U)' \to \mathscr{C}^1([t_0, T] \times X)'$ by the relation $\langle v, \mathscr{L}' \mu \rangle := \langle \mathscr{L} v, \mu \rangle$ for all $\mu \in \mathscr{M}([t_0, T] \times X)$ and $v \in \mathscr{C}^1([t_0, T] \times X)$. More explicitly, this operator can be expressed as

$$\mu \mapsto \mathscr{L}' \mu = -\frac{\partial \mu}{\partial t} - \sum_{i=1}^{n} \frac{\partial (f_i \mu)}{\partial x_i} = -\frac{\partial \mu}{\partial t} - \operatorname{div} f \mu,$$

where the derivatives of measures are understood in the weak sense, i.e., via their action on smooth test functions, and the change of sign comes from integration by parts. Equation (10.10) can be rewritten equivalently as $\langle v, \mu_T \rangle - \langle v, \mu_0 \rangle = \langle v, \mathscr{L}' \mu \rangle$, and since this equation should hold for all test functions $v \in \mathscr{C}^1([t_0, T] \times X)$, we obtain a linear PDE on measures $\mathscr{L}' \mu = \mu_T - \mu_0$ that we write as

$$\frac{\partial \mu}{\partial t} + \operatorname{div} f \mu + \mu_T = \mu_0. \tag{10.12}$$

This linear transport equation is classical in fluid mechanics, statistical physics, and analysis of PDEs. It is called the equation of conservation of mass, or the continuity equation, or the advection equation, or Liouville's equation. Under the assumption that the initial data $(t_0, x_0) \in \mathscr{A}$ and the control law $\omega_t \in \mathscr{P}(U)$ are given, the following result can be found, e.g., in [1851, Theorem 5.34] or [66].

Lemma 10.6 (Liouville PDE = Cauchy ODE). *There exists a unique solution to the Liouville PDE (10.12) which is concentrated on the solution of the Cauchy ODE (10.7), i.e., such that (10.9) and (10.11) hold.*

In our context of conic optimization, the relevance of the Liouville PDE (10.12) is its linearity in the occupation measures μ, μ_0, and μ_T, whereas the Cauchy ODE (10.7) is nonlinear in the state trajectory $x(t)$.

Example 10.7. Following Example 10.5, we explain (10.12) in the context of Bolza's problem. Consider the following:

$$\mu(dt,dx,du) = dt\, \delta_0(dx)\frac{\delta_{-1}(du)+\delta_1(du)}{2},$$

$$\mu_0(dt,dx) = \delta_0(dt)\,\delta_0(dx),$$

$$\mu_T(dt,dx) = \delta_1(dt)\,\delta_0(dx).$$

Let us show that these measures satisfy the Liouville PDE. For this purpose, given an arbitrary $v \in \mathscr{C}^1([0,1]\times\mathbb{R})$, we have

$$\left\langle v, \frac{\partial \mu}{\partial t}\right\rangle = -\int \frac{\partial v(t,x)}{\partial t}\mu(dt,dx,du) = -\int_0^1 \frac{\partial v(t,0)}{\partial t}dt = v(0,0) - v(1,0),$$

$$\langle v, \operatorname{div} f\,\mu\rangle = -\int \frac{\partial v(t,x)}{\partial x}u\,\mu(dt,dx,du) = -\int_0^1 \frac{\partial v(t,0)}{\partial x}\frac{(1-1)}{2}dt = 0.$$

Hence we have $\left\langle v, \frac{\partial \mu}{\partial t} + \operatorname{div} f\,\mu\right\rangle = v(0,0) - v(1,0) = \langle \mu_0, v\rangle - \langle \mu_1, v\rangle$. Since v was arbitrary, this shows that (10.12) is satisfied.   ∎

10.4.3 ▪ Primal LP on Measures

The cost in relaxed POCP (10.6) can therefore be written

$$\int_{t_0}^T \langle l(x(t),.),\omega_t\rangle dt + l_T(x(T)) = \langle l,\mu\rangle + \langle l_T,\mu_T\rangle,$$

and we can now define a relaxed optimal control problem as an LP problem in the cone of nonnegative measures:

$$\begin{aligned}
p^*(t_0,x_0) \quad := \quad &\min_{\mu,\mu_T} \quad \langle l,\mu\rangle + \langle l_T,\mu_T\rangle\\
&\text{s.t.} \quad \frac{\partial \mu}{\partial t} + \operatorname{div} f\,\mu + \mu_T = \delta_{t_0}\delta_{x_0},\\
&\quad\quad \mu \in \mathscr{M}_+([t_0,T]\times X\times U),\\
&\quad\quad \mu_T \in \mathscr{M}_+(\{T\}\times X_T),
\end{aligned} \tag{10.13}$$

where the minimization is w.r.t. the occupation measure μ (which includes the relaxed control ω_t; see (10.9)) and the terminal measure μ_T, for a given initial measure $\mu_0 = \delta_{t_0}\delta_{x_0}$, which is the right-hand side in the Liouville equation constraint.

Note that in LP problem (10.13), the infimum is always attained since the admissible set is (weak-star) compact and the functional is linear. However, since classical trajectories are a particular case of relaxed trajectories corresponding to the choice (10.9), the minimum in LP problem (10.13) is smaller than the minimum in relaxed POCP (10.6) (this latter one being equal to the infimum in POCP (10.2); recall Assumption 10.4), i.e.,

$$v^*(t_0,x_0) \geq p^*(t_0,x_0). \tag{10.14}$$

The following result, due to [1852], essentially based on convex duality, shows that there is no gap occurring when considering more general occupation measures than those concentrated on solutions of the ODE.

Lemma 10.8. *It holds that* $v^*(t_0, x_0) = p^*(t_0, x_0)$ *for all* $(t_0, x_0) \in \mathcal{A}$.

10.4.4 ▪ Dual LP on Functions

Primal measure LP problem (10.13) has a dual LP problem in the cone of nonnegative continuous functions:

$$d^*(t_0, x_0) \quad := \quad \sup_v \quad v(t_0, x_0)$$

$$\text{s.t.} \quad l + \frac{\partial v}{\partial t} + \text{grad}\, v \cdot f \in \mathscr{C}_+([t_0, T] \times X \times U), \qquad (10.15)$$

$$l_T - v(T, .) \in \mathscr{C}_+(X_T),$$

where maximization is with respect to a continuously differentiable function $v \in \mathscr{C}^1([t_0, T] \times X)$ that can be interpreted as a Lagrange multiplier of the Liouville equation in (10.13).

In general, the supremum in dual LP problem (10.15) is not attained, and weak duality with the primal LP problem (10.13) holds:

$$p^*(t_0, x_0) \geq d^*(t_0, x_0),$$

but it can be shown that there is actually no duality gap.

Lemma 10.9 (no duality gap). *It holds that* $p^*(t_0, x_0) = d^*(t_0, x_0)$ *for all* $(t_0, x_0) \in \mathcal{A}$.

10.5 ▪ Approximation Results

Primal LP problem (10.13) and dual LP problem (10.15) are infinite-dimensional conic problems. If we want to solve them with a computer, we invariably have to use discretization and approximation schemes. The aim of this section is to derive various approximation results that prove useful when designing numerical methods based on moment-SOS hierarchies.

10.5.1 ▪ Lower Bounds and Uniform Approximation of the Value Function

First, observe that there always exists an admissible solution for dual LP problem (10.15). For example, choose $v(t, x) := a + b(T - t)$ with $a \in \mathbb{R}$ such that $l_T(x) \geq a$ on X_T and $b \in \mathbb{R}$ such that $l(x, u) \geq b$ on $X \times U$. Moreover, by construction, any admissible function for dual LP problem (10.15) gives a global lower bound on the value function.

Lemma 10.10 (lower bound on value function). *If* $v \in \mathscr{C}^1([t_0, T] \times X)$ *is admissible for dual LP problem (10.15), then* $v^* \geq v$ *on* $[t_0, T] \times X$.

The relation between the dual LP problem (10.15) and the original optimal control problem (10.2) is given by the following result, which relates maximizing sequences to the value function.

Lemma 10.11 (maximizing sequence). *Given* $(t_0, x_0) \in \mathscr{A}$, *there is a sequence* $(v_k)_{k \in \mathbb{N}}$ *admissible for the dual LP problem* (10.15) *such that* $v^*(t_0, x_0) \geq v_k(t_0, x_0)$ *and* $\lim_{k \to \infty}$ $v_k(t_0, x_0) = v^*(t_0, x_0)$.

Building on these two results, we can describe more precisely how maximizing sequences of the dual LP problem (10.15) uniformly approximate the value function v^* along solutions of problem (10.6).

Theorem 10.12 (uniform convergence along relaxed trajectories). *For any sequence* $(v_k)_{k \in \mathbb{N}}$ *maximizing the dual LP problem* (10.15), *for any solution* (x, ω_t) *of relaxed POCP* (10.6), *and for any* $t \in [t_0, T]$, *it holds that*

$$0 \leq v^*(t, x(t)) - v_k(t, x(t)) \leq v^*(t_0, x_0) - v_k(t_0, x_0) \underset{k \to \infty}{\longrightarrow} 0.$$

It is important to notice that Theorem 10.12 holds for any trajectory realizing the minimum of POCP (10.6) and, therefore, for all of them simultaneously. In addition, these trajectories are identified with limiting trajectories of POCP (10.2) by Assumption 10.4.

10.5.2 ▪ Numerical Illustration

The results presented in Theorem 10.12 are related to properties of minimizing or maximizing elements, or sequences of elements for infinite-dimensional LO problems. From a practical point of view, it is possible to construct these sequences using the same numerical tools as in static polynomial optimization. On the primal side, this allows us to approximate the minimizing elements of measure LP problems with a converging hierarchy of moment SDP problems. On the dual side, we can construct numerically maximizing sequences of polynomial SOS certificates for the continuous function LP problems. The convergence properties that we investigated hold in particular for these solutions of the moment-SOS hierarchy.

We consider here the one-dimensional turnpike POCP that was presented in Example 10.2. For this problem, the infimum is attained at a unique optimal control that is piecewise constant. This result follows from Pontryagin's maximum principle (see [481, Section 22.2]). The optimal trajectory $t \mapsto x^*(t)$ starting at $(t_0, x_0) = (0, 0)$ is presented in Figure 10.1. The uniform convergence of approximate value functions $t \mapsto v_k(t, x^*(t))$ to the true value function $t \mapsto v^*(t, x^*(t))$ along this optimal trajectory, stated by Theorem 10.12, is illustrated in Figure 10.2. Moreover, the difference $t \mapsto v^*(t, x^*(t)) - v_k(t, x^*(t))$ is a decreasing function of time, which is essential in the proof of Theorem 10.12.

10.6 ▪ Optimal Control over a Set of Initial Conditions

Liouville equation (10.12) is used as a linear equality constraint in POCP (10.13) with a Dirac right-hand side as an initial condition. However, this right-hand side can be replaced by more general probability measures. The linearity of the constraint allows us to extend most of the results of the previous section to this setting. It leads to similar convergence guarantees regarding a (possibly uncountable) set of optimal control problems. These guarantees hold for solutions of a single infinite-dimensional LP problem.

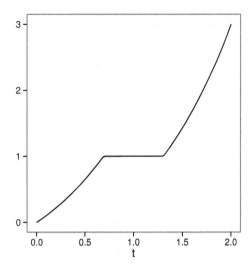

Figure 10.1. *Optimal trajectory* $t \mapsto x^*(t)$ *starting at* $(t_0, x_0) = (0, 0)$ *for the turnpike POCP* (10.4) *of Example* 10.2.

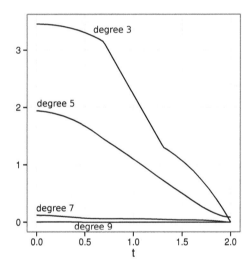

Figure 10.2. *Differences* $t \mapsto v^*(t, x^*(t)) - v_k(t, x^*(t))$ *between the actual value function and its polynomial approximations of increasing degree* $k = 3, 5, 7, 9$ *along the optimal trajectory* $t \mapsto x^*(t)$ *starting at* $(t_0, x_0) = (0, 0)$ *for the turnpike POCP* (10.4) *of Example* 10.2. *We observe uniform convergence along this trajectory, as well as time decrease of the difference, illustrating Theorem* 10.12.

10.6.1 ▪ Average Occupation Measure and Approximation Results

Suppose that we are given a set of initial conditions $X_0 \subset X$ such that $(t_0, x_0) \in \mathscr{A}$ for every $x_0 \in X_0$. Given a probability measure $\xi_0 \in \mathscr{P}(X_0)$, let

$$\mu_0(dt, dx) = \delta_{t_0}(dt)\xi_0(dx)$$

and consider the average value

$$\bar{v}^*(\mu_0) := \int_{X_0} v^*(t,x)\mu_0(dt,dx) = \langle v^*, \mu_0 \rangle, \tag{10.16}$$

where v^* is the value of POCP (10.2). Under Assumption 10.4, by linearity this value is equal to the value of POCP (10.13) with μ_0 as the right-hand side of the equality constraint, namely the primal averaged LP problem

$$\bar{p}^*(\mu_0) \quad := \quad \min_{\mu, \mu_T} \quad \langle l, \mu \rangle + \langle l_T, \mu_T \rangle$$

$$\text{s.t.} \qquad \frac{\partial \mu}{\partial t} + \operatorname{div} f \mu + \mu_T = \mu_0, \tag{10.17}$$
$$\mu \in \mathcal{M}_+([t_0, T] \times X \times U),$$
$$\mu_T \in \mathcal{M}_+(\{T\} \times X_T),$$

with dual averaged LP problem

$$\bar{d}^*(\mu_0) \quad := \quad \sup_v \quad \langle v, \mu_0 \rangle$$

$$\text{s.t.} \qquad l + \frac{\partial v}{\partial t} + \operatorname{grad} v \cdot f \in \mathscr{C}_+([t_0, T] \times X \times U), \tag{10.18}$$
$$l_T - v(T,.) \in \mathscr{C}_+(X_T).$$

The absence of a duality gap is justified in the same way as in Lemma 10.9. Moreover, Lemma 10.10 also holds, and, as in Lemma 10.11, we have the existence of maximizing lower bounds v_k such that

$$\lim_{k \to \infty} \langle v_k, \mu_0 \rangle = \langle \bar{v}^*, \mu_0 \rangle = \bar{p}^*(\mu_0) = \bar{d}^*(\mu_0).$$

Intuitively, primal LP problem (10.17) models the superposition of optimal control problems. The LP formulation allows us to express it as a single program over measures satisfying a transport equation. A relevant question here is the relation between solutions of averaged measure LP problem (10.17) and optimal trajectories of the original POCP (10.2). The intuition is that measure solutions of LP problem (10.17) represent the superposition of optimal trajectories of the relaxed POCP (10.6). These trajectories are themselves limiting trajectories of the original POCP (10.2). The superposition principle of [66, Theorem 3.2] allows us to formalize this intuition and to extend the result of Theorem 10.12 to this setting.

Theorem 10.13 (uniform convergence on support of optimal measure). *For any solution (μ, μ_T) of primal averaged LP problem (10.17), there are parameterized measures $\xi_t \in \mathscr{P}(X)$ (for the state) and $\omega_t \in \mathscr{P}(U)$ (for the control) such that $\mu(dt,dx,du) = dt\, \xi_t(dx)\omega_t(du)$, $\mu_0(dt,dx) = \delta_{t_0}(dt)\xi_{t_0}(dx)$, and $\mu_T(dt,dx) = \delta_T(dt)\xi_T(dx)$. In addition, if $(v_k)_{k \in \mathbb{N}}$ is a maximizing sequence for dual averaged LP problem (10.18), for any $t \in [t_0, T]$, it holds that*

$$0 \leq \int_X (v^*(t,x) - v_k(t,x))\xi_t(dx) \leq \int_X (v^*(t_0,x_0) - v_k(t_0,x_0))\xi_0(dx_0) \underset{k \to \infty}{\longrightarrow} 0.$$

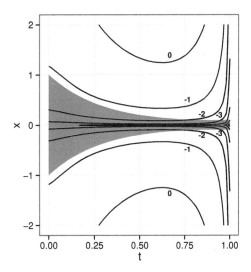

Figure 10.3. *Contour lines (at* $0, -1, -2, -3, \ldots$*) of the decimal logarithm of the difference* $(t, x) \mapsto v^*(t, x) - v_6(t, x)$ *between the actual value function and its polynomial approximation of degree six for LQR POCP (10.3) of Example 10.1. The dark area represents the set of optimal trajectories starting from* $x_0 \in X_0 = [-1, 1]$ *at time* $t_0 = 0$*. We observe that the difference is smaller in this area, as predicted by Theorem 10.13.*

A remarkable practical implication of this result is that maximizing sequences of averaged dual LP problem (10.18) approximate the value function of POCP (10.2) that is uniform in time and almost uniform in space along limits of optimal trajectories starting from X_0.

10.6.2 ▪ Numerical Illustration

As in Section 10.5.2, we illustrate the convergence result of Theorem 10.13 using Example 10.1, presented in Section 10.3.2. For each (t_0, x_0), the infimum is attained and the value of the problem can be computed by solving a Riccati differential equation. To illustrate Theorem 10.13, we are interested in the average value (10.16) for an initial measure μ_0 concentrated at time zero and uniformly distributed in space in $X_0 = [-1, 1]$. We approximate this value with primal and dual solutions of LP problem (10.17) and (10.18). The results are presented in Figure 10.3. The contour lines (in decimal logarithmic scale) represent the difference between the true value function $(t, x) \mapsto v^*(t, x)$ and a polynomial approximation of degree six $(t, x) \mapsto v_6(t, x)$. We also show the support of optimal trajectories starting from X_0. This illustrates the fact that the approximation of the value function is precise in this region, as stated by Theorem 10.13. It is noticeable that this is computed by a single LP problem and provides approximation guarantees on average over an uncountable set of optimal control problems.

Chapter 11

Truss Topology Design by Conic Linear Optimization

Michal Kočvara

11.1 ▪ Introduction

This chapter can be viewed as a complement to Chapter 2 (Truss Topology Design by Linear Optimization). We will use the same mechanical model of trusses and, whenever possible, the same notation. In Chapter 2, the truss topology design problem is formulated and solved as a linear optimization (LO) problem. In this chapter, we will introduce alternative formulations using conic linear optimization (CLO). In particular, we will present linear second-order cone optimization (SOCO) and linear semidefinite optimization (SDO) formulations of the minimum volume and minimum compliance problems. All formulations will be developed in the "primal" variables (bar cross-sectional areas) and the "dual" variables (displacements).

We will start with the nonlinear (and nonconvex) formulation of the basic truss topology problem, prove the existence of a solution, and show that the Lagrangian dual to this problem is a convex quadratically constrained quadratic problem. Then we introduce the SOCO formulations of the problem, both primal and dual, and the SDO formulations, again primal and dual. In the last section, we will demonstrate why we need these conic formulations when we already have the LO formulations from Chapter 2. In particular, we will show that by adding new important constraints to the basic problem, the conic formulations will prove to be very useful.

11.2 ▪ Truss Notation

By truss we understand an assemblage of pin-jointed uniform straight bars. The bars can carry axial tension and compression only when the truss is loaded at the joints. We denote by m the number of bars and by N the number of joints. The positions of the joints are collected in a vector y of dimension $\tilde{n} := dim \cdot N$, where dim is the spatial dimension of the truss. The material properties of bars are characterized by their Young's moduli E_i; the bar lengths are denoted by ℓ_i and the bar cross-sectional areas by a_i, $i = 1, \ldots, m$.

Let $dim = 2$, to shorten the notation. For the ith bar, let

$$r_i = \frac{1}{\ell_i}\left(-(y_1^{(k)} - y_1^{(j)}),\ -(y_2^{(k)} - y_2^{(j)}),\ y_1^{(k)} - y_1^{(j)},\ y_2^{(k)} - y_2^{(j)}\right)^T,$$

where $y^{(j)}$ and $y^{(k)}$ are the endpoints of the bar. The compatibility matrix is then defined by $R = (r_1, r_2, \ldots, r_m)$.

Let $f \in \mathbb{R}^{\tilde{n}}$ be a load vector of nodal forces. The response of the truss to the load f is measured by nodal displacements collected in a displacement vector $u \in \mathbb{R}^{\tilde{n}}$. Some of the displacement components may be restricted: a node can be fixed in a wall, and then the corresponding displacements are prescribed to be zero. The number of free nodes multiplied by the spatial dimension will be denoted by n, and we will assume that $f \in \mathbb{R}^n$ and $u \in \mathbb{R}^n$; see Figure 11.1.

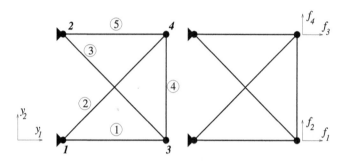

Figure 11.1. *A five-bar truss with four nodes, two of them fixed. Here $y^{(1)} = (0,0)$, $y^{(2)} = (0,1)$, $y^{(3)} = (1,0)$, $y^{(4)} = (1,1)$, $dim = 2$, $N = 4$, $\tilde{n} = 8$, and $n = 4$.*

In agreement with Chapter 2, we denote by

$$q_i = \frac{E_i a_i}{\ell_i} r_i^T u$$

the axial force in the ith bar; introduce bar stiffness matrices and assemble them in the global stiffness matrix of the truss

$$K(a) = \sum_{i=1}^{m} a_i K_i = \sum_{i=1}^{m} a_i \frac{E_i}{\ell_i} r_i r_i^T, \quad i = 1, \ldots, m;$$

and introduce the equilibrium equation

$$K(a)u = f. \tag{11.1}$$

Finally, to simplify the notation, we define $b_i := \sqrt{\frac{E_i}{\ell_i}} r_i$.

Assumption 11.1. $K(1)$ *is positive definite, and the load vector f is in the range space of $K(1)$.*

11.3 ▪ Nonlinear Optimization Formulation

The most natural objective is to minimize the volume of the structure. The minimal requirement on the optimal structure is that it should satisfy equilibrium equation (11.1). The lightest structure satisfying equilibrium tends to be no structure at all, so it is reasonable to introduce another constraint that would control the stiffness (or weakness) of the optimal structure. Commonly used is the *compliance* of the truss, $f^T u$, where f and u satisfy the equilibrium equation (11.1). The smaller the compliance, the smaller the displacement at the loaded nodes and thus the stiffer the truss.

11.3.1 ▪ Primal Problem

Let $\gamma \in \mathbb{R}$ and $\underline{a} \in \mathbb{R}^m, \bar{a} \in \mathbb{R}^m$ such that $\gamma > 0, 0 \le \underline{a}_i \le \bar{a}_i$, $i = 1,\dots,m$, and the following assumption is satisfied.

Assumption 11.2. *Let* $\bar{u} = K(\bar{a})^{-1}f$. *Then* $f^T \bar{u} < \gamma$.

The simplest meaningful truss design problem is the single-load topology optimization minimum volume problem

$$\min_{a \in \mathbb{R}^m,\, u \in \mathbb{R}^n} \quad \sum_{i=1}^m a_i \ell_i \tag{11.2}$$

$$\text{s.t.} \quad K(a)u = f,$$
$$f^T u \le \gamma,$$
$$\underline{a}_i \le a_i \le \bar{a}_i, \quad i = 1,\dots,m.$$

Now let $V \in \mathbb{R}$ be a given maximal volume satisfying the following assumption.

Assumption 11.3. *The maximum volume V must satisfy* $\sum_{i=1}^m \underline{a}_i \ell_i < V < \sum_{i=1}^m \bar{a}_i \ell_i$.

Alternatively to (11.2), we can maximize the stiffness of the truss (minimize compliance) subject to equilibrium and resource constraints:

$$\min_{a \in \mathbb{R}^m,\, u \in \mathbb{R}^n} \quad f^T u \tag{11.3}$$

$$\text{s.t.} \quad K(a)u = f,$$
$$\sum_{i=1}^m a_i \ell_i \le V,$$
$$\underline{a}_i \le a_i \le \bar{a}_i, \quad i = 1,\dots,m.$$

Theorem 11.4 [16]. *Problems (11.2) and (11.3) are equivalent. Any solution* (a^*, u^*) *of (11.2) is also a solution of (11.3) with* $V = \sum_{i=1}^m a_i^* \ell_i$. *Any solution* (a^*, u^*) *of (11.3) is also a solution of (11.2) with* $\gamma = f^T u^*$.

Problems (11.2) and (11.3) are nonlinear optimization (NLO) problems that are rather difficult to solve. This is because the equilibrium constraint does not satisfy the Mangasarian–Fromowitz constraint qualification required by most NLO algorithms.

11.3.2 ▪ Existence of Solution

In this section, we closely follow [214]. The minimum compliance problem (11.3) can be equivalently written as

$$\min_{a \in \mathbb{R}^m} \sup_{u \in \mathbb{R}^n} \quad 2f^T u - u^T K(a)u \tag{11.4}$$

$$\text{s.t.} \quad \sum_{i=1}^m a_i \ell_i \le V,$$
$$\underline{a}_i \le a_i \le \bar{a}_i, \quad i = 1,\dots,m.$$

Remark **11.1.** Formulation (11.4) is, in a sense, the basic and "most natural" formulation of the truss design problem, and it is also a general structural optimization problem. Searching equilibrium by minimization of the potential energy is more general than just the equilibrium equation, as it allows for more general physical laws (nonlinear material and geometry) and more general constraints (e.g., unilateral contact conditions). We can say that nature minimizes the potential energy, while the designer tries to bring this minimum as close to zero as possible.

Let us define the *compliance function* as

$$c(a) := \sup_{u \in \mathbb{R}^n} (2f^T u - u^T K(a) u).$$

Theorem 11.5. *The compliance function is convex and lower semicontinuous on* $\mathbb{R}^m$.

Proof. $c(\cdot)$ is a pointwise supremum of linear and thus closed functions. Hence it is convex and closed [1572, Theorems 5.5, 9.4] and so lower semicontinuous [1572, p. 52]. □

Now assume that $\underline{a}_i < a_i < \bar{a}_i$, $i = 1, \ldots, m$, which, in particular, means that $a > 0$ and thus $K(a)$ is invertible. Then the "sup" in the definition of the compliance is attained (Assumption 11.1), and we can write the compliance function as $c(a) = f^T K^{-1}(a) f$. Consider the following problem:

$$\min_{\alpha \in \mathbb{R}, \, a \in \mathbb{R}^m} \quad \alpha \tag{11.5}$$
$$\text{s.t.} \quad (a, \alpha) \in \mathrm{cl}\,\Omega,$$

where Ω is defined as

$$\Omega := \left\{ (a, \alpha) \in \mathbb{R}^m \times \mathbb{R} \mid f^T K^{-1}(a) f < \alpha, \ \sum_{i=1}^{m} a_i \ell_i \leq V, \ \underline{a}_i < a_i < \bar{a}_i, \ i = 1, \ldots, m \right\}.$$

In addition to Assumptions 11.1–11.3, we further assume that $\mathrm{int}\,\Omega \neq \emptyset$.

Proposition 11.6 [1572, Theorem 7.6]. *Let* $\varphi : \mathbb{R}^n \to \mathbb{R}$ *be a convex proper function, and let* $\alpha \in \mathbb{R}$, $\alpha > \inf \varphi$. *The convex level sets* $\{x \mid \varphi(x) \leq \alpha\}$ *and* $\{x \mid \varphi(x) < \alpha\}$ *then have the same closure and the same relative interior, namely*

$$\{x \mid (\mathrm{cl}\,\varphi)(x) \leq \alpha\}, \qquad \{x \in \mathrm{ri}(\mathrm{dom}\,\varphi) \mid f(x) < \alpha\},$$

respectively.

Corollary 11.7. *Let* $\varphi : \mathbb{R}^n \to \mathbb{R}$ *be a closed proper convex function, and let* $\omega \subset \mathbb{R}^n$ *be closed and convex. Then for* $\inf \varphi < \alpha < +\infty$, *one has*

$$\{x \mid \varphi(x) \leq \alpha, \ x \in \omega\} = \mathrm{cl}\{x \mid \varphi(x) < \alpha, \ x \in \mathrm{ri}\,\omega\}.$$

Hence problem (11.5) is equivalent to problem (11.4).

Theorem 11.8. *Problem (11.5) has at least one solution.*

Proof. $\mathrm{cl}\,\Omega$ is a closed convex set due to the convexity of the compliance function. We are thus minimizing a linear function over a closed convex set $\mathrm{cl}\,\Omega$. Furthermore, the level sets $\{(a, \alpha) \in \mathrm{cl}\,\Omega \mid \alpha < c\}$ are clearly bounded. Hence, (11.5) is solvable. □

Corollary 11.9. *Problem* (11.4) *has at least one solution.*

Above we have also proved the following proposition.

Proposition 11.10. *Let* $a^* \in \mathbb{R}^m$ *be a solution of the minimum volume problem* (11.2) *with* $\underline{a}_i = 0$, $i = 1,\dots,m$. *There exists a sequence* $\{a_k\}_{k=1}^{\infty}$, $a_k \in \mathbb{R}^m$, *with the following properties:*

- a_k *is a solution of* (11.2) *with* $\underline{a}_i = \varepsilon_k$, $i = 1,\dots,m$;
- $\varepsilon_k \to 0$ *and* $a_k \to a^*$ *as* $k \to \infty$.

11.3.3 ▪ Dual Problem (QCQO Formulation)

We now introduce a reformulation of (11.3) that has had a great impact on the numerical solution of truss design problems. It is the following quadratically constrained quadratic optimization (QCQO) problem:

$$\min_{u \in \mathbb{R}^n, \, \alpha \in \mathbb{R}, \, \underline{p} \in \mathbb{R}^m, \, \overline{p} \in \mathbb{R}^m} \quad \alpha V - f^T u - \underline{a}^T \underline{p} + \overline{a}^T \overline{p} \tag{11.6}$$

$$\text{s.t.} \quad \frac{1}{2} u^T K_i u \le \alpha \ell_i - \underline{p}_i + \overline{p}_i, \quad i = 1,\dots,m,$$

$$\underline{p} \ge 0,$$

$$\overline{p} \ge 0.$$

Theorem 11.11 [16]. *Problems* (11.3) *and* (11.6) *are equivalent in the following sense:*

(i) *If one problem has a solution, then the other problem also has a solution and the optimal objective values of the two problems are equal.*

(ii) *Let* $(u^*, \alpha^*, \underline{p}^*, \overline{p}^*)$ *be a solution to* (11.6). *Further, let* τ^* *be the vector of Lagrangian multipliers for the inequality constraints associated with this solution. Then* (u^*, τ^*) *is a solution of* (11.3). *Moreover,* $\underline{p}_i^* \overline{p}_i^* = 0$, $i = 1,\dots,m$.

(iii) *Let* (u^*, a^*) *be a solution of* (11.3). *Further, let* $\underline{r}^*$ *and* $\overline{r}^*$ *be the Lagrangian multipliers associated with the lower and upper bounds on* t, *respectively, and let* α^* *be the multiplier for the volume constraint. Then* $(u^*, \alpha^*, \underline{r}^*, \overline{r}^*)$ *is a solution of* (11.6).

11.4 ▪ SOCO Formulation

11.4.1 ▪ Primal SOCO Problem

We start with a simple but useful lemma that shows the relation between convex quadratic constraints and SOCO constraints.

Lemma 11.12. *Let* $x \in \mathbb{R}^n$, $t \in \mathbb{R}$, *and* $s \in \mathbb{R}$, $s > 0$. *Then*

$$\frac{x x^T}{s} \le t \quad \Longleftrightarrow \quad \left\| \begin{pmatrix} x \\ \frac{t-s}{2} \end{pmatrix} \right\|_2 \le \frac{t+s}{2}.$$

As (11.6) is a convex quadratic problem, it can now be immediately rewritten as a SOCO problem:

$$\min_{u\in\mathbb{R}^n,\ \alpha\in\mathbb{R},\ \underline{\rho}\in\mathbb{R}^m,\ \overline{\rho}\in\mathbb{R}^m} \quad \alpha V - f^T u - \underline{a}^T \underline{\rho} + \overline{a}^T \overline{\rho} \tag{11.7}$$

$$\text{s.t.} \quad \left\| \begin{pmatrix} \frac{\sqrt{2}}{2} b_i^T u \\ \frac{\alpha \ell_i - \underline{\rho}_i + \overline{\rho}_i - 1}{2} \end{pmatrix} \right\|_2 \leq \frac{\alpha \ell_i - \underline{\rho}_i + \overline{\rho}_i + 1}{2}, \quad i = 1,\ldots,m,$$

$$\underline{\rho} \geq 0,$$
$$\overline{\rho} \geq 0.$$

11.4.2 ▪ Dual SOCO Problem

Proposition 11.13. *The dual problem to (11.7) can be written as*

$$\min_{a\in\mathbb{R}^m,\ \tau\in\mathbb{R}^m,\ q\in\mathbb{R}^m} \quad \frac{1}{2}\sum_{i=1}^m \tau_i \tag{11.8}$$

$$\text{s.t.} \quad \sum_{i=1}^m a_i \ell_i = V,$$

$$\sum_{i=1}^m q_i b_i = f,$$

$$\left\| \begin{pmatrix} \sqrt{2} q_i \\ \frac{2a_i - \tau_i}{2} \end{pmatrix} \right\|_2 \leq \frac{2a_i + \tau_i}{2}, \quad i = 1,\ldots,m,$$

$$\underline{a} \leq a_i \leq \overline{a}, \quad i = 1,\ldots,m,$$

where a are the bar areas, q are the bar axial forces, and the objective function is equal to the compliance.

Proof. Let μ_i, ν_i be Lagrangian multipliers to the conic constraints, where $\mu_i = (\mu_{i,1}, \mu_{i,2}) \in \mathbb{R}^2$ and $\nu_i \in \mathbb{R}$, $i = 1,\ldots,m$. Further, let $x \in \mathbb{R}^m$ be the multiplier to the bound constraint on ρ. The Lagrangian dual to (11.7) reads as

$$\max_{\mu\in\mathbb{R}^{m\times2},\ \nu\in\mathbb{R}^m,\ \underline{x}\in\mathbb{R}^m,\ \overline{x}\in\mathbb{R}^m} \quad \frac{1}{2}\sum_{i=1}^m \mu_{i,2} - \nu_i$$

$$\text{s.t.} \quad \frac{1}{2}\sum_{i=1}^m (\mu_{i,2} + \nu_i)\ell_i = V,$$

$$\sum_{i=1}^m \mu_{i,1} \frac{\sqrt{2}}{2} b_i = f,$$

$$\|\mu_i\|_2 \leq \nu_i, \quad i = 1,\ldots,m,$$

$$\frac{1}{2}\mu_{i,2} + \frac{1}{2}\nu_i - \underline{x}_i = \underline{a}_i, \quad i = 1,\ldots,m,$$

$$\frac{1}{2}\mu_{i,2} + \frac{1}{2}\nu_i + \overline{x}_i = \overline{a}_i, \quad i = 1,\ldots,m,$$

$$\underline{x}_i \geq 0,\ \overline{x}_i \geq 0, \quad i = 1,\ldots,m.$$

Setting $q_i = \frac{\sqrt{2}}{2}\mu_{i,1}$, $a_i = \frac{1}{2}(\mu_{i,2} + \nu_i)$, and $\tau = \nu_i - \mu_{i,2}$, we see from the last three constraints that $\underline{a}_i \le a_i \le \overline{a}$. Consequently, we get (11.8). □

Similarly, we would get the minimum volume SOCO formulation just by replacing the objective in (11.8) by $\sum_{i=1}^{m} a_i \ell_i$ and the first constraint by $\sum_{i=1}^{m} \tau_i \ell_i \le \gamma$, where γ is an upper bound on compliance, as in (11.2).

11.5 ▪ SDO Formulation

11.5.1 ▪ Primal SDO Problem

The SDO formulation of the primal problem is based on the Schur complement theorem [211]. Because we allow the stiffness matrix to be singular, we need a minor generalization of the standard theorem. The proof can be found, e.g., in [18].

Proposition 11.14 [18]. *Let* $a \in \mathbb{R}^m$, $a \ge 0$, *and* $\gamma \in \mathbb{R}$ *be fixed. Then there exists* $u \in \mathbb{R}^n$ *satisfying*

$$K(a)u = f \quad and \quad f^T u \le \gamma$$

if and only if

$$\begin{pmatrix} \gamma & -f^T \\ -f & K(a) \end{pmatrix} \succeq 0.$$

Using this proposition, we get equivalent formulations of problems (11.2) and (11.3), respectively:

$$\min_{a \in \mathbb{R}^m} \quad \sum_{i=1}^{m} a_i \ell_i \tag{11.9}$$

$$\text{s.t.} \quad \begin{pmatrix} \gamma & -f^T \\ -f & K(a) \end{pmatrix} \succeq 0,$$

$$\underline{a}_i \le a_i \le \overline{a}_i, \quad i = 1, \dots, m,$$

and

$$\min_{a \in \mathbb{R}^m, \gamma \in \mathbb{R}} \quad \gamma \tag{11.10}$$

$$\text{s.t.} \quad \begin{pmatrix} \gamma & -f^T \\ -f & K(a) \end{pmatrix} \succeq 0,$$

$$\sum_{i=1}^{m} a_i \ell_i \le V,$$

$$\underline{a}_i \le a_i \le \overline{a}_i, \quad i = 1, \dots, m.$$

Theorem 11.15. *Problems (11.2) and (11.9) are equivalent. If* (a^*, u^*) *is a solution of (11.2), then* a^* *is a solution of (11.9). If* a^* *is a solution of (11.9), then there exists* $u^* \in \mathbb{R}^n$ *such that* $K(a^*)u^* = f$ *and* (a^*, u^*) *is a solution of (11.2). The same holds for problems (11.3) and (11.10).*

11.5.2 ▪ Dual SDO Problem

Let us now write down a dual to the SDO problem (11.9). It is the problem

$$\max_{W \in \mathbb{S}^{n+1},\, \overline{p} \in \mathbb{R}^m,\, \underline{p} \in \mathbb{R}^m} \left\langle \begin{pmatrix} -\gamma & f^T \\ f & 0 \end{pmatrix}, W \right\rangle - \sum_{i=1}^m \overline{p}_i \overline{a}_i + \sum_{i=1}^m \underline{p}_i \underline{a}_i \qquad (11.11)$$

$$\text{s.t.} \quad \left\langle \begin{pmatrix} 0 & 0 \\ 0 & K_i \end{pmatrix}, W \right\rangle - \overline{p}_i + \underline{p}_i = \ell_i, \quad i = 1, \dots, m,$$

$$W \succcurlyeq 0,$$

$$\overline{p}_i \geq 0, \; \underline{p}_i \geq 0, \quad i = 1, \dots, m.$$

Proposition 11.16. *There exists a solution* $(W^*, \overline{p}^*, \underline{p}^*) \in \mathbb{S}^{n+1} \times \mathbb{R}^m \times \mathbb{R}^m$ *of the dual SDO problem (11.11) such that the rank of* W^* *is one.*

Proof. Let us first show that both the primal and the dual SDO problems (11.9) and (11.11) satisfy the Slater condition. Indeed, by Assumption 11.2, $\gamma - f^T K^{-1}(\overline{a}) f > 0$, so, by the Schur complement theorem, $\overline{a}$ is a Slater point for (11.9). Now let $\widehat{W}$ be the identity matrix. Then $\widehat{W} \succ 0$ and $\langle \begin{pmatrix} 0 & 0 \\ 0 & K_i \end{pmatrix}, \widehat{W} \rangle = \text{trace}(K_i) > 0$. For any $i = 1, \dots, m$, we can now always find $\overline{p}_i \geq 0$ and $\underline{p}_i \geq 0$ satisfying $\text{trace}(K_i) - \overline{p}_i + \underline{p}_i = \ell_i$, so $\widehat{W}$ is a Slater point for (11.11). Hence, the assumptions of Theorem 9.7 from Chapter 9 are satisfied.

Let a^* be a solution of the primal problem (11.9). Denote by $S^* \in \mathbb{S}^{n+1}$ the primal slack matrix variable

$$S^* = \begin{pmatrix} \gamma & -f^T \\ -f & K(a^*) \end{pmatrix}.$$

From SDO duality, S^* is complementary to any solution W of the dual problem (11.11), $\langle W, S^* \rangle = 0$, and the pair satisfies the following rank condition:

$$\text{rank}(S^*) + \text{rank}(W) \leq n + 1.$$

If $\underline{a}_i > 0$, $i = 1, \dots, m$, the matrix $K(a^*)$ is positive definite and, due to Assumption 11.1, has a full rank n. Hence, by the above condition, the rank of any dual solution W is at most one. Excluding trivial solutions, the rank of W is then equal to one.

Assume now that $\underline{a}_i = 0$, $i = 1, \dots, m$, i.e., the matrix $K(a^*)$ can be rank deficient. Due to Proposition 11.10, there is a sequence of solutions $\{S_k\}$ to the primal problem with $(\underline{a}_i)_k = \varepsilon_k$, $\varepsilon_k \to 0$ as $k \to \infty$ such that $S_k \to S^*$. Associated with $\{S_k\}$ is a sequence of dual solutions $\{W_k\}$ such that any pair (S_k, W_k) satisfies the complementarity and thus the rank condition. Using the same argument as above, the rank of matrices W_k is one. Hence, there is a sequence of vectors $\{w_k\}$ such that $W_k = w_k w_k^T$ for each k and $w_k \to w^*$ as $k \to \infty$. By construction, $W^* = w^* w^{*T}$ is a rank-one matrix, which is attained due to the continuity of the Frobenius inner product. Finally, W^* is a solution to (11.11) because it satisfies the necessary and sufficient optimality conditions: due to the continuity of the Frobenius inner product, it is complementary to S^* and feasible in (11.11).   ☐

Analogously, we can write down a dual to the semidefinite formulation of the minimum compliance problem (11.10). It is the problem

$$\max_{W \in \mathbb{S}^{n+1}, \overline{p} \in \mathbb{R}^m, \underline{p} \in \mathbb{R}^m, \omega \in \mathbb{R}} \left\langle \begin{pmatrix} 0 & f^T \\ f & 0 \end{pmatrix}, W \right\rangle - \sum_{i=1}^m \overline{p}_i \overline{a}_i + \sum_{i=1}^m \underline{p}_i \underline{a}_i - \omega V \quad (11.12)$$

$$\text{s.t.} \quad \left\langle \begin{pmatrix} 0 & 0 \\ 0 & K_i \end{pmatrix}, W \right\rangle - \overline{p}_i + \underline{p}_i - \omega \ell_i = 0, \quad i = 1, \ldots, m,$$

$$W_{11} = 1,$$

$$W \succcurlyeq 0,$$

$$\omega \geq 0,$$

$$\overline{p}_i \geq 0, \ \underline{p}_i \geq 0, \quad i = 1, \ldots, m.$$

Proposition 11.17. *There exists a solution* $(W^*, \overline{p}^*, \underline{p}^*, \omega^*) \in \mathbb{S}^{n+1} \times \mathbb{R}^m \times \mathbb{R}^m \times \mathbb{R}$ *of the dual SDO problem* (11.12) *such that the rank of* W^* *is one.*

11.5.3 ▪ Closing the Circle

Theorem 11.18. *The dual SDO problem* (11.12) *is equivalent to the problem* (11.6):

(i) *Let* $(u^*, \alpha^*, \underline{p}^*, \overline{p}^*)$ *be a solution of* (11.6). *Then*

$$(W^+, \underline{p}^+, \overline{p}^+, \omega^+) := \left(u^* u^{*T}, \underline{p}^*, \overline{p}^*, \frac{2\alpha^*}{V} \right)$$

is a solution of (11.12).

(ii) *Let* $(W^*, \underline{p}^*, \overline{p}^*, \omega^*)$ *be a solution to* (11.12) *such that* rank $W^* = 1$. *Then there exists* $w^* \in \mathbb{R}^{n+1}$ *with* $W^* = w^* w^{*T}$ *and such that*

$$(u^+, \alpha^+, \underline{p}^+, \overline{p}^+) := \left(w^*_{2:n+1}, \frac{1}{2} \omega V, \underline{p}^*, \overline{p}^* \right)$$

is a solution of (11.6).

Proof. Let $(W, \overline{p}, \underline{p}, \omega) \in \mathbb{S}^{n+1} \times \mathbb{R}^m \times \mathbb{R}^m \times \mathbb{R}$ be a solution of (11.12). By the above proposition, assume that $W = \tilde{w}\tilde{w}^T$ with some $\tilde{w} \in \mathbb{R}^{n+1}$. Denote $\tilde{w} = \begin{pmatrix} c \\ w \end{pmatrix}$ with $c \in \mathbb{R}$. The dual SDO problem (11.12) can thus be written as

$$\max_{W \in \mathbb{S}^{n+1}, \overline{p} \in \mathbb{R}^m, \underline{p} \in \mathbb{R}^m, \omega \in \mathbb{R}} 2cf^T w - \sum_{i=1}^m \overline{p}_i \overline{a}_i + \sum_{i=1}^m \underline{p}_i \underline{a}_i - \omega V$$

$$\text{s.t.} \quad w^T K_i w - \overline{p}_i + \underline{p}_i = \omega \ell_i, \quad i = 1, \ldots, m,$$

$$c = 1,$$

$$\omega \geq 0,$$

$$\overline{p}_i \geq 0, \ \underline{p}_i \geq 0, \quad i = 1, \ldots, m,$$

which is just problem (11.6) with $\omega = 2\alpha/V$ and $w = u$. □

This closes the circle of equivalences.

11.6 ▪ Applications

All the formulations of the truss topology design problem introduced above can be seen as a nice exercise in conic optimization. When it comes to the numerical solution of the problem, however, they appear to be virtually useless. We know from Chapter 2 that the problem without upper bounds on a can be equivalently formulated as an LO problem, and the modern LO solvers will certainly beat linear SOCO and SDO solvers for problems of the same size. Introduction of upper bounds on a disallows the use of LO reformulation (see Chapter 2); however, the (convex) QCQO formulation (11.6) can still be efficiently solved by interior-point methods [990]. So is this chapter anything other than an exercise? Below we will demonstrate that the conic formulations of the problem can be extremely useful, as soon as we add more (important) constraints to the basic problem. Other problems benefiting from the conic formulation include problems with stability constraints [1079], multiple load problems [206], problems with local and global stress constraints, and related problems such as cable networks [1035] and free-material optimization [1082].

11.6.1 ▪ Vibration Constraints

We may be required to set an additional constraint on free vibrations of the optimal structure. The free vibrations are the squares of the eigenvalues of the following generalized eigenvalue problem:

$$K(a)w = \lambda(M(a)+M_0)w. \tag{11.13}$$

Here $M(a) = \sum_{i=1}^m a_i M_i$ is the so-called mass matrix that collects information about the mass distribution in the truss. The matrices M_i are positive semidefinite (PSD) and have the same sparsity structure as K_i. The nonstructural mass matrix M_0 is a constant, typically diagonal matrix with very few nonzero elements.

Low vibrations are dangerous and may lead to structural collapse. Hence we typically require the smallest free vibration to be bigger than some threshold, that is,

$$\lambda_{\min} \geq \overline{\lambda} \quad \text{for a given } \overline{\lambda} > 0,$$

where $\lambda_{\min}$ is the smallest eigenvalue of (11.13). This constraint can be equivalently written as the linear matrix inequality

$$K(a) - \overline{\lambda}(M(a)+M_0) \succcurlyeq 0, \tag{11.14}$$

which is to be added to the basic truss topology problem. As (11.14) is a linear matrix inequality in variable a, it is natural to add this constraint to the primal SDO formulation (11.9). We will thus get the following linear SDO formulation of the truss topology design with a vibration constraint:

$$\min_{a \in \mathbb{R}^m} \quad \sum_{i=1}^m \ell_i a_i$$

$$\text{s.t.} \quad \begin{pmatrix} \gamma & -f^T \\ -f & K(a) \end{pmatrix} \succcurlyeq 0,$$

$$K(a) - \overline{\lambda}(M(a)+M_0) \succcurlyeq 0,$$

$$\underline{a}_i \leq a_i \leq \overline{a}_i, \quad i = 1,\dots,m.$$

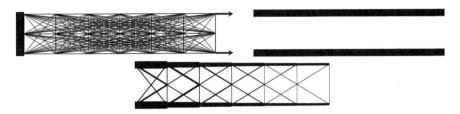

Figure 11.2. *A medium-size example (Example 11.19): initial layout (top left); optimal topology without (top right) and with (bottom) vibration constraints.*

Example 11.19. Consider a 7×3 nodal grid with ground structure, boundary conditions, and load as depicted in Figure 11.2, top left. The result of the standard minimum volume problem with no vibration constraints with $\gamma = 10$ is shown in Figure 11.2, top right—two independent horizontal bars. The volume of this structure is $V^* = 5.0$. Figure 11.2, bottom, shows the result of the problem with vibration constraints for $\gamma = 20$ and $\overline{\lambda} = 1.0 \cdot 10^{-3}$; the optimal structure has volume $V^* = 7.6166$. ∎

11.6.2 ▪ Problems with Integer Variables

In truss topology design we try to find an optimum structural design of the truss by finding the optimal cross-sectional areas of the bars. Often, from the manufacturing point of view, it is highly desirable that variables attain only few given discrete values, for instance, 0, 1, 2, 3. Then the problem becomes an optimization problem with integer variables.

As we are adding constraints on the bar areas a, we can only consider the primal formulations of the problem. Our basic formulation (11.2) is a nonlinear and nonconvex optimization problem, where the nonlinearity is due to equilibrium conditions. When searching for binary or integer design, the resulting problem is then a nonconvex mixed-integer nonlinear optimization problem. These problems are, typically, extremely difficult to solve, due to both nonconvexity and the integer nature of some of the variables. There have been many attempts to solve these problems, most of them based on heuristic optimization methods that cannot give any guarantees about the solution. A few articles have recently appeared in the literature that are based on a mathematical optimization approach to the problem and that deliver a guaranteed global minimum; see, e.g., [19, 1723].

However, we know from Section 11.4 that the basic problem is equivalent to a linear SOCO problem (11.8). Using that, we can reformulate the nonconvex mixed-integer nonlinear optimization problem as a conic linear problem with integer variables. As such, it is much easier to solve than the original formulation. In particular, we can directly apply available software, such as MOSEK or Gurobi, to its solution.

The problem formulation is obvious:

$$\min_{a \in \mathbb{R}^m, \, \tau \in \mathbb{R}^m, \, q \in \mathbb{R}^m} \quad \frac{1}{2} \sum_{i=1}^{m} \tau_i \tag{11.15}$$

$$\text{s.t.} \quad \sum_{i=1}^{m} a_i \ell_i = V,$$

$$\sum_{i=1}^{m} q_i b_i = f,$$

$$\left\| \begin{pmatrix} \sqrt{2}q_i \\ \frac{2a_i - \tau_i}{2} \end{pmatrix} \right\|_2 \le \frac{2a_i + \tau_i}{2}, \quad i = 1, \dots, m,$$

$$a_i \in \{0, 1, \dots, T\}, \quad i = 1, 2, \dots, m,$$

where $T > 0$ is a given integer.

Example 11.20 [1723]. Consider the minimum volume problem (11.15). The initial layout is shown in Figure 11.3. The dimensions are $m = 72$ and $n = 27$.

We have solved two instances of the problem, one with binary variables $a_i \in \{0, 1\}$ and compliance bound $\gamma = 50.0$ and one with integer variables $a_i \in \{0, 1, 2, 3\}$ and compliance bound $\gamma = 25.0$. The mixed-integer SOCO problem (11.15) was solved by Gurobi 5.62 with the default setting of parameters on an Intel Core i7-620M (2.66 GHz) processor with 4GB memory. The solution of the first problem (288 continuous and 72 binary variables) required visiting 3959 nodes and 14 seconds of CPU time. The optimal objective value of the relaxed problem ($a_i \in [0, 1]$) was 14.7342, and of the binary problem was 16.9853. To solve the second problem (288 continuous and 72 integer variables), Gurobi visited 25,273 nodes and needed 28 seconds of CPU time. The optimal objective value of the relaxed problem ($a_i \in [0, 3]$) was 23.3415, and of the integer problem was 24.6924.

The optimal solutions for the binary and integer problems, together with solutions of the relaxed problems, are shown in Figure 11.4. ∎

Another major advantage of this approach, apart from linearity, lies in the fact that we can easily add more conic constraints, such as the vibration constraint introduced in the previous section. This constraint amounts to a linear matrix inequality, so we may

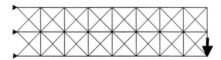

Figure 11.3. *Integer variables (Example 11.20): initial design.*

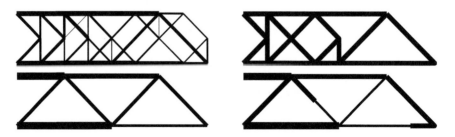

Figure 11.4. *Example 11.20: Top: binary variables $a_i \in \{0, 1\}$ and $\gamma = 50.0$, relaxed solution (left) and binary solution (right). Bottom: integer variables $a_i \in \{0, 1, 2, 3\}$ and $\gamma = 25.0$, relaxed solution (left) and integer solution (right).*

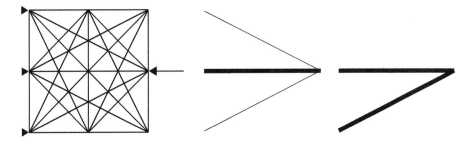

Figure 11.5. *Integer variables with vibration constraints (Example* 11.21*): initial design, relaxed solution, and binary solution.*

as well involve the linear SDO formulation of the truss problem (11.9). The resulting problem reads as

$$\min_{a \in \mathbb{R}^m} \quad \sum_{i=1}^{m} \ell_i a_i \tag{11.16}$$

$$\text{s.t.} \quad \begin{pmatrix} \gamma & -f^T \\ -f & K(a) \end{pmatrix} \succeq 0,$$

$$K(a) - \bar{\lambda}(M(a) + M_0) \succeq 0,$$

$$a_i \in \{0, 1, \dots, T\}, \quad i = 1, 2, \dots, m.$$

The drawback, as compared to (11.15), is that (11.16) is a mixed-integer linear SDO problem, and, at the time of writing, it is not supported by any "mainstream" optimization software, unlike mixed-integer linear SOCO. The next example was solved by the branch-and-bound algorithm implemented in YALMIP [1217]. The relaxations were solved by PENSDP [1081].

Example 11.21. We consider the minimum volume problem with binary variables and the vibration constraint (11.16). The initial layout is shown in Figure 11.5, left. The dimensions are $m = 36$ and $n = 12$. The upper bound on the compliance was chosen as $\gamma = 1.0$, and the bound on the smallest eigenfrequency is $\bar{\lambda} = 0.01$. To find the optimal binary solution, the branch-and-bound algorithm in YALMIP only needed to visit 149 nodes. The optimal objective value of the relaxed problem was 1.0471 (with $\max_i a_i = 0.9162$), and of the binary problem was 2.118. The optimal solutions for the relaxed and the binary problems are shown in Figure 11.5 (middle and right). ∎

Chapter 12

Applications of Conic Linear Optimization in Financial Engineering

Oleksandr Romanko and Helmut Mausser

12.1 ▪ Introduction

Optimization is one of the quantitative techniques used for financial decision making. Optimization problems in finance include *portfolio selection*, *risk management*, regression problems, pricing and hedging of derivatives, and asset liability management. A general overview of *financial optimization* or *optimization in finance* can be found in [514] and [1952].

Linear optimization (LO) models are widely used in finance due to the availability of intuitive and relatively simple formulations and computational attractiveness. The class of quadratic optimization (QO) problems gained importance among the financial community after Markowitz [1277, 1278] used it to conduct mean-variance analysis of investment portfolios. Even though conic linear (second-order conic and semidefinite) optimization problems are relatively new to the financial community, a number of financial problems can be solved using these techniques [514, 1177]. Mixed-integer optimization is required to solve financial problems involving value-at-risk (VaR) minimization, cardinality constraints, or fixed-plus-linear transaction cost. As most financial optimization models need to deal with uncertainty, robust optimization [203,684] and stochastic optimization (SO) [1940] are among the tools used by quantitative analysts. An added challenge is that many financial problems are large scale, yet require solutions in near real time.

Most financial models involve maximizing a performance measure, e.g., expected portfolio return, or minimizing a risk associated with a financial decision. As performance usually conflicts with risk, trade-offs between performance measures and risks need to be identified and explored. The trade-off curve is known in finance as an efficient frontier.

Let us consider a typical portfolio optimization problem. We denote the asset weights in a portfolio by x, i.e., x_i is the proportion of total funds invested in asset i, where $i = 1,\ldots,N$. The expected return vector for the assets is denoted by μ, where $\mu \in \mathbb{R}^N$. A typical financial risk management optimization problem can be formulated as

$$
\begin{aligned}
\max_{x \in \mathbb{R}^N} \quad & \mu^T x \\
\text{s.t.,} \quad & g_k(x) \leq \gamma_k \quad k = 1, \ldots, K, \\
& \sum_{i=1}^{N} x_i = 1, \\
& x \in \Omega.
\end{aligned}
\tag{12.1}
$$

In formulation (12.1), a performance measure $\mu^T x$, e.g., the expected portfolio return, is maximized subject to a constraint $g_k(x) \leq \gamma_k$ that a particular risk measure g_k does not exceed a prescribed bound γ_k and other operating constraints $x \in \Omega$ are satisfied. Constraints on asset holdings are usually assumed to be $\Omega = \{x \in \mathbb{R}^N : x \geq 0\}$ in the presence of no-short-sales restrictions. We can also solve problem formulations when the no-short-sales constraint $x \geq 0$ is relaxed and when arbitrary linear constraints $Ax \geq b$ on asset weights in the portfolio are specified. In this chapter, we assume that Ω is a set of linear constraints. A risk constraint $g_k(x) \leq \gamma_k$ can be specified for different risk measures, e.g., $\mathrm{Var}(x) \leq \gamma_k$ restricts portfolio variance and $\mathrm{TE}(x) \leq \gamma_k$ restricts portfolio tracking error (TE), while tail-risk constraints $\mathrm{VaR}_\alpha(x) \leq \gamma_k$ and $\mathrm{CVaR}_\alpha(x) \leq \gamma_k$ bound VaR and conditional value-at-risk (CVaR) at a quantile level α, where $\alpha \in (0,1)$.

Constraints on variance and TE are convex quadratic constraints that have a second-order cone (SOC) representation; see Section 12.2. If the distribution of asset returns is assumed to be normal, constraints on CVaR and VaR are also convex quadratic [1574]. When the normality assumption is relaxed and the distribution of asset returns is represented by a set of discrete scenarios, constraints on CVaR are linear [1574], while constraints on VaR require introducing binary variables that make VaR optimization a mixed-integer linear optimization (MILO) probem [1585].

In this chapter, we discuss conic linear optimization (CLO) formulations of financial optimization problems. In financial engineering and financial risk management, optimizing problems with conic linear constraints may present significant computational challenges. The complexity of modeling the uncertainty of future asset values is coupled with optimization challenges to compute stable and robust portfolios. Second-order cone optimization (SOCO) problems are characterized by having only continuous (no integer) variables in the problem as well as by the presence of an SOC objective function and/or constraints.

Most commercial and open-source solvers for SOCO allow specifying two types of expressions for SOCO constraints:

(1) $x^T Q x \leq y^2$, where $y \geq 0$ and $Q \succeq 0$ (SOC constraint), or

(2) $x^T Q x \leq y \cdot z$, where $y \geq 0$, $z \geq 0$, and $Q \succeq 0$ (rotated SOC constraint),

while other solvers require defining block-type variables. For instance, using block-type variables, the SOC constraint $x^T Q x \leq y^2$ can be defined as $Q^{1/2} x - u = 0$, $(y, u) \in \mathcal{K}$, where $\mathcal{K} = \{(x_0, x) \in \mathbb{R}^{N+1} : x_0 \geq \sqrt{x_1^2 + \cdots + x_N^2}\}$ is the SOC.

As semidefinite optimization (SDO) problems are not widely used in the financial industry yet, we do not discuss these models in detail. One example of an SDO problem in financial engineering is computing a positive semidefinite (PSD) covariance matrix that is close (in some norm) to a given estimated covariance matrix. Both SOCO and SDO are subclasses of CLO problems.

Here, we consider four types of financial SOCO problems:

- portfolio optimization problems with convex quadratic constraints on tracking error and/or variance (Section 12.2),
- minimizing or restricting the market impact costs of portfolio rebalancing [1337] (Section 12.3),
- robust mean-variance optimization problems with ellipsoidal uncertainty sets [423] (Section 12.4), and
- computing equal risk contribution (ERC) portfolios [1301] (Section 12.5).

12.2 • Portfolio Optimization Problems with Convex Quadratic Constraints

Multiobjective optimization is a natural tool for portfolio selection models as they involve minimizing one or several risk measures and maximizing (for the return) or minimizing (for the losses) a number of portfolio performance indicators. The mean-variance optimization problem discussed in this section is one example of a multiobjective problem in finance, where variance serves as a risk measure. However, variance is not the only risk measure that is used in practice.

One of the most famous portfolio management models that involve a risk-return trade-off is the mean-variance portfolio optimization problem introduced by Markowitz [1277]. The conflicting objectives in the Markowitz model are minimizing the variance (risk) and maximizing the expected return. Using variance to measure a portfolio's risk, as proposed by Markowitz, is central to finance in both a theoretical and a practical sense.

The variance-covariance matrix $Q \in \mathscr{S}^N$ and the expected return vector μ are usually estimated from historical returns or from a factor model. The following formulation of the mean-variance optimization problem maximizes the expected return $\mu^T x$ while constraining the variance to be at most V:

$$\begin{array}{ll} \max_{x \in \mathbb{R}^N} & \mu^T x \\ \text{s.t.} & x^T Q x \leq V, \\ & \sum_{i=1}^N x_i = 1, \\ & x \in \Omega. \end{array} \tag{12.2}$$

This problem, with the variance in the constraint, is a quadratically constrained optimization (QCO) problem that can be solved with a SOCO algorithm. The quadratic constraint $x^T Q x \leq V$ can be rewritten as the SOC constraint $x^T Q x \leq y^2$, $y = \sqrt{V}$.

Markowitz [1278] defined a portfolio to be efficient if for some fixed level of variance (risk) no other portfolio gives a larger expected return. Equivalently, an efficient portfolio can be defined as one for which at some fixed level of expected return no other portfolio gives a smaller variance (risk). The determination of the efficient portfolio frontier in the Markowitz mean-variance model is equivalent to solving the SOCO problem (12.2) for all values of $V \in [V_{\min}, V_{\max}]$, where $V_{\min}$ is the variance of a minimum variance portfolio and $V_{\max}$ is the variance of a maximum return portfolio.

The mean-variance optimization problem has a number of alternative formulations. A second formulation, where variance is minimized subject to the return constraint, is the following QO problem:

$$\min_{x \in \mathbb{R}^N} \quad x^T Q x$$
$$\text{s.t.} \quad \mu^T x \geq R,$$
$$\sum_{i=1}^{N} x_i = 1, \tag{12.3}$$
$$x \in \Omega,$$

where the lower bound for the required return is denoted by R.

Finally, minimizing the weighted sum of the variance and the expected return yields the following QO formulation [687]:

$$\min_{x \in \mathbb{R}^N} \quad -\mu^T x + \lambda x^T Q x$$
$$\text{s.t.} \quad \sum_{i=1}^{N} x_i = 1, \tag{12.4}$$
$$x \in \Omega.$$

The parameter $\lambda > 0$ in (12.4) is referred to as the investor's risk aversion parameter. Solutions of the optimization problem (12.4) for different values of λ also trace the efficient frontier in the mean-variance space. The mean-variance efficient frontier is known to be the graphical depiction of the Markowitz efficient set of portfolios and represents the boundary of the set of feasible portfolios that have the maximum return for a given level of risk.

The equivalence between the three mean-variance formulations, namely models (12.2), (12.3), and (12.4), is shown in [1106]. All three formulations are multiobjective optimization problems, where the conflicting objectives are maximizing expected return and minimizing variance. Using the standard deviation risk measure $\sqrt{x^T Q x}$ instead of variance in mean-variance formulations (12.2)–(12.4) would produce the same efficient frontier, but problem (12.4) becomes SOCO instead of QO,[11] while the other two problems can be cast back to their original formulations (12.2) and (12.3) by squaring the quadratic objective or the constraint.

TE is an alternative risk measure that gives rise to SOCO problems. TE is the standard deviation of the active return, which is the portfolio's return relative to a benchmark. As working with variance as the risk measure is more convenient mathematically, we utilize squared TE, $\text{TE}^2(x) = (x - \tilde{x})^T Q (x - \tilde{x})$, where x_i, $i = 1, \ldots, N$, are the asset weights in a portfolio, $\tilde{x}$ is the vector of asset weights of the benchmark, and Q is a covariance matrix of asset returns.

Minimizing or bounding TE in an optimization problem is required when the goal is to find a portfolio that tracks the returns of a benchmark, e.g., the S&P 500 index, as closely as possible. The *benchmark (index) tracking problem* consists of tracking a benchmark or index portfolio with another portfolio. There are a number of reasons to create a replicating portfolio that tracks the benchmark, e.g., the benchmark portfolio may not be tradable or may contain a large number of assets.

The mean-variance optimization problem with a (squared) TE constraint is a QCO problem:

$$\max_{x \in \mathbb{R}^N} \quad \mu^T x$$
$$\text{s.t.} \quad (x - \tilde{x})^T Q (x - \tilde{x}) \leq T,$$
$$\sum_{i=1}^{N} x_i = 1, \tag{12.5}$$
$$x \in \Omega.$$

---

[11]QO problems can be solved more efficiently algorithmically than QCO problems or SOCO problems due to the existence of simplex-type methods for QO. It is also easier to handle integer variables in QO formulations.

Problem (12.5) can be solved with a SOCO algorithm. Similarly, minimizing portfolio variance and simultaneously bounding portfolio TE, and minimizing portfolio tracking error and simultaneously bounding portfolio variance, are other examples of SOCO problem formulations. In all such cases we need to solve a multiobjective SOCO problem with an absolute risk measure (variance) and a relative risk measure (TE).

Additional SOCO formulations of mean-variance optimization problems that include constraints on systemic and specific risks, fund of funds with multiple active risk constraints, and constraints on risk that use more than one risk model are described in [1088]. These models follow the same layout as problems (12.2) and (12.5) but incorporate multiple covariance matrices. Another overview of SOCO mean-variance optimization formulations can be found in [421].

Means and variances that are used in a mean-variance framework may not be sufficient to quantify risk and return. Another, in many cases more attractive, possibility is to construct an approximate return distribution from a set of scenarios representing possible prices for financial instruments. Effectively, the scenario-based approach uses statistical sampling to model uncertainty. In contrast to mean-variance optimization, the scenario-based approach allows for general and discrete distributions.

Scenario-based models do not rely on any distributional assumptions about asset returns. Instead of using variance or TE as the only risk measure, like in mean-variance models, scenario-based models rely on tail-based risk measures such as VaR and CVaR. Unlike variance, which is a symmetric risk measure, tail-based risk measures are downside risk measures that focus on the tail of the distribution, i.e., on the worst possible returns.

The scenario-based CVaR optimization problem is an LO problem that can easily be incorporated into mean-variance optimization problems with variance or TE risk measures. The scenario-based VaR optimization model is a mixed-integer optimization problem, and in the presence of variance or TE constraints it becomes a mixed-integer SOCO (MISOCO) problem that is computationally difficult to solve. A heuristic algorithm that solves a sequence of SOCO CVaR problems instead of a single MISOCO VaR problem can offer significantly faster solution times [1585].

12.3 ▪ Transaction Costs with Market Impact

Section 12.2 assumed that assets could be traded freely. In practice, however, trading incurs transaction costs, which include not only brokerage fees but also so-called market impact costs. The latter are due to finite liquidity; as a larger quantity of an asset is bought (sold), the laws of supply and demand dictate that the price of the asset increase (decrease). Empirically, market impact costs are found to be approximated well by a power law function of the trading rate (see, for example, [56]). This type of cost model can be incorporated into a portfolio optimization problem through the use of SOC constraints. In the following, we describe the proportional market impact cost model introduced in [1337].

Consider an initial (pre-trading) portfolio $\bar{x}$ and a final (post-trading) portfolio x, where $\bar{x}_i$ and x_i denote the respective values of the asset i position divided by the initial portfolio value. The total market impact cost, expressed as a proportion of the initial portfolio value, is

$$c(x - \bar{x}) = \sum_{i=1}^{N} c_i(x_i - \bar{x}_i) = \sum_{i=1}^{N} m_i |x_i - \bar{x}_i|^{1+\nu}, \tag{12.6}$$

where $m_i \geq 0$, $i = 1,\ldots,N$, are parameters for each asset that can be estimated from historical data, and $v > 0$ is a rational number ($v = 0.5$ or $v = 1.5$ is commonly used in practice).

The initial wealth is normalized to be $\sum_{i=1}^{N} \bar{x}_i = 1$ and, consequently, $\bar{x}_i$ is the weight of asset i in the initial portfolio. With full investment while rebalancing, we require

$$\sum_{i=1}^{N} x_i + c(x - \bar{x}) = \sum_{i=1}^{N} \bar{x}_i = 1. \tag{12.7}$$

The equality constraint (12.7) is not convex, but it may be replaced by the convex and SOCO-representable inequality constraint

$$\sum_{i=1}^{N} x_i + c(x - \bar{x}) \leq 1. \tag{12.8}$$

When the market impact cost is modeled by (12.6), constraint (12.8) is equivalent to the family of constraints

$$\sum_{i=1}^{N} x_i + \sum_{i=1}^{N} m_i \cdot y_i \leq 1,$$
$$|x_i - \bar{x}_i|^{1+v} \leq y_i \qquad i = 1,\ldots,N. \tag{12.9}$$

In general, constraints of the form $|x_i - \bar{x}_i|^{1+v}$ for positive rational v can be represented as SOCO constraints [51]. For $v = 0.5$, it is straightforward to represent

$$|x_i - \bar{x}_i|^{3/2} \leq y_i$$

in terms of SOC constraints. We introduce an additional variable z_i satisfying

$$-z_i \leq x_i - \bar{x}_i \leq z_i, \qquad \frac{z_i^2}{\sqrt{z_i}} \leq y_i, \qquad z_i \geq 0,$$

and note that constraint $\frac{z_i^2}{\sqrt{z_i}} \leq y_i$ can be represented by the SOC constraints

$$z_i^2 \leq u \cdot y_i, \qquad u^2 \leq z_i, \qquad u \geq 0.$$

Note that the weight of asset i in the final portfolio is not x_i, but $x_i / \sum_{i=1}^{N} x_i$, as $\sum_{i=1}^{N} x_i$ is not necessarily equal to one. As a result, the final portfolio return variance is

$$\sigma^2(x) = \left(\frac{x}{\sum_{i=1}^{N} x_i}\right)^T Q \left(\frac{x}{\sum_{i=1}^{N} x_i}\right) = \frac{x^T Q x}{(\sum_{i=1}^{N} x_i)^2}, \tag{12.10}$$

and the final portfolio expected return constraint is

$$\mu(x) = \mu^T \left(\frac{x}{\sum_{i=1}^{N} x_i}\right) \geq R', \tag{12.11}$$

where R' is the expected return target for the final portfolio.

It is more intuitive to specify an expected return target, R, for the initial portfolio, i.e., the desired performance should be achieved net of transaction costs. Suppose that the value of the initial portfolio is Υ, so that a return of R corresponds to a monetary gain of $R \cdot \Upsilon$. If the final portfolio has value Υ', then obtaining the desired gain requires that $R' \cdot \Upsilon' = R \cdot \Upsilon$, and combining this with $\Upsilon' = \Upsilon \cdot \sum_{i=1}^{N} x_i$ leads equivalently to $R' = R / \sum_{i=1}^{N} x_i$. Thus, the expected return constraint (12.11) simplifies to $\mu^T x \geq R$.

The mean-variance optimization problem with market impact transaction cost constraint (12.8), where we maximize the expected return subject to constraining portfolio return variance (12.10), becomes

$$
\begin{aligned}
\max_{x \in \mathbb{R}^N, \, y \in \mathbb{R}^N} \quad & \mu^T x \\
\text{s.t.} \quad & x^T Q x \leq V \cdot \left(\sum_{i=1}^{N} x_i\right)^2, \\
& \sum_{i=1}^{N} x_i + \sum_{i=1}^{N} m_i \cdot y_i \leq 1, \\
& |x_i - \bar{x}_i|^{3/2} \leq y_i, \qquad\qquad i = 1, \ldots, N, \\
& x \in \Omega.
\end{aligned}
\tag{12.12}
$$

Formulation (12.12) can be solved as the following SOCO problem:

$$
\begin{aligned}
\max_{x \in \mathbb{R}^N, \, t \in \mathbb{R}, \, y \in \mathbb{R}^N, \, z \in \mathbb{R}^N, \, u \in \mathbb{R}} \quad & \mu^T x \\
\text{s.t.} \quad & x^T Q x \leq V \cdot t^2, \\
& \sum_{i=1}^{N} x_i = t, \\
& \sum_{i=1}^{N} x_i + \sum_{i=1}^{N} m_i \cdot y_i \leq 1, \\
& -z_i \leq x_i - \bar{x}_i \leq z_i, \qquad i = 1, \ldots, N, \\
& z_i \geq 0, \qquad\qquad\qquad\quad\; i = 1, \ldots, N, \\
& z_i^2 \leq u \cdot y_i, \qquad\qquad\quad i = 1, \ldots, N, \\
& u^2 \leq z_i, \qquad\qquad\qquad i = 1, \ldots, N, \\
& u \geq 0, \\
& x \in \Omega.
\end{aligned}
\tag{12.13}
$$

The quadratic mean-variance optimization problem with risk constraint (12.11) instead of the expected return constraint, while keeping market impact transaction costs in the model, becomes the fractional optimization problem

$$
\begin{aligned}
\min_{x \in \mathbb{R}^N, \, y \in \mathbb{R}^N} \quad & \frac{x^T Q x}{\left(\sum_{i=1}^{N} x_i\right)^2} \\
\text{s.t.} \quad & \mu^T x \geq R, \\
& \sum_{i=1}^{N} x_i + \sum_{i=1}^{N} m_i \cdot y_i \leq 1, \\
& |x_i - \bar{x}_i|^{3/2} \leq y_i, \qquad\qquad i = 1, \ldots, N,
\end{aligned}
\tag{12.14}
$$

which is SOCO representable with the change of variables $q = 1/(\sum_{i=1}^{N} x_i)$ and $\hat{x}_i = q \cdot x_i$; see [1337].

12.4 ▪ Robust Mean-Variance Optimization

One of the common criticisms of mean-variance optimization is its sensitivity to return estimates; i.e., small changes in the return estimates can result in big shifts of the portfolio weights x. Empirical tests by Chopra and Ziemba [464] showed that errors in return estimates are more important than errors in risk estimates. One of the solutions to this problem is *robust optimization*, which incorporates uncertainties into the optimization problem. For a review of robust optimization applied to portfolio management, we refer the reader to [684]. *Robust portfolio selection* allows us to reduce the sensitivity of an optimal portfolio to data perturbations, and, consequently, *robust optimization* [203] techniques can improve the robustness of portfolio selection models. Robust optimization puts uncertain data into a bounded uncertainty set and incorporates that uncertainty set into the optimization formulations. The idea is to optimize the worst-case outcome from the uncertainty set.

Let us recall formulations (12.2)–(12.4) of the Markowitz mean-variance model, where the set Ω of constraints is known with certainty, but the problem data, namely expected asset returns μ or covariances Q, are uncertain. As μ is an estimate of the expected return, we would like to consider a range of possible values of μ (uncertainty set). An uncertainty set can be defined in various ways, e.g., with intervals or with an ellipsoid. While it is also possible to define an uncertainty set for the covariance matrix Q, this leads to more complex nonlinear models with semidefinite constraints; see, e.g., [1213] and [821].

For robust portfolio optimization, we start with a model where return and covariance matrix information is given in the form of bounds (interval uncertainty sets) [514]:

$$\mathcal{U}_B = \{(\mu, Q) : \mu^L \leq \mu \leq \mu^U, \, Q^L \leq Q \leq Q^U, \, Q \succeq 0\}.$$

The robust optimization problem corresponding to formulation (12.4) minimizes the objective function in the worst-case realization of the input parameters μ and Q, i.e.,

$$\min_{\sum_{i=1}^{N} x_i = 1, \, x \in \Omega} \left\{ \max_{(\mu, Q) \in \mathcal{U}_B} -\mu^T x + \lambda x^T Q x \right\}. \tag{12.15}$$

In general, formulation (12.15) is a nonlinear optimization (NLO) problem with a semidefinite constraint $Q \succeq 0$ [821]. Without simplifying assumptions [1795], the computational complexity of solving large-scale problems with semidefinite constraints may prohibit using uncertainty sets for the covariance matrix Q estimate in practice. An interval uncertainty set for the expected return estimates μ translates into linear constraints in the deterministic reformulation of the optimization problem.

We consider a variant of robust portfolio selection problems proposed by Ceria and Stubbs [423]. In their model, instead of the uncertainty set being given in terms of bounds, they use an ellipsoidal uncertainty set $\mathcal{U}_E$. In [423] the authors assume that only r, the vector of estimated expected returns, is uncertain in the Markowitz model formulations (12.2)–(12.4). To consider the worst case of problem (12.4), it is assumed that the vector of true expected returns r is normally distributed and lies in the ellipsoidal set $\mathcal{U}_E$,

$$(r - \mu)^T \Theta^{-1} (r - \mu) \leq \varkappa^2,$$

with probability ρ, where μ is an estimate of the expected returns; Θ is the covariance matrix of the estimates of expected returns; and $\varkappa^2 = \chi_N^2 (1 - \rho)$, with χ_N^2 being the inverse cumulative distribution function of the chi-squared distribution with N degrees of freedom.

Let $\hat{x}$ be the optimal portfolio on the estimated frontier for a given target risk level. Then, the worst-case (maximal) difference between the estimated expected return and the actual expected return with the given portfolio $\hat{x}$ can be found by solving

$$\begin{aligned} \max_{r} \quad & (\mu - r)^T \hat{x} \\ \text{s.t.} \quad & (r - \mu)^T \Theta^{-1} (r - \mu) \leq \varkappa^2. \end{aligned} \tag{12.16}$$

As derived in [423], the optimal objective value of $(\mu - r)^T \hat{x}$ in (12.16) is equal to $\varkappa\sqrt{\hat{x}^T\Theta\hat{x}}$. So, the true expected return of the portfolio can be expressed as $r^T\hat{x} = \mu^T\hat{x} - \varkappa\sqrt{\hat{x}^T\Theta\hat{x}}$.

Now, formulation (12.4) transforms into the robust mean-variance portfolio selection problem

$$
\begin{aligned}
\min_{x\in\mathbb{R}^N} \quad & -\mu^T x + \varkappa\sqrt{x^T\Theta x} + \lambda x^T Q x \\
\text{s.t.} \quad & \sum_{i=1}^{N} x_i = 1, \\
& x \in \Omega,
\end{aligned}
\tag{12.17}
$$

where $\varkappa$ is the estimation error aversion and λ is the risk aversion. Matrix Θ is typically set to the identity matrix or a diagonal matrix with asset return variances on the diagonal. Problem (12.17) is a SOCO-representable problem; moreover, it is an optimization problem with two parameters, λ and $\varkappa$, that we need to set. If we look at problem (12.17) in the multiobjective sense, it is the problem of maximizing expected return, minimizing risk (variance of returns), and minimizing estimation error for the expected return.

We can rewrite problem (12.17) in the form

$$
\begin{aligned}
\min_{x\in\mathbb{R}^N, u\in\mathbb{R}, v\in\mathbb{R}} \quad & -\mu^T x + \varkappa u + \lambda v \\
\text{s.t.} \quad & x^T\Theta x \le u^2, \\
& x^T Q x \le v, \\
& \sum_{i=1}^{N} x_i = 1, \\
& x \in \Omega,
\end{aligned}
\tag{12.18}
$$

where parameters $\varkappa \ge 0$ and $\lambda \ge 0$ are fixed to appropriate values. The constraint $x^T Q x \le v$ can be expressed as the rotated SOC constraint $x^T Q x \le v \cdot y$, $y = 1$.

With known bounds on the variance, $x^T Q x \le V$, and on the size of the uncertainty set, $\sqrt{x^T\Theta x} \le \delta$, we can formulate a robust mean-variance optimization problem with variance and return estimation error bounds in the constraints:

$$
\begin{aligned}
\max_{x\in\mathbb{R}^N} \quad & \mu^T x \\
\text{s.t.} \quad & x^T Q x \le V, \\
& x^T\Theta x \le \delta^2, \\
& \sum_{i=1}^{N} x_i = 1, \\
& x \in \Omega,
\end{aligned}
\tag{12.19}
$$

where parameters $\delta \ge 0$ and $V \ge 0$ are fixed to appropriately chosen values.

12.5 ▪ Equal Risk Contribution Portfolios

Given undesirable effects related to expected return estimation issues described in Section 12.4, portfolio construction techniques that are only based on risk estimates have grown in popularity. For instance, ERC portfolios have an identical risk contribution from each asset or asset class, so that the portfolio is fully diversified from a risk perspective. In this section, we describe SOCO models that underlie the ERC approach and compare these to NLO models that are currently used to construct ERC portfolios.

Recall that $g(x)$ in (12.1) is some measure of the risk of portfolio x, and define $c_i(x)$ to be the risk contribution of asset i. By the definition of risk contribution, it holds that $\sum_{i=1}^{N} c_i(x) = g(x)$. If the portfolio's risk is measured by the variance of its return, then $g(x) = x^T Q x$ and $c_i(x) = x_i (Q x)_i$, where $(Q x)_i = \sum_{j=1}^{N} Q_{ij} x_j$. Using the standard deviation of the return as the risk measure yields $g(x) = \sqrt{x^T Q x}$ and $c_i(x) = \frac{x_i (Q x)_i}{\sqrt{x^T Q x}}$.

An ERC portfolio x^{ERC} satisfies $c_i(x^{\mathrm{ERC}}) = \frac{g(x^{\mathrm{ERC}})}{N}$ for $i = 1, \dots, N$. Since $x_i (Q x)_i = \frac{x^T Q x}{N}$ if and only if $\frac{x_i (Q x)_i}{\sqrt{x^T Q x}} = \frac{\sqrt{x^T Q x}}{N}$, it follows that the ERC portfolios for the variance and the standard deviation risk measures are the same. In the remainder of this section, we consider only the variance risk measure, but all results also apply to the standard deviation.

The computational complexity of finding an ERC portfolio depends on the problem characteristics, namely the covariance matrix Q and the set of feasible portfolios Ω. In [1255] and [1301], the authors consider long-only portfolios, i.e., $\Omega = \{x \in \mathbb{R}^N : x \ge 0\}$, because an ERC portfolio is guaranteed to exist in the long-only case. Removing the long-only restriction and/or adding other constraints complicates the problem of finding an ERC portfolio because it may not be unique even if Q is a positive-definite matrix, or the ERC portfolio may not even exist. In [1255], the authors show that if all correlations between assets are the same, then the ERC portfolio is computed by a simple formula.

When the correlations are different, the ERC portfolio must be found using numerical methods. In this case, [1255] proposes to solve

$$\min_{x \in \mathbb{R}^N} \quad \sum_{i=1}^{N} \sum_{j=1}^{N} \left(x_i (Q x)_i - x_j (Q x)_j \right)^2$$
$$\text{s.t.} \quad \sum_{i=1}^{N} x_i = 1, \qquad \qquad \qquad \qquad (12.20)$$
$$x_i \ge 0, \qquad i = 1, \dots, N.$$

As problem (12.20) minimizes the total squared differences between the risk contributions of all pairs of assets, the optimal value of problem (12.20) is zero.

In [1255] the authors also present an alternative approach to problem (12.20). The KKT optimality conditions imply that if y^* solves

$$\min_{y \in \mathbb{R}^N} \quad y^T Q y$$
$$\text{s.t.} \quad \sum_{i=1}^{N} \log(y_i) \ge \varsigma, \qquad \qquad \qquad (12.21)$$
$$y_i \ge 0, \quad i = 1, \dots, N,$$

then $y_i^* (Q y^*)_i = \lambda_\varsigma$ for $i = 1, \dots, N$, where λ_ς is the Lagrange multiplier of the sum of logarithms constraint. The ERC portfolio is obtained from an optimal solution of problem (12.21) by normalization, i.e., $x_i^{\mathrm{ERC}} = \frac{y_i^*}{\sum_{j=1}^{N} y_j^*}$, $i = 1, \dots, N$. Since problem (12.21) minimizes a lower-bounded convex function on a nonempty convex set, it must have a solution. Thus, it follows that a long-only ERC portfolio always exists, and it is unique if Q is positive definite.

Both problems (12.20) and (12.21) are general NLO problems. The way that an NLO problem is formulated can dramatically affect the performance of solution algorithms. In light of this, consider the following alternative to problem (12.20):

$$\min_{x \in \mathbb{R}^N, \tau \in \mathbb{R}} \quad \sum_{i=1}^{N} (x_i (Q x)_i - \tau)^2$$
$$\text{s.t.} \quad \sum_{i=1}^{N} x_i = 1, \qquad \qquad \qquad \qquad (12.22)$$
$$x_i \ge 0, \qquad i = 1, \dots, N.$$

Problem (12.22) minimizes the total squared differences between the risk contributions of all assets and some unconstrained variable τ. Using calculus, it can be shown that $\tau = \frac{x^T Q x}{N}$ is an optimal solution. Like problem (12.20), problem (12.22) attains a minimal value of zero when $x = x^{\text{ERC}}$, so these problems are effectively equivalent in the long-only case, when an ERC portfolio is known to exist. Computationally, formulation (12.22) is preferable to (12.20) because simplifying the objective function reduces the computational cost of calculating derivatives and results in fewer nonlinearities [131].

It turns out that it is possible to reformulate problems (12.20) and (12.21) so that they can be solved with a SOCO algorithm [1301]. The reformulations rely in part on transformations of constraints involving products of nonnegative variables into SOC constraints; see, e.g., [51] or [1214].

In [1301] the authors considered the expression

$$\sqrt{\frac{x^T Q x}{N}} - \sqrt{\min_{1 \le i \le N}\{x_i (Qx)_i\}}, \qquad (12.23)$$

which computes the difference between the square root of the average risk contribution and the square root of the smallest risk contribution. Equation (12.23) is zero if $x = x^{\text{ERC}}$ and positive otherwise. Thus, minimizing (12.23) is a suitable objective function for finding long-only ERC portfolios. Adapting problem (12.20) to use this objective function, we first introduce nonnegative variables $z_i = (Qx)_i$ for $i = 1,\ldots,N$. Next, we introduce nonnegative variables t and p and add the constraints

$$x^T Q x \le N p^2$$

and

$$x_i z_i \ge t^2, \quad i = 1,\ldots,N.$$

Note that p^2 is an upper bound for the average risk contribution, while t^2 is a lower bound for the individual risk contributions, so that $p \ge t$ always holds. Since $p = t$ for an ERC portfolio, our goal is to minimize $p - t$, which corresponds to (12.23). This leads to the following SOCO problem [1301]:

$$
\begin{aligned}
\min_{x \in \mathbb{R}^N, z \in \mathbb{R}^N, p \in \mathbb{R}, t \in \mathbb{R}} \quad & p - t \\
\text{s.t.} \quad & z_i = (Qx)_i, && i = 1,\ldots,N, \\
& x^T Q x \le N p^2, \\
& x_i z_i \ge t^2, && i = 1,\ldots,N, \\
& \textstyle\sum_{i=1}^N x_i = 1, \\
& x_i \ge 0, && i = 1,\ldots,N, \\
& z_i \ge 0, && i = 1,\ldots,N, \\
& p \ge 0,\ t \ge 0.
\end{aligned}
\qquad (12.24)
$$

Recalling problem (12.21), we can observe that

$$\sum_{i=1}^N \log(y_i) \ge \varsigma$$

is equivalent to

$$y_1 y_2 \cdots y_N \ge e^\varsigma.$$

Given real variables t and $s_k \geq 0$, $k = 1, \ldots, 2^K$, the constraint $s_1 s_2 \cdots s_{2^K} \geq t^{2^K}$ can be expressed as a set of $2^K - 1$ partial products, each of which is a rotated SOC constraint. As a result, we can derive an alternative formulation of problem (12.21) given in [1301] as

$$
\min_{y \in \mathbb{R}^{2^K}, \, t \in \mathbb{R}^{2^K - 1}} \quad y^T Q y
$$

$$
\begin{aligned}
\text{s.t.} \quad & y_{2(i-1)+1} \cdot y_{2i} \geq t_i^2, & i = 1, \ldots, 2^{K-1}, \\
& t_{2(j-1)+1} \cdot t_{2j} \geq t_{2^{K-1}+j}^2, & j = 1, \ldots, 2^{K-1} - 1, \\
& t_{2^K - 1} \geq e^{\varsigma / 2^K}, & \\
& y_i \geq 0, & i = 1, \ldots, N, \\
& y_i = 1, & i = N + 1, \ldots, 2^K, \\
& t_j \geq 0, & j = 1, \ldots, 2^K - 1,
\end{aligned}
\tag{12.25}
$$

where K is the smallest value satisfying $2^K \geq N$.

For ERC optimization problems with long-only constraints, SOCO formulations (12.24) and (12.25) are computationally attractive when compared to NLO (12.20), (12.21), and (12.22).

12.6 • Conclusions

During the last decade, CLO problem formulations have found novel applications in financial engineering. In this chapter, we have described a number of models that can be formulated as SOCO problems, namely portfolio optimization problems with convex quadratic constraints (variance and TE) and with market impact costs, robust mean-variance optimization problems, and optimization problems for computing ERC portfolios. As commercial SDO solvers were not available until recently, SDO models have been used in financial engineering less frequently than SOCO models. With ongoing advances in solution algorithms and software, especially for solving large-scale problems, we can expect to see greater use of CLO models by financial practitioners in the near future.

Part IV

Global Optimization

Chapter 13

Deterministic Global Optimization

Sergiy Butenko and Panos M. Pardalos

13.1 ▪ Introduction

Global optimization is a branch of mathematical optimization concerned with finding global optima of nonconvex, multiextremal problems. A typical global optimization problem has numerous local optima that are not global, which makes finding a globally optimal solution extremely difficult. In its most general form, it may involve a highly nonlinear, nonsmooth, difficult-to-evaluate objective function and nonconvex feasible region, with both continuous and discrete variables.

It is well known that convexity plays a central role in nonlinear optimization (NLO), and most classical optimization methods take advantage of the convexity assumptions for both the objective function and the feasible region. One of the fundamental properties of a convex problem is the fact that its local minimum is also its global minimum. Finding a global minimum of a smooth convex problem reduces to the task of detecting a point satisfying the KKT conditions. While this still requires solving a nonlinear system, the goal is clear (find a KKT point) and, under certain additional assumptions, can be efficiently achieved. In fact, all the classical methods in smooth NLO focused on the problem of determining a KKT (stationary) point. Such methods are referred to as *local methods*. Since the numerous available local methods converge to a stationary point, a natural question to ask is, can we find a large class of problems for which a stationary point is guaranteed to be a global minimizer? It appears that if such a class is required to be sufficiently wide to include linear functions and to be closed under the operations of addition of two functions and multiplication by a positive scalar, then it is exactly the class of smooth convex functions [1394]. Thus, the local methods are not expected to converge to a global minimum for nonconvex optimization problems.

The field of global optimization originated in 1964, when Hoang Tuy published his seminal paper on concave minimization [1798]. With advances in computer technologies, the field has expanded in many different directions, having become one of the most active and fruitful avenues of research in optimization. Hundreds of journal articles, edited volumes, research monographs, and textbooks have been devoted to the field.

The objective of this chapter is to briefly introduce deterministic global optimization. Its remainder is organized as follows. Section 13.2 describes some examples of typical global optimization problems, with emphasis on continuous formulations of discrete problems. Section 13.3 focuses on the computational complexity issues related to global optimization. Global optimality conditions are discussed in Section 13.4. Section 13.5 presents some basic ideas behind algorithmic approaches to solving global optimization problems and briefly reviews global optimization software. Finally, Section 13.6 concludes the chapter with pointers for further reading.

13.2 ▪ Examples of Global Optimization Problems

Many important tasks can be stated as global optimization problems. Consider, for example, a *fixed charge problem*, where one needs to decide on the level $x_i \geq 0$, $i = 1, \ldots, n$, of n given activities so that their overall cost $f(x) = \sum_{i=1}^{n} f_i(x_i)$ is minimized. In many applications, the cost function f_i for each activity $i = 1, \ldots, n$ has the form

$$f_i(x_i) = \begin{cases} 0 & \text{if } x_i = 0, \\ a_i + c_i(x_i) & \text{if } x_i > 0, \end{cases} \tag{13.1}$$

where $a_i > 0$ is the fixed setup cost and $c_i : \mathbb{R}_+ \to \mathbb{R}$ is a continuous concave function representing the variable cost of the activity.

One of the interesting directions in global optimization research deals with continuous (nonconvex) approaches to discrete optimization problems. Bridging the continuous and discrete domains may reveal surprising connections between problems of seemingly different natures and lead to new developments in both discrete and continuous optimization. Many integer linear optimization (LO) and integer quadratic optimization (QO) problems can be reformulated as QO problems in continuous variables. For example, a binary LO problem of the form

$$\min c^T x \text{ s.t. } Ax \leq b, \ x \in \{0, 1\}^n, \tag{13.2}$$

where A is a real $m \times n$ matrix, $c \in \mathbb{R}^n$, and $b \in \mathbb{R}^m$, is equivalent to the following *concave minimization* problem:

$$\min c^T x + \mu x^T (e - x) \text{ s.t. } Ax \leq b, \ 0 \leq x \leq e, \tag{13.3}$$

where e is the appropriate vector of ones and μ is a sufficiently large positive number used to ensure that the global minimum of this problem is attained only when $x^T (e - x) = 0$. Other nonlinear binary problems can be reduced to equivalent concave minimization problems in a similar fashion.

Next, we consider a quadratic unconstrained binary optimization (QUBO) problem,

$$\min f(x) = x^T Q x + c^T x \text{ s.t. } x \in \{0, 1\}^n, \tag{13.4}$$

where Q is a real symmetric $n \times n$ matrix. For a real μ, let $Q_\mu = Q + \mu I_n$, where I_n is the $n \times n$ identity matrix, and let $c_\mu = c - \mu e$. Then the above QUBO problem is equivalent to the following problem:

$$\min f_\mu(x) = c_\mu^T x + x^T Q_\mu x \text{ s.t. } x \in \{0, 1\}^n, \tag{13.5}$$

since $f_\mu(x) = f(x)$ for any x. If $\mu \le -\lambda_{\max}$, where $\lambda_{\max}$ is the largest eigenvalue of Q, then $f_\mu(x)$ is concave and $\{0, 1\}^n$ can be replaced with the unit hypercube $[0, 1]^n$ as the feasible region.

In yet another example of a continuous approach to combinatorial optimization, we consider the classical *maximum clique* and *maximum independent set* problems on graphs. Given a simple undirected graph $G = (V, E)$ with the set of vertices $V = \{1, \dots, n\}$ and the set of edges E, a clique C in G is a subset of pairwise-adjacent vertices, and an independent set I is a subset of pairwise-nonadjacent vertices of V. The maximum clique problem is to find a clique C of the largest size in G, and the clique number $\omega(G)$ is the size of a maximum clique in G. The maximum independent set problem is defined likewise, and the independence number is denoted by $\alpha(G)$.

Let A_G be the adjacency matrix of G and let e be the n-dimensional vector with all components equal to one. For a subset C of vertices, its characteristic vector x^C is defined by $x_i^C = 1/|C|$ if $i \in C$ and $x_i^C = 0$ otherwise for $i = 1, \dots, n$. Motzkin and Straus [1359] show that the global optimal value of the following quadratic program (Motzkin–Straus QP):

$$1 - \frac{1}{\omega(G)} = \max_{x \in S} x^T A_G x, \tag{13.6}$$

where $S = \{x \in \mathbb{R}^n : e^T x = 1, \ x \ge 0\}$ is the standard simplex in $\mathbb{R}^n$. Moreover, a subset of vertices $C \subseteq V$ is a maximum clique of G if and only if the characteristic vector x^C of C is a global maximizer of the above problem.

The Motzkin–Straus QP may have maximizers that are not in the form of characteristic vectors. Bomze [298] introduces the following "regularization" of the Motzkin–Straus formulation:

$$\max g(x) = x^T \left(A_G + \frac{1}{2} I_n \right) x \ \text{ s.t. } e^T x = 1, \ x \ge 0. \tag{13.7}$$

He shows that x^* is a local maximum of (13.7) if and only if it is the characteristic vector of a maximal-by-inclusion clique in the graph.

The maximum independent set problem can also be formulated as the problem of minimizing a quadratic function over a unit hypercube [1, 373, 907]:

$$\alpha(G) = \max_{x \in [0,1]^n} \left(e^T x - \frac{1}{2} x^T A_G x \right), \tag{13.8}$$

as well as the following global optimization problems [150, 151, 908]:

$$\alpha(G) = \max_{x \in [0,1]^n} \sum_{i \in V} x_i \prod_{j \in N(i)} (1 - x_j), \tag{13.9}$$

$$\alpha(G) = \max_{x \in [0,1]^n} \sum_{i \in V} \frac{x_i}{1 + \sum_{j \in N(i)} x_j} = \max_{x \in [0,1]^n} \sum_{i \in V} \frac{x_i}{1 + \sum_{j \in N(i)} x_j} - \sum_{(i,j) \in E} x_i x_j, \tag{13.10}$$

where $N(i)$ denotes the neighborhood of vertex i, which is the set of all vertices adjacent to i in G. Just like Bomze's regularization of the Motzkin–Straus formulation, the second formulation in (13.10) is characterized by a one-to-one correspondence between its local maxima and maximal-by-inclusion independent sets in G.

The examples above illustrate several important classes of global optimization problems, such as concave minimization, nonconvex QO, polynomial optimization, and fractional optimization. In addition, they can be shown to belong to the class of *difference of convex* (D.C.) functions. A function $f : S \to \mathbb{R}$, where $S \subseteq \mathbb{R}^n$ is a convex set, is called D.C. on S if it can be expressed in the form $f(x) = p(x) - q(x)$ for all $x \in S$, where p and q are convex functions on S.

13.3 ▪ Computational Complexity

Understanding a problem's complexity is crucial for developing mathematical optimization models and algorithms for solving the problem. While in discrete optimization establishing a problem's complexity from the NP-hardness perspective [787] is considered an important step to take before developing algorithms for the problem of interest, many important complexity questions pertaining to continuous optimization are still to be addressed. On the one hand, establishing NP-hardness may be very challenging and often requires sophisticated reductions from known hard problems (see, e.g., [958] and references therein). On the other hand, most problems in NLO are so difficult that proving their NP-hardness can be reasonably viewed as a mere formality. Nevertheless, a number of important complexity results related to global optimization have been established in the literature. In addition, alternative approaches to complexity analysis in continuous optimization have been developed, some of which aim to establish lower bounds on the running time of an algorithm belonging to a certain class of numerical methods [1392, 1394].

Next, we mention some of the known NP-hardness results concerning global optimization. Since our focus is on nonconvex problems, the first fundamental question to address is distinguishing between convex and nonconvex problems. In 1992, Pardalos and Vavasis [1450] posed the question of characterizing the computational complexity of checking whether a given multivariable polynomial is convex. The question was addressed in 2013 by Ahmadi et al. [34], who show that the problem is strongly NP-hard even when restricted to polynomials of degree four.

Convexity is easy to recognize for quadratic problems by, e.g., checking the sign of the eigenvalues of the Hessian. Since a general nonconvex optimization problem is at least as hard as a nonconvex QO problem, it is reasonable to focus on this special case when establishing the complexity of global optimization problems. Given an $n \times n$ real symmetric matrix Q; an $m \times n$ matrix A; and vectors $b \in \mathbb{R}^m, c \in \mathbb{R}^n$, consider the problem

$$\min f(x) = \frac{1}{2} x^T Q x + c^T x \text{ s.t. } Ax \geq b. \tag{13.11}$$

If Q is positive semidefinite (PSD), the problem is a convex QO problem, and every KKT point is a local minimum, which is also global. The problem can be solved in polynomial time using the ellipsoid method or an interior-point method. It turns out that as soon as Q has at least one negative eigenvalue, the problem becomes NP-hard [1449]. The Motzkin–Straus formulation (13.6) of the maximum clique problem and the quadratic formulation (13.8) for the maximum independent set problem show that QO remains NP-hard even when the feasible domain is restricted to the standard simplex or the unit hypercube in $\mathbb{R}^n$. As follows from the discussion in Section 13.2, the problem of minimizing a concave quadratic (or maximizing a convex quadratic) subject to linear constraints is NP-hard as well.

It should be pointed out that quadratic problems remain difficult even if instead of global optimization one resorts to a less ambitious goal, such as finding a stationary point or computing a local minimum. Indeed, the problem of checking the existence of a KKT point for a problem of minimizing a quadratic function subject to nonnegativity constraints has also been shown to be NP-hard by observing that the KKT conditions for this problem are given by the linear complementarity problem, which is NP-hard [958, 1444]. The problem of checking whether a given point is a local minimum of a given quadratic problem is also NP-hard [1448].

Even though nonconvex optimization is extremely hard in general, there exist nonconvex problems that can be solved efficiently. One example is the problem of minimizing a quadratic function over a sphere [1832]:

$$\min \frac{1}{2} x^T Q x + c^T x \quad \text{s.t.} \quad x^T x \leq 1. \tag{13.12}$$

Another example is the problem of maximizing the Euclidean norm $\|x\|^2 = \sum_{i=1}^{n} x_i^2$ over a rectangular parallelotope [292].

The last example of a nonconvex problem that is efficiently solvable that we mention here is the following *fractional linear optimization* problem [1831]:

$$\min (c^T x + \gamma)/(d^T x + \delta) \quad \text{s.t.} \quad Ax \geq b, \tag{13.13}$$

where the denominator $d^T x + \delta$ is assumed to be positive (or negative) over the feasible domain. The objective of this problem is pseudoconvex and may be minimized using the ellipsoid algorithm. Alternatively, the problem can be solved efficiently using Dinkelbach's algorithm [600, 1643].

13.4 ▪ Global Optimality Conditions

The classical optimality conditions, such as the KKT conditions mentioned above, are both necessary and sufficient for a global minimizer for convex problems. To take advantage of the classical optimality conditions in a more general setting, one may try to convexify nonconvex problems, leading to the concept of the *convex envelope*, defined as follows. Let S be a nonempty convex set in $\mathbb{R}^n$. For a lower semicontinuous function $f : S \to \mathbb{R}$, the convex envelope of $f(x)$ over S is a function $C_f(x)$ such that

  (i) C_f is a convex underestimation of f on S (that is, C_f is convex on S and $C_f(x) \leq f(x)$ for all $x \in S$), and

  (ii) for any convex underestimation h of f on S we have $h(x) \leq C_f(x)$ for all $x \in S$.

Geometrically, $C_f(x)$ is the function with epigraph coinciding with the convex hull of the epigraph of $f(x)$. With respect to global optima of the problem of minimizing $f(x)$ over S, it has the following properties:

$$\min_{x \in S} f(x) = \min_{x \in S} C_f(x), \tag{13.14}$$

$$\{x \in S : f(x) = f^*\} \subseteq \{x \in S : C_f(x) = f^*\}, \tag{13.15}$$

where $f^* = \min_{x \in S} f(x)$. These properties can be used to derive global optimality conditions for nonconvex problems. For example, the following result is proved in [944]: a point x^* is a global minimizer of a differentiable function $f : \mathbb{R}^n \to \mathbb{R}$ on $\mathbb{R}^n$ if and only if

$$\nabla f(x^*) = 0 \quad \text{and} \quad C_f(x^*) = f(x^*). \tag{13.16}$$

These conditions can be generalized to the nondifferentiable case in several ways [945].

Next we state global optimality conditions for the problem of minimizing the D.C. functions, utilizing the concept of ϵ-subdifferential, defined as follows. For an arbitrary function $\phi : \mathbb{R}^n \to \mathbb{R} \cup \{+\infty\}$, a point x^* where $\phi(x^*) < +\infty$, and $\epsilon \geq 0$, the ϵ-subdifferential $\partial_\epsilon \phi(x^*)$ of ϕ at x^* is the following set:

$$S = \{s \in \mathbb{R}^n : \phi(x) \geq \phi(x^*) + s^T(x - x^*) - \epsilon \text{ for all } x \in \mathbb{R}^n\}. \tag{13.17}$$

Hiriart-Urruty [945] shows that for a lower-semicontinuous convex function $g : \mathbb{R}^n \to \mathbb{R} \cup \{+\infty\}$ (with a nonempty domain) and a convex function $h : \mathbb{R}^n \to \mathbb{R}$, x^* is a global minimizer of $f = g - h$ on $\mathbb{R}^n$ if and only if

$$\partial_\epsilon h(x^*) \subseteq \partial_\epsilon g(x^*) \quad \text{for all } \epsilon > 0. \tag{13.18}$$

Deriving meaningful global optimality conditions for broad classes of problems is extremely challenging in general; however, restricting one's attention to specially structured cases sometimes yields interesting results. Consider, for example, a QUBO problem in the form

$$\min q(x) = \frac{1}{2} x^T Q x + c^T x \quad \text{s.t. } x \in B = \{-1, 1\}^n, \tag{13.19}$$

where Q is an $n \times n$ real symmetric matrix and $c \in \mathbb{R}^n$. Beck and Teboulle [182] use a continuous representation of this problem to derive global optimality conditions. Denote by $\lambda_{\min}(Q)$ the smallest eigenvalue of Q, $X = \text{diag}(x)$, and $e = [1, \ldots, 1]^T \in \mathbb{R}^n$. Then any global minimizer $x \in B$ satisfies

$$XQXe + Xc \leq \text{diag}(Q)e, \tag{13.20}$$

and any $x \in B$ satisfying

$$\lambda_{min}(Q)e \geq XQXe + Xc \tag{13.21}$$

is a global optimal solution of the above QUBO problem. Alternative optimality conditions for QUBO can be found in [1181]. The conditions are based on geometric arguments and are successfully used in a variable fixing procedure within a branch-and-bound framework.

Other special cases that have been studied in the literature in terms of global optimality conditions include indefinite and concave QO [547], nonconvex QO with quadratic constraints [996], polynomial optimization [391], convex maximization [642], and other problems with special structures [815, 995].

13.5 ▪ Algorithms and Software

We consider the following global optimization problem:

$$\min \ f(x) \tag{13.22}$$
$$\text{s.t. } \ g_i(x) \leq 0, \quad i = 1, \ldots, m, \tag{13.23}$$
$$x \in X, \tag{13.24}$$

where $X \subseteq \mathbb{R}^n$ and $f, g_i, i = 1, \ldots, m$, are real-valued functions, each defined on a domain containing X.

We assume that a global optimal solution x^* of problem (13.22)–(13.24) exists and denote by $f^* = f(x^*)$ the corresponding global optimal value of the objective function. Computing a global optimal solution exactly is nearly impossible for a general optimization problem; hence instead one may settle for the more realistic goal of finding a globally ϵ-optimal value of f and a corresponding feasible point. That is, the aim is to find a feasible point $x' \in X$ such that $f(x') \leq f^* + \epsilon$, where ϵ is a given small positive constant. We call an algorithm ϵ-convergent if it finds such a point in a finite number of iterations.

Depending on the information available about the functions involved in this formulation, we can define different classes of methods for solving the problem. In zero-order or derivative-free methods, we assume that there is an oracle that can be called to obtain $f(x)$ and $g_i(x)$ values for any $x \in X$. The first-order methods assume that the functions appearing in (13.22)–(13.24) are differentiable and, in addition to the function values, we can use the values of their gradients, $\nabla f(x), \nabla g_i(x), i = 1, \dots, m$, for any $x \in X$. First-order methods are also known as *gradient-based* methods. Similarly, the second-order methods assume availability of the Hessians $\nabla^2 f(x), \nabla^2 g_i(x), i = 1, \dots, m$, for any $x \in X$ in addition to the function and gradient values.

Most common deterministic approaches for solving global optimization problems are based on techniques involving some kind of *branch-and-bound* or *successive partitioning* procedures. In a branch-and-bound framework, *branching* is used to split the feasible region of the problem into smaller parts, whereas *bounding* schemes are applied to prune suboptimal branches and to prove optimality. The choice of specific branching and bounding strategies usually depends on the structure of the problem of interest. The key elements of a branch-and-bound algorithm are discussed next.

Branching (Partitioning)   A typical branch-and-bound scheme proceeds by splitting the problem's feasible region (or its relaxation) into a finite number of subsets in a systematic fashion. The set X (or its relaxation) is usually assumed to have a simple structure in the sense that it allows for an easy partition into subsets having similar structure. This is essential for devising efficient branching techniques. Popular choices of basic partition elements include hyperrectangles, simplices, and cones [957]. The choice of the next subset to partition at each step is an important part of the branching strategy.

Bounds   For each subset resulting from the partitioning, upper and lower bounds on the optimal objective function value are computed. The objective function value at any evaluation point constitutes an upper bound, which could also be enhanced using some heuristic strategies. The best evaluation point encountered in the process is recorded and updated whenever a better-quality solution is found. The methods used to obtain lower bounds include the reformulation-linearization technique (RLT) [1687], αBB approaches [28, 85], McCormick relaxations [1304, 1654], Lagrangian dual bounds [109, 201, 355, 966, 1384, 1802], semidefinite and other convex relaxations [300, 1004, 1241, 1929, 1960], bounds based on interval analysis [902, 1549], and Lipschitz bounds [904, 1494, 1730].

Pruning (Fathoming)   If for some subset the lower bound exceeds the currently best upper bound, the subset cannot contain an optimal solution and hence is eliminated from further consideration.

Termination   The algorithm terminates when the best upper bound is sufficiently close to the smallest upper bound among all the active subsets, yielding a global minimum or its close estimate.

Partitioning a continuous feasible region during the branching process is more complicated than dealing with discrete variables, since the number of partitions may be infinite and the partitions may overlap on the boundary. Hence, even establishing finite convergence is a nontrivial task [159].

Since a general global optimization problem is extremely hard to solve, many practical algorithms exploit certain structural properties or make additional assumptions regarding the problem of interest to make it more tractable. In the following two subsections, we illustrate this point by discussing some ideas behind algorithmic approaches for two important classes of global optimization problems: concave minimization and Lipschitz optimization.

13.5.1 ▪ Concave Minimization

Consider the problem (13.22)–(13.24); where X is a polytope; $f(x)$ is a concave function over X; and $g_i, i = 1, \ldots, m$, are convex over X. This problem is referred to as a *concave minimization* problem, or as a *concave optimization* problem when the feasible region is given by a polytope. As we have seen in Section 13.2, fixed-charge problems, integer optimization, and QUBO problems can be posed as concave programs. Other broad classes of problems that can also be transformed into concave minimization problems include, e.g., bilinear optimization problems, complementarity problems, and multiplicative programs.

Following the influential work by Tuy [1798] published in 1964, concave minimization became one of the focal points in the emerging field of global optimization during the next several decades. As a result, a number of algorithmic approaches have been devised for general concave minimization as well as for restricted subclasses of concave optimization. Most popular methods for solving concave optimization problems can be classified into three categories: enumerative methods, successive approximation (outer, inner, underestimation) methods, and branch-and-bound techniques.

Enumerative methods exploit the property that a global minimum of a concave function f over a polytope X is attained at a vertex of X, implying that the problem can be solved in a finite number of enumerations by ranking the extreme points of X. The first such approach was proposed by Murty in 1968 [1365]. To alleviate the difficulties associated with an excessively large number of extreme points, the enumerative approaches are often enhanced by introducing cutting planes, such as Tuy's concavity cuts [959, 1798], that allow one to eliminate some suboptimal parts of the feasible region.

Successive approximation techniques proceed by solving a sequence of simpler problems approximating the original, more complicated problem of interest. The approximation quality is successively improved until a solution to an approximation problem is guaranteed to be an exact or approximate solution to the original problem. Common techniques used for approximation purposes include outer approximation (OA), inner approximation, and successive underestimation.

Consider the concave minimization problem

$$\min \; f(x) \;\; \text{s.t.} \;\; x \in X, \tag{13.25}$$

where X is a nonempty, closed convex set. In OA methods, a sequence of sets X_k, $k = 1, 2, \ldots$, containing the feasible region X is constructed step by step, and the following approximating problem of (13.25) is considered at each step:

$$\min\ f(x)\ \text{ s.t. } x \in X_k. \tag{13.26}$$

If the computed optimal solution x^k of (13.26) is in X, it is also an optimal solution for (13.25), and the original problem is solved. Otherwise, we construct a new OA $X_{k+1} \subset X_k \backslash \{x^k\}$, which cuts off the infeasible point x^k, and repeat the process. Most often, the improved OA X_{k+1} is obtained using inequality cuts separating x^k from X. A much less common alternative approach is based on the notion of collapsing polytopes [686].

In the inner approximation approach, which is also known as the polyhedral annexation, the feasible region is approximated from inside by constructing an expanding sequence of approximating sets $X_1 \subset X_2 \subset \cdots \subset X$. In this case, the sequence $\{x^k : k \geq 1\}$ of optimal solutions to the corresponding approximating problems is monotonically decreasing and converges to the solution of the original problem from above [1799, 1800]. The successive underestimation methods approximate the problem of minimizing a concave function over a bounded polyhedron by replacing the original objective function with its underestimates at each iteration [685].

Just as in the case for global optimization problems in general, branch-and-bound methods represent the most common approach to solving concave minimization problems. The reader is referred to [222, 223, 959] for detailed surveys of concave minimization techniques.

13.5.2 ▪ Lipschitz Optimization

One of the simplest and most natural assumptions made concerning the structure of optimization problems is given by the *Lipschitz condition*, which bounds the variation of functions involved over the feasible domain. A function is called *Lipschitz* with Lipschitz constant L over X if

$$|f(x) - f(y)| \leq L\|x - y\| \quad \text{for all } x, y \in X. \tag{13.27}$$

The methods that take advantage of the Lipschitz assumption are referred to as Lipschitz optimization methods. The feasible region X is typically assumed to be (a subset of) an n-dimensional unit hypercube, and no smoothness assumptions are made for the objective function. The Lipschitz assumption does not make the problem easy. In fact, the problem remains NP-hard even when the objective is a quadratic function, as evident from formulation (13.8) for the independence number of a graph. It can also be shown that there always exists an instance of a Lipschitz optimization over a hypercube that requires at least $\lfloor L/2\epsilon \rfloor^n$ function evaluations to guarantee an ϵ-approximate solution, where the Lipschitz constant L is specified for the infinity norm [1394]. This bound on the number of evaluations can be achieved (asymptotically when $n \to \infty$) using an appropriate uniform grid to define evaluation points, implying that this naive strategy is optimal in the sense that it is unbeatable in the worst case. However, the Lipschitz assumption can be used to obtain algorithms that are much more effective in practice.

A key property of Lipschitz functions utilized in global optimization algorithms is the fact that for a given set $\{x_1, \ldots, x_k\}$ of evaluation points, the function

$$h_k(x) = \max_{i=1,\ldots,k} \{f(x_i) - L\|x - x_i\|\} \tag{13.28}$$

provides a lower bound on f over X, that is, $h_k(x) \leq f(x)$ for all $x \in X$. In particular, in the univariate case, h_k is a "saw-tooth" function and is easy to optimize. This observation forms the core of Piyavskii's algorithm [1494], which proceeds by adding a global minimizer x_{k+1} of h_k as the new evaluation point at the kth step of the algorithm, thus yielding an updated lower-bounding function h_{k+1}.

Piyavskii's algorithm can be generalized to the multivariate case. As in the univariate case, an underestimate of the objective is built and updated by adding evaluation points corresponding to the global minima of the lower-bounding function. Several algorithms for computing these global minima have been proposed in the literature [1340, 1495].

Alternatively, a multivariate Lipschitz optimization problem can be solved by reduction to a sequence of univariate problems using nested optimization [904, 1494]. Another way to reduce a multidimensional Lipschitz optimization problem to a univariate one is based on the concept of *space-filling curves*. A space-filling curve or *Peano curve* is a curve passing through every point of a multidimensional region. Such curves have been studied by mathematicians since the end of the 19th century. To exploit them in a global optimization context, we first use space-filling curves to transform the multidimensional optimization problem $\min\{f(x) : x \in [0,1]^n\}$ into the equivalent problem $\min\{f(x(t)) : t \in [0,1]\}$ of a single variable, where $x(t)$ is a Peano curve. Then we utilize a one-dimensional global optimization method to solve the resulting univariate problem. This methodological direction has been explored in depth for Lipschitz-continuous problems, as discussed in the monographs [1661, 1729, 1730]. The proposed techniques utilize the fact that if $f(x)$ is Lipschitz continuous with a constant L on $[0,1]^n$, then the one-dimensional function $f(x(t)), t \in [0,1]$, satisfies the following *Hölder condition*:

$$|f(x(t')) - |f(x(t''))| \leq 2L\sqrt{n+3}(|t'-t''|)^{1/n}, \quad t', t'' \in [0,1]. \qquad (13.29)$$

Strongin and Sergeyev [1730] provide a detailed account of the corresponding methodology. A more recent book by Sergeyev et al. [1661] gives a brief introduction to the topic and discusses the most recent related developments.

In addition to the approaches mentioned above, several branch-and-bound algorithms have been developed for multivariate Lipschitz optimization [681, 904, 1310, 1311, 1461, 1487]. Typically such algorithms utilize partitions of the feasible region into hyperrectangles; the objective function f is evaluated at the center of the current hyperrectangle to get the corresponding upper bound; branching rules are based on splitting the current hyperrectangle along its longest edge into several equal hyperrectangles, and the lower bound is based on the Lipschitz constant.

Despite their theoretical attractiveness, the early Lipschitz optimization methods suffered from several drawbacks in practice, such as the need to specify the Lipschitz constant, low speed of convergence, and high computational complexity in higher dimensions. In particular, their slow convergence can be explained by the high emphasis placed on global search, dictated by the large values of Lipschitz constants typically used to ensure a valid upper bound on the rate of change of the objective function. This motivated more recent approaches that take advantage of an alternative role given to the Lipschitz constant. One such method, DIRECT [1011, 1012], interprets the Lipschitz constant as a weighting parameter used in balancing between global and local search. Due to its simplicity and effectiveness, DIRECT and its variations have attracted considerable attention in the literature [711, 775, 1210, 1460].

13.5.3 ▪ Global Optimization Software

In the last two decades, we have witnessed significant progress in the development of global optimization software. What used to be "the road less traveled" 20 years ago [1444] is becoming an essential feature of state-of-the-art optimization software. Most modern optimization modeling systems and solvers include global optimization functionalities; see [1486] for a survey. Prominent global optimization solvers include αBB [28, 85], BARON (Branch and Reduce Optimization Navigator) [1611, 1754,1755], GloMIQO (Global Mixed Integer Quadratic Optimizer) [1331], and LGO (Lipschitz Global Optimizer) [1488], to name a few.

Some standard testbeds and problem generators have also been developed that can be used to evaluate the performance of computer implementations of global optimization algorithms. For example, the book by Floudas et al. [742] contains a large collection of various types of test instances for global optimization, including QO problems, minimum concave cost transportation problems, quadratically constrained problems, and polynomial problems. Gaviano et al. [790] propose a method for generating test functions with user-specified properties for global optimization. Websites [1339, 1399] maintaining lists of available global optimization software provide download links to numerous open-source packages. Rios and Sahinidis [1569] review algorithms and compare 22 derivative-free optimization solvers on 502 test problems; their problems and computational results are available at the website: http://archimedes. cheme.cmu.edu/?q=dfocomp.

13.6 ▪ Further Reading

Due to the limited scope of this work, many important topics dealing with global optimization have not been discussed, including D.C. and monotonic optimization [1797, 1801], stochastic global optimization and metaheuristic methods [817, 1943, 1974], bilevel and multilevel optimization [490,578,1322], multiobjective optimization [654, 1321], complementary problems [1440], polynomial optimization [923, 1131, 1133, 1691], methods based on interval analysis [902, 1045, 1550], and decomposition techniques [959]. The reader may consult the numerous textbooks [735, 932, 958, 959, 1215], research monographs [1447, 1461, 1646, 1687, 1730], edited volumes [741, 957, 1445, 1446, 1779], and comprehensive surveys of recent advances [119, 739] to learn more about theoretical, algorithmic, and applied aspects of the vibrant field of global optimization.

Chapter 14

De Novo Design of Protein-DNA Systems Using Global Optimization

James Smadbeck, George A. Khoury, and Christodoulos A. Floudas

14.1 ▪ Introduction

Computational protein design methods have seen notable successes in the last decade, using both deterministic and stochastic methods of design. Despite these successes, there are still a variety of difficulties in the application of well-established optimization methods to biologically relevant problems. Combinatorially speaking, designing on a biological scale can be challenging and time-consuming due to the NP-hard nature of the protein design problem [1481]. For this reason, careful definition of the optimization model used in such applications is necessary for success, especially when the aim is to find a globally optimal solution. By extending established quadratic assignment-like protein design methods to new applications, we can build on previous successes in a systematic way. Such extensions allow for the design of therapeutics for many disease targets previously unapproachable by global optimization–based design methods, while taking advantage of methods of improving optimization efficiency. Protein-DNA design represents an example of a design application with significant therapeutic potential where extensions of existing optimization models to the application would have a great impact.

Protein-DNA interactions are pervasive in nature, taking place in many aspects of cellular replication, transcription, and epigenetics [1583]. Determining the DNA binding and specificity properties of proteins that interact with DNA is important in the genomic era. This is especially true considering the large number of genomes being sequenced that are still in need of annotation. Beyond the increased knowledge of protein-DNA biology that a full understanding of protein-DNA interaction would afford, the ability to design protein-DNA specificity and binding has a wide range of applications in the development of cancer therapeutics and gene therapy strategies [82].

Recent studies strongly support the view that full structural information on both the protein and the DNA involved in the interaction must be present for full understanding of DNA recognition [1453]. For this reason, the development of a structure-based protein-DNA design model is important for such systems.

In this chapter, the process by which a global optimization–based model is extended to the design of protein-DNA systems is presented along with an application of the method to the design of gene therapeutics. The example demonstrates how the new method can produce physically relevant results that provide insight into the underlying complex biological process. The protein-DNA design method is a two-stage design method employing integer linear optimization (ILO) and all-atomistic validation. The first stage is a sequence selection optimization stage that takes in a protein-DNA structure and determines the optimal sequence for protein-protein and protein-DNA interactions. The second stage structurally validates the modified region of the protein loop using fold specificity and protein-DNA binding using the all-atomistic protein-DNA force field DDNA3 [1966]. The method is applied to the design of the prototype foamy virus (PFV) binding to a consensus DNA sequence with the aim being improved binding affinity. This is meant to demonstrate how existing optimization-based design models can be extended to new biological applications and how the method can be applied to the design of protein-DNA interactions.

14.2 ▪ Methods

The application of optimization methods to a combinatorially complex biological application such as protein-DNA interaction requires several careful considerations to find a global solution efficiently. Our original quadratic assignment–like sequence selection method was first developed by Klepeis et al. [1072, 1073]. It selects and ranks amino acid sequences according to their energies in the design template using an ILO model. This global optimization method does not rely on random mutations and is theoretically guaranteed to determine a global solution from the full sequence space. This is a major advantage of our approach compared to other approaches, which primarily use stochastic algorithms to search the sequence space.

Due to the structure of the ILO model, a binary interaction energy function must be used in the sequence selection method. The most accurate energy functions available for protein-protein and protein-DNA interaction calculations, however, are all-atomistic. For this reason, a two-stage method is used when applying the model to complex, biologically relevant problems. The first stage is the optimization stage, which produces a list of candidate solutions. The second stage utilizes more detailed energy functions to validate the candidate solutions and remove false positives.

This general two-stage approach to optimization-based protein design methods has been applied to a wide variety of problems with significant success. By extending a general design framework to a new application, necessary optimization efficiency is maintained, even when the new application may be significantly more complex than what the model was original developed for. Figure 14.1 gives an overview of the steps involved in the framework as it is applied in the protein-DNA application.

A design template, constraints, and a force field are inputted into an optimization-based sequence selection stage to produce a rank-ordered list of sequences that are validated using fold specificity and interaction energy calculations developed expressly for protein-DNA design. Notice the similarity in structure of this workflow to previously developed protein design methods that have demonstrated significant success [187, 188]. Details of each step in the workflow diagram are provided here.

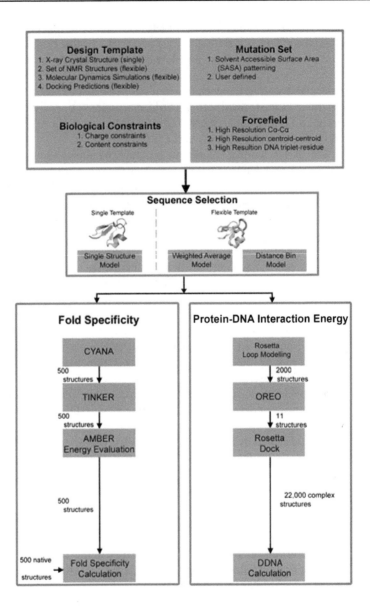

Figure 14.1. *The workflow diagram for the protein-DNA design method* [1700].

14.2.1 ▪ Input Parameters and Constraints

In any application of optimization to biological design, many of the features of the biological design space must be transformed or coarse-grained. These are defined as the input parameters for the de novo protein-DNA design framework. Since an ILO model is used in our design model, the design template, mutation constraints, and biological constraints must be produced in a linear form amenable to the standard ILO form.

The first design input is the design template. This is a three-dimensional protein structure that contains coordinates for all of the atoms in the protein. The structure can be rigid or flexible. Rigid templates are a set of fixed atom coordinates and are obtained from X-ray crystallography structures. Flexible templates can be a set of fixed atom coordinates or upper and lower bounds on the atom coordinates. These templates can be obtained from nuclear magnetic resonance (NMR) solution structures, molecular dynamics, or docking simulations. In all cases the three-dimensional protein structure is transformed into a set of binary interaction distances, which are input as a large two-dimensional matrix. This transformation allows for the easy use of binary interaction energy functions in the ILO form.

The design template is used to generate the allowed mutation set of the designed protein. This set defines which positions of the sequence can mutate and to what amino acids. The mutation set can be calculated by the solvent accessible surface area (SASA) of each residue in the template, which provides the model with which positions in the protein sequence define the structural core of the protein.

Biological constraints, in the form of charge or content bounds, can be defined using an extensive alignment to the native sequence. Further constraints may be manually defined through analysis of experimental data. This type of sequence analysis of proteins or domains with similar function can provide better constraint information. These constraints are represented as simple linear constraints on the summation of amino acid type across the protein being designed.

Unlike in previous applications of the quadratic assignment–like model, where only one binary energy function is needed, two distance-dependent force fields are necessary for protein-DNA design to define both the protein-protein interactions and the protein-DNA interactions. A high-resolution centroid-centroid force field [1544] was previously developed to define protein-protein interactions in ILO models for protein design. It was not developed for calculating protein-DNA interaction energy, however. For this reason, a similar protein-DNA potential [1208], based on the interaction of amino acid centroids and centroids of triplets of nucleotides, has been integrated into this design procedure for use in protein-DNA design. The force field, developed by Liu et al. [1208], operates in a similar manner to the centroid-centroid force field, whereby a coarse-graining of the experimental structure is made such that mutations in the structure can be treated without consideration of atomistic or rotameric changes. This is an important distinction, as this means the force field can be directly used in an ILO framework, e.g., at the sequence selection stage. This is an important example of how simple additions or changes to the input into a model can drastically increase the number of applications that a biological optimization method can be used for. Additionally, it is a good example of how the modularity of an optimization workflow can help ease the transition from one application to another. A graphical description of the potential energy calculation is given in Figure 14.2.

The primary difference between this protein-DNA force field and the amino acid force field is the fact that instead of pairwise centroid-centroid interactions, the DNA model treats each DNA centroid as a triplet set of nucleotides. So while in the protein-protein energy function the centroid of one amino acid is used as an interaction point with another centroid of an amino acid (Figure 14.2(a)), in the protein-DNA energy function, the centroid of up to three nucleotides is used as an interaction point with the centroid of an amino acid or another centroid of nucleotides (Figure 14.2(b)). This is an important consideration as the variety of nucleotide types (A,C,T,G) is one fifth the variety in types of amino acids. The use of DNA triplets ensures that the protein-DNA energy function is not too coarse grained in the DNA space for effective design.

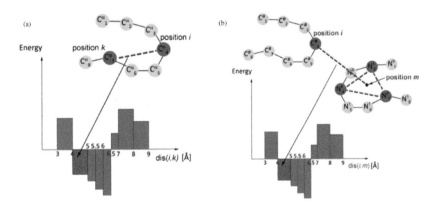

Figure 14.2. (a) *The original centroid-centroid force field.* (b) *The protein-DNA force field added for application to protein-DNA design problems* [1700].

14.2.2 ▪ Protein-DNA Sequence Selection

The original sequence selection method was first developed by Klepeis et al. [1072, 1073]. It selects and ranks amino acid sequences according to their energies in the design template using an ILO model. In detailing the method, several important constraints used to increase the computational efficiency of the model will be noted. These types of constraints, used to improve computational efficiency, are what allow for the use of a global optimization method for problems as complex as a protein-DNA application.

For the protein-DNA design problem, we aimed to develop an optimization model capable of designing a protein sequence that interacts with the other amino acids in the protein in a favorable way (i.e., does not disrupt the necessary protein structure) and favorably interacts with the specific DNA nucleotides to which the protein must bind. For the protein-DNA problem, the final form of the sequence selection optimization model, where only the protein sequence can be modified, is given in (14.1). This model is presented as a rigid template formulation for simplicity:

$$\min_{y_i^j, y_k^l} \sum_{i=1}^{n-1} \sum_{j=1}^{m_i} \left(\sum_{k=i+1}^{n} \sum_{l=1}^{m_k} \sum_{d=1}^{b_m} E_{jl}^d b_{ik}^d w_{ik}^{jl} + \sum_{p=1}^{n_d} \sum_{q=1}^{m_p} \sum_{d_d=1}^{b_d} E_{DNA_{jq}}^{d_d} b_{ip}^{d_d} y_i^j \hat{y}_p^q \right)$$

$$\text{s.t.} \quad \sum_{j=1}^{m_i} y_i^j = 1 \ \forall \ i,$$

$$\sum_{l=1}^{m_k} y_k^l = 1 \ \forall \ k,$$

$$\sum_{j=1}^{m_i} w_{ik}^{jl} = y_k^l \ \forall \ i, k > i, l,$$

$$\sum_{l=1}^{m_k} w_{ik}^{jl} = y_i^j \ \forall \ i, k > i, j,$$

$$y_i^j, y_k^l, w_{ik}^{jl} \in \{0,1\} \ \forall \ i, j, k > i, l.$$

(14.1)

The set $i = 1,\ldots,n$ defines the set of amino acid positions in the protein. The set $p = 1,\ldots,n_d$ defines the set of valid DNA triplet centroids in the DNA strand. At each position i, mutations are represented by $j\{i\} = 1,\ldots,m_i$, where $m_i = 20$ if position i is allowed to mutate to any of the 20 natural amino acids. At each DNA centroid position p, centroid identities are represented by $q\{p\} = 1,\ldots,m_p$, where $m_p = 34$ if all possible nucleotide triplets are allowed at position p. The alias sets $k \equiv i$ and $l \equiv j$, with $k > i$, are employed to represent all unique pairwise interactions. Binary variables y_i^j and y_k^l are introduced to model amino acid mutations. The y_i^j variable will assume a value of one if the model assigns amino acid j to position i, and a value of zero otherwise (similarly for y_k^l). The binary parameters $\hat{y}_p^q$ represent the valid DNA centroids and their triplet identities. The objective function represents the sum of all pairwise energy interactions in the design template for both protein-protein and protein-DNA interactions. Parameter E_{jl}^b is the energy interaction between amino acid type j and amino acid type l that interact within distance bin b. Parameter b_{ik}^d is equal to one if positions i and k interact within distance bin d. The energetic term only contributes to the overall objective function if both y_i^j and y_l^k are equal to one (and thus the linearization variable w_{ik}^{kl} is equal to one) and the parameter b_{ik}^d is equal to one. Note that the formulation does not allow for mutation to the DNA sequence, as the model is an example of what would be used in a protein-DNA binding specificity design problem.

The constraints involving the linearization variable w_{ik}^{kl} should be specially noted as they are used in what is termed second-order reformulation-linearization technique (RLT) constraints. As has been detailed in previous publications, RLT constraints are employed to efficiently solve for the global optimum solution for a complex biological application [771, 1073]. In such applications, if one does not formulate the model carefully, its combinatorial complexity can result in computational intractability. In the original formulation of the model, a simple McCormick relaxation technique was employed to linearize the model [1072]. RLT constraints were added to improve efficiency [1073] and quantitatively shown to improve the solve efficiency by several orders of magnitude [771]. Using lessons learned from previous applications is an important aspect of extending existing optimization models to new, more complex applications. Even though the original linearization technique was developed for a simpler biological application than protein-DNA design, it has proved sufficient to make the highly complex biological optimization model computationally tractable.

14.2.3 ▪ Protein-DNA Fold Specificity

To further validate and analyze the designed protein structures, a modified version of the fold specificity method was developed to assess the fold specificity of the designed regions of the protein [770, 772]. This method uses a constrained annealing simulation in CYANA [878, 879] to produce a set of initial loop models. A local AMBER energy minimization using TINKER [1501] is then performed on each model to produce a set of 500 final models, along with corresponding AMBER ff94 energy values [513]. Using these AMBER energy values, the fold specificity value is calculated as

$$F_{Spec} = \frac{\sum\limits_{i \in \text{New}} e^{-\beta E_i}}{\sum\limits_{i \in \text{Native}} e^{-\beta E_i}}, \tag{14.2}$$

where New is the set of models produced for the new sequence, Native is the set of models produced for just the modified region of the protein that interacts with the DNA strand, and E_i is the AMBER energy value calculated for model i. Physically, fold specificity assesses how energetically favorable it is for the designed sequences to adopt the target loop structure. The aim is to assess the specificity of the designed sequences for the target structure using a more detailed, atomistic potential energy than that used in the sequence selection stage.

14.2.4 • Protein-DNA Interaction Energy Metric

For a more detailed validation, a DNA-protein interaction energy metric, termed here K_{DNA}^*, was developed. For this calculation, a three-dimensional structure of each sequence bound to the desired DNA strand is needed. Since the designed sequences are derived from the quadratic assignment–like protein-DNA sequence selection method and/or protein-DNA fold specificity of the protein, in almost all cases the loop region of the protein will be the only area mutated. For this reason, the structure generation step of this validation focuses on this region of the protein, exploiting loop prediction methods developed in Rosetta [1140] expressly for this purpose. This kinetic loop modeling method [1260, 1261], which has been shown to generate better loop structures than fragment-based loop modeling, is used to generate the newly designed regions of the protein structure.

The proteins with modified loop structure from the kinematic loop modeling step are then clustered based on their ϕ and ψ angles using OREO [597, 598, 1738] to identify representative loop structures. This step is made computationally more efficient by isolating and clustering only the modified loop region of the generated structures.

Docking prediction is done using RosettaDock [541, 850]. For each sequence, each representative protein structure with modified binding loop is docked against the target DNA sequence. RosettaDock uses a Monte Carlo algorithm for low- and high-resolution docking movements of protein-DNA structures. Each docking run generates a large ensemble of docked structures.

RosettaDesign [1111] is then used to generate the final conformation ensemble based on the representative docked structures. The final ensemble size is 22,000 structures. This allows for a wide variety of docked structures to contribute to the overall energetic score of the designed sequence. This is more accurate than assuming similar docking structure to the native, as it is likely that the designed loop sequences are not identical to the native loop sequence.

Once the set of 22,000 protein-DNA docked structures is generated, they must be evaluated using a detailed, full-atomistic protein-DNA force field. Due to its excellent correlation with experimental protein-DNA interaction energies and good performance in identifying structural ensembles for energetic analysis, DDNA3 [1966] was chosen for use in our protein-DNA interaction energy metric calculation. The protein-DNA approximate binding affinity energy is calculated as follows:

$$K_{DNA}^* = \frac{\sum\limits_{b:E_b < E_{\text{cutoff}}} e^{-\beta E_b}}{N_{\text{cutoff}}}. \tag{14.3}$$

Here, the DDNA3 energy of a given complex, b, is given as E_b. These energies are used in a Boltzmann average calculation to produce an ensemble energy for all structures whose energies are lower than a selected energetic cutoff, E_{cutoff}. The ensemble energy is normalized by the total number of structures that satisfy the energetic requirement, N_{cutoff}, to produce the DNA-protein interaction energy metric.

It should be noted that the method used for the protein-DNA interaction energy calculation is fundamentally similar to methods used for approximate binding affinity calculations in previous publications [187, 188]. This is another example of how previous methods of validation can be extended to more complex biological applications, while maintaining many properties of the previous application method that made it a success. Through these previous applications it has been found that using such validation stages as fold specificity and the interaction energy calculation to remove false positives from the results of the optimization-based design is highly valuable.

14.3 ▪ Protein-DNA Design of PFV Loop DNA Specificity

In this section, the Protein-DNA design method described above is applied to the design of a PFV integrase enzyme with the aim of improving the binding specificity of the integrase loop for a consensus DNA sequence. This section is meant to demonstrate how an optimization model extended to a new biological application can be used to produce practical results with clear biological relevance.

The PFV is closely related to many viruses commonly used in retroviral gene therapy [82]. Like all viruses, the PFV life cycle includes the integration of viral DNA into the host DNA using integrase enzymes [523, 1169]. Through the replacement of viral DNA with therapeutic DNA, these retroviruses can be used as therapeutics for otherwise difficult to treat illnesses. Limitations of potential therapeutic applications of retroviruses have hindered their continued development for medical applications. Primarily the limitations involve the fact that many integrases integrate randomly into the host DNA [1044]. This can cause serious medical issues [887, 888], such as cancer [365, 366], through the integration of the DNA into important regulatory genes of the host genome. For this reason, PFV integrase represents a promising initial application of the protein-DNA design framework. With successful demonstration of the use of the method for PFV integrase design, one could hope to develop a de novo designed PFV integrase that integrates at a DNA position in the host genome that does not carry the same risk as nonspecific insertion.

14.3.1 ▪ Template Generation

The structure of the PFV integrase in the process of integration was recently published (PDB:3OS0 and PDB:3OS1 [1250]). This structure is shown in Figure 14.3(a). The triangle in Figure 14.3(a) highlights the binding pocket of the integrase, and the circle highlights the loop that is designed for DNA-binding specificity. It is known that PFV works by binding to the DNA strand of the host genome, involving a binding loop that is equivalently found in many integrase proteins (Figure 14.3(b) and (c)). This loop is targeted for the protein-DNA design application, as it is important for binding and represents a potential way in which binding specificity can be increased.

14.3.2 ▪ Mutation Constraints, Biological Constraints, and Force Fields

For the mutation constraints on the design positions, the relative SASA of the binding loop was analyzed. Three different design runs were performed. Run 1 designed using the standard mutation constraints. Run 2 designed using standard mutation constraints, without cysteine mutations. Run 3 kept native proline positions unmutated,

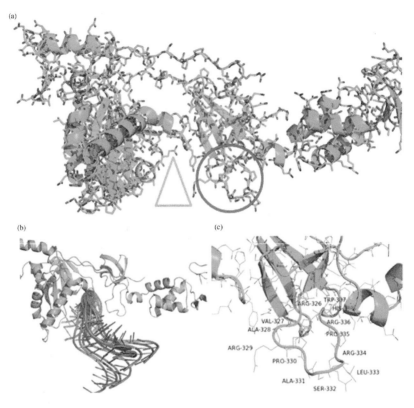

Figure 14.3. (a) *The structure of the PFV integrase (PDB:3OS0) in its bound state.* (b) *A series of alternative bound DNA structures based on RosettaDock.* (c) *Zoom in on the design loop of the integrase* [1700].

without cysteine mutations. The allowed mutations for a hydrophilic position were D, E, G, H, K, N, P, Q, R, S, and T. The allowed mutations for a hydrophobic position were A, F, I, L, M, V, W, Y, and C. Here, hydrophilic means a position exposed to the solvent in the template protein structure, and hydrophobic means a position in the interior of the protein structure. Proline and cysteine are often specially treated in optimization-based designs, as they are geometrically unique amino acids.

The biological constraints were added consistently to each design run. These constraints were determined through the collection of viral integrase proteins with similar sequences and function to PFV. A total of 15 integrase sequences with high sequence identity were found. The sequences of the loops were then analyzed for overall charge and the number of individual amino acid types. Upper and lower bounds on each amino acid type and total charge were calculated, with a minimum upper bound of one, and included in design runs 1–3 as constraints.

14.3.3 ▪ Protein-DNA Design Results

The optimization-based protein-DNA sequence selection method was used for runs 1–3 with input mutation constraints, biological constraints, and force field as detailed in the previous section. A full set of 500 globally optimal candidate sequences were solved for in each run, and fold specificity values were calculated for each. The rank-ordered lists of fold specificities were then used to select the top 50 sequences to be

Table 14.1. *Top sequences from runs 1–3 of the protein-DNA fold specificity and interaction energy validation. Modified from* [1700].

Run #	Name	Sequence	F_{Spec}	$K^*_{DNA,-35}$	$K^*_{DNA,-38}$
	PFV	RVARPASLRPRWH	1.0	−5.99	−0.80
Run 1	FSpec1_2	KYWPQNRTKFNCR	1.28	−11.13	8.46
	FSpec1_19	KYWPQNRRKFNCL	1.30	−9.42	−6.25
	FSpec1_20	AYWPQNRLKFRCR	1.29	−9.07	−5.34
	FSpec1_11	KYWPQNRLKINCR	1.33	−9.06	−6.39
	FSpec1_22	KYWPQNRLKFNCR	1.28	−7.53	−4.28
	FSpec1_24	KYWPQNRLKRNCL	1.27	−6.53	−2.60
	FSpec1_18	NYWPQNRSKFKCR	1.30	−6.37	−5.03
	FSpec1_21	NYWPQNRPKFKCR	1.28	−6.17	−2.57
Run 2	FSpec2_4	DYWPQNRRKRNFR	64.63	−10.90	−8.04
	FSpec2_7	NYWPQNRLKFRLR	41.10	−9.25	−5.02
	FSpec2_5	NYWPQNRLKIRTR	50.29	−8.48	−6.2
	FSpec2_11	NYWPQNRAKFRTR	35.84	−7.56	−3.26
Run 3	FSpec3_2	NYWDPSRRRPKFR	14.72	−8.83	−5.38
	FSpec3_3	NYWDPKRRQPKTR	6.09	−8.82	−4.45
	FSpec3_4	NYWDPKRRQKPFR	3.75	−6.37	−2.88

tested in the more rigorous protein-DNA interaction energy validation stage. In total, the top 25, 13, and 12 sequences from runs 1, 2, and 3, respectively, were selected for validation. This distribution was chosen because run 1 is the least constrained design run and could present a more challenging problem in identifying promising candidate loop designs, so more sequences may be needed to identify true positives. The data for the fold specificity validation stage are included in Table 14.1, which is rank ordered by the calculated protein-DNA interaction energy metrics.

By analyzing the sequences of the designed binding loop from runs 1–3, we can see several interesting trends. For runs 1–3, it can be observed that while the original sequence had a tryptophan (W) amino acid in position 12 of the loop, the designed sequences have a highly conserved tryptophan amino acid in position 3 of the loop. As a result, the tryptophan in position 12 is replaced by a cysteine (C) mutation in run 1 and a variety of other hydrophobic residues in runs 2 and 3. This variability confirms that the movement of the tryptophan was not due to favorability of other amino acids in that position, but rather favorability for tryptophan in position 3. This is interesting, as tryptophan residues are known to be found in protein-DNA interacting regions [1619]. Tryptophans are primarily associated with stabilizing the interacting regions so that polar amino acids can interact directly. We can see from the front of the loop sequence that we have a patterning of [polar]-[aromatic]-[aromatic]-[polar] in many of the design sequences. This result may be a reflection of the two parts of the objective energy function influencing the final design.

One can also see that there is a lot more variability in the designed sequences on the areas of the loop away from the DNA-binding pocket (positions 6–13) in comparison to the amino acids directly in contact with the DNA-binding pocket (positions 1–5). This suggests that the amino acids that contribute the most to interaction with the DNA sequence contribute the most to the energetics of the designed sequence, such that they do not change as easily to alternative mutations. This is a promising result as it suggests that the design is primarily interested in the protein-DNA interacting amino acids and can identify amino acid mutations that strongly interact with the specific DNA strand included in the design. These sequences represent viable sequences for experimental validation. The several promising properties of the designed loop sequences demonstrate the use of the method.

14.4 ▪ Conclusion

To the best of our knowledge this method constitutes the first quadratic assignment–like protein design method for protein-DNA design, and one of very few methods that address the problem in general. Its application to the design of protein loops for the increased specificity of protein-DNA binding is one of many applications to which a method that deals with protein-DNA interactions could be applied. This application is particularly important for the future of gene therapy, as gene therapy techniques require DNA specificity of binding to make them safe for clinical use. While the application area is more complex than what was used in the development of many of our optimization-based protein design tools, they can be extended in a straightforward manner to produce promising results.

Chapter 15

Global Optimization: Optimal Power Flow Problem

Vitaliy Krasko and Steffen Rebennack

15.1 ▪ Introduction

The optimal power flow (OPF) problem seeks to optimize the operation of an electric power system subject to network power flow constraints and system operating limits. It was first introduced in 1962 by Carpentier [407] as an extension of the economic dispatch (ED) problem. Carpentier's seminal contribution is the inclusion of power flow equations as constraints. These equations govern electric power flow and are derived from the application of three physical laws: Ohm's law, Kirchoff's current law, and Kirchoff's voltage law. They yield a steady-state alternating current (AC) electrical network.

Over the past five decades, OPF has become a very actively researched subfield of electrical engineering and operations research. The literature on OPF is vast. We point the reader to the textbooks [1551, 1901, 1979] and some classical as well as recent literature surveys [265, 408, 755, 756, 968, 1344, 1345]. For a novice in OPF with a sound background in optimization, we also refer to the primer [753].

In general, any power system optimization problem that includes the power flow equations as equality constraints can be considered an OPF problem. Hence the term OPF encompasses a broad range of formulations seeking to optimize the cost of energy generation, system reliability, power loss, load shedding, environmental impact, or other operational or planning objective. Additional constraints beyond the power flow equations can include bus voltage limits, power generation limits, branch flow limits, and demand constraints, among others. OPF can be applied to any planning horizon, from long-term capacity expansion planning with a planning horizon spanning several decades, to short-term power dispatch with a minute-to-minute planning horizon [753]. OPF problems generally seek to determine voltage, current, and injected power (both real and reactive) throughout the system. In this chapter, we focus on the

Table 15.1. *Notation used in this chapter.*

Indices	$i, k \in \mathbf{N}$	set of system buses (nodes)
	$i \in \mathbf{G}$	set of generation buses (supply nodes)
	$ik \in \mathbf{L}$	set of transmission lines and other connecting equipment (branches)
	$ik \in \mathbf{H}$	set of branches with controllable phase-shifting transformers
	$ik \in \mathbf{K}$	set of branches with controllable tap-changing transformers
Variables	P_i^G	generator (supply) real power at bus i
	Q_i^G	generator (supply) reactive power at bus i
	V_i	voltage magnitude at bus i
	δ_i	voltage angle at bus i
	E_i	real component of complex voltage at bus i
	F_i	imaginary component of complex voltage at bus i
	ρ_{ik}	phase shift of phase-shifting transformer in branch ik
	T_{ik}	tap ratio of tap-changing transformer in branch ik
Parameters	P_i^L	load (demand) real power at bus i
	Q_i^L	load (demand) reactive power at bus i
	G_{ik}	conductance (real component) of ikth element of bus admittance matrix
	B_{ik}	susceptance (imaginary component) of ikth element of bus admittance matrix
	Y_{ik}	magnitude of ikth element of bus admittance matrix

classical formulation that extends the ED problem, following Carpentier [407, 408] and Dommel and Tinney [610]. To enhance readability, the nomenclature used in this chapter is summarized in Table 15.1.

15.2 ▪ Problem Statement

The classical formulation of the OPF problem [610] seeks to minimize the total cost of electricity generation in the system by optimally controlling real and reactive power at the generator buses (electricity supply nodes). It is assumed that the transmission network is defined and no additional generators or other changes are considered on this planning horizon. Given are the following:

1. a network of nodes, $i \in \mathbf{N}$, and arcs, $ik \in \mathbf{L} \subset \mathbf{N} \times \mathbf{N}$, where the nodes represent electrical buses and the arcs represent transmission lines;

2. a subset of supply nodes $\mathbf{G} \subset \mathbf{N}$ representing generator buses;

3. lower and upper limits on voltage magnitudes, V_i, and voltage angles, δ_i, for each system bus, representing proper system operation;

4. real and reactive power demands P_i^L and Q_i^L for the nongenerator buses $i \in \mathbf{N} \setminus \mathbf{G}$;

5. lower and upper limits on real power generation, P_i^G, and reactive power generation, Q_i^G, for each generator bus which can vary from generator to generator;

6. cost functions $C_i(P_i^G)$ for each generator bus; and

7. parameters to calculate the bus admittance matrix, including line resistance, line reactance, and any bus or line shunt susceptance and conductance, where the admittance matrix essentially measures how easily current can flow in the lines.

The data associated with OPF problems are typically distributed in two common formats: IEEE Common Data Format [1903], and MATPOWER Case Format [1983].

15.3 ▪ Formulation

The classical formulation is:

$$\min \sum_{i \in G} C_i \left(P_i^G \right) \tag{15.1}$$

$$\text{s.t. } P_i^G - P_i^L = P_i(V, \delta) \qquad \forall i \in \mathbf{N}, \tag{15.2}$$

$$Q_i^G - Q_i^L = Q_i(V, \delta) \qquad \forall i \in \mathbf{N}, \tag{15.3}$$

$$P_i^{G,\min} \leq P_i^G \leq P_i^{G,\max} \qquad \forall i \in \mathbf{G}, \tag{15.4}$$

$$Q_i^{G,\min} \leq Q_i^G \leq Q_i^{G,\max} \qquad \forall i \in \mathbf{G}, \tag{15.5}$$

$$V_i^{\min} \leq V_i \leq V_i^{\max} \qquad \forall i \in \mathbf{N}, \tag{15.6}$$

$$\delta_i^{\min} \leq \delta_i \leq \delta_i^{\max} \qquad \forall i \in \mathbf{N}. \tag{15.7}$$

The cost of operating generator $i \in \mathbf{G}$ is expressed as a function, C_i, of its real output power, P_i^G. This cost function is typically convex quadratic or a piecewise-linear approximation [265,753]. Constraints (15.4) and (15.5) enforce bounds on real and reactive power generation at generator buses. These vary depending on the type of generator and the level of unit commitment (UC). Constraints (15.6) and (15.7) enforce bounds on voltage magnitude and voltage angle, ensuring that equipment operating limits are met.

The independent decision variables are the real and reactive power injections at all controllable generator buses, P_i^G and Q_i^G for all $i \in \mathbf{G}$. To avoid an underdetermined system with multiple solutions, the voltage magnitude and angle are fixed at one bus. This is done because it is possible to have multiple solutions that all have the same values for real and reactive generation and only differ in the voltages and phase angles, producing the same objective function value. This fixed bus is referred to as the slack or swing bus and its voltage magnitude is typically set to one and its angle to zero. The remaining voltage angles and magnitudes make up the state (dependent) variables, which are related to the independent variables through the power flow equations [1901].

The power flow equations are enforced in constraints (15.2) and (15.3), where $P_i(V, \delta)$ and $Q_i(V, \delta)$ can be expressed as

$$P_i(V, \delta) = V_i \sum_{k=1}^{N} V_k \left(G_{ik} \cos(\delta_i - \delta_k) + B_{ik} \sin(\delta_i - \delta_k) \right), \tag{15.8}$$

$$Q_i(V, \delta) = V_i \sum_{k=1}^{N} V_k \left(G_{ik} \sin(\delta_i - \delta_k) - B_{ik} \cos(\delta_i - \delta_k) \right) \tag{15.9}$$

or, alternatively,

$$P_i(V, \delta) = V_i \sum_{k=1}^{N} V_k Y_{ik} \cos(\delta_i - \delta_k - \theta_{ik}), \tag{15.10}$$

$$Q_i(V, \delta) = V_i \sum_{k=1}^{N} V_k Y_{ik} \sin(\delta_i - \delta_k - \theta_{ik}). \tag{15.11}$$

Equations (15.8) and (15.9) use the admittance matrix in *polar coordinates*, while (15.10) and (15.11) express it in *rectangular coordinates*, and both sets of equations use voltage in *polar coordinates*. Alternatively, a quadratic (more precisely, bilinear) formulation expressing both admittance and voltage in rectangular coordinates has been derived as [1781]

$$P_i(V,\delta) = \sum_{k=1}^{N} G_{ik}(E_i E_k + F_i F_k) + B_{ik}(F_i E_k - E_i F_k),　\qquad (15.12)$$

$$Q_i(V,\delta) = \sum_{k=1}^{N} G_{ik}(F_i E_k - E_i F_k) - B_{ik}(E_i E_k + F_i F_k).　\qquad (15.13)$$

The most significant advantage of (15.12)–(15.13) over (15.8)–(15.9) and (15.10)–(15.11) for optimization is the replacement of trigonometric terms via bilinear expressions. Such a formulation is preferred by current global optimization software. In addition, there are two numerical advantages: the Hessian matrix of a quadratic is constant and only needs to be evaluated once, and the Taylor series expansion of the quadratic function terminates at the second-order term without error [1780]. The first advantage simplifies the application of Newton's method to the KKT conditions, while the second makes higher-order interior-point methods more efficient. There are several disadvantages [753] as well stemming from the indirect expression of voltage magnitude because voltage must be expressed indirectly as $V_i = \sqrt{E_i^2 + F_i^2}$. Constraint (15.6) is no longer a simple variable limit and instead becomes a functional inequality constraint of the form

$$V_i^{\min} \le \sqrt{E_i^2 + F_i^2} \le V_i^{\max}　\quad \forall i \in \mathbf{N}.　\qquad (15.14)$$

If any voltage magnitudes are fixed, this formulation requires the addition of an equality constraint for each fixed magnitude rather than the removal of a variable. We note that all three representations are equivalent, i.e., (15.1)–(15.7) with (15.8)–(15.9) is equivalent to (15.1)–(15.7) with (15.10)–(15.11), which is equivalent to (15.1)–(15.5), (15.7), (15.14) with (15.12)–(15.13).

Regardless of which of the three formulations above is used, the feasible region, defined by (15.2)–(15.7), is a nonconvex set. Practitioners typically refer to the AC power flow equations as being "highly" nonconvex. Thus, (15.1)–(15.7) is a *global optimization* problem.

15.4 ▪ Variants and Extensions

OPF encompasses a broad range of power system optimization problems. A simple extension to the classical OPF formulation is to consider phase-shifting and tap-changing transformers [753]. This requires introducing a phase shift variable, ρ_{ik}, for each branch that contains such a transformer $ik \in \mathbf{H}$, and similarly introducing a tap ratio variable, T_{ik} for all $ik \in \mathbf{K}$. Since these transformers have an effect on the admittance matrix, the right-hand sides of power flow equations (15.2) and (15.3) become functions of ρ and T: $P_i(V,\delta,\rho,T)$ and $Q_i(V,\delta,\rho,T)$. Additionally, limits on phase angles and tap ratios must be enforced through the following constraints:

$$\rho_{ik}^{\min} \le \rho_{ik} \le \rho_{ik}^{\max},　\qquad (15.15)$$

$$T_{ik}^{\min} \le T_{ik} \le T_{ik}^{\max}.　\qquad (15.16)$$

Besides the addition of various power system equipment, there are several well-known extensions and variations on the classical OPF formulation, including *security-constrained economic dispatch* (SCED), *security-constrained unit commitment* (SCUC), *optimal reactive power flow* (ORPF), and *reactive power planning* (RPP) problems. The rest of this section focuses on these problems.

The SCED extension [61] seeks a power system operation that is robust to a set of predefined contingencies, where a contingency is defined as an event that removes one or more generators or transmission lines from the system. It is a restriction to the classical OPF formulation that ensures system operation during a contingency event by redistributing system power flows and voltages. In others words, when a contingency event occurs, taking out one or more generators or transmission lines and reducing system connectivity, the state variables readjust and keep the solution feasible without readjusting the independent decision variables. This requires that the slack bus power injections not be fixed.

SCED is more difficult to solve than the classical OPF formulation because each contingency event is essentially represented by a distinct system with its own distinct admittance matrix. This means that each contingency requires its own set of constraints, including the power flow equations, and its own set of state variables. The SCED is an essential tool for many transmission system operators, and a linearized formulation has been used to price electricity in several markets, including PJM, New England, and California [401].

The general ED problem differs from the UC problem in that it takes the set of operational generators as given, while UC optimally turns them on and off. Thus, UC has to consider multiple time periods because it takes time to start up and shut off power generators. When the power flow equations are included in the UC problem, it is called SCUC. In practical applications, UC and SCUC are used to plan on a horizon of weeks to several days ahead, while ED and SCED are used to plan one or two days or even hours ahead.

The SCUC problem [767] is a large-scale, mixed-integer nonlinear optimization (MINLO) problem with many binary variables. Additional constraints can include ramp rate limits, spinning and operating reserve requirements, fuel constraints, and emission limits, among others. It is NP-hard [868] and very difficult to solve, often leading practitioners to linearize it or otherwise relax the formulation [753].

Some OPF variants focus on reactive power flow. The ORPF problem essentially considers the classical OPF problem (or OPF extension such as SCED or SCUC), where real power outputs are given and reactive power outputs are optimally adjusted to minimize total power losses in the system. Only the slack bus has a real power generation that is not fixed, and minimizing this value minimizes the total system power loss [753]. With large power systems, OPF and its extensions are often linearized in a way that ignores reactive power. To get a full solution, the solution to the linearized OPF problem can be used to fix real power injections in the ORPF and then solve for the reactive power by minimizing system losses, thereby reducing the variable space [506]. ORPF can also be used to respond to changes in electricity load without changing real power outputs.

The RPP problem is an extension to the ORPF problem that considers a set of possible new reactive power sources, such as capacitor banks and static VAR compensators [1961].

15.5 ▪ Solution Methods

Various deterministic optimization methods have been used to solve OPF problems, including gradient methods, Newton's and quasi-Newton's methods, sequential linear optimization, sequential quadratic optimization, interior-point methods, and Benders decomposition, among others [755]. None of these methods can guarantee global optimality for a nonconvex problem such as the OPF problem because the KKT conditions are not sufficient. Various nondeterministic metaheuristics have been applied to OPF as well. Some of these, including chaos optimization, genetic algorithms, particle swarm optimization, simulated annealing, and tabu search, are known to asymptotically converge to the globally optimal solution [756]. In practice, however, the required computation time for global convergence may not be reasonable.

The rest of this section will present several solution methods using a small example.

15.5.1 ▪ Example Data and Formulation

Our test system comes from a recent paper [354] that compares local OPF solutions in multiple test cases. These test cases can also be found at [353] in MATPOWER format. The selected case is a nine-bus system with three generators and nine transmission lines.

In Figure 15.1, buses 1–3 are generators with limits on real power generation, P_i^G, and reactive power generation, Q_i^G; buses 5, 7, and 9 have nonzero real power demands P_i^L and reactive power demands Q_i^L.

The cost of running each of the three respective generators is expressed as a polynomial function of its real power output. These functions are added together to formulate the objective function (15.1) as follows:

$$\min \quad 0.11\left(P_1^G\right)^2 + 5P_1^G + 150 \tag{15.17}$$
$$+\, 0.085\left(P_2^G\right)^2 + 1.2P_2^G + 600$$
$$+\, 0.1225\left(P_3^G\right)^2 + P_3^G + 335.$$

To express the power flow constraints (15.2) and (15.3), it is necessary to calculate the bus admittance matrix, which represents nodal admittance (a measure of how easily current can flow) between buses in the system. The admittance for each branch ik can be expressed in terms of its real part, called the conductance, G_{ik}, and its imaginary part, called the susceptance, B_{ik}, both of which appear in the power flow constraints. First, the resistance R_{ik} and reactance X_{ik} provided in the data set are used to calculate series conductance g_{ik} and series susceptance b_{ik} for each branch using the following equations:

$$g_{ik} = \frac{R_{ik}}{R_{ik}^2 + X_{ik}^2} \quad \forall i, k,$$
$$b_{ik} = \frac{-X_{ik}}{R_{ik}^2 + X_{ik}^2} \quad \forall i, k.$$

Next, the conductance $G_{i,k}$ and susceptance $B_{i,k}$ are calculated using $g_{i,k}$ and $b_{i,k}$ from above, along with the branch shunt susceptance g_{ik}^{sh} (also called the line-charging susceptance) given in the data set. Ignoring phase-shifting transformers, tap-changing transformers, bus shunt susceptance, and shunt conductance for both branches and buses (since these are not present in the nine-bus test system), the equations are

$$G_{ii} = \sum_{k:ik\in N}\left(g_{ik} + \frac{g_{ik}^{sh}}{2}\right) + \sum_{k:(ki)\in N}\left(g_{ki} + \frac{g_{ki}^{sh}}{2}\right),$$
$$G_{ik} = -\sum_{k:ik\in N} g_{ik} - \sum_{k:(ki)\in N} g_{ki}, \qquad\qquad i \neq k,$$

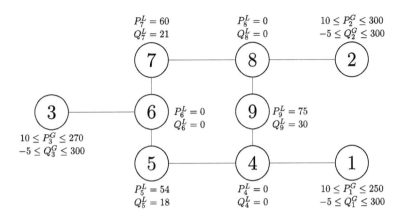

Figure 15.1. *The nine-bus test system.*

$$B_{ii} = \sum_{k:ik\in N} b_{ik} + \sum_{k:(ki)\in N} b_{ki},$$

$$B_{ik} = -\sum_{k:ik\in N} b_{ik} - \sum_{k:(ki)\in N} b_{ki}, \qquad\qquad i \neq k.$$

The power flow constraints are expressed here in polar coordinates following (15.8) and (15.9). As mentioned in Section 15.3, an equivalent problem can be formulated following (15.10)–(15.11) or (15.12)–(15.13). The real power flow constraints (15.2) are

$$P_1^G = 1736.11 V_1 V_4 \sin(\delta_1 - \delta_4), \tag{15.18}$$

$$P_2^G = 1600 V_2 V_8 \sin(\delta_2 - \delta_8), \tag{15.19}$$

$$P_3^G = 1706.48 V_3 V_6 \sin(\delta_3 - \delta_6), \tag{15.20}$$

$$0 = 330.74(V_4)^2 \tag{15.21}$$
$$+ 1736.11 V_4 V_1 \sin(\delta_4 - \delta_1)$$
$$- 194.22 V_4 V_5 \cos(\delta_4 - \delta_5) + 1051.07 V_4 V_5 \sin(\delta_4 - \delta_5)$$
$$- 136.52 V_4 V_9 \cos(\delta_4 - \delta_9) + 1160.41 V_4 V_9 \sin(\delta_4 - \delta_9),$$

$$-54 = 322.42(V_5)^2 \tag{15.22}$$
$$- 194.22 V_5 V_4 \cos(\delta_5 - \delta_4) + 1051.07 V_5 V_4 \sin(\delta_5 - \delta_4)$$
$$- 128.20 V_5 V_6 \cos(\delta_5 - \delta_6) + 558.82 V_5 V_6 \sin(\delta_5 - \delta_6),$$

$$0 = 243.71(V_6)^2 \tag{15.23}$$
$$+ 1706.48 V_6 V_3 \sin(\delta_6 - \delta_3)$$
$$- 128.20 V_6 V_5 \cos(\delta_6 - \delta_5) + 558.82 V_6 V_5 \sin(\delta_6 - \delta_5)$$
$$- 116.71 V_6 V_7 \cos(\delta_6 - \delta_7) + 978.43 V_6 V_7 \sin(\delta_6 - \delta_7),$$

$$-60 = 277.22(V_7)^2 \tag{15.24}$$
$$- 115.51 V_7 V_6 \cos(\delta_7 - \delta_6) + 978.43 V_7 V_6 \sin(\delta_7 - \delta_6)$$
$$- 161.71 V_7 V_8 \cos(\delta_7 - \delta_8) + 1369.80 V_7 V_8 \sin(\delta_7 - \delta_8),$$

$$0 = 280.47(V_8)^2 \tag{15.25}$$
$$+ 1600 V_8 V_2 \sin(\delta_8 - \delta_2)$$

$$- 161.71 V_8 V_7 \cos(\delta_8 - \delta_7) + 1369.80 V_8 V_7 \sin(\delta_8 - \delta_7)$$
$$- 118.76 V_8 V_9 \cos(\delta_8 - \delta_9) + 597.51 V_8 V_9 \sin(\delta_8 - \delta_9),$$
$$-75 = 255.28(V_9)^2 \tag{15.26}$$
$$- 136.52 V_9 V_4 \cos(\delta_9 - \delta_4) + 1160.41 V_9 V_4 \sin(\delta_9 - \delta_4)$$
$$- 118.76 V_9 V_8 \cos(\delta_9 - \delta_8) + 597.51 V_9 V_8 \sin(\delta_9 - \delta_8).$$

The reactive power flow constraints (15.3) are

$$Q_1^G = 1736.11(V_1)^2 - 1736.11 V_1 V_4 \cos(\delta_1 - \delta_4), \tag{15.27}$$
$$Q_2^G = 1600(V_2)^2 - 1600 V_2 V_8 \cos(\delta_2 - \delta_8), \tag{15.28}$$
$$Q_3^G = 1706.48(V_3)^2 - 1706.48 V_3 V_6 \cos(\delta_3 - \delta_6), \tag{15.29}$$
$$0 = 3930.89(V_4)^2 \tag{15.30}$$
$$- 1736.11 V_4 V_1 \cos(\delta_4 - \delta_1)$$
$$- 194.22 V_4 V_5 \sin(\delta_4 - \delta_5) - 1051.07 V_4 V_5 \cos(\delta_4 - \delta_5)$$
$$- 136.52 V_4 V_9 \sin(\delta_4 - \delta_9) - 1160.41 V_4 V_9 \cos(\delta_4 - \delta_9),$$
$$-18 = 1584.09(V_5)^2 \tag{15.31}$$
$$- 194.22 V_5 V_4 \sin(\delta_5 - \delta_4) - 1051.07 V_5 V_4 \cos(\delta_5 - \delta_4)$$
$$- 128.20 V_5 V_6 \sin(\delta_5 - \delta_6) - 558.82 V_5 V_6 \cos(\delta_5 - \delta_6),$$

$$0 = 3215.39(V_6)^2 \tag{15.32}$$
$$- 1706.49 V_6 V_3 \cos(\delta_6 - \delta_3)$$
$$- 128.20 V_6 V_5 \sin(\delta_6 - \delta_5) - 558.82 V_6 V_5 \cos(\delta_6 - \delta_5)$$
$$- 115.51 V_6 V_7 \sin(\delta_6 - \delta_7) - 978.43 V_6 V_7 \cos(\delta_6 - \delta_7),$$
$$-21 = 2330.32(V_7)^2 \tag{15.33}$$
$$- 115.51 V_7 V_6 \sin(\delta_7 - \delta_6) - 978.43 V_7 V_6 \cos(\delta_7 - \delta_6)$$
$$- 161.71 V_7 V_8 \sin(\delta_7 - \delta_8) - 1369.80 V_7 V_8 \cos(\delta_7 - \delta_8),$$
$$0 = 344.56(V_8)^2 \tag{15.34}$$
$$- 16.00 V_8 V_2 \cos(\delta_8 - \delta_2)$$
$$- 161.71 V_8 V_7 \sin(\delta_8 - \delta_7) - 1369.80 V_8 V_7 \cos(\delta_8 - \delta_7)$$
$$- 118.76 V_8 V_9 \sin(\delta_8 - \delta_9) - 597.51 V_8 V_9 \cos(\delta_8 - \delta_9),$$
$$-30 = 1733.82(V_9)^2 \tag{15.35}$$
$$- 136.52 V_9 V_4 \sin(\delta_9 - \delta_4) - 1160.41 V_9 V_4 \cos(\delta_9 - \delta_4)$$
$$- 118.76 V_9 V_8 \sin(\delta_9 - \delta_8) - 597.51 V_9 V_8 \cos(\delta_9 - \delta_8).$$

The left-hand sides of the real and reactive power flow equations only contain a variable for the three generator buses; the other six buses have a zero or negative constant representing demand. It is important to note that while the coefficients are approximated to two decimal places here, they actually have many more decimal places—this rounding can have an impact on numerical results. The next set of constraints, (15.4) and (15.5), enforce limits on real and reactive power output for the generator buses:

$$10 \le P_1^G \le 250, \tag{15.36}$$

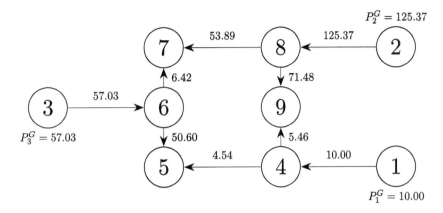

Figure 15.2. *Global solution to the nine-bus test system: real power flows.*

$$10 \le P_2^G \le 300, \tag{15.37}$$
$$10 \le P_3^G \le 270, \tag{15.38}$$
$$-5 \le Q_2^G \le 300, \tag{15.39}$$
$$-5 \le Q_1^G \le 300, \tag{15.40}$$
$$-5 \le Q_3^G \le 300. \tag{15.41}$$

The last two sets of constraints, (15.6) and (15.7), enforce bounds on bus voltage magnitudes and voltage angles:

$$0.9 \le V_i \le 1.1 \quad \forall i \in \{1, \ldots, 9\}, \tag{15.42}$$
$$0° \le \delta_i \le 360° \quad \forall i \in \{1, \ldots, 9\}. \tag{15.43}$$

The full problem is to minimize objective function (15.17) subject to constraints (15.18)–(15.43). After solving the problem, it is possible to compute the real and reactive power flows P_{ik} and Q_{ik} from bus i to bus k using the following equations:

$$P_{ik} = g_{ik}(V_i)^2 - g_{ik} V_i V_k \cos(\delta_i - \delta_k) - b_{ik} V_i V_k \sin(\delta_i - \delta_k), \tag{15.44}$$
$$Q_{ik} = -b_{ik}(V_i)^2 - g_{ik} V_i V_k \sin(\delta_i - \delta_k) + b_{ik} V_i V_k \cos(\delta_i - \delta_k). \tag{15.45}$$

Note that P_{ik} should be approximately equal to $-P_{ki}$, and the transmission line power loss in branch ik is equal to $P_{ik} + P_{ki}$.

Solving this program to global optimality returns an objective value of 3087.84, but there are also three local optima with objective values of 3398.03, 4246.48, and 4265.15, respectively [354, 1764]. The real power generation and flows from the globally optimal solution are illustrated in Figure 15.2. Note that the flows are expressed in terms of the origin bus as opposed to the destination bus (i.e., the flows include line losses). The costs of running the three generators are 211.00, 2086.46, and 790.40, respectively.

To compare several solvers, the nine-bus example was solved in polar form as formulated above and in rectangular form (15.12)–(15.13) using GAMS. Two versions of GAMS are used in conjunction with the global solvers to illustrate recent advances in the field. The computational results of GAMS version 24.2, released May 2014, are displayed in Table 15.2, and the results of GAMS version 24.0, released December 2012, are displayed in Table 15.3. The time limit was set to 10 minutes, error to 1%, and seed

Table 15.2. *Solving the nine-bus example with GAMS 24.2, released May 2014 (2.13 GHz, 2.93 GB RAM).*

Solver	Form	Solution time	Objective	Bounds
CONOPT [632]	Polar	< 1 sec	3087.84	
	Rectangular	< 1 sec	locally infeasible	
IPOPT [1858]	Polar	< 1 sec	4246.48	
	Rectangular	< 1 sec	locally infeasible	
PATHNLP [603]	Polar	< 1 sec	3087.84	
	Rectangular	< 1 sec	locally infeasible	
SNOPT [801]	Polar	< 1 sec	4246.48	
	Rectangular	< 1 sec	locally infeasible	
BARON* [336]	Polar	-	-	-
	Rectangular	10 min	3087.84	[2761.06, 3087.84]
COUENNE* [193]	Polar	10 min	3087.84	[1776.51, 3087.84]
	Rectangular	10 min	3087.84	[2667.81, 3087.84]
GLOMIQO* [1331]	Polar	-	-	-
	Rectangular	3 sec	3087.84	[3087.84, 3087.84]
LINDOGLOBAL* [1200]	Polar	10 min	3087.86	[1248.00, 3087.86]
	Rectangular	10 min	3087.84	[2203.55, 3087.84]
SCIP* [11]	Polar	-	-	-
	Rectangular	10 min	3087.84	[1188.75, 3087.84]

Table 15.3. *Solving the nine-bus example with GAMS 24.0, released December 2012.*

Solver	Form	Solution time	Objective	Bounds
BARON* [1756]	Polar	-	-	-
	Rectangular	10 min	3087.84	[2757.47, 3087.84]
COUENNE* [193]	Polar	10 min	3087.84	[1789.11, 3087.84]
	Rectangular	10 min	4264.85	[2725.07, 4264.85]
GLOMIQO* [1331]	Polar	-	-	-
	Rectangular	10 min	3087.84	[2783.52, 3087.84]
LINDOGLOBAL* [1200]	Polar	10 min	3100.64	[1225.46, 3100.64]
	Rectangular	10 min	-	[1188.75, ∞]
SCIP* [11]	Polar	-	-	-
	Rectangular	10 min	3087.84	[1188.75, 3087.84]

to 3141. Global solvers are marked with (*) in Table 15.2. We observe that all local solvers are challenged by the rectangular formulation but solve the polar formulation very quickly. In contrast, the global solvers tend to do better with the rectangular formulation. Notice that none of the solvers besides the relatively new GLOMIQO could prove optimality and that the computed lower bounds are still very weak after 10 minutes of computational time. Also notice that the bounds are generally weaker with the older GAMS version. While GLOMIQO solved this nine-bus system to optimality with GAMS version 24.2, a more realistic power system has hundreds or thousands of buses and remains challenging. The slow computation is the main reason that a local solver is likely to be used in practice. In our example, the local solvers find a solution almost immediately. CONOPT and PATHNLP actually find the global optimum (without proving that it is the global optimum), while IPOPT and SNOPT only compute a locally optimal solution. With such large optimality gaps after a few minutes of computation, most global solvers do not provide a practical advantage for the problem the way it is set up here.

15.5.2 ▪ Linear Approximation with Direct Current Power Flow

The inherent nonconvexity of the power flow equations implies that convergence to a globally optimal solution is not guaranteed. Furthermore, the large size of modern power systems means that finding a local optimum can also be too computationally expensive in practice. Because of this, OPF problems are often simplified and linearized so that linear optimization (LO) techniques can be used to solve them.

The most frequently used method for simplifying the nonlinear OPF is called the direct current (DC) power flow linearization, named so because the resulting equations resemble DC power flow. This method is frequently used in commercial and industry applications [755] because the methods are relatively simple and fast, solutions are noniterative and unique, contingency analysis is sped up dramatically, and the required data are minimal and relatively easy to obtain, among other reasons [1431, 1726].

The DC OPF problem relies on several simplifying assumptions [753, 1431, 1979]: (1) all bus voltage magnitudes are equal to one; (2) the difference in voltage angles between adjacent buses is small enough that $\sin(\delta_i - \delta_k) \approx \delta_i - \delta_k$ and $\cos(\delta_i - \delta_k) \approx 1$; (3) branch resistances are zero, meaning that there is no power loss in transmission; and (4) reactive power is ignored entirely. These assumptions reduce the constraints to a set of linear equations. In the classical OPF problem, power flow equation (15.3) is removed and the right-hand side of (15.2) is reduced to [755]

$$P_i(\delta) = \sum_{k=1}^{N} B_{ik}(\delta_i - \delta_k). \tag{15.46}$$

In our nine-bus example, reactive power flow constraints (15.27)–(15.35) are removed from the formulation and constraints (15.18)–(15.26) are reduced to the following:

$$P_1^G = 1736.11(\delta_1 - \delta_4), \tag{15.47}$$

$$P_2^G = 1600(\delta_2 - \delta_8), \tag{15.48}$$

$$P_3^G = 1706.48(\delta_3 - \delta_6), \tag{15.49}$$

$$0 = 1736.11(\delta_4 - \delta_1) + 1051.07(\delta_4 - \delta_5) + 1160.41(\delta_4 - \delta_9), \tag{15.50}$$

$$-54 = 1051.07(\delta_5 - \delta_4) + 558.82(\delta_5 - \delta_6), \tag{15.51}$$

$$0 = 1706.48(\delta_6 - \delta_3) + 558.82(\delta_6 - \delta_5) + 978.43(\delta_6 - \delta_7), \tag{15.52}$$

$$-60 = 978.43(\delta_7 - \delta_6) + 1369.80(\delta_7 - \delta_8), \tag{15.53}$$

$$0 = 1600(\delta_8 - \delta_2) + 1369.80(\delta_8 - \delta_7) + 597.51(\delta_8 - \delta_9), \tag{15.54}$$

$$-75 = 1160.41(\delta_9 - \delta_4) + 597.51(\delta_9 - \delta_8). \tag{15.55}$$

Solving the DC OPF problem with the nine-bus test system (while keeping the original quadratic objective function) returns an objective function value of 2733.55, which is less than the globally optimal solution to the AC OPF and therefore represents an infeasible solution to the AC OPF. However, in this particular case the infeasibility mostly comes from the lower bounds on reactive power flow (15.39)–(15.41). By relaxing these lower bounds from −5 to −40 in the original AC formulation and allowing real power flow to deviate by at most ±1 from its optimal DC value, it is possible to obtain a feasible solution to the AC OPF problem with an objective function value of 2753.04. The same solution is obtained without fixing the power flow injections. This means that if this particular AC OPF problem had relaxed lower limits on reactive power flow, the DC formulation would return a solution in the neighborhood of the global optimum.

While the DC OPF problem can be solved quickly, there is no guarantee that the solution will be optimal, or even feasible, for the original AC formulation in general. Our nine-bus test system happens to be such an example, but there are other cases of small systems where the two solutions are exactly the same, and others where the DC solution is completely wrong (e.g., the DC formulation returns a unique solution while the AC formulation is infeasible). With realistically large systems, it is difficult to quantify the error introduced through the DC formulation a priori [1431].

15.5.3 ▪ Convex Outer Approximation via McCormick Envelopes

An LO formulation based on McCormick envelopes [1304] has been proposed for the OPF problem [572]. This method is a classical convex relaxation that takes an outer approximation (OA) of the feasible region and thus provides a lower bound on the optimal objective function value of the original problem.

For example, the real power flow constraint (15.18) for bus 1 can be expressed in rectangular coordinates (following (15.12)) as

$$P_1^G = 1736.11 (F_1 E_4 - E_1 F_4). \tag{15.56}$$

The two bilinear terms in the equation above are then reformulated using auxiliary variables $W_{1,4}^{fe}$ and $W_{1,4}^{ef}$ as follows:

$$P_1^G = 1736.11 \left(W_{1,4}^{fe} - W_{1,4}^{ef} \right), \tag{15.57}$$

$$W_{1,4}^{fe} \geq u_4^E F_1 + u_1^F E_4 - u_1^F u_4^E, \tag{15.58}$$

$$W_{1,4}^{fe} \leq l_4^E F_1 + u_1^F E_4 - l_1^F u_4^E, \tag{15.59}$$

$$W_{1,4}^{fe} \leq u_4^E F_1 + l_1^F E_4 - u_1^F l_4^E, \tag{15.60}$$

$$W_{1,4}^{fe} \geq l_4^E F_1 + l_1^F E_4 - l_1^F l_4^E, \tag{15.61}$$

$$W_{1,4}^{ef} \geq u_1^E F_4 + u_4^F E_1 - u_4^F u_1^E, \tag{15.62}$$

$$W_{1,4}^{ef} \leq l_1^E F_4 + u_4^F E_1 - l_4^F u_1^E, \tag{15.63}$$

$$W_{1,4}^{ef} \leq u_1^E F_4 + l_4^F E_1 - u_4^F l_1^E, \tag{15.64}$$

$$W_{1,4}^{ef} \geq l_1^E F_4 + l_4^F E_1 - l_4^F l_1^E, \tag{15.65}$$

where $\left(l_i^E, u_i^E \right)$ and $\left(l_i^F, u_i^F \right)$ are (lower, upper) bounds on E_i and F_i, respectively. In the nine-bus test system these are both effectively equal to $[-1.1, 1.1]$.

The approximation accuracy is dictated by the tightness of the bounds on the continuous variables that form the bilinear terms. Unfortunately, the bounds in the nine-bus test system are not very tight and in general power system equipment has a wide range of operation. Solving the nine-bus test system with McCormick envelopes returns an optimal objective function value 1188.75, which is not feasible in the original OPF problem but provides a weak lower bound. In fact, this objective function value reflects the cost of enforcing lower bounds on real power generation (15.36)–(15.38) without actually meeting demand.

Such a weak lower bound resulting from the McCormick relaxations is typical for OPF problems [793]. Notice that the lower bound computed for the rectangular formulation by SCIP is the same as that obtained by the McCormick relaxation.

15.5.4 ▪ Convex OA via Piecewise-Linear Envelopes

While the classical McCormick envelopes [1304] have low accuracy for this problem due to weak bounds, it is possible to extend this idea to an arbitrary degree of accuracy by expressing the envelopes as piecewise functions. The idea is to divide the domain between the lower and upper bounds into multiple sections and to apply the envelope relaxation for each section. By increasing the number of breakpoints in the envelopes, it is possible to approximate the original function to an arbitrary degree of accuracy. However, this comes at the cost of computational complexity because binary decision variables and other auxiliary variables enter the formulation and the number of constraints increases by a multiple.

There are many different ways to formulate piecewise-linear envelopes around bilinear terms, and these formulations can exhibit very different numerical behavior [835, 1887]. In this section, two such mixed-integer linear optimization (MILO) formulations are used to approximate the nine-bus system.

The first formulation comes from [835] (formulation nf4l). The piecewise-linear approximation is defined by the following. Let $W_{i,k} := F_i E_k$, and define $m \in \mathbf{M}$ as the set of breakpoints in the approximation. Then replace the bilinear terms with their respective auxiliary variables and add constraints:

$$\sum_m \lambda^F_{i,m} = 1 \qquad \forall i, \qquad (15.66)$$

$$F_i = \sum_m \left(f^L_{i,m} \lambda^F_{i,m} + \delta^F_{i,m} \right) \qquad \forall i, \qquad (15.67)$$

$$0 \le \delta^F_{i,m} \le \left(f^U_{i,m} - f^L_{i,m} \right) \lambda^F_{i,m} \qquad \forall m, i, \qquad (15.68)$$

$$E_k = l^E_k + \sum_m \gamma^{FE}_{i,k,m} \qquad \forall i, k, \qquad (15.69)$$

$$0 \le \gamma^{FE}_{i,k,m} \le \left(u^E_k - l^E_k \right) \lambda^F_{i,m} \qquad \forall m, i, k, \qquad (15.70)$$

$$W_{i,k} = l^E_k + \sum_m \left(f^L_{i,m} \gamma^{FE}_{i,k,m} \right) + \delta^W_{i,k} \qquad \forall i, k, \qquad (15.71)$$

$$\delta^W_{i,k} \ge \sum_m \left(u^E_k - l^E_k \right) \gamma^{FE}_{i,k,m}, \qquad (15.72)$$

$$+ \left(u^E_k - l^E_k \right) \sum_m \left(\delta^F_{i,m} - \left(u^E_k - l^E_k \right) \lambda^F_{i,m} \right) \qquad \forall i, k,$$

$$\delta^W_{i,k} \le \left(u^E_k - l^E_k \right) \sum_m \delta^F_{i,m} \qquad \forall i, k, \qquad (15.73)$$

$$\delta^W_{i,k} \le \sum_m \left(u^E_k - l^E_k \right) \gamma^{FE}_{i,k,m} \qquad \forall i, k, \qquad (15.74)$$

$$\lambda^F_{i,m} \text{ binary} \qquad \forall i, m. \qquad (15.75)$$

We have introduced constants $f_{i,m}^L, f_{i,m}^U$, which are the bounds on F_i when it is in partition m. These are equal to the following:

$$f_{i,m}^L = l_k^F + \frac{\left(u_i^F - l_i^F\right)(m-1)}{M}, \tag{15.76}$$

$$f_{i,m}^U = l_k^F + \frac{\left(u_i^F - l_i^F\right)m}{M}. \tag{15.77}$$

We have also introduced variables $\lambda_{i,m}^F$, $\delta_{i,m}^F$, $\gamma_{i,k,m}^{FE}$, and $\delta_{i,k}^W$. Note that $\lambda_{i,m}^F = 1$ if F_i is in partition m, meaning that $f_{i,m}^L \le F_i \le f_{i,m}^U$, and zero otherwise. This is enforced by constraints (15.66)–(15.67) and (15.75). The variable $\delta_{i,m}^F$ is used to represent the deviation of F_i from its lower bound in the partition, $f_{i,m}^L$, if F_i is in partition m, and is equal to zero otherwise. This is enforced by constraints (15.67)–(15.68). The variable $\gamma_{i,k,m}^{FE}$ is similarly used to represent the deviation of E_k from its absolute lower bound, l_k^E, if F_i is in domain m, and is equal to zero otherwise. This is enforced by constraints (15.69)–(15.70). The variable $\delta_{i,k}^W$ is essentially used to model the aggregate of the bilinear term $\delta_{i,m}^F \delta_{k,m}^E$ and is defined with constraints (15.71)–(15.74).

Note that to reformulate term $E_i E_k$, it is necessary to add variables $\lambda_{i,m}^E$, $\delta_{k,m}^E$, and $\gamma_{i,k,m}^{EE}$, with constants $e_{i,m}^L$ and $e_{i,m}^U$; to reformulate $F_i F_k$, it is necessary to add variable $\gamma_{i,k,m}^{FF}$, similar to the above. Each bilinear term will also require its own auxiliary variables, $W_{i,k}$ and $\delta_{i,k}^W$, to represent it. To represent the OPF problem as a MILO problem, the bilinear terms in the objective function must also be reformulated.

The nine-bus example has been solved using formulation (15.66–15.75) with bilinear reformulations using several values of M, and the numerical results are reported in Table 15.4. The same M was used for all bilinear terms; the optimality gap was set to 1% and the time limit to 10 hours. We observe that while the bounds become more accurate with more breakpoints, it takes a lot of breakpoints to begin to approach an optimal solution to the original problem. Although the problem is now linear, this computational advantage is offset by the addition of new binary variables and becomes more computationally expensive than the original.

Table 15.4. *Solving the nine-bus example with piecewise-linear envelopes with CPLEX, GAMS 24.1.3 (3.5 GHz, 32 GB RAM).*

M	Binary variables	Continuous variables	Constraints	Solution time	Objective	Bounds
10	210	3,391	3,970	00:05:51	1247.68	[1234.68, 1247.68]
20	420	6,241	6,610	00:46:42	1390.97	[1377.07, 1390.97]
30	630	9,091	9,250	05:03:23	1431.24	[1416.93, 1431.24]
40	840	11,941	11,890	10:00:00	1436.21	[1455.08, 1436.21]
50	1,050	14,791	14,530	08:24:15	1470.68	[1455.98, 1470.68]
60	1,260	17,641	17,170	10:00:00	1476.13	[1212.60, 1476.13]
70	1,470	20,491	19,810	10:00:00	1482.99	[1188.75, 1482.99]

An alternative formulation that uses breakpoints for both variables in the bilinear term is defined as follows. Let $W_{i,k} := F_i E_k$, and define $m \in \mathbf{M}$ as the set of breakpoints in the approximation for variables F and $n \in \mathbf{N}$ the set of breakpoints in the approximation for E. Then replace the bilinear terms with their respective auxiliary variables and add constraints:

$$\sum_m \lambda_{i,m}^F = 1 \qquad \forall i, \qquad (15.78)$$

$$\sum_n \lambda_{k,n}^E = 1 \qquad \forall k, \qquad (15.79)$$

$$\sum_m f_{i,m}^L \lambda_{i,m}^F \leq F_i \leq \sum_m f_{i,m}^U \lambda_{i,m}^F \qquad \forall i, \qquad (15.80)$$

$$\sum_n f_{k,n}^L \lambda_{k,n}^E \leq E_k \leq \sum_n f_{k,n}^U \lambda_{k,n}^E \qquad \forall k, \qquad (15.81)$$

$$W_{i,k} \geq \left(\sum_n f_{k,n}^U \lambda_{k,n}^E \right) F_i \qquad (15.82)$$
$$+ \left(\sum_m f_{i,m}^U \lambda_{i,m}^F \right) E_k - \left(\sum_n f_{k,n}^U \lambda_{k,n}^E \right) \left(\sum_m f_{i,m}^U \lambda_{i,m}^F \right) \qquad \forall i,k,$$

$$W_{i,k} \leq \left(\sum_n f_{k,n}^L \lambda_{k,n}^E \right) F_i \qquad (15.83)$$
$$+ \left(\sum_m f_{i,m}^U \lambda_{i,m}^F \right) E_k - \left(\sum_n f_{k,n}^L \lambda_{k,n}^E \right) \left(\sum_m f_{i,m}^U \lambda_{i,m}^F \right) \qquad \forall i,k,$$

$$W_{i,k} \leq \left(\sum_n f_{k,n}^U \lambda_{k,n}^E \right) F_i \qquad (15.84)$$
$$+ \left(\sum_m f_{i,m}^L \lambda_{i,m}^F \right) E_k - \left(\sum_n f_{k,n}^U \lambda_{k,n}^E \right) \left(\sum_m f_{i,m}^L \lambda_{i,m}^F \right) \qquad \forall i,k,$$

$$W_{i,k} \geq \left(\sum_n f_{k,n}^L \lambda_{k,n}^E \right) F_i \qquad (15.85)$$
$$+ \left(\sum_m f_{i,m}^L \lambda_{i,m}^F \right) E_k - \left(\sum_n f_{k,n}^L \lambda_{k,n}^E \right) \left(\sum_m f_{i,m}^L \lambda_{i,m}^F \right) \qquad \forall i,k,$$

$$\lambda_{i,m}^F \quad \text{binary} \qquad \forall i,m, \qquad (15.86)$$

$$\lambda_{k,n}^E \quad \text{binary} \qquad \forall k,n. \qquad (15.87)$$

In this formulation, $\lambda_{i,m}^F = 1$ if and only if F_i is in partition m, meaning that $f_{i,m}^L \leq F_i \leq f_{i,m}^U$. This relationship and a similar one for E_k are enforced by constraints (15.78–15.81) and (15.86–15.87). Constraints (15.82–15.85) correspond to the piecewise McCormick envelopes.

To solve the piecewise McCormick formulation described above with a linear solver, we use the standard (exact and linear) reformulation for the product of binary and continuous variables ($\lambda_{k,n}^E F_i$ and $\lambda_{i,m}^F E_k$) as well as the product of binary variables ($\lambda_{k,n}^E \lambda_{i,m}^F$). For example, $\lambda_{k,n}^E F_i$ is replaced with an equivalent auxiliary variable $W_{k,n,i}^{\lambda,E}$, and the following constraints are added:

$$W_{k,n,i}^{\lambda,E} \leq u_i^F \lambda_{k,n}^E, \qquad (15.88)$$

$$W_{k,n,i}^{\lambda,E} \geq F_i - u_i^F (1 - \lambda_{k,n}^E), \qquad (15.89)$$

$$W_{k,n,i}^{\lambda,E} \leq F_i. \qquad (15.90)$$

For a product of binary variables, $\lambda_{k,n}^E \lambda_{i,m}^F$ is replaced with the equivalent auxiliary variable $W_{i,k,m,n}^\lambda$, and the following constraints are added:

$$W_{i,k,m,n}^\lambda \leq \lambda_{k,n}^E, \tag{15.91}$$

$$W_{i,k,m,n}^\lambda \leq \lambda_{i,m}^F, \tag{15.92}$$

$$W_{i,k,m,n}^\lambda \geq \lambda_{k,n}^E + \lambda_{i,m}^F - 1. \tag{15.93}$$

The OPF example problem of this chapter was reformulated as a MILO problem as described above and solved using the CPLEX solver. The optimality gap was set to 1% and the time limit to 10 hours. Numerical results are displayed in Table 15.5. We observe that this formulation requires fewer breakpoints to achieve accuracy similar to the results of Table 15.4, but the computation time required to reach an optimal solution to the original problem is still prohibitive. The second formulation requires fewer binary and continuous variables but more constraints.

Table 15.5. *Solving the nine-bus example with piecewise-linear envelopes with CPLEX, GAMS 24.1.3 (3.5 GHz, 32 GB RAM).*

M	Binary variables	Continuous variables	Constraints	Solution time	Objective	Bounds
2	42	1,909	5,961	00:00:22	1188.75	[1188.75, 1188.75]
3	63	3,451	10,587	00:01:13	1195.20	[1188.75, 1195.20]
4	84	5,479	16,671	04:18:36	1335.59	[1323.15, 1335.59]
5	105	7,993	24,213	10:00:00	1231.32	[1188.75, 1231.32]
6	126	10,993	33,213	09:31:39	1682.16	[1665.75, 1682.16]

15.5.5 • Lagrangian Relaxation

A Lagrangian relaxation can be applied to the OPF problem. In the absence of strong duality, the solution of this relaxation provides a lower bound to the optimal objective function of the original problem. With the rectangular formulation (15.12)–(15.13), the Lagrangian can be formed by dualizing constraints (15.2)–(15.3) while keeping the bounding constraints. A Lagrangian can be defined by the following:

$$\min \sum_{i \in G} C_i \left(P_i^G \right) + \alpha^P \sum_{i \in N} e_i^P + \alpha^Q \sum_{i \in N} e_i^Q \tag{15.94}$$

$$\text{s.t. } e_i^P \geq P_i^G - P_i^L - G_{ik}(E_i E_k + F_i F_k) + B_{ik}(F_i E_k - E_i F_k) \qquad \forall i, \tag{15.95}$$

$$e_i^P \leq -P_i^G + P_i^L + G_{ik}(E_i E_k + F_i F_k) - B_{ik}(F_i E_k - E_i F_k) \qquad \forall i, \tag{15.96}$$

$$e_i^Q \geq Q_i^G - Q_i^L - G_{ik}(F_i E_k - E_i F_k) - B_{ik}(E_i E_k + F_i F_k) \qquad \forall i, \tag{15.97}$$

$$e_i^Q \leq -Q_i^G + Q_i^L + G_{ik}(F_i E_k - E_i F_k) + B_{ik}(E_i E_k + F_i F_k) \qquad \forall i, \tag{15.98}$$

$$(15.4\text{--}15.5), (15.14). \tag{15.99}$$

The e_i^P and e_i^Q variables represent the violations of the original real and reactive power flow constraints (15.2)–(15.3) at bus i. Constraints (15.95)–(15.98) define these variables. The α^P and α^Q parameters are penalty terms for violating the real and reactive power flow constraints, respectively. The Lagrangian relaxation can be solved iteratively to find the optimal solution to the original problem. This is accomplished by iteratively increasing the penalty terms until the violations e_i^P and e_i^Q equal zero in the Lagrangian solution, indicating that the original power flow constraints are satisfied.

The example of this chapter has been formulated as (15.94)–(15.99) and solved with several values of α^P and α^Q, with results displayed in Table 15.6. In Table 15.6 the second and third columns are the sums of real and reactive power flow constraint violations, respectively, while the third column displays a sum of all the power flow constraint violations multiplied by the penalty terms. We observe that as the penalty terms are increased, the violations of the original power flow constraints decrease and the objective function increases until it reaches the solution to the original problem. The solution times displayed in Table 15.6 were achieved with the GLOMIQO solver; other current solvers are significantly slower for this problem. In practice, a more sophisticated algorithm may be used. We refer the interested reader to [1473], for example.

Table 15.6. *Solving the nine-bus example Lagrangian with GLOMIQO, GAMS 24.1.3 (3.5 GHz, 32 GB RAM).*

α^P, α^Q	$\sum_{i \in N} e_i^P$	$\sum_{i \in N} e_i^Q$	Total penalty	Objective	Solution time
5	142.38	0	711.91	1264.11	00:00:02
10	82.06	0	820.55	1735.05	00:00:02
20	0	8.77	175.34	2779.91	00:00:08
30	0	6.89	206.64	2828.02	00:00:02
40	0	2.33	93.05	2991.65	00:00:05
50	0	0	0	3087.42	00:00:01

15.5.6 • Convex Reformulation with Sufficient Strong Duality Condition

Lavaei and Low [1136] provide a sufficient condition that guarantees a zero duality gap. This means that strong duality holds and the global optimum of the primal problem can be computed from the solution of the dual problem. Furthermore, they prove that the dual of the classical OPF problem can be formulated as a convex semidefinite program and can therefore be solved efficiently. Although the OPF problem is NP-hard in general [1136, 1449], the subset of the problem instances that satisfy this condition are equivalent to the convex dual.

The condition essentially states that the dual has a solution, and a certain semidefinite matrix derived from the solution has a zero eigenvalue of multiplicity two. This condition is met for several IEEE benchmark systems as long as a small resistance (as small as 10^{-5} per unit) is added to the transformers. The common simplification of assuming zero resistance for transformers violates this condition. If the condition is not met, solving the semidefinite program still provides a lower bound.

The convex semidefinite program can be derived as follows [1136]. Let $e_1, e_2, \ldots, e_N$ be the standard basis vectors in $\mathbf{R}^N$, where $N = |\mathbf{N}|$. For every system bus, let $M_i \in \mathbf{R}^{2N \times 2N}$ be a diagonal matrix with its (i, i) and $(i+N, i+N)$ entries equal to one and all other entries equal to zero. Note that the symbol * below denotes the conjugate transpose operator, Y is the complex admittance matrix, and Re and Im represent the real and imaginary operators, respectively. Next define

$$Y_i := e_i e_i^* Y, \tag{15.100}$$

$$\mathbf{Y}_i := \frac{1}{2} \begin{bmatrix} Re\{Y_i + Y_i^T\} & Im\{Y_i^T - Y_i\} \\ Im\{Y_i - Y_i^T\} & Re\{Y_i + Y_i^T\} \end{bmatrix}, \tag{15.101}$$

$$\bar{\mathbf{Y}}_i := \frac{-1}{2} \begin{bmatrix} Im\{Y_i + Y_i^T\} & Re\{Y_i - Y_i^T\} \\ Re\{Y_i^T - Y_i\} & Im\{Y_i + Y_i^T\} \end{bmatrix}. \tag{15.102}$$

The dual of the classic OPF problem can be defined as

$$\text{max} \qquad \sum_{i \in N} \left(\lambda_i^{\min} P_i^{\min} - \lambda_i^{\max} P_i^{\max} \lambda_i P_i^L \right) \tag{15.103}$$

$$+ \bar{\lambda}_i^{\min} Q_i^{\min} - \bar{\lambda}_i^{\max} Q_i^{\max} + \bar{\lambda}_i Q_i^L + u_i^{\min} \left(V_i^{\min} \right)^2 \tag{15.104}$$

$$- u_i^{\max} (V_i^{\max})^2 \Big) + \sum_{i \in G} (c_i^0 - r_{i2}) \tag{15.105}$$

$$\text{s.t.} \qquad \sum_{i \in N} \left(\lambda_i \mathbf{Y}_i + \bar{\lambda}_i \bar{\mathbf{Y}}_i + \mathbf{u}_i M_i \right) \succeq 0, \tag{15.106}$$

$$\begin{bmatrix} 1 & r_{i1} \\ r_{i1} & r_{i2} \end{bmatrix} \succeq 0 \quad \forall i \in G, \tag{15.107}$$

where parameters c_i^0, c_i^1, and c_i^2 are coefficients in the polynomial damage function of generator i. The decision variables are $\lambda_i^{\min}, \lambda_i^{\max}, \bar{\lambda}_i^{\min}, \bar{\lambda}_i^{\max}, u_i^{\min}, u_i^{\max}$ for all buses $i \in N$, and r_{i1} and r_{i2} for all generator buses $i \in G$. Further, we define:

$$\lambda_i := \begin{cases} -\lambda_i^{\min} + \lambda_i^{\max} + c_i^1 + 2\sqrt{c_i^2} r_{1k} & \text{if } i \in G, \\ -\lambda_i^{\min} + \lambda_i^{\max} & \text{otherwise,} \end{cases} \tag{15.108}$$

$$\bar{\lambda}_i := -\bar{\lambda}_i^{\min} + \bar{\lambda}_i^{\min}, \tag{15.109}$$

$$\bar{u}_i := -\bar{u}_i^{\min} + \bar{u}_i^{\min}. \tag{15.110}$$

Although the sufficient condition is met for several benchmark systems, another work has presented counterexamples where the condition is not met and where semidefinite optimization (SDO) fails to provide a physically meaningful solution to the original OPF problem [1167].

15.5.7 ▪ Semidefinite Optimization—Moment Sum of Squares

An SDO method known as moment sum of squares (SOS) has been used to solve OPF problems to global optimality [1014]. This method works when the sufficient condition of [1136] is not met, at the cost of higher runtime. Numerical experiments show that the size is often a quadratic function of the number of buses in the network, whereas it is a linear function of the number of buses with the method of [1136] when the sufficient condition is met [1014].

15.5.8 ▪ Spatial Branch-and-Bound

Another global solution method to the OPF problem relies on the spatial branch-and-bound algorithm [829, 1473]. In general, the branch-and-bound algorithm can be used to solve a nonconvex minimization problem by iteratively creating a convex relaxation that is relatively easy to solve. The solution to this relaxation provides a lower bound on the optimal objective function value of the original problem, while a local search can be used to find an upper bound. If the difference between the lower and upper bounds is below the tolerance, the algorithm terminates; otherwise it recursively partitions the feasible region and repeats the above procedure until convergence.

The OPF branch-and-bound method relies on either the semidefinite relaxation from [1136] (assuming the sufficient condition for a global solution is not satisfied) or a Lagrangian relaxation following [1473] to compute the lower bound. The upper bound is then computed using a local nonlinear programming (NLP) solver.

15.6 ▪ Future Research Directions

Solution methods for OPF problems are of very high scientific interest. Basically every optimization solution and modeling method has been applied to OPF problems. Nevertheless, OPF problems were never a focus of the global optimization community. The available off-the-shelf global optimization solvers are very challenged by this problem. Currently, they can solve only academic toy problems of systems comprising a few buses and branches (in more or less acceptable computational times). Real power systems are at least two orders of magnitude larger. In contrast, nonlinear optimization (NLO) algorithms are very efficient for OPF problems and are used in practice.

Thus, there is a huge potential for performance improvement for global optimization algorithms. We highlight three promising directions.

First, OPF problems have a very unique and special structure: they are tightly constrained, and the difference in the objective function values of the worst and the best feasible solutions is relatively small [754]. Global optimization algorithms need to take advantage of this structure.

Second, the convex relaxations currently used by global optimization solvers happen to be very weak for OPF problems. We need tight convexification procedures tailored to these problems; the positive semidefinite (PSD) relaxations seem to be very promising in this context.

The third suggestion concerns the power flow equations. Many practical MILO problems contain AC power flow equations, which make them nonconvex and computationally intractable. Thus, an approximation formulation via MILO that is both tight and computationally tractable is of very significant practical interest. Recent advances in piecewise-linear approximations [1027, 1556, 1557] and their representation [1848] have not been fully utilized for the OPF problem, while they are a success story for another network transmission problem: gas pipeline transportation [768, 1970].

15.7 ▪ Conclusions

In this chapter, we have introduced the optimal power flow (OPF) problem from the perspective of global optimization. A simple illustrative example of the classical formulation taken from the literature was used to explore several reformulations and solution methods. Several GAMS solvers were also used to solve the problem, illustrating the difficulty of solving the OPF problem to global optimality even with a small nine-bus example. Despite all of the different global optimization methods that have been applied to the OPF problem, the ability to make use of the global solution in a practical setting, with a transmission system of hundreds or thousands of buses, remains elusive for the short-term (e.g., minute-to-minute) planning horizon.

Chapter 16

The Pooling Problem

Sreekanth Rajagopalan and Nikolaos V. Sahinidis

16.1 ▪ Introduction

The pooling problem has its roots in the refining industry, where the requirement for blending of raw materials (crudes) becomes an integral part of planning. Here, the raw materials of different grades often need to be stored in a limited number of storage tanks, blended to achieve a certain target quality specification, and dispatched to further downstream processes or customers. The pooling problem aims at identifying the right mixing proportions toward minimizing the overall cost of satisfying the demands. The multicommodity flow nature of the pooling problem results in a multiextremal nonlinear optimization (NLO) problem. Alfaki and Haugland [49] showed that the pooling problem is NP-hard, even when the number of pools is fixed.

Before we continue with this brief exposition on pooling problems, we should understand the difference between the blending problem and the pooling problem to help acknowledge the challenges posed in solving the latter. First, both problems have a given set of input nodes (raw material grades), which need to be mixed, and a set of output nodes (targeted grades). While the blending problem addresses the mixing of different proportions of material from the inputs at every given output node, the pooling problem, comprising a set of transshipment nodes (intermediate pools) in addition, addresses a two-stage mixing scenario—one at every pool node, and the other at every output node. Figure 16.1 depicts the schematics of the two problems. Since the in-flow material rate and the out-flow material rate along with their qualities are to be determined for every pool node, the pooling problem possesses nonlinearity in material balance across the pool nodes.

The pooling problem is encountered frequently in mixing operations in chemical industries. However, each pooling node could be a process unit (e.g., a distiller in the case of crude refining) and, thus, may contain additional process model equations relating input to output in addition to mixing. For example, in the work of Amos et al. [69], the pooling node is a crude distillation unit. Although the sulfur mixing rules are additive, the density of fractions at the output depends on temperature.

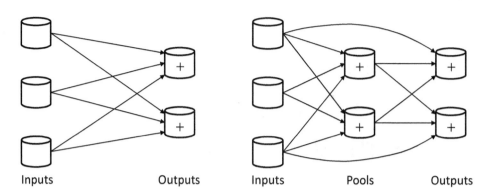

Figure 16.1. *Schematics of blending (left) and pooling (right). Pooling has intermediate nodes where inputs are mixed prior to final mixing at the outputs.*

Thus, additional cut-point temperature variables and constraints are specified, further increasing the nonlinearities in the model. Another instance of the pooling problem in industrial applications is gasoline blending. DeWitt et al. [589] detail the impact on savings upon employing a nonlinear model in their blending procedure to produce gasoline blends from different feedstock qualities. Their blending optimizer technology OMEGA evolved into a decision system across all Texaco refineries after initial annual savings to the tune of $30 million in the 1990s. Recently, Furman and Androulakis [773] developed a mixed-integer nonlinear programming (MINLP)-based approach to extend the pooling problem in gasoline blending to include reformulated gasoline requirements introduced by the Environmental Protection Agency (EPA) to reduce emissions in vehicles. Here, a model to blend and produce boutique fuels and designer fuels is included as additional quality constraints using disjunctive programming techniques.

The pooling problem has attracted considerable attention in the chemical engineering and operations research literature. For other recent surveys of the pooling problem and its applications, we refer the reader to [885, 1023, 1090, 1334]. In the current chapter, we summarize old and recent developments, while providing references to other sources for any related details. The remainder of this chapter is organized as follows. Section 16.2 describes the widely recognized P-formulation and its variants. Section 16.3 is an overview of early solution techniques and local methods. Section 16.4 is a discussion on global solution techniques. Finally, Section 16.5 summarizes the current state of the art in computation.

16.2 ▪ Formulation

Table 16.1 lists the variables and parameters used throughout this chapter. First, we develop the objective and constraints for the most intuitive formulation—the P-formulation, variations of which are discussed subsequently.

16.2.1 ▪ Problem Statement

Given are the following:

1. a network of nodes comprising a set of inputs I, pools L, and outputs J, and directed edges—connections between the nodes $(I \times L) \cup (L \times J) \cup (I \times J)$;

Table 16.1. *Notation used in this chapter.*

Indices	$i \in I$	raw materials or inputs, $I = \{1, \dots, |I|\}$
	$j \in J$	blends or outputs, $J = \{1, \dots, |J|\}$
	$k \in K$	pool qualities, $K = \{1, \dots, |K|\}$
	$l \in L$	pools, $L = \{1, \dots, |L|\}$
Variables	x_{il}	total flow of material from ith input to lth pool
	y_{lj}	total flow of material from lth pool to jth output
	z_{ij}	total flow of material from ith input to jth output
	p_{lk}	kth quality of material from lth pool
	q_{il}	ratio of total flow of material from ith input to lth pool to total flow of material into lth pool
	t_{il}	ratio of total flow of material from lth pool to jth output to total flow of material out of lth pool
Parameters	c_i	unit cost of raw material or input i
	d_j	price of final blend or output j
	A_i	capacity/availability of raw material or input i
	D_j	capacity/demand for final blend or output j
	C_l	capacity of lth pool
	Q_{ik}	kth quality of input i
	P^L_{jk}	required minimum (lower bound) kth quality at output j
	P^U_{jk}	required maximum (upper bound) kth quality at output j
	$\underline{C}^{IL}_{il}$	lower bound of flow from ith input to lth pool
	$\underline{C}^{IJ}_{ij}$	lower bound of flow from ith input to jth output
	$\underline{C}^{LJ}_{lj}$	lower bound of flow from lth pool to jth output
	$\overline{C}^{IL}_{il}$	upper bound of flow from ith input to lth pool
	$\overline{C}^{IJ}_{ij}$	upper bound of flow from ith input to jth output
	$\overline{C}^{LJ}_{lj}$	upper bound of flow from lth pool to jth output

2. a set K of qualities specified for material at the input, and their requirements at the outputs, typically as lower and upper bounds, where a perfect mixing rule is followed, i.e., the kth quality at the output of any pool l is $p_{lk} = \frac{\sum_{i \in I} Q_{ik} x_{il}}{\sum_{j \in J} y_{lj}}$; and

3. the capacities of nodes and edges, along with per unit flow cost for edges.

The objective is to identify all the flows and the pool qualities to meet demand and quality requirements at all outputs, while minimizing overall cost.

16.2.2 ▪ P-formulation

The P-formulation (also known as the *flow model* or the *concentration model*) was proposed by Haverly [921, 922]. Here, the decision variables are the flows between the nodes in the pooling problem and the pool qualities at the pool nodes. The formulation is as follows.

16.2.2.1 ▪ Objective Function

The net cost to be minimized is the total cost of input material minus the total profit from sales:

$$\min \quad \sum_{i \in I} c_i \sum_{l \in L} x_{il} - \sum_{j \in J} d_j \sum_{l \in L} y_{lj} - \sum_{i \in I} \sum_{j \in J} (d_j - c_i) z_{ij}. \tag{16.1}$$

16.2.2.2 • Constraints

$$\sum_{l \in L} x_{il} + \sum_{j \in J} z_{ij} \leq A_i \quad \forall\, i \in I, \tag{16.2}$$

$$\sum_{l \in L} y_{lj} + \sum_{i \in I} z_{ij} \leq D_j \quad \forall\, j \in J, \tag{16.3}$$

$$\sum_{i \in I} x_{il} - \sum_{j \in J} y_{lj} = 0 \quad \forall\, l \in L, \tag{16.4}$$

$$\sum_{i \in I} Q_{ik} x_{il} - p_{lk} \sum_{j \in J} y_{lj} = 0 \quad \forall\, k \in K,\, l \in L, \tag{16.5}$$

$$\sum_{l \in L} (P_{jk}^L - p_{lk}) y_{lj} + \sum_{i \in I} (P_{jk}^L - Q_{ik}) z_{ij} \leq 0 \quad \forall\, k \in K,\, j \in J, \tag{16.6}$$

$$\sum_{l \in L} (p_{lk} - P_{jk}^U) y_{lj} + \sum_{i \in I} (Q_{ik} - P_{jk}^U) z_{ij} \leq 0 \quad \forall\, k \in K,\, j \in J, \tag{16.7}$$

$$\sum_{i \in I} x_{il} \leq C_l \quad \forall\, l \in L, \tag{16.8}$$

$$x_{il} \in \left[\underline{C}_{il}^{IL}, \overline{C}_{il}^{IL}\right], \quad y_{lj} \in \left[\underline{C}_{lj}^{LJ}, \overline{C}_{lj}^{LJ}\right], \quad z_{ij} \in \left[\underline{C}_{ij}^{IJ}, \overline{C}_{ij}^{IJ}\right] \tag{16.9}$$

$$\forall\, i \in I,\, l \in L,\, j \in J,$$

$$x_{il}, y_{lj}, z_{ij}, p_{lk} \geq 0 \quad \forall\, i \in I,\, l \in L,\, j \in J,\, k \in K. \tag{16.10}$$

Constraints (16.2) and (16.3) ensure availability of raw materials at the inputs and satisfaction of demand or capacity at the outputs, respectively. Constraints (16.4) and (16.5) represent the overall mass balances and individual quality mass balances across the pools, respectively. Constraints (16.6) and (16.7) represent the lower bound and upper bound quality requirements at the outputs. The various capacities at the nodes and on the flows are represented by constraints (16.8) and (16.9).

16.2.3 • Q-formulation

Ben-Tal et al. [201] derived the Q-formulation (also known as the *proportional model*) by defining a q_{il} variable as follows, effectively replacing $x_{il}\ \forall\, i \in I,\, l \in L$ in the objective and all the constraints:

$$q_{il} = \frac{x_{il}}{\sum_{j \in J} y_{lj}} \quad \forall\, i \in I,\, l \in L \iff x_{il} = q_{il} \sum_{j \in J} y_{lj} \quad \forall\, i \in I,\, l \in L. \tag{16.11}$$

While q_{il} replaces x_{il}, it introduces nonlinearities in the form of $q_{il} y_{lj}$. Both x_{il} and p_{lk} are eliminated with the introduction of q_{il}, and q_{il} belongs to a simplex, whereas p_{lk} belongs to a hypercube. Once q_{il} is introduced, the mass balance constraint (16.4) and the quality balance constraint (16.5) simplify to

$$\sum_{i \in I} q_{il} \sum_{j \in J} y_{lj} - \sum_{j \in J} y_{lj} = 0 \implies \sum_{i \in I} q_{il} = 1 \quad \forall\, l \in L, \tag{16.12}$$

$$\sum_{i \in I} Q_{ik} q_{il} \sum_{j \in J} y_{lj} - p_{lk} \sum_{j \in J} y_{lj} = 0 \implies p_{lk} = \sum_{i \in I} Q_{ik} q_{il} \quad \forall\, k \in K,\, l \in L. \tag{16.13}$$

Objective (16.1) and constraints (16.2)–(16.3) and (16.6)–(16.10), with x_{il} and p_{il} replaced according to (16.11) and (16.13), and constraint (16.12) complete the Q-formulation.

16.2.4 ▪ PQ-formulation

The PQ-formulation is nothing but the Q-formulation with the following constraint in addition:

$$\sum_{i\in I} q_{il} y_{lj} = y_{lj} \quad \forall\, l \in L, j \in J. \tag{16.14}$$

Quesada and Grossmann [1535] derived (16.14) using reformulation-linearization techniques (RLTs) [1686]—by multiplying constraint (16.12) by y_{lj}—in the context of global optimization of bilinear chemical process networks. Clearly, constraint (16.14) is redundant as far as the feasible region of Q-formulation is concerned. However, the linear relaxation of the PQ-formulation provides a tighter lower bound than that of the P-formulation [1754].

16.2.5 ▪ TP- and STP-formulations

Following the idea of the Q-formulation and the subsequent PQ-formulation, Alfaki and Haugland [49] define proportion variables t_{lj}, similar to q_{il}, based on flow between a pool and its connected outputs (T refers to the outputs or *terminal* nodes in [49]). The TP-formulation is derived similarly to PQ, with the following relations instead of (16.11) and (16.14):

$$t_{lj} = \frac{y_{lj}}{\sum_{i\in I} x_{il}} \quad \forall\, j \in J, l \in L \quad \Longleftrightarrow \quad y_{lj} = t_{lj}\sum_{i\in I} x_{il} \quad \forall\, j \in J, l \in L, \tag{16.15}$$

$$\sum_{j\in J} t_{lj} x_{il} = x_{il} \quad \forall\, l \in L, i \in I. \tag{16.16}$$

Constraint (16.16) is redundant and derived using RLT, similar to (16.14) in the PQ-formulation. By introducing (16.15), we can show that the objective functions in both PQ and TP are the same, and so are the node capacity constraints, (16.2), (16.3), and (16.8), and the quality requirement constraints, (16.6) and (16.7).

While the numerical comparison of lower bounds from linear relaxations of TP and PQ formulations shows that neither dominates the other, the STP-formulation (S referring to the inputs or *sources* in [49]), which utilizes constraints (16.11) and (16.14) in addition to those in the TP-formulation, is a stronger variant [49]. The linear relaxation of STP inherits all the constraints in the linear relaxation of PQ (and TP), and, thus, the lower bound on the objective provided by the STP-relaxation is at least as tight as that of PQ (and TP) [49].

16.2.6 ▪ Variants and Extended Formulations

Other extensions to the above formulations exist in the literature. Audet et al. [111] propose the *generalized pooling problem*, where additional edges between pool nodes are considered. Here, a formulation analogous to Q-formulation, i.e., based on proportions, leads to quadratically constrained quadratic programs (QCQP) instead of a bilinear program when extended naturally. To address this, Audet et al. [111] derive a hybrid model where the proportion model (based on q variables) is followed for pools that receive flow only from inputs, and the flow model (based on p variables) is followed for all other pools. Recently, Alfaki and Haugland [48] introduced the *multicommodity flow formulation* for the generalized pooling problem, an extension of the PQ-formulation for the generalized pooling problem, whose linear relaxation is stronger than that of the hybrid model. Boland et al. [296, 297] study some special cases of pooling problems in coal and iron ore mining industry supply chains—a special case of the generalized pooling problem that penalizes deviations from target qualities, and the supply chain extension of the standard pooling problem with hard demand constraints and penalization of deviations from target qualities.

Generalizations in terms of the choice of the number of pools and the interconnectivity of nodes in the network are modeled through MINLPs, as proposed by Meyer and Floudas [1319] and Karuppiah and Grossmann [1041]. Extended pooling problems are proposed that involve additional constraints, for example, emission constraints by Misener et al. [1334] and Furman and Androulakis [773]. A recent work of Li et al. [1187] presents a stochastic pooling problem formulation for natural gas production network design and operation, wherein in addition to interconnections between pools, a two-stage recourse formulation is developed based on bilevel optimization—a first stage to decide connectivity between nodes, and a second stage to plan the operation under uncertainty. Kolodziej et al. [1090] introduce a multiperiod scheduling version of the pooling problem to represent time-varying blending systems, formulated as a nonconvex MINLP that is solved using a mixed-integer linear programming (MILP) approximation based on discretization. Lotero et al. [1224] model a multiperiod blending problem with time-varying supply and demand using generalized disjunctive programming techniques. An MILP-MINLP decomposition technique is presented to demonstrate computational performance over general purpose state-of-the-art solvers. Gupte et al. [885] propose a time-indexed pooling problem to include discrete decisions and planning over a horizon to meet demands at the outputs.

16.3 ▪ Local Solution Methods and Heuristics

An early approach to solving pooling problems was based on recursive application of linear programming (LP) [921]. When the quality variables p_{lk} are fixed, the problem results in an LP problem that can be solved easily. Postsolving, p_{lk} are calculated using flow rates from the LP solution, and corrections are applied to fix p_{lk} for the successive iteration. Haverly [921, 922] formalizes this approach and discusses the quality of the solutions it provides—global, local, or failure of convergence. Besides the lack of convergence guarantee (even to a local solution), this method becomes computationally expensive for large problems.

Lasdon et al. [1128] applied generalized reduced gradient and successive linear programming (SLP) techniques—using Taylor approximation of nonlinear terms based on the choice of an initial set of flow rates (and qualities calculated subsequently). An efficient and successful implementation of SLP was later proposed by Palacios-Gomez et al. [1437] and Baker and Lasdon [136]. Greenberg [851] utilizes computational geometry methods to calculate ranges of flow rates and qualities for sensitivity analysis of pooling problems to diagnose infeasibilities that may result in SLP subproblems.

Aggarwal and Floudas [30] introduce a primal-dual technique utilizing decomposition of bilinear terms $p_{lk}y_{lj}$, resulting in a sequence of LP problems. In general, the method cannot guarantee global optimality [1613].

More recently, Audet et al. [111] employ the alternate heuristic (ALT), sometimes in combination with a variable neighborhood search (VNS) heuristic, to kick-start their QCQP algorithm [118] for the generalized pooling problem. ALT solves a sequence of LPs or convex nonlinear programming (NLP) problems resulting from fixing either of the variables in a bilinear or a biconvex term, subsequently alternating with iterations. In a VNS scheme, at each iteration, the solution from ALT is perturbed in a neighborhood of increasing size in an effort to obtain a better extremum. Recently, Almutairi and Elhedhli [57] proposed a Lagrangian approach by dualizing the nonlinear constraints in the P- and PQ- formulations. Almutairi and Elhedhli [57] propose two heuristics, similar to ALT, to obtain feasible solutions for the Lagrangian subproblems. On applying these heuristics, equal or better lower bounds to 14 out of 15 problems with respect to that by Adhya et al. [25] were obtained [57].

16.3.1 ▪ Heuristics and Approximation Schemes Based on MILP Discretization

In this section, we present heuristics for the pooling problem based on discretization of variables participating in the bilinear terms. In the first method, introduced by Pham et al. [1472], a complete graph is assumed for the network. That is, any input is connected to all the pools, and any output is connected to all the pools. Though this is typically the case in any general pooling problem, the methods discussed previously also hold for problems where only a subset of these connections exist.

Pham et al. [1472] follow a three-step approach. First, the pool qualities p_{lk} are discretized into N values for every kth quality. The discretization will be based on the same bounds P_{lk}^L and P_{lk}^U denoted previously. Then, we have $P_{lk}^L = p_{lk}^1 \leq \cdots \leq p_{lk}^N = P_{lk}^U$ discrete values for any kth quality of any lth pool. Thus, instead of $|L|$ pools, $L' = \{1, 2, \ldots, N^k\}$ hypothetical pools are considered with qualities belonging to $\times_{l \in L}(p_{lk}^1, \ldots, p_{lk}^N)$, and the problem reduces to an LP problem (blending problem).

Once discretized, the second step is to apply integer cuts to the new set of pools. Clearly, to obtain a valid solution, we need to restrict the number of pools in the solution to the original $|L|$. Thus, $N^k - |L|$ pools are excluded from the solution through the following integer cut:

$$\sum_{l \in L'} b_l \leq |L|. \tag{16.17}$$

This results in an MILP via introduction of a binary variable b_l for each pool. Then, we have constraints (16.5)–(16.7) as a set of $3N$ constraints with p_{lk} fixed to each of the $p_{lk}^n, n = 1, \ldots, N$, values. Additionally, constraints (16.8)–(16.9) have a binary b_l multiplied by the capacities.

Since the formulation considers a finite set of values for the continuous-quality variables, the solution is an approximation to the original problem. However, this MILP restriction of the pooling problem provides a valid upper bound in a rigorous global optimization approach or, in quite a few cases, a good-quality solution. One may choose to discretize each quality independently as N_l discrete-quality vectors, resulting in $N_1 \times N_2 \times \cdots \times N_k$ pools instead of N^k.

Finally, a convex hull search technique is proposed for problems resulting in a large MILP due to several qualities and discretization points. Given a set of input-quality vectors Q_{ik} for all the inputs, the quality of any pool lies in the convex hull of this k-dimensional space. Thus, this technique sufficiently reduces the problem size by only considering pools in the convex hull instead of the entire hypercube in $\mathbb{R}^k$ with edges $[P_{lk}^L, P_{lk}^U]$. Since the (perfect) mixing rule (16.5) is a natural convex combination of inputs, discretizing q_{il} provides the required pools in the convex hull. That is, for any pool l, the matrix $\tilde{Q}_l = [q_i^1, \ldots, q_i^m, \ldots, q_i^{M_l}]$ represents the M_l assigned combinations of input flow proportions, any mth column vector q_i^m being such that $\sum_{i \in I} q_i^m = 1$.

Using these M_l combinations, discretizations for p_{lk} can be calculated from the mixing rule for the lth pool. Thus, one can discretize the flow proportions instead.

Alfaki and Haugland [47] derive an equivalent convex hull search discretization through an extended graph representation. Pool l in the original network is duplicated to result in M_l pools such that the proportional flows assigned to each of these pools are consistent, i.e., $\sum_{i \in I} q_{il} = 1$. This indeed gives the matrix $\tilde{Q}_l$ described previously. Here, the constraint that only one of the duplicates (including the original parent pool) is active in the solution is imposed as the integer cut for any original pool in the problem. Thus, this results in an MILP restriction as well.

Recently, Dey and Gupte [590] proved that, in the worst case, the ratio of the optimal solution to the lower bound via any MILP relaxation is at most $|J|$ (number of outputs) and that this bound is tight. Their analysis led to the first polynomial-time $|J|$-approximation algorithm for the pooling problem.

16.4 · Global Solution Methods

A widely used global optimization strategy is branch-and-bound. The local solution methods discussed in the previous section, then, serve as an upper-bounding technique. Most work in the literature focuses on deriving strong lower bounds to expedite the branch-and-bound algorithm. In this section, we survey some relaxations of the pooling problem for lower bounding—typically relaxations based on convex/concave envelopes, strengthening of relaxations in a branch-and-cut scheme via RLT, Lagrangian relaxations, and MILP-based relaxation or discretization.

16.4.1 · Envelope-Based Relaxations

Foulds et al. [747] introduce a linear relaxation of the pooling problem via McCormick convex/concave envelopes [1304]. To illustrate, the P-formulation can be linearized by introducing $u_{lkj} = p_{lk} y_{lj}$. Thus, we have the following constraints in place of (16.5)–(16.7):

$$\sum_{i \in I} Q_{ik} x_{il} - \sum_{j \in J} u_{lkj} = 0 \quad \forall k \in K, l \in L, \tag{16.18}$$

$$\sum_{l \in L} P_{jk}^L y_{lj} - \sum_{l \in L} u_{lkj} + \sum_{i \in I} (P_{jk}^L - Q_{ik}) z_{ij} \leq 0 \quad \forall k \in K, j \in J, \tag{16.19}$$

$$\sum_{l \in L} u_{lkj} - \sum_{l \in L} P_{jk}^U y_{lj} + \sum_{i \in I} (Q_{ik} - P_{jk}^U) z_{ij} \leq 0 \quad \forall k \in K, j \in J. \tag{16.20}$$

In addition, the following envelopes of bilinear terms relax u_{lkj} based on bounds on p_{lk} and y_{lj}:

$$\left. \begin{array}{ll} u_{lkj} \geq y_{lj}^L p_{lk} + p_{lk}^L y_{lj} - p_{lk}^L y_{lj}^L & \forall l \in L, k \in K, j \in J \\ u_{lkj} \geq y_{lj}^U p_{lk} + p_{lk}^U y_{lj} - p_{lk}^U y_{lj}^U & \forall l \subset L, k \subset K, j \subset J \\ u_{lkj} \leq y_{lj}^L p_{lk} + p_{lk}^U y_{lj} - p_{lk}^U y_{lj}^L & \forall l \in L, k \in K, j \in J \\ u_{lkj} \leq y_{lj}^U p_{lk} + p_{lk}^L y_{lj} - p_{lk}^L y_{lj}^U & \forall l \in L, k \in K, j \in J \end{array} \right\}. \tag{16.21}$$

Similarly, the TP-formulation can be relaxed by introducing u_{lkj}, while the Q- and PQ-formulations can be relaxed by introducing $v_{ilj} = q_{il} y_{lj}$, and the TP- and STP-formulations can be relaxed by defining both u_{lkj} and v_{ilj}.

Proposition 16.1 [1754]. *The PQ-formulation when relaxed using bilinear envelopes produces a tighter lower bound than relaxations of P and Q using similar bilinear envelopes.*

Proposition 16.2 [49]. *The lower bound on the optimal objective function value provided by the STP-relaxation is at least as tight as those provided by the PQ- and TP-relaxations.*

16.4.2 ▪ Lagrangian Relaxations

Adhya et al. [25] introduce a Lagrangian lower-bounding approach by dualizing all constraints (16.2)–(16.8) in the P-formulation. They show that the Lagrangian subproblem can be decomposed and reduced to solving two LP and $|L|$ MILP problems. Also, the strength of Lagrangian relaxation is shown to be the same as that of linear relaxation using bilinear envelopes when $k = 1$, whereas the same relaxation may provide a strictly stronger lower bound when $k > 1$ (multiple qualities). Later, Tawarmalani and Sahinidis [1754] showed that the linear relaxation of PQ-formulation is at least as tight as the Lagrangian relaxation by Adhya et al. [25].

Ben-Tal et al. [201] used Lagrangian relaxation to close the duality gap in a branch-and-bound algorithm for the Q-formulation. The algorithm partitions the q variables over $\sum_{i \in I} q_{il} = 1$, which represents a unit simplex in $|I|$ dimensions for each pool l. On the other hand, for a partitioning algorithm based on p_{lk} for the P-formulation, p_{lk} belong to a hypercube in $|L||K|$ dimensions, the size of which depends on minimum and maximum input pool qualities Q_{ik}.

16.4.3 ▪ MILP—Piecewise Relaxation

The MILP-based relaxation, at a higher level, can be thought of as a linear relaxation applied over small subintervals of continuous variables present in nonlinear terms. Here, the subintervals are typically derived a priori to the formulation by partitioning the variable space between its bounds. The number of subintervals, then, determines the number of binaries introduced.

For example, the quality vector p_{lk} can be normalized (for every kth quality) as $\tilde{p}_{lk} = \frac{p_{lk} - P_k^L}{P_{lk}^U - P_{lk}^L}$, where P_k^L and P_k^U are the respective minimum and maximum of all input values for the kth quality for the lth pool. The normalized qualities can be partitioned into N subintervals whose bounds are determined by the points $0, \frac{1}{N}, \frac{2}{N}, \ldots,$ $\frac{n-1}{N}, \frac{n}{N}, \ldots, 1$, where the nth subinterval is $[\frac{n-1}{N}, \frac{n}{N}]$. If b_{lk}^n is the binary representing the nth interval for the kth quality in the lth pool, we have $\tilde{p}_{lk} \in [\frac{n-1}{N} b_{lk}^n, \frac{n}{N} b_{lk}^n]$. More generally, if p_{lk} is partitioned based on points $P_{lk}^L = p_{lk}^0 \leq p_{lk}^1 \leq \cdots \leq p_{lk}^n \leq p_{lk}^{n-1} \leq \cdots \leq p_{lk}^{N_l} = P_{lk}^U$, we have the following for the nth subinterval:

$$p_{lk} \geq p_{lk}^{n-1} b_{lk}^n \quad \text{and} \quad p_{lk} \leq p_{lk}^n b_{lk}^n \quad \forall n = 1, \ldots, N_l. \tag{16.22}$$

To force p_{lk} to be in one of the subintervals, we require

$$\sum_{n=1}^{N_l} b_{lk}^n = 1. \tag{16.23}$$

Thus, any relaxation of a constraint given by $g(\ldots, p_{lk}; p_{lk}^0, p_{lk}^{N_l}) \leq 0$ is replaced by a set of N_l constraints given by

$$g(\ldots, p_{lk}; p_{lk}^{n-1}, p_{lk}^n) \le M(1 - b_{lk}^n) \quad \forall n = 1, \ldots, N_l \qquad (16.24)$$

and (16.23), where the value of M is sufficiently large.

Meyer and Floudas [1319] model the water treatment network problem as an MINLP problem using the generalized pooling problem as the superstructure. Using the above piecewise relaxation technique and RLT, an MILP problem is obtained as the lower-bounding problem. Wicaksono and Karimi [1887] discuss different piecewise MILP underestimators and overestimators for bilinear programs. Floudas et al. [835, 1329, 1333] compare some of these MILP relaxation schemes, along with their own proposed schemes, for the pooling problem. Misener et al. [1334] utilize a technique introduced by Vielma and Nemhauser [1848] to keep a logarithmic number of binary variables in modeling disjunctions.

16.5 • State of the Art in Computation

In this final section, we present some computational advancements achieved through the global solution strategies discussed in the previous section. The standard benchmark pooling problems are Haverly 1–3 [921,922], Foulds 2–5 [747], Ben-Tal 4–5 [201], Adhya 1–4 [25], and RT [111]. These problems contain up to 11 inputs, 16 outputs, 8 pools, and 6 qualities. While Haverly, Foulds, and Ben-Tal 4 contain a single quality attribute, the others contain multiple qualities. Other library problems include Alfaki–Haugland [49] (35 problems with 20 large-scale instances), Meyer–Floudas [1319] (plant test cases), EPA problems [773], and some recently reported instances [885].

16.5.1 • Linear Relaxations and Lagrangian

A comparison between the lower bounds obtained from linear relaxations and Lagrangian relaxations of the P-, Q-, and PQ-formulations of the standard benchmark problems is provided by Tawarmalani and Sahinidis [1754] and Almutairi and Elhedhli [57]. The linear relaxation of the PQ-formulation provided the tightest bound among all in the RT2 and Adhya 1–4 problems, and the same bound as that of P in the rest, with a 6.6% improvement in the lower bound on average. The Lagrangian relaxation of PQ provided even better results, with almost 10% improvement over its linear relaxation counterpart, and within a 2.2% optimality gap from the global solution. However, the linear relaxation of the PQ-formulation is a natural lower-bounding scheme in a typical general purpose solver. Through an effective branching strategy on the bilinears in PQ, the benchmark Haverly, Foulds, and Ben-Tal problems were solved at the root node, and the Adhya problems were solved within two seconds using BARON [1611, 1754]. Alfaki and Haugland [49] show that linear relaxation of the STP-formulation can further improve the lower bound with respect to that of PQ in two problems—Ben-Tal 4 (1.51%) and RT2 (8.39%).

16.5.2 • MILP Relaxations

Gounaris et al. [835] compare 15 different relaxation schemes from three different classes—big-M, convex combination, and incremental cost—on the standard benchmark problems and an EPA1 problem with emission constraints. Though the overall relaxation strength for any bilinear term is the same for all of these schemes, the number of variables (binary and continuous) and constraints introduced to represent convex and concave envelopes in subintervals varies between the schemes. Gounaris et al.

[835] show that the quality of MILP relaxations in general improves with increased discretization. As the number of subintervals increases, the relaxation lower bound tends toward the global optimum [835]. However, the trends were not monotonic for RT2 and EPA1 problems for the p-variant of P-formulation. Nonuniform partitioning performed better in terms of overall relaxation strength. Overall, the P-formulation outperformed the Q-formulation, while PQ effectively closed the gap for the Ben-Tal 4 problem at the root with a coarse and uniform partitioning consisting of just two subintervals. In terms of computational effort, Gounaris et al. [835] propose a few schemes that are efficient at solving all of the benchmark problems within a few seconds in a global branch-and-bound procedure.

16.5.3 ▪ Quality of MILP Restriction

Even though MILP discretization and restriction methods do not necessarily guarantee global solutions, they may still provide good-quality solutions, especially for large problems. Pham et al. [1472] show that, through a suitable discretization scheme, solution qualities to the tune of $< 2\%$ with respect to the global optimum can be achieved, with an average quality of 1% in Adhya problems, and very small errors in other benchmark problems. Although the number of discretized pools was an order 100 more than that of the original specifications for Adhya problems, the solver times were an order of magnitude less than that of the global solver [1472]. Here, both the MILP discretization and the original P-formulation (NLP) were solved using LINGO [1648].

Alfaki and Haugland [47] compare continuous and discrete formulations for a set of 20 large-scale problems [49] up to a few thousand variables and nonlinear terms, and 15 other arbitrary instances with 8–60 inputs, 6–15 pools, 6–50 outputs, and 4–40 qualities, with 57–1451 edges or connections. Better upper bounds were obtained in the discrete model, in 21/35 instances in one hour using the CPLEX solver [47]. CPLEX, however, was unable to prove optimality of the MILP problem in 19/21 of these instances. While the continuous model delivered better upper bounds in 5 of the remaining problems, both methods find the global solution in the remaining 9 problems. In the same resource time, BARON found global solutions to 14/35 of the NLP problems.

16.6 ▪ Conclusions

The pooling problem literature has evolved over time in conjunction with nonconvex optimization algorithms. In most chemical engineering applications, especially those pertaining to the oil and gas industry, the pooling problem can often be identified as a subproblem—in both upstream and downstream processing, with extensions to complex target specification models. The solution strategies are largely tied to the formulation of the problem. For example, a generalized pooling problem formulation is more amenable to a water treatment facility, where recycling is advantageous for high-throughput processing. On the other hand, a standard pooling problem is more suitable for blending operations in refineries. In terms of solution strategies, with the current state-of-the-art global solvers, a PQ- or STP-formulation is more flexible, with standard reformulations, lower-bounding techniques, and bound-tightening schemes. While short of a cutting-edge global solver, a piecewise MILP-based relaxation scheme is attractive and may offer high-quality solutions.

Part V

Nonlinear Optimization

Chapter 17

Nonlinear Optimization Algorithms

Andreas Wächter

17.1 ▪ Introduction

This part of the book is concerned with nonlinear optimization (NLO) problems, also referred to as nonlinear programming (NLP) problems, that can be stated as

$$\min_{x \in \mathbb{R}^n} \quad f(x) \tag{17.1a}$$

$$\text{s.t.} \quad g(x) \leq 0, \tag{17.1b}$$

$$\phantom{\text{s.t.}} \quad h(x) = 0, \tag{17.1c}$$

where the objective function $f : \mathbb{R}^n \to \mathbb{R}$, as well as the constraint functions $g : \mathbb{R}^n \to \mathbb{R}^m$ and $h : \mathbb{R}^n \to \mathbb{R}^l$, are usually assumed to be at least once continuously differentiable. NLO with discrete variables is discussed in Part VI.

NLO problems arise in a large variety of scientific, engineering, and financial disciplines in which phenomena are inherently nonlinear and cannot be approximated sufficiently well with linear functions. The following chapters present particular problems from chemical (Chapter 18), aerospace (Chapter 19), and environmental engineering (Chapter 20).

The goal of this chapter is to give practitioners an overview of some basic concepts and algorithms that have proved to be effective for challenging applications. A number of high-quality general purpose software packages for the solution of (17.1) are available, some of which will be mentioned throughout the chapter. Furthermore, in many situations, specialized algorithms and implementations tailored to specific problem characteristics and particular applications have been developed, but they are not discussed in this chapter in much depth.

The material in the chapter is presented from a broad perspective, to provide some background that might help the reader make informed choices of a suitable algorithm given a particular setting: to decide either which existing software to use or which type of method might be worth exploring in more detail.

Interested readers who want to deepen their understanding in practical NLO techniques will find more detailed treatment in textbooks such as [242, 261, 724, 803, 1402, 1604]. In the interest of brevity, only a few additional selected references are given, mainly to give initial pointers for aspects that are not covered in standard textbooks. Detailed information about the software mentioned in this chapter can be found on the Decision Tree for Optimization Software website [1339], which gives a comprehensive list of available optimization codes in general. Another valuable online resource for numerical optimization methods is the *NEOS Guide* [1393], which gives a detailed overview of different types of optimization problem classes and algorithms and also provides free online use for a number of optimization codes.

This chapter is structured as follows. In Section 17.2 we discuss properties of (17.1) that should be taken into account during the modeling of particular applications and for the choice of the most suitable algorithm. Section 17.3 summarizes some theoretical background, and Section 17.4 introduces basic algorithmic concepts that are common to most NLO algorithms. Particular classes of algorithms are discussed in Section 17.5, and Section 17.6 briefly comments on variations of (17.1) that require special treatment.

17.2 ▪ Problem Characteristics

17.2.1 ▪ Convex versus Nonconvex Problems

A (globally) optimal solution to the NLO problem (17.1) is a point x_* for which the objective function value is as small as that for any other feasible point, i.e., $f(x) \geq f(x_*)$ for all $x \in F$, where $F = \{x : g(x) \leq 0 \text{ and } h(x) = 0\}$ is the feasible set.

This chapter concentrates on algorithms that aim at finding a *local solution* to (17.1), i.e., a point x_* for which $f(x) \geq f(x_*)$ for all $x \in F$ in a sufficiently small neighborhood of x_*. These methods can address problems with a large number of variables and constraints (up to millions). In many practical applications, a local solution is sufficient; after all, a solution that cannot be improved easily by minor changes is better than a merely feasible point or a heuristic solution. In addition, in many applications a reasonable guess of the optimal solution can be exploited to steer algorithms to a "good" local solution.

In the lucky situation where the objective function and the feasible region are convex, every local minimizer is a global minimizer, and any convergent method is guaranteed to return the best solution. When it is not possible to model one's application as a convex problem, it is often a good idea to consider different equivalent formulations that might lead to "a lesser degree of nonconvexity" (a vague phrase), even if this increases the problem size. This may result in convergence to local solutions with better objective function values. To give a simple example for a convex reformulation, consider the nonconvex constraint $\Pi_{i=1}^{N} p_i \geq \bar{p}$, which might express that the probability of N independent events occurring at the same time should be at least $\bar{p}$. This can be modeled equivalently by the convex constraint $-\sum_{i=1}^{N} \log(p_i) \leq -\log(\bar{p})$. Also note that the Hessian of the new constraint is diagonal, while the Hessian of the original one is dense.

There are methods that are guaranteed to compute globally optimal solutions for nonconvex problems (see Part IV of this book). These typically exploit particular structures that might not be present in the problem of interest, or they are based on some kind of enumeration scheme that requires a significant computational effort. As a consequence, they are often only practical for problems of small or moderate size and complexity.

We mention in passing that there are also randomized algorithms, such as simulated annealing or genetic algorithms, with varying convergence guarantees to global optima. However, they are beyond the scope of this chapter and are practical only for problems with a small number of variables.

17.2.2 ▪ Availability of Derivatives

The methods discussed in this chapter utilize pointwise information of the problem functions, specifically, values and derivatives at given points. No global representation of the problem, in the form of formulas or an expression tree, is required. To make predictions of the functions around a particular point, the methods work with "local models" of the nonlinear functions based on Taylor expansions. This requires that the objective and constraint functions be at least differentiable; otherwise, the predictions made by these models might be valid only in a very small neighborhood around the current iterate, or the model cannot even be computed. Furthermore, the availability of second derivatives can often result in better models and improve the convergence speed.

Depending on the particular modeling situation, first- and second-order derivatives might be easily available, for example, when the objective and constraint functions are given as formulas. When a modeling language (such as AIMMS, AMPL, or GAMS) is used, the practitioner does not need to worry about derivatives because they are computed by the modeling software and provided to the optimization codes. Often, however, function values are calculated by some computer code that can be as complicated as the numerical solution of a system of PDEs. Sometimes, automatic differentiation techniques [853] (not to be confused with symbolic differentiation) or specialized methods, such as adjoint computations for differential equations, can be used to compute derivatives in these cases. Chapter 19 presents examples of this approach for functions defined by the numerical solution of PDE constraints. In any case, for the methods discussed in this chapter, sufficiently accurate first-order derivatives are indispensable; note that care must be taken when derivatives are approximated by finite differences because their accuracy depends crucially on the size of the perturbation. On the other hand, the availability of second derivatives is less crucial because techniques exist that approximate second-order derivative information, if it is not available, to accelerate convergence; see Section 17.4.1.

Even when the problem functions are not differentiable, it is sometimes possible to reformulate the original problem as a smooth problem. For example, $\min |f(x)|$ is equivalent to $\min t$ s.t. $f(x) \leq t, -f(x) \leq t$. Alternatively, one may smooth the function at nondifferentiable points (e.g., $|f(x)| \approx \sqrt{x^2 + \epsilon^2} - \epsilon$). There is, however, usually a trade-off between the accuracy by which the smoothed function approximates the nonsmooth one and the degree of nonlinearity that is introduced into the problem in the form of high curvature. Several methods are specifically designed for nonsmooth problems (see, e.g., [361, 1039]), but they are beyond the scope of this chapter.

When derivatives are not available, derivative-free methods (see [502] and Part X) can be employed.

17.2.3 ▪ Other Aspects

It is useful to distinguish several subclasses of the NLO problem (17.1). *Unconstrained optimization* deals with the minimization of an objective function in the absence of constraints (i.e., $F = \mathbb{R}^n$). The category of *constrained optimization* problems can be divided loosely into problems that have only bound constraints, only equality constraints, only inequality constraints, or mixed constraints. The best choice of algorithm depends on the particular problem class, in addition to other factors, which we discuss next.

The number of variables and constraints is certainly an important consideration, but it is not the only characteristic that determines the computational difficulty of an NLO problem; some small but highly nonlinear and nonconvex problems can be much more difficult to solve than larger but "less nonlinear" problems.

Because all methods discussed here require the solution of linear systems, the characteristics of the derivative matrices play a crucial role. Often, these matrices are sparse, and existing linear algebra software that exploits the sparsity structure in a direct factorization can be used, even when the number of variables and constraints is on the order of millions. Therefore, it might be worthwhile to seek a formulation of an optimization problem that leads to sparse derivative matrices, even if the number of variables and constraints increases.

In some situations, e.g., when the constraints are derived from the discretization of a three-dimensional PDE, the direct factorization of the sparse constraint Jacobian would be very expensive due to a significant increase in the number of nonzero elements in the factors (fill-in). In this case, iterative linear solvers are preferred that only require products of the matrix with a vector. This "matrix-free" approach is also important in case the storage of the full derivative matrices is prohibitive, or when the computation of all matrix entries would require a lot of time (e.g., when the constraints involve the integration of differential equations). A challenge is then that the solutions of the linear systems can be computed only approximately, and the optimization algorithm must be designed to handle inexact steps robustly.

Another important factor is the time required for computing values and derivatives of the problem functions. When the evaluation of problem functions is the computational bottleneck (e.g., because it involves a complicated numerical procedure such as integration), a method that requires few iterations is preferred; in contrast, when function evaluations are cheap, the speed at which steps are computed in the method is more crucial. An example here is that active-set sequential quadratic programming (SQP) methods (see Section 17.5.1) typically require fewer iterations than interior-point methods (see Section 17.5.2), but they solve a subproblem by an algorithm with combinatorial complexity in each iteration. Active set methods can become prohibitively slow when there are many inequality constraints, whereas the performance of interior-point methods is much less affected by the number of inequality constraints.

Finally, the number of *degrees of freedom* influences the best choice of an optimization algorithm. In some cases, only a few optimization variables can be chosen freely, whereas the others are determined implicitly by the equality constraints (17.1c). Then, a *reduced-space method* that works with a projection of the problem into this smaller space can often be very efficient. However, the projection usually destroys sparsity, and, as a consequence, these methods do not perform well when the reduced space is large.

17.3 ▪ Optimality Conditions

The numerical NLO algorithms discussed here are guided by the search for a point that satisfies a mathematical characterization of optimal solutions to (17.1) in the form of optimality conditions that involve the derivatives of the problem functions.

These optimality conditions are based on Taylor approximations of the functions; therefore, smoothness of the functions is crucial.[12] The most important of these conditions are the first-order necessary optimality conditions. These essentially state that, at an optimal solution x_*, no feasible *descent direction* exists that would improve a first-order Taylor approximation of the objective function.

In the unconstrained case, the approximation of the objective function at a point x_* becomes $f(x_* + \alpha d) \approx f(x_*) + \alpha \nabla f(x_*)^T d$, where d is a direction and $\alpha \geq 0$ a step size. Requiring that no descent direction exist implies that the gradient has to be zero at an optimal solution.

The analogue for the constrained case asserts that, at a feasible point x_*, there is no descent direction for the linearization of the objective that maintains feasibility for a linearization of the constraints, i.e., there exists no $\bar{d} \in L = \{d : g(x_*) + \nabla g(x_*)^T d \leq 0 \text{ and } h(x_*) + \nabla h(x_*)^T d = 0\}$ such that $\nabla f(x_*)^T \bar{d} < 0$. This negative condition is equivalently stated in the form of the *KKT conditions*: there exist *Lagrangian multipliers* $\lambda_*^g \in \mathbb{R}^m$ and $\lambda_*^h \in \mathbb{R}^l$ so that

$$\nabla f(x_*) + \nabla g(x_*)\lambda_*^g + \nabla h(x_*)\lambda_*^h = 0, \tag{17.2a}$$
$$g(x_*) \leq 0, \tag{17.2b}$$
$$h(x_*) = 0, \tag{17.2c}$$
$$g_j(x_*)\lambda_{*,j}^g = 0 \qquad \text{for all } j = 1, \ldots, m, \tag{17.2d}$$
$$\lambda_*^g \geq 0. \tag{17.2e}$$

Defining the Lagrangian function $\mathscr{L}(x, \lambda^g, \lambda^h) = f(x) + g(x)^T \lambda^g + h(x)^T \lambda^h$, the first equation can compactly be written as $\nabla_x \mathscr{L}(x_*, \lambda_*^g, \lambda_*^h) = 0$. Geometrically, this means that the projection of the negative objective gradient onto L is zero. Condition (17.2d) is called a *complementarity condition*; it states that either a particular inequality constraint is *active* (i.e., it is satisfied with equality), or its corresponding Lagrangian multiplier is zero. Intuitively it makes sense that an inequality constraint that remains strictly satisfied for all small changes in x cannot play a role in local optimality conditions. It is important to note that (17.2) are only necessary conditions for x^* to be a local minimizer.

Most NLO solvers aim at finding points that satisfy the KKT conditions. However, recall that the KKT conditions are based on an approximation L of the feasible region derived from the linearization of the constraint functions. If this approximation represents the true feasible region F sufficiently well, the KKT conditions are indeed optimality conditions for a local solution of NLO. Conditions that guarantee that this is the case are called *constraint qualifications (CQs)*. An example is the *linear-independence constraint qualification (LICQ)*, which states that the gradients of the constraints active at x_* (including all equality constraints) are linearly independent. A looser constraint qualification is the *Mangasarian–Fromovitz constraint qualification (MFCQ)* [1402]. To summarize this as a theorem: if x_* is a local solution to NLO problem (17.1) at which a CQ holds, then there exist $\lambda_*^g \in \mathbb{R}^m$ and $\lambda_*^h \in \mathbb{R}^l$ so that (17.2) hold.

---

[12]Throughout this chapter, we assume for simplicity that all functions are twice Lipschitz continuously differentiable, even though many results hold under weaker assumptions. The notation $\nabla f(x)$ stands for the gradient of f as a column vector evaluated at x, and $\nabla^2 f(x)$ is the Hessian matrix of $f(x)$ at x. For a vector-valued function c, the matrix $\nabla c(x)$ denotes the transpose of its Jacobian at x, i.e., the columns of $\nabla c(x)$ correspond to the gradients of c_i at x.

There are also a number of sufficient optimality conditions that typically involve the second derivatives of the problem functions. An example of such a second-order sufficient optimality condition (SSOC) is the following: if $x*$, λ_*^g, and λ_*^h satisfy (17.2), *strict complementarity* holds (i.e., $-g_j(x_*) + \lambda_{*,j}^g > 0$ for all $j = 1,\ldots,m$), and $w^T \nabla_{xx}^2 \mathcal{L}(x_*, \lambda_*^g, \lambda_*^h)w > 0$ for all $w \in L \setminus \{0\}$, then x_* is a local minimizer of NLO problem (17.1).

From a theoretical standpoint, most NLO methods are only proved to converge to points that satisfy the KKT conditions (under certain assumptions pertaining to smoothness and boundedness of the problem functions, as well as CQs). Since these are only necessary conditions, a method might converge to other points that satisfy the KKT conditions and might only be saddle points or even maxima. Even so, most algorithms take into account the objective function explicitly when choosing a new iterate (see Section 17.4.3), so that they are usually attracted to local minima. Some methods explicitly exploit directions of negative curvature to escape maxima or saddle points.

It is important to note that in the absence of a CQ, no multipliers satisfying (17.2) might exist at an optimal solution, and trying to find KKT points will fail. Furthermore, some methods require that LICQ hold, a relatively strong CQ that is often not satisfied in practice. Optimization codes might still converge when these assumptions do not hold. However, their practical performance is then often less reliable or slow, or the returned point might not even be a stationary point. It is important to keep this in mind when formulating an optimization model and, for example, to avoid constraints that are redundant or whose gradients might become zero.

Termination tests in software implementations of NLO algorithms typically require that some first-order optimality measure, such as the residual of the KKT conditions (17.2), be smaller than a specified tolerance. Specifically for nonconvex problems, the returned point might only be a stationary point, not a local minimum.

For very ill-conditioned or badly scaled problems, the solver might not be able to satisfy the regular termination tests. Sophisticated software implementations include a number of often undocumented heuristics that attempt to prevent failure when the method appears to break down. For example, the code might stop at a point if the difference between the iterates becomes very small. But in this case there is no guarantee that this point is close to a local minimizer. In general, when the return message indicates that the default termination criteria are not satisfied, the return point is often still a "reasonable" solution, but the user is advised to look at it somewhat critically, perform some sanity checks, and attempt to write the optimization model in some different way.

17.4 ▪ Fundamental NLO Algorithm Concepts

To facilitate the description of the NLO algorithms in Section 17.5, we first introduce a few common concepts. We start with techniques for unconstrained optimization, where one seeks to find a local minimizer of

$$\min_{x \in \mathbb{R}} f(x), \tag{17.3}$$

and extend them to the constrained setting.

Even if the original problem has constraints, it is sometimes advantageous to eliminate variables by exploiting the fact that equality constraints implicitly define some of the variables as a function of others. Problem function evaluations then become more expensive, but this strategy reduces the problem size and makes it easier to satisfy highly nonlinear constraints. This approach is used in Chapter 19, where the governing equations $R(x, y(x)) = 0$ are solved to find the states $y(x)$ as a function of the design variables x.

In some applications, there are several objective functions that each have their individual merit. Multiobjective optimization (discussed in Section 20.4 and used for the examples in Sections 18.3, 19.4, and 20.4) deals with the exploration of the trade-off between the different goals.

17.4.1 ▪ Newton's Method and Quasi-Newton Approximations

The negative of the gradient of the objective, $-\nabla f(x_k)$, is the direction of steepest descent at an iterate x_k. Therefore, it is tempting to use the *steepest descent* method, which, at each iterate x_k, minimizes the objective along the direction $d_k = -\nabla f(x_k)$. However, this usually leads to inefficient algorithms (with linear convergence rate at best) due to potential zigzagging behavior (see, e.g., [1402]). In contrast, Newton's method exhibits fast local convergence properties. Here, the original nonlinear objective function (17.3) is replaced by its second-order Taylor approximation at the current iterate x_k,

$$\min \ f(x_k) + \nabla f(x_k)^T d + \tfrac{1}{2} d^T B_k d \quad (\approx f(x_k + d)), \tag{17.4}$$

where $B_k = \nabla^2 f(x_k)$ is the Hessian matrix of $f(x)$ calculated at x_k. From the optimality conditions in the previous section, it follows that the optimal solution d_k of this quadratic problem can be computed by solving the linear system $B_k d_k = -\nabla f(x_k)$, as long as B_k is positive definite.

At an iterate x_k, the basic Newton algorithm generates the new iterate as the minimizer of the quadratic approximation, i.e., $x_{k+1} = x_k + d_k$, where d_k is the optimal solution of (17.4). Like its counterpart for the solution of nonlinear systems of equations, it converges to the optimal solution x_* with a quadratic convergence rate if $\nabla^2 f(x_*)$ is nonsingular, and if the initial iterate x_0 is sufficiently close to x_*. However, if x_0 is far from x_*, then the sequence x_k might not converge. Even worse, away from x_*, the Hessian $\nabla^2 f(x_k)$ might not be positive definite, and the quadratic problem (17.4) might be unbounded and the new iterate not well defined.

When second derivatives are not available, the Hessian matrix can be approximated using a *quasi-Newton* approach. Here, an estimate $B_k \approx \nabla^2 f(x_k)$ is maintained and updated in each iteration based on gradient information $\nabla f(x_k)$ and $\nabla f(x_{k-1})$ that has been computed already. The most widely used updating scheme is the BFGS formula [1402], which ensures that B_k is always positive definite, so that the model (17.4) is never unbounded. (For a nonconvex objective function, some safeguards are required.) Under certain assumptions, a local superlinear convergence rate can be proved. *Limited-memory* versions exist for some quasi-Newton methods that avoid the explicit storage of the dense B_k matrix but can only achieve a theoretical linear convergence rate at best.

17.4.2 ▪ Line-Search and Trust-Region Globalization Techniques

As just discussed, the basic Newton method is not guaranteed to converge unless it is started close to the optimal solution. To ensure *global convergence*, i.e., convergence from any starting point (not to be confused with the concept of finding a global optimum), two mechanisms as follows are typically used that carefully control which points can be accepted as new iterates:

- *Line-search methods* compute a search direction d_k from the solution of (17.4) and explore a sequence of trial points $x_k + \alpha d_k$ for different step sizes $\alpha \in (0, 1]$ (using a backtracking procedure $\alpha = 2^{-i}$, $i = 0, 1, 2, \ldots$, for example) until the corresponding objective function value is sufficiently smaller than $f(x_k)$.
- *Trust-region methods* [496] solve a variation of (17.4) in which the constraint $\|d\| \leq \Delta_k$ is added with a given parameter $\Delta_k > 0$. This quantity is updated during the iterations and is called the trust-region radius because it determines the area around the current iterate in which the Taylor approximation is trusted to be a sufficiently good approximation of the true function. If the step d_k, obtained as the optimal solution from the trust-region problem, leads to a sufficient reduction in the objective function, the trial point is accepted as the new iterate and the current trust-region radius is potentially increased. Otherwise, the trial point is rejected and a new radius Δ_{k+1} less than Δ_k is chosen because the model does not represent the true objective function sufficiently well within the current trust region. A new trial step is then computed from the updated trust-region subproblem.

Comparing the two, line-search methods typically require less work per iteration because a search direction is obtained simply by solving a linear system, and it is used to explore several trial points, whereas each trial point in a trust-region approach is generated by the solution of an optimization problem. On the downside, line-search methods require that B_k be positive definite, which is not guaranteed for the choice $B_k = \nabla^2 f(x_k)$ away from a local solution. As a remedy, modifications of the true Hessian have to be made (e.g., by a modified Cholesky factorization or by adding a multiple of the identity matrix), or a suitable quasi-Newton method has to be used. In contrast, trust-region methods handle indefinite B_k matrices, such as the true Hessian matrices, in a natural way that often leads to fewer iterations. Typically, trust-region methods are preferred if ill-conditioned or highly nonlinear problems are solved.

17.4.3 ▪ Merit Functions and Filters

The globalization techniques just described have to decide whether a trial point should be accepted as a new iterate. In the case of unconstrained optimization (17.3), it is natural to require a (sufficient) decrease in the objective function $f(x)$. The situation is more complicated when constraints are present, because then the algorithm has to balance two usually competing goals: (a) reduce the objective function and (b) find a feasible point. (Most algorithms allow intermediate iterates to be infeasible and achieve feasibility only in the limit.) *Merit functions* combine these goals into a single function, and progress in the algorithm is then measured in terms of the decrease in that function.

One such function is the *exact penalty function*

$$\psi_\rho(x) = f(x) + \rho \theta(x), \tag{17.5}$$

with $\theta(x) = \|\max(g(x), 0)\| + \|h(x)\|$; here, the max operator is taken componentwise. A point x is feasible for (17.1) if and only if $\theta(x) = 0$. This penalty function explicitly balances the two goals and weights their relative importance via a penalty parameter $\rho > 0$. In theory, any norm can be chosen in the definition of $\theta(x)$, but often the ℓ_1-norm is preferred because it is less sensitive to outliers. This function is called *exact* because it can be shown, under certain regularity assumptions, that a local minimizer x_* of (17.1) with KKT multipliers λ_*^g and λ_*^h corresponds to a local minimizer of $\psi_\rho(x)$ provided that ρ is larger than the dual norm of $(\lambda_*^g, \lambda_*^h)$. Because λ_*^g and λ_*^h are not known in advance, many methods update the penalty parameter over the course of the iterations based on an estimate of the optimal multipliers. The newer *penalty-steering* methods [377] update the penalty parameter by comparing the (predicted) reduction in infeasibility made by the optimization step to the reduction obtained by a step that focuses solely on minimizing the constraint violation $\theta(x)$. A related line of recent research addresses the question of how infeasibility of the NLO problem (17.1) can be detected quickly [374].

Another frequently used merit function is the augmented Lagrangian function

$$\mathcal{L}_\rho^{\mathrm{aug}}(x, \lambda^g, \lambda^h) = f(x) + \tfrac{\rho}{2}\left\|\max\left\{0, g(x) + \tfrac{\lambda^g}{\rho}\right\}\right\|_2^2 + \tfrac{\rho}{2}\left\|h(x) + \tfrac{\lambda^h}{\rho}\right\|_2^2, \qquad (17.6)$$

where λ^g and λ^h are estimates of the optimal multipliers and $\rho > 0$ is a penalty parameter. In contrast to $\psi_\rho(x)$, this function is continuously differentiable in x. Again, under certain regularity assumptions, a local minimizer x_* of (17.1) with corresponding multipliers λ_*^g and λ_*^h is a local minimizer of $\mathcal{L}_\rho^{\mathrm{aug}}(\,\cdot\,, \lambda_*^g, \lambda_*^h)$ provided that $\rho > 0$ is sufficiently large.

A different approach is taken by a *filter method* [1402], which judges desirability of trial points by considering the two goals separately: with each point x, we associate the pair $(f(x), \theta(x))$. In its most basic form, a filter method stores some $(f(x_k), \theta(x_k))$ pairs from previous iterates (in a "filter") and accepts a new iterate if it sufficiently improves the objective or the constraint violation compared to all points in the filter. A globally convergent filter method has a number of additional ingredients, such as a restoration phase that temporarily ignores the objective function and focuses on minimizing $\theta(x)$ only.

17.5 • Algorithm Frameworks

17.5.1 • Sequential Quadratic Programming Methods

SQP methods [293] can be thought of as an extension of Newton's method for the unconstrained problem (17.3) to the constrained case. Similar to the unconstrained case, iterates are computed as solutions of a subproblem (17.4) that involves a quadratic approximation of the objective function. In addition, the nonlinear constraints in (17.1b)–(17.1c) are replaced by their linear Taylor approximation. More precisely, at an iterate x_k, a step d_k is computed as the optimal solution of the *quadratic optimization problem* (QP)

$$\min \quad f(x_k) + \nabla f(x_k)^T d + \tfrac{1}{2}d^T B_k d \qquad (17.7a)$$

$$\text{s.t.} \quad \nabla g(x_k)^T d + g(x_k) \le 0, \qquad (17.7b)$$

$$\nabla h(x_k)^T d + h(x_k) = 0. \qquad (17.7c)$$

Here, B_k is the Hessian of the Lagrangian function evaluated at x_k for a current estimate of multipliers λ_k^g and λ_k^h; i.e., $B_k = \nabla^2 f(x_k) + \sum \lambda_{k,i}^g \nabla^2 g_i(x_k) + \sum \lambda_{k,i}^h \nabla^2 h_i(x_k)$. As before, B_k can also be an approximation of the true Hessian, using a quasi-Newton approach. A new iterate x_{k+1} is then computed based on the step d_k. The multiplier iterates λ_{k+1}^g and λ_{k+1}^h are updated based on the optimal multipliers from (17.7) or are estimated in some other way (e.g., as least-squares solutions of (17.2a)).

To ensure global convergence in SQP methods, both line-search and trust-region options are used, where the acceptance of trial points relies on the different merit functions or the filter method discussed in the previous section. For the trust-region approach, the QP (17.8) is augmented by the constraint $\|d\| \leq \Delta_k$.

17.5.1.1 ▪ Equality-Constrained NLO Problems

Before handling the general formulation (17.1), we first consider the case where no inequality constraints (17.1b) are present. Then, the QP solved in each iteration is

$$\min \quad f(x_k) + \nabla f(x_k)^T d + \tfrac{1}{2} d^T B_k d \tag{17.8a}$$

$$\text{s.t.} \quad \nabla h(x_k)^T d + h(x_k) = 0. \tag{17.8b}$$

It follows from the KKT conditions that a stationary point of (17.8) can be computed from the linear system

$$\begin{bmatrix} B_k & A_k^T \\ A_k & 0 \end{bmatrix} \begin{pmatrix} d_k \\ \tilde{\lambda}_{k+1}^h \end{pmatrix} = - \begin{pmatrix} \nabla f(x_k) \\ h(x_k) \end{pmatrix}. \tag{17.9}$$

Here, $A_k = \nabla h(x_k)^T$ and $\tilde{\lambda}_{k+1}^h = \lambda_k^h + d_k^{\lambda^h}$, where $d_k^{\lambda^h}$ is the Newton step for the multiplier values. It is not difficult to see that the solution of this linear system is actually the Newton step for KKT conditions corresponding to the *original* NLO problem (17.1a) and (17.1c) at the iterate (x_k, λ_k^h), as long as B_k is the Hessian of the Lagrangian function; this observation is the motivation behind this particular choice of B_k. As a consequence, the SQP method enjoys the fast local convergence rate of Newton's method.

Unless the evaluation of the problem functions and their derivatives is very time-consuming, the solution of the QP (17.8) is typically the computational bottleneck. In some cases, the linear system (17.9) is solved directly using a sparse factorization method for symmetric indefinite linear systems as it is implemented in sophisticated numerical linear algebra packages. A different (often called *reduced-space*) approach decomposes the step $d_k = v + w$ into a component v toward feasibility that aims at satisfying $A_k v + h(x_k) = 0$ and a component w that aims at improving the objective function while keeping the (linearized) constraint violation unchanged.

There are different ways to perform this decomposition. Here, we highlight a particular option that is natural for many engineering applications. Often, the optimization variables can be partitioned into "free variables" x^f (such as design or control variables) and "dependent variables" x^d (such as state variables), say $x = (x^d, x^f)$. The dependent variables are uniquely determined by the equality constraints (17.1c) once the free variables x^f have been chosen. Take, as an example, a physical system with controllable quantities x^f (e.g., input flow rates). For a given control $\bar{x}^f$, the state x^d of the system (described by quantities such as flow rates and pressures) is determined by energy and mass balances, etc., which corresponds to a nonlinear system of equations $h(\cdot, \bar{x}^f) = 0$. Then, when the constraint Jacobian is partitioned according to the variables, $A = [C_k \ N_k]$, the submatrix C_k corresponding to the state variables is usually nonsingular. To exploit the decomposition in a line-search algorithm, recall that the solution of the QP (17.8) is obtained from the linear system (17.9). (The situation is more complicated in the trust-region case, due to the additional constraint for the step size.) After defining $Y_k = \begin{bmatrix} I \\ 0 \end{bmatrix}$ and $Z_k = \begin{bmatrix} -C_k^{-1} N_k \\ I \end{bmatrix}$, we can compute the solution $d_k = v + w$ of (17.9) from

$$p_Y = -[A_k Y_k]^{-1} h(x_k) = -C_k^{-1} h(x_k), \tag{17.10a}$$

$$p_Z = -[Z_k^T B_k Z_k]^{-1} Z_k^T (\nabla f(x_k) + B_k Y_k p_Y), \tag{17.10b}$$

with $v = Y_k p_Y$ and $w = Z_k p_Z$. The matrix $Z_k^T B_k Z_k$ is commonly called the *reduced Hessian matrix* and can be proved to be positive semidefinite at a local optimum of the NLO problem. If needed, the QP multipliers can be calculated from $\hat{\lambda}_{k+1}^h = -[A_k Y_k]^{-T}$ $(Y_k^T \nabla f(x_k) + Y_k^T B_k d_k)$.

Note that $p_Y = -C_k^{-1} h(x_k)$ is the Newton step for $h(\,\cdot\,, \bar{x}^f) = 0$. Because this quantity is required if one computes the state x^d for the system with a fixed control $\bar{x}^f$ by Newton's method, efficient techniques for computing p_Y are often already implemented. For example, if $h(x)$ arises from a discretization of a PDE, C_k could be the discretized operator matrix for which iterative linear solvers with efficient preconditioners are available. Then, the reduced-space approach makes it possible to extend efficient solution techniques for the simulation of a physical system to the optimization of that system. Often, C_k has special characteristics, such as an almost block diagonal structure, that can be exploited.

If iterative linear solvers are employed for the solution of (17.9), it is beneficial not to require highly accurate solutions to the linear system. To make this work, the optimization method must be adjusted to account for inexact steps that satisfy specific accuracy requirements (see, e.g., [528, 924, 1982]).

In this discussion, we implicitly assumed that the reduced Hessian in (17.10b) is positive definite since otherwise the equality-constrained QP (17.8) would be unbounded. For nonconvex problems, the reduced true Lagrangian Hessian $Z_k^T \nabla_{xx}^2 \mathcal{L}(x_k, \lambda_k^h) Z_k$ might be indefinite away from an optimal solution. Consequently, similar to the unconstrained case, care must be taken when true second-order information is used. When the reduced Hessian $Z_k^T B_k Z_k$ is explicitly available in a decomposition approach, positive-definiteness can be verified explicitly (e.g., via a Cholesky factorization). Alternatively, when the search direction is computed from the linear system (17.9) directly, inertia-revealing factorization methods provide the number of negative eigenvalues of the indefinite matrix in (17.9). The reduced Hessian is positive definite if the matrix in (17.9) is nonsingular and the number of negative eigenvalues is equal to the number of equality constraints, m. If there are more negative eigenvalues, a suitable modification, such as $B_k = \nabla_{xx}^2 \mathcal{L}(x_k, \lambda_k^h) + \delta I$ with $\delta \geq 0$, can be found by trial and error.

In a situation where second-order derivatives are not available or too expensive to compute, maintaining a BFGS approximation of the Hessian is an option. Because the true Hessian $\nabla_{xx}^2 \mathcal{L}(x_*, \lambda_*^h)$ might not be positive definite in the full space $\mathbb{R}^n$ even for a local solution that satisfies SSOC, its approximation via the original BFGS update can be inefficient. In practice, variants of the BFGS methods are employed nevertheless, using a modification of the update formula or simply skipping updates if necessary. However, this can lead to slow linear local convergence or even failure when B_k becomes very ill-conditioned. Faster convergence can be achieved for variants in which the reduced Hessian $Z_k^T \nabla_{xx}^2 \mathcal{L}(x_k, \lambda_k^h) Z_k$ is approximated directly in a decomposition approach.

17.5.1.2 ▪ General NLO Problems

We now turn our attention to the general case (17.1), which involves inequality constraints. In the SQP framework, the QP from which the steps are computed then includes the linearization of those constraints; see (17.7). Before discussing how this QP is solved, let us point out an important property. It can be shown that, if x_* is a local solution of (17.1) satisfying LICQ and SSOC and if x_k is sufficiently close to x_*, then the linearized constraints that are active in QP (17.7) are exactly those that correspond to the original constraints in (17.1) that are active at the optimal solution x_*. In other words, close to x_*, the steps are generated as if the inactive constraints were dropped from the original problem, and as if all remaining constraints were posed with equality. Therefore, the general SQP method behaves like the SQP method for equality constraints discussed above, and fast local convergence can be achieved.

17.5.1.3 • Active Set QP Solvers

Finding the optimal solution of the general QP (17.7) is considerably more difficult than solving the equality-only QP (17.8). If the optimal *active set*, the set of constraints that hold with equality at the solution of (17.7), is known, the inactive constraints can be ignored and the solution can again be computed from a linear system,

$$\begin{bmatrix} B_k & A_{\mathscr{W}_k}^T \\ A_{\mathscr{W}_k} & 0 \end{bmatrix} \begin{pmatrix} d_{\mathscr{W}_k} \\ \lambda \end{pmatrix} = -\begin{pmatrix} \nabla f(x_k) \\ c_{\mathscr{W}_k} \end{pmatrix}, \tag{17.11}$$

where the rows of $A_{\mathscr{W}_k}$ consist of the gradients of the constraints that are in the active set, say $\mathscr{W}_k$, and $c_{\mathscr{W}_k}$ are the corresponding right-hand sides.

The challenge lies in the combinatorial problem of identifying the optimal active set. *Active set QP solvers*, as they are usually employed within SQP methods, update a *working set* $\mathscr{W}_k$, a guess of the optimal active set, within "inner" (or "minor") QP iterations. Many variants are available (e.g., primal, dual, parametric), but they all typically change the working set only by one constraint per inner iteration. As a consequence, converging to the optimal working set without prior knowledge often requires many iterations and a significant amount of time. This approach is only practical for a moderate number of inequality constraints, up to a few thousand, depending on the particular circumstances. On the upside, because the optimal working set for (17.7) does not change much during the later "outer" (or "major") iterations of the SQP method, active set QP solvers can begin the solution of a new QP from the previous optimal working set and then solve the new QP in very little time. This *warm-start* feature can be exploited further when a sequence of nonlinear problems (17.1) with few data changes has to be solved, e.g., in online optimal control; then a good initial working set is known even in the first SQP iteration for a new NLO problem. Interior-point QP solvers (see Section 17.5.2) usually do not perform well within an SQP method, because they are difficult to warm-start.

The iterate $d_{\mathscr{W}_k}$ corresponding to the working set $\mathscr{W}_k$ is the solution of the linear system (17.11). Many of the approaches discussed earlier for the solution of (17.9) can also be used here. However, because the working set changes only by one constraint at a time, it would be very inefficient to solve the linear system from scratch in each inner iteration. Instead, active set solvers typically maintain some kind of factorization that can be updated. For example, in the reduced-space approach, a change in the working set corresponds to adding or removing a row and a column of C_k and $Z_k^T B_k Z_k$.

One problematic situation occurs when the constraints of the QP render the subproblem infeasible (particularly in the presence of the trust-region constraint $\|d\| \leq \Delta_k$) and no step can be computed, even when the original NLO problem is feasible. In the decomposition approach, this can be handled by allowing the feasibility component v to make only partial progress toward linearized feasibility (as in the KNITRO code). Alternatively, the "flexible mode" (as in the SNOPT code) modifies the NLO objective and adds the constraint violation $\theta(x)$ with a large weight.

17.5.1.4 ▪ General Purpose Solvers

A number of robust and efficient implementations of SQP methods are available. Examples of general purpose solvers are FILTERSQP (a trust-region method using filter globalization that exploits exact second derivatives) and SNOPT (a line-search method based on a variant of the augmented Lagrangian merit function that approximates the reduced Hessian matrix via limited-memory BFGS).

A different active set approach (sequential linear optimization with equality-constrained QPs) with the potential of addressing larger NLO problems has been developed [375, 457]. Here, in each iteration of the SQP method, a working set $\mathcal{W}_k$ is determined by solving a linear optimization problem; this can usually be done much faster than solving QP (17.7). The optimization step is then obtained by solving the linear system (17.11). This method is available as the "active set" option in the KNITRO solver.

17.5.2 ▪ Interior-Point Methods

Because of the combinatorial complexity of the QP solver, the active set SQP approach is usually only efficient up to a moderate problem size (a few thousand inequality constraints). Interior-point methods [745], also known as barrier methods, avoid the explicit selection of a working set and are capable of solving very large problems. But they typically require more (outer) iterations and are less efficient than SQP methods for small and medium-sized problems, particularly when function evaluations are expensive.

Most interior-point methods solve a reformulation of (17.1) where general inequality constraints are replaced by equality constraints and nonnegative slack variables:

$$\min \quad f(x) \tag{17.12a}$$
$$\text{s.t.} \quad h(x) = 0, \tag{17.12b}$$
$$x \geq 0. \tag{17.12c}$$

The crucial difference between these and active set SQP methods is that the inequality constraints (17.12c), which are responsible for the combinatorial complexity, are replaced by a *barrier term* that is added to the objective function:

$$\min \quad \phi_\mu(x) \equiv f(x) - \mu \sum_{i=1}^{n} \log(x_i) \tag{17.13a}$$
$$\text{s.t.} \quad h(x) = 0. \tag{17.13b}$$

Here, $\mu > 0$ is the *barrier parameter*. The logarithm term in (17.13a) goes to infinity as $x > 0$ approaches the boundary of the region defined by the bound constraints (17.12c). It acts as a barrier for the bound constraints in the sense that the optimal solution of (17.13) is "pushed away" from the boundary of the nonnegative orthant. The influence of the barrier term is determined by the size of the barrier parameter, and it can be shown, under certain regularity assumptions such as LICQ and SSOC, that a local solution of the original NLO problem (17.12) can be approached by local solutions $x_*(\mu)$ of (17.13) as $\mu \to 0$.

This observation is the foundation of the classical barrier algorithm that solves barrier problems (17.13) for a sequence of barrier parameters $\{\mu_l\}$ with $\mu_l \to 0$. Because we are not interested in a highly accurate solution of (17.13) for a large value of μ_l, the corresponding barrier problem is solved only approximately, to a tolerance usually proportional to μ_l. After an update of μ_l, the solution of the new barrier problem is started from the final iterate for the previous barrier problem. Because (17.13) has no inequality constraints, it can be solved efficiently with the SQP method for equality constraints described in Section 17.5.1. However, there are a few crucial considerations.

Even though (17.13) is formally written without inequality constraints, the presence of the logarithm in the objective $\phi_\mu(x)$ implicitly restricts the iterates x_k to be strictly positive. To deal with this, an interior-point method ensures that x_k is always strictly in the interior of the nonnegative orthant (i.e., $x_k > 0$), justifying the term "interior-point method." As a consequence, interior-point methods are difficult to warm-start: if the starting point is chosen too close to the boundary, the initial steps often leave the nonnegative orthant and might have to be cut back significantly. This typically leads to much poorer performance compared to an active set approach. Nevertheless, in some cases sufficiently accurate approximate solutions of a perturbed problem can be computed quickly using sensitivity information [1492].

Defining $X = \mathrm{diag}(x)$ and $e = (1,\ldots,1)^T$ allows us to write down the gradient $\nabla\phi_\mu = \nabla f(x) - \mu X^{-1} e$ and the Hessian $\nabla^2\phi_\mu(x) = \nabla^2 f(x) + \Sigma$ with $\Sigma = \mu X^{-2}$ of the barrier objective function. These are unbounded when a component of x approaches (its potentially optimal value of) zero. As a result, the matrix in (17.9) becomes increasingly ill-conditioned as the method converges to an optimal solution with an active bound constraint; the matrix is not even well defined in the limit. Luckily, this ill-conditioning is essentially benign for the direct factorization of the matrix in (17.9), and highly accurate solutions can be computed. However, the use of an iterative linear solver for systems that involve the barrier Hessian is challenging and requires careful preconditioning.

To achieve fast local convergence, modern interior-point methods for NLO follow the primal-dual approach originally developed for linear optimization. By introducing new variables $z = \mu X^{-1} e$, the KKT conditions for the barrier problem (17.13) can be written equivalently as

$$\nabla f(x) + \nabla h(x)\lambda - z = 0, \tag{17.14a}$$
$$h(x) = 0, \tag{17.14b}$$
$$XZe = \mu e, \tag{17.14c}$$

where $Z = \mathrm{diag}(z)$. If $\mu = 0$ in (17.14) and $x, z \geq 0$, then x satisfies the KKT conditions for the original NLO problem (17.12), with multipliers λ and z. From this point of view, we may interpret the barrier approach as a homotopy method for solving the *primal-dual equations* (17.14) as $\mu \to 0$. Because the Jacobian of the primal-dual equations is nonsingular at an optimal solution of (17.12) at which LICQ and SSOC hold, one can devise update rules for μ that, overall, lead to local superlinear convergence. As before, global convergence is promoted with line-search or trust-region mechanisms.

17.5.2.1 ▪ General Purpose Solvers

Line-search versions of the primal-dual interior-point framework are implemented in IPOPT and LOQO (using merit functions and filters), and the trust-region globalization mechanism is used in IPFILTER (filter) and KNITRO (merit function). The

KNITRO code uses a step decomposition where the linear system in the reduced space is solved with the iterative conjugate gradient method. All of these codes are based on the primal-dual framework, and some give the option to approximate the Hessian of the Lagrangian function with the limited-memory BFGS formula.

17.5.3 ▪ Other Methods

Augmented Lagrangian methods are based on the function (17.6). In an outer loop, they update the penalty parameter ρ and the multiplier estimates λ^g and λ^b, and for each such choice, the function $\mathscr{L}_\rho^{\text{aug}}(\,\cdot\,, \lambda^g, \lambda^b)$ is optimized (approximately) in an inner loop, using a suitable method for unconstrained optimization (e.g., in the ALGENCAN solver). The LANCELOT code works with the formulation (17.12) and solves a bound-constrained subproblem. The advantage of augmented Lagrangian methods is that the Newton steps for the inner loop can be computed with an iterative linear solver method, so that only vector products with Hessian matrices are required.

Also based on the augmented Lagrangian function is the MINOS code, which solves a sequence of linearly constrained subproblems using a reduced-space method with quasi-Newton approximations of the reduced Hessian. The CONOPT solver is based on a *generalized reduced gradient* method, which attempts to keep the iterates close to the feasible region.

17.6 ▪ Extensions

Mathematical programs with complementarity constraints (MPCCs) are NLO problems that include complementarity constraints of the form $x_1, x_2 \geq 0$ and $x_1 x_2 = 0$. Even though CQs fail to hold, some standard NLO solvers (particularly active set methods) may still work satisfactorily in practice [728]. However, special treatment of these constraints is often necessary. Section 18.2.2 presents several reformulations that are compared in Section 18.3.2.

Some applications call for conic constraints (such as second-order cone (SOC) constraints or requiring semi-positive-definiteness of matrices) that cannot be cast into the standard NLO formulation (17.1). The PENNON solver augments the interior-point framework described above and employs different barrier terms that can handle these conic constraints.

Chapter 18

Optimization of Distillation Systems

Lorenz T. Biegler

18.1 ▪ Introduction

Distillation is the most widely used unit operation for the separation of chemical species. Because it has no moving parts and scales easily and economically to all production levels, distillation is one of the most important tasks in petroleum refining and chemical processing. On the other hand, distillation is highly energy intensive and can consume from 40% to 90% of the total energy in a typical chemical or petrochemical process. Overall, distillation operations require about a quarter of the energy consumed by the U.S. manufacturing sector and about 3% of the total energy consumed in the United States. As a result, optimization of distillation processes is essential for their profitability. Moreover, because the input data for distillation models, including feed conditions, product demands, and ambient conditions, change frequently, it becomes important to update the optimal response to these changes to ensure successful operation of these systems.

Moreover, distillation optimization problems are a rich source of challenges that apply to current topics in nonlinear optimization (NLO). First, optimization models for distillation form a block tridiagonal equation structure that scales in a number of ways. The number of diagonal blocks corresponds to the number of equilibrium stages or trays, typically from 10 to 200; the size of each block is proportional to the number of chemical species, ranging from 2 to 1,000. Second, phase equilibrium, which can encompass various elements of nonideal thermodynamics, will govern the degree of nonlinearity. Optimization requires the satisfaction of product specifications as well as adjustment of heat addition and cooling, internal flow rates, and the number of equilibrium stages. Third, distillation columns operate over a limited range, as operations require equilibrium of both liquid and vapor phases on each tray. Hence, the need to model disappearing phases on distillation trays leads to interesting mathematical programs with complementarity constraints (MPCCs). Finally, there are interesting ways to combine distillation column sections, so the problem structure evolves from a block tridiagonal set of equations to a blocked structure with arbitrary off-diagonal interconnections. These multicolumn structures lead to a challenging interplay of feasible operation and trade-offs in performance and energy requirements.

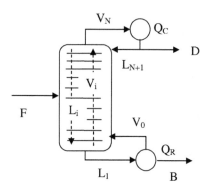

Figure 18.1. *Distillation column example* [261].

This study considers the optimal design of distillation systems using NLO formulations and state-of-the-art NLO algorithms. The basic model is considered in the next section along with an extension of this model to consider possible phase disappearance on a tray through complementarity conditions. This is illustrated by two distillation case studies in Section 18.3. Finally, Section 18.4 concludes this study with a brief discussion of problem extensions to multicolumn systems and dynamic optimization of distillation systems.

18.2 ▪ Distillation Optimization Models

There is a broad literature devoted to distillation modeling, process synthesis, and optimization. Textbooks that deal with fundamental concepts, models, and design strategies include [606, 1877]. Comprehensive reviews and case studies of distillation optimization can be found in [729, 862, 863, 1233]. To deal with the modeling and optimization tasks, we begin with the following basic model of a single distillation column.

18.2.1 ▪ Basic Distillation Model

We consider the distillation column shown in Figure 18.1 with N stages. In an actual distillation column, each physical tray provides well-mixed contact of internal liquid and vapor streams to approach phase equilibrium (i.e., boiling), which corresponds to an equilibrium stage. Although perfect equilibrium stages are rarely realized in practice, ideal stage models are still the most widely considered, with any deviations from equilibrium addressed through tray efficiencies or additional mass transfer relations. Through a cascade of equilibrium stages, countercurrent flow of liquid and vapor enriches low-boiling components in the top product and high-boiling components in the bottom product. Two heat exchangers, the top condenser and the bottom reboiler, act as sources for the condensed liquid vapor (L_{N+1}) and boiled-up vapor (V_0), respectively. The process feed contains chemical components given by the set $\mathscr{C} = \{comp_1, comp_2, \ldots, comp_n\}$. The column is specified to recover most of a low-boiling component (the light key) in the top product and most of a high-boiling component (the heavy key) in the bottom product. As seen in Figure 18.1, we assume a total condenser and a partial reboiler, and that the liquid and vapor phases are in equilibrium at each stage. The corresponding distillation column model is constructed below using the MESH (mass-equilibrium-summation-heat) equations:

Total mass balances:

$$B + V_0 - L_1 = 0, \tag{18.1}$$
$$L_i + V_i - L_{i+1} - V_{i-1} = 0, \ i \in [1,N], i \notin \mathscr{S},$$
$$L_i + V_i - L_{i+1} - V_{i-1} - F = 0, \ i \in \mathscr{S},$$
$$L_{N+1} + D - V_N = 0.$$

Component mass balances for each $j \in \mathscr{C}$:

$$B x_{0,j} + V_0 y_{0,j} - L_1 x_{1,j} = 0,$$
$$L_i x_{i,j} + V_i y_{i,j} - L_{i+1} x_{i+1,j} - V_{i-1,j} y_{i-1,j} = 0, \ i \in [1,N], i \notin \mathscr{S},$$
$$L_i x_{i,j} + V_i y_{i,j} - L_{i+1} x_{i+1,j} - V_{i-1} y_{i-1,j} - F x_{F,j} = 0, \ i \in \mathscr{S},$$
$$(L_{N+1} + D) x_{N+1,j} - V_N y_{N,j} = 0,$$
$$x_{N+1,j} - y_{N,j} = 0.$$

Heat balances:

$$B H_B + V_0 H_{V,0} - L_1 H_{L,1} - Q_R = 0, \tag{18.2}$$
$$L_i H_{L,i} + V_i H_{V,i} - L_{i+1} H_{L,i+1} - V_{i-1} H_{V,i-1} = 0, \ i \in [1,N], i \notin \mathscr{S},$$
$$L_i H_{L,i} + V_i H_{V,i} - L_{i+1} H_{L,i+1} - V_{i-1} H_{V,i-1} - F H_F = 0, \ i \in \mathscr{S},$$
$$V_N H_{V,N} - (L_{N+1} + D) H_{L,D} - Q_C = 0.$$

Summation, enthalpy, and equilibrium relations:

$$\sum_{j=1}^{m} y_{i,j} - \sum_{j=1}^{m} x_{i,j} = 0, \ i = 0, \ldots, N, \tag{18.3}$$
$$y_{i,j} - K_{i,j}(T_i, P, x_i) x_{i,j} = 0, \ j \in \mathscr{C}, i = 0, \ldots, N,$$
$$H_{L,i} = \varphi_L(x_i, T_i), \ H_{V,i} = \varphi_V(y_i, T_i), \ i = 1, \ldots, N,$$
$$H_B = \varphi_L(x_0, T_0), \ H_F = \varphi_L(x_F, T_F), \ H_{N+1} = \varphi_L(x_{N+1}, T_{N+1}),$$

where

i	tray index number starting from reboiler ($= 0$),
P	pressure in the column,
$\mathscr{S} \in \{1, \ldots, N\}$	set of feed tray locations in column,
F	feed, molar flow rate,
L_i/V_i	molar flow rate of liquid/vapor leaving tray i,
T_i	temperature of tray i,
H_F	feed enthalpy,
$H_{L,i}/H_{V,i}$	enthalpy of liquid/vapor leaving tray i,
x_F	feed mole fraction,
$x_{i,j}$	mole fraction component j in liquid leaving tray i,
$y_{i,j}$	mole fraction component j in vapor leaving tray i,
$K_{i,j}$	nonlinear vapor/liquid equilibrium constant,
φ_V/φ_L	nonlinear vapor/liquid enthalpy function,
D/B	distillate/bottoms, molar flow rate,
Q_R/Q_C	heat load on reboiler/condenser.

In addition, the model includes a set of thermodynamic relations to compute equilibrium constants, $K_{i,j}(T_i,P,x_i)$, and the enthalpy of saturated liquid and saturated vapor, $H_{L,i} = \varphi_L(x_i,T_i)$, $H_{V,i} = \varphi_V(y_i,T_i)$, as functions of composition, pressure, and temperature. Depending on the nonideality of the mixture thermodynamics, these relations are often quite complex and may constitute the majority of equations in the model. Often quantities such as $K_{i,j}, H_{L,i}, H_{V,i}$ are not given as explicit formulas but computed by callable procedures for $K_{i,j}(T_i,P,x_i)$, $H_{L,i} = \varphi_L(x_i,T_i)$, $H_{V,i} = \varphi_V(y_i,T_i)$. These provide function values and, in some packages, derivatives with respect to their inputs.

For the optimization problem, internal flow rates, tray pressures, process feeds and their phase states, products, and side draw-offs serve as continuous decision variables, while the number of stages is an integer variable. For fixed feed, tray number, and pressure, the flow rates for V_0 and L_{N+1} can be chosen as decisions, and all of the other quantities can be calculated from equations (18.1)–(18.4). Constraints are also specified on product flows and compositions, and a typical optimization problem is given as

$$\min \quad \Phi(Q_R, Q_C, D, B) \tag{18.4a}$$
$$\text{s.t.} \quad (18.1)-(18.4), \tag{18.4b}$$
$$g(x_1, x_{N+1}, D, B) \le 0, \tag{18.4c}$$
$$L_i, V_i, T_i \ge 0 \ , i = 1,\ldots,N+1, \tag{18.4d}$$
$$D, Q_R, Q_C \ge 0, \tag{18.4e}$$
$$y_{i,j}, x_{i,j} \in [0,1], j \in \mathscr{C}, \ i = 1,\ldots,N+1, \tag{18.4f}$$

where the objective may incorporate energy costs and product sales. Solution of this optimization problem is often difficult, and initialization and solution strategies have been detailed in [729, 1233]. On the other hand, a key issue is that equilibrium of liquid and vapor phases, described by (18.4), occurs over only a limited range. Fortunately, this range can be extended by allowing dry or vaporless trays, which leads to a realistic (albeit undesired) relaxation of column operation. A straightforward modification of (18.4) and the addition of complementarity constraints, and solution of an MPCC, allow this extension to be realized.

18.2.2 ▪ MPCC Formulation

The equilibrium equations (18.4) arise from the first-order optimality conditions of the following optimization problem:

$$\min_{l_j, v_j} \quad G(T,P,l,v) = \sum_{j \in \mathscr{C}} l_j \, \tilde{G}_j^L + \sum_{j \in \mathscr{C}} v_j \, \tilde{G}_j^V \tag{18.5}$$
$$\text{s.t.} \quad \sum_{j \in \mathscr{C}} l_j \ge 0, \quad \sum_{j \in \mathscr{C}} v_j \ge 0,$$
$$l_j + v_j = m_j^T > 0, j \in \mathscr{C},$$

where $G(T,P,l,v)$ is the total Gibbs free energy:

$$\tilde{G}_j^L = \tilde{G}_j^{ig}(T,P) + RT \, \ln(f_j^L),$$
$$\tilde{G}_j^V = \tilde{G}_j^{ig}(T,P) + RT \, \ln(f_j^V).$$

$\bar{G}_j^{ig}$ is the ideal gas free energy per mole for component j, f_j^L and f_j^V are the mixture liquid and vapor fugacities for component j, l_j and v_j are the moles of component j in the liquid and vapor phase, R is the gas constant, and m_j^T are the total moles of component j for the given mixture. Applying the Gibbs–Duhem relation [261] allows the first-order KKT conditions to be simplified as

$$RT \ln(f_j^V / f_j^L) - \alpha_V + \alpha_L = 0, \tag{18.6a}$$

$$0 \le \alpha_L \perp \sum_{j \in \mathscr{C}} l_j \ge 0, \quad 0 \le \alpha_V \perp \sum_{j \in \mathscr{C}} v_j \ge 0, \tag{18.6b}$$

$$l_j + v_j = m_j^T, \ j \in \mathscr{C}, \tag{18.6c}$$

where $q \perp r$ is the complementarity operator that forces orthogonality of vectors q and r. Moreover, defining $f_j^V = \phi_j^V(T,P,y)y_j$, $f_j^L = \phi_j^L(T,P,x)x_j$, and $K_j = \phi_j^L / \phi_j^V$, where ϕ_j^L and ϕ_j^V are fugacity coefficients, leads to

$$y_j = \exp\left(\frac{\alpha_V - \alpha_L}{RT}\right) K_j x_j.$$

Defining $\beta = \exp(\frac{\alpha_V - \alpha_L}{RT})$ leads to the following complementarity system:

$$y_j = \beta K_j(T,P,x)x_j, \quad -s_L \le \beta - 1 \le s_V, \tag{18.7a}$$

$$0 \le L \perp s_L \ge 0, \quad 0 \le V \perp s_V \ge 0. \tag{18.7b}$$

In this manner, phase existence can be determined within the context of an MPCC. If a slack variable (s_L or s_V) is positive, either the corresponding liquid or the corresponding vapor phase is absent and $\beta \ne 1$ relaxes the phase equilibrium condition, as required in (18.6).

The constraints (18.7) replace the equilibrium equations (18.4) above and render problem (18.4) as the MPCC

$$\min \Phi(\mathbf{x},\mathbf{y},\mathbf{z}) \quad \text{s.t. } c(\mathbf{x},\mathbf{y},\mathbf{z}) = 0, \ g(\mathbf{x},\mathbf{y},\mathbf{z}) \le 0, \ 0 \le \mathbf{x} \perp \mathbf{y} \ge 0, \tag{18.8}$$

where $\perp$ is the complementarity operator that forces orthogonality of the nonnegative vectors $\mathbf{x}$ and $\mathbf{y}$, with $\mathbf{z}$ as the remaining variables. However, MPCCs should not be solved directly as NLO problems, because the optimality conditions of (18.8) do not satisfy the constraint qualifications (CQs). As a result, MPCC reformulations are needed that allow standard NLO solvers to be applied. Three typical reformulations are

$$\min \Phi(\mathbf{x},\mathbf{y},\mathbf{z}) \quad \text{s.t. } c(\mathbf{x},\mathbf{y},\mathbf{z}) = 0, \ g(\mathbf{x},\mathbf{y},\mathbf{z}) \le 0, \ \mathbf{x},\mathbf{y} \ge 0, \ x_i y_i = \epsilon, \tag{18.9}$$

$$\min \Phi(\mathbf{x},\mathbf{y},\mathbf{z}) \quad \text{s.t. } c(\mathbf{x},\mathbf{y},\mathbf{z}) = 0, \ g(\mathbf{x},\mathbf{y},\mathbf{z}) \le 0, \ \mathbf{x},\mathbf{y} \ge 0, \ x_i y_i \le \epsilon, \tag{18.10}$$

$$\min \Phi(\mathbf{x},\mathbf{y},\mathbf{z}) + \rho \mathbf{x}^T \mathbf{y} \quad \text{s.t. } c(\mathbf{x},\mathbf{y},\mathbf{z}) = 0, \ g(\mathbf{x},\mathbf{y},\mathbf{z}) \le 0, \ \mathbf{x},\mathbf{y} \ge 0, \tag{18.11}$$

where $\epsilon > 0$ is a relaxation parameter and $\rho > 0$ is a penalty parameter. These formulations are used in the following distillation case studies.

18.3 ▪ Distillation Case Studies

We consider two case studies based on the single model formulated as an MPCC, (18.4) with the modification (18.7), given above. While both cases formulate relatively small

NLO problems, they illustrate the challenges posed by distillation optimization. The first case study deals with a ternary system with N fixed and only two degrees of freedom for optimization. This case study shows how the MPCC handles disappearing phases and leads to interesting optimal solutions. The second case study extends this problem to deal with the optimization of the number of stages with a continuous variable relaxation of integer variables.

18.3.1 ▪ Column Optimization with Phase Changes

This column optimization case deals with optimal separation of a ternary mixture, $\mathscr{C} = \{\text{n-hexane, n-heptane, n-nonane}\}$, described in [1028]. The column has 20 trays, is operated at a pressure of 6 bar, and has negligible pressure drop across the column. A saturated liquid feed of 1 kmol/s enters on the 10th stage with mole fractions $x_F = [0.46, 0.08, 0.46]^T$. The column is modeled using (18.4) with the modification (18.7) and thermodynamics based on the Soave–Redlich–Kwong (SRK) cubic equation of state (CEOS). The constants for physical and thermodynamic properties are retrieved from the Aspen Plus process simulator [104].

Both top (n-hexane) and bottom (n-nonane) products are expected to contain small amounts of intermediate-boiling n-heptane. The objective of the optimization formulation is to minimize the reboiler duty (major operating cost) while satisfying the purity constraints for the top and/or bottom products. Note that this case study is similar to [1540], where ideal thermodynamics was used. As derived in [1028], additional inequality constraints were imposed to handle the selection of cubic roots in the CEOS and the penalty formulation (18.11) with $\rho = 50{,}000$.

To demonstrate the treatment of disappearing phases, we consider three cases differing in the minimum purity specification for top and/or bottom products. The product purity constraints and the optimal reboiler duty for these cases are given in Table 18.1. In all three cases, the product purity constraints are active at the optimal solution, and the reboiler duty increases with a higher purity specification. In Case 1, both liquid and vapor are present at all stages, and their corresponding flows are shown in Figure 18.2(a). Case 2 has a purity specification only on the bottom product. Since medium purity is acceptable in this case, the reboiler load is lower than in Case 1. Also, there is no specification on the top product, and the minimum reboiler duty solution has trays above the feed with no liquid outlet, as seen in Figure 18.2(b).

In Case 3, the feed composition itself satisfies the product purity constraint, and no further separation operation is required (trivial solution). The optimizer finds this solution by setting the reboiler duty to zero; liquid feed flows down the column and exits through the reboiler without any change in temperature or composition. All stages above the feed have no liquid and vapor flows, as shown in Figure 18.2(c).

The optimization formulation for this case study has 681 variables and 742 constraints. All three cases could be reliably solved using the NLO solver CONOPT [630, 632] with less than 1.5 CPUs (Intel Dual Core 2.4GHz processor, 2GB RAM). The same starting point was used for all cases, i.e., an Aspen Plus simulation with

Table 18.1. *Product purity constraints for the three cases of distillation column optimization.*

Case	Product purity constraints	Optimal reboiler duty (MW)
1	$x_{N+1,\text{n-hexane}} \geq 0.9,\ x_{1,\text{n-nonane}} \geq 0.9$	28.14
2	$x_{1,\text{n-nonane}} \geq 0.8$	19.337
3	$x_{1,\text{n-nonane}} \geq 0.45$	0.0

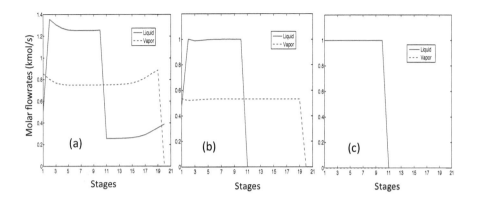

Figure 18.2. (a) *Liquid and vapor flows in the column for Case 1 in Table 18.1: high top and bottom recoveries;* (b) *liquid and vapor flows in the column for Case 2: medium bottom recovery;* (c) *liquid and vapor flows in the column for Case 3: low bottom recovery. Reprinted with permission from Elsevier* [1028].

distillate flow of 0.5 kmol/s and reflux ratio of 2. This indicates that the proposed MPCC formulation is robust; optimal solutions with three distinct flow schemes are found from the same starting point.

18.3.2 ▪ Optimal Design of Distillation Columns

Optimal determination of the distillation stage number is an essential part of column design; this has been widely addressed using a mixed-integer nonlinear optimization approach [863]. Instead, the NLO formulation in [172, 1125, 1540] is presented here, which uses a differentiable distribution function (DDF) to direct all feed, reflux, and intermediate product streams to the column trays. As shown in Figure 18.3(a), streams for the feed and the reflux are fed to all trays as dictated by two discretized Gaussian distribution functions. Note that the grayed area in Figure 18.3(a) consists only of vapor traffic, and, consequently, each tray model must include complementarities (18.7) that allow for disappearance of the liquid phase. We choose two additional continuous optimization variables: N_f, the feed location, and N with $N \geq N_f$. We also specify feed and reflux flow rates based on the value of the DDF at that tray, given by

$$F_i = \frac{\exp\left(\frac{-(i-N_f)^2}{\sigma_f}\right)}{\sum\limits_{j\in\{1,\dots,N\}} \exp\left(\frac{-(j-N_f)^2}{\sigma_f}\right)}, \quad R_i = \frac{\exp\left(\frac{-(i-N)^2}{\sigma_t}\right)}{\sum\limits_{j\in\{1,\dots,N\}} \exp\left(\frac{-(j-N)^2}{\sigma_t}\right)}, \quad i \in \{1,\dots,N\},$$

where $\sigma_f, \sigma_t = 0.5$ are parameters in the distribution. Finally, we modify the overall mass, energy, and component balances in Section 18.3.1 by allowing feed and reflux flowrates on all trays $i \in \mathscr{S}$.

This modification enables the placement of feeds, sidestreams, and number of trays in the column to be continuous variables in the DDF. On the other hand, this approach leads to upper trays with no liquid flows, which requires the complementarity constraints (18.7). The resulting model is used to determine the optimal number of trays, reflux ratio, and feedtray location for a benzene/toluene separation, using ideal

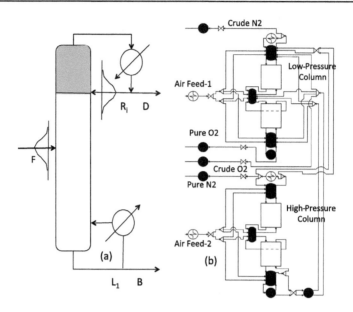

Figure 18.3. (a) *Diagram of distillation column showing feed and reflux flows distributed according to the DDF. The grayed section of the column is above the primary reflux location and has negligible liquid flows.* (b) *Double-column superstructure for air separation with column sections (rectangles) and exchangers for heating (closed circles) and cooling (open circles).*

thermodynamics. Again, three MPCC formulations, modeled in GAMS and solved with CONOPT, were considered for these cases:

- penalty formulation (PF), (18.11), with $\rho = 1,000$;
- relaxed formulation (RF), (18.10), a strategy that was solved in two NLO stages with $\epsilon = 10^{-6}$ followed by $\epsilon = 10^{-12}$; and
- Nonlinear Complementarity Problem (NCP) formulation using the Fischer–Burmeister function equivalent to (18.9), solved in three NLO stages with $\epsilon = 10^{-4}$ followed by $\epsilon = 10^{-8}$ and $\epsilon = 10^{-12}$.

The binary column has a maximum of 25 trays, its feed (F) is 100 mol/s of a 70%/30% mixture of benzene/toluene, and distillate flow (D) is specified to be 50% of the feed. The reflux ratio is allowed to vary between 1 and 20, the feedtray location varies between 2 and 20, and the total stage number varies between 3 and 25. The objective function for the benzene-toluene separation minimizes

$$\text{objective} = wt \cdot D \cdot x_{D,\text{Toluene}} + wr \cdot r + wn \cdot N,$$

where $r = L_{N+1}/D$ is the reflux ratio; D is distillate flow; $x_{N+1,\text{Toluene}}$ is the toluene mole fraction; and wt, wr, and wn are the weighting parameters for each term; these weights allow the optimization to trade off product purity, energy cost, and capital cost. The column optimizations were initialized with 21 stages, a feedtray location at the seventh stage, and a reflux ratio at 2.2. Temperature and mole fraction profiles were initialized with linear interpolations based on the top and bottom product properties. The resulting GAMS models consists of 353 equations and 359 variables for the RF formulation (18.10) and NCP formulations, and 305 equations and 361 variables for the PF formulation (18.11). The following cases were considered:

Table 18.2. *Distillation optimization: comparison of MPCC formulations. All computations were performed on a Pentium 4 with 1.8 GHz and 992 MB RAM. The asterisk (*) indicates that CONOPT terminated at a feasible point with negligible improvement of the objective function, without satisfying KKT tolerance.*

Cases	1	2	3
iter-PF (18.11)	356	395	354
iter-RF (18.10)	538	348	370
iter-NCP (18.9)	179	119	254
CPUs-PF (18.11)	7.188	7.969	4.875
CPUs-RF (18.10)	14.594	7.969	7.125
CPUs-NCP (18.9)	7.906	3.922	9.188
obj-PF (18.11)	9.4723	6.8103	3.0053
obj-RF (18.10)	9.5202*	6.8103*	2.9048*
obj-NCP (18.9)	11.5904*	9.7288*	2.9332*
reflux-PF (18.11)	1.619	4.969	1.485
reflux-RF (18.10)	1.811	4.969	1.431
reflux-NCP (18.9)	1.683	2.881	1.359
N-PF (18.11)	6.524	5.528	8.844
N-RF (18.10)	6.645	5.528	9.794
N-NCP (18.9)	9.437	9.302	9.721

- Case 1 *(wt = 1, wr = 1, wn = 1)*: This represents the base case with equal weights for toluene in distillate, reflux ratio, and stage count. As seen in the results in Table 18.2, the optimal solution has an objective function value of 9.4723 with intermediate values of r and N. This solution is found quickly by the PF formulation (18.11). On the other hand, the RF formulation (18.10) terminates close to this solution, while the NCP formulation (18.9) terminates early with poor progress.
- Case 2 *(wt = 1, wr = 0.1, wn = 1)*: In this case, less emphasis is given to energy cost. As seen in the results in Table 18.2, the optimal solution now has a lower objective function value of 6.8103 along with a higher value of r and a lower value of N. This is found quickly by both the PF and the RF formulations, although only the former satisfies the convergence tolerance. On the other hand, the NCP formulation (18.9) again terminates early with poor progress.
- Case 3 *(wt = 1, wr = 1, wn = 0.1)*: In contrast to Case 2, less emphasis is now given to the stage number. As seen in the results in Table 18.2, the optimal solution now has an objective function value of 2.9048 with lower reflux values and higher values of N. This is found quickly by the (18.10) formulation. Although (18.10) does not satisfy the convergence tolerance, the optimum could also be verified by (18.11). On the other hand, (18.11) quickly converges to a slightly different solution, which it identifies as a local optimum, while the NCP formulation requires more time to terminate with poor progress.

Along with the numerical study in Section 18.3.1, these three cases demonstrate that optimization of detailed distillation column models can be performed efficiently with MPCC formulations. In particular, it can be seen that the penalty formulation (18.11) represents a significant improvement over the NCP and the RF formulations in terms of both iterations and CPU seconds.

18.4 ▪ Extensions of Distillation Models

The previous sections describe NLO models with complementarity constraints for single distillation columns. Although restricted to small models, they demonstrate the potential for significantly improved performance with distillation optimization. Extensions of these models to much larger and complex systems build directly on these models and have also been applied to challenging real-world applications. These extensions can be classified in the following areas.

First, nonideal phase equilibrium requires additional consideration because the corresponding minimization problem (18.5) is generally nonconvex with few exceptions (e.g., ideal thermodynamics and Wilson liquid activity correlations). In particular, the widely used cubic equation of state [622, 1028] requires additional inequalities to isolate correct cubic roots to define K_{ij} values correctly. Moreover, highly nonideal cases may require (18.5) to be solved *globally*. While significant work has been done in this direction (see [1305]), global optimization for each distillation stage is still prohibitive, and first-order conditions, with constraints and initialization strategies as heuristics, are common modeling practice.

Second, multicolumn systems lead to more complex integration of column sections and heat exchangers. This is especially important in air separation systems, where nitrogen (N_2) and oxygen (O_2) are separated via distillation at cryogenic temperatures. This is typically done with the coupled low- and high-pressure columns shown in Figure 18.3(b). Feed air is compressed, cooled, and split between the two columns (Air Feed-1 and Air Feed-2), and each column is divided into two sections (tray stacks). High-purity N_2 is produced at the top of the high-pressure (HP) column. Its bottom product, crude O_2, is throttled to low pressure and recycled to the low-pressure (LP) column and/or used as product. The LP column has crude N_2 and the high-purity O_2 as top and bottom products, respectively. Cooling for the LP column is supplied by the high-purity N_2 product stream from the HP column, after it is throttled to ambient pressure. For this process, the energy required for air compression is the dominant cost, and heat integration is essential to minimize this energy cost. As a result, the two columns are tightly heat integrated to increase energy efficiency. This is done by matching heat duties of the heating and cooling exchangers in Figure 18.3(b), with the result that product streams cool the feed streams through multistream heat exchangers, and the condenser of the HP column supplies heat to the reboiler of the LP column. Optimal design of the air separation unit is formulated as an optimization model using extended versions of the MPCCs in Section 18.2. Here, pressures and stream flow rates are optimized as well and management of phase changes is needed in the exchangers as well as on column stages. As a result of the more complex superstructure and incorporation of CEOS, the NLO model, using the PF reformulation, requires careful initialization through nested solution of simpler distillation models (including shortcut columns and ideal thermodynamics). The overall modeling and solution strategy has been built into a six-step automated framework on the GAMS platform with typical problem sizes of over 16,000 variables and 15,000 equalities. Solved with CONOPT, the overall framework requires about 15 CPU minutes (Dual Core 2.4 GHz Intel Xeon processors). As discussed in [621, 622], quite efficient air separation units result from this optimization strategy.

Finally, the steady-state distillation model extends naturally to dynamic distillation systems. Dynamic operation of these units includes batch distillation for low-volume, high-value products; dynamic transitions between steady-state operations; and optimal rejection of disturbances for process control. Here, the dynamic optimization model is derived by (a) adding liquid and vapor holdup at each stage for the steady-state mass and heat equations and (b) supplying additional equations to describe tray hydraulics to compute pressure drops. The resulting dynamic model is generally stiff and quite expensive to solve. An effective model reduction to reduce stiffness is to neglect vapor holdup, which is 2 to 3 orders of magnitude less than liquid holdup.

The resulting differential-algebraic system leads to an index-2 DAE, which upon reformulation leads to a much cheaper model for optimization. Optimization of dynamic distillation systems has been studied for the start-up and operation of batch distillation columns while monitoring phase appearance and disappearance [1541]. Applied to a batch column with 12 stages, the time-discretized NLP problem consists of 22,226 variables and over 2,100 degrees of freedom. In addition, optimal control of high-purity separations with liquid phase disappearance in reboilers [714] has been formulated for the dynamic operation of a propane/propylene separation with 158 trays and over 48,000 discretized variables. Finally, a nonlinear model predictive control (MPC) strategy for an air separation unit was developed in [965]. For a double-column model with 80 stages and over 1500 DAEs, the resulting NLO with over 117,000 variables and 240 degrees of freedom was solved in about 200 CPUs. To accommodate the larger problem sizes for dynamic distillation, IPOPT was used as the NLO solver.

In all of these cases, the sparse, structured optimization formulations derived in Section 18.2 effectively track the phase condition and lead to efficient and robust performance for the solution of large-scale NLO problems.

Chapter 19

Multidisciplinary Design Optimization of Aerospace Systems

Joaquim R. R. A. Martins[13]

19.1 ▪ Introduction

Aerospace engineering has been at the forefront both in modeling and design optimization due to the demand for high performance. The objective of this chapter is to show how numerical optimization has been useful in the design of aerospace systems and to give an idea of the challenges involved.

In the design of aerospace systems, mass is especially critical for performance. In aircraft applications, every unit of mass must be compensated with a corresponding increase in lift. This additional lift incurs additional drag and thus thrust, which in turn increases the required amount of fuel, leading to a further increase in total mass. Thus, one unit increase in structural or payload mass can lead to up to a four-unit increase overall due to the additional fuel required. In space applications, there is a high cost associated with mass due to the large amount of potential and kinetic energy required to put spacecraft into orbit. Since there is such a high premium on mass, it is economically advantageous to invest in mass reduction through research and development in the modeling and design process. The application of numerical optimization to the design of aerospace systems is thus a natural research and development area that has been fruitful, with the potential for further improvements.

Aerospace systems require the integration of multiple disciplines for a successful design. In aircraft design, for example, aerodynamics, structures, propulsion, and control all come into play. Any design changes within one discipline impact the other disciplines, sometimes in significant and nonintuitive ways. Thus it is important to couple all disciplines when modeling the performance of aerospace systems. In addition, the design optimization must be performed by considering the design variables in all disciplines simultaneously to make sure that the true multidisciplinary optimum is found.

---

[13]I would like to thank John T. Hwang, Graeme J. Kennedy, Gaetan K. W. Kenway, and Zhoujie Lyu for their contributions to the numerical results presented in this chapter.

The field of multidisciplinary design optimization (MDO) was born out of this necessity, as aerospace engineers invested in the application of numerical optimization to the design of multidisciplinary aerospace systems. One of the first applications was wing design, where the coupling of aerodynamics and structures is particularly important [890]. Since then, the application of MDO has been extended to a wide range of other engineering systems [1292].

We start the remainder of this chapter by giving (in Section 19.2) a background on the types of optimization programs that are typically found in aerospace applications (objective functions, design variables, constraints, and models). We then present three different applications: aerodynamic shape optimization applied to wing design (Section 19.3); simultaneous optimization of aerodynamic shape and structure for this same application (Section 19.4); and, finally, a satellite case that includes both the design optimization of the satellite and the optimization of its operations (Section 19.5).

19.2 ▪ Background

19.2.1 ▪ Computational Models

The computational models used in aerospace engineering applications range from simple explicit algebraic expressions to the solution of nonlinear PDEs in three dimensional domains, which could also be time-dependent. We denote the computational models by the residual of a system of k governing equations $R(x, y(x)) = 0$, where x are fixed parameters and y are the k state variables that are implicitly determined by solving the governing equations. The number of state variables in the applications presented herein is up to $\mathcal{O}(10^6)$. These governing equations are usually solved using specialized solvers that were developed to solve the physics at hand.

One of the challenges of MDO is the solution of coupled systems of governing equations that pertain to different physical models (e.g., fluid versus structure). While we might be tempted to use a single solver for the coupled system, specialized segregated solvers are usually used for practical reasons. In Sections. 19.3–19.5 we specify the governing equations and solvers used in the respective applications.

19.2.2 ▪ Optimization Problem

Most problems in the design optimization of aerospace systems are nonlinear optimization (NLO) problems of the same form as presented in Chapter 17, equation (17.1), where both the objective functions f and constraints g, h are usually nonlinear and nonconvex. The design variables can be continuous or discrete, but we restrict ourselves to continuous variables in the applications presented in this chapter.

The governing equations $R(x, y(x)) = 0$ introduced in Section 19.2.1 could be considered as equality constraints in the NLO problem, where y would also be a vector of optimization variables. However, as previously mentioned, we tend to use specialized solvers to solve the governing equations separately. At each optimization iteration, the solver finds the y that satisfies $R(x, y(x)) = 0$ for a given x, and thus $y(x)$ is effectively an implicit function of x. The dimension of the states equals that of the governing equations, and there is usually a unique solution for this problem. Once the governing equations are satisfied and y is computed, we can evaluate the objective function and the constraints that are not governing equations. Hence this is an optimization problem that is solved in a reduced space, which can be expressed as

$$\min_{x \in \mathbb{R}^n} \quad f(x, y(x))$$

$$\text{s.t.} \quad g(x, y(x)) \leq 0,$$

$$h(x, y(x)) = 0,$$

where y satisfies $R(x, y(x)) = 0$ for each x. While we could consider a full-space approach where the optimizer is responsible for satisfying the governing equations, it is unlikely that a general purpose nonlinear optimizer would be efficient enough in the solution of nonlinear PDEs with $\mathcal{O}(10^6)$ state variables. Furthermore, the implementation of a full-space method is intrusive and requires an extensive development effort.

19.2.3 ▪ Optimization Algorithms

There are two fundamental choices when it comes to optimization algorithms: derivative-free or derivative-based methods. The design optimization applications presented herein are afflicted by two compounding challenges: large numbers of design variables ($\mathcal{O}(10^2)$ variables or more), and a high cost of evaluating the objective and constraints (which involves the solution of governing equations with $\mathcal{O}(10^6)$ state variables).

Since the number of iterations required by derivative-free methods does not scale well with the number of optimization variables, we choose to use a derivative-based method. Given the efficiency of derivative-based methods, it is possible to address the two compounding challenges mentioned above, provided we can compute the required derivatives efficiently (addressed in Section 19.2.4).

Derivative-based methods are not without some drawbacks. The two main drawbacks are (1) convergence to local (as opposed to global) optima and (2) sensitivity to discontinuities in the design space. When it comes to local versus global minima, most engineers are satisfied by having their designs improved relative to the initial design. Furthermore, in the problems considered herein, it is impossible to guarantee convergence to a global minimum even when using a global optimizer. Discontinuities of the objective and constraints with respect to the design variables can be caused either by discontinuities in y with respect to x due to the solution of the governing equations $R(x, y(x)) = 0$, or by discontinuities in $f(x, y(x))$, $g(x, y(x))$, or $h(x, y(x))$ with respect to x. In our experience, there are few cases where there are discontinuities that have an underlying physical justification, and many discontinuities are introduced in the modeling. Thus, we need to minimize the introduction of such discontinuities when developing solvers and constructing the objective and constraint functions. Even when there are physical discontinuities, these can be smoothed out without loss of fidelity in the physical model (see Section 19.5).

There are a number of derivative-based NLO packages available. We use SNOPT [802], an active set sequential quadratic programming (SQP) algorithm suitable for general nonlinear constrained problems (see Section 17.5.1). To facilitate the use of SNOPT and its integration with the various solvers considered herein, we use the Python interface pyOpt [1465], which also enables the use of other optimizers with few modifications to the main program.

19.2.4 ▪ Computing Gradients Efficiently

Given the choice of a derivative-based optimizer, the efficiency of the overall optimization hinges on an efficient evaluation of the gradients of the functions f, g, and h

with respect to the design variables. A number of methods are available for evaluating derivatives of PDE systems: finite differences, the complex-step method, algorithmic differentiation (forward or reverse mode), and analytic methods (direct or adjoint) [1291]. The computational cost of these methods is either proportional to the number of optimization variables or proportional to the number of functions being differentiated.

Since we have a large number of design variables, the best option is reverse mode algorithmic differentiation or the adjoint method. In our applications, we tend to use a hybrid method that combines the adjoint method with algorithmic differentiation (both forward and reverse modes).

We now derive the adjoint method for evaluating the derivatives of a function of interest, $f(x, y(x))$ (which in our case are the objective function and constraints), with respect to the design variables x. The only independent variables are x, since y is determined implicitly by the solution of the governing equations, $R(x, y(x)) = 0$, for a given x. Using the implicit function theorem, the gradient of f with respect to x is

$$\frac{\mathrm{d}f}{\mathrm{d}x} = \frac{\partial f}{\partial x} + \frac{\partial f}{\partial y}\frac{\mathrm{d}y}{\mathrm{d}x}. \tag{19.1}$$

A similar expression can be written for the Jacobian of R,

$$\frac{\mathrm{d}R}{\mathrm{d}x} = \frac{\partial R}{\partial x} + \frac{\partial R}{\partial y}\frac{\mathrm{d}y}{\mathrm{d}x} = 0 \quad \Rightarrow \quad \frac{\partial R}{\partial y}\frac{\mathrm{d}y}{\mathrm{d}x} = -\frac{\partial R}{\partial x}. \tag{19.2}$$

We can now solve this linear system to evaluate the gradients of the state variables with respect to the optimization variables. Substituting this solution into the evaluation of the gradient of f (19.1) yields

$$\frac{\mathrm{d}f}{\mathrm{d}x} = \frac{\partial f}{\partial x} - \frac{\partial f}{\partial y}\left[\frac{\partial R}{\partial y}\right]^{-1}\frac{\partial R}{\partial x}. \tag{19.3}$$

The adjoint method consists of factorizing the Jacobian $\partial R/\partial y$ with $\partial f/\partial y$, i.e., we solve the adjoint equations

$$\left[\frac{\partial R}{\partial y}\right]^{T}\psi = -\frac{\partial f}{\partial y}, \tag{19.4}$$

where $\psi \in \mathbb{R}^{k}$ is the adjoint vector. We can then substitute the result into the total gradient equation (19.1) to get the required gradient,

$$\frac{\mathrm{d}f}{\mathrm{d}x} = \frac{\partial f}{\partial x} + \psi^{T}\frac{\partial R}{\partial x}. \tag{19.5}$$

The partial derivatives in the equations above are inexpensive to evaluate, since they do not require the solution of the governing equations. The computational cost of evaluating gradients with the adjoint method is independent of the number of design variables, but dependent on the number of functions of interest.

19.3 ▪ Aerodynamic Shape Optimization

Among the many parameters that an aircraft designer must decide on, the parameters determining the shape of the wing are among the most crucial and complex. Small changes have a large effect on both drag and lift, and a large number of parameters are required to reduce the drag as much as possible.

The typical wing aerodynamic shape optimization is to minimize the drag coefficient (C_D) by varying the design variables x (which consist of shape design variables and angle of attack) subject to a lift coefficient constraint, a moment constraint, and a number of geometric constraints. In our example, we have

$$\min_{x \in \mathbb{R}^n} \quad C_D(x, y(x)) \tag{19.6a}$$

$$\text{s.t.} \quad C_L(x, y(x)) = 0.5, \tag{19.6b}$$

$$C_M(x, y(x)) \geq -0.17, \tag{19.6c}$$

$$g_{\text{geo}}(x) \leq 0. \tag{19.6d}$$

The geometric constraints g_{geo} are required to prevent the wing from becoming too thin. There is a strong incentive to decrease the thickness of the wing cross-sections to reduce the drag at high subsonic speeds, but a thin wing makes for a heavy structure. Since in this case the structure is not considered in the modeling of the optimization, geometric constraints are required.

The solution of the PDE governing the flow is implicit in this formulation: as previously mentioned, we use a reduced-space approach, where the governing equations are solved for a given shape x at each optimization iteration. In these examples, we use the ADflow flow solver [1816], a finite volume, cell-centered multiblock solver for the Reynolds-averaged Navier–Stokes (RANS) equations. The discrete adjoint method for the RANS equations was developed by forming (19.4) and (19.5), where the partial derivatives are implemented by performing algorithmic differentiation in the relevant parts of the original code [1246].

We solve the RANS equations in the three-dimensional domain surrounding the wing, using a mesh with 450,560 cells, resulting in over 2.2 million state variables. The evaluation of the drag, lift, and moment coefficients consists of the numerical integration of the flow pressure distribution on the surface of the aircraft. The wing geometry is taken from the Common Research Model (CRM) aircraft configuration [1830]. The shape design variables (x) are the vertical positions of 768 points that control a free-form deformation volume, which allows for fine control of the wing airfoil shapes. We enforce 750 thickness constraints distributed on the wing to be greater than or equal to 25% of the initial thickness at the respective locations. The internal volume is also constrained to be greater than or equal to the initial volume. The reason for this volume constraint is that the wing also serves as the main fuel tank for the aircraft.

To demonstrate the robustness of the optimization process, we solve the optimization problem (19.6) starting with a wing that is a random perturbation of the CRM wing. The solution of this design optimization problem is shown in Figure 19.1. The upper left quadrant shows a planform view of the wing with contours of the pressure coefficient, C_p. The left wing is the initial design, where we can see that the C_p distribution is erratic due to the random shape. The wing on the right side is the optimized one, where we can see that the C_p contours tend to be parallel to the spanwise direction of the wing. Below the planform view, we can see a front view of the wing showing the shock surface (in yellow) and the spanwise distributions of lift, wing geometric twist, and thickness-to-chord ratio. All the red lines refer to the starting wing, while the blue lines refer to the optimized one. The two columns on the right show the airfoil cross-sectional shapes corresponding to the positions denoted by the respective letters

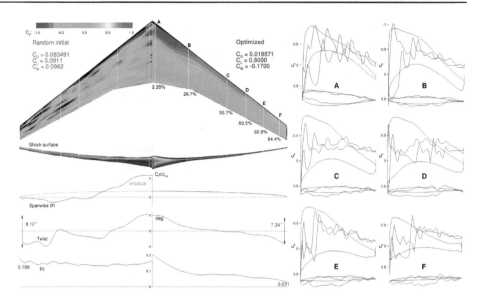

Figure 19.1. *The optimization starts from a random geometry (left/red) and converges to an optimal wing (right/blue). Originally appeared in* [1245] *and* [1053].

shown on the wing planform view. The C_p distribution is shown above each airfoil. We can see the random nature of the airfoil shapes (in red) and the corresponding C_p distributions. The smooth optimized airfoils (in blue) result in a shock-free solution with a much lower drag.

More details on this work, as well as other design optimization cases (including a case with 144 million state variables, multiple flight conditions, and different thickness constraints), and a more thorough analysis of the results are presented by Lyu et al. [1245].

19.4 ▪ Aerostructural Design Optimization

In the previous example, we performed aerodynamic shape optimization of an aircraft wing, where we had to impose thickness constraints to prevent the wing from becoming too thin and thus infeasible from the structural point of view. In this section, we include a structural model coupled to the same aerodynamic model of the previous section to perform aerostructural design optimization. This enables the optimizer to evaluate trades between aerodynamic and structural performance, provided we define an objective function representative of the overall aircraft performance. With aerostructural optimization it is possible to eliminate the geometric thickness constraints and to include wing planform shape (e.g., sweep and span). In addition, coupling the aerodynamic and structural models is essential for predicting the performance of even moderately flexible wings, since the structural deflections have a strong effect on the aerodynamic loads and vice versa.

For the aerostructural design optimization problem, we choose to use a linear combination of fuel burn (FB) and take-off gross weight (TOGW). FB affects mainly the aircraft operating cost, while TOGW affects mainly the aircraft acquisition cost.

Airlines look for a balance between these two metrics, and this balance depends on each particular case. We state the optimization problem as follows:

$$\min_{x \in \mathbb{R}^n} \quad \beta \mathrm{FB}(x, y(x)) + (1-\beta)\mathrm{TOGW}(x, y(x)) \tag{19.7a}$$

$$\text{s.t.} \quad L(x, y(x)) = n_i W(x, y(x)), \quad i = 1, 2, 3, \tag{19.7b}$$

$$C_M(x, y(x)) \geq -0.17, \tag{19.7c}$$

$$\mathrm{KS}_{\mathrm{struct}}(x, y(x)) \leq 1, \tag{19.7d}$$

$$g_{\mathrm{geo}}(x) \leq 0, \tag{19.7e}$$

where $\beta \in [0, 1]$ defines the balance between the performance metrics. The design variables consist of angle of attack, airfoil shape, planform shape (sweep, span, and taper), and structural thicknesses. KS (Kreisselmeier–Steinhauser) is an aggregation function representing the failure of the structure. Since the structural failure originally includes a large number of constraints, and there is no efficient way of computing the Jacobian of a large number of constraints with respect to a large number of design variables, we aggregate all the constraints into a few KS functions so that we can use the adjoint method to evaluate the gradients of KS [1055, 1503]. Three different flight conditions are considered for which the lift must be equal to the weight ($L = W$): $n = 1$ corresponds to the cruise condition, which is used to evaluate the objective function, while $n = 2.5$ and $n = -1$ are maneuver conditions, which are used to enforce the structural failure constraints.

The aerostructural model consists of the RANS aerodynamic model described in the previous example coupled to a structural finite element model of the wing. The structural solver is a parallel direct solver that can also evaluate derivatives using the adjoint method [1054]. The coupled system is solved using a Newton–Krylov method. The adjoint method described in Section 19.2.4 can also be implemented to the coupled problem to obtain all the derivatives required in the aerostructural design optimization problem [1055].

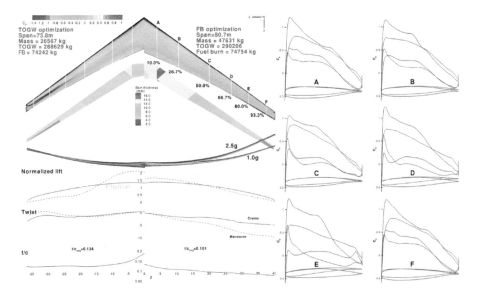

Figure 19.2. *Comparison of two wing aerostructural design optimization problems: TOGW minimization (left/red) and FB minimization (right/blue).*

Figure 19.2 shows the result of solving the aerostructural design optimization problem for $\beta = 0$ (left wing) and $\beta = 1$ (right wing), which corresponds to minimizing TOGW and FB, respectively. As in Figure 19.1, we show the contours of the pressure coefficient on a planform view of the wing in the upper left quadrant, but now we add another planform view below showing the wing box structural thickness distribution. The front view now shows the actual deflected shapes of the wings for both the cruise and the maneuver conditions.

As we can see, the FB minimization converged to a larger span than the TOGW minimization. This is because FB minimization trades off a further decrease in drag for a penalty in structural weight. Detailed examples of wing aerostructural optimization can be found in Kenway and Martins [1056, 1057].

19.5 ▪ Optimization of a Satellite's Design and Operation

The previous two examples focused on wing design optimization, starting with one discipline (aerodynamics) in the first example, and then coupling this to a second discipline (structures) to perform MDO. The solution of this problem was enabled by the coupled adjoint method, which evaluates gradients of the coupled system efficiently. We can take this idea further in many more disciplines. In this section we present a satellite MDO problem with seven disciplines: orbit dynamics, attitude dynamics, cell illumination, temperature, solar power, energy storage, and communication. Many of these disciplines include functions with discontinuities and nonsmooth regions that are addressed to enable the numerically exact computation of derivatives for all the modeled variables. The wide-ranging time scales in the design problem, spanning from 30 seconds to 1 year, are captured through a combination of multipoint optimization and the use of a small time step in the analyses.

This satellite design optimization problem can be stated as follows:

	Function/variable	Description	Quantity
$\min_{x \in \mathbb{R}^n}$	$\sum_{i=1}^{6} D_i$	Data downloaded	
w.r.t.	$0 \leq I_{\text{solar}} \leq 0.4$	Solar panel current	$300 \times 12 \times 6$
	$0 \leq \gamma \leq \pi/2$	Roll angle profile	300×6
	$0 \leq P_{\text{comm}} \leq 25$	Communication power	300×6
	$0 \leq \text{cell} \leq 1$	Cell versus radiator	84
	$0 \leq \alpha_{\text{fin}} \leq \pi/2$	Fin angle	1
	$0 \leq \alpha_{\text{ant}} \leq \pi$	Antenna angle	1
	$0.2 \leq \text{SOC}_i \leq 1$	Initial state of charge	6
		Total	25292
s.t.	$I_{\text{bat}} - 5 \leq 0$	Battery charge constraint	6
	$-10 - I_{\text{bat}} \leq 0$	Battery discharge constraint	6
	$0.2 - \text{SOC} \leq 0$	Battery capacity constraint	6
	$\text{SOC} - 1 \leq 0$	Battery capacity constraint	6
	$\text{SOC}_f - \text{SOC}_i = 0$	Battery state of charge periodicity constraint	6
		Total	30

The objective is to maximize the data downloaded from the satellite to ground stations by varying both satellite design parameters (cell and radiator allocation, fin angle, and antenna angle) and satellite control variables (these are variables that vary in time: solar panel current, roll angle, and communication power). The total number of design variables is 25,292. All the design constraints are related to limitations of the battery, and the governing equations for all disciplines are solved separately as before, i.e., we use a reduced-space approach. The crucial combination of numerical methods that makes the solution of this large-scale nonlinear problem tractable is the combination of SNOPT, a Newton method for the solution of the coupled governing equations, and a coupled adjoint for evaluating the derivatives efficiently. Hwang et al. [972] provide much more detail on this optimization problem and the governing equations for each discipline.

19.6 ▪ Conclusion

The strong incentive to invest in improving the performance of aerospace systems, together with the complex multidisciplinary nature of these systems, led to the development of high-fidelity MDO. Because these design optimization problems involve large numbers of design variables, our approach was to use a derivative-based optimizer in conjunction with adjoint methods, which efficiently evaluates the derivatives for such problems.

The three examples presented here are not meant to span the wide variety of MDO problems in aerospace, but they show how it is possible to solve large-scale design optimization problems by combining a derivative-based optimizer, efficient methods for solving coupled systems of governing equations, and a coupled adjoint for evaluating the derivatives efficiently. Furthermore, these approaches to MDO can be generalized and applied to the design optimization of other engineering systems that involve multiple disciplines or components.

Chapter 20

Nonlinear Optimization for Building Automation

Victor M. Zavala

20.1 ▪ Introduction

Automation is a critical technology that enables *sustainable buildings*. It has been estimated that air-conditioning in commercial buildings is responsible for up to 25% of the total energy consumption in the United States. Automation can thus reduce indirect carbon emissions by reducing total energy and by reducing demand peaks because this prevents the construction of power plants [1295]. Automation also enables demand elasticity, which is key in accommodating intermittent renewable power in the power grid. Finally, automation enables healthy and productive environments through improved monitoring of air quality and comfort.

The heating, ventilation, and air-conditioning (HVAC) system is the physical system conditioning the building. The HVAC system comprises a large number of equipment units and material and energy resources that are monitored and controlled by a building management system (BMS). The traditional architecture sends set-points to low-level, single-loop controllers such as thermostats that try to *track* the set-points closely. The determination of set-points by operators or logical rules does not usually capture the interaction between multiple control loops and introduces a disconnect between high-level economic performance and zone-level control performance (e.g., set-point tracking) [263]. Because of these limitations, the automation industry has shifted its interest to optimization-based BMS systems. These are also known as model predictive control (MPC) systems [936, 1247, 1424, 1950]. These systems use dynamic models to predict the interactions between HVAC variables and local zone conditions. With these predictive capabilities, they can directly optimize high-level economic objective functions and enforce zone-level comfort and air-quality constraints. Optimization-based systems use optimal control formulations that are cast as large-scale nonlinear optimization (NLO) problems and solved in real time.

In this chapter, we argue that modern NLO solvers enable the consideration of powerful automation systems capable of exploiting *building-wide nonlinear variable interactions*. This is important because a significant number of control studies still use linearized models [1424]. In addition, we discuss how NLO solvers can be used to perform off-line *multiobjective studies*.

The chapter is structured as follows. In Section 20.2 we motivate the need for optimization-based automation systems. In Section 20.3 we present a typical physical building model and operational constraints. In Section 20.4 we present a multi-objective optimal control formulation to explore the trade-offs between comfort and energy use and we demonstrate the performance of different control strategies. In Section 20.5 we describe computational strategies for the solution of optimal control problems (OCPs), and the chapter closes in Section 20.6 with final remarks.

20.2 • Building Operations and Sources of Inefficiency

The HVAC system comprises **four essential control** systems (see Figure 20.1). **(I) The temperature** in a given zone is sensed by a thermostat. Each thermostat is in a closed loop with an air-volume box that supplies air at a given supply temperature. This loop keeps the zone temperature at a given set-point by rejecting heat loads generated by humans and interactions with the external environment. The temperature of the supply air is in a closed loop with the air-handling unit (AHU), and its set-point has a strong influence on energy use [1950]. **(II) Relative humidity** (RH) is controlled centrally by sensing RH in the return duct. This is in a closed loop with the AHU. Moisture removal from the air involves cooling down the air stream to a low temperature by a cooling coil to condense water. The air stream is then heated up (if necessary) by a heating coil to reach the desired supply temperature. Note that this process is inefficient but limited by thermodynamics. If the humidity is low, the AHU will add moisture to the supply air. **(III) Pressure** control is essential to prevent air infiltration from the surroundings [1870], which affects air quality and energy efficiency. Commercial buildings typically operate under *positive pressure*, which means that the internal pressure of the building is kept *slightly* higher than the atmospheric pressure. Pressure control is normally performed by sensing pressure in the outer building layer and having this in a closed loop with the exhaust air flow. **(IV) Air quality** control is necessary to deal with carbon dioxide and volatile organic compounds (responsible for odors). These contaminants are generated by occupants and are removed by outside ventilation control. The simplest control strategy consists of estimating the total number of occupants and

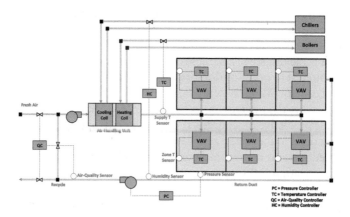

Figure 20.1. *Schematic representation of building HVAC system showing four main control systems.*

using this to set a minimum ventilation rate. In a more advanced control setting, a CO_2 sensor in the return duct is maintained in a closed loop with the recycle system. This is a mixing system that combines contaminated air with fresh air from the outside.

In a traditional control architecture, the four control systems operate in isolation (each one trying to reach its goal) and a central operator tries to orchestrate them. The objectives of the operator are to minimize energy and to ensure that the building conditions are comfortable. To do so, the operator determines operating conditions (set-points) for the four control systems. Temperature and RH set-points are typically set to a combination that ensures comfortable conditions. Typical values are 70°F and 50%. The CO_2 concentration is kept close to atmospheric concentration (approximately 400–700 ppm), but regulations allow concentrations as high as 1,000 ppm. Some buildings do not have ventilation (air-quality) control, so operators typically overventilate the building to ensure air quality. Supply temperature is set to 50–55°F for cooling and 100–110°F for heating. Pressure is normally set 1–2 pascals above atmospheric.

While the four control systems are operated in isolation, they are in fact *physically coupled*. The disconnect between the control systems yields inefficient operations because the systems tend to fight each other. This disconnect also has another major implication from an energy efficiency standpoint: *variable interactions* provide an opportunity to gain *control flexibility*. An interesting interaction is that of RH and temperature control. *RH measures the ability of the air to absorb moisture*. At high RH, humans feel warmer (even at mild temperature) because the air cannot easily absorb the generated perspiration. The ability to absorb moisture by the air is *not only a function of moisture present* in the air but also a strong *nonlinear* function of temperature (see equation (20.3)). This is because higher temperatures increase the void spaces between molecules in the air. Consequently, manipulating temperature can aid the RH control system, *which only relies on moisture removal/addition in the AHU*.

Capturing interactions between the four (or more) control systems is by no means trivial. This requires *building-wide models* capable of *predicting* the impact of the variables of one system on the variables of another system. This is the key concept behind MPC. MPC systems use building-wide models coupled to NLO solvers to minimize energy and ensure comfort by overseeing the interaction between the different control systems of the building. In the following sections, we describe the components of an MPC system.

20.3 ▪ Physical Model

The building model considered here is presented in detail in [1949]. This is a simplified model that captures the most basic variable interactions.

20.3.1 ▪ Material Balances

The total and individual component mass balances in the building are given by

$$\frac{dm(\tau)}{d\tau} = \rho \cdot (q^{in}(\tau) - q^{out}(\tau)), \tag{20.1a}$$

$$V \cdot \frac{dC_i(\tau)}{d\tau} = q^{in}(\tau) \cdot C_i^{in}(\tau) - q^{out}(\tau) \cdot C_i(\tau) + n_{oc}(\tau) \cdot G_i, \tag{20.1b}$$

where $i \in \{CO_2, H_2O\}$ are subindices denoting species, $\tau \in [0, N]$ denotes time, and $q^{in}(\cdot), q^{out}(\cdot)$ are inlet and outlet zone flows. The number of occupants is $n_{oc}(\tau)$ and the pollutant generation is given by $n_{oc}(\tau) \cdot G_i$. The balances in the recycler and in the AHU are

$$q^{out}(\tau) + q^{amb}(\tau) = q^{ex}(\tau) + q^{in}(\tau), \tag{20.2a}$$

$$C_i(\tau) \cdot q^{out}(\tau) + C_i^{amb}(\tau) \cdot q^{amb}(\tau) = C_i(\tau) \cdot q^{ex}(\tau) + C_i^m(\tau) \cdot q^{in}(\tau), \tag{20.2b}$$

$$m_i^{rm}(\tau) = q^{in}(\tau) \cdot C_i^{in} - q^{in}(\tau) \cdot C_i^m(\tau). \tag{20.2c}$$

The symbols $q^{amb}(\cdot)$ and $q^{ex}(\cdot)$ denote the ambient flow (brought from outdoors) and exhaust flow (sent to outdoors), respectively. The removal rates in the AHU are denoted as m_i^{rm}. The relationship between the zone pressure, mass, and temperature is given by the ideal gas law. The RH is given by

$$RH(\tau) = 100 \cdot \frac{C_{H_2O}(\tau)}{C_{H_2O}^{sat}(\tau)}, \quad \log_{10}(C_{H_2O}^{sat}(\tau)) = 8.07131 - \frac{1730.63}{T(\tau) - 39.73}. \tag{20.3}$$

Here, $C_{H_2O}^{sat}(\tau)$ is the saturation density. The concentration of CO_2 in parts per million, $ppmV_{CO_2}(\tau)$, is computed from $C_{CO_2}(\tau)$ and the ideal gas law.

20.3.2 ▪ Energy Balances

The dynamic energy balance of the zone and in the recycler are

$$\frac{m(\tau)}{\rho} \cdot \frac{dT(\tau)}{d\tau} = q^{in}(\tau) \cdot T^{in}(\tau) - q^{out}(\tau) \cdot T(\tau)$$

$$- \frac{U^w \cdot A^w}{\rho \cdot c_p} \cdot (T(\tau) - T^{amb}(\tau)) + \frac{n_{oc}(\tau) \cdot Q^{oc}}{\rho \cdot c_p}, \tag{20.4a}$$

$$q^{out}(\tau) \cdot T(\tau) + q^{amb}(\tau) \cdot T^{amb}(\tau) = q^{ex}(\tau) \cdot T(\tau) + q^{in}(\tau) \cdot T^m(\tau). \tag{20.4b}$$

These equations describe the dynamics of the building temperature due to accumulation/losses of energy resulting from interactions with the environment and the generation of heat from the occupants. Here, $T(\cdot)$ is the zone temperature, $T^{in}(\cdot)$ is the air-supply temperature, $T^{amb}(\cdot)$ is the ambient temperature, and $T^m(\cdot)$ is the mixed-stream temperature entering the AHU. U^w denotes the global heat-transfer coefficient of the building envelope and A^w is the heat-transfer area. The heat gains generated from the occupants are given by $n_{oc}(\tau) \cdot Q^{oc}$. The latent, sensible, and total energy consumed in the AHU are given by

$$Q^{lat}(\tau) = h^{lat} \cdot m_{H_2O}^{rm}(\tau), \tag{20.5a}$$

$$Q^{sens}(\tau) = q^{in}(\tau) \cdot \rho \cdot c_p \cdot (T^{in}(\tau) - T^m(\tau)), \tag{20.5b}$$

$$Q^{hvac}(\tau) = |Q^{lat}(\tau)| + |Q^{sens}(\tau)|. \tag{20.5c}$$

We reformulate the absolute value terms in (20.5) to a differentiable form by using dummy variables.

20.4 ▪ Predictive Control

Consider the following multiobjective optimal control formulation:

$$\min_{u(\cdot)} \quad \{\Phi_e(x(\cdot), y(\cdot), u(\cdot)), \, \Phi_c(x(\cdot), y(\cdot), u(\cdot))\} \tag{20.6a}$$

$$\text{s.t. } \frac{dx}{d\tau} = f(x(\tau), y(\tau), u(\tau), d(\tau)), \ \tau \in [0, N], \quad (20.6\text{b})$$

$$0 = g(x(\tau), y(\tau), u(\tau), d(\tau)), \ \tau \in [0, N], \quad (20.6\text{c})$$

$$x(0) = x_{init}, \quad (20.6\text{d})$$

$$0 \leq h_a(x(\tau), y(\tau), u(\tau), d(\tau)), \ \tau \in [0, N], \quad (20.6\text{e})$$

$$0 \leq h_c(x(\tau), y(\tau), u(\tau), d(\tau)), \ \tau \in [0, N]. \quad (20.6\text{f})$$

System (20.6b)–(20.6c) is a set of DAEs. $x(\tau)$ denotes the dynamic states, which are variables whose dynamics are described by differential equations (e.g., temperature $T(\cdot)$, species concentrations $C_i(\tau)$, and air mass $m(\tau)$); $y(\tau)$ are the algebraic states, which are variables whose reactions are "instantaneous" in time (e.g., relative humidity $RH(\cdot)$); $u(\tau)$ are the control trajectories (sensitive heat load $Q^{sens}(\cdot)$, latent heat load $Q^{lat}(\cdot)$, ambient flow q^{amb}, and exhaust flow $q^{ex}(\cdot)$); and $d(\tau)$ are the system disturbance trajectories (ambient temperature $T^{amb}(\cdot)$, ambient concentrations $C_i^{amb}(\cdot)$, and number of occupants $n_{oc}(\cdot)$). The inequality constraints have been partitioned into two sets: $h_a(\cdot)$ are the air-quality constraints and $h_c(\cdot)$ are comfort constraints to be described in Section 20.4.1. For additional constraints, please refer to [1949]. The objective functions $\Phi_e(\cdot), \Phi_c(\cdot)$ will be described in Section 20.4.2.

20.4.1 ▪ Constraints

20.4.1.1 ▪ Air Quality Constraints

We consider constraints on air quality of the form

$$ppmV_{CO_2}(\tau) \leq ppmV_{CO_2,U}. \quad (20.7)$$

The use of such constraints assumes that CO_2 concentration can be measured or inferred. In the absence of such information, we can also impose constraints on the total ambient air flow rate (ventilation) as a function of the number of occupants.

20.4.1.2 ▪ Thermal Comfort Constraints

Thermal comfort is the result of the heat exchange between the body and the environment. The comfort index used in the ASHRAE 55-2004 and the ISO 7730 standards is the predicted mean vote (PMV) [103]. The PMV is measured on a scale of seven values defined as follows: +3 hot, +2 warm, +1 slightly warm, 0 neutral, −1 slightly cool, −2 cool, −3 cold. On the PMV scale, maximum comfort is achieved at a value of zero. Acceptable comfort conditions are typically in the PMV range $[−0.5, +0.5]$. Two different comfort zones are shown in Figure 20.2. The predicted percentage dissatisfied (PPD) index represents the percentage of people that are expected not to be comfortable in a given environment. PMV and PPD can be calculated from

$$PMV(\tau) = (0.303 \cdot \exp(-0.036 \cdot M_m) + 0.028) \cdot L(\tau), \quad (20.8\text{a})$$

$$PPD(\tau) = 100 - 95 \cdot \exp\left(-0.033 \cdot PMV(\tau)^4 - 0.218 \cdot PMV(\tau)^2\right). \quad (20.8\text{b})$$

Here, $L(\tau)$ is the thermal load of the body and is a complex nonlinear function of temperature and RH, and M_m is a constant. For more details, see [1346]. We consider PMV constraints of the form

$$-PMV_U \leq PMV(\tau) \leq PMV_U. \quad (20.9)$$

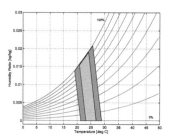 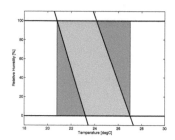

Figure 20.2. *Comfort zone visualization in a psychometric chart (left). Overestimation of the PMV feasible region using temperature and RH bounds (right). Reprinted with permission from Elsevier* [1346].

We can also impose constraints on PPD, but previous experience demonstrates that this is equivalent to imposing constraints on PMV [1346]. Imposing PMV constraints is rigorous because it fully defines the physical comfort region. This, however, comes at the expense of additional computational complexity because of the high nonlinearity. We can avoid this by using simple temperature and humidity constraints of the form

$$T_L \leq T(\tau) \leq T_U, \quad RH_L \leq RH(\tau) \leq RH_U. \tag{20.10}$$

The upper and lower bounds can be chosen as the corners of the comfort region for a given desired range of PMV. This approach, however, can significantly overestimate the comfort region. This can be seen in Figure 20.2.

We can exploit the close-to-polyhedral shape of the comfort region to avoid the need for PMV constraints and, at the same time, to avoid overestimating the comfort region. We determine the coordinates of the corners of the polyhedron and approximate the region using the set of linear inequality constraints

$$RH_1 \leq RH(\tau) \leq RH_2, \tag{20.11a}$$

$$RH(\tau) \geq RH_2 + \frac{RH_2 - RH_1}{T_1^{left} - T_2^{left}}\left(T(\tau) - T_1^{left}\right), \tag{20.11b}$$

$$RH(\tau) \leq RH_2 + \frac{RH_2 - RH_1}{T_1^{right} - T_2^{right}}\left(T(\tau) - T_1^{right}\right). \tag{20.11c}$$

20.4.2 ▪ Objective Functions and Multiobjective Concepts

The objective function $\Phi_e(\cdot)$ represents the HVAC energy demand accumulated over the time horizon $[0, N]$:

$$\Phi_e(\cdot) = \int_0^N Q^{hvac}(\tau)d\tau. \tag{20.12}$$

$\Phi_c(\cdot)$ represents a comfort objective. Note that comfort can be enforced by using the constraint (20.6f) or by combining the economic and comfort terms in a single objective function. From an implementation standpoint, constraints are preferred because it is often difficult to weight objectives. From an analysis standpoint, however, a multiobjective interpretation can help us evaluate the effects of comfort on energy use, as we describe next.

When *objectives are conflicting*, we cannot simultaneously minimize them. Instead, the solution is expressed in terms of the so-called Pareto front. The Pareto front can be computed by using the so-called ϵ-constrained method or the weighting method. Here, one of the objectives is moved into the constraints and is relaxed by a constant parameter ϵ. This gives the following canonical problem:

$$\text{min} \quad \Phi_e(\cdot) \tag{20.13a}$$

$$\text{s.t.} \quad (20.6b)-(20.6e), \quad \Phi_c(\cdot) \leq \epsilon. \tag{20.13b}$$

We use the short notation $\Phi_e(\cdot) = \Phi_e(x(\cdot), y(\cdot), u(\cdot))$ and $\Phi_c(\cdot) = \Phi_c(x(\cdot), y(\cdot), u(\cdot))$. The relaxation factor ϵ is varied in the range $[\epsilon^L, \epsilon^U]$, and the solution of problem (20.13) for a given ϵ gives a point on the Pareto front. When strong duality holds, one can show that a solution of the weighted problem

$$\text{min} \quad (1-\omega) \cdot \Phi_e(\cdot) + \omega \cdot \Phi_c(\cdot) \tag{20.14a}$$

$$\text{s.t.} \quad (20.6b)-(20.6e) \tag{20.14b}$$

with $\omega \in [0, 1]$ is a Pareto solution of the multiobjective optimization problem [1321].

20.4.3 ▪ Control Strategies

In this section, we present different optimal control formulations that seek to balance energy and occupant comfort. In Section 20.4.4, we will demonstrate that the rigorous use of economic objective functions and comfort constraints gives the best performance, but this comes at the expense of increasing computational complexity due to higher nonlinearity.

20.4.3.1 ▪ Maximum Comfort Tracking Strategy

The simplest strategy to enforce comfort consists of penalizing quadratic deviations from a given set-point:

$$\Phi_c(\cdot) = \frac{1}{N} \int_0^N w_T (T(\tau) - T^c)^2 + w_{RH}(RH(\tau) - RH^c)^2 d\tau. \tag{20.15}$$

Here, T^c and RH^c are temperature and RH set-points, respectively, and $w_T, w_{RH} > 0$ are weighting factors. A tracking controller moves as fast as possible to a given set-point. Typically, the set-point is set to a point of maximum comfort (PMV $= 0$). This typically occurs around a temperature of 22°C and an RH of 50%. We refer to this strategy as the *maximum comfort tracking* strategy. This strategy is attractive because of its simplicity and is widely used in industry. Solving the ϵ-constrained problem (20.13) by relaxing (20.15) is equivalent to allowing the tracking controller to deviate from the point of maximum comfort. As we will show in Section 20.4.4, this can result in significant inefficiency and high comfort volatility.

20.4.3.2 ▪ Set-Back Relaxation Strategy

A control strategy widely used in practice to reduce energy consumption is *set-back relaxation*. Because a tracking control architecture is typically in place, the energy can be decreased by setting the set-points T^c, RH^c to values of relaxed comfort. These set-points are also known as set-back conditions. For instance, we might think of increasing the temperature set-point from 23°C during occupied times to 25°C during unoccupied times in summer. We will refer to this strategy as *set-back relaxation*. While intuitive, this strategy offers limited efficiency potential because of the inherent interactions between the control systems and because of the tracking structure of the controller.

20.4.3.3 ▪ PMV Constrained Strategy

We now consider a building-wide optimization strategy that minimizes total energy $\Phi_e(\cdot)$ and imposes PMV constraints to enforce comfort (20.9). We construct a Pareto front to analyze the effect of relaxing PMV bounds on energy consumption by solving the following problem:

$$\min \quad \Phi_e(\cdot) \tag{20.16a}$$
$$\text{s.t.} \quad (20.6b)–(20.6e), (20.9), \tag{20.16b}$$

where we define $\epsilon := PMV^U$ in (20.9), and this bound is progressively relaxed. We refer to this strategy as the *PMV constrained* strategy. One can show that the *PMV constrained* strategy is equivalent to constructing the Pareto front by solving the ϵ-constrained problem (20.13) using the worst-case comfort metric

$$\Phi_c^{PMV,max}(\cdot) = \max_{t \in [0,N]} |PMV(\tau)|.$$

20.4.3.4 ▪ Comfort Overrelaxation Strategy

We consider a building-wide strategy that minimizes energy and replaces the PMV constraints (20.9) from (20.16) with (20.10). This strategy is practical because it avoids the incorporation of the nonlinear comfort model. As we have argued, however, this overestimates the feasible region imposed by PMV. Because of this, we refer to this strategy as *comfort overrelaxation*.

20.4.3.5 ▪ Polyhedral Constraints Strategy

To avoid the overrelaxation of the comfort region, we use the polyhedron (20.11). We replace the PMV constraints (20.9) from (20.16) with (20.11). This strategy more tightly approximates the comfort region.

20.4.4 ▪ Case Studies

We demonstrate the benefits of building-wide optimization over traditional tracking control architectures. To do so, we perform year-round receding-horizon simulations with real ambient data. We then perform a multiobjective optimization analysis to demonstrate that exploiting variable temperature and RH interactions results in much higher efficiency. Details can be found in [1346, 1949].

20.4.4.1 ▪ Year-Round Analysis

We first analyze the performance of the *set-back relaxation strategy*. The temperature and RH set-points under occupied conditions were set to 23°C and 50%. During unoccupied conditions, we *relax* the set-points by letting them be the ambient temperature and RH conditions. We consider this extreme case because engineering intuition suggests that zero energy consumption would be needed during unoccupied times.

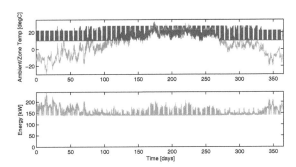

Figure 20.3. *Temperature and energy profiles for the set-back relaxation strategy. Reprinted with permission from* [1949]. *Copyright* 2013 *American Chemical Society.*

The year-round policies are presented in Figure 20.3. In the top panel we present the ambient temperature (light gray) and the building internal temperature (dark blue). In the bottom panel we present the total energy consumption. The total energy consumption for this case was 1,530 MWh. Note that a large amount of energy is wasted during unoccupied periods because the tracking system tries to stay at the temperature and RH set-points without realizing that it can shut down the entire ventilation system to do so. This is a fundamental design issue of a tracking system that we wish to emphasize here. This simulation also highlights that *relaxing set-points* does not necessarily yield high efficiency.

We now consider a *building-wide strategy* for energy minimization using temperature and RH bounds. The comfort zone in occupied mode is assumed to be 21–27°C for temperature and 30–50% for humidity, and these are imposed as constraints of the form (20.10). During unoccupied mode, the temperature and humidity bounds were relaxed to 10–30°C and 20–70%, respectively. Note that this corresponds to the *overrelaxation* strategy. The year-round energy consumption in this scenario was 230 MWh. This represents an 85% *improvement over the tracking strategy*. The optimal profiles are presented in Figure 20.4. As can be seen, the building-wide strategy *drops the energy consumption to almost zero* during unoccupied conditions. This is the direct result of exploiting the interactions between all variables.

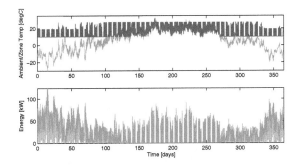

Figure 20.4. *Temperature and energy profiles for building-wide MPC strategy. Reprinted with permission from* [1949]. *Copyright* 2013 *American Chemical Society.*

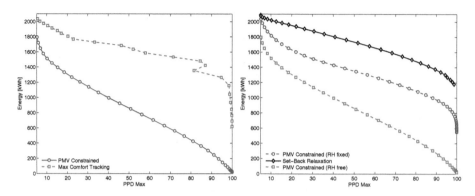

Figure 20.5. *Pareto fronts for different control strategies. Reprinted with permission from Elsevier* [1346].

20.4.4.2 ▪ Multiobjective Analysis

We now explore the energy flexibility of building-wide and tracking control architectures as comfort conditions are relaxed. We consider a multiobjective OCP with a time horizon of 72 hours. We first compare the building-wide *PMV constrained* strategy (minimizes total energy and imposes direct constraints of PMV) against the *maximum comfort tracking* strategy (relaxes comfort by allowing excursions of temperature and relative humidity from the set-point). The Pareto fronts are presented in the left panel of Figure 20.5. The tracking strategy consistently uses more energy and exhibits high comfort volatility as conditions are relaxed. This illustrates that tracking is highly inefficient and difficult to tune to give appropriate comfort relaxation, thus motivating the use of rigorous physical models and economic objective functions.

We now compare the building-wide *PMV constrained* strategy against the *set-back relaxation* strategy. We fix the RH set-point RH^c to a value of 50% and find the temperatures T^c at which PMV values in the range $[0, 3]$ are obtained. This corresponds to temperature set-points in the range $[26, 33]°$C. To perform a fair comparison, we also fix the RH to a value of 50% for the building-wide strategy. To evaluate the additional flexibility that RH offers, we also compare the performance of the building-wide strategy by allowing RH to be free. The results are presented in the right panel of Figure 20.5. We can see that set-back relaxation is not competitive. Moreover, allowing RH to be free yields significant energy savings. This illustrates that exploiting variable interactions is key.

20.5 ▪ Computational Considerations

In all of the presented results, we approximate the solution of the infinite-dimensional OCPs by using a full discretization scheme (direct transcription). This transforms the OCP into a large-scale and highly sparse NLO problem that embeds the discretized DAEs as equality constraints. For details of different discretization schemes, please refer to [259, 261]. To obtain our results, we implement the discretized OCPs in AMPL to obtain exact first and second derivatives.[14] We solve the NLO problems implemented in AMPL using the interior-point solver IPOPT with default options [1859].

---

[14]The user can access some of these AMPL models at http://homepages.cae.wisc.edu/~zavalatejeda/publications.html.

Table 20.1. *Computational and energy performance of different control strategies.*

PMV region	Strategy	$\Phi_e(\cdot)$ [kWh]	Max PPD [%]	Iterations
	PMV	1,495	10.2	76
[-0.5,+0.5]	Overrelax.	1,222	46.8	23
	Polyhedral	1,493	10.8	15
	PMV	469	76.1	51
[-2.0,+2.0]	Overrelax.	161	95.9	11
	Polyhedral	429	79.3	32

We choose IPOPT because it enables the exploitation of sparsity and because of the robustness of its globalization strategy (filter line-search). Other solvers, such as KNITRO, IPFilter, CONOPT, and FilterSQP, also offer similar capabilities.

In Table 20.1, we present the number of NLO iterations required for different optimal control strategies. Note that the use of economic objectives and the incorporation of the PMV comfort model introduces significant nonlinearities, but this can be handled well by IPOPT. Overrelaxation provides much better computational performance, but energy and comfort are overestimated. The polyhedral strategy provides tighter energy and comfort estimates than overrelaxation and improved computational performance over PMV constraints because it avoids the use of the highly nonlinear PMV constraints.

20.6 ▪ Conclusions

NLO solvers are a key enabling technology for building automation. We have argued how the ability to exploit nonlinear variable interactions through advanced NLO solvers can provide significant energy savings and improved occupant comfort satisfaction.

Part VI

Mixed-Integer Nonlinear Optimization

Chapter 21

State of the Art in Mixed-Integer Nonlinear Optimization

Mustafa R. Kılınç and Nikolaos V. Sahinidis

This chapter provides a comprehensive overview of solution techniques for mixed-integer nonlinear optimization (MINLO), an area that represents one of the most general modeling paradigms in optimization. The reader is assumed to be familiar with the topics covered in Parts I–V of this book, in particular, mixed-integer linear optimization (MILO) and global optimization.

21.1 ▪ Introduction

A mixed-integer nonlinear programming (MINLP) problem is expressed as

$$\begin{aligned} \min \quad & f(x) \\ \text{s.t.} \quad & g(x) \le 0, \\ & x \in X, \\ & x_i \in \mathbb{Z} \quad \forall i \in I, \end{aligned} \qquad (21.1)$$

where $X = \{x \mid x \in \mathbb{R}^n, Ax \le b\}$ is a bounded polyhedral subset of $\mathbb{R}^n$. We assume that some finite lower bounds l and upper bounds u on the values of the variables are known and that bound constraints $l \le x \le u$ are included among the linear constraints $Ax \le b$ defining X. The functions $f : [l,u] \to \mathbb{R}$ and $g : [l,u] \to \mathbb{R}^m$ are twice continuously differentiable functions. The set $I \subseteq \{1,\dots,n\}$ is the index set of integer variables. In most applications, some of the integer variables are restricted to 0 or 1, i.e., $x_i \in \{0,1\} \ \forall i \in B \subseteq I$, where B is the index set of binary variables, in which case, (21.1) is called 0-1 MINLP.

MINLP is one of the most general class of problems containing both nonlinear programming (NLP) (when $I = \emptyset$) and mixed-integer linear programming (MILP) as special cases. There are two important special cases of MINLP that draw special attention. One special case is mixed-integer quadratically constrained programming (MIQCP) [356, 358], where all nonlinear functions in the objective and constraints are quadratic. Another special case is mixed-integer second-order cone programming (MISOCP) [51, 225], where the objective function is linear, and nonlinear constraint functions have the form $g_j(x) = \|A_j x + b_j\|_2 - c_j^T x - d_j$. The scope of this Part VI will be general MINLP problems.

Problem (21.1) is nonconvex for two reasons: the presence of integer variables and the presence of nonconvex functions in the objective and constraints. This most general form is referred to as *nonconvex* MINLP. As a special case, if the functions f and g are all convex functions, then we say that (21.1) is a *convex* MINLP. Even though both kinds of MINLP are NP-hard in general [1034], convex MINLP problems are usually easier to solve than nonconvex ones. In fact, nonconvex MINLP problems are even undecidable [994] when X is unbounded. We refer the reader to the surveys [929, 1096] for details on different cases.

There are several comprehensive books [734, 735, 1754] and good surveys on MINLP, including surveys of algorithms [192, 305, 356, 865, 929], surveys of software [192, 305, 372, 544], a review of MINLP and generalized disjunctive programming [860], and a collection of articles from a recent IMA workshop on MINLP [1160].

There exist a wide variety of algorithms for solving MINLP problems. Most of them use a form of the *branch-and-bound* method [1124]. The underlying concept behind branch-and-bound is to create a *relaxation* of (21.1) by enlarging *the feasible set* of (21.1), the set of points satisfying the constraints of (21.1) and/or underestimating its objective function value. A relaxation is generally much easier to solve to global optimality than the MINLP problem and gives a valid *lower bound* on the *optimal value* of the MINLP problem. The most straightforward way of getting a relaxation is to relax the integrality constraints from $x_i \in \mathbf{Z}$ to $x_i \in \mathbf{R}\ \forall i \in I$. This procedure yields the *continuous relaxation* of (21.1) and is given by

$$
\begin{aligned}
\min \quad & f(x) \\
\text{s.t.} \quad & g(x) \leq 0, \\
& x \in X.
\end{aligned}
\tag{21.2}
$$

Relaxed constraints are enforced either by refining/tightening the relaxation or by *branching*, where the relaxation is divided into two or more subproblems. A feasible solution to (21.1) provides an *upper bound,* and branch-and-bound terminates whenever lower and upper bounds are within an *optimality tolerance.* Tighter relaxations provide better lower bounds, but there is generally a trade-off between the tightness and easiness of a relaxation. Algorithms for solving MINLP problems differ in how they create relaxations and enforce relaxed constraints.

Some MINLP problems allow for equivalent formulations that lead to tighter relaxations without increasing the complexity of the problem. As a rule of thumb, convex formulations should be preferred over nonconvex ones, and linear formulations over nonlinear ones. Many MINLP solvers fail to solve a problem because of nondifferentiable functions; thus, if possible, they should be avoided in problem formulation. We refer the reader to the book [1894] for useful modeling tricks with integer variables and to Belotti et al. [192] for some simple formulation tricks for nonlinear functions.

The remainder of this chapter is structured as follows. In Sections 21.2 and 21.3, we present methods for solving convex and nonconvex MINLP problems, respectively. In Section 21.4, we discuss how these methods can be improved by advanced techniques and implementation considerations. The state of the art in software for MINLP and

computational experiments is given in Section 21.5. Finally, we mention some application areas in Section 21.6.

21.2 ▪ Convex Mixed-Integer Nonlinear Programming

In this section, we focus on convex MINLP problems, i.e., the case where the functions
f and g are convex. Thus, the continuous relaxation, (21.2), is a convex NLP problem.
Since all local optima of the continuous relaxation (21.2) are also global optima, local
NLP solvers can be used for its solution. This fact immediately leads to an algorithm
for solving MINLP problems based on the branch-and-bound method, which is called
nonlinear branch-and-bound [542, 883].

21.2.1 ▪ Nonlinear Branch-and-Bound

The algorithm starts by solving (21.2), whose solution value provides a lower bound
on the optimal value of the MINLP problem. If this continuous relaxation is infeasible,
then the MINLP problem is also infeasible. If the solution of the continuous relaxation
is feasible and all integer variables take integer values, then the relaxation solution
also solves the MINLP problem. Otherwise, the domain is divided into two or more
subproblems by refining the bounds on the integer variables. Given a solution $\bar{x}$ to
(21.2) such that $\bar{x}_i \notin \mathbb{Z}, i \in I$, two new subproblems can be created by adding the
bound $x_i \leq \lfloor \bar{x}_j \rfloor$ to one subproblem and $x_i \geq \lceil \bar{x}_i \rceil$ to the other subproblem. Note that
$\bar{x}$ is excluded in both subproblems:

Continuing in this manner yields a *branch-and-bound tree* such that nodes correspond to subproblems and edges correspond to branching decisions; the root node
corresponds to the continuous relaxation (21.2). A branch-and-bound tree node is
uniquely defined by the bounds on the integer variables and corresponds to the following subproblem:

$$\begin{aligned}
\min \quad & f(x) \\
\text{s.t.} \quad & g(x) \leq 0, \\
& x \in X, \\
& l_i \leq x_i \leq u_i \quad \forall i \in I.
\end{aligned} \tag{21.3}$$

We will refer to this subproblem as $NLP(l, u)$. After branching on x_j at node
$NLP(l, u)$, two *child nodes* are created. Bounds of child nodes are initialized from
their *parent node*, $(l^-, u^-) \leftarrow (l, u)$ and $(l^+, u^+) \leftarrow (l, u)$, and updated by modify-
ing the bound on the branching variable as $u_j^- \leftarrow \lfloor \bar{x}_j \rfloor$ and $l_j^+ \leftarrow \lceil \bar{x}_j \rceil$. The root node
corresponds to $NLP(-\infty, \infty)$ since no additional bounds are enforced by branching.

We can discard the subtree rooted at any node if the bound obtained by relaxation
indicates that it cannot contain an optimal solution. *Fathoming* of a node occurs when
one of the following rules applies: by infeasibility if the subproblem is infeasible, by
bound if the lower bound obtained by solving $NLP(l, u)$ exceeds the current known
upper bound (U), or by integrality if the subproblem provides a feasible integral solution. In this last case, if the optimal value of the subproblem is less than the current
upper bound U, then the best feasible solution x^* is updated with the subproblem solution along with the current upper bound. The tree search continues until all nodes
are fathomed, at which point x^* is the optimal solution. The formal definition of the
nonlinear branch-and-bound algorithm is given in Algorithm 21.1.

ALGORITHM 21.1. **Nonlinear Branch-and-Bound for MINLP.**

0. **Initialize.**
   Choose an optimality tolerance $\epsilon > 0$, $\mathcal{H} \leftarrow NLP(-\infty, \infty)$. $U = \infty$.
   $x^* \leftarrow$ NONE.
1. **Terminate?**
   Is $\mathcal{H} = \emptyset$? If so, terminate the algorithm. Otherwise, go to step 2.
   If x^* is NONE, then the problem is infeasible. Otherwise, x^* is optimal.
2. **Node Selection.**
   Choose and remove a problem $NLP(l, u)$ from $\mathcal{H}$.
3. **Evaluate.**
   Solve $NLP(l, u)$ and let $\bar{x}$ be its optimal solution.
4. **Fathom.**
   If the problem is infeasible, fathom by infeasibility and go to step 1.
   If $f(\bar{x}) \geq U - \epsilon$, fathom by bound and go to step 1.
   If $\bar{x}$ is integral, let $U \leftarrow f(\bar{x})$, $x^* \leftarrow \bar{x}$ and fathom by integrality and go
   to step 1.
5. **Branch.**
   Select index j such that $\bar{x}_j \notin \mathbb{Z}$ and $j \in I$.
   Set $(l^-, u^-) \leftarrow (l, u)$, $(l^+, u^+) \leftarrow (l, u)$, $u_j^- \leftarrow \lfloor \bar{x}_j \rfloor$ and $l_j^+ \leftarrow \lceil \bar{x}_j \rceil$.
   Add $NLP(l^-, u^-)$ and $NLP(l^+, u^+)$ to $\mathcal{H}$.

21.2.2 ▪ Polyhedral Relaxations

One drawback of nonlinear branch-and-bound is that it requires solution of an NLP subproblem at each branch-and-bound node, which can be prohibitive for some problems. Another class of methods has been developed to use polyhedral relaxations to employ robust and efficient linear programming (LP) and MILP solvers. Since the minimizer of (21.2) may lie in the strict interior of the feasible region, it may not be separated by linear inequalities. However, we can reformulate (21.1) by introducing an auxiliary variable z as the objective function and adding the constraint $f(x) \leq z$. We obtain the following equivalent MINLP:

$$
\begin{aligned}
\min \quad & z \\
\text{s.t.} \quad & f(x) \leq z, \\
& g(x) \leq 0, \\
& x \in X, \\
& x_i \in \mathbb{Z} \quad \forall i \in I.
\end{aligned}
\tag{21.4}
$$

Since f and g are convex and differentiable functions, their first-order Taylor series approximations can be used to relax nonlinear constraints by replacing them by a set of linear inequalities. Consider any point $\hat{x}$ where f and g are defined. Then, the following *linearizations* are valid for MINLP, where $\nabla f(\hat{x})$ and $\nabla g(\hat{x})$ are the gradients of f and g, respectively, at point $\hat{x}$:

$$
f(\hat{x}) + \nabla f(\hat{x})^T (x - \hat{x}) \leq z,
\tag{21.5a}
$$

$$
g(\hat{x}) + \nabla g(\hat{x})^T (x - \hat{x}) \leq 0.
\tag{21.5b}
$$

Then an MILP relaxation of (21.4) is given by

$$
\begin{aligned}
\min \quad & z \\
\text{s.t.} \quad & f(\hat{x}) + \nabla f(\hat{x})^T (x - \hat{x}) \leq z, \quad \hat{x} \in K, \\
& g(\hat{x}) + \nabla g(\hat{x})^T (x - \hat{x}) \leq 0, \quad \hat{x} \in K, \\
& x \in X, \\
& x_i \in \mathbb{Z} \quad \forall i \in I
\end{aligned}
\tag{21.6}
$$

where K is the set of linearization points.

A feasible solution $\tilde{x}$ to (21.6) satisfies the integrality constraints but not necessarily the nonlinear constraints. To find a feasible solution to (21.1), one can solve the following subproblem:

$$\begin{aligned} \min \quad & f(x) \\ \text{s.t.} \quad & g(x) \leq 0, \\ & x \in X, \\ & x_i = \tilde{x}_i \quad \forall i \in I, \end{aligned} \tag{21.7}$$

where all of the integer variables are fixed to corresponding values at $\tilde{x}$. If (21.7) is feasible, then the optimal value of (21.7) provides an upper bound to problem (21.1). Otherwise, to find a certificate of infeasibility, most NLP solvers will automatically solve a feasibility subproblem of the form

$$\begin{aligned} \min \quad & \sum_{j=1}^{m} w_j g_j(x)^+ \\ \text{s.t.} \quad & x \in X, \\ & x_i = \tilde{x}_i \quad \forall i \in I, \end{aligned} \tag{21.8}$$

where $g_j(x)^+ = \max\{g_j(x), 0\}$ for some weights $w_j > 0$. This problem can be interpreted as the minimization of the constraint violation. Valid linearizations generated at an optimal solution to (21.8) are violated by $\tilde{x}$ [726].

21.2.3 ▪ Outer Approximation

First proposed by Duran and Grossmann [643], the outer approximation (OA) algorithm starts by solving (21.6) with an empty set ($K = \emptyset$) of linearization points. Since (21.6) is a relaxation of MINLP, it gives a valid lower bound on the optimal solution value of (21.1). Its optimal solution satisfies the integrality constraints and is used to construct the NLP subproblem (21.7). If (21.7) is feasible, the optimal solution value to (21.7) yields an upper bound to MINLP. The optimal solution of (21.7), $\hat{x}$, is added to the set K of linearization points. The linearizations about $\hat{x}$ eliminate the current solution from being optimal to (21.6) unless $\hat{x}$ is optimal for MINLP. If (21.7) is infeasible, the feasibility subproblem (21.8) provides linearizations that exclude $\hat{x}$ from the feasible region of (21.6). The algorithm iterates until the lower and upper bounds are within a specified optimality tolerance ϵ. If X is bounded, then OA terminates in a finite number of steps [726]. The OA algorithm is described in Algorithm 21.2.

ALGORITHM 21.2. **Outer Approximation.**

0. **Initialize.**
   $U \leftarrow +\infty$. $L \leftarrow -\infty$. $x^* \leftarrow$ NONE. $K \leftarrow \emptyset$. Choose an optimality tolerance $\epsilon > 0$.
1. **Terminate?**
   Is $U - L < \epsilon$ or (21.6) infeasible? If so, terminate the algorithm. Otherwise, go to step 2.
   If x^* is NONE, then the problem is infeasible. Otherwise, x^* is optimal.
2. **Lower Bound**
   Solve (21.6) and let $(\bar{z}, \tilde{x})$ be its optimal solution. $L \leftarrow \bar{z}$.
3. **Upper Bound**
   Solve (21.7) and let $\hat{x}$ be the optimal (or minimally infeasible) solution. Is $\hat{x}$ feasible for (21.7) and $f(\hat{x}) < U$? If so, $x^* \leftarrow \hat{x}$ and $U \leftarrow f(\hat{x})$.
4. **Refine**
   $K \leftarrow K \cup \{\hat{x}\}$. Go to 1.

21.2.4 ▪ LP/NLP-Based Branch-and-Bound

Introduced by Quesada and Grossmann [1534], LP/NLP-based branch-and-bound (LP/NLP-BB) is an extension of the OA algorithm. LP/NLP-BB improves OA by avoiding solving a sequence of MILP relaxations (21.6). Instead, LP/NLP-BB solves the continuous relaxation of (21.6), where the integrality requirements are further relaxed and enforced by branching. We denote this LP subproblem at any node as $LP(l, u)$, defined by the bounds on the integer variables and corresponding to the following subproblem:

$$\begin{aligned}
\min \quad & z \\
\text{s.t.} \quad & f(\hat{x}) + \nabla f(\hat{x})^T (x - \hat{x}) \leq z, \quad \hat{x} \in K, \\
& g(\hat{x}) + \nabla g(\hat{x})^T (x - \hat{x}) \leq 0, \quad \hat{x} \in K, \\
& x \in X, \\
& l_i \leq x_i \leq u_i \quad \forall i \in I,
\end{aligned} \tag{21.9}$$

where K is the set of linearization points and the problem is initialized by the optimal solution of the continuous relaxation (21.2).

The branch-and-bound algorithm for solving (21.6) is interrupted to solve NLP subproblem (21.7) whenever an integer solution is found. Then, the linearizations generated at the solution of NLP subproblem (21.7) are added to (21.6). Unlike regular branch-and-bound, a node cannot be pruned by integrality, and the LP relaxation (21.9) at this node must be re-solved. Then, the tree search continues with the updated MILP relaxation (21.6).

Since the MILP master problem is dynamically updated in a single branch-and-bound tree, the LP/NLP-BB algorithm closely resembles the branch-and-cut method for MILP. An outline of the LP/NLP-BB method is given in Algorithm 21.3.

ALGORITHM 21.3. The LP/NLP-BB Algorithm.

0. **Initialize.**
   Choose an optimality tolerance $\epsilon > 0$, $\mathcal{H} \leftarrow LP(-\infty, \infty)$. $U = \infty$.
   $x^* \leftarrow$ NONE. Let $\tilde{x}$ be the optimal solution of (21.2). $K \leftarrow \{\tilde{x}\}$.

1. **Terminate?**
   Is $\mathcal{H} = \emptyset$? If so, terminate the algorithm. Otherwise, go to step 2.
   If x^* is NONE, then the problem is infeasible. Otherwise, x^* is optimal.

2. **Node Selection.**
   Choose and remove a problem $LP(l, u)$ from $\mathcal{H}$.

3. **Evaluate.**
   Solve $LP(l, u)$, and let $\bar{x}$ be its optimal solution.

4. **Fathom.**
   If $LP(l, u)$ is infeasible, fathom by infeasibility and go to step 1.
   If $f(\bar{x}) \geq U - \epsilon$, fathom by bound and go to step 1.

5. **Refine.**
   If $\bar{x}_I$ is integer, solve (21.7) and let $\hat{x}$ be the optimal (or minimally infeasible) solution.
   Is $\hat{x}$ feasible for (21.7) and $f(\hat{x}) < U$? If so, $x^* \leftarrow \hat{x}$ and $U \leftarrow f(\hat{x})$.
   Let $K \leftarrow K \cup \{\hat{x}\}$ and go to step 3 .

6. **Branch.**
   Select index j such that $\bar{x}_j \notin \mathbb{Z}$ and $j \in I$.
   Set $(l^-, u^-) \leftarrow (l, u)$, $(l^+, u^+) \leftarrow (l, u)$, $u_j^- \leftarrow \lfloor \bar{x}_j \rfloor$ and $l_j^+ \leftarrow \lceil \bar{x}_j \rceil$.
   Add $LP(l^-, u^-)$ and $LP(l^+, u^+)$ to $\mathcal{H}$ and go to step 1.

21.2.5 ▪ Additional Methods

Generalized Benders decomposition (GBD) [795] is another method developed for solving convex MINLP problems and is very similar to the OA algorithm. Using Benders decomposition, the linearizations generated at the optimal solution of (21.7) can be aggregated to the Benders cut

$$f(\hat{x}) + \sum_{i \in I} \mu_i(x_i - \hat{x}_i) \leq z,$$

where μ_i is the dual multiplier of the bound $x_i = \bar{x}_i$ in the optimal solution of (21.7). Instead of linearizations, Benders cuts are added to the MILP relaxation (21.6).

The extended cutting plane (ECP) method of [1883] is another variant of OA. The ECP method does not solve NLP subproblems (21.7); instead, it generates linearizations at the optimal solution of the MILP relaxation (21.6). The method alternates between solving the MILP relaxation and generating linearizations. At each step, a small number of violated linearizations are added to (21.6), and the algorithm terminates when the optimal solution of (21.6) satisfies all linearizations.

Abhishek et al. [2] use linearizations from nonintegral points, such as the LP relaxation solution and Benders cuts, to accelerate the convergence of the LP/NLP-BB algorithm. Bonami et al. [302] proposed a hybrid method of OA and nonlinear branch-and-bound that alternates between LP and NLP relaxations for bounding.

The nonlinear branch-and-bound can be accelerated by inexact NLP subproblem solves. Borchers and Mitchell [315] use the sequential quadratic programming (SQP) method to solve NLP subproblems. At each iteration of SQP, it is checked whether convergence to a nonintegral solution has been detected. In this case, the solution is interrupted and they proceed with branching. Leyffer [1175] uses linearizations at each SQP step to obtain valid implicit lower bounds. Mahajan et al. [1253] propose using a single SQP problem generated at the root node throughout the branch-and-bound tree.

21.3 ▪ Nonconvex Mixed-Integer Nonlinear Programming

Nonconvex MINLP is much harder than convex MINLP because it contains nonconvexities of integer variables and nonconvex functions in the objective and constraints. Hence, even when integrality requirements are relaxed, the feasible region may be nonconvex. MINLP methods utilize factorable programming, spatial branch-and-bound, and domain propagation and bound-tightening techniques, which we describe next.

21.3.1 ▪ Factorable Programming

One way of dealing with nonconvex functions is to replace each of them with a convex underestimating function, i.e., convex functions $\check{f}(x)$ such that $\check{f}(x) \leq f(x)$ and $\check{g}(x)$ such that $\check{g}(x) \leq g(x) \ \forall \ x \in [l, u]$. For an equality constraint $h(x) = 0$, one also needs a concave overestimating function $\hat{h}(x)$, i.e., $\hat{h}(x) \geq h(x) \ \forall \ x \in [l, u]$, in addition to the convex underestimating function $\check{h}(x)$. The tightest possible under- and overestimators, namely convex and concave envelopes of the functions, reduce the number of subproblems required to be solved in branch-and-bound.

If a function $f(x)$ is *separable*, i.e., if it can be written as a sum of univariate functions ($f(x) = \sum_{i=1}^{n} f_i(x_i)$), then an overestimator of f can be obtained by summing up overestimators of the individual univariate functions. For factorable MINLP problems, the functions can be expressed as recursive compositions of sums and products of univariate functions. Relaxations for such problems can be obtained via transformation to separable programs through the introduction of additional variables and constraints. For example, the function $f(x_1, x_2, x_3) = \sqrt{x_1} \log(x_2) x_3^3$ can be written equivalently as

$$x_4 = \log(x_2),$$
$$x_5 = x_1 x_4,$$
$$x_6 = x_3^3,$$
$$x_7 = x_5 x_6,$$
$$x_8 = \sqrt{x_7},$$

where $x_8 = f(x_1, x_2, x_3)$. Then, over- and underestimators of the relatively simple functions $\log(x_2), x_1 x_4, x_3^3$, and $\sqrt{x_7}$ suffice to construct a relaxation.

Factorable MINLP problems allow a reformulation of (21.1) as follows [1304, 1703, 1754]:

$$
\begin{aligned}
\min \quad & x_{n+q} \\
\text{s.t.} \quad & x_k = \mathcal{V}_k(x) \quad \forall k \in \{n+1, n+2, \ldots, n+q\}, \\
& x \in X, \\
& x_i \in \mathbb{Z} \quad \forall i \in I,
\end{aligned}
\tag{21.10}
$$

where $\mathcal{V}_k$ is an operator of univariate or bivariate functions. The reformulation contains a set of q new auxiliary variables. By convention, the last auxiliary variable denotes the objective function. The lower and upper bounds on auxiliary variables can be deduced from the operator $\mathcal{V}_k$ and the bounds on the arguments of $\mathcal{V}_k$.

Once the reformulation (21.10) is formed, a polyhedral relaxation can be formed by linear under- and overestimators of operators $\mathcal{V}_k$. For example, for the product of two variables, $x_k = x_i x_j$, with bounds $l_i \leq x_i \leq u_i$ and $l_j \leq x_j \leq u_j$, McCormick [1304] proposed two linear inequalities, $x_k \geq l_j x_i + l_i x_j - l_i l_j$ and $x_k \geq u_j x_i + u_i x_j - u_i u_j$, as the convex envelope, and $x_k \leq l_j x_i + u_i x_j - u_i l_j$ and $x_k \leq u_j x_i + l_i x_j - l_i u_j$, as the concave envelope. We refer the reader to Horst and Tuy [959] and Tawarmalani and Sahinidis [1754] for under- and overestimators and envelopes for other common expressions.

Often, convex relaxations are further relaxed with polyhedral relaxations [1755, 1756], thus facilitating their solution through efficient and reliable LP techniques.

21.3.2 ▪ Spatial Branch-and-Bound

The most widely used method to solve nonconvex MINLP problems is *spatial branch-and-bound*. Once a polyhedral relaxation of (21.10) is formed by linear under- and overestimators, an optimal solution to the corresponding LP problem is found. This solution may not be feasible for (21.10) even if it satisfies the integrality requirements. In this case, the domain of continuous variables must be partitioned. This type of branching is called spatial branching. For a continuous variable x_i whose current domain is $[l_i, u_i]$, a value β, such that $l_i < \beta < u_i$, is chosen, and two subproblems are created where the domain of x_i is partitioned as $[l_i, \beta]$ in one branch and $[\beta, u_i]$ in the other. The reduced domain in subproblems improves the under- and overestimators that involve variable x_i. Thus, the branching value β is chosen such that the relaxation solution becomes infeasible in both of the subproblems.

The performance of branch-and-bound strongly depends on the choice of branching variable x_i and branching point β [193, 1754]. Tawarmalani and Sahinidis [1754] propose violation transfer as a branching variable selection. Violation transfer chooses the variable x_i that has the largest total effect on the violation of nonconvex constraints of $\mathcal{V}_k(x)$ that involve variable x_i. The branching point is typically chosen as a convex combination of the current relaxation solution and the midpoint of the domain of the variable [1682]. While midpoint selection provides finite termination, the current relaxation solution is used to speed up convergence in practice.

21.3.3 ▪ Domain Propagation and Bound Tightening

A major improvement over spatial branch-and-bound is the introduction of the branch-and-reduce algorithm [1608, 1609], in which one attempts to reduce the domains of variables in addition to the reductions that occur due to branching. There are two types of bound-tightening methods: feasibility and optimality based. Feasibility-based bound tightening (FBBT) works by inferring tighter bounds without eliminating any feasible solutions, whereas in optimality-based bound tightening (OBBT), feasible solutions might be eliminated as long as they are not superior to the incumbent. With reduced variable domains, tighter under- and overestimators can be generated to further improve the relaxations.

The domain of a variable can be tightened by minimizing and maximizing this variable over the constraint set with the additional requirement of an upper bound on the objective function equal to the best-known feasible solution. This approach requires solution of two nonconvex NLP subproblems for each variable. Even solving these maximization and minimization problems over the polyhedral relaxation might be expensive.

Using interval arithmetic operations, domain reduction can be done more efficiently at the expense of possibly arriving at looser domains. Considering only one nonlinear constraint $x_k = \mathcal{V}_k(x)$ of (21.10) at a time, bounds of variables can be easily tightened. For example, if $x_k = x_i x_j$ with $1 \le x_i \le 6$, $2 \le x_j \le 4$, and $0 \le x_k \le 4$, then $l_i = 1$ and $l_j = 2$ imply a tighter lower bound on x_k, $l'_k = l_i l_j = 2 > 0$, and $u_k = 4$ and $l_j = 2$ imply an upper bound on x_i, $u'_i = l_k / l_j = 2 < 6$. This *bounds propagation* is repeated on different nonlinear constraints until no more bounds can be tightened. Propagation over a single linear constraint can also be done very efficiently [339]. If one of the linear constraints defining X is in the form of $\sum_{j=1}^n a_j x_j \le b$, the following bounds are derived on each variable x_h:

$$\text{for } a_h > 0, \quad x_h \le \frac{1}{a_h}\left(b - \sum_{j \ne h}\min\{a_j l_j, a_j u_j\}\right),$$

$$\text{for } a_h < 0, \quad x_h \ge \frac{1}{a_h}\left(b - \sum_{j \ne h}\min\{a_j l_j, a_j u_j\}\right).$$

FBBT methods are practically fast but may not converge in finite time. Belotti et al. [191] propose a polynomial-time method for computing the limit point of FBBT via solution of a single LP problem.

One OBBT method is *reduced-cost bound tightening* [1609]. Suppose a convex relaxation of (21.1) has been solved with an optimal value of L and optimal solution $\bar{x}$, where x_i is at its lower bound l_i with a positive reduced cost μ_i. Then, a valid upper bound on x_i is $u'_i = l_i + \frac{U - L}{\mu_i}$, where U is the value of the best feasible solution so far. Reduced-cost bound tightening can only be done for variables that are at their lower or upper bounds. If not, a fictitious bound can be added and relaxation could be solved to tighten bounds of such variables. This method is called *probing* and is used in MILP [1631] and MINLP [1609] problems on integer or continuous variables.

In the branch-and-reduce algorithm, before a relaxation is solved at a branch-and-bound node, its constraints are checked to see whether the domain of any variables can be reduced via FBBT. After the relaxation is solved to optimality, sensitivity information is used to check whether the domain of any variable can be reduced via OBBT.

21.3.4 ▪ Stronger Relaxations of Structured Sets

The factorable MINLP problem given in (21.10) makes it easy to find a relaxation since we only need relaxations of univariate or bivariate functions. However, the relaxation obtained may be weak and can lead to a very large branch-and-bound tree. One common strategy is to study specific structured sets that consider multiple constraints of the problem simultaneously to find a tighter relaxation.

One structure that appears in many MINLP problems is quadratic functions. A reformulation of quadratic functions can be done by defining the additional variables $X_{ij} = x_i x_j$ and replacing all quadratic and bilinear terms with X_{ij}. Then, all nonlinearities in the quadratic functions are moved to a single constraint of the form $X = xx^T$, where X is the $n \times n$ matrix containing all the X_{ij} variables. Then, McCormick inequalities yield a polyhedral relaxation of the nonconvex constraints $X = xx^T$.

For 0-1 MIQCP problems, Adams and Sherali [23] propose generating valid constraints by the reformulation-linearization technique (RLT). In RLT, each linear constraint of the problem, say $\sum_{j=1}^{n} a_j x_j \leq b$, multiplied by the binary variable x_k gives the quadratic inequality $\sum_{j=1}^{n} a_j x_j x_k \leq b x_k$. Then, this inequality is linearized by substituting quadratic terms with the auxiliary variables X to get the valid linear inequality $\sum_{j=1}^{n} a_j X_{jk} \leq b x_k$. Since $x_k^2 = x_k$ for a binary variable x_k, variable X_{kk} can be replaced with x_k to improve the relaxation further. Another round of RLT can be applied to already generated inequalities by introducing variables representing products of three variables. Further multiplying constraints with variables leads to a hierarchy of increasingly stronger relaxations [24]. The drawback of this approach is that problem size grows exponentially. RLT has been extended to other classes of MINLP problems beyond 0-1 MIQCP. We refer the reader to the book [1687] for a full treatment of RLT.

Another general technique for generating stronger relaxations of $X = xx^T$ is via semidefinite programming (SDP), which is a convex optimization problem. Note that X is a symmetric positive definite matrix; thus the nonconvex constraint $X = xx^T$ can be relaxed as the convex constraint $X - xx^T \succeq 0$, which leads to an SDP relaxation. These relaxations obtained by using RLT and SDP do not dominate each other. Thus, combining both may yield a stronger relaxation [88]. A tight LP relaxation of the SDP constraint is also possible via a cutting-plane approach [1531, 1688]. We refer the reader to [161] for a review and comparison of SDP relaxations.

Another line of research constructs the convex and concave envelopes of nonconvex functions with polyhedral envelopes [1568, 1747]. Explicit characterizations of polyhedral envelopes are given for bilinear [41], trilinear [1317], edge-concave [1318], and multilinear functions [160, 1237, 1685]. Tawarmalani et al. [1751] derived convex and concave envelopes for supermodular functions with polyhedral concave envelopes over hyperrectangles. Separation of polyhedral envelopes simplifies to an LP problem that can be solved efficiently [160].

Nonpolyhedral convex envelopes of lower semicontinuous functions were first studied by Tawarmalani and Sahinidis [1753]. Tawarmalani and Sahinidis [1752] give the convex and concave envelopes of the ratio function and convex envelope of a function $f(x, y)$ over a box such that f is concave in x and convex in y. Jach et al. [983] give convex envelopes for indefinite and $(n-1)$-convex functions. Khajavirad and Sahinidis [1059] study the convex envelope of lower semicontinuous functions whose generating set is the union of a finite number of compact sets. They also give convex envelopes for products of convex and componentwise concave functions [1058].

21.3.5 ▪ Additional Methods

When f and g are twice differentiable, another way of dealing with nonconvexities is by writing f and g as differences of convex functions. Given $f(x)$, one can find convex functions p and q such that $f(x) = p(x) - q(x)$ [916]. Then, one would only need to deal with the concave function $-q(x)$, which is much easier to relax. Androulakis et al. [85] propose forming a convex relaxation by adding a quadratic term to nonconvex functions:

$$f_\alpha(x) = f(x) + \alpha \sum_{i=1}^{n} (l_i - x_i)(u_i - x_i), \tag{21.11}$$

which underestimates function f over the box $[l, u]$. If α is chosen large enough, then $f_\alpha(x)$ becomes a convex function because the quadratic term dominates in the Hessian of the function. Thus, α should be bigger than negative one half of the most negative eigenvalue of the Hessian of $f(x)$ over the box $[l, u]$. On the other hand, as α increases, the quality of underestimation worsens. Choosing the best α is not an easy task if $f(x)$ is not a quadratic function since the Hessian of $f(x)$ is not a constant. One has to carry out interval arithmetic on the Hessian to ensure convexity. Novak and Vigerske [1407] estimate the Hessian of nonquadratic functions via sampling. Thus, their method is an approximate method for nonquadratic MINLP problems.

The branch-and-bound algorithm that uses (21.11) is called αBB [85]. The algorithm is enhanced by the use of tighter and more specialized over- and underestimators for certain specific functions [26, 28]. αBB is extended to the mixed-integer case by Adjiman et al. [27]. The advantage of the αBB method over factorable programming reformulation is that no additional variables are required.

Other methods employ decomposition techniques. For nonconvex NLP, Ben-Tal et al. [201] propose partitioning the variables into two sets so that fixing the variables in the first set leads to a convex NLP subproblem. Then, a lower bound on the NLP problem is obtained by a Lagrangian dual. A branch-and-bound method employed in the space of variables in the first set leads to a convergent algorithm that reduces the duality gap at each step. Ben-Tal et al. [201] tested the method on the pooling problem, where the only nonlinearities are bilinear constraints. They chose to fix one of the variables in each bilinear term so that the subproblem becomes an LP problem after fixing.

Another example of decomposition techniques is nonconvex generalized Benders decomposition (NGBD), an extension of GBD to the nonconvex case [1188]. Nonconvex NLP subproblems are solved via global optimization solvers. NGBD has been successfully applied to stochastic MINLP problems [1188].

21.3.6 ▪ Piecewise-Linear Approximations and Relaxations

In previous subsections, we described exact methods for solving nonconvex MINLP problems. Another idea is to replace nonlinear functions with piecewise-linear functions so that an approximate solution to the MINLP problem can be found by solving the resulting MILP approximations [1280]. To create a piecewise-linear approximation, a set of breakpoints in the domain of a function is chosen where the approximation will be exact. Then, the nonlinear function is approximated linearly in the segment between two adjacent breakpoints.

Modeling piecewise-linear functions leads to either sets of binary variables or special-ordered sets of type 2 (SOS2) [176]. An SOS2 is an ordered set of variables where at most two of the variables can be nonzero and the nonzero variables are adjacent. Beale and Tomlin [176] propose a specialized branching rule for SOS2 variables. The number of binary variables required for piecewise-linear functions is equal to the number of segments. An improved approximation can be obtained by including more segments at the expense of increased solution time for the approximate MILP problem. Vielma and Nemhauser [1848] propose a way of modeling piecewise-linear functions where the number of binary variables required is logarithmic in the number of linear segments.

Bergamini et al. [228] propose replacing nonlinear functions with piecewise-linear under- and overestimators. This ensures that all feasible points of the MINLP problem remain feasible for the MILP problem. Bergamini et al. [228] partition the domain of one of the variables of the bilinear term into disjoint segments. Then, for each segment, McCormick inequalities are used separately as convex and concave envelopes. The relaxations created by the piecewise-linear functions are used within the OA algorithm to solve nonconvex problems with bilinear and concave functions to global optimality. Hasan and Karimi [917] suggest partitioning both variables in a bilinear term for a tighter relaxation. Castro [415] suggests tightening the bounds on the other variable before generating McCormick envelopes for each segment.

Leyffer et al. [1176] calculate the approximation error of the piecewise-linear function in each segment and shift the piecewise-linear function so that it over- and underestimates the nonlinear function. Instead of branching on introduced SOS2 variables, they branch on the original variables and refine the piecewise-linear approximation each time a new subproblem is formed by branching.

Piecewise-linear relaxations have been used within spatial branch-and-bound for solving nonconvex MINLP and MIQCP problems [1041, 1176, 1330]. We refer the reader to [792, 1047, 1847] for further details on piecewise-linear functions for solving MINLP problems.

21.4 ▪ Enhancements

A naive implementation of branch-and-bound for both convex and nonconvex MINLP problems may lead to huge search trees. To speed up convergence, many techniques have been proposed, including presolve, reformulations, primal heuristics, and cutting planes.

21.4.1 ▪ Presolve, Branching, and Reformulations

The goal of presolve is to create a tighter and smaller relaxation of the problem that will likely result in a smaller search tree. Basic methods, such as like removing redundant rows and removing fixed variables, reduce the problem size without removing any feasible points from the relaxation. Other methods reformulate the problem to get tighter relaxations.

Proposed by Savelsergh [1631] for MILP problems, coefficient reduction reformulates the constraints by reducing the coefficient of a variable if the constraint becomes loose after fixing the integer variable. Application of coefficient reduction to the case of the MINLP problem resulted in dramatic performance improvements in solution times [192].

Constraint disaggregation is another reformulation method for MINLP problems. Tawarmalani and Sahinidis [1756] consider the set of the form $S = \{x \in \mathbb{R}^n \mid f(x) = h(g(x)) \leq 0\}$, where $g : \mathbb{R}^n \to \mathbb{R}^p$ is a smooth and convex function and $h : \mathbb{R}^p \to \mathbb{R}$ is smooth, convex, and nondecreasing function. These conditions imply that $f(x)$ is a convex function. Thus, an equivalent reformulation can be formed as $S_r = \{(x, y) \in \mathbb{R}^n \times \mathbb{R}^p \mid h(y) \leq 0, g(x) \leq y\}$ by introducing extra variables y. The polyhedral relaxation of S_r is tighter than the polyhedral relaxation of S generated by the same set of linearization points. Hijazi et al. [940] give a worst-case example for which the OA method takes an exponential number of iterations, whereas the number of iterations is linear if the OA method is applied to the reformulation after constraint disaggregation is applied. Computational results on different solvers show that significant improvements can be attained by constraint disaggregation [1061].

The size of the branch-and-bound tree mainly determines the solution time of a problem. Two crucial components that affect the size of the tree are branching variable and node selection. Since a node is fathomed whenever the relaxation bound of the node is above the current upper node, one would want to select the branching variable that increases the relaxation bound the most.

In the MILP case, the consensus is to estimate the change in the relaxation bound after branching. Once both estimates for branching up and down are available, these can be combined to compute a score for each candidate variable. Strong branching performs the branching on each candidate variable and solves the subproblems to compute changes in the relaxation bound. Even for convex MINLP problems, it can be prohibitive to solve two subproblems for each candidate variable. One approach is to solve an approximate quadratic program around the current relaxation solution to obtain the estimates [306]. Pseudocost branching keeps the history of the effect of past branching decisions and utilizes this information by averaging the bound changes over the history. Tawarmalani and Sahinidis [1755] propose a violation transfer mechanism for deciding branching variables. Reliability branching [13] is a hybrid method that utilizes strong branching early during the search tree and then switches to pseudocost branching once reliable estimates are available. Belotti et al. [193] extend reliability branching to the case of continuous variable selection in spatial branch-and-bound.

Another difficulty arises if MINLP formulations have symmetry. To find an optimal solution quickly, symmetry in MINLP formulations is recognized and exploited. One method of dealing with symmetry is to reformulate the problem by adding symmetry-breaking constraints [1192]. These constraints eliminate some symmetric solutions, leaving at least one optimal solution. Another way is to modify the branch-and-bound algorithm so that only one node is solved from all of the nodes that correspond to symmetric subproblems [1274, 1430].

21.4.2 ▪ Primal Heuristics

Primal heuristic methods are designed to find good feasible solutions quickly without any optimality guarantee. For harder problems, finding a good solution instead of solving the problem to optimality is acceptable. Primal heuristics accelerate branch-and-bound by quickly identifying a feasible solution that can be used for pruning. Bound-tightening methods also become more effective with a better upper bound.

The most straightforward way of finding feasible solutions is the rounding heuristic. Given a solution to a relaxation of (21.1), one can round fractional-valued integer variables. After all integer variables are fixed to integral values, an NLP subproblem (21.7) is solved to search for a feasible solution to (21.1). Nannicini and Belotti [1376] propose using MILP relaxations to increase the chance of finding a feasible solution after rounding. Another fixing-based heuristic is due to [240] and is called *undercover*. After fixing a minimal number of variables to reduce the problem to an MILP problem, this MILP restriction is solved either exactly or heuristically.

Diving heuristics conduct a depth-first search of the branch-and-bound tree with the aim of finding a feasible leaf node [304]. A quadratic programming relaxation could be solved to speed up the diving process [1253].

Bonami et al. [303] extend the idea of the feasibility pump in the context of MILP [715] to MINLP problems. The main idea of the feasibility pump is to invoke an alternating sequence of NLP and MILP solves that leads to a feasible solution for (21.1). The MILP relaxation is obtained by using linearizations of the nonlinear constraints at the solution of the NLP subproblems, and the objective function of MILP is the L_1 distance to the solution of last the NLP subproblem. The NLP subproblem is the continuous relaxation (21.2) where the objective function is replaced by the L_2 distance to the solution of the last MILP problem. In another variant, D'Ambrosio et al. [543] propose replacing the MILP relaxation with a convex MINLP relaxation for nonconvex MINLP problems.

The RENS (relaxation-enforced neighborhood search) [239] heuristic searches for a feasible solution around the continuous relaxation of (21.1). If $\bar{x}$ is the continuous relaxation solution, the domain of integer variable x_i is restricted to $[\lfloor \bar{x}_i \rfloor, \lceil \bar{x}_i \rceil]$ for all $i \in I$. In the RINS (relaxation-induced neighborhood search) heuristic [546], the integer variables in a continuous relaxation ($\bar{x}$) are fixed if they match the current best feasible solution (x^*), i.e., $x_i = \bar{x}_i$ if $\bar{x}_i = x_i^* \ \forall \ i \in I$. This idea was extended to convex MINLP problems by Bonami and Gonçalves [304] where once the integer variables are fixed as above, the resulting MINLP problem is solved by the LP/NLP-BB algorithm. In the *local branching* heuristic [716], only a few binary variables are allowed to differ from the best feasible solution. A constraint $\sum_{i \in B} |x_i - x_i^*| \leq k$ is added to restrict the search space, where k is chosen between 10 and 20. Local branching is extended to MINLP by [1377]. Finally, Liberti et al. [1194] integrate variable neighborhood search, local branching, and SQP into their RECIPE heuristic.

21.4.3 ▪ Cutting Planes

One way of enforcing nonconvex constraints is by branching either on integer or nonconvex continuous variables. Another way is to strengthen the convex relaxation of an MINLP problem via valid inequalities, which is called *cutting planes*.

Cutting planes were a major reason for the vast improvement in MILP solution technology and state-of-the-art MILP solvers. We refer the reader to [494] for a recent survey on cutting planes for MILP. Far fewer cutting planes have been developed for MINLP.

For the case of MISOCP, generalizations of cutting-plane techniques for MILP have been proposed. For example, mixed-integer rounding cuts [1273], Chvátal–Gomory cuts [474], and lift-and-project cuts [146, 1226, 1686] are extended to MISOCP by [106, 424], and [424, 1342], respectively.

In the context of MINLP, the main developments are in the area of *disjunctive programming*. In his seminal work, Balas [143] studies the convex hull of the union of finitely many polyhedra and gave a compact representation of the convex hull. Later, Ceria and Soares [422] extend these results to general convex sets.

Given a solution $\bar{x}$ to a convex relaxation of (21.1) such that $\bar{x}_i \notin \mathbb{Z}, i \in I$, instead of branching, a disjunction on the integer variable x_i can be created as $x_i \le \lfloor \bar{x}_i \rfloor \vee x_i \ge \lceil \bar{x}_i \rceil$. A valid inequality for both disjunctions that is violated by $\bar{x}$ can be added to the relaxation to cut off $\bar{x}$. This valid inequality is called a disjunctive cut.

In the context of MILP, Balas et al. [146] propose a lift-and-project procedure that solves a cut-generating linear problem (CGLP) whose objective is to maximize the violation of $\bar{x}$ by the generated cut. This CGLP has twice the size of the LP relaxation when a two-term disjunction is imposed. Later, Stubbs and Mehrotra [1731] extended this method to the case of convex MINLP. Unfortunately, the separation problem of [1731] has twice the number of variables and constraints as the original problem and is not differentiable at some points. Addressing this issue, an LP-based iterative separation method was recently proposed in [1061, 1062]. Also recently, Bonami [301] suggest solving a cut-generating convex problem in the space of the original variables, while only doubling the number of nonlinear constraints. Frangioni and Gentile [750] study 0-1 convex MINLP, where one side of the disjunction is a point and gives explicit characterization of the convex hull of the union. Kılınç et al. [1063] propose valid cutting planes that use the information gathered during strong branching.

In the context of factorable nonconvex MINLP, Belotti [190] creates disjunctive cuts by means of CGLP that exploit disjunctions on continuous variables, i.e., $x_i \le \beta \vee x_i \ge \beta$. Júdice et al. [1016] develop disjunctive cuts on complementarity constraints of the form $x_i x_j = 0$, where a valid disjunction is $x_i = 0 \vee x_j = 0$. Saxena et al. [1637, 1638] derive cutting planes using disjunctions of the form $(\sum_{j=1}^n a_j \le b) \vee (\sum_{j=1}^n a_j \ge b)$ for nonconvex MIQCPs. We refer the reader to [308] for a recent review of disjunctive cuts for MINLP problems.

In a different setting where all disjunctions belong to orthogonal subspaces, Tawarmalani et al. [1750] explicitly characterize the convex hull of the union of the disjunctions.

21.5 • Software for MINLP

There are many software packages for modeling and solving MINLP problems. In this section, we briefly survey the currently available solvers and modeling packages that provide interfaces to these solvers.

Algebraic modeling languages are programming languages that are tailored to specifying and solving optimization models with a syntax similar to mathematical notation. Modeling languages do not have solver capabilities but provide interfaces to the actual MINLP solvers. Modeling languages allow the use of abstract entities such as sets and indices, which makes it easy to formulate large instances of optimization models. Some modeling languages are also flexible enough to allow the user to implement complicated algorithms and provide effective preprocessing techniques to simplify the problem.

The basic functionalities of modeling languages include reading the model and data for the problem provided by the user, translating the model into a form readable by the solvers, providing the evaluation of, first and second derivatives of nonlinear functions at given points, and returning the solution found by the solver in a human-readable format.

The most common and popular modeling software packages are GAMS [343], AIMMS [283], and AMPL [749]. AIMMS provides built-in algorithms for OA and

Table 21.1. *Convex MINLP solvers.*

Solver	Algorithms	Interfaces
α-ECP	ECP	GAMS
BONMIN	NLP-BB, OA, LP/NLP-BB	AMPL, GAMS, MATLAB, Opti Toolbox
DICOPT	OA	GAMS
FilMINT	LP/NLP-BB	AMPL
KNITRO	NLP-BB	AIMMS, AMPL, GAMS
MILANO	NLP-BB, OA	MATLAB
MINLPBB	NLP-BB	AMPL, MATLAB
MINOTAUR	NLP-BB	AMPL
SBB	NLP-BB	GAMS
Xpress-SLP	NLP-BB, OA	Xpress-MOSEL

Table 21.2. *Nonconvex MINLP solvers.*

Solver	Algorithms	Interfaces
ANTIGONE	spatial-BB	GAMS
BARON	spatial-BB	AIMMS, AMPL, GAMS, MATLAB, Opti Toolbox, YALMIP
Couenne	spatial-BB	AMPL, GAMS
LaGO	spatial-BB	GAMS
LindoGlobal	spatial-BB	GAMS, LINGO
OQNLP	heuristic	GAMS, MATLAB
SCIP	spatial-BB	AMPL, GAMS, Opti Toolbox

LP/NLP-BB built on its modeling language. TomLab [951], YALMIP [1217], and Opti Toolbox [527] are MATLAB-based modeling languages that provide interfaces to MINLP solvers within MATLAB. Another modeling package is Pyomo [915], which is available for Python.

The NEOS server [535] provides free access to a collection of state-of-the-art optimization software, including most of the software listed below. Users submit optimization problems through a web interface, and problems are solved remotely and free of charge to the user.

The MINLP solvers can be divided into those for convex and nonconvex MINLPs. A recent survey [372] provides a nearly complete list of available MINLP solvers. Tables 21.1 and 21.2 list publicly available convex and nonconvex MINLP solvers, respectively. The tables also include the implemented algorithms and available modeling language interfaces. A detailed discussion of these solvers follows.

21.5.1 ▪ Convex MINLP solvers

- α-ECP [1884] implements the ECP method and depends on external MILP solvers that are available within GAMS. α-ECP has been extended to ensure global optimality for pseudoconvex MINLP problems.

- Bonmin [302] is an open-source solver available from COIN-OR [1225] that implements NLP-BB, OA, and LP/NLP-BB algorithms and a hybrid method of OA and NLP-BB. It has been built on the MILP solver CBC [744] and can use FilterSQP [725] and IPOPT [1859] as NLP solvers. It implements disjunctive cutting planes for convex MINLP and different primal heuristics, such as feasibility pump, diving, and RINS.

- Dicopt [1077] implements the OA algorithm that ensures global optimality only for convex MINLP problems. It can be used as a heuristic for nonconvex MINLP problems since linearizations of nonconvex functions are allowed to move away from their support point through an augmented penalty function [1853]. Termination criteria are based on deterioration of the objective function value in NLP subproblems. Dicopt depends on MILP and NLP solvers that are available within GAMS.

- FILMINT [2] implements an LP/NLP-BB method and was built on the MILP solver MINTO [1389] and the NLP solver FilterSQP. Different strategies for linearization point selection have been incorporated for faster convergence. FILMINT also includes disjunctive cutting planes for convex MINLP and a feasibility pump heuristic.

- KNITRO [376] was initially designed as an NLP solving, later evolving to an MINLP solver that implements an NLP-BB algorithm.

- MILANO [224] is a MATLAB-based solver that implements NLP-BB and OA algorithms. It uses LOQO [1827] to solve NLP subproblems where the main focus is to develop efficient warm-start strategies for interior-point methods.

- MINLPBB [1174] implements an NLP-BB algorithm and uses FilterSQP for NLP subproblems.

- MINOTAUR [1325] implements NLP-BB and can do inexact NLP solves for faster convergence. It depends on external MILP and LP solvers and can use FilterSQP and IPOPT as NLP solvers.

- SBB [369] implements an NLP-BB algorithm and depends on external NLP solvers that are available within GAMS.

- Xpress-SLP [706] implements three algorithms: MIP within SLP (resembles the OA algorithm), SLP within MIP (resembles the NLP-BB algorithm), and SLP then MIP (first solves an NLP relaxation, then a MILP relaxation, and finally an NLP relaxation with fixed integer variables). Xpress-Mosel [488,707] can be used to model problems, and NLP subproblems are solved via a built-in solver based on SLP (successive linear programming).

21.5.2 ▪ Nonconvex MINLP Solvers

- ANTIGONE [1332] ensures global optimality by spatial branch-and-bound and utilizes MILP relaxations for bounding.

- BARON [1611,1756] implements a spatial branch-and-bound algorithm. The algorithm is enhanced by various bound-tightening methods and convexification techniques for structured functions [160]. It can use LP, MILP, and NLP relaxations for bounding. It uses external MILP, LP, and NLP solvers and comes with links to a variety of such solvers, including COIN-OR solvers.

- COUENNE [193] is an open-source solver available from COIN-OR that implements a spatial branch-and-bound algorithm. It also implements disjunctive cutting planes for nonconvex MINLP problems, a feasibility pump heuristic, and different branching schemes. It uses CBC as the MILP solver and IPOPT as the NLP solver.

- LaGO [1407] implements a spatial branch-and-bound algorithm and is available from COIN-OR. Nonconvex nonquadratic functions are approximated by sampling. As a result, global optimality is ensured only for convex MINLP and

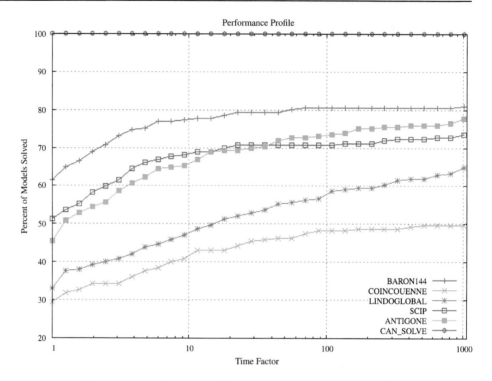

Figure 21.1. *Performance profile comparing nonconvex MINLP solvers.*

nonconvex MIQCP problems. It depends on an external LP solver and uses IPOPT as the NLP solver.

- LINDOGlobal [1200] implements a branch-and-cut algorithm that utilizes LP and NLP relaxations for bounding. The LINGO modeling suite [1647] can be used to model a variety of problems. It can be run as a multistart heuristic by disabling global solver components. The NLP relaxations are solved by CONOPT [631].

- OQNLP [1803] implements a multistart search heuristic for MINLP problems and solves NLP subproblems with fixed integer variables.

- SCIP first started as an MILP solver [11] and evolved to MINLP [1849]. It ensures global optimality by spatial branch-and-bound that utilizes LP relaxations for bounding. It includes various primal heuristics and bound-tightening methods.

21.5.3 ▪ Test Problems and Computational Experiments

There exist many libraries of test problems in AMPL and GAMS formats. These include the GAMS MINLPLib World [370], the MacMINLP collection [1324], the CMU-IBM Cyber-Infrastructure for MINLP collaborative site [483], and the CMU-IBM Open Source MINLP Project [484]. Recently, a new version of MINLPLib has been published with the addition of many publicly available test problems [1323]. Computational experiments conducted on this test [1850] demonstrate that the nonconvex MINLP area has progressed rapidly over the last few years due largely to developments in BARON.

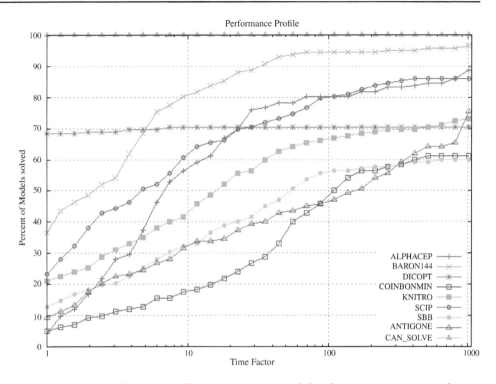

Figure 21.2. *Performance profile comparing convex and three best nonconvex MINLP solvers.*

We provide our own computational experiments that compare convex and nonconvex MINLP solvers separately. To compare different solvers, we use GAMS Performance Tools [371] to create performance profiles [607] in which a solver is counted as having solved a problem if it finds the best objective value among all solvers on an instance.

For testing nonconvex MINLP solvers, we used MINLPLib [370] test problems and considered solvers that are available in GAMS and guarantee global optimality. Namely, the following solvers were tested: ANTIGONE, BARON, Couenne, LindoGlobal, and SCIP. The resulting performance profiles are provided in Figure 21.1. For this test, the solvers ANTIGONE, BARON, and SCIP clearly dominate Couenne and LindoGlobal. BARON slightly outperforms ANTIGONE and SCIP.

For the next test, we compared convex MINLP solvers available in GAMS. We also included the three best nonconvex MINLP solvers from the previous test. For this computational experiment, we used the CMU-IBM test set [484], which contains only convex MINLP problems. The performance profiles are provided in Figure 21.2. For the problems that it can solve, Dicopt tends to be very fast, whereas BARON can solve many more instances and clearly outperforms all other solvers.

21.6 ▪ MINLP Applications

Given the generality and flexibility of MINLP, many diverse and important practical problems are naturally modeled as MINLP problems. The field of chemical engineering gives rise to many applications, including integrated design and control of chemical processes [731], multiproduct batch plant design [1553, 1834], synthesis and retrofit design problems [643, 1635, 1792], pooling problems [1328], the simultaneous design and control of a process [1244], and heat-exchanger networks [1930]. Other sources for MINLP applications include network design problems [243, 313, 313, 315, 334, 769, 1041, 1118, 1285], layout design problems [414, 1635], portfolio optimization [264, 750, 1007], supply chain [1163], energy systems [53, 266, 1536], facility location [876], minimizing the environmental impact of utility plants [666], service sectors [665], and cutting stock problems [911].

The remainder of Part VI is devoted to applications in air traffic management (Chapter 22), energy systems (Chapter 23), and process systems (Chapter 24).

Chapter 22

Mixed-Integer Nonlinear Optimization in Air Traffic Management: Aircraft Conflict Avoidance

Sonia Cafieri[15]

22.1 ▪ Introduction

Air traffic management (ATM) represents a domain of emerging and challenging applications of mixed-integer nonlinear programming (MINLP). A number of problems arising in ATM lead naturally to optimization problems whose efficient and reliable solution constitutes a key ingredient of air traffic safety [576]. The air traffic level currently attained in Europe is on the order of tens of thousands of flights per day, and it is expected to grow further on the world scale during the next 20 years. Increasing levels of traffic raise the problem of increasing the capacity of air sectors by better managing the air traffic. This requires increasing the level of automation in ATM, as pointed out in the context of the major projects Single European Sky ATM Research (SESAR) [1667] in Europe and Next Generation Air Transportation System (NextGen) [906] in the United States, which are aimed at designing future ATM systems. Aircraft conflict detection and resolution in en route flights, and the related problem of conflict-free aircraft trajectory planning, are prominent examples of problems that urgently need to be addressed to ensure a higher level of automation in ATM, and consequently more efficiency and safety in air traffic. These problems still deserve investigation of both the identification of suitable mathematical models and the development of efficient and reliable solution methods and algorithms. Mixed-integer nonlinear optimization formulations appear particularly suitable, as they allow us to simultaneously consider continuous as well as discrete decision-making variables and model the complex nonlinear processes characterizing ATM systems.

[15]Financial support provided by the French National Research Agency (ANR) through grant ANR 12-JS02-009-01 "ATOMIC" is gratefully acknowledged.

In the following, we focus on aircraft conflict detection and resolution for en route flights. The aim of this chapter is to present and discuss the use of mixed-integer optimization for this real-life application, with an emphasis on MINLP modeling.

Aircraft conflict avoidance is described in Section 22.2, and the main approaches based on MINLP are recalled. In Section 22.3, the main ingredients and issues of mathematical MINLP modeling are discussed. Section 22.4 presents a brief overview of solution approaches. Section 22.5 draws a few conclusions.

22.2 • Aircraft Conflict Avoidance

Aircraft sharing the same airspace are said to be potentially in *conflict* when they are too close to each other according to their predicted trajectories, i.e., their relative horizontal and vertical distances do not satisfy two given safety-distance constraints. More precisely, the standard separation norms in the en route airspace are 5 NM horizontally and 1000 ft vertically (1 NM (nautical mile) = 1852 m; 1 ft (foot) = 0.3048 m). One can then imagine an aircraft as the center of a cylinder of 5-NM radius and 1000-ft height: it is conflict free if no other aircraft enters this protection volume. When a loss of separation occurs, aircraft have to be separated by performing suitable maneuvers. Aircraft conflict resolution, also referred to as *aircraft deconfliction*, is the problem of providing, starting from an initial configuration given by aircraft trajectories (positions, heading angles) and velocities, a new conflict-free configuration. In general, a selected portion of the airspace is observed over a given time horizon, and then the process is restarted on the next time window. We consider the case of en route cruise flights at a tactical level, i.e., potential conflicts are resolved a few minutes before the loss of separation potentially occurs.

The main challenge is to propose mathematical formulations that can model the complex choices characterizing the target problem without assuming any unrealistic simplifying hypotheses, and that can be solved by efficient algorithms. The underlying problem is an optimization problem, as conflict avoidance may be performed that deviates as little as possible from the original aircraft flight plan, i.e., that minimizes the impact of the separation maneuvers on the flight efficiency.

Modeling aircraft conflict avoidance strictly depends on the separation maneuver chosen to solve conflicts. The most common way to achieve separation, representing the separation maneuver usually exploited by air traffic controllers, is based on changing either the trajectory (heading) or the flight level of the aircraft involved in a conflict. Alternatively, conflict avoidance can be performed through aircraft velocity regulation, keeping the predicted trajectories. The European aeronautical project ERASMUS [311] in recent years promoted the idea of *subliminal control*, according to which velocity regulation should be performed so that aircraft speeds are modified only in a small range (namely, from −6% to +3%) around the original speed. Subliminal speed control is considered promising in view of a future, more automated ATM system, thanks to its limited impact on the workload of air traffic controllers.

Various models and solution strategies have been proposed for the underlying optimization problem. A survey up to the year 2000 can be found in [1110]. Since then, numerous other approaches have been introduced, and mixed-integer linear and nonlinear optimization appear to be a powerful framework for mathematically modeling aircraft deconfliction. In 2002, Pallottino et al. [1438] and Richards and How [1566] first proposed geometrical constructions to model aircraft separation using either velocity or trajectory changes. They obtained mixed-integer linear programming (MILP) models that, even though characterized by quite stringent hypothesis, have the advantage of being solvable by any state-of-the-art MILP solver. A mixed-integer linear model for conflict resolution by velocity and altitude changes is proposed in [58]. The main drawback of this approach is that altitude changes are not preferred because they are uncomfortable for passengers, as well as being fuel consumptive. In [59], a non-linear model is developed starting from the geometrical construction in [1438]. The authors propose a sequential linear solution approach for its solution. The same authors develop a variable neighborhood search heuristic in [60]. In [1836] a model is proposed based on speed control and flight-level assignments for conflict resolution over predefined routes, while the authors in [467] propose an MINLP problem coming from a combination of velocity control and heading angle control methods. In [388], an MINLP model based on velocity changes is proposed, as well as deterministic solution approaches. The speed regulation strategy is also the basis of mixed-integer linear and nonlinear models proposed in [1562, 1563] for the related problem of minimizing potential conflicts. A number of contributions also come from the related problem of conflict-free trajectory planning (i.e., finding trajectories such that aircraft are separated all along their path, a priori in long-term strategical planning or in tactical phases of flights). In [425, 1578], mixed-integer models are proposed for trajectory-planning problems that are solved through heuristic approaches. Interesting MINLP problems also come from mixed-integer optimal control models for conflict-free trajectory planning and conflict resolution, where typical discretization steps lead to the solution of MINLP problems; see, e.g., [309].

22.3 ▪ MINLP Modeling

As for many real-life applications, aircraft conflict avoidance is quite challenging in that a successful model should exhibit a good trade-off between being adherent to realistic constraints and amenable to computational treatment.

The focus of this section is on MINLP formulations. We provide an overview of the main formulation elements, pointing out different possible choices and discussing some modeling issues. We mainly refer to the MINLP model in [388], and variants therein.

We assume that aircraft are represented by points moving at constant speed along linear trajectories at the same altitude and can instantaneously change their course and speed. Thus, the focus is on two-dimensional space, and only horizontal separation has to be ensured.

The choice of separation maneuvers to solve conflicts directly affects the choice of decision variables. Thus, the main decision variables in a conflict avoidance model may include aircraft velocities, heading angles, or flight levels, according to the chosen separation maneuvers. These variables are in general continuous, typically bounded because of operational constraints: aircraft velocity cannot be reduced to zero and is limited by the aircraft engine, and trajectories presenting sharp turns may not be feasible for the aircraft. Referring to modeling based on velocity regulation [388], we consider, as main decision variables, speed variations q_i for each aircraft $i \in A$, bounded in the interval $[-6\%, +3\%]$ of the original aircraft speed v_i, according to the ERASMUS directives. The actual aircraft speed is then $q_i v_i$. A few authors consider, alternatively, instants when aircraft cross intersection points as the main continuous variables [1563]. Integer, and in particular binary, variables usually come from the need to express combinatorial decisions characterizing the problem: typically, the choice among possible scenarios or that of an order for aircraft to perform their maneuver or to arrive at a given point of the airspace. In that sense, the problem can naturally be modeled as a mixed-integer optimization problem.

Different cost functions can be identified as the objective of the optimization in the considered framework. In general, the aim is to achieve deconfliction so that aircraft deviate as little as possible from their original flight plan. In modeling based on speed regulation, this corresponds to minimizing aircraft speed changes; when trajectory changes (heading) are performed, the aircraft's flight on a deviated trajectory has to be minimized. This is in turn related to the minimization of time delays due to deconflicting maneuvers. On the other hand, minimization of fuel consumption, associated with the changes imposed on aircraft to ensure their separation, is important for airline companies and is valued in the context of a sustainable environment. In our example using speed regulation, we minimize the speed variations over the set of all aircraft:

$$\min \sum_{i \in A} (q_i - 1)^2. \tag{22.1}$$

This shows that one can reasonably model the objective using functions with "desirable" properties, like convexity, and in general that are amenable to computational treatment. The main difficulties in fact come from modeling the aircraft separation condition, as discussed below.

The constraints that definitely characterize the aircraft conflict avoidance problem are separation constraints. They have to be expressed for pairs of aircraft, so their number rapidly increases with the number, n, of aircraft. These constraints, in the general form $g(t) \geq 0 \; \forall t$, are in principle nonconvex constraints expressed by relations involving state variables, such as positions and velocities, and depending on time t, although they are often reformulated. The aircraft separation between two aircraft i and j at time t is expressed by the condition

$$\|\mathbf{x}_{ij}^r(t)\| \geq d \qquad \forall t, \tag{22.2}$$

where d is the minimum required separation distance (5 NM for en route flights), $\mathbf{x}_{ij}^r(t) = \mathbf{x}_i(t) - \mathbf{x}_j(t)$ is a vector representing the relative distance between aircraft i and j, and we use the Euclidean norm. Notice that a few already mentioned approaches (see [1438]) express separation in a different way, looking at the geometry of the problem when the trajectories are intersecting straight lines. Such approaches require a few hypotheses and lead to "OR" constraints to take into account different possible configurations. In the following, we use (22.2) to express separation. A reformulation can be provided [388] to find an expression not explicitly dependent on time, and numerically treatable. We can assume that uniform motion laws apply, so that the relative distance of aircraft i and j is expressed as the sum of their relative initial position and the product of their relative speed $\mathbf{v}_{ij}^r$ by the time:

$$\mathbf{x}_{ij}^r(t) = \mathbf{x}_{ij}^{r0} + \mathbf{v}_{ij}^r t \qquad \forall t,$$

which, substituting into (22.2) and squaring, gives

$$\|\mathbf{v}_{ij}^r\|^2 \, t^2 + 2(\mathbf{x}_{ij}^{r0} \cdot \mathbf{v}_{ij}^r) \, t + (\|\mathbf{x}_{ij}^{r0}\|^2 - d^2) \geq 0 \qquad \forall t. \tag{22.3}$$

The study of (22.3) and its associated equation can provide interesting insights into modeling aircraft separation. Let us consider the equation associated with (22.3). It is a quadratic equation in t. Its graph is a parabola that, as $||\mathbf{v}_{ij}^r||^2 > 0$, has a minimum point and opens upward. The discriminant Δ is defined as

$$\Delta = (\mathbf{x}_{ij}^{r0} \cdot \mathbf{v}_{ij}^r)^2 - ||\mathbf{v}_{ij}^r||^2 (||\mathbf{x}_{ij}^{r0}||^2 - d^2). \tag{22.4}$$

If $\Delta < 0$, there are no solutions of the quadratic equation and aircraft are not in conflict, while if $\Delta \geq 0$, there are two solutions, eventually coincident (entry and exit points from the protection zone). If the two roots t' and t'' are both negative, then the conflict is over (it happened in the past). So, there is no conflict when the discriminant is negative or when the discriminant is positive and the two roots are negative (see [386,797]). Binary variables and disjunctive constraints can then be introduced to model separation, taking into account these possible scenarios. Also, one can look at the sign of the scalar product $\mathbf{x}_{ij}^{r0} \cdot \mathbf{v}_{ij}^r$ to infer which kind of angle the vectors form. When the scalar product $\mathbf{x}_{ij}^{r0} \cdot \mathbf{v}_{ij}^r$ is negative, then aircraft flying on straight-line trajectories are converging, potentially generating a conflict, while they are diverging when the product is positive. This last condition can be used in conjunction with other, such as the condition on Δ [386].

Again considering (22.3), we can observe that, by differentiation, the minimum occurs at

$$t_{ij}^m = -\frac{\mathbf{v}_{ij}^r \cdot \mathbf{x}_{ij}^{r0}}{||\mathbf{v}_{ij}^r||^2}. \tag{22.5}$$

So, t_{ij}^m represents the time of closest separation for aircraft i and j: it is the worst case, and aircraft are farther apart for times greater than t_{ij}^m. Thus, by substituting into (22.3) we obtain

$$||\mathbf{x}_{ij}^{r0}||^2 - \frac{(\mathbf{v}_{ij}^r \cdot \mathbf{x}_{ij}^{r0})^2}{||\mathbf{v}_{ij}^r||^2} - d^2 \geq 0, \tag{22.6}$$

which represents a separation condition in the considered time window $(0, T)$. In [388], separation is imposed through condition (22.6) when $t_{ij}^m \in (0, T)$ and $||\mathbf{x}_{ij}^{r0}|| \geq d$, $||\mathbf{x}_{ij}^{r0} + \mathbf{v}_{ij}^r T|| \geq d$ (separation at the initial time and at the time horizon). Through the above transformation, we have obtained nonlinear expression (22.6), but no longer depending on time. Notice that constraint (22.6) has to be imposed, for each i and j, when $t_{ij}^m > 0$. This condition requires us to introduce a binary variable, and corresponding constraints (y_{ij} and constraints (22.11) in the formulation below), to check this condition and impose separation when this variable takes the value one. Finally, we obtain the following MINLP formulation:

$$\min \quad \sum_{i \in A} (q_i - 1)^2 \tag{22.7}$$

$$\text{s.t.} \quad ||\mathbf{v}_{ij}^r|| = ||\mathbf{v}_i q_i - \mathbf{v}_j q_j|| \qquad \forall i, j \in A, i < j, \tag{22.8}$$

$$y_{ij} \left(||\mathbf{x}_{ij}^{r0}||^2 - \frac{(\mathbf{v}_{ij}^r \cdot \mathbf{x}_{ij}^{r0})^2}{||\mathbf{v}_{ij}^r||^2} - d^2 \right) \geq 0 \quad \forall i, j \in A, i < j, \tag{22.9}$$

$$t_{ij}^m = -\frac{(\mathbf{v}_{ij}^r \cdot \mathbf{x}_{ij}^{r0})}{||\mathbf{v}_{ij}^r||^2} \qquad \forall i, j \in A, i < j, \tag{22.10}$$

$$t_{ij}^m (2y_{ij} - 1) \geq 0 \qquad \forall i, j \in A, i < j, \tag{22.11}$$

$$||\mathbf{x}_{ij}^{r0}||^2 \geq d^2 \qquad \forall i, j \in A, i < j, \tag{22.12}$$

$$\|\mathbf{x}_{ij}^{r0} + \mathbf{v}_{ij}^{r} \, T\|^2 \geq d^2 \qquad\qquad \forall i,j \in A, i < j, \qquad (22.13)$$

$$y_{ij} \in \{0,1\} \qquad\qquad \forall i,j \in A, i < j, \qquad (22.14)$$

$$0.94 \leq q_i \leq 1.03 \qquad\qquad \forall i \in A \qquad (22.15)$$

where bounds on q_i take into account the limitation on speed variation for a subliminal control. We remark that most of the constraints, (22.8)–(22.14), are on pairs of aircraft, so there are $n(n-1)/2$ of each of these constraints. The main nonlinearities come from squares and products of continuous variables and of binary and continuous variables, for which linearizations like Fortet's linearization can be computed [746].

It is worth noting that this kind of model, like most of the models in the literature, considers only one maneuver per aircraft, and maneuvers performed simultaneously by all aircraft in the airspace to be deconflicted. A more flexible model can be developed, on the one hand allowing different kinds of maneuvers, combining for example velocity and heading changes (see, e.g., [467]), and on the other hand modeling different scenarios where aircraft perform their maneuvers at different times instead of simultaneously. Evidently, new (mainly integer) variables and constraints have to be added to the formulation to accommodate these new model features.

To model the problem so that separation maneuvers are not performed simultaneously, we introduce for each aircraft i a time t_i^1 and a time t_i^2 when the aircraft can modify its speed and respectively go back to its original speed. These times represent new (continuous) variables of the problem. As no conditions are imposed on the order of execution of separation maneuvers, and consequently on the order of times t^1 and t^2 for each pair of aircraft i and j, the idea is to handle different possible time configurations for pairs of aircraft. These configurations, for a pair $i,j \in A$, correspond to all possible permutations of t^1 and t^2 for the two aircraft of the pair (six configurations per pair). For example, suppose that aircraft i is the first to start its maneuver (changing its speed from the initial v_i to $v_i q_i$), then j starts flying with modified speed, and then i ends its maneuver while j is still flying with modified speed. The sequence of times is then $t_i^1, t_j^1, t_i^2, t_j^2$, with $0 \leq t_i^1 \leq t_j^1 \leq t_i^2 \leq t_j^2 \leq T$. We then introduce, for each pair $i,j \in A$, binary variables z_{ij}^{ℓ} to identify the configuration ℓ holding ($\ell \in \{1,\dots,6\}$), i.e., the sequence of times. For example, z_{ij}^1 identifies the first configuration:

$$z_{ij}^1 = \begin{cases} 1 & \text{if } t_i^1 \leq t_j^1 \quad \text{and} \quad t_j^1 \leq t_i^2 \quad \text{and} \quad t_i^2 \leq t_j^2, \\ 0 & \text{otherwise.} \end{cases}$$

Then, the idea is to deal with the different time windows, in each configuration, where each aircraft flies either with its original (known) speed or with a changed speed. For example, in the first configuration, the time windows are from zero to t_i^1, from t_i^1 to t_j^1, from t_j^1 to t_i^2, from t_i^2 to t_j^2, and from t_j^2 to T, and aircraft i flies with a modified speed in the second and third time windows. As we consider instantaneous speed changes, aircraft speeds are piecewise constant in time windows.

Dealing with time windows means imposing separation conditions like (22.6) in each time window, for each pair of aircraft, in each possible configuration. This makes the number of variables and constraints increase significantly with respect to the above model. First, constraints have to be added to the model to identify, for each pair $i,j \in A$, which is the current time configuration (i.e., which is the order of times t^1 and t^2 for i and j). Then, for each time window, further constraints express the size of the time window, the initial position and speed of each aircraft, and finally the separation condition. The reader is referred to [388] for the complete detailed model. Here we focus on constraints expressing time configurations, as they can be formulated in different ways, thus showing the interest in reformulations in mathematical optimization [1193] in the considered context. Following the definition, the first time configuration for a pair $i, j \in A$ is expressed by

$$z_{ij}^1(t_i^1 - t_j^1) \le 0, \qquad z_{ij}^1(t_j^1 - t_i^2) \le 0, \qquad z_{ij}^1(t_i^2 - t_j^2) \le 0,$$

and similarly for the other five configurations, for each $i, j \in A$. To eliminate the nonlinearities given by the products between continuous and binary variables (t and z, respectively), these are easily reformulated using big-M constraints:

$$
\begin{aligned}
t_i^1 &\le t_j^1 + M(1 - z_{ij}^1), & t_j^1 &\le t_i^2 + M(1 - z_{ij}^1), & t_i^2 &\le t_j^2 + M(1 - z_{ij}^1), \\
t_j^1 &\le t_i^1 + M(1 - z_{ij}^2), & t_i^1 &\le t_j^2 + M(1 - z_{ij}^2), & t_j^2 &\le t_i^2 + M(1 - z_{ij}^2), \\
t_i^1 &\le t_j^2 + M(1 - z_{ij}^3), & t_j^2 &\le t_i^1 + M(1 - z_{ij}^3), & t_i^1 &\le t_j^2 + M(1 - z_{ij}^3), \\
t_j^1 &\le t_i^2 + M(1 - z_{ij}^4), & t_i^2 &\le t_j^1 + M(1 - z_{ij}^4), & t_i^1 &\le t_i^2 + M(1 - z_{ij}^4), \\
t_i^1 &\le t_j^1 + M(1 - z_{ij}^5), & t_i^1 &\le t_j^2 + M(1 - z_{ij}^5), & t_j^2 &\le t_i^2 + M(1 - z_{ij}^5), \\
t_j^1 &\le t_i^1 + M(1 - z_{ij}^6), & t_i^1 &\le t_j^2 + M(1 - z_{ij}^6), & t_j^2 &\le t_i^2 + M(1 - z_{ij}^6).
\end{aligned}
$$

Taking into account all pairs $i, j \in A$, this gives $18n(n-1)/2$ constraints. Interestingly, the value of constant M, usually difficult to choose, in this case can be chosen using the time horizon T, which represents an upper bound on the length of all time intervals. The above constraints can be further reformulated, again using a big-M approach, where the constant M can be chosen using the time horizon T, but using a different formulation with new variables p_{ij} (one variable for each pair i, j and for each time window) [387]:

$$
\begin{aligned}
-M + M(z_{ij}^1 + z_{ij}^3 + z_{ij}^5) &\le p_{ij}^1 - t_i^1 \le M - M(z_{ij}^1 + z_{ij}^3 + z_{ij}^5), \\
-M + M(z_{ij}^2 + z_{ij}^4 + z_{ij}^6) &\le p_{ij}^1 - t_j^1 \le M - M(z_{ij}^2 + z_{ij}^4 + z_{ij}^6), \\
-M + M(z_{ij}^2 + z_{ij}^6) &\le p_{ij}^2 - t_i^1 \le M - M(z_{ij}^2 + z_{ij}^6), \\
-M + M(z_{ij}^1 + z_{ij}^5) &\le p_{ij}^2 - t_j^1 \le M - M(z_{ij}^1 + z_{ij}^5), \\
-M + M z_{ij}^3 &\le p_{ij}^2 - t_i^2 \le M - M z_{ij}^3, \\
-M + M z_{ij}^4 &\le p_{ij}^2 - t_j^2 \le M - M z_{ij}^4, \\
-M + M(z_{ij}^1 + z_{ij}^2) &\le p_{ij}^3 - t_i^2 \le M - M(z_{ij}^1 + z_{ij}^2), \\
-M + M z_{ij}^3 &\le p_{ij}^3 - t_j^1 \le M - M z_{ij}^3, \\
-M + M z_{ij}^4 &\le p_{ij}^3 - t_i^1 \le M - M z_{ij}^4, \\
-M + M(z_{ij}^5 + z_{ij}^6) &\le p_{ij}^3 - t_j^2 \le M - M(z_{ij}^5 + z_{ij}^6), \\
-M + M(z_{ij}^1 + z_{ij}^2 + z_{ij}^3) &\le p_{ij}^4 - t_j^2 \le M - M(z_{ij}^1 + z_{ij}^2 + z_{ij}^3), \\
-M + M(z_{ij}^4 + z_{ij}^5 + z_{ij}^6) &\le p_{ij}^4 - t_i^2 \le M - M(z_{ij}^4 + z_{ij}^5 + z_{ij}^6),
\end{aligned}
$$

with

$$0 = p_{ij}^0 \le p_{ij}^1 \le p_{ij}^2 \le p_{ij}^3 \le p_{ij}^4 \le p_{ij}^5 = T.$$

Taking into account all pairs $i, j \in A$, this gives $12n(n-1)/2$ constraints, which also give the sizes of time intervals (directly obtained by p_{ij} values), which in the prior formulation requires adding specific constraints.

Summarizing, MINLP formulations for the aircraft conflict avoidance problem are generally characterized by

- a number of variables and constraints that are growing rapidly with the number n of aircraft and generally leading to large-scale problems;
- integer, and in particular binary, variables, due to the combinatorial nature of the problem;
- continuous variables often bounded on the basis of operational constraints, which restricts their degree of freedom; and
- nonlinearities appearing in the objective(s) and constraints.

In some cases, we can reduce the size of the problem via suitable pre-processing, identifying pairs of aircraft whose trajectories remain separated regardless of other separation maneuvers in the airspace. However, the number of integer variables depends on the generality and flexibility of the chosen model. The nonlinearities mainly arise from the modeling of separation conditions and the logical choices (using binary variables). Thus, they appear in the form of products of continuous as well as continuous and binary variables, of trigonometric functions when angles have to be decided, and of "OR" constraints. This clearly affects the complexity of the solution process, and reveals the importance of reformulations and of the choice of an appropriate solver.

22.4 • Solution Approaches

While a thorough discussion of solution approaches for aircraft deconfliction via MINLP is beyond the scope of this chapter, we recall in this section the main approaches and issues related to the numerical solution of the considered application.

Deterministic approaches to compute a global solution, mainly based on branch-and-bound methods, are applied, e.g., in [58, 309, 388, 467, 1438]. On simpler models with linear formulations, solutions are efficiently obtained using state-of-the-art solvers [58, 1438]. The global exact solution is evidently more difficult to compute for complex mixed-integer nonlinear models, mainly due to the nonconvexity of the region described by aircraft separation constraints. Note that, as these constraints are indexed on all pairs of aircraft, their number rapidly grows with the number n of aircraft. Furthermore, note that for the considered application a feasible solution is not guaranteed to exist. This is especially related to the tight bounds that are often imposed on decision variables (speeds, angles) because of aerodynamical and operational limitations, thus restricting the freedom to find a feasible solution. Results are obtained for small to medium-scale instances in moderate computing time [309, 388], using global optimization engines like Couenne. These results demonstrate that the aforementioned MINLP formulations behave reasonably well for the considered application and are amenable to solution by general purpose solvers for MINLP. The proposed reformulation of the separation condition to avoid the dependence on time, on the one hand, enables us to avoid a time discretization, and on the other hand is flexible enough to potentially be used in different models (e.g., also based on aircraft angle modifications). In other cases, like in [59], the proposed complex model cannot be solved by general purpose MINLP solvers, and a sequence of linear approximations based on Taylor polynomials is used instead.

Alternatively, some authors resort to heuristics [60, 425, 1578] and solve large instances, though obtaining feasible (if any) but not guaranteed global optimal solutions.

More specifically, a variable neighborhood search (VNS) metaheuristic is proposed in [60], while a simulated annealing and, respectively, a genetic algorithm tailored to the problem, are used in [425, 1578] for the related problem of trajectory planning. Finally, in [388], besides the exact solution of the whole MINLP, a heuristic is proposed based on exact solutions of subproblems that are represented by clusters of conflicting aircraft. These kinds of hybrid solution strategies appear to be promising and deserve further development.

22.5 ▪ Conclusion

We have presented an application of MINLP arising in ATM, namely aircraft conflict avoidance for en route flights. We primarily focused on modeling aspects, specific to the considered application, emphasizing the role of MINLP. It appears that the considered application is challenging and leaves room for further interesting developments using mixed-integer optimization.

Current research is specifically addressed to efficiently solving large-scale instances, and to incorporating in the problem formulation objectives that are relevant for a sustainable environment [1835], as well as the uncertainty affecting the aircraft motion (see, e.g., [962, 1563]).

Chapter 23

Short-Term Planning of Cogeneration Energy Systems via Mixed-Integer Nonlinear Optimization

Leonardo Taccari, Edoardo Amaldi, Aldo Bischi, and Emanuele Martelli

23.1 ▪ Introduction

Combined heat and power (CHP) plants, also called *cogeneration power plants*, are energy systems composed of a network of units that convert primary energy (fossil fuels) into electricity and useful heat to meet the demand for electric power and heat of a set of users at certain temperature levels. As in a cascade process, primary energy is converted into electric power through a thermodynamic cycle, and the heat discharged by the cycle is used to satisfy the users' heat demand. Thanks to the improved integration of these heat flows, CHP plants achieve remarkable savings in primary energy and in CO_2 emissions with respect to noncogenerative plants at both large [329] and small [1087] scales. Therefore, several European and North American countries have recently adopted incentive policies to strongly favor CHP plants as well as combined cooling, heat, and power (CCHP) plants that also cogenerate refrigeration power.

Due to its practical relevance, the optimization of cogeneration systems has received growing attention during the last decade. The problems addressed range from design optimization and long-term tactical planning of energy plants to short-term operational planning, where the components of the energy systems are considered in greater detail.

In short-term operation planning, given a set of cogeneration units and other possible generation and heat storage units, one has to determine for each time period t of a given time horizon which units must be switched on/off, the value of their operating variables (e.g., input fuel), and the amount of stored energy to minimize an objective function (e.g., the total operating costs), while satisfying the demands of electric, thermal, and refrigeration power. In addition, electrical energy can be sold to/purchased from the power grid, and the price of electrical energy can vary hourly in deregulated markets. Since the different cogeneration units (e.g., multiple CHP gas turbines) can be independently controlled (e.g., switched on/off) and the performance curves of several cogeneration units are nonlinear due to the significant efficiency decrease at partial loads, the short-term operational planning of a cogeneration system is a mixed-integer nonlinear optimization (MINLO) problem.

In this chapter, we describe a basic version of the problem, present a MINLO formulation, and summarize some computational results obtained with state-of-the-art mixed-integer nonlinear programming (MINLP) global solvers, namely BARON [1756], SCIP [11], and Couenne [189], on some realistic instances of small to medium size and complexity. As we shall see, even small and relatively simple instances of the short-term cogeneration systems planning problem can be very challenging.

23.2 ▪ Related Work

In the literature, two main approaches are adopted to tackle short-term operational planning of cogeneration systems: a data-driven one and one based on first principles, which differ in the way the behavior of the (co)generation units is modeled. In both cases, combinatorial constraints can be included to model the commitment of the units and related constraints on ramp rates, start-up/shut-down costs, etc.

In data-driven *black-box* approaches, the behavior of the energy systems is described with approximate models obtained from experimental data. This is a rather common approach that gives considerable flexibility with respect to the level of accuracy in the description of the system. It is possible to consider an explicit approximation of the performance curves of the units in the system; see, e.g., [280] and [1978], which consider linear and nonlinear models approximated via piecewise-linear functions. An alternative is to consider only the space of the output variables (heat and electric power), projecting out the input variables (fuel, consumed electricity), either considering linear costs [95] or a convex hull representation in the power-heat-cost space, as done in [1117] and [466]. These representations typically lead to linear optimization (LO) models, or mixed-integer linear optimization (MILO) ones if discrete unit operation modes and other nonconvexities are accounted for [1257].

In first-principles thermodynamic approaches, the system is decomposed into simpler components with well-known performance curves, and mass/energy balance equations are imposed to determine the plant operating points. This kind of approach can be adopted, for example, for complex CHP steam cycles and combined cycles with multiple operating variables and highly complex behaviors. Examples are [648] and [1338], where the behavior of each component is described starting from specific thermodynamic relations.

The cogeneration system operational planning problem can be considered as a variant of what is known in the power systems community as the unit commitment (UC) problem [1436]. UC consists of determining when to start up and shut down the power plants, and how much each committed unit should generate to meet the demand, while typically minimizing a quadratic cost function. Successful approaches for UC include dynamic programming (DP) for simple cases and Lagrangian methods for more complex, large-scale problems (see, e.g., [752, 1589]). In recent years, growing attention has been devoted to mathematical optimization approaches (see, e.g., [317, 751]) due to substantial advances in mixed-integer optimization theory and practice.

The core structure of the problem we address here, i.e., production planning with the presence of storage, also has several similarities to the multi-item, multimachine lot-sizing problem with bounded inventory and minimum lot size. For an extensive account of lot-sizing problems and in particular of polyhedral approaches, see, e.g., [1497] and the references therein.

23.3 ▪ Cogeneration Energy Systems

Depending on the actual application and setting, real-world cogeneration systems can substantially vary in terms of number and types of units as well as in terms of scale, ranging from small-scale plants (applications with <50 kW fuel input) to large-scale ones (industrial applications with >100 MW fuel input). Here, we consider cogeneration energy systems involving the following types of cogeneration units:

- one-degree-of-freedom cogeneration units that simultaneously generate electric and thermal power, e.g., gas turbines, internal combustion engines, back-pressure steam cycles, fuel cells, and
- two-degrees-of-freedom cogeneration units that simultaneously generate electric and thermal power (depending on two operating variables), for instance, gas turbines with supplementary firing in the heat recovery section, steam cycles with extraction-condensing turbine, combined cycles with supplementary firing in the heat recovery steam generator (HRSG), and back-pressure bottoming cycle.

Cogeneration systems may also include generation units such as

- boilers (i.e., one-degree-of-freedom units generating only heat from fuel),
- compression heat pumps (i.e., one-degree-of-freedom units generating only heat from electricity),
- compression chillers (i.e., one-degree-of-freedom units generating only refrigeration power from electricity), and
- absorption chillers (i.e., one-degree-of-freedom units generating only refrigeration power from heat).

The above-mentioned types of units allow us to account for a wide variety of cogeneration systems involving units with multiple degrees of freedom (two or more operating variables) and different sizes.

Figure 23.1 gives a schematic representation of a cogeneration system comprising multiple cogeneration and generation units as well as networks for the distribution of electric power, refrigeration power, and high- and low-temperature thermal power. For instance, the HT heat network allows us to model a steam network for an industrial heat user, while the LT heat network accounts for a district heating network. Storage tanks can be connected to the heat networks as well as to the refrigeration power network. The electric power generated by the units can be used to fulfill the customers' demands and, at the same time, drive the compression heat pumps and compression chillers and satisfy the electricity needs of the absorption chillers. Electric power can be sold to/purchased from the electric grid. The HT and LT heat networks are interconnected to have the possibility of downgrading the high-temperature heat to the low-temperature heat network. Finally, thermal power in excess can, if needed, be dissipated through a dedicated heat exchanger.

23.4 ▪ The Basic Problem and Its Peculiarities

The basic version of the short-term cogeneration system planning problem we consider is defined as follows.

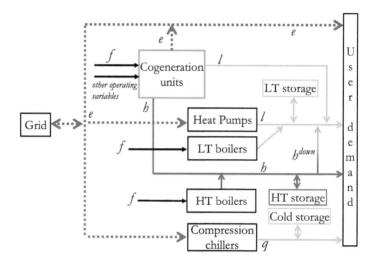

Figure 23.1. *Schematic representation of a CCHP network connecting the (co)generation units with the storage tanks, the electric grid, and the users. Red and yellow arrows represent, respectively, the high-(h) and low-temperature (l) thermal power flows, light blue arrows represent the refrigeration power flows (q), blue dotted arrows represent the electric power (e), and black arrows represent the fuel (f) consumed by each unit.*

We are given the following:

- a cogeneration system as described above, including CCHP cogeneration units with possibly other generation units and heat storage tanks with fixed capacity;
- time-dependent demands for low- and high-temperature thermal and electric power;
- a time-dependent price of electricity; and
- time-dependent ambient temperatures.

From this information, we must determine, for each time period $t \in \mathcal{T}$, the schedule that minimizes the total operating costs while satisfying the given demands.

We adopt a data-driven approach and consider nonlinear performance curves derived from data, either obtained experimentally or provided by the manufacturer, that make a good approximation of the behavior of each unit. The unit performance curves can be time varying, as the ambient temperature affects performance. Units can cogenerate electric power, thermal power, and refrigeration power starting from fuel, electricity, or heat. We also account for the start-up phase of some units, which may incur a significant energy penalty due to their warm-up phase. Additional logical constraints can also be included to limit the number of start-up operations.

Let us now briefly summarize the main peculiarities of the short-term cogeneration system planning problem.

Compared to classical UC problems, a CCHP system includes the generation not only of electric power but also of other commodities, such as thermal power (at different temperature levels) and refrigeration power. Note also that cogeneration units produce both electric power and thermal power simultaneously, and the cost cannot be considered simply as a quadratic function of the production level. Unlike in the usual plant-level approach of UC, single components of the cogeneration systems are considered in greater detail. Their interdependence is also crucial, because commodities can be converted (with some caveats): high-temperature heat can easily be converted to low-temperature heat (not the opposite), electricity can be used to generate thermal power via a heat pump, and so on. A crucial feature of our problem is also the possibility of storing thermal energy from one time period to the next. The storage can be accessed by multiple units, effectively making a decomposition harder.

Compared to the classical multi-item, multimachine lot-sizing problem with bounded inventory, our operational planning problem involves a nonconvex objective function and complex interdependence between the "items," and it does not include an explicit inventory cost. Moreover, stored energy is subject to losses, as stored thermal energy decreases naturally with time.

23.5 ▪ MINLP Formulation

23.5.1 ▪ Sets and Parameters

$\mathcal{T}$: set of time periods (hours)
$\mathcal{U}$: set of all generation units
$\mathcal{F}$: set of units consuming fuel
$\mathcal{E}$: set of units consuming electricity
$\mathcal{C}$: set of units that generate refrigeration
$\mathcal{H}$: set of units that generate high-temperature heat
$\mathcal{L}$: set of units that generate low-temperature heat
$\mathcal{G}$: set of units that generate electricity
c_i^{OM}: hourly operation and maintenance cost for unit $i \in \mathcal{U}$ [€]
c_i^{δ}: start-up cost for unit $i \in \mathcal{U}$ [€]
c_i^{f}: unit cost of fuel consumed by unit $i \in \mathcal{F}$ [€/kWh]
b_t: unit price of electricity bought from the grid at time t [€/kWh]
p_t: unit price of electricity sold to the grid at time t [€/kWh]
$F_{it}^{min}, F_{it}^{max}$: minimum and maximum fuel input for unit $i \in \mathcal{F}$ at time t [kWh]
$E_{it}^{min}, E_{it}^{max}$: minimum and maximum electricity input for unit $i \in \mathcal{E}$ [kWh]
N_i: maximum number of start-ups for unit $i \in \mathcal{U}$
U, V, W: capacity of low-/high-temperature heat and refrigeration storage [kWh]
α, β, γ: heat loss parameters of thermal and refrigeration storage systems
$D_t^{low}, D_t^{high}, D_t^{cold}, D_t^{e}$: demand for low- and high-temperature heat, refrigeration power, electricity at time t [kWh]

23.5.2 ▪ Decision Variables

f_{it}: fuel consumed by unit $i \in \mathcal{F}$ in period t [kWh]
y_{it}: secondary fuel consumed by unit $i \in \mathcal{F}$ with postfiring [kWh]
x_{it}: extraction valve opening percentage for combined cycle units [%]
e_{it}^{cons}: electricity consumed by unit $i \in \mathcal{E}$ in period t [kWh]
e_{it}^{gen}: electricity generated by unit $i \in \mathcal{G}$ in period t [kWh]
l_{it}: low-temperature heat generated by unit $i \in \mathcal{L}$ in period t [kWh]

h_{it}: high-temperature heat generated by unit $i \in \mathcal{H}$ in period t [kWh]

h_t^{down}: high-temperature heat downgraded to low-temperature heat in period t [kWh]

q_{it}: refrigeration energy generated by unit $i \in \mathcal{C}$ in period t [kWh]

e_t^-: electricity sold to the grid in period t [kWh]

e_t^+: electricity bought from the grid in period t [kWh]

u_t: high-temperature thermal energy stored at the beginning of period t [kWh]

v_t: low-temperature thermal energy stored at the beginning of period t [kWh]

w_t: refrigeration energy stored at the beginning of period t [kWh]

z_{it}: binary variable, on/off status of unit i in period t

δ_{it}: binary start-up variable ($\delta_{it} = 1$ if unit i is switched on at beginning of period t)

23.5.3 ▪ MINLP model

Using the above-mentioned sets, parameters, and decision variables, the basic version of the short-term cogeneration systems planning problem can be formulated as the following MINLP problem:

$$\min \quad \sum_{t \in \mathcal{T}} \left(\sum_{i \in \mathcal{U}} c_i^{OM} z_{it} + \sum_{i \in \mathcal{U}} c_i^\delta \delta_{it} + \sum_{i \in \mathcal{F}} c_i^f f_{it} + b_t e_t^+ - p_t e_t^- \right) \tag{23.1}$$

$$\text{s.t.} \quad \sum_{i \in \mathcal{G}} e_{it}^{gen} - \sum_{i \in \mathcal{E}} e_{it}^{cons} + e_t^+ - e_t^- = D_t^e \qquad \forall t \in \mathcal{T}, \tag{23.2}$$

$$\sum_{i \in \mathcal{H}} h_{it} - h_t^{down} + \left(u_t - \frac{u_{t+1}}{1-\alpha} \right) \geq D_t^{high} \quad \forall t \in \mathcal{T}, i \in \mathcal{U}, \tag{23.3}$$

$$\sum_{i \in \mathcal{L}} l_{it} + h_t^{down} + \left(v_t - \frac{v_{t+1}}{1-\beta} \right) \geq D_t^{low} \quad \forall t \in \mathcal{T}, i \in \mathcal{U}, \tag{23.4}$$

$$\sum_{i \in \mathcal{C}} q_{it} + \left(w_t - \frac{w_{t+1}}{1-\gamma} \right) \geq D_t^{cold} \qquad \forall t \in \mathcal{T}, i \in \mathcal{U}, \tag{23.5}$$

$$z_{it} F_{it}^{min} \leq f_{it} \leq z_{it} F_{it}^{max} \qquad \forall t \in \mathcal{T}, i \in \mathcal{F}, \tag{23.6}$$

$$z_{it} E_{it}^{min} \leq e_{it}^{cons} \leq z_{it} E_{it}^{max} \qquad \forall t \in \mathcal{T}, i \in \mathcal{E}, \tag{23.7}$$

Performance constraints described in (23.13) (below),

$$\sum_{t \in \mathcal{T}} \delta_{it} \leq N_i \qquad \forall t \in \mathcal{T}, i \in \mathcal{U}, \tag{23.8}$$

$$\delta_{it} \geq z_{it} - z_{it-1} \qquad \forall t \in \mathcal{T}, i \in \mathcal{U}, \tag{23.9}$$

$$e_{it}^{gen}, h_{it}, l_{it}, q_{it}, h_t^{down}, e_t^+, e_t^- \geq 0 \qquad \forall t \in \mathcal{T}, i \in \mathcal{U}, \tag{23.10}$$

$$0 \leq u_t \leq U, \; 0 \leq v_t \leq V, \; 0 \leq w_t \leq W \qquad \forall t \in \mathcal{T}, \tag{23.11}$$

$$\delta_{it} \in \{0,1\}, z_{it} \in \{0,1\} \qquad \forall t \in \mathcal{T}, i \in \mathcal{U}. \tag{23.12}$$

The aim is to minimize the operational costs minus the revenue obtained by selling extra electricity to the grid. In the objective function (23.1), we consider unit-dependent fuel costs c_i^f. Start-up penalties c_i^δ account for the extra cost due to the warm-up phase. The fixed cost c_i^{OM} accounts for operation and maintenance costs proportional to the number of working hours. It can include the cost of staff needed to operate and maintain the unit, or machine deterioration costs.

Constraints (23.2) are balance equations for electricity. The net amount of electric power, either generated or bought, must satisfy the demand D_t^e for period t. Note that some units generate electric power, while others consume electricity. It is necessary to separate energy that is purchased from the power grid, e_t^+, from energy that is sold, e_t^-, since their price is different. Constraints (23.3) are balance constraints for high-temperature heat. The requirement D_t^{high} for period t must be covered by the generated high-temperature heat and/or by that which is available in the storage (u_t). High-temperature heat can be downgraded to low-temperature heat. Thermal energy can be stored in the tank for the next period, as long as the capacity U is not saturated. Accordingly, the stored energy at the beginning of the following period will be

$$u_{t+1} = \min\left\{U, (1-\alpha)\left(\sum_{i \in \mathcal{H}} h_{it} - h_t^{down} + u_t - D_t^{high}\right)\right\},$$

where $\alpha \in [0, 1)$ is the constant deterioration rate for high-temperature heat. We assume that thermal energy in excess can be dissipated with no additional costs. Similarly, constraints (23.4) are balance constraints for low-temperature heat, where we also include the high-temperature heat that has been downgraded to low-temperature heat. Constraints (23.5) are balance constraints for the refrigeration units. Constraints (23.6) and (23.7) ensure that the operating variables for a unit i (fuel, consumed electricity) are within the technical minimum and maximum. Constraints (23.13), which model the nonlinear behavior of the generation units, linking operating and output variables, are described in detail in the next paragraph. Constraints (23.8) and (23.9) limit the number of start-ups in the time horizon. Finally, constraints (23.10)–(23.12) impose lower and upper bounds and integrality for variables z_{it}.

23.5.4 ▪ Nonlinear Performance Constraints

The performance of each unit $i \in \mathcal{U}$ is described in terms of a nonlinear function $g_{it}(\cdot)$ for each period t that maps one or more operating variables (fuel, consumed electricity, supplementary fuel) to an output variable (low- or high-temperature heat, refrigeration power, electric power). The performance curves, which are usually continuous and nondecreasing, are often nonconvex (sometimes even nondifferentiable) and time-varying due to the nonnegligible temperature effect in each period t. In addition, if unit i is off, its output has to be zero. Thus, the corresponding output variables are semicontinuous. The performance constraints for the generation units can then be expressed as inequalities of the form

$$0 \le \zeta \le z_{it} g_{it}(\vec{\theta}), \tag{23.13}$$

where $\vec{\theta}$ is the vector of input variables and ζ is an output variable.

In the case of generation or cogeneration units with one degree of freedom, each performance curve, g_{it}, is a function of one variable (θ is scalar). For instance, given a high-temperature auxiliary boiler, the output variable is the high-temperature thermal power h_{it}, while the only operating variable is fuel, f_{it}. The feasible region for h_{it} will be $\{0\} \cup [g_{it}(F_{it}^{min}), g_{it}(F_{it}^{max})]$.

In the case of cogeneration units with more degrees of freedom, the performance curves are functions of two or more operating variables ($\vec{\theta}$ is a vector). Two examples are combined cycles with extraction valve regulation (23.14) and gas turbines with post-firing (23.15):

$$l_{it} \leq z_{it} g_{it}^l(f_{it}, x_{it}),$$
$$h_{it} \leq z_{it} g_{it}^h(f_{it}, x_{it}), \qquad\qquad (23.14)$$
$$e_{it}^{gen} \leq z_{it} g_{it}^e(f_{it}, x_{it}),$$

$$l_{it} \leq z_{it} g_{it}^l(f_{it}, y_{it}),$$
$$h_{it} \leq z_{it} g_{it}^h(f_{it}, y_{it}), \qquad\qquad (23.15)$$
$$e_{it}^{gen} \leq z_{it} g_{it}^e(f_{it}, y_{it}),$$

where the operating variables are the fuel quantity f_{it}, the valve opening percentage $x_{it} \in [0, 0.4]$ for the combined cycle (23.14), and the supplementary fuel y_{it} for the gas turbine (23.15). The variable y_{it} has a positive cost that must be added to the objective function and must satisfy an additional technical constraint $y_{it} \leq a + d f_{it}$, with $a, d \geq 0$.

Note that constraints (23.13) are rather general. They can be adapted to several types of generation and cogeneration units, as long as it is possible to model their behavior as nonlinear functions of one or more operating variables.

As to the properties of the performance curves, it is worth pointing out that the units cannot always be classified a priori according to their convexity/concavity because it depends not only on the type of unit but also on the control strategy implemented by the manufacturer. For example, while boilers and heat pumps typically have concave performance curves, the performance curve of gas turbines is often neither concave nor convex, and even nonsmooth, due to a change of control strategy occurring at about 60% of the load.

23.5.5 ▪ Extensions

The model can be extended with additional features and constraints. For example, it is possible to introduce other temperature levels for thermal power. In some cases, it is desirable to model the transition states of the generation units more accurately with ramp-up and ramp-down constraints. Technical constraints regarding temperature limits, mutually exclusive units, and minimum and maximum up-/down-time are also common in similar problems. Moreover, in large-scale systems, the topological aspect of the distribution network, for both heat and power, could be taken into account.

23.6 ▪ Computational Experiments

Given the wide variety of cogeneration systems, ranging from small to large scale, we consider two scenarios: a domestic application (scenario 1) with a few small cogeneration units, and an industrial application (scenario 2) with a considerable number of larger cogeneration units. Six instances generated from the scenarios are available from [1745].

23.6.1 ▪ Scenario 1

The first scenario is a microcogeneration system designed to provide thermal power, refrigeration power, and electricity to a 2,000-m² building. The building has the following

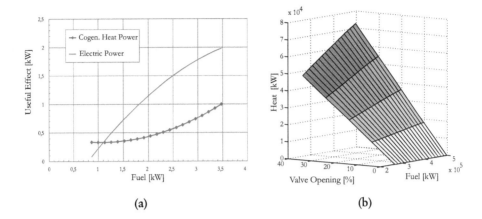

(a)   (b)

Figure 23.2. (a) *Useful effect (electric and thermal power) of a fuel cell unit as a function of consumed fuel. Note that the concavity is different: at larger loads the thermal efficiency increases, while the electrical efficiency decreases. The performance curves are derived from the data of a commercially available machine.* (b) *Heat as a function of fuel and extraction valve opening percentage for a natural gas combined cycle. Heat production is zero when the valve is closed, and it increases with fuel when the valve is opened. Data were obtained via simulation with the dedicated software THERMOFLEX* [1766].

power requirements: high-temperature thermal power (hot water above 60 °C) for domestic hot water, low-temperature thermal power (hot water at 35–45 °C) for heating, refrigeration power for air-conditioning during the summer, and electric power. The cogeneration system is made of the following units:

- a solid oxide fuel cell (SOFC) using natural gas to cogenerate up to 30 kW and 15 kW of, respectively, electric and thermal power;
- a heat pump (HP) using electric power to generate low-temperature heat by "pumping" heat from ambient temperature up to 35–45 °C, which generates about 130 kW at nominal conditions but is very sensitive to ambient temperature;
- an auxiliary boiler (AB) burning natural gas to generate up to 100 kW of high-temperature heat; and
- a thermal storage system to store up to 100 kWh of heat energy.

Figure 23.2(a) shows the performance curves of the SOFC units, i.e., the useful effects, heat, and electric power as a function of the fuel input. Because the thermal and electric requests may have independent time profiles, the heat storage tank is essential to allow the cogeneration system to generate extra electric power (to be sold to the grid) when the selling price is higher, without wasting the cogenerated heat (which will be stored and used when needed). The auxiliary boiler is included in the system mainly as a backup, and it is capable of fulfilling the peak requirements of both high- and low-temperature heat.

23.6.2 ▪ Scenario 2

The second scenario is a large-scale cogeneration system providing heat to a district heating network. The requirement is thermal power at one level of temperature, about

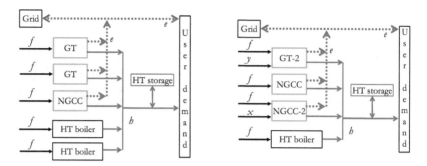

Figure 23.3. *Representation of instances 2-a (left) and 2-b (right).*

90 °C, while all of the electricity produced is sold to the electric grid. The cogeneration system includes one or more of the following units:[16]

- gas turbines (GT) with heat recovery, burning natural gas to generate up to about 10 MW of heat and 5.5 MW of electricity;
- gas turbines (GT-2) with supplementary firing and heat recovery, burning natural gas to generate up to about 40 MW of heat and 11 MW of electricity;
- natural gas combined cycles (NGCC) with a bottoming back-pressure steam turbine, burning natural gas to generate up to 30 MW of heat and 45 MW of electricity;
- natural gas combined cycles (NGCC-2) with a bottoming extraction-type steam turbine, burning natural gas to generate up to about 70 MW of heat and 30 MW of electricity;
- auxiliary boilers (AB) burning natural gas to generate up to about 40 MW of heat; and
- a thermal storage system to store up to 50 MWh of heat energy.

The thermal power requirements are fulfilled by well-established CHP units, such as gas turbines and combined cycles, with the help of auxiliary boilers. This scenario includes cogeneration units with two degrees of freedom, namely, gas turbines (GT-2) with postfiring injection and combined cycles with an extraction condensing steam turbine (NGCC-2). In GT-2, it is possible to burn supplementary fuel to increase the amount of heat that can be recovered from the exhaust gases (23.15). In NGCC-2, the amount of cogenerated heat and electric power is a function of the consumed fuel and the opening of a steam extraction valve, as described in (23.14). Opening the valve reduces the electric power efficiency and increases the amount of recovered heat, while closing the valve drives heat production to zero (see Figure 23.2(b)) but provides larger electric output.

23.6.3 ▪ Selected Results

In Table 23.1, the type and number of (co)generation units contained in each instance are specified. The performance curves are obtained by fitting experimental or simulated data with quadratic functions.

---

[16]We report nominal values at an ambient temperature of 15°C.

Table 23.1. *Type of units included in each instance. Input and output variables for each unit are indicated in the second row. For instance, unit NGCC produces heat power h and electric power e from fuel f.*

	HP $f \to l$	AB $f \to h$	SOFC $f \to h,e$	NGCC $f \to h,e$	NGCC-2 $f,x \to h,e$	GT $f \to h,e$	GT-2 $f,y \to h,e$	number of units
1-a	1	1	1	-	-	-	-	3
1-b	2	2	2	-	-	-	-	6
2-a	-	2	-	1	-	2	-	5
2-b	-	2	-	1	1	-	2	6
2-c	-	4	-	4	-	4	-	12
2-d	-	4	-	2	1	2	2	11

For scenario 1, we consider two instances. For scenario 2, we consider four different unit configurations. In instances 2-a and 2-b (see Figure 23.3 for a schematic representation), heat demand is relatively low. In instances 2-c and 2-d, heat requirements are higher, and more units are necessary to fulfill them. Instances 2-b and 2-d also include cogeneration units with two degrees of freedom. In short-term planning, it is common to face instances with a time horizon ranging from a few hours to several weeks. Here we consider a time horizon of two days, with 48 one-hour periods. When taking into account annual economic incentives, we may even be forced to consider one-year time horizons (see, e.g., [281]).

For full details on this set of instances in AMPL (**.nl**) and GAMS (**.gms**) format, the reader is referred to [1745].

Computational experiments were performed with commercial and open-source global solvers for generic MINLP on an Intel Xeon with E3125@3.30GHz CPUs and 16 GB of RAM. Tests were carried out on a single core, with no memory limit, a time limit of two hours, and a relative optimality gap tolerance of 0.01%. We report results obtained with BARON 14.4 [1756], Couenne 0.4 [189], and SCIP 3.1.0 [11], called from GAMS 24.4.

Starting from formulation (23.1)–(23.12), the MINLP model used in the computational experiments is strengthened by adding simple lower and upper bounds to all the decision variables. For concave g_{it}, constraint (23.13) can be easily convexified, for example, with a big-M reformulation. When g_{it} is convex and univariate, we also add a valid inequality that approximates the convex hull of the (nonconvex) region defined by the performance constraints.[17] Note that there has been recent work on strategies to further tighten similar MINLP formulations, such as those based on perspective reformulations (see, e.g., [751, 877]). From a practical point of view, scaling is also essential, since the original data often contain values that are several orders of magnitude apart (e.g., generated energy with respect to cost coefficients), leading to numerical difficulties or, sometimes, even infeasible or suboptimal solutions.

Table 23.2 summarizes a selection of computational results. As usual, the gap is computed as $100 \frac{|best-LB|}{\min\{|best|,|LB|\}}$. Observe that, in some instances of scenario 2, where all of the generated power is sold to the grid, it is sometimes possible to find a solution with a negative cost, i.e., a net revenue.

---

[17]For a unit i, the function g_{it} is approximated by connecting, in the input-output space (see, e.g., Figure 23.2(a)), the extreme point $(F_{it}^{max}, g_{it}(F_{it}^{max}))$ either with the origin $(0,0)$ or with the point $(F_{it}^{min}, g_{it}(F_{it}^{min}))$.

Table 23.2. *Selected results obtained with three MINLP global solvers. We report the best valid upper and lower bounds found within the time limit of two hours and the relative gap %. Certified optima within the tolerance are in bold. The symbol – indicates that not even a feasible solution was found. Note that ∞ is used when the gap is meaningless because the two bounds have different signs.*

	BARON				SCIP				Couenne			
	time	best	LB	gap	time	best	LB	gap	time	best	LB	gap
1-a	298.3	**21.48**	**21.48**	0.01	50.5	**21.48**	**21.48**	0.00	7200	25.54	15.88	60.83
1-b	7200	25.65	24.68	3.94	7200	–	24.49	∞	7200	32.58	18.04	80.63
2-a	502.3	**24.58**	**24.58**	0.01	58.4	**24.58**	**24.58**	0.00	7200	24.58	20.81	18.13
2-b	7155.2	**−16.27**	**−16.28**	0.01	7200	−16.27	−16.32	0.27	*Infeasible solution*			
2-c	7200	86.78	38.28	126.67	7200	–	39.76	∞	7200	–	23.11	∞
2-d	7200	265.86	42.98	518.53	7200	–	44.49	∞	7200	45.75	13.97	227.52

It is worth noting that the difficulty of even small instances can vary substantially depending on the structure of the cogeneration system. For example, instances with a few generation units that dominate the others in terms of efficiency are significantly easier, which is quite reasonable.

BARON certifies optimality (within the set tolerance) for three of the six instances and finds the best-known solution for four of them. SCIP has a similar behavior. It solves 1-a and 2-a rather easily, both within a minute, and achieves good bounds for 2-b, though it does not close the gap. For instance 1-b, SCIP is not able to find even a single feasible solution. Instances 2-c and 2-d are very challenging for both BARON and SCIP, which are unable to find satisfactory solutions within the time limit.

Couenne is usually not competitive with the other two solvers. For instance 2-b, it even gives an infeasible solution. We assume that this is due to numerical errors or to a bug. Interestingly, for instance 2-d Couenne provides a good solution that neither BARON nor SCIP is able to find.

23.7 ▪ Concluding Remarks

The above selected computational results, for two relatively simple scenarios of a basic version of the short-term operation planning problem with a two-day time horizon, indicate that even small instances can be very challenging to solve to optimality. The additional constraints and (binary) variables needed to account for longer time horizons and capture other typical features, such as ramp-up/down constraints or minimum and maximum unit up-time requirements, often make the mixed-integer nonlinear programming (MINLP) models corresponding to real-world applications even harder. A natural alternative approach consists of approximating the nonlinear performance curves with piecewise-linear functions; see, e.g., [280] and [1978]. Although the resulting approximate mixed-integer linear programming (MILP) models can generally be solved more efficiently than their MINLP counterparts, the advantages and disadvantages of the two approaches still need to be investigated for complex energy systems involving several interacting cogeneration units with nonlinear performance curves.

Chapter 24

Review of Mixed-Integer Nonlinear Optimization and Generalized Disjunctive Programming Applications in Process Systems Engineering

Francisco Trespalacios and Ignacio E. Grossmann[18]

24.1 ▪ Introduction

In this chapter, we present some of the applications of mixed-integer nonlinear programming (MINLP) and generalized disjunctive programming (GDP) in process systems engineering (PSE). For a comprehensive review of mixed-integer nonlinear optimization (MINLO), we refer the reader to the work by Belotti et al. [192]. Bonami et al. [305] review convex MINLP algorithms, and software in more detail. Tawarmalani and Sahinidis [1754] describe global optimization theory, algorithms, and applications. Grossmann [860] reviews nonlinear mixed-integer and disjunctive programming techniques. A systematic method for deriving MINLP models through GDP is provided by Grossmann and Trespalacios [866]. For a detailed review of MINLP solvers, we refer the reader to the work by Bussieck and Vigerske [372] and D'Ambrosio and Lodi [545]. Burer and Letchford [356] present a survey on applications, algorithms, and software specifically focused on nonconvex MINLP.

One of the main concepts in optimization in PSE is the idea of superstructure. A superstructure contains most (or all) of the alternatives a system can have. Superstructures are defined by the modeler, and different superstructures can be derived for the same process. The mathematical optimization approach seeks to find the optimal configuration of the proposed alternatives that are embedded in the superstructure. For a

[18]The authors would like to acknowledge financial support from the Center for Advanced Process Decision-making (CAPD).

more comprehensive description of superstructures, we refer the reader to the work by Grossmann et al. [864]. The online library of problems www.minlp.org contains several PSE problems and includes a detailed description of the model, application, and MINLP problem formulation. In most PSE problems, the discrete variables are binary. The general MINLP model is as follows:

$$
\begin{aligned}
\min \quad & z = f(x,y) \\
\text{s.t.} \quad & h(x,y) = 0, \\
& g(x,y) \le 0, \\
& x \in \mathbb{R}^n, \\
& y \in \{0,1\}^p.
\end{aligned}
\tag{24.1}
$$

Note that the constraint functions $f(\cdot)$, $h(\cdot)$, and $g(\cdot)$ are typically linear in y ($f(x,y) = r(x) + c^T y$, $g(x,y) = h(x) + By$) [860].

Discrete-continuous optimization problems are typically formulated as mixed-integer linear or nonlinear programs (MILP/MINLP). An alternative modeling framework is GDP [1546]. GDP models are formulated using continuous variables, Boolean variables, disjunctions, and logic propositions. The GDP model is as follows:

$$
\begin{aligned}
\min \quad & f(x) \\
\text{s.t.} \quad & g(x) \le 0, \\
& \bigvee_{i \in D_k} \begin{bmatrix} Y_{ki} \\ r_{ki}(x) \le 0 \end{bmatrix}, && k \in K, \\
& \bigvee_{i \in D_k} Y_{ki}, && k \in K, \\
& \Omega(Y) = \textit{True}, \\
& x^{lo} \le x \le x^{up}, \\
& x \in \mathbb{R}^n, \\
& Y_{ki} \in \{\textit{True, False}\}, && k \in K, i \in D_k.
\end{aligned}
\tag{24.2}
$$

In (24.2), the objective is a function of the continuous variables $x \in \mathbb{R}^n$. The global constraints ($g(x) \le 0$) must be satisfied regardless of the discrete decisions. There are disjunctions ($k \in K$) in (24.2), and each one contains disjunctive terms ($i \in D_k$). The disjunctive terms in each disjunction are linked by an "or" operator ($\vee$). A Boolean variable (Y_{ki}) and a set of constraints ($r_{ki}(x) \le 0$) are assigned to each disjunctive term. Exactly one disjunctive term in each disjunction must be enforced ($\bigvee_{i \in D_k} Y_{ki}$). A Boolean variable takes a value of *True* ($Y_{ki} = \textit{True}$) when a disjunctive term is active, and the corresponding constraints ($r_{ki}(x) \le 0$) are enforced. When a term is not active ($Y_{ki} = \textit{False}$), its corresponding constraints are ignored. The logic constraint $\Omega(Y) = \textit{True}$ represents the relations between the Boolean variables in propositional logic.

The GDP representation facilitates the modeling process while keeping the logic structure of the problem [866]. GDP models can be reformulated as MILP (MINLP) models or solved with specialized GDP algorithms [1783]. The GDP-to-MILP reformulations are the big-M (BM) [1390], multiple-parameter big-M (MBM) [1784], and Hull reformulation (HR) [1161]. Alternatively to direct MINLP reformulations, special techniques can help to improve the performance in solving GDP problems. Two GDP solution methods that have proved useful in some applications are the disjunctive branch-and-bound [180, 1161] and the logic-based outer approximation (OA) [1793].

Different types of models have been used for different PSE applications. Table 24.1, adapted from Biegler and Grossmann [262], provides an overview of the different types of models used in the different PSE applications. The different models in Table 24.1

Table 24.1. *Applications of mathematical programming in PSE.*

	LP	MILP/LGDP	QP, LCP	NLP	MINLP/GDP
Process synthesis					
Process Flowsheet		✓		✓	✓
Reactor Networks	✓			✓	✓
Separations		✓			✓
Heat Exchanger Networks	✓	✓		✓	✓
Water Networks	✓	✓		✓	✓
Operations					
Planning	✓	✓			✓
Scheduling	✓	✓			✓
Real-time Optimization	✓		✓	✓	
Process control					
Linear MPC	✓		✓		
Nonlinear MPC				✓	
Hybrid		✓		✓	✓
Molecular computing		✓			✓

are linear programming (LP), mixed-integer linear programming (MILP), linear generalized disjunctive programming (LGDP), quadratic programming (QP), linear complimentary programming (LCP), nonlinear programming (NLP), MINLP, and GDP problems. This chapter presents a review of the applications that require the use of MINLP and nonlinear GDP models.

Kallrath [1026] presents a review of MILP and MINLP in planning and design problems in the process industry. The work presents an overview of the different characteristics and models of the process industry. It also gives an overview of MINLP methods and of optimization under uncertainty. The review work presents in detail five real-world case studies in which optimization was used successfully.

The organization of this chapter follows the framework presented in Table 24.1. Section 24.2 presents a review of MINLP applications in process synthesis: the subsections correspond to process flowsheet synthesis, reactor networks, separation, heat exchanger networks (HENs), and water networks. Section 24.3 presents applications of MINLP in planning and scheduling. Sections 24.4 and 24.5 present MINLP applications in control and in molecular computing, respectively. Finally, Section 24.6 presents some concluding remarks.

24.2 ▪ Process Synthesis

In this section, we summarize the main applications of MINLP and GDP in process synthesis. Grossmann et al. [864] provide a comprehensive review of the development of optimization models and methods in process synthesis.

Real process models can sometimes be highly nonlinear and nonconvex. Furthermore, the models may require the use of ODEs or PDEs. Because of the complexity of real models, it may be impractical to solve synthesis problems with rigorous models in some cases. Therefore, process synthesis models can be classified into three levels: aggregated, shortcut, and rigorous.

Aggregated models focus on a high-level perspective of the synthesis problem and do not consider detailed superstructures. Aggregated models considerably simplify the synthesis problem. Their solution provides a "target" for a metric in the system and does not provide a detailed solution to the problem. For example, the transshipment model for HENs [1443] is an LP problem that provides the minimum cost for utilities. However, it does not provide a solution to the HEN problem (i.e., how many heat exchangers are needed, and which streams belong to each heat exchanger). Aggregated models are typically used as a first step in the sequential optimization of synthesis problems.

The majority of applications that optimize process synthesis make use of shortcut models. These models typically include a detailed superstructure but use simplified equations to describe real processes. These equations can be based on the simplified behavior of systems or on surrogate models [1868] (also known as metamodels or reduced-order models (ROMs)). Some examples of simplified behavior are assuming ideal systems [1933], engineering correlations (such as the Antoine equation for vapor pressure), and input-output relations. Surrogate models can be built by reducing the dimensionality of complex systems [89] (i.e., ROMs) or by building models through simulated or experimental data [521, 564].

The use of rigorous models in a detailed flowsheet optimization is typically computationally expensive. Applications in which rigorous models are used for flowsheet optimization have been reported (for example, in distillation column design [171]). Rigorous models in detailed superstructures yield large-scale nonconvex MINLP problems.

24.2.1 ▪ Process Flowsheet Synthesis

MINLP has been widely used for process flowsheet synthesis. The problem seeks to obtain the optimal configuration of a process contained in a given superstructure [857–859]. The superstructure contains alternative units, interconnections, and process properties. The selection of units and their interconnections is modeled using binary variables, while the properties of the process (flow, concentration, etc.) are modeled with continuous variables. Alternative superstructure representations of processes have also been propose [764, 765, 1441, 1702].

Process flowsheet synthesis is one of the areas where GDP has been most successful. Raman and Grossmann [1545] propose a GDP model for process flowsheet synthesis. Disjunctive programming techniques for the optimization of process systems with discontinuous investment costs and multiple size regions are presented by Türkay and Grossmann [1791]. These authors [1792] also present logic-based MINLP algorithms for the optimal synthesis of process networks. Later, Yeomans and Grossmann [1932] formulate with GDP the two fundamental superstructures for process systems (state task network and state equipment network). GDP methods have been shown to improve solution times for linear, convex, and nonconvex process flowsheet synthesis problems [1601, 1602, 1636, 1834, 1932].

The accurate modeling of a process flowsheet, including the detailed formulations of each unit, normally yields a large-scale nonconvex MINLP. Two general frameworks are normally used to tackle these problems: decomposition techniques [183, 540, 1078, 1104] and surrogate models [931, 1675, 1900, 1914]. There are many processes in which flowsheet optimization has been applied over the last 30 years. The most recent work has focused mainly on bioenergy systems and biorefineries [155, 1029, 1064, 1286, 1287, 1363, 1624, 1757, 1758, 1867], polygeneration systems [452, 1204, 1205], and gas-to-liquid complexes [669].

Figure 24.1 presents a process flowsheet superstructure for the hydrodealkylation (HDA) of toluene process [1078]. The given superstructure contains 192 different flowsheets. The figure shows the possibility of selecting alternative connections and units, such as adiabatic or isothermal reactors, membranes, absorber, distillation columns, flash units, compressors, heat exchangers, and valves.

The MINLP problem presented in Figure 24.1 is highly nonconvex. For illustration, we present the following equations that describe the simplified behavior of the adiabatic reactor presented in Figure 24.1(a):

$$F_{h2}^{IN} \geq 5(F_{ben}^{IN} + F_{tol}^{IN} + F_{dip}^{IN}),$$

$$k = 6.3 * 10^{10} \exp(-26167/100Tr),$$

$$1 - X_{tol} = (1/(1 + 0.372kV(F_{tol}^{IN})^{0.5}(F_{tot}^{IN})^{-1.5}))^2,$$

$$1 - \xi = 0.0036(1 - X_{tol})^{-1.544},$$

$$R_{tol} = X_{tol}F_{tol}^{IN},$$

$$F_{tol}^{OUT} = F_{tol}^{IN} - R_{tol}, \qquad (24.3)$$

$$F_{ben}^{OUT} = F_{ben}^{IN} + \xi R_{tol},$$

$$F_{dip}^{OUT} = F_{dip}^{IN} + 0.5R_{tol} + 0.5(F_{ben}^{IN} - F_{ben}^{OUT}),$$

$$F_{h2}^{OUT} = F_{h2}^{IN} - R_{tol} + (F_{dip}^{OUT} - F_{dip}^{IN}),$$

$$F_{ch4}^{OUT} = F_{ch4}^{IN} + R_{tol},$$

$$501R_{tol} = Cp^{OUT}F_{tot}^{OUT}T_{tot}^{OUT} - Cp^{IN}F_{tot}^{IN}T_{tot}^{IN}.$$

In (24.3), F_j^{IN} represents the inlet flow of $j = h2, ch4, ben, tol, dip, tot$, where $h2$ is hydrogen, $ch4$ is methane, ben is benzene, tol is toluene, dip is diphenyl, and tot is total. F_j^{OUT} represents the outlet flow. T_{tot}^{IN} and T_{tot}^{OUT} are the inlet and outlet temperatures of the streams. k is the reaction kinetics constant, Tr is the reactor temperature, X_{tol} is the conversion of toluene, V is the volume of the reactor, ξ is the selectivity in the reaction to benzene, R_{tol} is the consumption rate of toluene, and Cp^{OUT} is an approximate heat capacity. Notice that (24.3) is a simplified model of a single unit of the process superstructure. It is easy to see that flowsheet optimization with rigorous models can become computationally intractable.

24.2.2 ▪ Reactor Network

This problem seeks to optimize the configuration of a reactor system for a given feed and a given set of reactions. There are two main mathematical programming approaches for this problem: superstructure optimization and targeting. A review of both methods is provided by Hildebrandt and Biegler [941].

Superstructure optimization methods require the modeler to postulate a superstructure that represents possible configurations for the reactor network. This structure is formulated as an MINLP problem and solved with optimization tools. Achenie and Biegler [8] postulate a superstructure NLP model. The continuous model allows the selection of the network structure, reactor type, and amount of heat addition. This model uses a dispersion coefficient to determine the reactor type. Later, the same authors postulate an alternative NLP problem that uses the recycle ratio as the determinant of reactor type [9]. Kokossis and Floudas [1084–1086] present a superstructure of continuous stirred-tank reactors (CSTRs) and plug flow reactors (PFRs).

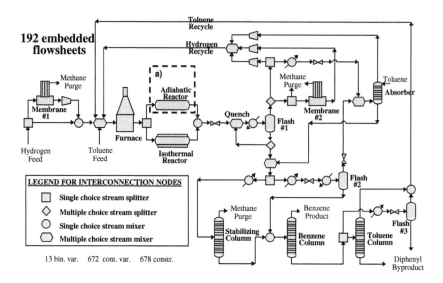

Figure 24.1. *Process flowsheet superstructure for the HDA process.*

In the MINLP problem, the PFRs are modeled as a series of CSTRs, which eliminates the use of differential equations in the model. Smith and Pantelides [1702] present a reaction and separation network with detailed unit operation models. Esposito and Floudas [678] make use of global optimization tools to solve a nonconvex MINLP problem in which the PFRs of the superstructure are modeled using DAEs.

The targeting method seeks to identify the maximum possible performance. A reactor network that meets this criterion is then determined. Horn [955] introduces the concept of the "attainable region." The attainable region is the convex hull of possible concentrations for a given feed and reaction scheme. The geometrical concepts that allow the derivation of the attainable region, and further extensions to higher dimensions, were developed mainly by Glasser, Hildebrandt, Feinberg, and Crowe [697, 809, 942, 943]. Burri et al. [363] apply the Infinite DimEnsionAl State-space (IDEAS) framework to construct the attainable region of reactors. This framework is also used by Zhou and Manousiouthakis [1977] and Davis et al. [562] to find the attainable region of nonideal reactors and of non-steady-state networks of batch reactors. There are two main downsides to the attainable region technique. The first one is that since it is a graphical method, it can handle at most three dimensions. The second one is that every time that the conditions for the reactor problem change, the region must be recalculated. To solve this issue and improve the performance of the method, Biegler et al. [137–139, 1119, 1590] develop hybrid methods that combine the attainable region with optimization techniques.

24.2.3 ▪ Single Distillation Columns

The simplest distillation design problem is to select the feed tray location in a distillation column with a given number of trays. The superstructure of this problem is postulated by Sargent and Gaminibandara [1628] for ideal mixtures, and extended later for azeotropic cases [1627] (see Figure 24.2(a)). Viswanathan and Grossmann [1854] present the superstructure of a distillation column to optimize not only the feed tray but also the number of trays in the column (see Figure 24.2(b)). The model also

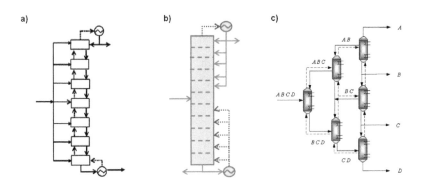

Figure 24.2. *Superstructures for (a) feed tray location, (b) feed tray location and number of plates, and (c) distillation sequence for a four-component zeotropic mixture.*

allows the possibility of multiple feeds. This model was later incorporated into more comprehensive superstructures that incorporated thermodynamic considerations [171, 1702] and thermally coupled distillation columns and dividing wall columns [638].

One of the main issues of the tray-by-tray MINLP model of a column is the large number of nonconvex constraints in the model. Yeomans and Grossmann [1933] propose a GDP model for the tray-by-tray representation of the column, as well as for the sequence superstructure. They propose a logic-based OA approach, which considerably reduces the number of nonconvex constraints in each iteration. Based on the tray-by-tray GDP model, Barttfeld et al. [167] present a computational comparison between the MINLP and GDP models using a heuristic algorithm, showing the advantages in solution times of the GDP model.

For the feed location problem, let $I = \{1, \ldots, N\}$ denote the set of trays. Let $R = \{1\}$ denote the tray that corresponds to the reboiler, $C = \{N\}$ the one that corresponds to the condenser, and $S = \{2, \ldots, N - 1\}$ the set of trays within the column. For a set of feeds $k \in K$ and components $j \in J$, let $F^k, T^k, p^k, v^k, h^k, z_j^k$ denote, respectively, the molar flow rate, temperature, pressure, vapor fraction, molar specific enthalpy, and mole fractions of the feeds. let P_1 be the molar flow rate of the distillate and P_2 the molar flow rate at the bottom of the column. For each tray $i \in I$, let p_i denote its pressure and T_i its temperature. Let $L_i, x_{ij}, h_i^L, f_{ij}^L$ denote, respectively, the molar flow rate, molar fractions, molar specific enthalpy, and fugacity of the liquid leaving tray i. Similarly, let $V_i, y_{ij}, h_i^V, f_{ij}^V$ denote the corresponding variables for the vapor leaving tray i.

The fugacity and molar specific enthalpy are functions of the corresponding variables. These functions depend on the thermodynamic model that is used and therefore vary in complexity. For this problem, consider the following general functions:

$$\begin{aligned}
h_i^L &= h_i^L(T_i, p_i, x_{ij}), \\
h_i^V &= h_i^V(T_i, p_i, y_{ij}), \\
f_{ij}^L &= f_{ij}^L(T_i, p_i, x_{ij}), \\
f_{ij}^V &= f_{ij}^V(T_i, p_i, y_{ij}).
\end{aligned} \qquad (24.4)$$

The MINLP problem to optimize the feed location is as follows.

The objective is to minimize the cost of utilities in the reboiler and condenser (considering costs w_{reb} and w_{cond}):

$$\min \quad w_{reb}q_{reb} + w_{cond}q_{cond} \tag{24.5}$$

s.t.

Phase equilibrium:

$$f_{ij}^L = f_{ij}^V \quad i \in I, j \in J; \tag{24.6}$$

Material balances:

$$
\begin{aligned}
V_{N-1}y_{N-1,j} &= (L_N + P_1)x_{ij}, & j \in J, \\
L_i x_{ij} + V_i y_{ij} &= L_{i+1}x_{i+1,j} + V_{i-1}y_{i-1,j} + \sum_{k \in K} f_i^k z_j^k, & i \in S, j \in J, \\
P_2 x_{1j} + V_i y_{1j} &= L_2 x_{2j}, & j \in J, \\
\sum_{j \in J} x_{ij} &= 1, & i \in I, \\
\sum_{j \in J} y_{ij} &= 1, & i \in I;
\end{aligned}
\tag{24.7}
$$

Energy (heat) balances:

$$
\begin{aligned}
(L_N + P_N)h_N^L - V_{N-1}h_{N-1}^V &= q_{cond}, \\
L_i h_i^L + V_i h_i^V &= L_{i+1}h_{i+1}^L + V_{i-1}h_{i-1}^V + \sum_{k \in K} f_i^k h^k, & i \in S, \\
P_2 h_1^L + V_i h_1^V - L_2 h_2^L &= q_{reb}, & j \in J;
\end{aligned}
\tag{24.8}
$$

Feed location:

$$
\begin{aligned}
f_i^k &\leq F^k z_i^k, & k \in K, i \in S, \\
\sum_{i \in S} f_i^k &= F^k, & k \in K, \\
\sum_{i \in S} z_i^k &= 1, & k \in K, \\
z_i^k &\in \{1, 0\}, & k \in K, i \in S;
\end{aligned}
\tag{24.9}
$$

Pressure profile ($p_{con}, p_{top}, p_{bot}, p_{reb}$ are given):

$$
\begin{aligned}
p_N &= p_{con}, \\
p_{N-1} &= p_{top}, \\
p_2 &= p_{bot}, \\
p_1 &= p_{reb}, \\
p_{i-1} - 2p_i + p_{i+1} &= 0, \quad 3 \leq i \leq N-2.
\end{aligned}
\tag{24.10}
$$

Constraints (24.4), (24.6), (24.7), and (24.8) are often referred to as MESH equations (material balance, equilibrium, summation and heat balance).

In the GDP by Yeomans and Grossmann [1933], the MESH equations are enforced for (a) the fixed plates (i.e., the feed plate, top plate, and bottom plate) and (b) the selected plates. The plates that are not selected behave as a simple bypass. The high-level formulation, without explicitly including all the detailed equations, is as follows:

$$\min \ COST$$

s.t. *Cost equations,*

MESH equations of fixed trays,

$$\begin{bmatrix} Y_i \\ MESH \ equations \ for \ tray \ i \\ stg_i = 1 \end{bmatrix} \lor \begin{bmatrix} \neg Y_i \\ x_{ij} = x_{i+1,j} \ j \in J \\ y_{ij} = y_{i-1,j} \ j \in J \\ L_i = L_{i+1} \\ V_i = V_{i-1} \\ T_i = T_{i-1} \\ stg_i = 1 \end{bmatrix}, \quad i \in S, \quad (24.11)$$

$$ntray = \sum_{i \in S} stg_i,$$

$$\Omega(Y) = True,$$

where Ω represents the logic relations between trays (for example, $Y_i \rightarrow Y_{i-1}$ when i is above the feed tray to reduce symmetry). Problem (24.11) requires the use of more binary variables than the feed location problem. Also, the *cost equations* in (24.11) typically involve nonconvex constraints that correspond to capital cost. The nonconvexities arise in the estimation of the column diameter, the heat exchange area of the reboiler and condenser, and the concave cost correlations.

24.2.4 ▪ Distillation Sequences

The separation problem was originally defined almost 50 years ago by Rudd and Watson [1598]. For the MINLP and GDP applications, Yeomans and Grossmann [1932] characterize two major types of superstructure representation: state-task-network (STN) (see also Sargent and Gaminibandara [1628] and Sargent [1627]) and state equipment network (SEN) (see also Smith [1701]). They describe both representations using GDP models.

For sharp split columns, the superstructure can be modeled as an MINLP problem. The GDP formulation for the STN model is straightforward [378]. In the case where each column is assigned to each of the split separation tasks, the model reduces to the MILP superstructure proposed by Andrecovich and Westerberg [84]. The SEN structure is not as straightforward, but it is numerically more robust [378, 1933]. Novak et al. [1406] used this representation before it was formalized by Yeomans and Grosmann [1932].

To develop more efficient distillation sequences, heat integration between different separation tasks can be considered. Paules and Floudas [743, 1462] propose an MINLP model for heat-integrated distillation sequences, and Raman and Grossmann [1545] propose a disjunctive representation.

Thermally coupled systems seek to reduce the inherited inefficiencies due to the irreversibility during the mixing of streams at the feed, top, and bottom of the column. The design and control of these columns are complicated, but they were made possible by the design concept of "thermodynamically equivalent configuration" [31, 33, 380, 1019] and the improvement in control strategies [1663–1666, 1896]. For zeotropic mixtures, Agrawal [32] characterized a subset of the possible configurations named basic configurations. This concept helps to reduce the search space of feasible configurations, since nonbasic configurations normally have higher overall costs [382, 806, 807]. Algorithms for finding the basic configurations [103], and mathematical representations of these configurations, have been proposed [979, 1670]. In terms of GDP, logic rules that include all the basic column configurations are developed by Caballero and Grossmann [379, 381, 382].

Figure 24.2(c) presents the distillation sequence superstructure for a four-component zeotropic mixture, presented by Sargent and Gaminibandara [1628].

24.2.5 ▪ HENs

A comprehensive HEN review with annotated bibliography is provided by Furman and Sahinidis [774]. A more recent review on the developments of HENs is provided by Morar and Agachi [1347]. Following the classification proposed by Furman and Sahinidis [774], the work in HENs is divided into sequential or simultaneous synthesis methods.

The sequential synthesis method decomposes the design of the HEN problem typically into three simpler sequential problems, but it does not guarantee the global optimum of the HEN problem. The first problem seeks to minimize the utility costs. The second one seeks to minimize the number of heat exchanger units while satisfying the minimum utility cost previously found. The last problem minimizes the network cost subject to the minimum number of units found. The first problem (minimizing utility usage/cost) can be an LP [420, 999, 1443], an MILP [782, 783], or an MINLP [781] problem. The second problem is an MILP problem [420, 874, 875, 1443]. The most common model for the third problem that minimizes the network cost is an NLP problem [738].

In simultaneous synthesis, the HEN problem is optimized without decomposing it. One of the first MINLP models was presented by Yuan et al. [1942], describing a superstructure of the network. This MINLP model, however, does not allow the splitting or mixing of streams. The work by Ciric and Floudas [476, 477, 736, 737] presents an MINLP model that optimizes all of the costs of a HEN problem without the need for decomposition. Yee, Grossmann, and Kravanja [1930, 1931] present a linearly constrained MINLP formulation of a multistage superstructure that allows any pair of hot and cold streams to exchange heat in every stage. The superstructure of this model for two possible stages is shown in Figure 24.3. This model was extended to include flexibility [440, 1095, 1842], detailed exchanger design models [758, 1662], and isothermal streams involving phase change [1500].

The simultaneous optimization of process flowsheet and heat integration is addressed by Duran and Grossmann [644] through a set of nonlinear inequalities. Particular applications of heat integration with optimization of distillation columns and distillation sequences were addressed earlier in this section. Grossmann et al. [867] present a GDP model for simultaneous flowsheet optimization and heat integration. Recently, Navarro-Amorós et al. [1380] extended this GDP model to heat integration with variable temperatures.

24.2.6 ▪ Utility Systems

Petroulas and Reklaitis [1469] propose a mathematical optimization model for the design of utility systems, based on an LP model and a dynamic programming (DP) approach. An MILP model is formulated by Papoulias and Grossmann [1442], which

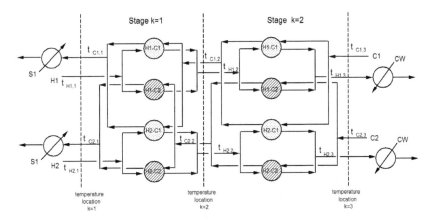

Figure 24.3. *Superstructure of multiple stages with potential heat exchangers.*

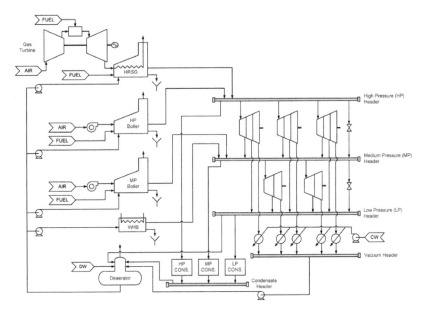

Figure 24.4. *Superstructure for utility plants.*

assumes linear capital costs with fixed charge and linear energy balance. Bruno et al. [351] propose a refined version of this model by including nonlinear functions, presenting the first MINLP model for the design of utility systems. The superstructure of this MINLP is shown in Figure 24.4. This MINLP problem involves nonlinearities to represent the cost of equipment and the plant performance in terms of enthalpies, entropies, and efficiencies. The model considers steam properties at specific pressures, so it cannot simultaneously optimize the operating conditions of the steam levels. Savola et al. [1634] present a modified MINLP model that allows the simultaneous optimization of pressure levels by using correlations that depend on both pressure and temperature.

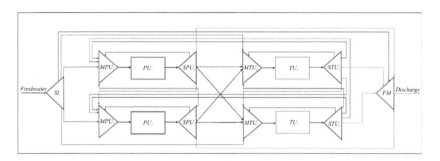

Figure 24.5. *Superstructure of integrated water network with two process and two treat-ment units.*

24.2.7 ▪ Water Networks

A comprehensive review of water network design methods with literature annota-tions is provided by Jeżowski [997] (see also the review by Bagajewicz [125]). The optimization-based methods for water network design are based on superstructures and encompass most (or all) feasible solutions for the network. Four types of wa-ter networks are normally modeled. The network of water-using processes is called a water-using network (WUN). The second type is the wastewater treatment network (WWTN). These two networks are subsystems of the third type of network: the total water network (TWN) [1875, 1876]. Finally, the fourth type is the complete water net-work (CWN), which is a TWN with pretreatment units [126]. The models in water networks typically seek to optimize the configuration of the network and unit oper-ating conditions, satisfying the water quantity and quality requirements in the WUN, with water discharges within environmental restrictions.

The superstructure-based optimization of the WUN is relatively simple, due to the fixed number of water-using units. However, the use of alternative water treatment units in WWTN, TWN, and CWN makes the problem more complex. In particular, the mixing of streams yields bilinear terms (which are nonconvex), and the operating constraints in each of the treatment units may be nonconvex. In addition, the units and interconnections are represented by binary variables. The combination of these constraints and variables results in a nonconvex MINLP problem. The WWTN alone has been shown to be a very difficult problem, equivalent to the generalized pooling problem [1319].

Figure 24.5 presents the superstructure of a water network [38] with one source, one discharge, two water-using processes, and two treatment units. It can be seen from the figure that all possible connections between the water-using units and water treat-ment units are considered, including recycle streams.

The first MINLP model for the simultaneous global optimization of water net-works was presented by Karuppiah and Grossmann [1042], which is derived from a GDP formulation. This model was extended by Ahmetović and Grossmann [38] to include all feasible connections between one or multiple sources of water of different quality, water-using processes, and wastewater treatment operations. Since the MINLP problem that arises in water networks is difficult to solve, five strategies are normally used to solve it. The first one is the linearization of constraints, typically used for the WUN. The first linearization was developed for a single-contaminant problem [130], and its optimality constraints were derived by Savelski and Bagajewicz [1632]. The model was later extended to multiple contaminants [129], though the conditions are different [1633], and it requires a specialized branching approach. Other linearization techniques for WUN with regeneration processes [1097], a near-optimal single-stage method for WUN with multicontaminant [1864], and WWTN [1227] have been presented. The second approach is to find good solutions (locally optimal) through good initial points. A common practice is to fix the outlet concentrations at the maximum values [1178,1759]. The third approach is to use sequential optimization, originally presented by Takama et al. [1746]. Some of the sequential optimization approach methods include a relaxed NLP model [625], dividing and sequentially reducing concentration intervals for WUN [1804], an MILP-LP technique for TWN [873], and an LP-NLP approach for the WWTN [938]. The fourth strategy is to solve the MINLP problem by using metaheuristic algorithms [998, 1137, 1788]. The last approach is to solve the MINLP problem using rigorous global optimization techniques [38, 1042, 1947].

Dong et al. [611, 612], Li et al. [1184, 1185], and Zhou et al. [1976] have recently proposed MINLP models that incorporate the design of HENs with water networks, based on the state-space superstructure proposed by Bagajewicz et al. [127, 128] (see also [39] for recent work on simultaneous synthesis of water and HENs). Ruiz and Grossmann [1600] present a simplified GDP formulation of the WWTN as an illustration of the bound-tightening technique.

24.3 ▪ Planning and Scheduling

Process planning and scheduling have been areas of much interest for industry and research over the past decades [910, 1024, 1484, 1497]. Although the boundary between planning and scheduling is sometimes not clear, the time scale distinguishes between the two. Planning normally involves medium-term (e.g., years, quarters, months, or weeks) decisions, such as assignment of production tasks to facilities and transportation logistics. Scheduling defines short-term decisions, such as assignment of tasks to units, sequencing of tasks in each unit, and determining starting and ending times for the execution of tasks. A review of planning and scheduling in the process industry is provided by Kallrath [1024]. Models and techniques to integrate planning and scheduling have also received attention in recent years [1272]. In this section we present some of the main MINLP applications for planning and for scheduling.

24.3.1 ▪ Planning

Planning models are usually not as detailed as scheduling models, so most planning models are represented using MILP models [1497]. However, refinery planning is one of the few areas where nonlinear models have been used. Moro et al. [1355] present an NLP planning model for diesel production refineries, extending it later to petroleum refineries Pinto and Moro, [1491]. Zhang at al. [1956] presented one of the first MINLP models for refinery planning, which simultaneously optimizes liquid flows and allocation of utilities (hydrogen, and steam and power flows). They deal with the nonlinear terms by using piecewise-linear approximations. Elkamel et al. [667] present an MINLP multiobjective model for refinery planning, seeking to maximize profit while reducing CO_2 emissions. Neiro and Pinto [1385] present an MINLP implementation for multiperiod refinery planning models using empirical relations. Alattas et al. [43] present a multiperiod MINLP model that utilizes a more precise representation of the crude distillation unit, using the model introduced by Geddes [791]. Shah et al. [1668] provide a literature review of approaches that address the problem of enterprise-wide optimization in the petroleum industry.

24.3.2 ▪ Scheduling

Many scheduling models assume process simplifications and can be modeled as MILP problems [740, 910, 1092, 1314]. Castro and Grossmann [417] present the traditional MILP formulations as a derivation of GDP models. Many applications, however, require the use of nonlinear constraints to better represent the process.

There are two non-process-related cases that typically introduce nonlinearities to the traditional MILP representations. The simplest one is when cyclic schedules with infinite horizon are considered [416, 1489, 1614, 1615, 1669]. In this case, the objective function is divided by the length of the cycle time, yielding a linear fractional objective function. The rest of the model is represented using a traditional MILP representation. Specialized algorithms have been developed to address this problem [1496, 1937]. The second case is when a cyclic scheduling model includes average inventory equations [1489]. The additional constraints in this case are nonconvex and may require the use of a global optimization method for its solution [959, 1614, 1615].

In terms of process-related nonlinearities, blending equations are probably the most common type that appear in scheduling models [1315]. These constraints make use of nonconvex bilinear terms, leading to multiple local solutions. To solve this type of scheduling model to global optimality, rigorous global optimization techniques are used [1090, 1328, 1329]. The work by Gupte et al. [884] presents a comprehensive review of different models for the pooling problem. The work analyzes the linear relaxations of the different models when using McCormick envelopes, and it presents discretization strategies to approximate the MINLP problems by MILP problems.

One of the most complex nonconvexities in scheduling arises in the modeling of dynamic processes. The dynamic, discrete, and continuous behavior of these systems gives rise to mixed-integer dynamic optimization (MIDO). This type of problem typically occurs in the scheduling and control of polymerization reactors [732, 1335, 1409, 1519, 1763]. The most common approach for solving MIDO problems is to discretize the system of differential equations through orthogonal collocation, yielding large-scale MINLP models. Biegler [260] provides an overview of solution strategies for dynamic optimization.

24.4 ▪ Process Control

Many processes in chemical engineering require the modeling of nonlinear systems, especially when representing dynamic behavior. Additionally, discontinuity is expected in the operation and control of a process [166]. The dynamic models with mixed discrete and continuous variables are called hybrid systems (which are MIDO problems in general). Start-up and shut-down operations [1541], batch systems [1423], and grade transitions [1519, 1763] are some of the main applications of optimization of hybrid systems. Several hybrid formulations have been developed and implemented in recent years [122, 194, 368, 1230, 1354, 1422, 1737, 1944]. A comprehensive review of modeling, simulation, sensitivity analysis, and optimization of hybrid systems is provided by Barton and Lee [165]. Morari and Barić [1348] present a review paper on developments in the control of hybrid systems.

GDP has also been relevant in the development of models and solution methods in hybrid systems. Oldenburg and Marquardt [1422] develop a GDP model of "fixed alternative sequences," formalizing modeling approaches proposed by several authors [165, 1410]. The authors solve the GDP model using a modified version of the logic-based OA, showing the efficiency of the algorithm for finding global solutions.

Ruiz-Femenia et al. [1603] present a logic-based OA algorithm for solving optimal control problems.

24.5 ▪ Molecular Computing

A collection of works on computer-aided design is provided by Achenie et al. [10]. The review paper on advances in global optimization by Floudas and Gounaris [739] provides several applications of global optimization in molecular computing.

From a mathematical programming perspective, one of the first works on molecular computing was developed by Macchietto et al. [1248] and Odele and Macchietto [1412]. The MINLP model they proposed was based on group contribution, accounting for the presence or absence of a group in a molecule. Churi and Achenie [471–473] further refine this model by incorporating some information on how the groups are connected to each other in the molecule. Naser and Fournier [1378] propose an alternative MINLP representation and develop a heuristic method for solving the problem. All of these MINLP models and methods are solved using local tools, so they do not guarantee the global solution. Joback and Stephanopoulos [1005, 1006] propose the use of interval arithmetic techniques, providing bounds on the properties of aggregate molecules, overcoming the multiple local optima issue.

Several applications of MINLP have been used in molecular design. Some of them are design of solvents [471, 1248, 1428, 1429, 1493, 1698, 1699, 1873, 1874], refrigerants [472, 473, 647, 1617, 1618], polymers [393, 1269–1271, 1813], and pharmaceutical products [1547, 1692].

24.6 ▪ Concluding Remarks

As can be seen from the review of this chapter, MINLP and GDP techniques have had wide applicability in many problems in PSE. This has allowed not only a more systematic approach to the problems, but also one where optimal or near-optimal solutions can be obtained. However, despite many successful applications, the global optimization of MINLP and GDP problems still remains a major challenge due to the limited problem size that can be handled with current methods.

Part VII

Robust Optimization

Chapter 25

Robust Optimization Methods

Timothy C. Y. Chan and Philip Allen Mar

25.1 ▪ Introduction

Uncertainty in the parameters of an optimization problem can have a large impact on the quality of the resulting solutions, even rendering such solutions infeasible if unexpected scenarios arise [212]. Robust optimization is one method for optimizing in the presence of uncertainty. The last 15 years have seen an explosion in activity in robust optimization driven by initial work done to explicitly reformulate several different optimization models under set-based uncertainty as tractable mathematical optimization problems. The success of robust optimization can be attributed to its computational tractability and its ability to produce quality solutions for a large class of problems. From a modeling viewpoint, robust optimization conveniently does not require complete knowledge of the probability distribution of the uncertain parameters.

Although tractable, robust optimization is inherently conservative because it focuses on worst-case scenarios, possibly to the detriment of non-worst-case scenarios. Thus, much research has focused on addressing the inherent conservatism of robust solutions [196, 257, 367, 717, 819, 973, 1595, 1767]. Some of these models, in some special cases, have been shown to provide good solutions in multistage problems [248, 249, 251].

An early demonstration of the applicability of robust optimization in engineering was in robust truss design [208]. Since then, the value of robust optimization and its various generalizations has been demonstrated in many other problems, for example, in optics, [254, 278, 1403]; in chemical engineering [1162, 1189]; in wireless network design [1166, 1180]; in circuit design [1920]; in transportation, in problems in air traffic control [246], railway timetabling [475, 1195], vehicle routing [442, 836, 1742], and network design [107, 1361], as well as autonomous aerial vehicle mission planning [680] and emergency response [200]; in energy, in the unit commitment (UC) problem [253, 1001, 1913] and energy systems optimization and design [895, 1290, 1451, 1629] and in minimizing the energy cost of water distribution [833]; in finance, in portfolio optimization under uncertainty [310, 439, 575, 811, 821, 871, 1532]; in operations management, in supply chain management [204] and airline revenue management

[275]; and in healthcare, in problems in cancer therapy [255, 321, 427, 429, 469, 762, 831, 1183, 1268] and emergency response facility location [430].

Other comprehensive reviews of robust optimization go into more depth on a wide range of technical topics [213, 245, 1189], highlight recent advances in robust optimization [776], or cover some practical advice for application to real-world problems [832]. This chapter aims to provide a brief overview of topics in the robust optimization literature, highlighting the main results while leaving detailed derivations to their original sources.

25.2 ▪ Robust Optimization Formulations

25.2.1 ▪ Setup

We start with the following general optimization problem over decisions $\mathbf{x} \in \mathbb{R}^n$:

$$
\begin{aligned}
\min_{\mathbf{x} \in \mathbb{R}^n} \quad & f_0(\mathbf{x}; \mathbf{s}) \\
\text{s.t.} \quad & f_i(\mathbf{x}; \mathbf{s}) \le 0 \quad \forall i \in I,
\end{aligned}
\tag{25.1}
$$

where f_0 is the cost function and f_i are the constraint functions, and $\mathbf{s}$ represents the problem data and is fixed, and the index set I is finite. We refer to formulation (25.1) as the *nominal problem* — in this problem, there is no uncertainty in the data. Robust optimization is concerned with the problem that arises when the parameters $\mathbf{s}$ are uncertain and belong to some *uncertainty set* $\mathcal{U}$. We call $\mathbf{s} \in \mathcal{U}$ a *scenario*, and the particular scenario that actually occurs is called the *realized scenario*. The decision maker seeks a solution that is robust, i.e., still feasible, no matter what scenario is realized in the uncertainty. In other words, the decision maker seeks a solution to the following *robust problem* corresponding to the above nominal problem:

$$
\begin{aligned}
\min_{\mathbf{x} \in \mathbb{R}^n} \quad & \max_{\mathbf{s} \in \mathcal{U}} f_0(\mathbf{x}; \mathbf{s}) \\
\text{s.t.} \quad & f_i(\mathbf{x}; \mathbf{s}) \le 0 \quad \forall i \in I, \forall \mathbf{s} \in \mathcal{U}.
\end{aligned}
\tag{25.2}
$$

A solution to formulation (25.2) is called a *robust solution*. A robust solution optimizes the objective for the worst-case realization of $\mathbf{s} \in \mathcal{U}$, subject to the constraint that no matter which scenario $\mathbf{s} \in \mathcal{U}$ is realized, the constraints $f_i(\mathbf{x}; \mathbf{s}) \le 0$ are satisfied. Generally, $\mathcal{U}$ is an uncountable set, which means the robust problem has an uncountably infinite number of constraints. Thus, much effort is focused on reformulating problems of the form (25.2) into deterministic problems of finite size, which is known as a *robust counterpart*. Note that some papers in the literature call (25.2) the robust counterpart.

Without loss of generality, we can focus on uncertainty solely in the constraints [212], since (25.2) is equivalent to

$$
\begin{aligned}
\min_{\mathbf{x} \in \mathbb{R}^n, t \in \mathbb{R}} \quad & t \\
\text{s.t.} \quad & f_0(\mathbf{x}; \mathbf{s}) \le t \quad \forall \mathbf{s} \in \mathcal{U}, \\
& f_i(\mathbf{x}; \mathbf{s}) \le 0 \quad \forall i \in I, \forall \mathbf{s} \in \mathcal{U}.
\end{aligned}
\tag{25.3}
$$

Thus, to keep the exposition streamlined, we will consider problems with linear cost functions and uncertainty only in the constraints. We will focus on two major categories of robust problems in the subsequent sections: problems with uncertainty in linear constraints and problems with uncertainty in nonlinear constraints.

25.2.2 ▪ Problems with Uncertainty in Linear Constraints

In this section, we focus on robust linear optimization problems of the form

$$
\begin{aligned}
\min_{x \in \mathbb{R}^n} \quad & c'x \\
\text{s.t.} \quad & Ax \le b \quad \forall A \in \mathcal{U},
\end{aligned}
\tag{25.4}
$$

where $A \in \mathbb{R}^{m \times n}$, $b \in \mathbb{R}^m$, and $c \in \mathbb{R}^n$. Without loss of generality [209], we may consider the case where there is only uncertainty in A and where the uncertainty is row-wise, where $a_i \in \mathbb{R}^n$, $i \in I$, are the rows of A, and $b_i \in \mathbb{R}$ are the components of b:

$$
\begin{aligned}
\min_{x \in \mathbb{R}^n} \quad & c'x \\
\text{s.t.} \quad & a_i'x \le b_i \quad \forall a_i \in \mathcal{U}_i, \forall i \in I.
\end{aligned}
\tag{25.5}
$$

Here, the uncertainty sets for each row are the projections of the original uncertainty set into the appropriate spaces containing the a_i vectors. Thus, we need only ensure each constraint is satisfied for all scenarios that impact that constraint.

25.2.2.1 ▪ Polyhedral Uncertainty Set

The simplest case is when the uncertainty set is polyhedral [209, 212, 245]. Let $\mathcal{U}_i = \{a_i : C_i a_i \le d_i\}$, where $C_i \in \mathbb{R}^{m_i \times n}$, where m_i is the number of constraints in the definition of the uncertainty set of $\mathcal{U}_i$, and $d_i \in \mathbb{R}^{m_i}$. The constraint $a_i'x \le b_i \ \forall a_i \in \mathcal{U}_i, \forall i \in I$, can be written as

$$
\max_{a_i \in \{a_i \,:\, C_i a_i \le d_i\}} a_i'x \le b_i.
\tag{25.6}
$$

Noting the connection with duality, the above problem can be reformulated into the following linear optimization (LO) problem, where $q_i \in \mathbb{R}^{m_i}$ are appropriately defined dual vectors for the maximization problem in (25.6):

$$
\begin{aligned}
\min_{x \in \mathbb{R}^n, q_i \in \mathbb{R}^{m_i}} \quad & c'x \\
\text{s.t.} \quad & q_i'd_i \le b_i \quad \forall i \in I, \\
& C_i'q_i = x \quad \forall i \in I, \\
& q_i \ge 0_{m_i} \quad \forall i \in I.
\end{aligned}
\tag{25.7}
$$

25.2.2.2 ▪ Ellipsoidal Uncertainty Set

Next, we consider uncertainty sets that take the form of an ellipsoid [209, 245, 332]:

$$
\mathcal{U} := \left\{ A \in \mathbb{R}^{m \times n} : A = P^{(0)} + \sum_{l=1}^{L} u_l P^{(l)} \text{ and } u'u \le 1 \right\},
$$

where $\mathbf{P}^{(l)} \in \mathbb{R}^{m \times n}$ for $l = 0,\ldots,L$ are given matrices and $\mathbf{u} \in \mathbb{R}^L$. Ellipsoidal uncertainty sets form a fairly general class of uncertainty sets and can be used to approximate many different types of convex sets [209]. Let $\mathbf{r}_i^{(l)} \in \mathbb{R}^n$ be the ith row of $\mathbf{P}^{(l)}$, where $\mathbf{r}_i^{(l)}$ is in the form of a column vector, and define

$$\mathbf{R}_i := \left(\begin{array}{cccc} | & | & & | \\ \mathbf{r}_i^{(1)} & \mathbf{r}_i^{(2)} & \cdots & \mathbf{r}_i^{(L)} \\ | & | & & | \end{array} \right) \in \mathbb{R}^{n \times L}.$$

For example, $\mathbf{r}_1^{(0)} + \mathbf{R}_1 \mathbf{u}$ denotes the first row of $\mathbf{P}^{(0)} + \sum_{l=1}^{L} u_l \mathbf{P}^{(l)}$ in the form of a column vector. Under this type of uncertainty set, the linear problem (25.4) can be transformed into an equivalent second-order cone optimization (SOCO) problem [209], which can be solved using interior-point methods [332]:

$$\begin{aligned} \min_{\mathbf{x} \in \mathbb{R}^n} \quad & \mathbf{c}'\mathbf{x} \\ \text{s.t.} \quad & \|\mathbf{R}_i'\mathbf{x}\|_2 \le b_i - (\mathbf{r}_i^{(0)})'\mathbf{x} \quad \forall i \in I. \end{aligned} \tag{25.8}$$

25.2.2.3 • Cardinality-Constrained Uncertainty Set

An alternative uncertainty set was proposed in [257], where the decision maker can control the level of conservatism of the robust solution by tuning the uncertainty set. In this cardinality-constrained uncertainty set, the number of entries of $\mathbf{A}$ that can deviate from their respective nominal values is bounded by some constant Γ. Let matrix $\mathbf{A} \in \mathbb{R}^{m \times n}$, $I = \{1,\ldots,m\}$, and $J = \{1,\ldots,n\}$.

Suppose the uncertain coefficients a_{ij} of $\mathbf{A}$ can take values in $[a_{ij}^{(0)} - \epsilon_{ij}, a_{ij}^{(0)} + \epsilon_{ij}]$, where $a_{ij}^{(0)}$ is the nominal value and ϵ_{ij} is the maximum deviation from the nominal value. The uncertainty set limits the number of coefficients a_{ij} in each row i that can deviate from their nominal values. The parameter Γ_i is used to denote the "budget" of deviations: up to $\lfloor \Gamma_i \rfloor$ coefficients are allowed to deviate completely within their respective intervals from the nominal value $a_{ij}^{(0)}$, and one coefficient, a_{ij_*}, is allowed to deviate only by $(\Gamma_i - \lfloor \Gamma_i \rfloor)\epsilon_{ij_*}$. Let $K(J_i, \Gamma_i) := \{S_i \cup \{t_i\} : S_i \subseteq J_i, |S_i| = \lfloor \Gamma_i \rfloor, t_i \in J_i \backslash S_i\}$, where J_i is the set of coefficients in row i of $\mathbf{A}$ that are subject to uncertainty and defined for $0 \le \Gamma \le |J_i|$. The robust problem is

$$\begin{aligned} \min_{\mathbf{x}, \mathbf{y} \in \mathbb{R}^n} \quad & \mathbf{c}'\mathbf{x} \\ \text{s.t.} \quad & \sum_{j \in J} a_{ij}^{(0)} x_j + \max_{S_i \cup \{t_i\} \in K(J_i, \Gamma_i)} \left\{ \sum_{j \in S_i} \epsilon_{ij} y_j + (\Gamma_i - \lfloor \Gamma_i \rfloor)\epsilon_{it_i} y_{t_i} \right\} \le b_i \quad \forall i \in I, \\ & -y_j \le x_j \le y_j \quad \forall j \in J, \\ & \mathbf{l} \le \mathbf{x} \le \mathbf{u}, \\ & \mathbf{y} \ge \mathbf{0}_n \end{aligned} \tag{25.9}$$

for some lower and upper bounds $\mathbf{l}$ and $\mathbf{u} \in \mathbb{R}^n$. In the original paper, the term

$$\max_{S_i \cup \{t_i\} \in K(J_i, \Gamma_i)} \left\{ \sum_{j \in S_i} \epsilon_{ij} y_j + (\Gamma_i - \lfloor \Gamma_i \rfloor)\epsilon_{it_i} y_{t_i} \right\} \tag{25.10}$$

is first shown to have the same optimal value as an LO problem (Proposition 1 in [257]). Then, the resulting LO problem is dualized and reinserted into (25.9). The full robust problem then finally takes the form of an LO problem:

$$\min_{x,y\in\mathbb{R}^n, z\in\mathbb{R}^m, p_i\in\mathbb{R}^{|J_i|}} \quad c'x$$

$$\text{s.t.} \quad \sum_{j\in J} a_{ij}^{(0)} x_j + z_i\Gamma_i + \sum_{j\in J_i} p_{ij} \le b_i \quad \forall i \in I,$$

$$z_i + p_{ij} \ge \epsilon_{ij} y_j \quad \forall j \in J_i, \forall i \in I,$$

$$-y \le x \le y, \tag{25.11}$$

$$p_{ij} \ge 0 \quad \forall j \in J_i, \forall i \in I,$$

$$l_j \le x_j \le u_j \quad \forall j \in J,$$

$$y \ge 0_n,$$

$$z \ge 0_m.$$

The parameter Γ_i can be varied by the decision maker to control the conservatism of the robust solution. We remark that this cardinality-constrained uncertainty set is a special case of a polyhedral uncertainty set.

25.2.2.4 ▪ Conic Linear Problem with Convex Uncertainty Set

A recent paper [830] uses results from [181] to provide a general method for dealing with general constraintwise convex uncertainty sets. Suppose that we have constraints of the form $Ax \le b$ and matrix $A \in \mathbb{R}^{m\times n}$. Let $I = \{1,\ldots,m\}, J = \{1,\ldots,n\}$, and suppose each constraint vector a_i is contained in the uncertainty set $\mathcal{U}_i := \{a_i : g_{i,k}(a_i) \le 0 \,\forall k \in K\}$ for some finite index set K, where $g_{i,k}$ are closed convex real-valued functions and the decision variables x are in a closed convex cone $\mathcal{K} \subset \mathbb{R}^n$. The robust problem is of the form

$$\min_{x\in\mathbb{R}^n} \quad c'x$$
$$\text{s.t.} \quad a_i'x \le b_i \quad \forall a_i \in \mathcal{U}_i, \forall i \in I, \tag{25.12}$$
$$x \in \mathcal{K}.$$

If $\mathcal{K} = \mathbb{R}^n_+$ and $g_{i,k}$ are linear functions, we recover the situation in (25.4). As stated in [830], by using the results in [181], we find that (25.12) has the same optimal value as the following convex optimization problem, which is equivalent to the dual of (25.12), where $\mathcal{K}^*$ is the dual cone of $\mathcal{K}$ and $y \in \mathbb{R}^I$ and $v_i \in \mathbb{R}^n$:

$$\max_{v_i\in\mathbb{R}^n, y\in\mathbb{R}^I} \quad -b'y$$

$$\text{s.t.} \quad \sum_{i\in I} v_i + c \in \mathcal{K}^*,$$

$$y_i g_{i,k}\left(\frac{v_i}{y_i}\right) \le 0 \quad \forall i \in I, \forall k, \tag{25.13}$$

$$y \ge 0_{|I|},$$

provided that Slater's condition is satisfied.[19] Here, the KKT vector associated with the constraint $\sum_{i\in I} v_i + c \in \mathcal{K}^*$ forms a robust optimal solution to (25.12). Furthermore, if (25.13) is bounded, then an optimal solution can be directly recovered for the

---

[19]See [332] for more on dual cones and Slater's condition.

original robust problem. Note that (25.13) is convex because $y_i g_{i,k}\left(\frac{\mathbf{v}_i}{y_i}\right)$ are perspective functions. One advantage of this method is that it does not require taking the conjugate of $g_{i,k}$.

25.2.2.5 ▪ Remark

Note that the derivations for the robust counterparts in this section, except for the conic linear constraints with convex uncertainty sets, also hold for when the decision variable $\mathbf{x}$ is integer valued. This is because all the derivations are essentially focused on dualizing a constraint as an optimization problem whose decision variable is the coefficient vector $\mathbf{a}$, with $\mathbf{x}$ held constant.

25.2.3 ▪ Problems with Uncertainty in Nonlinear Constraints

25.2.3.1 ▪ Conic Constraints with Ellipsoidal Uncertainty Set

The SOCO methods discussed here were originally presented in [310]. Consider the following model:

$$\begin{aligned} \min_{\mathbf{x}\in\mathbb{R}^n} \quad & \mathbf{c}'\mathbf{x} \\ \text{s.t.} \quad & \|A\mathbf{x}+\mathbf{f}\|_2 \le \mathbf{b}'\mathbf{x}+d \quad \forall(A,\mathbf{f},\mathbf{b},d)\in\mathcal{U}, \end{aligned} \tag{25.14}$$

where $A\in\mathbb{R}^{m\times n}$, $\mathbf{f}\in\mathbb{R}^m$, $\mathbf{b}\in\mathbb{R}^n$, and $d\in\mathbb{R}$.
The uncertainty set $\mathcal{U}$ is defined as

$$\left\{(A,\mathbf{f},\mathbf{b},d) \mid (A,\mathbf{f},\mathbf{b},d)=(A^{(0)},\mathbf{f}^{(0)},\mathbf{b}^{(0)},d^{(0)})+\sum_{l=1}^L u_l(A^{(l)},\mathbf{f}^{(l)},\mathbf{b}^{(l)},d^{(l)}) \ \& \ \mathbf{u}'\mathbf{u}\le 1\right\}.$$

Then, [310] states that the robust problem can be approximated by the following semidefinite optimization (SDO) problem, called the approximate robust counterpart:

$$\begin{aligned} \min_{\mathbf{x},\lambda,\mu} \quad & \mathbf{c}'\mathbf{x} \\ \text{s.t.} \quad & \left(\begin{array}{c|c|c} v[\mathbf{x}]-\mu-\lambda & (w[\mathbf{x}])' & -(r[\mathbf{x}])' \\ \hline w[\mathbf{x}] & \lambda\mathbf{I}_L & -(U[\mathbf{x}])' \\ \hline -r[\mathbf{x}] & -U[\mathbf{x}] & \mu\mathbf{I}_m \end{array}\right) \succeq 0, \\ & \lambda\ge 0, \\ & \mu\ge 0, \end{aligned} \tag{25.15}$$

where the functions $v[\mathbf{x}]$, $w[\mathbf{x}]$, $r[\mathbf{x}]$, and $U[\mathbf{x}]$-are defined as

$$\begin{aligned} v[\mathbf{x}] &:= (\mathbf{b}^{(0)})'\mathbf{x}+d^{(0)}, & w[\mathbf{x}] &:= \tfrac{1}{2}\big((\mathbf{b}^{(l)})'\mathbf{x}+d^{(l)}\big)'_{l=1,\dots,L}, \\ r[\mathbf{x}] &:= \tfrac{1}{2}\big(A^{(0)}\mathbf{x}+\mathbf{f}^{(0)}\big), & U[\mathbf{x}] &:= \tfrac{1}{2}\big(A^{(l)}\mathbf{x}+\mathbf{f}^{(l)}\big)_{l=1,\dots,L}, \end{aligned} \tag{25.16}$$

where $U[\mathbf{x}]\in\mathbb{R}^{m\times L}$ is a matrix with $\tfrac{1}{2}\big(A^{(l)}\mathbf{x}+\mathbf{f}^{(l)}\big)$ as its columns.

25.2.3.2 ▪ Nonlinear Constraints with General Uncertainty Set

A recent paper [199] outlines a general method for deriving robust counterparts for problems with nonlinear constraints under uncertainty through concepts in convex analysis. We reproduce the main result from that paper here and recommend the reader look to the original paper for an extensive list of robust counterparts for a wide variety of problems. Consider the following mathematical optimization problem:

$$\min_{\mathbf{x}\in\mathbb{R}^n} \quad \mathbf{c}'\mathbf{x}$$
$$\text{s.t.} \quad f(\mathbf{a},\mathbf{x})\leq 0 \quad \forall \mathbf{a}\in\mathcal{U}, \qquad (25.17)$$

where f is a nonlinear function, concave in $\mathbf{a}\in\mathbb{R}^m$ for each $\mathbf{x}\in\mathbb{R}^n$. Furthermore, $\mathcal{U}$ is an uncertainty set dependent on a compact and convex set $\mathcal{X}\subset\mathbb{R}^L$ with zeros in the relative interior of $\mathcal{X}$, and $\mathcal{U}:=\{\mathbf{a}=\mathbf{a}^0+\mathbf{A}\mathbf{u}:\mathbf{u}\in\mathcal{X}\}$ for some nominal value $\mathbf{a}^0$ and matrix $\mathbf{A}\in\mathbb{R}^{m\times L}$. The main result of [199, Theorem 2] reformulates (25.17) without the $\forall \mathbf{a}\in\mathcal{U}$ quantifier. This theorem states that, if the nominal value $\mathbf{a}^0$ is regular,[20] then the vector $\mathbf{x}$ is a feasible solution to (25.17) if and only if $\mathbf{x}$ satisfies

$$(\mathbf{a}^0)'\mathbf{v}+\sup_{\mathbf{u}\in\mathcal{X}}\mathbf{v}'\mathbf{A}\mathbf{u}-f_*(\mathbf{v},\mathbf{x})\leq 0 \qquad (25.18)$$

for some $\mathbf{v}\in\mathbb{R}^m$, where $f_*(\cdot,\mathbf{x})$ is the partial concave conjugate, defined as

$$f_*(\mathbf{v},\mathbf{x}):=\inf_{\mathbf{z}\in\mathrm{dom}(f(\cdot,\mathbf{x}))}\left(\mathbf{v}'\mathbf{z}-f(\mathbf{z},\mathbf{x})\right). \qquad (25.19)$$

The rest of [199] focuses on calculating $\sup_{\mathbf{u}\in\mathcal{X}}\mathbf{v}'\mathbf{A}\mathbf{u}$ and $f_*(\mathbf{v},\mathbf{x})$ for different classes of "primitive" uncertainty sets $\mathcal{X}$ and constraints f and so is a good comprehensive resource for finding tractable robust counterparts for a large class of problems.

25.2.4 ▪ Uncertainty Set Parameterizations

In many of the previous formulations, we are given a set $\mathcal{U}$, which may be a polyhedron or an ellipsoid, and we are to reparameterize it so that it is of the form

$$\mathcal{U}:=\left\{\mathbf{v}:\mathbf{v}^{(0)}+\sum_{l=1}^{L}u_l\mathbf{v}^{(l)}\text{ and }\mathbf{u}\in\mathcal{X}\right\}. \qquad (25.20)$$

Here we write out how to explicitly find this reparameterization (also discussed briefly in [205]). Suppose $\mathbf{v}\in\mathbb{R}^n$. If $\mathcal{U}$ is a polyhedron, then it can be written in the form $\mathcal{U}=\{\mathbf{v}:\mathbf{A}\mathbf{v}\leq\mathbf{b}\}$. Then by choosing $L:=n$ and $\mathbf{v}^{(0)}=\mathbf{0}_n$, and choosing $\mathbf{v}^{(l)}$ to be an orthonormal basis in $\mathbb{R}^n$, any $\mathbf{v}\in\mathcal{U}$ can be rewritten as $\sum_{l=1}^{L}u_l\mathbf{v}^{(l)}$ for some choice of $\mathbf{u}=(u_1,\ldots,u_L)$, and $\mathbf{A}\mathbf{v}\leq\mathbf{b}\iff\sum_{l=1}^{L}u_l\mathbf{A}\mathbf{v}^{(l)}\leq\mathbf{b}$. Thus,

$$\mathcal{X}:=\left\{\mathbf{u}:(\mathbf{A}\mathbf{v}^{(1)},\ldots,\mathbf{A}\mathbf{v}^{(L)})\mathbf{u}\leq\mathbf{b}\right\},$$

where $(\mathbf{A}\mathbf{v}^{(1)},\ldots,\mathbf{A}\mathbf{v}^{(L)})$ is the matrix with $\mathbf{A}\mathbf{v}^{(l)}$ as its columns, so that $\mathcal{X}\subset\mathbb{R}^n$ is a polyhedron and we obtain the uncertainty set $\mathcal{U}$ in the form of (25.20).

If $\mathcal{U}$ is an ellipsoid, then it can be written in the form $\mathcal{U}=\{\mathbf{v}:\mathbf{v}'\mathbf{X}\mathbf{v}\leq 1\}$, where $\mathbf{X}$ is a positive definite matrix. Since positive definite matrices are invertible, and $\mathbf{X}$ is an $n\times n$ symmetric matrix, we should choose $L:=n$ and $\mathbf{v}^{(l)}$ to be the eigenvectors of $\mathbf{X}$, which form an orthonormal basis for $\mathbb{R}^n$. So for any $\mathbf{v}\in\mathcal{U}$, there is a representation $\mathbf{v}=\sum_{l=1}^{L}u_l\mathbf{v}^{(l)}$ for some $\mathbf{u}$, and then $\mathbf{v}'\mathbf{X}\mathbf{v}\leq 1\iff\sum_{l=1}^{L}\lambda^{(l)}u_l^2\leq 1$, where $\lambda^{(l)}$ is the strictly positive eigenvalue of $\mathbf{v}^{(l)}$, due to orthonormality of $\mathbf{v}^{(l)}$, so that by defining $\mathcal{X}:=\left\{\mathbf{u}:\sum_{l=1}^{L}\lambda^{(l)}u_l^2\leq 1\right\}$ we obtain the set $\mathcal{U}$ in the form of (25.20).

---

[20]$\mathbf{a}^0$ is regular if for all $\mathbf{x}$, we have $\mathbf{a}^0\in\mathrm{ri}(\mathrm{dom}(f(\cdot,\mathbf{x})))$, that is, in the relative interior of the domain of the constraint function. The domain of a concave function is defined as the set $\{\mathbf{a}:f(\mathbf{a},\mathbf{x})>-\infty\}$, denoted $\mathrm{dom} f(\cdot,\mathbf{x})$.

25.2.5 ▪ Pareto Robust Optimality

Consider the specialized robust optimization (maximization) problem with uncertainty in the cost function only:

$$\max_{\mathbf{x} \in X} \min_{\mathbf{p} \in \mathcal{U}} \mathbf{p}'\mathbf{x}, \tag{25.21}$$

with some bounded polyhedral feasible region X and bounded polyhedral uncertainty set $\mathcal{U}$. The way (25.21) is formulated, any robust optimal solution optimizes for the worst-case realization of $\mathbf{p} \in \mathcal{U}$ but does not take into account the quality of this solution if some non-worst-case $\mathbf{p} \in \mathcal{U}$ occurs. Thus, while all robust optimal solutions will have the same objective value for their respective worst-case $\mathbf{p}$, it is not true that all robust optimal solutions will perform equally with respect to the other $\mathbf{p} \in \mathcal{U}$. Borrowing concepts from economics, the paper [973] aims to address this issue through defining a *Pareto robust optimization paradigm*.

The main contribution of this [973] is the development of a methodology that finds Pareto robust optimal solutions, that is, robust optimal solutions for which there are no other feasible solutions that perform at least well as in all scenarios $\mathbf{p} \in \mathcal{U}$ *and* strictly better in at least one scenario $\mathbf{p}$. The paper also outlines ways to find such Pareto robust optimal solutions and illustrates their usefulness through a few examples. While the Pareto robust paradigm is primarily concerned with cost uncertainty, it raises the important issue of being aware of the existence of multiple robust solutions. In many applications, there may be several robust optimal solutions, and some may confer more advantage to the decision maker than others.

25.3 ▪ Robust Optimization with Recourse

Recourse-based robust optimization draws inspiration from two-stage stochastic models. A standard "single-stage" robust solution needs to be identified and implemented before any uncertainty is realized. However, some problems naturally involve multiple stages, where decisions can be made both immediately and in future stages, after some uncertainty has been resolved. In this paradigm, the decision maker can construct a less conservative first-stage solution and make adjustments later [205].

Variants of robust optimization with recourse involve augmenting a first-stage robust solution in response to realized scenarios. The "method of augmentation" is one way of classifying some of the methods of recourse-based robust optimization. Some methods consider optimizing over recourse robust solutions that are functions of the uncertain parameters, as in the case of affine adjustable robust optimization and its extensions [197, 205, 248, 252, 451], which consider affine and polynomial functions of the uncertainty, and the closely related finite adaptable robust optimization [246, 249], which considers piecewise-constant functions of the uncertainty. Other methods consider a framework in which algorithms are designed before a scenario is revealed that modify solutions to "recover" them to feasibility after the scenario is revealed [475, 1195].

25.3.1 ▪ Fully Adjustable Robust Solutions

We focus on the adjustable robust methodologies discussed in [205, 245]. Adjustable robust solutions are mainly concerned with distinguishing the "here-and-now," $\mathbf{x}_h$, decisions and the "wait-and-see," $\mathbf{x}_w$, decisions. Suppose $\mathbf{x}_h \in H$ and $\mathbf{x}_w \in W$, where H and W are subsets of the Euclidean space. The first type denotes the variables that must be optimized before the scenario is revealed, and the second type denotes the variables that are assigned after the scenario is revealed. First, consider the standard robust optimization problem with a rearrangement of our indices so that $\mathbf{x} := (\mathbf{x}_h, \mathbf{x}_w)$, $\mathbf{A} := (\mathbf{A}_h, \mathbf{A}_w)$, and $\mathbf{c} := (\mathbf{c}_h, \mathbf{c}_w)$, with $\mathbf{A} \in \mathbb{R}^{m \times n}$ and $\mathbf{c} \in \mathbb{R}^n$, and $(\mathbf{A}_h, \mathbf{A}_w), (\mathbf{c}_h, \mathbf{c}_w)$ in the appropriate spaces. The robust problem without adjustability is

$$\min_{\mathbf{x}_h \in H, \mathbf{x}_w \in W} \quad \mathbf{c}_h' \mathbf{x}_h + \mathbf{c}_w' \mathbf{x}_w$$
$$\text{s.t.} \quad \mathbf{A}_h \mathbf{x}_h + \mathbf{A}_w \mathbf{x}_w \leq \mathbf{b} \quad \forall (\mathbf{A}_h, \mathbf{A}_w) \in \mathcal{U}, \tag{25.22}$$

and the corresponding robust solution does not distinguish between here-and-now and wait-and-see variables. For ease of exposition, we use $\mu := (\mathbf{A}_h, \mathbf{A}_w)$ to denote the uncertainty ($\mathbf{A}$ matrix). The key feature of an adjustable robust solution is that $\mathbf{x}_w$ is allowed to be different for each $(\mathbf{A}_h, \mathbf{A}_w)$. In other words, the wait-and-see decision can be a function of the uncertainty. We denote this vector-valued function as χ_w to distinguish it from the vector $\mathbf{x}_w$, so that for a particular μ, we have $\mathbf{x}_w = \chi_w(\mu)$, so that χ_w belongs to the space of vector-valued functions of μ, which we denote as $\mathcal{W}$:

$$\min_{\mathbf{x}_h \in H, \chi_w \in \mathcal{W}} \quad \mathbf{c}_h' \mathbf{x}_h + \max_{\mu \in \mathcal{U}} \mathbf{c}_w' \chi_w(\mu)$$
$$\text{s.t.} \quad \mathbf{A}_h \mathbf{x}_h + \mathbf{A}_w \chi_w(\mu) \leq \mathbf{b} \quad \forall \mu := (\mathbf{A}_h, \mathbf{A}_w) \in \mathcal{U}. \tag{25.23}$$

This is the fully adjustable robust optimization problem. The solution $(\mathbf{x}_h, \chi_w)$ that minimizes this problem is an adjustable robust solution. This problem is generally difficult because χ_w is a function of the uncertainty, and so we are optimizing over a (vector-valued) function space, rather than a finite vector space.

The adjustable robust optimization paradigm is more general than standard robust optimization and hence can theoretically provide better solutions when applied correctly (cf. [199, Section 10]). However, there is an important case where they are equivalent, as outlined in [205]. One of these conditions amounts to having the uncertainty be constraintwise. Although the conditions of robust optimization and adjustable robust optimization turn out to be fairly stringent, it is important to check these conditions, otherwise effort may be wasted in trying to solve the fully adjustable robust problem. Practically speaking, this means that when solving a problem in the adjustable robust paradigm, if possible, the uncertainty set should not be constraintwise.

25.3.2 ▪ Min-Max-Min Models

The fully adjustable robust optimization problem (25.23) is also known as the two-stage adaptable optimization problem [248, 249]. Since $\chi_w(\mu)$ depends on the uncertain parameter μ, we can transform the fully adjustable problem (25.23) into a min-max-min problem:

$$\min_{\mathbf{x}_h \in H} \max_{\mu \in \mathcal{U}} \min_{\mathbf{x}_w \in W(\mathbf{x}_h, \mu)} f(\mathbf{x}_h, \mu, \mathbf{x}_w), \tag{25.24}$$

where H is the feasible region for the here-and-now decision, μ is the uncertain parameter, and $W(\mathbf{x}_h, \mu)$ is the feasible region of the wait-and-see decisions, based on the values of $\mathbf{x}_h$ and μ. In [45], they discretize and enumerate $\mathcal{U}$ and use a decomposition algorithm to solve (25.24), where the subproblem solves the inner max-min problem.

From an application viewpoint, military applications related to interdiction have been considered in this min-max-min framework [45, 346], in a max-min framework [347], and specifically on Markovian networks [886]. In the defender-attacker-defender framework, the first stage represents actions of the defender to harden his/her defenses to attack; the uncertainty realization represents actions taken by the attacker to cause as much damage as possible; then the second-stage decision represents actions by the defender to minimize the impact of the attack. This min-max-min framework has also been considered in a healthcare facility location setting [430].

25.3.3 ▪ Affine Adjustable Robust Solutions

Following the development in [205], we can assume without loss of generality that (25.23) can be rewritten so that the cost function is independent of the wait-and-see decisions, and so we can have the simpler form

$$
\begin{aligned}
\min_{\mathbf{x}_b \in H, \chi_w \in \mathscr{W}} \quad & \mathbf{c}'\mathbf{x}_b \\
\text{s.t.} \quad & \mathbf{A}_b\mathbf{x}_b + \mathbf{A}_w\chi_w(\mu) \leq \mathbf{b} \quad \forall \mu := (\mathbf{A}_b, \mathbf{A}_w) \in \mathscr{U}.
\end{aligned}
\tag{25.25}
$$

The trade-off for the generality of adjustable robust solutions is that the adjustable robust problem is generally intractable [205]. Thus, instead of considering all possible functions of the uncertainty μ, we consider the restricted case where $\chi_w(\mu) := \mathbf{W}\mu + \mathbf{w}$ is a linear function of μ for some $\mathbf{W}$ matrix and $\mathbf{w}$ vector in the appropriate spaces. This problem is known as the affine adjustable robust counterpart (AARC). With this simplification, and under a Slater-like condition and structural assumptions on the feasible region, the paper [205] explicitly reformulates (25.25) into a conic quadraticly optimization or SDO problem (Theorem 3.2) in the case when $\mathbf{A}_w$ is independent of uncertainty. For the case of an uncertain recourse matrix and a *primitive* ellipsoidal uncertainty set centered at the origin, the robust problem can be reformulated to an SDO problem. Details of this formulation are given in [205]. The paper [248] also reviews the effectiveness of affine adjustable robust solutions for problems with right-hand side uncertainty.

25.3.3.1 ▪ Remark

A few direct extensions to affine adjustable robust optimization have been proposed [197,451]. In [197], AARC is extended by controlling constraint violations outside the uncertainty set in addition to the standard constraint requirements. In its most general form, the model given in [197] can be seen as an extension to the general adjustable robust counterpart. In [451], AARC is extended by introducing an extended uncertainty set on which the recourse variables are dependent in an affine way. The extended uncertainty set allows more flexibility in the recourse variables. In the multistage context, AARC has been shown to be optimal for restricted types of multistage robust problems [251]. The logical extension of affine functions to polynomial functions of the uncertainty is also used in analyzing multistage problems [252]. Other forms of decision rules have also been proposed for stochastic optimization (SO) and distributionally robust optimization (DRO) problems [450, 796, 820].

25.3.4 ▪ Finite Adaptability

An alternative method of adjustable robust optimization considers piecewise-constant, rather than affine, functions of the uncertainty, providing a novel geometric interpretation. We shall focus on the methods in [246, 404]. To have a piecewise-constant function over the uncertainty set is the same as having a partition of the uncertainty set and individually solving the standard robust problem on each piece in the partition. Each subset is a smaller uncertainty set, so the corresponding robust solution is less constrained than the standard robust solution over the entire set. Furthermore, since this is a partition, there is always a robust solution prepared for any scenario that might be realized. The problem is then to choose the best partition of the uncertainty set $\mathcal{U}$ into $\mathcal{U}_1 \cup \cdots \cup \mathcal{U}_K$, with robust optimal solutions for each $\mathcal{U}_k$ denoted as $\mathbf{x}_w^k$ [246]:

$$\min_{\mathcal{U}=\mathcal{U}_1\ldots\cup\mathcal{U}_K} \left[\begin{array}{cc} \min_{\mathbf{x}_h\in H,\mathbf{x}_w^1,\ldots,\mathbf{x}_w^K\in W} & \mathbf{c}'\mathbf{x}_h + \max\{\mathbf{d}'\mathbf{x}_w^1, \mathbf{d}'\mathbf{x}_w^2, \ldots, \mathbf{d}'\mathbf{x}_w^K\} \\ \text{s.t.} & \mathbf{A}_h\mathbf{x}_h + \mathbf{A}_w\mathbf{x}_w^k \le \mathbf{b} \quad \forall(\mathbf{A}_h, \mathbf{A}_w) \in \mathcal{U}_k, \forall k = 1, \ldots, K \end{array} \right]. \tag{25.26}$$

Even in the case of a two-element partition (i.e., $K = 2$), the adaptability problem is, "in general, NP-hard" [246], though under some strict conditions it is possible to compute approximations to the solution. Note, however, that even in the case of a nonoptimal partition, choosing one of the robust solutions of a piece of the partition forms a better solution than the standard robust solution using the entire uncertainty set (provided it is a true partition). It should be noted that between finite adaptability and affine adjustability neither is superior to the other; there are cases where finite adaptability performs better than affine adjustability and vice versa [246]. Finite adaptability can be a good approximation to the fully adjustable optimization problem [249]. More recently, finite adaptability has been applied to two-stage robust binary optimization problems [900] as well as to multistage robust mixed-integer optimization problems [247, 1507].

25.3.5 ▪ Myopic Adaptive Reoptimization

A different approach to multistage robust optimization takes a myopic view toward recourse, leveraging new observations of the uncertainty at every stage. The idea is that updated observations of the uncertainty allow us to refine our beliefs about the uncertainty set and perhaps shrink it if uncertain realizations stabilize over time. To make the approach tractable and simple to implement, an open-loop approach is considered [429]. Consider the following robust LO problem, which is solved in stage t of the multistage problem:

$$\begin{array}{cl} \min_{\mathbf{x}\in\mathbb{R}^n} & \mathbf{c}'\mathbf{x}^{(t)} \\ \text{s.t.} & \mathbf{A}\mathbf{x}^{(t)} \le \mathbf{b} \quad \forall \mathbf{A} \in \mathcal{U}^{(t)}, \end{array} \tag{25.27}$$

where $\mathcal{U}^{(t)}$ is a polyhedron. After implementing the solution $\mathbf{x}^{(t)}$ (which is prepared only with knowledge of $\mathcal{U}^{(t)}$), a realization of the uncertain parameter vector is observed: $\mathbf{A}^{(t)}$. Using this $\mathbf{A}^{(t)}$, we update the uncertainty set $\mathcal{U}^{(t+1)}$ from $\mathcal{U}^{(t)}$ for use in the next stage according to $\mathcal{U}^{(t+1)} \leftarrow (1-\alpha)\mathcal{U}^{(t)} + \alpha\mathbf{A}^{(t)}$ for some parameter $\alpha \in [0,1]$. A larger choice of α means that the uncertainty set will shrink more quickly and eventually try to follow $\mathbf{A}^{(t)}$ (i.e., more weight is given to recent observations over more historical observations).

Although the approach is open loop, it was shown in [429] that, under some fairly mild conditions, if the sequence of uncertain realizations $\mathbf{A}^{(t)}$ converges to some limit $\mathbf{A}^{(*)}$, the sequence of optimal robust solutions $\mathbf{x}^{(t,*)}$ converges to a solution $\mathbf{x}^*$, which is an optimal solution to the following problem:

$$\begin{array}{cl} \min_{\mathbf{x}\in\mathbb{R}^n} & \mathbf{c}'\mathbf{x} \\ \text{s.t.} & \mathbf{A}^{(*)}\mathbf{x} \le \mathbf{b}. \end{array} \tag{25.28}$$

It was shown that in a healthcare application, if the observations $\mathbf{A}^{(t)}$ are stable over time, the myopic adaptive reoptimization outperforms a comparable nonadaptive robust optimization approach [1268]. The advantage of this method is that the myopic adaptive reoptimization methodology is tractable as long as the single-stage optimization is tractable.

25.4 ▪ Distributionally Robust Optimization

DRO can be viewed as a generalization of both robust and stochastic optimization and has become a very active field of research in recent years [390, 575, 640, 677, 820, 899, 1567, 1740, 1889, 1915, 1985]. Classical SO requires complete knowledge of the probability distribution, but such knowledge is often unavailable or difficult to obtain. On the other hand, in classical robust optimization, no assumptions are made on the distributional information, which may result in overconservative solutions, and any distributional information that is available beyond the support is effectively wasted [1889]. DRO, also known as minimax SO [640, 1567], addresses these issues and unifies both stochastic and robust optimization in one paradigm.

DRO takes the idea of an uncertainty set, where uncertain parameters may lie, and generalizes it to an *ambiguity* set, where the uncertain distributions of the uncertain parameters may lie. We present the formulation given in [575]. Suppose we want to optimize the cost function $c(\mathbf{x}, \mathbf{u})$, where $\mathbf{x}$ is the decision vector in some feasible set X and $\mathbf{u}$ is an uncertain parameter. If we knew that the distribution of $\mathbf{u}$ is exactly the distribution $D_{\mathbf{u}}$, then we could use SO to solve

$$\min_{\mathbf{x} \in X} \mathbb{E}_{D_{\mathbf{u}}}[c(\mathbf{x}, \mathbf{u})]. \tag{25.29}$$

Now suppose that even the distribution $D_{\mathbf{u}}$ is uncertain but is known to lie in the ambiguity set $\mathcal{D}$. Then the distributionally robust problem is given by

$$\min_{\mathbf{x} \in X} \max_{D_{\mathbf{u}} \in \mathcal{D}} \mathbb{E}_{D_{\mathbf{u}}}[c(\mathbf{x}, \mathbf{u})]. \tag{25.30}$$

We refer interested readers to the references listed above for approaches to solve (25.30).

25.5 ▪ Conclusion

In this chapter, we reviewed some of the foundational results in modern robust optimization, starting with the robust counterparts of LO problems under various types of uncertainty, and closing with modeling approaches to multistage robust optimization. We also presented some modern variants of robust optimization, including affine adjustable robust optimization, finite adaptability, and DRO.

Other important aspects of robust optimization have not been covered in this chapter, for example, the application of robust optimization to stochastic dynamic optimization [980, 1888]. Another important topic not covered in this chapter is the construction of appropriate uncertainty sets, for example, using data-driven methods [244, 250]. Another recent topic that has not been discussed in this review is the stability and continuity properties of robust optimization, which have practical implications when implementing robust optimization computationally [428, 429, 525, 662, 1341, 1740, 1882].

Chapter 26

Robust Optimization in Radiation Therapy

Albin Fredriksson

26.1 ▪ Introduction

Radiation therapy is the medical use of ionizing radiation. It is used to treat nearly two thirds of the cancer patients in the United States [68], either alone or in combination with surgery or chemotherapy. Radiation kills cells by damaging the cellular DNA. A curative treatment requires administration of a sufficiently high dose to the tumor to eradicate the clonogenic cancer cells to an extent that results in permanent tumor control. The amount of dose delivered to the surrounding healthy tissues must at the same time be restricted for the treatment not to result in adverse effects. Radiation therapy treatment planning aims to strike the right balance between the probability of tumor control and the probability of complications due to the treatment.

In intensity-modulated radiation therapy (IMRT), the dose is delivered by external beams that are incident to the patient from multiple directions. The fluences over the cross-sections of the beams are modulated. This enables the superposition of the beam doses to conform closely to the target while avoiding the nearby healthy organs. The most common form of IMRT is delivered in the form of high-energy photon beams. An illustration of a photon-mediated IMRT plan is shown in Figure 26.1. For a review of IMRT treatment planning, see Bortfeld [320]. An alternative treatment modality is intensity-modulated proton therapy (IMPT). A key difference between proton and photon beams is that proton beams have a controllable range with a sharp increase in the dose deposition at the end of the range—the Bragg peak. The possibility of controlling the range of the protons provides an additional degree of freedom compared to photon beams. Disadvantages are that the proton equipment is more costly and that protons are more sensitive to errors. For a review of IMPT treatment planning, see Schwarz [1653].

Because the DNA of healthy cells is repaired to a higher extent than that of cancer cells, dividing the treatment into fractions increases the chance of complication-free tumor control. Thus, treatments are typically divided into 30–40 fractions. When the patient is positioned for treatment in a fraction, there is an inevitable risk that the patient position relative to the beams differs from what was planned. The conventional way of handling this uncertainty is to deliver a high dose to an enlarged target volume, while trying to protect enlarged regions encompassing the sensitive organs. Such a use

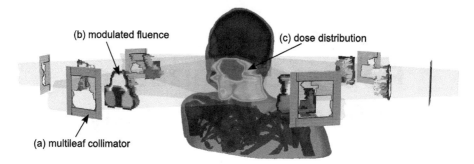

Figure 26.1. *An IMRT plan for a head and neck case. (a) Each beam is delivered as a sequence of beam profiles shaped by a multileaf collimator made of a blocking material. (b) The superposition of beam profiles enables modulation of the fluence from a single direction. (c) The superposition of doses from multiple beams yields a three-dimensional modulated dose* [760].

of margins of course aggravates the conflicts between the high- and low-dose regions. More elaborate methods that explicitly take the effects of possible errors into account during the treatment planning can yield plans that are robust to errors while delivering lower doses than margin-based plans. Many of these methods are based on robust optimization.

For photon-mediated IMRT, robust optimization methods are usually employed to dispense with margins and instead better exploit the fact that the patient geometry changes under and between treatment fractions: if the target doses in some fractions become too low, this can be compensated by higher doses in other fractions. This can lead to solutions that are almost as robust as margin-based plans, but that deliver lower total doses.

Proton beams are highly dependent on the density of the traversed medium. If the patient is slightly misaligned or the patient anatomy has changed between planning and delivery of the treatment, the delivered dose distribution might be deformed compared to the planned one. Because errors can result in deformed dose distributions, margins do not provide the intended robustness for IMPT plans. Methods for robust IMPT planning have therefore generally been oriented toward compensating for the lack of functioning margins by ensuring that the worst-case outcomes become as beneficial as possible. This can lead to solutions that are more robust than margin-based IMPT plans while at the same time delivering lower doses to healthy structures.

26.2 ▪ Treatment Planning and Optimization

The goal of treatment planning is to find plans that have high probability of curing the patients. Computed tomography (CT) images provide a three-dimensional representation of the patient geometry that guides the planning. The boundaries of the regions of interest (ROIs) are delineated on the CT images. The ROIs are typically the cancerous regions that are to be irradiated and the healthy organs—the organs at risk (OARs)—that are to be spared.

An IMRT treatment depends on tens of thousands of variables representing the settings of the treatment machine, such as settings for the collimating leaves that shape the beam profile, shown as (a) in Figure 26.1. Determining all of these manually would be impractical. Therefore, optimization is used to find settings that result in high-quality treatment. From the optimization variables, the resulting dose absorbed by the patient, measured in gray, is calculated. The dose is scored on the basis of an objective function that measures the quality of the dose. For dose computation and optimization to be made possible, the patient geometry is discretized into a grid of volume elements called voxels.

For brevity, the following exposition will concern a patient having one target, with voxels indexed by the set $\mathcal{T}$, and one OAR, with voxels indexed by the set $\mathcal{O}$. The set of all voxels in the irradiated region is denoted by $\mathcal{V}$.

26.3 ▪ Uncertainties

Among the many sources of uncertainty that can affect radiation therapy treatment are errors in the alignment of the patient during the CT image acquisition [1819], erroneously delineated ROIs [713], changes in the ROI geometries [341], errors in the positioning of the patient relative to the beams [969], and uncertainties in biological parameters such as the dose-response relations for various endpoints [1880]. The effects of motion on photon-mediated IMRT plans have been reviewed by Webb [1879], and the effects on IMPT plans by Lomax [1218].

To great generality, the uncertainties can be modeled as a random variable S picking an error scenario from the set $\S$ of possible scenarios. The scenario set must generally be discretized to become fit for use in tractable optimization problems, and it will henceforth be assumed that the number of elements in $\S$, denoted by $|\S|$, is finite.

The expected value of a random variable Y under a probability measure π is denoted by $\mathbb{E}_\pi[Y]$ and is given by

$$\mathbb{E}_\pi[Y] = \sum_{s \in \S} \pi_s y_s,$$

where y_s is the value that Y takes under scenario s (in the present case of finitely many scenarios, π is a probability mass function). The a priori probability measure, prescribing the historically measured or assumed probability to each scenario, is denoted by p. The standard deviation is only considered under p and is denoted by $\sigma(Y)$.

26.4 ▪ Scenario Doses

The dose distribution $d(x; s)$, illustrated by (c) in Figure 26.1, is a function of the optimization variable vector x and of the scenario s. All methods that are reviewed in the following section rely on a linear relationship between the dose distribution and the optimization variables, i.e.,

$$d(x; s) = D(s)x$$

for some matrix $D(s)$, which typically has a nonconvex dependency on s. The dose $d_i(x; s)$ to voxel i is then given by $d_i(x; s) = D_i(s)x$, where $D_i(s)$ is the ith row of $D(s)$. A common setting in which linearity holds is when the variable vector x represents a discretization of the beam fluences into area elements, illustrated by (b) in Figure 26.1. Each column of the matrix $D(s)$ is then the vector of voxel dose depositions that results when the fluence is set to unity in one of the beam fluence area elements. In IMPT, such a linear representation corresponds to deliverable machine settings, whereas in photon-mediated IMRT, the beam fluences are only indirectly controllable via superposition of multileaf collimator shapes; see (a) in Figure 26.1. A conversion step from beam fluences into deliverable machine settings is thus required when the linear representation is used.

26.5 ▪ Robust Optimization Methods Used in Treatment Planning

This section surveys robust optimization methods that have been used to generate robust treatment plans. The methods are presented in a distilled form that highlights their distinguishing features. Natural variations and extensions are possible, e.g., emphasizing the nominal scenario by including nominal dose deviation penalties in the objective, or emphasizing the reduction of total dose by including terms penalizing any dose to $\mathcal{V}$. Methods such as stochastic optimization have also been used to achieve robustness [1812].

26.5.1 ▪ Distributional Robustness via Linear Optimization

Lung cancer treatments are affected by intrafraction motion—patient motion that occurs during the delivery of the treatment [1126]. The effect on the dose of intrafraction motion is an averaging of the doses $d(x;s)$ delivered during different phases of the motion, modeled by the scenario s, weighted by the proportion of time, p_s, spent in each phase. The use of a margin that covers the position of the target over all phases can yield fully robust target coverage. However, some phases can be more favorable than others. If the optimization algorithm is informed about how the patient geometry changes over the phases, it can find a plan that yields a sufficient dose to the target while reducing the dose to healthy tissues.

Because intrafraction motion results in an averaging of the phase doses, optimizing an averaged dose distribution may seem advisable. Doing so, however, results in heterogeneous doses, the robustness of which is highly dependent on the averaging realized during the delivery of the treatment being the same as that considered during the planning [426, 759, 1812].

Chan et al. [426] address the problem that the time spent in each phase can differ from the times assumed during the treatment planning. They introduce a method in which the fraction of time spent in each phase is considered uncertain and only known to lie within some bounds. The fraction of time π_s that is spent in phase s is bounded by $a_s \leq \pi_s \leq b_s$, where the bounds are such that $0 \leq a_s \leq p_s \leq b_s \leq 1$ holds for the a priori distribution p. An illustration of this is shown in Figure 26.2. The uncertainty set $\mathcal{U}$ of possible fractions of time spent in the phases is defined by

$$\mathcal{U} = \left\{ \pi \in \mathbb{R}^{|S|} : a \leq \pi \leq b, \sum_{s \in \mathcal{S}} \pi_s = 1 \right\}.$$

They then minimize the total expected dose delivered under the a priori distribution p subject to constraints requiring the expected dose to each target voxel to exceed a reference dose level under all distributions π from $\mathcal{U}$. This is formulated mathematically as

$$\begin{aligned} \min_{x \geq 0} \quad & \sum_{i \in \mathcal{V}} \mathbb{E}_p[d_i(x;S)] \\ \text{s.t.} \quad & \mathbb{E}_\pi[d_i(x;S)] \geq \delta_i \quad \forall \pi \in \mathcal{U}, \quad \forall i \in \mathcal{T}. \end{aligned} \tag{26.1}$$

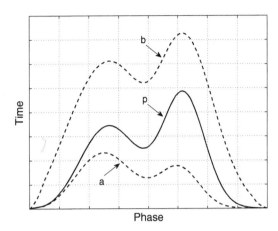

Figure 26.2. *An illustration of the a priori distribution p of time spent in each phase s from the set S and the bounds a and b on the true distribution π.*

This formulation introduces robustness by the constraints for all π in $\mathcal{U}$. When $a \neq b$, this problem has infinitely many constraints because $\mathcal{U}$ contains infinitely many points. However, the objective and the constraints are linear, and the uncertainty set $\mathcal{U}$ is polyhedral, so there exists a reformulation of (26.1) that is a linear optimization (LO) problem with finitely many constraints (see Ben-Tal and Nemirovskii [209] and Bertsimas and Sim [257]). To derive the tractable reformulation, the first observation to make is that for each $i \in \mathcal{T}$, if the constraint holds for the worst π in $\mathcal{U}$, then it holds for all of them. This yields the equivalence

$$\mathbb{E}_\pi[d_i(x;S)] \geq \delta_i \quad \forall \pi \in \mathcal{U} \quad \Leftrightarrow \quad \min_{\pi \in \mathcal{U}} \left\{ \sum_{s \in S} \pi_s d_i(x;s) \right\} \geq \delta_i, \tag{26.2}$$

where we use $\mathbb{E}_\pi[d_i(x;S)] = \sum_{s \in S} \pi_s d_i(x;s)$. By strong duality for LO, the value of the minimum in (26.2) equals the value of its dual

$$\max_{\lambda_i, \mu_i, \nu_i} \left\{ \begin{array}{l} \lambda_i + a^T \mu_i - b^T \nu_i : \\ \lambda_i + \mu_{i,s} - \nu_{i,s} \leq d_i(x;s) \quad \forall s \in S \\ \mu_i, \nu_i \geq 0 \end{array} \right\}, \tag{26.3}$$

where λ_i is a scalar, whereas μ_i and ν_i are vectors of length $|S|$. The value of any $\lambda_i + a^T \mu_i - b^T \nu_i$ satisfying the constraints in (26.3) is bounded from above by the minimum in (26.2), so the maximum operation can be discarded and the tractable counterpart of (26.1) can be formulated as

$$\begin{array}{ll} \min_{x, \lambda, \mu, \nu} & \sum_{i \in \mathcal{Y}} \mathbb{E}_p[d_i(x;S)] \\ \text{s.t.} & \lambda_i + a^T \mu_i - b^T \nu_i \geq \delta_i \quad \forall i \in \mathcal{T}, \\ & \lambda_i + \mu_{i,s} - \nu_{i,s} \leq d_i(x;s) \quad \forall s \in S, \quad \forall i \in \mathcal{T}, \\ & x, \mu, \nu \geq 0, \end{array} \tag{26.4}$$

which is an LO problem with finitely many constraints.

The plans resulting from this method provide target coverage under all distributions π in $\mathcal{U}$ of times spent in each phase. At the same time, they have been found to deliver a substantially lower total dose than plans achieved using margins [321, 426].

Chan et al. [427] generalize model (26.1). Instead of constraining each target voxel to receive the dose δ under all distributions π in $\mathcal{U}$, the conditional value-at-risk (CVaR) [99, 1574], or mean tail loss, of the dose distribution in the target volume is constrained to receive at least the dose δ. It should be noted that the CVaR is not used to achieve robustness, but rather as a means of controlling the dose distribution in the geometric structures of the patient. Two methods are then used to introduce robustness: either the expectation of the CVaR over the phases is constrained, or the CVaR of the expected dose over the phases is constrained. For these two methods, the expectation is taken with respect to an uncertain probability distribution π that is only known to lie within a set $\mathcal{U}$. The constraint in (26.1) is thus modified to, respectively,

$$\mathbb{E}_{\pi}\left[\text{CVaR}_{\beta}(d_{\mathcal{T}}(x;S))\right] \geq \delta \quad \forall \pi \in \mathcal{U} \text{ and}$$
$$\text{CVaR}_{\beta}\left(\mathbb{E}_{\pi}[d_{\mathcal{T}}(x;S)]\right) \geq \delta \quad \forall \pi \in \mathcal{U},$$

where $d_{\mathcal{T}}$ is the vector of target voxel doses and CVaR_{β} is given by

$$\text{CVaR}_{\beta}(d_{\mathcal{T}}) = \min_{\zeta}\left\{\zeta - (1-\beta)^{-1}|d_{\mathcal{T}}|^{-1}\sum_{d \in d_{\mathcal{T}}}(d - \zeta)_{+}\right\},$$

where $(\cdot)_{+} = \max\{\cdot, 0\}$. Moreover, upper CVaR constraints are introduced for the target as well as the OARs. Similarly to the derivation of (26.4), strong duality for LO can be used to derive computationally tractable LO problems with robust CVaR constraints.

Chan and Mišić [429] and Mar and Chan [1268] generalize the model (26.1) in another direction, viz., to combine robust and adaptive radiation therapy. In adaptive radiation therapy, information regarding the uncertainties is acquired during the treatment. After a number of fractions, the treatment plan can be reoptimized, with the effects of errors that have occurred taken into account. When robust optimization is combined with adaptive planning, the acquired information about the uncertainties is used to reoptimize the plan with an updated uncertainty set $\mathcal{U}$. The updating rule from Chan and Mišić [429] is given by

$$\mathcal{U}_{k} \leftarrow (1-\alpha)\mathcal{U}_{k-1} + \alpha p_{k-1},$$

where $\mathcal{U}_{k}$ is the uncertainty set in fraction k and p_{k} is the probability distribution observed in fraction k. The parameter α specifies how fast the method should adapt to new information. The combination of robustness and adaptive planning was shown to substantially improve plan quality compared to robust optimization without adaptation in the presence of well-behaved breathing motion [429] as well as less well-behaved patterns with drift [1268].

26.5.2 ▪ Probabilistic Robustness via Second-Order Cone Optimization

During the course of a radiation therapy treatment, the patient geometry changes. The interfraction motion—the geometric changes occurring between treatment fractions—occurs because the patient might be positioned differently during each fraction, and the patient anatomy changes due to organ motion, tumor shrinkage, and weight loss.

The total dose delivered to the patient is the sum of n fraction doses $d(x;S_{j})$ for $j = 1,\ldots,n$, where the random variable S_{j} selects the scenario in fraction j. Under the assumption that $S_{1},\ldots,S_{n}$ are independent and identically distributed, distributed like S, and using the fact that the number of fractions n is fairly large (typically 30–40), the central limit theorem implies that the total dose to a given voxel is approximately normally distributed.

An approach that has been used to find plans that are robust to interfraction motion optimizes toward high probability that the dose to each voxel, considered individually, is above or below the desired dose level. For an OAR voxel i, achieving a low probability ϵ that the total dose exceeds some dose level $\delta^{\mathcal{O}}$ can be formulated as the individual chance constraint

$$\mathbb{P}\left(\sum_{j=1}^{n} d_i(x; S_j) > \delta^{\mathcal{O}}\right) \leq \epsilon \tag{26.5}$$

(the true goal of a joint chance constraint over $\mathcal{O}$ has not yet been considered in a rigorous setting, but preliminary work is given in Fredriksson et al. [763]). When the total dose $\sum_{j=1}^{n} d_i(x; S_j)$ is assumed to be normally distributed, it is fully characterized by its expected value $\mu = n\mathbb{E}_p[d_i(x;S)]$ and standard deviation $\sigma = \sqrt{n}\sigma(d_i(x;S))$, and (26.5) is equivalent to

$$\mathbb{P}\left(Z > (\delta^{\mathcal{O}} - \mu)/\sigma\right) \leq \epsilon \quad \Leftrightarrow \quad (\delta^{\mathcal{O}} - \mu)/\sigma \geq z_{1-\epsilon},$$

where Z denotes a standard normal deviate and $z_{1-\epsilon}$ is chosen such that $\mathbb{P}(Z > z_{1-\epsilon}) = \epsilon$. This expression and a similar one requiring a target voxel to achieve a lower probability than ϵ of falling below the level $\delta^{\mathcal{T}\text{low}}$ are equivalent to, respectively,

$$\mu + z_{1-\epsilon}\sigma \leq \delta^{\mathcal{O}} \quad \text{and} \quad \mu - z_{1-\epsilon}\sigma \geq \delta^{\mathcal{T}\text{low}}. \tag{26.6}$$

The left-hand sides of these expressions can be used to define modified dose distributions

$$\begin{aligned} d_i^{\text{high}}(x) &= n\mathbb{E}_p[d_i(x;S)] + z_{1-\epsilon}\sqrt{n}\sigma(d_i(x;S)), \\ d_i^{\text{low}}(x) &= \left(n\mathbb{E}_p[d_i(x;S)] - z_{1-\epsilon}\sqrt{n}\sigma(d_i(x;S))\right)_+, \end{aligned} \tag{26.7}$$

where $n\mathbb{E}_p[d_i(x;S)]$ and $\sqrt{n}\sigma(d_i(x;S))$ have been substituted for μ and σ, respectively, and the positive part is taken for $d_i^{\text{low}}(x)$ because dose cannot be negative.

Chu et al. [469] relax the constraints (26.6) into objectives by penalizing the worst violation of these constraints over the voxels for each structure. They solve the problem

$$\begin{aligned} \min_{x \geq 0} \quad & w^{\mathcal{T}\text{low}} \max_{i \in \mathcal{T}}\left\{\left(\delta^{\mathcal{T}\text{low}} - d_i^{\text{low}}(x)\right)_+\right\} \\ & + w^{\mathcal{T}\text{high}} \max_{i \in \mathcal{T}}\left\{\left(d_i^{\text{high}}(x) - \delta^{\mathcal{T}\text{high}}\right)_+\right\} \\ & + \sum_{s \in \mathbb{S}} w_s^{\mathcal{T}} \max_{i \in \mathcal{T}}\left\{\left(\delta_s^{\mathcal{T}} - d_i(x;s)\right)_+\right\} \\ & + w^{\mathcal{O}} \max_{i \in \mathcal{O}}\left\{\left(d_i^{\text{high}}(x) - \delta^{\mathcal{O}}\right)_+\right\}, \end{aligned} \tag{26.8}$$

where $w^{\mathcal{T}\text{low}}$, $w^{\mathcal{T}\text{high}}$, $w_s^{\mathcal{T}}$, and $w^{\mathcal{O}}$ are penalty weights for failing to reach the minimum target dose level $\delta^{\mathcal{T}\text{low}}$, for exceeding the maximum target dose level $\delta^{\mathcal{T}\text{high}}$, for failing to reach the minimum target dose level $\delta_s^{\mathcal{T}}$ in scenario s, and for exceeding the maximum OAR dose level $\delta^{\mathcal{O}}$, respectively. The nonlinear optimization (NLO) problem (26.8) can be rearranged into a second-order cone optimization (SOCO) problem, which is more practical to solve: the maxima and positive parts in (26.8) can be avoided by the introduction of auxiliary linear variables and constraints; see, e.g., Vanderbei [1823]. The modified dose distributions contain standard deviation terms that require second-order cone (SOC) constraints. Substituting auxiliary variables y^{low} and y^{high} for $d_i^{\text{low}}(x)$ and $d_i^{\text{high}}(x)$, combined with constraints of the form $y_i^{\text{low}} \le d_i^{\text{low}}(x)$ and $y_i^{\text{high}} \ge d_i^{\text{high}}(x)$, reduces the objective to a linear function. The constraints are equivalent to

$$z_{1-\epsilon}\sqrt{n}\|RA_i x\|_2 \le n\mathbb{E}_p[d_i(x;S)] - y_i^{\text{low}} \text{ and}$$

$$z_{1-\epsilon}\sqrt{n}\|RA_i x\|_2 \le y_i^{\text{high}} - n\mathbb{E}_p[d_i(x;S)],$$

where the matrix A_i consists of the rows $D_i(s)$ of the dose matrix for all scenarios s in S and $R = P^{1/2}(I - ep^T)$, in which $P = \text{diag}(p)$, $I \in \mathbb{R}^{|S| \times |S|}$ is the identity matrix, and $e \in \mathbb{R}^{|S|}$ is the vector of all ones. See Chu et al. [469] for more details concerning the reformulation of (26.8) into a SOCO problem.

The formulation of the SOC constraints indicates a relationship to robust optimization with ellipsoidal uncertainty sets (see Ben-Tal and Nemirovski [209]). Indeed, the problem may be interpreted in terms of robust LO. The dose to voxel i is then modeled to be given by $a_i^T x$ for some uncertain vector a_i from the ellipsoidal uncertainty set $\mathcal{U}_i$, given by

$$\mathcal{U}_i = \left\{ n\mathbb{E}_p[D_i(S)] + W_i u_i : u_i^T u_i \le 1 \right\}.$$

For each voxel i from the set $\mathcal{O}$ of OAR voxels, the problem is subject to constraints

$$a_i^T x \le \delta^{\mathcal{O}} \quad \forall a_i \in \mathcal{U}_i \quad \Leftrightarrow \quad n\mathbb{E}_p[d(x;S)] + \max_{u:u^T u \le 1}\left\{u^T W_i^T x\right\} \le \delta^{\mathcal{O}}.$$

The maximum over u equals $\|W_i^T x\|_2$. Thus, when $z_{1-\epsilon}\sqrt{n}RA_i$ has been substituted for W_i^T, this constraint is seen to be equivalent to (26.6).

Ólafsson and Wright [1421] use a model similar to (26.8) but consider uncertainty in the Monte Carlo dose computation in addition to organ motion. They solve the problem

$$\begin{aligned}
\min_{x \ge 0} \quad & \sum_{i \in \mathcal{T}} w_i^{\mathcal{T}\text{low}}\left(\delta_i^{\mathcal{T}} - n\mathbb{E}_p[d_i(x;S)]\right)_+ \\
& + \sum_{i \in \mathcal{T}} w_i^{\mathcal{T}\text{high}}\left(n\mathbb{E}_p[d_i(x;S)] - \delta_i^{\mathcal{T}}\right)_+ \\
& + \sum_{i \in \mathcal{O}} w_i^{\mathcal{O}}\left(d_i^{\text{high}}(x) - \delta_i^{\mathcal{O}}\right)_+ \\
\text{s.t.} \quad & d_i^{\text{low}}(x) \ge \delta_i^{\mathcal{T}\text{low}} && \forall i \in \mathcal{T}, \\
& d_i^{\text{high}}(x) \le \delta_i^{\mathcal{T}\text{high}} && \forall i \in \mathcal{T}.
\end{aligned} \qquad (26.9)$$

Like (26.8), this problem can be equivalently formulated as an SOCO problem.

The plans optimized with probabilistic robustness requirements were found to provide similar robustness with respect to target coverage as planning with margins (when evaluated against the assumed probability distribution p), while reducing the total dose and the doses to the OARs [469, 1421].

26.5.3 ▪ Worst-Case Robustness via NLO

For margins to provide robustness against errors, the effects of the errors must be well approximated by rigid translations of the dose distribution. Otherwise, there is

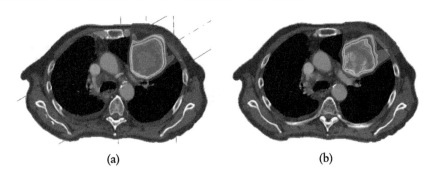

(a)   (b)

Figure 26.3. *Plan optimized to deliver dose to an enlarged target region. (a) The planned dose distribution covers the target. (b) The dose distribution resulting when the patient is misaligned exhibits a too-low dose to the target.*

no guarantee that the dose distribution will cover the target when an error has oc-curred. The doses of IMPT treatments often deform drastically as a result of errors. Figure 26.3 shows the effects of a setup error on the dose distribution of a plan reached by margin-based planning. This sensitivity to errors has motivated the development of methods that generate plans that are robust in the sense intended by margin-based planning. Such methods do not exploit intra- or interfraction motions, like the meth-ods described in Sections 26.5.1 and 26.5.2. Instead, they aim for target coverage and OAR sparing under the, say, 95% most probable error scenarios that can occur, even if the same error occurs in each fraction. Similarly, margins are specified to provide a given probability of target coverage.

Fredriksson et al. [762] consider a nonlinear treatment plan optimization problem and minimize the penalty of the objective function under the worst-case scenario. The problem is formulated as a minimax problem of the form

$$\min_{x \geq 0} \quad \max_{s \in \S} \left\{ w^{\mathcal{T}} f^{\mathcal{T}}(d(x;s)) + w^{\mathcal{O}} f^{\mathcal{O}}(d(x;s)) \right\}, \qquad (26.10)$$

where $f^{\mathcal{T}}$ and $f^{\mathcal{O}}$ are nonlinear functions penalizing deviation from the planning goals of the target and OAR doses, respectively, such as those described by Oelfke and Bort-feld [1413]. The maximum in the objective is not continuously differentiable, which can result in problems with convergence for gradient-based optimization solvers. This disadvantage is alleviated when the problem is posed in epigraph form, i.e., when an auxiliary variable λ is substituted for the objective, and this variable is constrained by $\lambda \geq w^{\mathcal{T}} f^{\mathcal{T}}(d(x;s)) + w^{\mathcal{O}} f^{\mathcal{O}}(d(x;s))$ for all scenarios s in $\S$. Chen et al. [447] use a similar method in a multiobjective optimization framework but apply the maximum over $\S$ to each objective constituent $f^{\mathcal{T}}$ and $f^{\mathcal{O}}$ individually. In Fredriksson [759], a formulation similar to (26.10) is used, but instead of minimax, distributional robust-ness, including CVaR, is used and the framework is extended to also take into account random setup errors with uncertain probability distribution.

It has been empirically found that the worst scenarios are typically at the bound-ary of the uncertainty region [413, 431, 759], i.e., that the worst scenarios are those corresponding to the largest deviations from the planning image. This observation can be used to reduce the computational cost of solving (26.10) by removing nonboundary scenarios from $\S$.

Minimizing the worst-case penalty has been found to result in more robust target coverage than heuristic methods using margins, uniform beam doses, and planning

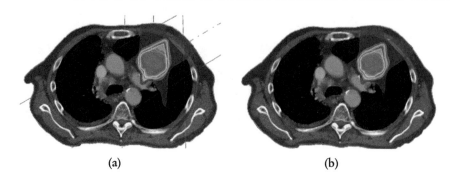

(a)   (b)

Figure 26.4. *Plan optimized to deliver dose to the target in the worst-case scenario. (a) The planned dose distribution covers the target. (b) The dose distribution resulting when the patient is misaligned still covers the target.*

with the densities in the target region overridden, while at the same time yielding lower doses to healthy structures [762]. Figure 26.4 shows the effects of a setup error on the dose distribution of a plan reached by worst-case optimization according to (26.10).

26.5.4 ▪ Voxelwise Worst-Case Robustness via LO and NLO

The concept of worst-case dose distributions is introduced by Lomax et al. [1219] as a tool to evaluate the robustness of IMPT plans. They define modified dose distributions d^{low} and d^{high} according to

$$d_i^{\text{low}}(x) = \min_{s\in\S}\{d_i(x;s)\} \quad \text{and} \quad d_i^{\text{high}}(x) = \max_{s\in\S}\{d_i(x;s)\}.$$

How much a treatment plan might deteriorate is bounded by how well d^{low} meets the minimum target dose goals and how well d^{high} meets the maximum OAR dose goals. Optimization of the worst-case dose distributions was first considered by Unkelbach et al. [1811] and later by Pflugfelder et al. [1471]. The optimization problem takes the form

$$\min_{x\geq 0} \quad w^{\mathcal{T}\text{low}} f^{\mathcal{T}\text{low}}(d^{\text{low}}(x)) + w^{\mathcal{T}\text{high}} f^{\mathcal{T}\text{high}}(d^{\text{high}}(x)) + w^{\mathcal{O}} f^{\mathcal{O}}(d^{\text{high}}(x)), \quad (26.11)$$

where $f^{\mathcal{T}\text{low}}$ penalizes deviations from the desired minimum target dose, while $f^{\mathcal{T}\text{high}}$ and $f^{\mathcal{O}}$ penalize deviations from the desired maximum target and OAR doses, respectively. For the problem to be convex, $f^{\mathcal{T}\text{low}}$ should be nonincreasing and convex, whereas $f^{\mathcal{T}\text{high}}$ and $f^{\mathcal{O}}$ should be nondecreasing and convex, (see, e.g., Boyd and Vandenberghe [332, Section 3.2.4]).

Unkelbach et al. [1811] solve (26.11) in the form of a robust LO problem. They consider errors that are independent per beam, which results in a scenario set $\S$ of exponential size in the number of beams. By using LO duality in a way similar to transforming (26.1) into (26.4), they reduce the number of constraints used to specify d^{low} and d^{high} from an exponential to a polynomial number.

Pflugfelder et al. [1471] optimize the worst-case dose distributions in an NLO setting. They apply the penalty functions to d^{low} and d^{high} directly, without introducing auxiliary variables and constraints to alleviate the discontinuous gradients that result from the minima and maxima. These discontinuities have not been observed to pose problems in practice. A plausible explanation is that the number of voxels is generally large, so when one voxel is affected by a discontinuity, many others are not.

The nonlinear functions $f^{\mathcal{T}\text{high}}$, $f^{\mathcal{T}\text{low}}$, and $f^{\mathcal{O}}$ that have been employed in the literature often penalize dose deviations linearly or quadratically, i.e., are of the form

$$f^{\mathcal{T}\text{low}}(d^{\text{low}}) = \sum_{i\in\mathcal{T}}\left(\delta^{\mathcal{T}\text{low}} - d_i^{\text{low}}\right)_+^q \quad \text{and} \quad f^{\mathcal{O}}(d^{\text{high}}) = \sum_{i\in\mathcal{O}}\left(d_i^{\text{high}} - \delta^{\mathcal{O}\text{high}}\right)_+^q,$$

where q is 1 or 2 (and $f^{\mathcal{T}\text{high}}$ has the same form as $f^{\mathcal{O}}$, but for the target). This hints at the close relationship between the voxelwise worst-case formulation (26.11) and the probabilistic formulations (26.8) and (26.9). The main difference between these formulations lies in the choice of risk measure used to define the modified dose distributions d^{low} and d^{high}: the probabilistic formulations use mean-deviation risk measures, whereas the voxelwise worst-case formulation uses a max-norm risk measure. See Ruszczyński and Shapiro [1607] for details concerning general optimization of risk measures.

Optimization of the worst-case dose distributions has been found to result in more robust target coverage and higher doses to healthy structures than optimization of the nominal dose distribution (without a margin) [1471, 1811]. When compared to conventional planning with margins, it resulted not only in more robust target coverage, but also in lower doses to healthy structures [1207]. The worst-case dose distribution has, however, been found to be overly conservative for robustness evaluation and for optimization under certain dose criteria [413,761].

26.6 ▪ Conclusion

Robust optimization has great potential for improving radiation therapy treatments. For photon-mediated IMRT, robust optimization methods like those in Sections 26.5.1 and 26.5.2 can yield treatment plans that exploit the structure of the changes in patient geometry during the treatment. Although margin-based plans can provide more robust target coverage, they do so by extending the high-dose region, which is arguably a crude means of increasing robustness. The cost of doing so is exhibited by the robust optimization methods, which provide almost as robust target coverage but reduce the OAR and total doses substantially.

For IMPT, where margins are often insufficient to generate robust plans, more advanced methods are indispensable. Here, robust optimization not only provides lower doses to the healthy structures but moreover provides more robust target coverage than conventional methods. The robust optimization methods that have been applied to IMPT, including those in Sections 26.5.3 and 26.5.4, have been constructed to achieve the type of robustness that margins are intended to provide. There is likely even more to be gained by methods that take into account that patient anatomy changes during the treatment.

In conclusion, the success of robust optimization methods in radiation therapy points to the benefits of (a) including additional information during the optimization and (b) utilizing the information in relevant ways. Still, robust optimization of radiation therapy treatment plans remains an active area of research; there is more information to be utilized and further ways to do so.

Chapter 27

Robust Optimization in Electric Power Systems

X. Andy Sun

27.1 ▪ Introduction

Electric power systems, acclaimed as the "supreme engineering achievement of the 20th century," are one of the most complex human-made systems [505]. As an example, the U.S. power grid has approximately 170,000 miles of high-voltage (voltage at or above 200 kV) transmission lines and almost 5,000 generating units with capacity of at least 50 MW [1405]. In an engineering system of such scale and complexity, uncertainty abounds. More specifically, uncertainties in generation, consumption, and unexpected failure of electric equipment have to be carefully considered in the operation of power systems. It is amazing that the reliability and security of power supply is held to such a high standard that flipping on a switch and expecting the lights to shine at any time has almost become a part of the subconscious of modern society. This is a great achievement, made possible by the sound engineering design and the tremendous efforts of the electricity industry.

However, it becomes increasingly challenging to uphold such a high standard of reliability, as power systems around the world experience fast and fundamental change. In particular, the output of generators, which traditionally has been very predictable, is becoming more and more difficult to predict. The key driving force behind this change is the large-scale integration of wind and solar power generation into the power grids, both of which are highly intermittent, correlated in time and space, and stochastic in nature [1336]. At the same time, the demand side is also becoming more and more intelligent and responsive, with smart-grid technologies enabling electricity consumers to change their consumption in real time. All of these changes have significant implications on power systems at multiple levels.

Figure 27.1 shows a picture of the major decision-making problems involved in power systems operation. The problems are categorized from a temporal perspective into real-time operations (minute-to-minute dispatch), daily scheduling (day-ahead commitment), mid-term maintenance planning (seasonal or annual schedules), and long-term investment planning for generation and transmission systems [1466]. In an

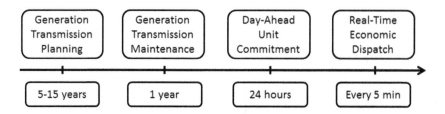

Figure 27.1. *Major decision-making problems in power systems operation. The time arrow points to the start of real-time operation. The typical decision horizon of each problem is given under each problem.*

electricity market environment, the independent system operator (ISO) is responsible for scheduling dispatch and daily commitment, as well as coordinating maintenance schedules between generation and transmission owners [1671]. The ISO would also conduct a long-term planning study and make technical recommendations for generation and transmission expansion. At the same time, all of these decision-making problems involve numerous stakeholders and affect millions of consumers. Mathematical optimization is heavily relied upon for reaching consistent, efficient, and optimal outcomes. It is not an exaggeration to say that modern power systems are built on rigorous and sound optimization models and efficient solution algorithms.

Traditionally, the above decision-making problems are solved by deterministic optimization models [1902]. That is, the optimization models consider a problem of fixed and known parameters. For example, in the short-term, economic dispatch (ED) and unit commitment (UC), the generation capabilities, the exact demand at each of the future periods, and the conditions of the transmission lines and generators are assumed to be known. The resulting solutions are therefore feasible and optimal for the planned or forecast scenario. Such a deterministic approach has been successful at maintaining power system reliability and security under the traditional conditions. However, as the uncertainties in power systems multiply due to the growing penetration of renewable resources, traditional approaches are becoming inadequate for multilevel decision-making problems. The industry is actively seeking new approaches to dealing with the growing uncertainties (see, e.g., [253]).

Recently, robust optimization has been developed into a rich and practical methodology for decision making under uncertainty [203]. In the past few years, a flurry of research activity has introduced robust optimization into the field of power systems [253, 1001, 1003, 1967]. In a short amount of time, interesting models have been proposed and promising results have been shown to demonstrate the power of robust optimization in almost every category of the decision problems depicted in Figure 27.1. The industry has taken the lead in supporting and pursuing collaborations with academic researchers. The present chapter aims to give a concise overview of the recent advances in the field, however, the review is bound to be incomplete due to the fast growth of this area.

This chapter first focuses on the UC problem in the day-ahead operation, which is a building block for many other decision-making problems in power systems; then we discuss real-time and long-term planning. In particular, Section 27.2 introduces the two-stage robust optimization based UC model, solution methods, and some computational results. Section 27.3 presents several extensions to this fundamental model. Section 27.4 discusses some recent progress on robust optimization models for real-time operation and long-term planning. Section 27.5 closes the chapter with some discussion of future directions.

27.2 ▪ Two-Stage Adaptive Robust Optimization for the Security-Constrained Unit Commitment Problem

In this section, we first present the deterministic security-constrained unit commitment (SCUC) model, which is used by most of the system operators for day-ahead scheduling. Then, a fundamental SCUC model based on two-stage adaptive robust optimization is discussed.

27.2.1 ▪ Deterministic Security-Constrained Unit Commitment

UC is a process of determining the on and off status of generation units (primarily thermal units, i.e., coal, nuclear, geothermal, and natural gas power plants) and their production levels for next-day operation. The scheduling horizon is usually 24 hours or 36 hours, with an hourly interval. The on and off decision is called the UC decision. The production level is called the dispatch decision. The production of the generation units is scheduled to meet the forecast demand in the power network, satisfying various physical and operational constraints. To ensure a certain level of security, the system operator usually requires the UC and dispatch solutions to be feasible for any one failure of a generation unit and/or a transmission line. Such constraints are called $N-1$ security constraints. A typical deterministic SCUC model with transmission security constraints is presented below [1739]:

$$\min_{x,u,v,p} \sum_{t \in \mathcal{T}} \sum_{i \in \mathcal{G}} \left(f_i^t(x_i^t, u_i^t, v_i^t) + c_i^t(p_i^t) \right) \tag{27.1a}$$

$$\text{s.t.} \quad x_i^{t-1} - x_i^t + u_i^t \geq 0 \quad \forall i \in \mathcal{G}, t \in \mathcal{T}, \tag{27.1b}$$

$$x_i^t - x_i^{t-1} + v_i^t \geq 0 \quad \forall i \in \mathcal{G}, t \in \mathcal{T}, \tag{27.1c}$$

$$x_i^t - x_i^{t-1} \leq x_i^\tau \quad \forall \tau \in [t+1, \min\{t + \text{MinUp}_i - 1, T\}],$$
$$t \in [2, T], i \in \mathcal{G}, \tag{27.1d}$$

$$x_i^{t-1} - x_i^t \leq 1 - x_i^\tau \quad \forall \tau \in [t+1, \min\{t + \text{MinDw}_i - 1, T\}],$$
$$t \in [2, T], i \in \mathcal{G}, \tag{27.1e}$$

$$\sum_{i \in \mathcal{G}} p_i^t = \sum_{j \in \mathcal{D}} \bar{d}_j^t \quad \forall t \in \mathcal{T}, \tag{27.1f}$$

$$p_i^t - p_i^{t-1} \leq RU_i x_i^{t-1} + SU_i u_i^t \quad \forall i \in \mathcal{G}, t \in \mathcal{T}, \tag{27.1g}$$

$$p_i^{t-1} - p_i^t \leq RD_i x_i^t + SD_i v_i^t \quad \forall i \in \mathcal{G}, t \in \mathcal{T}, \tag{27.1h}$$

$$-f_{l,k}^{\max} \leq a_{l,k}^\top(p^t - d^t) \leq f_{l,k}^{\max} \quad \forall t \in \mathcal{T}, l \in \mathcal{C}_k, k \in \mathcal{L}, \tag{27.1i}$$

$$p_i^{\min} x_i^t \leq p_i^t \leq p_i^{\max} x_i^t \quad \forall i \in \mathcal{G}, t \in \mathcal{T}, \tag{27.1j}$$

$$x_i^t, u_i^t, v_i^t \in \{0,1\} \quad \forall i \in \mathcal{G}, t \in \mathcal{T}. \tag{27.1k}$$

The UC decisions include binary variables x_i^t, u_i^t, v_i^t, where $x_i^t = 1$ if generator i is on at time t, and $x_i^t = 0$ otherwise; $u_i^t = 1$ if generator i is turned on from the off state at time t; and $v_i^t = 1$ if generator i is turned off at time t. The dispatch decision is p_i^t of generator $i \in \mathcal{G}$ at time $t \in \mathcal{T}$, where $\mathcal{G}$ is the set of generators and $\mathcal{T}$ is the set of time periods in the decision horizon.

The fixed cost $f_i^t(x_i^t, u_i^t, v_i^t)$ of each generator includes start-up and shut-down costs and other fixed costs. The variable cost $c_i^t(p_i^t)$ is usually approximated by a convex piecewise-linear function of the active power output p_i^t. The forecast demand $\bar{d}_j^t$ is the load at bus j, time t. Constraints (27.1b) and (27.1c) represent logic relations between on and off status and the turn-on and turn-off actions. Constraints (27.1d) and (27.1e) restrict the minimum up and down times for each generator. Constraint (27.1f) enforces systemwide energy balance in each time period. Constraints (27.1g)–(27.1h) limit the rate of production changes over a single period, where RU_i and RD_i are limits for ramp-up and ramp-down rates when the generator is already running, and SU_i and SD_i are ramping limits when generator i is just starting up and shutting down. Constraint (27.1i) expresses the power flow in the transmission lines as a linear function of power production and load in the entire system, where the coefficients of the linear function, $a_{l,k}$, are called the shift factors of line l, and the index k represents the kth contingency, i.e., when the line k is tripped offline, $\mathcal{C}_k$ is the set of remaining lines. Constraint (27.1j) represents the physical limits on the production level of each generator.

The above SCUC model (27.1) is the basic formulation for a UC model. An important aspect that is not presented in (27.1) is the system reserve and associated constraints [154]. The reliability of the power supply is so critical for the function of modern society that system operators pay extremely careful attention to ensure the system has enough generation capacity. The industry practice is to use so-called reserves, generation resources that are online or can be quickly brought online to respond to any demand surge or generation outages. The systemwide reserve requirement is usually predetermined by certain rules, such as a percentage of the forecasted peak load plus the largest online generator's capacity. Then, the reserve levels of individual generators are cooptimized with the UC and dispatch decisions [271,272,828]. In short, reserve is an engineering way to cope with uncertainties. In the following, we will present robust optimization–based SCUC models, which in a sense rigorously quantify uncertainties and replace or reduce the ad hoc reserve requirement.

27.2.2 ▪ Adaptive Robust UC Model with Net Load Uncertainty

As discussed in Section 27.1, the main sources of uncertainties in the day-ahead UC include demand uncertainty, renewable generation uncertainty, and unexpected failures of transmission lines and generators. In this subsection, we first present a fundamental robust optimization model that considers the uncertainty in the net load, where the net load is demand minus renewable generation. Then, we present several generalizations. Consider the following robust UC model [253, 1001, 1003, 1967]:

$$\min_{\mathbf{x}} \left\{ c(\mathbf{x}) + \max_{\mathbf{d} \in \mathcal{D}} \min_{\mathbf{p} \in \Omega(\mathbf{x},\mathbf{d})} b(\mathbf{p}) \right\} \tag{27.2}$$

$$\text{s.t. } \mathbf{x} \in \mathcal{F},$$

where $\mathbf{x}$ is the vector of commitment-related decisions, $\mathbf{p}$ is the vector of dispatch variables, and $\mathbf{d}$ is the vector of net load in the network; $\mathcal{D}$ is a set that describes the region of possible net load; and $\Omega(\mathbf{x},\mathbf{d})$ is the feasible region of the dispatch problem, defined by (27.1f)–(27.1j). More compactly, we can write it as $\Omega(\mathbf{x},\mathbf{d}) = \{\mathbf{p} : \mathbf{Hp}+\mathbf{Ed} \leq \mathbf{h}, \mathbf{Ax}+\mathbf{Bp} \leq \mathbf{g}\}$, where the first linear inequality represents the constraints (27.1f) and (27.1i), involving the dispatch $\mathbf{p}$ and demand $\mathbf{d}$, and the second linear inequality represents the constraints (27.1g), (27.1h), and (27.1j). The set $\mathcal{F}$ is the feasible region of the commitment decisions $\mathbf{x}$, which is defined by (27.1b)–(27.1e).

Formulation (27.2) is a two-stage fully adaptive robust optimization model. The commitment decision $\mathbf{x}$ is the first-stage decision, made before the realization of the uncertain net load $\mathbf{d}$, whereas the dispatch decision $\mathbf{p}$ is the second-stage decision made to respond to each specific realization of **d**. That is, the dispatch solution **p**(**b**) as a function of **b** fully adapts to any **b**. The solution of (27.2) is a UC decision **x** that is feasible, and therefore robust, for any possible realization of the uncertain net load **d**.

The *uncertainty set* $\mathscr{D}$ describes the ranges of possible net load **d**. We can also impose some constraints on the total variations of the uncertainty in $\mathscr{D}$ and use it to control the conservativeness of the robust model. The following so-called budgeted uncertainty set is widely used in the literature. For each time t,

$$\mathscr{D}^t(\bar{\mathbf{d}}^t, \hat{\mathbf{d}}^t, \Delta^t) := \left\{ \mathbf{d}^t : \sum_{i \in \mathcal{N}_d} \frac{|d_i^t - \bar{d}_i^t|}{\hat{d}_i^t} \leq \Delta^t, d_i^t \in [\bar{d}_i^t - \hat{d}_i^t, \bar{d}_i^t + \hat{d}_i^t] \forall i \right\}, \quad (27.3)$$

where $\mathcal{N}_d$ is the set of nodes that have uncertain net load, $\mathbf{d}^t = (d_i^t, i \in \mathcal{N}_d)$ is the vector of uncertain net load at time t, $\bar{d}_i^t$ is the nominal value of the net load of node i at time t, $\hat{d}_i^t$ is the deviation from the nominal net load, and the interval $[\bar{d}_i^t - \hat{d}_i^t, \bar{d}_i^t + \hat{d}_i^t]$ is the range of the uncertain d_i^t. The inequality in (27.3) controls the deviation of total net load from the nominal value. The parameter Δ^t is the "budget of uncertainty." With $\Delta^t = 0$, the uncertainty set $\mathscr{D}^t = \{\mathbf{d}^t\}$ is a singleton, corresponding to the nominal deterministic case. As Δ^t increases, the size of the uncertainty set $\mathscr{D}^t$ enlarges. This means that larger total deviation from the expected net load is considered, so that the resulting robust UC solutions are more conservative and the system is protected against a higher degree of uncertainty. With $\Delta^t = N_d$, where N_d is the total number of uncertain net loads, $\mathscr{D}^t$ equals the entire hypercube defined by the intervals for each d_i^t.

The uncertainty set in (27.3) is defined independently for each time period. A budget constraint over all the time periods can also be added to limit the variation of net loads over the entire planning horizon. The basic structure of the budgeted uncertainty sets remains the same. In Section 27.4, we will introduce another type of uncertainty sets that model the dynamics of uncertainty parameters between time periods and locations.

27.2.3 • Solution Method to Solve the Adaptive Robust Model

The two-stage robust optimization model (27.2) is general and fundamental. The complexity of solving such a model comes from two sources. One is the discrete nature of the first-stage decision **x**. The other is the max-min structure in the second stage. The latter is more fundamental in the sense that even if the first-stage decision is continuous with convex region $\mathscr{F}$, the second-stage problem can still be computationally challenging. To see this, we can reformulate the second-stage max-min problem by the strong duality of linear optimization (LO), assuming linear cost functions:

$$\max_{\mathbf{d} \in \mathscr{D}} \min_{\mathbf{p}} \left\{ \mathbf{b}^\top \mathbf{p} : H\mathbf{p} + E\mathbf{d} \leq \mathbf{h}, A\mathbf{x} + B\mathbf{p} \leq \mathbf{g} \right\}$$

$$= \max_{\mathbf{d}, \mu, \eta} \ \mathbf{d}^\top E^\top \mu - \mathbf{h}^\top \mu - (\mathbf{g} - A\mathbf{x})^\top \eta \quad (27.4a)$$

$$\text{s.t. } H^\top \mu + B^\top \eta = \mathbf{b}, \ \mu, \eta \geq 0, \quad (27.4b)$$

$$\mathbf{d} \in \mathscr{D}. \quad (27.4c)$$

Notice that the objective function (27.4a) has a bilinear term $\mathbf{d}^\top E^\top \mu$, and the feasible region is composed of two separate polyhedrons for (μ, η) and **d**, respectively. By the strong duality of LO, the above procedure can also be reversed so that a bilinear optimization problem with two separate polyhedral feasible regions can be reformulated as a max-min problem as in the second stage of (27.2). In general, this type of bilinear optimization problem is NP-hard to solve [195], which indicates that the second-stage problem of (27.2) is computationally challenging, independent of the first-stage problem.

However, it is interesting to note that the optimal objective value of the max-min problem (27.4), denoted as $R(\mathbf{x})$, is a convex function of the first-stage decision $\mathbf{x}$. This suggests that, modulo the complexity of the second-stage problem, the overall two-stage problem (27.2) may be reasonably solvable by a Benders decomposition type algorithm, which is developed in [253, 1001, 1003].

Another key property of the second-stage problem is that the worst-case net load is always an extreme point of the uncertainty set $\mathscr{D}$. This follows from the well-known property of bilinear optimization problems with separate polyhedron sets [1094]. With this observation, the two-stage robust model (27.2) can be rewritten as

$$\min_{\mathbf{x}\in\mathscr{F},z,\mathbf{p}} \quad c(\mathbf{x})+z \tag{27.5}$$

$$\text{s.t. } z \geq \mathbf{b}^\top \mathbf{p}_k \quad \forall k=1,\ldots,m,$$
$$\mathbf{p}_k \in \Omega(\mathbf{x},\mathbf{d}_k) \quad \forall k=1,\ldots,m,$$

where $(\mathbf{d}_1,\ldots,\mathbf{d}_m)$ is the set of extreme points of the polyhedron $\mathscr{D}$. The number of extreme points may be exponential in the dimension of $\mathscr{D}$, which is the case with the budget uncertainty (27.3). This presents an ideal situation to apply constraint generation on (27.5). Since new variables $\mathbf{p}_k$ are also generated with the new constraints, such an algorithm is formally proposed in [1951] with the name "column-and-constraint generation" (see also [1869]). A similar procedure is also proposed in [253] as a heuristic to speed up the Benders decomposition, where the worst-case extreme points of $\mathscr{D}$ together with the associated dispatch constraints are added to the first-stage problem in each iteration of the Benders algorithm. For details, please refer to [253, 1003, 1951]. The recent work in [1142] proposes efficient algorithms to deal with transmission constraints in (27.2), which dynamically include critical transmission lines.

27.2.4 ▪ Computational Study

The two-stage robust SCUC model (27.2) with the budgeted uncertainty sets (27.3) has been applied to the day-ahead scheduling of the ISO of New England's power system [253]. Table 27.1 compares average dispatch and total costs between the two-stage robust SCUC and the current practice of deterministic UC with reserve adjustment [253].

From this table, we can see that both dispatch and total costs are reduced by the two-stage robust UC model. Also, it is worth noticing that the lowest average cost is achieved at uncertainty budget $\Delta^t = 0.5\sqrt{N_d}$, which results in an uncertainty set that is much smaller than the box uncertainty set of the net load intervals in (27.3). The computational study also shows that the robust UC model can significantly reduce the variability of the production cost by more than an order of magnitude [253].

27.3 ▪ Extensions to Two-Stage Robust UC Models

In this section, we present several recent extensions to the two-stage robust SCUC model (27.2).

Table 27.1. *The average dispatch and total costs of the two-stage robust UC problem and the deterministic UC problem with reserve for normally distributed net load $\Delta^t/\sqrt{N_d} = 0.5, 1, \ldots, 3$ and $\hat{d}_j^t = 0.1\bar{d}_j^t$. © 2013 IEEE. Reprinted with permission from [253].*

| budget | Two-stage robust | | Reserve adjustment | |
$\Delta^t/\sqrt{N_d}$	dispatch cost (M\$)	total cost (M\$)	dispatch cost (M\$)	total cost (M\$)
0.5	16.9195	18.6050	18.1855	19.6837
1.0	16.9650	18.6688	17.4907	18.9942
1.5	16.9815	18.7365	17.3027	18.8006
2.0	17.0297	18.7937	17.7403	19.2415
2.5	17.0586	18.8366	17.6567	19.1618
3.0	17.0745	18.8526	18.0804	19.5889

27.3.1 ▪ Security-constrained UC with Corrective Actions

The UC model (27.1) finds a commitment and dispatch solution that is feasible for any one transmission line contingency. This is called $N-1$ SCUC with preventive actions. A more stringent requirement is to prepare for any loss of k transmission lines, i.e., $N-k$ SCUC. Furthermore, a more flexible way to respond to contingencies is to allow the commitment and dispatch solution to change using the so-called SCUC with corrective actions. It is easy to see that $N-k$ SCUC with corrective actions is a robust optimization model with decisions adaptive to contingency uncertainties. The seminal work [1728] presents such a robust UC model for SCUC with generation contingency, i.e., any k online generators may experience unexpected failures in each period, and remaining generators can be recommitted and redispatched. A subsequent work [1869] proposes an elegant robust UC model that considers both generation and transmission line contingencies in a power network. A similar two-stage robust UC model is proposed in [1727], which considers both generation and transmission contingencies and reserve scheduling. Notice that net load uncertainty is not explicitly considered in these models.

27.3.2 ▪ Taming the Conservativeness

The adaptive robust UC model (27.2) minimizes the sum of the commitment cost and the worst-case second-stage dispatch cost. As shown in [253], the balance between the conservativeness and the robustness of the resulting UC solution can be controlled by the budget constraints in (27.3). A nice discussion on the modeling choices of uncertainty sets and their implications on the conservativeness of the UC solutions is provided in [869]. To further tame the conservativeness, alternative objective functions are considered. For example, a regret optimization–based approach is proposed in [1002], and a hybrid model minimizing the expected cost and the worst-case cost is proposed in [1964].

The two-stage minimax regret robust UC model can be presented abstractly as $\min_{x\in\mathcal{F}} \{\text{Reg}(x) : \Omega(x,d) \neq \emptyset \ \forall d \in \mathcal{D}\}$, where the regret $\text{Reg}(x)$ of the first-stage commitment decision x is defined as

$$\text{Reg}(x) = \max_{d\in\mathcal{D}} \left\{ \min_{p\in\Omega(x,d)} \{c(x)+b(p)\} - Q(d) \right\}. \tag{27.6}$$

That is, the regret of $\mathbf{x}$ is the worst-case difference between the total production cost of choosing $\mathbf{x}$ before knowing $\mathbf{d}$ (i.e., $\min_{\mathbf{p} \in \Omega(\mathbf{x},\mathbf{d})} \{c(\mathbf{x}) + b(\mathbf{p})\}$) and the total production cost $Q(\mathbf{d})$ of a UC with the foresight of demand realization, where $Q(\mathbf{d})$ is defined as $Q(\mathbf{d}) = \min_{\mathbf{x},\mathbf{p}} \{c(\mathbf{x}) + b(\mathbf{p}) : \mathbf{x} \in \mathscr{F}, \mathbf{p} \in \Omega(\mathbf{x},\mathbf{d})\}$.

The inner LO $\min_{\mathbf{p} \in \Omega(\mathbf{x},\mathbf{d})} \{c(\mathbf{x}) + b(\mathbf{p})\}$ can be dualized to a maximization problem, just as we did in (27.4). The term $-Q(d)$ in (27.6) can also be easily rewritten as a maximization problem, given $Q(d)$ is defined by a minimization problem. Therefore, Reg($\mathbf{x}$) in (27.6) can be reformulated as a standard max-min problem. The solution method developed in Section 27.2.3 can be applied to solve the resulting two-stage robust model.

27.3.3 ▪ From Two-Stage to Multistage Robust UC Models

The two-stage robust UC model (27.2) ignores nonanticipativity in the dispatch, which dictates that the dispatch decision at hour t should only depend on realizations of uncertainty up to t. In a recent work [1222], the authors show that respecting the nonanticipativity condition in the dispatch process is crucial for managing a system with restricted ramping capability, which has become a limiting factor in today's power systems with growing penetration of intermittent generation. The paper proposes a multistage robust UC model, which utilizes decision rules, in particular, affine policy, for the multistage dispatch process and develops efficient cutting-plane algorithms to solve the resulting large-scale robust optimization problem. For the first time, multistage robust UC problems can be solved for power networks of more than 2,000 buses within a realistic time framework for applications in day-ahead electricity markets. Furthermore, on realistic test cases, the simplified affine policy proposed in the paper performs surprisingly well—usually within 1% of the true optimal multistage policy.

27.4 ▪ Robust Optimization in Real-Time Operation and Long-Term Planning

Robust optimization models are also proposed for real-time dispatch and long-term planning problems. In the following, we outline some of the recent works in these areas.

A static robust optimization model is proposed for the look-ahead ED problem [1912], where novel statistical models for wind forecast are incorporated into the robust ED model. In [1765], a static robust ED model is proposed for managing system ramping capability, which is shown to outperform the recent development in deterministic look-ahead ED with ramping products [1381].

In [1221], a new two-stage robust optimization model with a new type of dynamic uncertainty sets is proposed for the multiperiod dispatch problem. The dynamic uncertainty sets for wind power incorporate linear autoregression models of wind speeds at neighboring wind farms, so the temporal and spatial correlations of wind speeds are captured. It is shown in [1221] that the robust ED model with dynamic uncertainty sets can Pareto dominate the performance of the robust ED model with the traditional budgeted uncertainty set (27.3) in both the average and variability of the operational costs.

The fundamental two-stage robust optimization model (27.2) is applied to the real-time dispatch of automatic generation control (AGC) units in [1971], where the first-stage decision is the dispatch of normal generation units to satisfy a nominal demand and then the second-stage problem dispatches automatic generation units to respond to demand fluctuations. An affine policy–based robust optimization model is proposed for AGC dispatching in [981], where the second-stage dispatch decision of the AGC units is assumed to be an affine function of the uncertain load. as an approximation to the fully adaptive policy in (27.2), which makes the second-stage problem easier to solve. Affine decision rules are also proposed for managing reserves in the power system [1878].

The two-stage robust optimization model (27.2) has also been applied to the long-term transmission network expansion planning problem in [982], where uncertainties in renewable generation and loads are considered, and to the generation expansion planning problem in [573]. Adaptive robust models using affine decision rules are proposed for capacity expansion planning in [1313].

27.5 ▪ Closing Remarks

This chapter gave a brief review of some of the recent developments in applying robust optimization methodology to power system operation and planning under uncertainty. Many interesting directions, such as the integration of robust models into the electricity markets, multistage optimization for handling systems with high penetration of wind and solar power, and long-term investment, are important questions that are open for further investigation.

Chapter 28

Robust Wind Farm Layout Optimization

Peter Y. Zhang, Jim Y. J. Kuo, David A. Romero, Timothy C. Y. Chan, and Cristina H. Amon

28.1 ▪ Introduction

The number of commercial-scale wind farms and total installed capacity have grown dramatically over the past two decades [1958]. Relying on an essentially free but intermittent energy source and spanning large areas of land or water, wind farms face unique challenges in their design, installation, and operations. Two types of optimization problems arise naturally from this context: (a) optimization of aggregated energy output by strategic turbine placement and (b) optimization of electrical dispatch policies to reduce the economic impact of wind intermittency. In this chapter, we apply robust optimization to the former and discuss the potential benefits on both total energy output and variability in energy output, which can impact dispatch policies.

Site selection and wind turbine layout optimization are important steps in large-scale wind farm engineering projects. The characteristics of a good site include close proximity to the grid, abundance of wind resources, and smoothness of the surface [1267, Chapter 9]. The layout of wind turbines is also a nontrivial task due to the stochasticity of wind (direction and speed) and the complex interaction of wakes propagating through the wind farm. The energy production of a wind turbine, given stable operating conditions, is roughly proportional to the cube of the ambient wind speed [1267, Chapter 3]. Turbines convert the kinetic energy of wind into electrical energy, resulting in wind slowing down and becoming turbulent immediately behind the turbine. This wake region then expands and regains speed as it propagates further, until it encounters other turbines downstream.

Current literature on wind farm layout optimization focuses on the location of turbines within a selected farm considering the aforementioned complex wake interactions. The majority of the approaches in the literature use heuristics such as genetic algorithms [267, 443, 668, 827, 837, 1112, 1358, 1559, 1968] and simulated annealing [268, 438, 937, 1570], among others [267, 465, 1434, 1866]. Mathematical optimization approaches exist as well [617, 1794, 1959]. In addition to energy output, other relevant factors in the layout problem include the noise impact of wind farms [1114] and the layout of connection infrastructure [826], which is especially important in the context of offshore wind farms due to the high cost of foundation construction and cables [1267, Chapter 11].

To date, little consideration has been given to distributional uncertainties that may affect the quality of a layout. The uncertainties we refer to should be distinguished from the terminology used in existing papers. In previous work, uncertainty refers to the stochasticity of wind: wind comes from different directions at different speeds over time. Distributional uncertainty acknowledges such stochasticity but further assumes that the underlying distribution itself may be uncertain [575]. For example, wind measurement is subject to errors. Errors can be due to imperfect equipment setup and data collection, which may result in up to a 60° discrepancy between the measured wind direction and the corresponding true wind direction [170]. Uncertainties can come from the fact that the historical wind distribution is not necessarily the same as the wind distribution that will be realized in the future. For more detailed descriptions on the concepts of robust optimization and distributionally robust optimization (DRO), readers can review the robust optimization chapter in this book (Chapter 25).

This chapter focuses on the uncertainties resulting from wind measurement and forecasting, and we allocate one set of numerical experiments to each of the two uncertainty types described above. More specifically, we take into consideration the distributional uncertainty of wind. The distribution of wind refers to the frequencies of different wind *states*, and every state refers to a unique velocity (speed and direction). We are mainly interested in developing a wind farm layout optimization model that considers such long-term wind state uncertainty. Historical data are imperfect because wind farm developers usually cannot obtain long-term historical data at the exact geographical location of the wind farm—they usually correlate the long-term data from nearby weather stations with short-term local measurements to estimate the long-term wind information at the candidate site [1684, 1787].

In addition to minimizing power fluctuations due to wind uncertainties, there is a need to reduce the power fluctuations due to changing wind directions for grid integration [1455]. Some wind farms, such as the Horn Rev offshore wind farm in Denmark, are laid out in a structured manner. As a result, their power production is sensitive to changes in wind direction [1504].

In this chapter, we develop the first robust optimization model for wind farm layout design. We show that with simple wake expansion and combination models, we can formulate the wind farm layout design problem as a mixed-integer linear programming (MILP) problem. We then develop a robust optimization model based on this nominal problem considering the long-term wind resource uncertainty.

28.2 ▪ Wake Model

The energy generation of a wind farm is significantly less than the total power generated by individual turbines operating in isolation [164]. Wake is one of the primary contributors to energy loss [1267, Chapter 11]. Due to this wake effect, turbines located downstream in the wake "shadow" of upstream turbines will generate less energy than they would in the free stream. Therefore, incorporating wake effects is crucial for optimal wind farm layout [1776].

One of the most commonly used wake models in the literature is the Jensen model [992]. It assumes a linearly expanding wake diameter and uniform velocity profile within the wake. With these assumptions, because of momentum conservation, the result is that the velocity deficit decreases with the distance behind the rotor [992].

The velocity a distance x downstream from a turbine is

$$u(x) = U_\infty \left[1 - \frac{2a}{(1 + \alpha \frac{x}{r_1})^2} \right], \tag{28.1}$$

where U_∞ is the wind speed in the free stream, and the induction factor a (the fractional decrease in wind speed between the free stream and directly after the turbine) can be determined from the thrust coefficient, C_T, of a turbine using $C_T = 4a(1-a)$. $\alpha = 0.5/\ln(z/z_0)$ is the entrainment constant (which dictates the angle of wake expansion), z is the hub height, and z_0 is the surface roughness. Finally, $r_1 = r_r \sqrt{(1-a)/(1-2a)}$ is the wake radius, where r_r is the rotor radius of the turbine.

The interactions between wakes after multiple turbines increase the difficulty of the problem as they involve complex turbulence phenomena. The wind speed after multiple turbines is typically assumed to be a summation of the power deficit of the turbines upstream, which is the approach we take in this chapter [617]. The total instantaneous power, P_{tot}, a wind farm can generate can be estimated by the effective wind speed, u_i, at each turbine i, assuming N turbines [635]:

$$P_{tot} \propto \sum_{i=1}^{N} u_i^3. \tag{28.2}$$

28.3 ▪ Formulations

28.3.1 ▪ Nominal Formulation

We begin by formulating the nominal problem, which assumes the wind resource distribution is fixed. We discretize the wind farm into a grid with n cells, indexed by the set I, with each grid cell capable of hosting one turbine at its center. The primary decision variables are x_i, $i \in I$, which are binary variables indicating whether or not a turbine is placed in cell i.

Under each wind state d, the wind farm produces at one power level. We do not consider the transient behavior of the wind from when wind changes from one state to another; we only focus on its steady-state performance. The steady-state performance of a wind farm is characterized by the undisturbed turbine production at the state minus the loss due to wake interaction. More concretely, coefficient $q_{ijd} \geq 0$ refers to the (normalized) power loss due to location i being downstream of location j in wind state d [617]. Let $\mathcal{D}$ be the set of wind states, p_d be the probability of wind state d, and w_d be the wind power at state d without wake loss. Then we have the problem

$$\max \quad \sum_{d \in \mathcal{D}} \sum_{i \in I} \left(w_d x_i - \sum_{j \in I} q_{ijd} y_{ij} \right) p_d \tag{28.3a}$$

$$\text{s.t.} \quad \sum_{i \in I} x_i = K, \tag{28.3b}$$

$$x_i + x_j - y_{ij} \leq 1, \quad i, j \in I, \tag{28.3c}$$

$$0 \leq y_{ij} \leq 1, \quad i, j \in I, \tag{28.3d}$$

$$x_i \in \{0, 1\}, \quad i \in I. \tag{28.3e}$$

Variables y_{ij} are auxiliary variables used to linearize the quadratic terms $x_i x_j$. The computational performance of this model (or equivalent forms of this model) has been studied in [616–618, 1959]. Formulation (28.3) aims to maximize the energy production of K turbines. Some variations of this objective have been used in the literature. Kusiak and Song [1113] employed expected energy production as their objective function. Mosetti et al. [1358] and Grady [837] approached the problem with a weighted sum of energy production and cost per kilowatt-hour of energy produced and solved their models using genetic algorithms. Réthoré et al. [1559] proposed a multifidelity optimization framework and included a broad set of system costs, such as turbine foundations. Kwong et al. [1114] considered energy maximization and noise minimization at the same time and generated the Pareto frontiers of this biobjective optimization problem using genetic algorithms.

28.3.2 ▪ Robust Formulation

With the nominal formulation, a wind farm layout is optimized under the assumption that the vector **p**, which describes the wind resource distribution, is known with certainty. In reality, **p** is subject to significant uncertainty. For example, some current wind resource prediction models contain error up to 10% in wind speed and 60° in direction [170]. Thus, the realized wind regime may result in a substantially different amount of wind power production than assumed, increasing the financial risk of the project.

The robust problem incorporates uncertainty in the wind resource. We assume that the true wind resource distribution is only known to lie in a polyhedral uncertainty set $\mathscr{P} = \{\mathbf{p} \mid \mathbf{p} \in [\underline{\mathbf{p}}, \overline{\mathbf{p}}], \mathbf{1}'\mathbf{p} = 1\}$:

$$\max_{\tilde{\mathbf{p}} \in \mathscr{P}} \min \sum_{d \in \mathscr{D}} \sum_{i \in I} \left(w_d x_i - \sum_{j \in I} q_{ijd} y_{ij} \right) \tilde{p}_d \tag{28.4a}$$

$$\text{s.t.} \quad \sum_{i \in I} x_i = K, \tag{28.4b}$$

$$x_i + x_j - y_{ij} \leq 1, \quad i, j \in I, \tag{28.4c}$$

$$0 \leq y_{ij} \leq 1, \quad i, j \in I, \tag{28.4d}$$

$$x_i \in \{0, 1\}, \quad i \in I. \tag{28.4e}$$

Formulation (28.4) can be rewritten as the equivalent mathematical optimization using linear optimization (LO) duality [245]:

$$\max \sum_{d \in \mathscr{D}} \left(\sum_{i \in I} \left(w_d x_i - \sum_{j \in I} q_{ijd} y_{ij} \right) p_d - \left(\overline{p}_d r_d + \underline{p}_d s_d \right) \right) \tag{28.5a}$$

$$\text{s.t.} \quad \sum_{i \in I} x_i = K, \tag{28.5b}$$

$$s_d - r_d + t = \sum_{i \in I} \left(w_{id} x_i - \sum_{j \in I} q_{ijd} y_{ij} \right) \quad \forall d \in \mathscr{D}, \tag{28.5c}$$

$$x_i + x_j - y_{ij} \leq 1, \quad i, j \in I, \tag{28.5d}$$

$$0 \leq y_{ij} \leq 1, \quad i, j \in I, \tag{28.5e}$$

$$x_i \in \{0, 1\}, \quad i \in I, \tag{28.5f}$$

$$r_d, s_d \geq 0 \quad \forall d \in \mathscr{D}. \tag{28.5g}$$

The term $\sum_{d\in\mathscr{D}}(\overline{p}_d r_d + \underline{p}_d s_d)$ in the objective characterizes the price, in terms of lower energy production, that we pay for robustness.

28.4 ▪ Results and Discussion

Next we compare the performance of the nominal and robust models. The annual wind probability distribution in 2005 at the Vancouver International Airport was used in this study, shown in Figure 28.1 [670]. Wind directions are discretized into 36 evenly spaced directions. The wind directions are measured counterclockwise, with 0° corresponding to east. This distribution is bimodal, with the dominant wind directions coming from 90° and 290°.

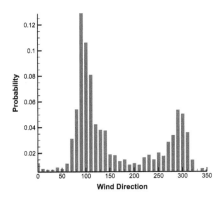

Figure 28.1. *Probability distribution of the wind regime at Vancouver International Airport in 2005.*

To take into account the effects of changing wind directions, we assume the same wind speed for each direction and consider the layout of 30 turbines in a 10 × 10 grid. Due to the dominant north-south wind directions in the given wind regime, the nominal solution results in the turbines being arranged in four rows (Figure 28.2). The layout efficiency, defined as the ratio of power production considering wake losses to that without losses is shown in Figure 28.3 for each wind direction. The nominal layout efficiency curve has a peak to trough difference of 42%.

Next we use formulation (28.5) to create robustly optimized layouts. We consider two different ways to generate the uncertainty set $\mathscr{P}$. Uncertainty set A is generated by shifting the probability distribution up to 50° to the left and the right and taking the maximum and minimum envelopes, as shown in Figure 28.4. Uncertainty set B is generated in a similar manner but considers shifting the probability distribution up to 90° to the right, as shown in Figure 28.5. The purpose is to capture the peaks and troughs of the efficiency curve in Figure 28.3, as the peaks and the troughs are within −50° to +50° and +90°.

Two different robust solutions are generated using the two uncertainty sets and model (28.5). Figure 28.6 shows the layout corresponding to uncertainty set A. Note that the turbines are more spread out than in the nominal layout. Figure 28.7 shows the layout corresponding to uncertainty set B.

The layout efficiency for each of the three layouts for every 10° is plotted in Figure 28.8. The two robust layouts experience peak to trough layout efficiency differences of 13.7% and 19.4%, respectively. We also used the annual wind probability

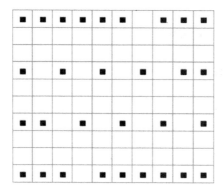

Figure 28.2. *Nominal layout.*

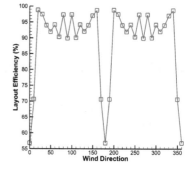

Figure 28.3. *Layout efficiency of nominal layout as a function of dominant wind direction.*

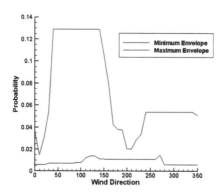

Figure 28.4. *Uncertainty set A.*

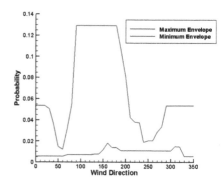

Figure 28.5. *Uncertainty set B.*

distribution from 2005 to 2012 to evaluate nominal and robust layouts on a longer time scale. It can be seen from Table 28.1 that the price of robustness is about a 1–2% reduction in average energy production compared to the nominal.

The hourly wind data in a one-week period in both 2005 and 2012 were used to assess the stability of energy production on smaller time scales, to capture the effects of sudden changes in wind direction, independent of wind speed. Figure 28.9 shows the hourly power production from one week in 2012. It is evident that in the nominal layout, power production can fluctuate greatly from hour to hour, which is undesirable for grid integration. On the other hand, the robust layouts produce much smoother energy flow at a cost of slightly lower maximum energy output.

What remains to be addressed is whether the slightly lower average production of the robust layouts is compensated for by the improved stability of the energy output.

Table 28.1. *Annual layout efficiency based on historic wind data.*

Layout	2005	2006	2007	2008	2009	2010	2011	2012	Average
Nominal	91.5%	91.3%	91.4%	91.6%	91.4%	91.7%	91.3%	91.7%	91.4%
Robust-A	89.1%	89.4%	89.2%	89.2%	89.3%	89.1%	89.1%	89.2%	89.2%
Robust-B	90.5%	90.3%	90.3%	90.2%	90.2%	90.3%	90.1%	90.3%	90.3%

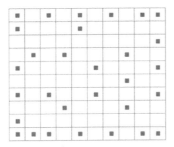
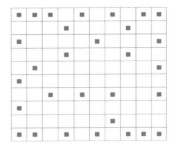

Figure 28.6. *Robust-A layout.*   **Figure 28.7.** *Robust-B layout.*

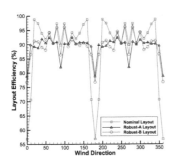
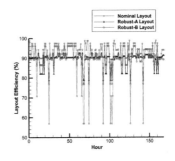

Figure 28.8. *Robust and nominal layout efficiency as a function of wind direction.*

Figure 28.9. *Power production of a wind farm under nominal and robust layouts in one week of 2012.*

To address this question, we performed a set of simulations to assess the economic impact of (hourly) wind energy variability based on the operations of typical wholesale electricity markets [1196]. In such markets, an energy supplier proposes a schedule of energy production one day in advance to the day-ahead market, so that the regulator can balance the supply and demand in the market. But since the realized energy production is not always the same as the proposed schedule for intermittent sources such as wind, the supplier has to pay a premium to compensate for the discrepancy between the proposed schedule and the actual energy production by purchasing and selling in the real-time electricity market. The simulations capture the difference in annual wind energy sales between the nominal and the robust-B layouts. For each layout, we calculate the sales revenue based on the sum of three values: (1) revenue from energy sales, (2) penalty paid for the hourly deficit between actual energy production and the energy committed, and (3) salvage revenue of the overproduction of energy. The penalty is defined as the unit price (normalized by real-time market price) under which the wind energy supplier must pay for an hourly deficit. The salvage value is defined as the (normalized) unit price at which the wind energy supplier can sell the excess energy. Hourly deficit and excess energies are defined as the negative and positive components of the difference between actual and proposed energy production. The penalties and salvage values are based on [1196]. The values of the salvage and deficit penalty vary by jurisdiction and can be substantial [79, 1581].

The simulation results are presented in Figure 28.10. The horizontal axis is the revenue given a nominal layout (million dollars), and the vertical axis represents the revenue of the robust-B layout. Each point represents one simulated year, based on

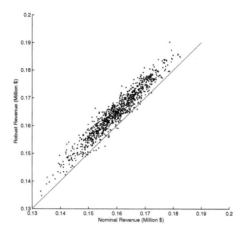

Figure 28.10. *Penalty for underproduction (deficit) is 50% of the market price, while the salvage value of excess energy production is zero. Each dot represents one instance of simulation: the annual revenues (in millions of dollars) based on a wind distribution randomly drawn from the uncertainty set.*

Table 28.2. *Percent difference between average (over 1,000 simulations) annual revenues given the robust-B layout and the nominal layout. Positive values indicate a higher revenue from the robust layout.*

	Salvage value = 0	Salvage value = 50%	Salvage value = 100%
Deficit penalty = 0	1.06	−0.13	−0.52
Deficit penalty = 50%	3.14	0.31	−0.29
Deficit penalty = 100%	18.05	0.94	−0.12

randomly generated wind forecast data and realized wind data. For each year, both data sets are based on a randomly drawn wind regime from the uncertainty set. Points above the diagonal represent cases where the robust layout generated more revenue. The plot shows a case with 50% penalty and zero salvage value, meaning that the wind energy supplier must pay 50% more than the real-time market price to "generate" the deficit (by utilizing storage or purchasing from another supplier), while the excess energy has zero value. Under this experimental setup, we observe that the revenue from the robust-B layout is almost always higher than the nominal layout revenue in the 1,000 simulations.

We also conducted a sensitivity analysis (Table 28.2) to study how the financial impact of wind energy variability depends on the salvage and penalty values. Naturally, when the penalty for deficit is high and the salvage value of overproduction is low, the market favors low-variability energy suppliers, like the robust layout. Thus, the design of a wind farm can play an important role in mitigating the financial impact of energy variability.

28.5 ▪ Conclusion

In this chapter, we presented the first study that considers distributional uncertainty in the wind farm layout optimization problem and assessed the potential improvement of robust optimization in minimizing the corresponding fluctuations in power production. We demonstrated that robust solutions can reduce power fluctuations at the price of a modest reduction in overall annual energy production. However, optimizing the layout of a wind farm is important not only for total energy production on long time scales but also for power fluctuations on small time scales, particularly when the market penalizes deviations from proposed output schedules. Future research could explore different types of uncertainty sets for the wind direction probability vector $\mathbf{p}$, such as the ones pointed out by [198]. Such uncertainty sets arise naturally when the parameters are elements of a probability vector, often leading to tractable reformulations.

Part VIII

Stochastic Optimization

Chapter 29

Stochastic Optimization

Shabbir Ahmed

29.1 ▪ Introduction

Consider the generic optimization problem

$$\min_{x}\{F(x,\xi)\colon\ x\in X(\xi)\subseteq\mathbb{R}^{n}\}, \tag{29.1}$$

where x is the decision vector, $\xi\in\mathbb{R}^{d}$ is a vector of problem parameters or data required to describe the objective and constraints, $F:\mathbb{R}^{n}\times\mathbb{R}^{d}\mapsto\mathbb{R}$ is the objective function, and $X(\xi)$ is a constraint set, e.g., of the form $X(\xi):=\{x:\ G(x,\xi)\le 0\}$, where $G:\mathbb{R}^{n}\times\mathbb{R}^{d}\mapsto\mathbb{R}^{m}$ is the constraint function. Note that the objective and constraints of (29.1) depend on the data ξ, and it is typically assumed that these data are completely specified before we can solve the problem. However, in many practical settings, the data are unknown because of imprecise or uncertain information about the future, and decisions have to be made prior to knowing the data realization and hence the full consequences of the decisions. Depending on the structure of the underlying optimization problem and the uncertainty involved, a variety of approaches may be applicable for such problems of optimization under uncertainty.

In the paradigm of stochastic optimization (SO), the uncertain parameters in an optimization problem are modeled as random variables. In this case the objective and constraint functions of (29.1) are random valued, so we must specify appropriate criteria for determining the optimality and feasibility of a given solution. Typical SO models consider optimizing the expected value of $F(x,\xi)$ while guaranteeing feasibility with a prespecified probability. The resulting models can be computationally very challenging and require powerful optimization methods integrated with sampling-based approximations.

In this chapter, we discuss a few common classes of SO approaches. Our presentation is limited to SO models and methods built on the mathematical optimization framework, for example, when the underlying deterministic optimization problem can be treated using linear optimization (LO) methodology. This approach is often referred to as *stochastic programming*. We do not discuss many other important SO approaches, such as simulation optimization, stochastic dynamic programming (DP), Markov decision processes, and stochastic control. The remainder of this chapter is organized as follows. We begin with the classical static SO model involving an expectation objective in Section 29.2. In Section 29.3, we discuss extensions of static SO models to the discrete-time, finite-horizon, dynamic setting. In Section 29.4, we discuss chance constrained stochastic optimization (CCSO) models for dealing with uncertain constraint data. Finally, we conclude by briefly mentioning some extensions of basic SO models in Section 29.5.

29.2 ▪ Static Stochastic Optimization

Let us begin with a simple supply chain engineering example.

Example 29.1. A firm has I distribution centers (DCs) and J retail locations for its product. For the next year, the firm needs to decide how much of the product to stock at the DCs, and how to ship the product to the retail locations to satisfy demand at minimum cost. The following data are available:

c_i = per unit stocking cost at DC i,
b_{ij} = per unit shipping cost from DC i to retail location j,
d_j = demand at retail location j,

where $i \in \{1,\dots,I\}$ and $j \in \{1,\dots,J\}$ are indices used for DCs and retail locations, respectively. This problem can be formulated as the following linear programming (LP) problem:

$$
\begin{aligned}
\min_{u,v} \quad & \sum_{i=1}^{I} c_i u_i + \sum_{i=1}^{I} \sum_{j=1}^{J} b_{ij} v_{ij} \\
\text{s.t.} \quad & \sum_{j=1}^{J} v_{ij} \le u_i, \quad i = 1,\dots,I, \\
& \sum_{i=1}^{I} v_{ij} \ge d_j, \quad j = 1,\dots,J, \\
& u_i, v_{ij} \ge 0, \quad i = 1,\dots,I, \ j = 1,\dots,J.
\end{aligned}
\tag{29.2}
$$

In the above LP problem, u_i denotes the stock at DC i and v_{ij} denotes the shipment from DC i to retail location j. The objective function is the total cost of stocking and shipping, the first set of constraints requires that the total shipment from a DC cannot exceed the stock available, the second set of constraints requires that the total shipment to a retail location should meet the demand, and the final set of constraints enforces the nonnegativity of the decision variables.

Suppose that the demand at each retail location is uncertain when the stocking and shipping decisions are made but can be reasonably modeled as a random variable with a known probability distribution, for example, using historical information. What happens, then, if the shipment is not sufficient to meet the demand? In such situations there are typically stockout penalties, and a reasonable approach is to determine stocking and shipping levels to minimize the total stocking and shipping costs and the *expected* stockout penalty. This gives rise to the following SO version of problem (29.2):

$$
\begin{aligned}
\min_{u,v} \quad & \sum_{i=1}^{I} c_i u_i + \sum_{i=1}^{I} \sum_{j=1}^{J} b_{ij} v_{ij} + \mathbb{E}\left[\sum_{j=1}^{J} \Psi_j \left(\tilde{d}_j - \sum_{i=1}^{I} v_{ij} \right) \right] \\
\text{s.t.} \quad & \sum_{j=1}^{J} v_{ij} \le u_i, \quad i = 1,\dots,I, \\
& u_i, v_{ij} \ge 0, \quad i = 1,\dots,I, \ j = 1,\dots,J.
\end{aligned}
\tag{29.3}
$$

Above, $\Psi_j(t) := q_j \cdot \max\{0, t\}$ is the stockout penalty function; i.e., the stockout at location j is penalized at a cost of q_j per unit. We have used $\tilde{d}$ to denote the random demand vector, and the expectation in the objective of (29.3) is with respect to the distribution of $\tilde{d}$. ∎

Problem (29.3) is an example of the following static SO problem:

$$\min_x \left\{ \mathbb{E}[F(x,\tilde{\xi})] : x \in X \subseteq \mathbb{R}^n \right\}. \tag{29.4}$$

To see this, note that x denotes the vector of stocking and shipping decisions (u,v); the random vector $\tilde{\xi}$ denotes the demand vector $\tilde{d}$; the function $F(x,\xi)$ denotes the total stocking, shipping, and penalty cost for a given decision x and realization ξ of the random vector; and the constraint system X denotes the constraints on u and v.

Problem (29.4) is a stochastic version of the generic optimization problem (29.1) where the uncertainty only affects the objective function. The constraint data are deterministic. The decision dynamics of the model are as follows. The decision x is made *here and now*, i.e., prior to the realization of the random vector $\tilde{\xi}$, and then a realization ξ is observed for which the value of the objective function is $F(x,\xi)$. The goal is to find an x to minimize the expected value of the objective. The distribution of $\tilde{\xi}$ is assumed to be completely known beforehand and remains unaffected by the decision vector x.

The static SO problem (29.4) is the most widely studied class of SO problems. The classical newsvendor or single-period inventory model is a familiar example. In this model, the newsvendor buys x units of newspapers in the morning to meet the demand ξ during the day and pays a shortage or underage cost of a and a surplus or overage cost of b, i.e., $F(x,\xi) = a \cdot \max\{0, \xi - x\} + b \cdot \max\{0, x - \xi\}$. It can be shown that the optimal buy quantity that minimizes expected costs is given by $x^* = F^{-1}(a/(a+b))$, where F^{-1} is the inverse cumulative distribution function (cdf) of the random variable $\tilde{\xi}$. Here we have an example of SO where an analytical solution is available. In general, however, problem (29.4) can be very challenging and must be dealt with using sophisticated approximation and algorithmic methods.

The primary difficulty of the SO problem (29.4) is the expectation operation in the objective function. Expectation preserves many of the desirable properties for optimization. For example, if $F(x,\xi)$ is convex and/or differentiable in x for all ξ, then $\mathbb{E}[F(x,\xi)]$ is a convex and/or differentiable function of x. If the distribution of ξ is finite with a modest-sized support, then the expectation reduces to a sum and (29.4) is a standard optimization problem with properties similar to its deterministic counterpart (29.1). In general, however, just evaluating the expectation objective for a given solution x can require computing a high-dimensional integral or a very large sum, which is most often impossible to do exactly, and we must resort to approximations.

Next we discuss two common sampling-based approximation methods for (29.4). We limit our discussion to the case where $F(x,\xi)$ is convex in x for all ξ; $\mathbb{E}[F(x,\xi)]$ is well defined and finite for all x; and X is a nonempty, compact, and convex set. Under these assumptions, the SO problem (29.4) involves minimizing a well-defined, albeit difficult to evaluate, convex function over a convex set, and hence is a convex optimization problem.

The first approach is *stochastic approximation* (SA). The classical SA method is similar to the usual deterministic subgradient method applied to the convex optimization problem (29.4), where a sampled subgradient is used. More precisely, starting from an initial solution $x^0 \in X$, it generates candidate solutions by the formula

$$x^{k+1} = \Pi_X(x^k - \alpha_k \hat{G}(x^k)), \quad k = 0, 1, 2, \ldots,$$

where $\Pi_X(y)$ is the projection of y on the feasible region X, $\{\alpha_k\}$ is a sequence of step sizes, and $\hat{G}(x^k)$ is a sampled subgradient of $\mathbb{E}[F(x,\xi)]$ at x^k. In the simplest case, such a sampled subgradient can be obtained by computing a subgradient of F at x^k for a single sampled realization ξ^k of $\tilde{\xi}$, i.e., $\hat{G}(x^k) \in \partial F(x^k,\xi^k)$. There is an abundance of literature on various conditions on the problem structure, the step size rules, and the subgradient sampling scheme, under which the iterates converge to an optimal solution of the SO problem.

The second approach is *sample average approximation* (SAA). In this approach, the SO problem (29.4) is approximated as

$$\min_x \left\{ \frac{1}{N}\sum_{i=1}^N F(x,\xi^i): \; x \in X \subseteq \mathbb{R}^n \right\}, \tag{29.5}$$

where $\{\xi^1,\ldots,\xi^N\}$ is an independent and identically distributed or Monte Carlo sample of $\tilde{\xi}$. For a given sample, (29.5) is a deterministic optimization problem that can be solved using standard optimization approaches. Under mild conditions it can be shown that for a sufficiently large sample size N, an optimal solution to the approximate problem (29.5) is a very good solution to the true SO problem (29.4) with high probability. Moreover, the average of the optimal values of several instances of (29.5) corresponding to different samples estimates a lower bound on the optimal value of (29.4), which can then be used to assess the quality of a given candidate solution.

The SAA scheme is applicable in much more general settings than the convex case to which SA is restricted. As long as there is an effective deterministic optimization method for the sampled problem (29.5), which may be large scale, the SAA method can be applied. On the other hand, for certain classes of convex problems, specialized SA schemes tailored to the structure of the problem can perform significantly better than SAA.

29.3 • Dynamic Stochastic Optimization

The SO model described in the previous section is static in the sense that it is assumed that the decisions are fully specified before the uncertainty is realized. In this section, we discuss SO models accommodating dynamic decision making, where additional actions, sometimes called recourse actions, are available after and in response to the uncertainty realized. We motivate such dynamic SO models using a multiperiod extension of Example 29.1.

Example 29.2. The firm is now planning for the next T years. The decision dynamics are as follows. Each year t, the firm receives stock at the DCs *before* the demand is observed and then ships to the retailers *after* the demand is known. Any leftover inventory gets carried over to period $t+1$. There is a holding cost associated with the inventory. The deterministic problem can be formulated as follows:

$$\min_{u,v,w} \quad \sum_{t=0}^T \left\{ \sum_i (c_i^t u_i^t + h_i^t w_i^t) + \sum_{i,j} b_{ij}^t v_{ij}^t \right\}$$

$$\text{s.t.} \quad w_i^t = w_i^{t-1} + u_i^{t-1} - \sum_j v_{ij}^t \; \forall \, i, t,$$

$$\sum_i v_{ij}^t \geq d_j^t \; \forall \, j, t, \tag{29.6}$$

$$w_i^t, \, u_i^t, \, v_{ij}^t \geq 0 \; \forall \, i, j, t,$$

$$w_i^0 = 0, \; u_i^T = 0 \; \forall \, i.$$

We have introduced inventory variables w_i^t to denote the end-of-year-t inventory at DC i with a holding cost of h_i^t per unit. We have also introduced superscript t on the stocking and shipment decisions and cost parameters for year t. The vector u^t is the stock ordered in period t that becomes available in period $t+1$ for shipment. For notational simplicity, we have suppressed the ranges of the indices i and j. The objective of the above LP problem is to minimize total stocking, holding, and shipping cost. The first set of constraints enforces inventory balance, the second set of constraints enforces demand satisfaction, and the third set of constraints enforces nonnegativity. The nonnegativity of the inventory variables imposes the constraint that the total shipment of a product from a DC in period t cannot exceed the sum of the inventory at the end of period $t-1$ and the stock ordered in period $t-1$. The final constraint enforces boundary conditions. The current year is denoted by $t=0$, and it is assumed that $w_i^0 = 0$ for all i, i.e., there is no initial on-hand inventory. The only decision for the current period is the stocking orders u_i^0 for all i that will become available in period $t=1$. No stock is ordered in the final period, T. The decision dynamics of the problem are illustrated in Figure 29.1.

Decide $u^0 \rightarrow$ Observe $d^1 \rightarrow$ Decide $u^1, v^1 \rightarrow \cdots \rightarrow$ Decide $u^{t-1}, v^{t-1} \rightarrow$ Observe $d^t \rightarrow$ Decide $u^t, v^t \rightarrow \cdots \rightarrow$ Decide $u^{T-1}, v^{T-1} \rightarrow$ Observe $d^T \rightarrow$ Decide v^T

Figure 29.1. *Decision dynamics for Example 29.2.*

Suppose now that the demand is uncertain and its evolution is modeled as a stochastic process $\{\tilde{d}^t\}$. Since we are interested in minimizing expected costs, the problem can be formulated as follows:

$$\min_{u^0} \quad \sum_i c_i^0 u_i^0 + \mathbb{E}_{\tilde{d}^1}[Q_1(u^0, w^0, \tilde{d}^1)]$$
$$\text{s.t.} \quad u_i^0 \geq 0 \ \forall \ i, \qquad\qquad\qquad\qquad (29.7)$$
$$w_i^0 = 0 \ \forall \ i,$$

where the cost of period t is captured in the functions Q_t, which are defined as

$$Q_t(u^{t-1}, w^{t-1}, d^t) := \min_{u^t, v^t, w^t} \sum_i (c_i^t u_i^t + h_i^t w_i^t) + \sum_{i,j} b_{ij}^t v_{ij}^t$$
$$+ \mathbb{E}_{\tilde{d}^{t+1}|d^t}[Q_{t+1}(u^t, w^t, \tilde{d}^{t+1})]$$

$$\text{s.t.} \quad w_i^t = w_i^{t-1} + u_i^{t-1} - \sum_j v_{ij}^t \ \forall \ i,$$
$$\sum_i v_{ij}^t \geq d_j^t \ \forall \ j,$$
$$w_i^t, u_i^t, v_{ij}^t \geq 0 \ \forall \ i,j, \qquad\qquad (29.8)$$

for $t = 1, \ldots, T-1$, and

$$Q_T(u^{T-1}, w^{T-1}, d^T) := \min_{v^T, w^T} \sum_i h_i^T w_i^T + \sum_{i,j} b_{ij}^T v_{ij}^T$$
$$\text{s.t.} \quad w_i^T = w_i^{T-1} - \sum_j v_{ij}^T \ \forall \ i,$$
$$\sum_i v_{ij}^T \geq d_j^T \ \forall \ j, \qquad\qquad (29.9)$$
$$w_i^T, v_{ij}^T \geq 0 \ \forall \ i,j.$$

The future-stage cost functions (29.8) and (29.9) encode the optimization problem to be solved in each period depending on the decisions made earlier and the information observed. The expectation operation in period t is with respect to the random demand corresponding to period $t+1$ conditioned on the observed demand of period t.   ∎

Next we present a general formulation for dynamic SO models. We restrict our discussion to models with linear objectives and constraints. Let us start with a canonical multiperiod (or multistage) deterministic LP problem with $T+1$ periods or stages, with $t=0$ denoting the current period:

$$
\begin{aligned}
\min_{x_0, x_1, \dots, x_T} \quad & \sum_{t=0,\dots,T} c_t^{\top} x_t \\
\text{s.t.} \quad & B_t x_{t-1} + A_t x_t = b_t \quad \forall\, t = 1, \dots, T, \\
& x_t \in X_t \qquad\qquad\quad \forall\, t = 0, \dots, T,
\end{aligned}
\tag{29.10}
$$

where X_t is a set of linear constraints on x_t, perhaps with integrality restrictions on some of the components. Note that there are two sets of constraints — one set linking decisions across stages, and the other set only involving variables within a stage. Without loss of generality, we assume that the temporal constraints only link successive stages since we can model the dependence on an earlier stage by introducing additional information-carrying variables.

The data required in stage t are $\xi_t = (c_t, B_t, A_t, b_t, X_t)$, where with some notational abuse we have used X_t to denote the data for the constraints in X_t. For $t \geq 1$, let us denote the feasible region of the stage-t problem by $S_t(x_{t-1}, \xi_t)$, which depends on the decision in stage $t-1$ and the information ξ_t available in stage t. That is, $S_t(x_{t-1}, \xi_t) := \{x_t \in X_t : A_t x_t = b_t - B_t x_{t-1}\}$. Suppose now that the data $(\tilde{\xi}_1, \dots, \tilde{\xi}_T)$ evolve according to a known stochastic process. We use $\tilde{\xi}_t$ to denote the random data vector in stage t and ξ_t to denote a specific realization. The decision dynamics is as follows: in stage t we first observe the data realization ξ_t and then take an action x_t depending on the previous stage decision x_{t-1} and the observed data to optimize the expected future cost. A formulation for this dynamic SO problem is

$$
\begin{aligned}
\min_{x_0 \in X_0} \Bigg\{ c_0^{\top} x_0 + \mathbb{E}_{\tilde{\xi}_1} \Bigg[\min_{x_1 \in S_1(x_0, \xi_1)} \Big\{ c_1(\xi_1)^{\top} x_1 \\
+ \mathbb{E}_{\tilde{\xi}_2 | \xi_1} \Big[\min_{x_2 \in S_2(x_1, \xi_2)} \Big\{ c_2(\xi_2)^{\top} x_2 + \cdots \Big\} \Big] \Big\} \Bigg] \Bigg\}.
\end{aligned}
\tag{29.11}
$$

Note that for stages $t = 2, \dots, T$ we used $c_t(\xi_t)$ to denote the cost coefficient corresponding to the data ξ_t, and $\mathbb{E}_{\tilde{\xi}_t | \xi_{t-1}}$ to denote that the expectation operation in stage t is with respect to the *conditional* distribution of $\tilde{\xi}_t$ given realization ξ_{t-1} in stage $t-1$. If the data are *stagewise independent*, this reduces to the usual expectation $\mathbb{E}_{\tilde{\xi}_t}$. The above class of problems are called multistage stochastic linear programming (MSSLP) problems.

An alternative formulation for the above MSSLP problem uses the notion of *cost-to-go functions* as in DP. The problem in stage zero (and the overall problem) is

$$
(P_0): \quad \min\{c_0^{\top} x_0 + \mathbb{E}_{\tilde{\xi}_1}[Q_1(x_0, \tilde{\xi}_1)] : x_0 \in S_0\},
\tag{29.12}
$$

where the cost-to-go functions or value functions corresponding to stages $t = 1, \ldots, T$ are recursively defined as values of optimization problems:

$$(P_t): \quad Q_t(x_{t-1}, \xi_t) = \min\{c_t^\top(\xi_t)x_t + \mathbb{E}_{\tilde{\xi}_{t+1}|\xi_t}[Q_{t+1}(x_t, \tilde{\xi}_{t+1})] : x_t \in S_t(x_{t-1}, \xi_t)\},$$
$$(29.13)$$

with $Q_{T+1} \equiv 0$.

We will assume throughout that for all $t = 1, \ldots, T$ the optimization problem (P_t) is feasible for any feasible x_{t-1} and realized ξ_t. The problem (P_T) in stage T is an LO; thus for any ξ_T the function $Q_T(x_{T-1}, \xi_T)$ is the value function of an LP problem with x_{T-1} appearing as a right-hand side parameter, and therefore is convex in x_{T-1}. Then, for any ξ_{T-1} the function $Q_{T-1}(x_{T-2}, \xi_{T-1})$ is the value function of a convex optimization problem where x_{T-2} appears on the right-hand side of the linear constraints. Thus, $Q_{T-1}(x_{T-2}, \xi_{T-1})$ is convex in x_{T-2}. Going back in this way we note that the function Q_t in stage t is convex with respect to x_{t-1}. Thus the overall problem (P_0) is a convex optimization problem.

While (P_0) looks identical to the static formulation (29.4) we have seen before, the key difference is that for given x_0 and realization ξ_1, a single evaluation of $Q_1(x_0, \xi_1)$ requires solving another SO problem (with just one less period). This recursive feature makes the problem very difficult even when the number of stages is just modestly large.

A first-stage solution x_0 of (P_0) is well defined, but all successive-stage solutions are functions of previous-stage solutions and data realized, i.e., $x_t(x_{t-1}, \xi_t)$. Such a solution is called a *policy*, i.e., a mapping from the state (x_{t-1}, ξ_t) to an action x_t. If there are only finitely many realizations of the uncertain data, then we can encode the policy/solution as a table. In the general case, an encoding of such a policy is given if we have a description of the function $\mathcal{Q}_{t+1}(x_t, \xi_t) := \mathbb{E}_{\tilde{\xi}_{t+1}|\xi_t}[Q_{t+1}(x_t, \tilde{\xi}_{t+1})]$ as

$$x_t(x_{t-1}, \xi_t) \in \arg\min\{c_t^\top(\xi_t)x_t + \mathcal{Q}_{t+1}(x_t, \xi_t) : x_t \in S_t(x_{t-1}, \xi_t)\}.$$

Since $\mathcal{Q}_{t+1}(x_t, \xi_t)$ is hard to describe, typically we have to resort to approximations to derive approximate policies.

Let us now assume that the stochastic data $\{\xi_1, \ldots, \xi_T\}$ have a finite distribution. In this case, the evolution of the data can be represented as a *scenario tree*, denoted by $\mathcal{T}$, with T levels corresponding to the T stages. The set of nodes in stage t is denoted by $\mathcal{S}_t$. The root node in the current stage $(t = 0)$ is labeled 0, i.e., $\mathcal{S}_0 = \{0\}$. Each node n in stage $t \geq 1$ has a unique ancestor node $a(n)$ in stage $t - 1$. We denote the stage corresponding to node n by $t(n)$. The set of children of a node n is denoted by $\mathscr{C}(n)$. The set of nodes on the (unique) path from node zero to node n (including node n) is denoted by $\mathscr{P}(n)$. A node $n \in \mathcal{S}_t$ represents a state of the world in stage t and corresponds to the data sequence $\{\xi_m = (c_m, B_m, A_m, b_m, X_m)\}_{m \in \mathscr{P}(n)}$. The *total probability* associated with node n is p_n, and this is the probability of realization of the $t(n)$ period data sequence $\{\xi_m\}_{n \in \mathscr{P}(n)}$. For $m \in \mathcal{T} \setminus \{0\}$ and $n = a(m)$, we let $q_{nm} = p_m/p_n$ be the *conditional* probability of transiting from node n to node m. Each node in the last stage, $\mathcal{S}_T$, or in other words a path from the root node zero to a leaf node in stage T, corresponds to a realization of the data for the full planning horizon (all T periods) and is called a *scenario*. Let S denote the set of all scenarios represented in $\mathcal{T}$. The data in stage t corresponding to scenario $s \in S$ are $(c_t^s, B_t^s, A_t^s, b_t^s, X_t^s)$, and the associated probability is p_s. Note that any two scenarios or root-to-leaf paths share all nodes up to a stage. We say two scenarios s and s' are *indistinguishable* at period t if the paths corresponding to s and s' pass through the same set of nodes up to stage t. Let $\mathcal{I}_t$ denote the collection of pairs of scenarios that are indistinguishable at stage t.

Next we describe two formulations based on the scenario tree description explained above.

The first formulation is called the *nodal formulation*. Here we index the decision variables corresponding to the nodes of the scenario tree $\mathcal{T}$. Upon observing the data in node n, a decision x_n in this node is made. The formulation is then as follows:

$$
\begin{aligned}
\min_x \quad & \sum_{n \in \mathcal{T}} p_n c_n^\top x_n \\
\text{s.t.} \quad & B_n x_{a(n)} + A_n x_n = b_n \quad \forall\, n \in \mathcal{T} \setminus \{0\}, \\
& x_n \in X_n \qquad\qquad\quad \forall\, n \in \mathcal{T}.
\end{aligned}
\tag{29.14}
$$

The second formulation, called the *scenario formulation*, is as follows:

$$
\begin{aligned}
\min_x \quad & \sum_s p_s \left(\sum_t c_t^{s\top} x_t^s \right) \\
\text{s.t.} \quad & B_t^s x_{t-1}^s + A_t^s x_t^s = b_t^s \quad \forall\, t \geq 1, s, \\
& x_t^s \in X_t^s \qquad\qquad\quad\ \forall\, t, s, \\
& x_t^s = x_t^{s'} \qquad\qquad\quad\ \forall\, (s, s') \in \mathcal{I}_t, \forall\, t.
\end{aligned}
\tag{29.15}
$$

The last set of constraints consists of *nonanticipativity* constraints that enforce the requirement that stage t decisions for two scenarios that are indistinguishable at stage t must be identical.

Both the nodal and scenario formulations are large-scale LP problems that are amenable to decomposition methods.

We can write a value function or DP-based formulation of the nodal formulation (29.14) as

$$
(P_0): \quad
\begin{aligned}
\min_{x_0} \quad & c_0^\top x_0 + \sum_{m \in \mathcal{C}(0)} q_{0m} Q_m(x_0) \\
\text{s.t.} \quad & x_0 \in X_0,
\end{aligned}
$$

where for all $n \in \mathcal{T} \setminus \{0\}$,

$$
(P_n): \quad
\begin{aligned}
Q_n(x_{a(n)}) := \quad \min_x \quad & c_n^\top x_n + \sum_{m \in \mathcal{C}(n)} q_{nm} Q_m(x_n) \\
\text{s.t.} \quad & A_n x_n = b_n - B_n x_{a(n)}, \\
& x_n \in X_n.
\end{aligned}
$$

The *nested decomposition method* for the nodal formulation uses iteratively refined piecewise-linear approximations of the cost-to-go functions Q_m in each node problem (P_n). The dual variables of the node LP problem (P_n) are used in the approximation of Q_n in the ancestor node $a(n)$, and the primal solutions are passed on to the children nodes $m \in \mathcal{C}(n)$ as candidate points, where the approximation of Q_m is to be further refined. In this way, each node subproblem (P_n) is decoupled from other subproblems.

In the scenario formulation (29.15), relaxing the nonanticipativity constraints decomposes the problem into $|S|$ instances of the deterministic multiperiod problem (29.10). Such a relaxation is achieved by penalizing the linear nonanticipativity constraints by Lagrange multipliers in the objective function. For a given set of Lagrange multipliers, each scenario problem can be solved independently. The Lagrange multipliers can be updated using standard subgradient or related techniques used for optimizing the Lagrangian dual. This scheme is called *scenario decomposition*.

Particularly important types of dynamic SO problems are two-stage problems where $T = 1$. The decomposition methods outlined above are very effective for such problems and have been extensively applied.

In describing the optimization approaches above, we assumed a finite scenario tree description of the underlying stochastic process. How to build such a finite but modest-sized scenario tree from a stochastic process description or historical observations of the problem data is a major technical question. Possibilities include discretizing the data to match moments or some other measure of similarity between the stochastic process and the scenario tree or using Monte Carlo sampling to build the scenario tree.

29.4 ▪ Chance Constrained Stochastic Optimization

In this section, we consider SO problems where the data in the constraints are random.

Example 29.3. Let us return to Example 29.1. Suppose the demand is uncertain and the firm needs to plan its stocking and shipping decisions now to guarantee with a probability of $(1-\epsilon)$ that the demand will be satisfied:

$$\min_{u,v} \sum_{i=1}^{I} c_i u_i + \sum_{i=1}^{I}\sum_{j=1}^{J} b_{ij} v_{ij}$$

$$\text{s.t.} \quad \sum_{j=1}^{I} v_{ij} \le u_i, \quad i = 1,\dots,I,$$

$$\Pr\left[\sum_{i=1}^{J} v_{ij} \ge \tilde{d}_j, \ j = 1,\dots,J\right] \ge 1-\epsilon,$$

$$u_i, \ v_{ij} \ge 0, \quad i = 1,\dots,I, \ j = 1,\dots,J, \tag{29.16}$$

where the probability of the demand constraint being satisfied is with respect to the distribution of $\tilde{d}$. ∎

Example 29.3 is an SO problem where constraints involving random variables are required to be satisfied with a prespecified probability. A generic formulation of such *chance constrained* (CC) or *probabilistically constrained* SO problems is

$$\min_x \left\{F(x) : \Pr[G(x,\tilde{\xi}) \le 0] \ge 1-\epsilon, \ x \in S\right\}, \tag{29.17}$$

where $S \subset \mathbb{R}^n$ represents a deterministic feasible region, $F : \mathbb{R}^n \to \mathbb{R}$ represents the objective to be minimized, $\tilde{\xi}$ is a random vector, $G : \mathbb{R}^n \times \mathbb{R}^d \to \mathbb{R}$ is a constraint function, and $\epsilon \in (0,1)$ is a given risk parameter (significance level). To simplify the presentation, we assume, without loss of generality, that the constraint function G in (29.17) is scalar valued. Of course, a number of constraints, $G_j(x,\xi) \le 0, j = 1,\dots,m$, can be equivalently replaced by one constraint, $G(x,\xi) := \max_{1\le j\le m} G_j(x,\xi) \le 0$. Formulation (29.17) seeks a decision vector x from the feasible set S that minimizes the function $F(x)$ while satisfying the chance constraint $G(x,\tilde{\xi}) \le 0$ with probability at least $1-\epsilon$. It is assumed that the probability distribution of $\tilde{\xi}$ is known.

There are two primary sources of difficulty for CC problems:

1. In general, for a given $x \in S$, computing $\Pr[G(x,\tilde{\xi}) \le 0]$ accurately, i.e., checking whether x is feasible to (29.17), can be hard. In multidimensional situations, this involves calculation of a multivariate integral that typically cannot be computed with high accuracy.
2. The feasible region defined by a chance constraint generally is not convex even if $G(x,\tilde{\xi})$ is convex in x for every possible realization of $\tilde{\xi}$. This implies that even if checking feasibility is easy, optimization of the problem can be very difficult.

In light of the above difficulties, existing approaches for CCSO can be classified as follows. First are the approaches for problems where both difficulties are absent, i.e., the distribution of $\tilde{\xi}$ is such that checking feasibility is easy, and the resulting feasible region is convex. A classical example of this case is when $G(x, \tilde{\xi}) = v - \tilde{\xi}^\top x$ and $\tilde{\xi}$ has a multivariate normal distribution with mean μ and covariance matrix Σ. Then for $\epsilon \in (0, 0.5)$,

$$\left\{ x \in \mathbb{R}^n : \Pr\!\left[\tilde{\xi}^\top x \geq v\right] \geq 1 - \epsilon \right\} = \left\{ x \in \mathbb{R}^n : v - \mu^\top x + z_\epsilon \sqrt{x^\top \Sigma x} \leq 0 \right\},$$

where $z_\epsilon = \Phi^{-1}(1 - \epsilon)$ is the $(1 - \epsilon)$-quantile of the standard normal distribution. In this case, under convexity of S and F, the CC problem reduces to a deterministic convex optimization problem.

The second class of approaches is for problems where only the second difficulty is absent, i.e., the feasible region of the chance constraint is guaranteed to be convex. The best-known example of this case is when $G(x, \tilde{\xi}) = \tilde{\xi} - Ax$, where A is a deterministic matrix and $\tilde{\xi}$ has a *log-concave* distribution. In this case the chance constraint feasible set is convex. However, it may still be difficult to compute $\Pr\!\left[G(x, \tilde{\xi}) \leq 0\right]$ exactly. Solution methods in this class are primarily based on classical nonlinear optimization (NLO) techniques adapted with suitable approximations of the chance constraint function and its gradients. The third class of approaches is for problems where the first difficulty is absent, i.e., computing $\Pr\!\left[G(x, \tilde{\xi}) \leq 0\right]$ is easy, e.g., when $\tilde{\xi}$ has a finite distribution with a modest number of realizations. In this case the feasible region is typically nonconvex and so optimization may be difficult. A number of approaches based on integer and global optimization have been developed for this class of problems. When both difficulties are present, various approximation approaches are used. The common theme in many of these approximation approaches is that a convex approximation of the nonconvex chance constraint is designed that yields solutions that are feasible, or at least highly likely to be feasible, to the original problem. Thus the difficulty of checking feasibility as well as optimization over a nonconvex set is avoided. Unfortunately, often, the solutions produced by these approaches are quite conservative.

For the remainder of this section, we discuss an approximation approach for CC problems using Monte Carlo sampling and integer optimization. In this scheme we generate an independent and identically distributed sample $\{\xi^1, \ldots, \xi^N\}$ from $\tilde{\xi}$ and consider the problem

$$\min_x \left\{ F(x) : (1/N) \sum_{i=1}^N \mathbb{I}_+[G(x, \tilde{\xi}^i)] \leq \gamma, \ x \in S \right\}, \tag{29.18}$$

where $\mathbb{I}_+[a] = 1$ if $a > 0$ and is zero otherwise. Note that (29.18) is an approximation of the CC problem (29.17) where we have replaced the original distribution with the empirical distribution corresponding to the generated sample and the risk level ϵ by γ. Problem (29.18) can be recast as

$$\min_{x,z} \left\{ F(x) : \sum_{i=1}^N z_i \leq \gamma N, \ x \in S, \ G(x, \xi^i) \leq M_i z_i, \ z_i \in \{0,1\} \ \forall \ i = 1, \ldots, N, \right\}, \tag{29.19}$$

where M_i is a large positive number such that $M_i \geq \max_{x \in S} G(x, \xi^i)$ for all i. Note that if the binary variable z_i is zero, then the constraint $G(x, \xi^i) \leq 0$ corresponding to the realization i in the sample is enforced. On the other hand, $z_i = 1$ does not pose any restriction on $G(x, \xi^i)$. The cardinality constraint $\sum_{i=1}^{N} z_i \leq \gamma N$ requires that at least $(1 - \gamma)N$ of the N constraints $G(x, \xi^i) \leq 0$ for $i = 1, \ldots, N$ are enforced.

The sampled approximation (29.19) is a deterministic optimization problem, albeit involving binary variables. Assuming we can solve this problem, it can be shown that when $\gamma < \epsilon$, for a sufficiently large sample size N, any feasible solution to (29.19) is feasible to the true CC problem (29.17) with very high probability. Furthermore, when $\gamma \geq \epsilon$, i.e., the risk level is higher in the sampled approximation than in the true problem, then the optimal value of the sampled problem (29.19) can estimate a lower bound to the optimal value of the true problem (29.17) and thus can be used to provide a quality guarantee for a candidate solution.

In the linear setting, i.e., F, G, and S are defined by linear functions of x, problem (29.19) is a mixed-integer linear programming (MILP) problem. If N is not too large, this MILP problem can be effectively solved by standard solvers. For large N, however, the MILP problem can be quite difficult. The difficulty is primarily due to the fact that the continuous relaxation of (29.19) (obtained by dropping the integrality restriction on the z variables) provides a weak relaxation and hence slows down the branch-and-bound algorithm that is the workhorse of MILP solvers. This difficulty can be alleviated by strengthening the formulation (29.19) by tightening coefficients M_i and/or adding cuts that chop off parts of the feasible region of the continuous relaxation. Such improved formulations have tighter continuous relaxation gaps and can serve to significantly cut down solution times.

29.5 ▪ Extensions

In this chapter, we have briefly discussed two classical SO frameworks, one involving the optimization of the expectation objective and the other involving random constraints that are required to hold with a prespecified probability. There are many important extensions to the models discussed here. We conclude by briefly mentioning a few of these.

In the standard SO model (29.4), we assumed that the set X is convex and the function $F(x, \xi)$ is convex in x for all ξ. If we have integrality restrictions on some of the decision variables, then we lose convexity. Such *discrete* SO or *stochastic integer* problems are extremely challenging. The difficulty is significantly exacerbated in dynamic SO models such as (29.11). Integrality restrictions in the variables of a stage destroy the convexity and even the continuity of the underlying cost-to-go function. Accordingly, the nested decomposition scheme that is built on the convex DP formulation is no longer applicable. The scenario decomposition scheme, which uses Lagrangian duality, is also not suitable since there remains a duality gap. Integer variables are commonly required for modeling many real-world constraints, and deterministic integer optimization is a prevalent and valuable tool. Accordingly, there has been a great deal of very significant work on extending the developments in deterministic integer optimization to the stochastic setting.

An expectation objective is risk neutral and does not capture the tail behavior of the random objective, which may be much more important to the decision maker. *Risk-averse* SO models replace the expectation operator in the classical model (29.4) with a risk measure ρ:

$$\min_{x}\{\rho[F(x, \tilde{\xi})]\colon x \in X\}.$$

A risk measure maps the random cost $F(x, \tilde{\xi})$ to a real number. A suitable risk measure could be expected utility, a weighted combination of expected value and a variability measure, probability of cost exceeding a threshold, etc. Incorporating risk aversion in dynamic SO models is significantly more difficult than in the static model since often the DP structure breaks down. There is a large body of recent research addressing the computational tractability of such models and their capability of dealing with risk preferences adequately.

A common criticism of SO models is regarding the availability of an accurate distribution of the uncertain parameters. There has been a lot of theoretical work on the sensitivity of the optimal value and the set of optimal solutions of an SO problem with respect to perturbations to the distribution. A typical analysis provides a bound on the change in the optimal value or the solution set as a function of the change in the distribution. *Distributionally robust* SO models optimize the worst-case expectation with respect to a plausible family of distributions:

$$\min_{x \in X} \max_{P \in \mathbb{P}} \mathbb{E}_P[F(x, \tilde{\xi})],$$

where $\mathbb{P}$ is a family of probability distributions associated with the probability space of $\tilde{\xi}$, and $\mathbb{E}_P$ denotes the expectation with respect to the distribution P. The distribution family $\mathbb{P}$ can be specified, for example, by imposing constraints on the moments of the random vector $\tilde{\xi}$ or as a reasonable neighborhood of a nominal distribution. Optimality conditions, reformulations, and solution approaches for the above min-max problem for various types of distribution families have been investigated extensively. It turns out that there is an interesting relationship between the risk-averse and distributionally robust models. Certain reasonable risk measures ρ (called coherent) can be represented as $\rho[\tilde{Z}] = \max_{P \in \mathbb{P}} \mathbb{E}_P[\tilde{Z}]$, where $\tilde{Z}$ is a random variable, for an appropriately defined distribution family $\mathbb{P}$ for $\tilde{Z}$. Thus, these two types of models are in some sense equivalent.

Distributionally robust CC models have also been studied. These take the form

$$\min_x \left\{ F(x) : \min_{P \in \mathbb{P}} \Pr[G(x, \tilde{\xi}) \leq 0] \geq 1 - \epsilon, \ x \in S \right\},$$

where the chance constraint is defined with respect to the worst-case distribution from a given family of distributions. Most work in this area has focused on reformulations of the chance constraint into a suitable deterministic counterpart.

29.6 ▪ Bibliographical Notes

The origins of the SA method for static expectation optimization can be traced back to the 1950s with the pioneering works of Robbins and Monro [1571] and Kiefer and Wolfowitz [1060]. A number of modifications to the classical SA approach have recently been proposed; see, e.g., [1123, 1391] and references therein. While various versions of SAA had been proposed previously, a detailed methodological description appeared in Kleywegt et al. [1074]. Some computational studies of the SAA approach applied to various classes of SO problems are [1201, 1626, 1844]. A computational comparison of modern versions of SA and SAA for various classes of SO problems is reported in [1391]. Early works on dynamic SO or, more precisely, stochastic linear programming with recourse, are due to Beale [175] and Dantzig [548]. The nested decomposition method for two-stage stochastic linear optimization is due to Van Slyke and Wets [1820] and was extended to multistage stochastic linear optimization by Birge [277]. A decomposition approach that integrates sampling within the algorithm, called *stochastic decomposition*, was developed by Higle and Sen [939]. The scenario decomposition approach obtained by relaxing nonanticipativity constraints originated with the work of Rockafellar and Wets [1576]. One of the first CC models was proposed in Charnes et al. [434]. A sampling and integer optimization–based method for chance constraints was developed by Luedtke and Ahmed [1236]. For surveys on theory and algorithms for stochastic integer optimization, see, e.g., [36, 1650, 1658, 1659]. Some references on risk aversion in SO are [35, 1470, 1605, 1606]. One of the first distributionally robust SO problems was studied in the context of the newsvendor problem by Scarf [1640]. Distributional robustness in more general SO models has been studied in, e.g., [639, 1677, 1679]. For the equivalence of risk-averse with coherent risk measures and distributionally robust models, see [1606]. Distributionally robust CC models have been studied in, e.g., [390, 575]. There are a number of excellent textbooks with extensive coverage of the material discussed here. We mention a few of these. Kall and Wallace [1022] offer a great introduction to recourse models and basic decomposition algorithms. Birge and Louveaux [279] have a broad coverage of various types of models and algorithms, including those for stochastic integer optimization. Shapiro et al. [1678] offer a very deep treatment of important theoretical aspects of various types of convex SO models and chance constraints with an emphasis on sampling methods and risk aversion. Prékopa [1522] has extensive coverage of CC models. Finally, Wallace and Ziemba [1865] include a number of recent applications of SO.

Chapter 30

Application of Stochastic Optimization to Chemical Engineering

Zukui Li and Christodoulos A. Floudas

30.1 ▪ Introduction

The operational planning of a chemical plant typically entails determining both the aggregate and the daily production profiles for a time horizon between one and three months. The projected aggregate production profile can be used to ascertain the raw material requirements for the operational planning time horizon, and the daily production profile supplied by the operational planning level can be used in production scheduling of the day-to-day operations of the plant. The production profiles should represent a tight upper bound on the production capacity of the plant so that resources are not misallocated.

In traditional planning models, it is generally assumed that the system parameters are deterministic. However, given the duration of the operational planning time horizon under consideration, the robustness of the planning decisions would be greatly enhanced if the uncertain nature of some system parameters were factored into the analysis. Specifically, taking into account the uncertain nature of customer demand when making the pertinent operational planning decisions would aid in maximizing the realized profit for the plant in question by concurrently minimizing product inventory, as well as maximizing the level of customer demand satisfaction.

There exist alternative approaches for dealing with demand uncertainty at the operational planning level that capture the stochastic nature of the various demand parameters. Petkov and Maranas [1468] apply chance constraint optimization within a multiperiod planning and scheduling framework for multiproduct batch plants with uncertain demand profiles due at the end of each period. Ahmed and Sahinidis [37] develop a two-stage stochastic planning under uncertainty model that minimizes the actual cost of the first-stage model and the expected cost of the second-stage model while also minimizing the variance of the expected cost for the stochastic inner problem. Gupta et al. [881, 882] formulate a two-stage stochastic supply chain planning model with an outer deterministic production site component and an inner recourse supply chain management component that depends on the realization of customer demand. Wu and Ierapetritou [1906] present a hierarchical approach to planning and scheduling under uncertainty. This approach models uncertainty by means of multistage optimization, where the overall horizon is discretized into time periods based on the given system parameters' degrees of certainty, which vary inversely with the time horizon. Colvin and Maravelias [492] develop a multistage stochastic optimization (SO) model for clinical trial planning in new drug development. You and Grossmann [1938] utilize chance constraint optimization when considering the design of responsive supply chains under demand uncertainty. You et al. [1939] develop a two-stage stochastic linear optimization (LO) approach for the risk management of a global supply chain at the planning level, and their studies indicate that the explicit consideration of system uncertainty can lead to cost reduction. Despite the aforementioned contributions to the field, planning under uncertainty remains a challenging problem according to the reviews by [1837] and [1190].

Robust optimization belongs to an important methodology for dealing with optimization problems with data uncertainty. One of the earliest papers on robust counterpart optimization is the work of [1717], which considers simple perturbations in the data and aims at reformulating the original LO problem such that the resulting solution is feasible under all possible perturbations. Other authors [209, 210, 257, 661, 662, 988, 1189, 1191, 1199, 1840, 1841] extend the framework of robust counterpart optimization and include sophisticated solution techniques with nontrivial uncertainty sets describing the data.

Value-at-risk (VaR) and conditional value-at-risk (CVaR) [99, 1415, 1574, 1575] seek to guard against unfavorable realization of uncertain parameters by going beyond expectation evaluation when expressing the uncertainty of system parameters. CVaR represents a more pessimistic loss level when compared to VaR, so it is often the preferable representation of loss. CVaR features the important property of maintaining convexity regardless of what type of probability distribution is being utilized, while VaR exhibits inherent theoretical and algorithmic difficulties within an optimization model when the probability distribution is not normal or lognormal because the resulting problem becomes nonconvex [1107]. With the convexity property and the more pessimistic level of risk evaluation, CVaR poses a potentially attractive alternative to other uncertainty approaches.

In this chapter, the operational planning under uncertainty problem is addressed with robust optimization and CVaR optimization techniques. The operational planning problem under investigation is introduced first in Section 30.2, followed by the application of robust optimization (Section 30.3) and the CVaR method (Section 30.4). Finally, a comparison between the two methods is made and the chapter is summarized (Section 30.5).

30.2 ▪ Problem Statement

The industrial plant under investigation is a multipurpose and multiproduct batch plant that can produce hundreds of different products [986, 987, 1838]. The operational planning problem is as follows [1840]. The plant produces both made-to-order (bulk) and made-to-stock (packed) goods. For the made-to-order products, the customer supplies the plant with intermediate demand due dates, which specify on a given day the amount of a product that needs to be sent to market. The supplied intermediate demand due dates have uncertainty associated with both the requested amount of product and the day of demand realization, and both of these uncertain characteristics need

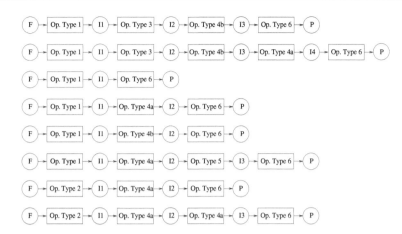

Figure 30.1. *Plant STN. Reprinted with permission from* [1840]. *Copyright* 2009 *American Chemical Society.*

to be explicitly modeled at the operational planning level. For the made-to-stock goods, the customer supplies the plant with weekly demand totals that should be satisfied by the end of the given week; however, there is uncertainty associated with the amount of demand requested at the end of each week. The plant in question has many different types of operations, and the state-task network (STN) for the plant can be seen in Figure 30.1, where the rectangles denote the various unit operations occurring within the plant and the circles denote the different states, with F, I, and P standing for feed, intermediate, and product, respectively. Given the plant and product characteristics as outlined by [986, 987] and the customer-supplied demand distribution, the objective is to determine the plant's daily production profile. For the detailed plant characteristics, the reader is directed to the work of [986, 987]. The nominal planning model was developed by [1838] and later extended to multisite production and distribution networks by [1839]. It contains the following major constraints.

Equation 30.1 calculates the total time required to undergo the activated batch processing tasks taking place within reactor u on day d:

$$\sum_{s \in S_p \cap S_u} [Fixedtime_{us} \cdot y(u,s,d)] + Cleanup \cdot \sum_{s \in S_p \cap S_u} y(u,s,d) = T^u(u,d) \ \forall u, \forall d.$$

(30.1)

For the first day, the total amount of time that reactor u can be utilized is bounded by the total available processing time for reactor u on day d:

$$T^u(u,d) \leq T_{ud} \ \forall u, \forall d | d = 1.$$

(30.2)

To allow for the carryover of unutilized reactor processing time, equation (30.3) replaces equation (30.2) for all reactors and days after the initial production day d:

$$T^u(u,d') \leq T_{ud'} + \sum_{d=1}^{d=d'-1} [T_{ud} - T^u(u,d)] \ \forall u, \forall d' | d' > 1.$$

(30.3)

Equation (30.3) allows reactor u to run for a duration of time equal to the total available processing time for a given intermediate day plus any unutilized reactor time that has accumulated. The implementation of equation (30.3) eliminates unnecessary downtime between reactor tasks due to the discretization of the time horizon. To ensure that all reactors are efficiently utilized, equation (30.4) is enforced:

$$\sum_d T^u(u,d) + slt(u) \geq H \; \forall u. \tag{30.4}$$

Equation (30.5) provides bounds on the size of a batch processing task for state s taking place within reactor u on day d:

$$Capmin_{us} \cdot y(u,s,d) \leq tot1(u,s,d) \leq Capmax_{us} \cdot y(u,s,d) \; \forall u, \forall s | s \in S_p, \forall d. \tag{30.5}$$

The total amount of state s produced on day d is equal to the summation of all the individual batches of state s produced on day d:

$$\sum_{u \in U_s} tot1(u,s,d) = tot(s,d) \; \forall s | s \in S_p, \forall d \tag{30.6}$$

Equations (30.7) and (30.8) represent the daily underproduction and overproduction demand constraints, respectively. The aforementioned constraints enforce that the amount of bulk state s produced on a given day d should be equivalent to the demand for the given state. The bulk demand constraints allow for the carryover of outstanding demand, as well as inventory from one day to the next:

$$\sum_{d'=1}^{d'=d} tot(s,d') + sl1a(s,d) \geq \sum_{d'=1}^{d'=d} r_{s,d'} \; \forall s | s \in S_p \cap S_p^b, \forall d, \tag{30.7}$$

$$\sum_{d'=1}^{d'=d} tot(s,d') - sl1b(s,d) \leq \sum_{d'=1}^{d'=d} r_{s,d'} \; \forall s | s \in S_p \cap S_p^b, \forall d. \tag{30.8}$$

Equation (30.9) expresses the inventory area term, which is the area under the inventory versus time curve for each bulk state s. The parameter Hst_d represents the plant operating time for day d and is set to 24 hours for every day within the operational planning horizon:

$$Area(s,d) = sl1b(s,d) \cdot Hst_d \; \forall s | s \in S_p \cap S_p^b, \forall d. \tag{30.9}$$

Equation (30.10) provides a lower bound on the production for packed state s by requiring that the amount of state s produced between the initial production day and the last day of week w be greater than or equal to the aggregate demand for state s within that same time period:

$$\sum_{d \leq term\_w_w} tot(s,d) + sl2a(s,w) \geq \sum_{d \leq term\_w_w} r_{s,d} \; \forall s | s \in S_p \cap S_p^p, \forall w, \tag{30.10}$$

where $term\_w_w$ is the last day of week w. Equation (30.11) provides an upper bound on packed product production that allows for some overproduction of packed states while still providing a valid upper bound on the plant production levels:

$$\frac{5}{6} \cdot \sum_{initial\_m_m \leq d \leq term\_m_m} tot(s,d) \leq \sum_{initial\_m_m \leq d \leq term\_m_m} r_{s,d}$$

$$\forall s | s \in S_p \cap S_p^p, \forall m, \tag{30.11}$$

where $initial\_m_m$ and $term\_m_m$ are the first and last days of month m, respectively. The objective function (30.12) penalizes the activation of all the slack variables present within the model while concurrently attempting to maximize the projected gross profit generated from the anticipated sale of the bulk and packed products scheduled to be produced within the time horizon of interest:

$$\min \left[\alpha \cdot \sum_{s \in S_p \cap S_p^b} \sum_d sl1a(s,d) \cdot price_s \right.$$
$$+ \beta \cdot \left(\sum_{s \in S_p \cap S_p^b} \sum_d Area(s,d) + \sum_{s \in S_p \cap S_p^p} \sum_d sl2a(s,w) \cdot price_s \right) \quad (30.12)$$
$$\left. + \gamma \cdot \sum_u slt(u) - \sum_{s \in S_p} tot(s,d) \cdot price_s \right].$$

30.3 ▪ Robust Optimization Method

In the robust optimization framework [1840], bulk products have two forms of demand uncertainty. These need to be taken into account if the production profile generated from the proposed operational planning model is to be robust with regard to bulk product demand uncertainty. Both the realized date and the required amount associated with each bulk product demand due date parameter, $r_{s,d}$, are uncertain. The actual date of demand realization is considered to follow a discrete uniform distribution with bounds of plus or minus one day from the nominal day. The realized amount of demand related to each demand due date parameter follows a user-specified probability distribution. To model concurrently both the amount and day realization, each demand due date parameter is given a unique order designation, or, and the order has a 33% chance of being realized by the previous day ($r\_or_{s,d-1,or}$), a 66% chance of being realized by the nominal day ($r\_or_{s,d,or}$), and a 100% chance of being realized by the following day ($r\_or_{s,d+1,or}$). Each possible realization of the customer order has a realized amount that follows the user-specified probability distribution (30.13), where ϵ stands for an uncertainty level and $\xi_{s,d}$ represents the random variable having a distribution that might be state and/or time specific:

$$r\_or_{s,d-1,or} = (1 + \epsilon \xi_{s,d-1}) r_{s,d},$$
$$r\_or_{s,d,or} = (1 + \epsilon \xi_{s,d}) r_{s,d}, \quad (30.13)$$
$$r\_or_{s,d+1,or} = (1 + \epsilon \xi_{s,d+1}) r_{s,d}.$$

To take into account the possibility of a customer order being realized on the previous, nominal, or following day, the daily demand constraints for the bulk products (30.7) need to be changed into the form seen in (30.14). The total demand for a bulk state s up to a given day d has now been modified to take into account the possibility of alternative order due date realization without double counting the same customer order:

$$\sum_{d'=1}^{d'=d} tot(s,d') + sl1a(s,d) \geq \sum_{or} \sum_{d'=1}^{d'=d} [r\_or_{s,d',or} - r\_or_{s,d'-1,or}(1 - last_{s,d'-1,or})]$$

$$\forall s | s \in S_p \cap S_p^b, \forall d, \quad (30.14)$$

where the parameter $last_{s,d,or}$ takes on a value of one if day d is the last day that the customer order or for state s could be realized. Equation (30.15) represents the probabilistic underproduction demand constraints. It is important to note that the term x, which stands for the reliability level of a probabilistic constraint being satisfied, has been replaced with the term $(1 - \prod_{or|prob_{s,d,or}>0} prob_{s,d,or})$, and in doing so, the level of constraint reliability is now directly related to the likelihood of the sum of the pertinent orders being realized by the day in question, where $prob_{s,d,or}$ is the probability of an individual order (or) for state s being realized by day d. The constraint reliability level now explicitly takes into account the discrete uniform distribution of demand due date realization, and, hence, the robust optimization framework now allows for the coupling of the uncertain due date with the uncertain demand amount within standard probabilistic constraints, which can in turn be reformulated into their deterministic robust counterparts:

$$Pr\left\{\sum_{d'=1}^{d'=d} tot(s,d')+sl1a(s,d)<\sum_{or}\sum_{d'=1}^{d'=d}[\tilde{r}\_or_{s,d',or}-\tilde{r}\_or_{s,d'-1,or}(1-last_{s,d'-1,or})]\right.$$

$$\left. -\delta \max\left[1,\left(\sum_{or}\sum_{d'=1}^{d'=d} r\_or_{s,d',or}-r\_or_{s,d'-1,or}(1-last_{s,d'-1,or})\right)\right]\right\}$$

$$\leq \left(1-\prod_{or|prob_{s,d,or}>0} prob_{s,d,or}\right) \forall s|s \in S_p \cap S_p^b, \forall d. \qquad (30.15)$$

Using the robust optimization framework, the generic robust bulk product underproduction demand constraints can be written as seen in equation (30.16). The function f depends on the nominal values of the uncertain right-hand side parameters, as well as the underlying probability distribution that these aforementioned stochastic elements follow:

$$\sum_{d'=1}^{d'=d} tot(s,d')+sl1a(s,d) \geq \sum_{or}\sum_{d'=1}^{d'=d}[r\_or_{s,d',or}-r\_or_{s,d'-1,or}(1-last_{s,d'-1,or})]$$

$$+\epsilon f(\lambda_1, r\_or_{s,d',or})-\delta \max\left[1,\left(\sum_{or}\sum_{d'=1}^{d'=d} r\_or_{s,d',or}-r\_or_{s,d'-1,or}(1-last_{s,d'-1,or})\right)\right]$$

$$\forall s|s \in S_p \cap S_p^b, \forall d. \qquad (30.16)$$

The robust operational planning for the normal distribution case includes all of the constraints encompassing the nominal operational planning along with the normal distribution deterministic robust counterpart demand constraints for both the bulk and packed product classes.

To demonstrate the viability of the robust optimization approach in dealing with demand uncertainty for the operational planning problem, robust operational planning for a normal distribution case has been applied to the plant in question for a three-month time horizon. The time horizon was divided into seven planning periods, each having a duration of approximately two weeks, to generate a tiering scheme for the standard deviation, σ_d, used to characterize the normal random variables utilized in the development of the normal distribution case. In other words, the standard deviations of the normal random variables used in part to express the true realization of the demand due date parameters for both the bulk and packed products varied in a positive and linear fashion with the time horizon's planning periods:

$$\sigma_d = 0.05 + 0.025(1-p) \ \forall d, \ \forall p \mid initial\_p_p \leq d \leq term\_p_p, \qquad (30.17)$$

where p stands for planning period and $initial\_p_p$ and $term\_p_p$ are the first and last days within the given planning period, respectively. For the bulk products, the due date realization was assumed to follow a uniform discrete distribution with bounds of plus or minus one day from the nominal day, while the due date was considered to be deterministic for the packed products. All of the 531 supplied demand due date parameters ($r_{s,d}$) were considered to be uncertain, with the average supplied customer order for bulk and packed products being 19.208 mu and 21.667 mu, respectively. The robust planning model was implemented using GAMS 22.5 and solved with CPLEX on a 3.2 GHz Linux workstation. A relative optimality tolerance equal to 1% was used as the termination criterion along with a five-hour time limit and an integer solution limit of 200 for the given planning model. The executed model contained 101,131 equations, 94,783 continuous variables, and 41,490 binary variables.

Figure 30.2 shows the planning period aggregate production totals for the bulk products at both the planning and scheduling levels. The figure demonstrates that robust operational planning provides a tight upper bound on the production capacity of the plant due to the relatively small and primarily positive difference between the planning level's aggregate production totals and the scheduling level's production totals for bulk products. The Gantt chart (Figure 30.3) for the three-month time horizon also demonstrates that the 13 bottleneck reactors present within the plant were highly utilized, providing further evidence that robust operational planning accurately captures the true production capacity of the plant in question.

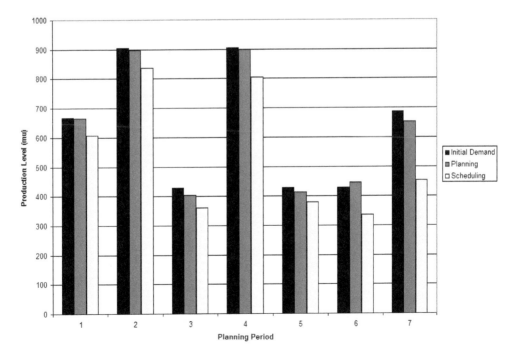

Figure 30.2. *Aggregate bulk product planning period production totals. Reprinted with permission from* [1840]. *Copyright* 2009 *American Chemical Society.*

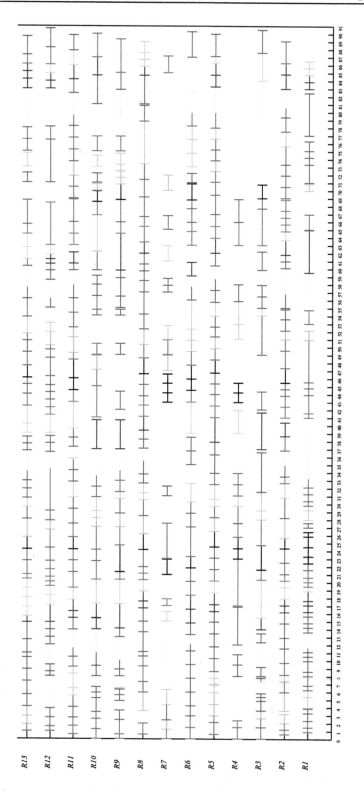

Figure 30.3. *Gantt chart for 13 bottleneck reactors for three-month time horizon.*

30.4 ▪ CVaR Method

In this section, we apply the CVaR method to the operational planning problem [1841]. We use the CVaR-based model to obtain a candidate production profile and acceptable loss function variable values. A larger, independent set of demand due date parameter scenarios is considered, and the averages and standard deviations of the CVaR demand constraints' left-hand side are calculated using the larger set of demand due date parameter scenario values along with the obtained production profile and loss function variable values. A z transformation is performed for each of the CVaR demand constraints by subtracting the average left-hand side value of the CVaR demand constraint from the original right-hand side representing the user's degree of risk aversion, and the given quantity is divided by the standard deviation of the left-hand side for the CVaR demand constraint. If the z-transformed quantity is large, then it is unlikely in a probabilistic sense that the given CVaR demand constraint would be violated by the production profile and acceptable loss function variables values currently under evaluation. As a termination criterion, the z-transformed quantity for each of the CVaR demand constraints needs to be greater than or equal to two, indicating that the right-hand side is two standard deviations greater than the average of the left-hand side. If the given termination criterion is not satisfied, then the CVaR operational planning model is solved again with another finite set of demand due date parameter scenarios. However, the right-hand side values of those CVaR demand constraints corresponding to the z-transformed quantities that were less than two are reduced to encourage the generation of a production profile and acceptable loss variable values that will satisfy the termination criterion when considering the original right-hand side values and a larger, independent set of demand due date parameter scenarios.

For a given scenario (sc), a demand due date parameter scenario, $r\_sc_{s,d,sc}$, is generated by simultaneously sampling the date distribution and amount distribution associated with each demand due date parameter $(r_{s,d})$. Having generated all of the requisite scenarios for each demand due date parameter, the following bulk product daily underproduction and overproduction CVaR demand constraints can be generated. The bulk product underproduction loss function is written weekly to stress the importance of satisfying customer orders when compared to violating inventory restrictions. The state-specific terms found within the bulk product underproduction loss function (30.18) are weighted by the respective price of the bulk product to emphasize the importance of meeting customer demand for profitable products. The underproduction loss function for a given week (w) and scenario (sc) takes on a positive value when the total weighted production of bulk product is less than the total weighted demand for the scenario, and the overproduction loss function for a given month m and scenario sc takes on a positive value when the total production is greater than the total demand for the scenario:

$$f_{bulk\_under}(w, sc) = \sum_{s \in S_p \cap S_p^b} \sum_{initial\_w_w \leq d \leq term\_w_w} price_s \cdot (r\_sc_{s,d,sc} - tot(s,d)) \ \forall w, \forall sc,$$

(30.18)

where $f_{bulk\_under}(w, sc)$ and $f_{bulk\_over}(m, sc)$ are the bulk product underproduction and overproduction loss functions for a given scenario (sc), respectively.

The bulk underproduction loss function is used in conjunction with the bulk underproduction acceptable loss variables $(\xi_{bulk\_under}(w))$ to define the loss function underproduction variables $(z_{bulk\_under}(w, sc))$, as shown in equation (30.19). If z_{bulk_under}

(w, sc) takes on a positive value, then the plant underproduced beyond the acceptable threshold for a given week (w) and scenario (sc):

$$z_{bulk\_under}(w, sc) \geq \sum_{s \in S_p \cap S_p^b} \sum_{initial\_w_w \leq d \leq term\_w_w} price_s \cdot (r\_sc_{s,d,sc} - tot(s,d))$$

$$- \xi_{bulk\_under(w)} \quad \forall w, \forall sc. \tag{30.19}$$

Having defined the requisite loss function variables $(z_{bulk\_under}(w, sc))$, the encapsulated CVaR constraints for the underproduction demand constraints can be written in the form:

$$\sum_w \xi_{bulk\_under(w)} + \frac{1}{(1-\omega)|Sc|} \cdot \sum_w \sum_{sc} z_{bulk\_under}(w, sc) \leq \delta \cdot \sum_{s \in S_p \cap S_p^b} \sum_d price_s \cdot r_{s,d}.$$

$$\tag{30.20}$$

The left-hand side of each encapsulated CVaR constraint represents the scenario-based approximation of the CVaR for the particular loss function, and the right-hand side is the acceptable loss threshold specified by the user, which partially depends on the historical demand distribution, the price of each of the products under investigation, and the user-specified parameter δ, which can take on values between zero and one and is used to express the user's level of risk aversion. Parameter ω represents a probabilistic confidence level with bounds between zero and one. Equation (30.19) replaces the nominal daily underproduction demand constraint (30.7) within the operational planning model to take into account bulk product demand uncertainty by means of CVaR theory. The averages, standard deviations, and z-transformed quantities for the CVaR demand constraints can be calculated using

$$LHS\_bulk\_under_{sc} = \sum_w \xi_{bulk\_under}(w) + \frac{1}{(1-\omega)} \sum_w \max \Bigg[0,$$

$$\left(\sum_{s \in S^p \cap S_p^b} \sum_{initial\_w_w \leq d \leq term\_w_w} price_s \cdot (r\_sc_{s,d,sc} - tot(s,d)) - \xi_{bulk\_under}(w) \right) \Bigg].$$

$$\tag{30.21}$$

The CVaR method has been applied to the same problem. One thousand scenarios were used in the application of the CVaR operational planning, and 60,000 scenarios were utilized within the implemented sample average approximation (SAA) algorithm to ensure that the final planning level production profile represents a solution that is a valid, feasible upper bound. The planning model was implemented using GAMS 22.5 and solved with CPLEX on a 3.2 GHz Linux workstation. A relative optimality tolerance equal to 10% was used as the termination criterion along with a five-hour time limit and an integer solution limit of 200 for the given planning model. The executed model contained 113,199 equations, 111,858 continuous variables, and 41,490 binary variables. CVaR operational planning was applied six times within the presented SAA algorithm before a valid, feasible upper bound was determined, and Table 30.1 shows the algorithm's progression. Each time a given CVaR demand constraint did not meet the aforementioned termination criterion, the right-hand side of the given constraint was reduced by 1% from its previous value before executing CVaR operational planning again. The conservative level of 1% was chosen to maintain model feasibility for every iteration of the SAA algorithm.

Table 30.1. *SAA algorithm progress. Reprinted with permission from* [1841]. *Copyright* 2009 *American Chemical Society.*

	Pack_Under	Pack_Over	Bulk_Under	Bulk_Over
		Iteration 1		
Avg.:	93711.8	979.8	39648.9	595.8
Std.:	0.0	121.9	9249.2	6.0
z:	∞	−0.035	−0.278	12.028
		Iteration 2		
Avg.:	87464.3	917.7	7115.9	587.7
Std.:	0.0	130.1	7095.5	5.9
z:	∞	0.444	4.223	13.583
		Iteration 3		
Avg.:	87464.6	847.7	17723.9	592.9
Std.:	54.6	124.0	7894.1	4.6
z:	114.422	1.030	2.424	16.370
		Iteration 4		
Avg.:	87464.3	787.8	22316.1	591.2
Std.:	0.0	128.7	11130.0	5.8
z:	∞	1.458	1.326	13.205
		Iteration 5		
Avg.:	87484.1	712.2	32260.7	596.2
Std.:	489.9	116.1	8988.0	4.8
z:	12.711	2.269	0.536	14.982
		Iteration 6		
Avg.:	87473.0	713.4	6854.6	589.6
Std.:	334.2	127.0	7213.9	5.4
z:	18.670	2.063	4.190	14.507

Table 30.2. *CVaR and robust production/profit difference comparison. Reprinted with permission from* [1841]. *Copyright* 2009 *American Chemical Society.*

	Robust		CVAR	
		Production		
	Bulk	Packed	Bulk	Packed
Avg.	85.887	195.104	73.565	212.247
Std.	56.252	91.626	75.536	146.362
		Profit		
	Bulk	Packed	Bulk	Packed
Avg.	4334.94	19433.67	4802.42	25667.90
Std.	3156.45	8593.24	4886.68	19341.54

Table 30.2 compares the planning and scheduling planning period production and profit totals for the CVaR operational planning and the robust operational planning applications, and it demonstrates that even though both planning approaches provide acceptable upper bounds on plant production capacity, robust operational planning on average provides a tighter upper bound on packed product production, while CVaR operational planning on average provides a tighter upper bound on bulk production.

30.5 ▪ Summary

The robust optimization technique and the scenario-based CVaR approach presented in this chapter represent two different ways of explicitly addressing demand due date and amount uncertainty for the operational planning problem. The robust optimization technique maintains the same functional form for the underproduction and overproduction demand constraints without any aggregation. On the other hand, the scenario-based CVaR approach requires aggregation of the demand constraints on both the time and state levels; however, as evidenced by the viable production profile generated by CVaR operational planning, the chosen levels of demand aggregation helped to maintain computational tractability while still allowing for a high level of solution quality. The presented scenario-based CVaR approach can adapt to any probability distribution under consideration without having to change model constraints, while the application of the robust optimization technique is distribution dependent, requiring model reformulation. A potentially detrimental characteristic of CVaR operational planning is the increased computational time caused by the application of the sample average approximation algorithm, which is used to ensure that the generated production profile represents a solution that is a valid, feasible upper bound. On the other hand, the robust operational planning model only needs to be applied once to generate the desired production profile, which can then be supplied to the scheduling level. The planning and scheduling optimization results indicate that both proposed planning models are viable alternative approaches.

Chapter 31

On the Marginal Value of Water for Hydroelectricity

Andy Philpott[21]

31.1 ▪ Introduction

This chapter discusses optimization models for computing prices in perfectly competitive wholesale electricity markets that are dominated by hydroelectric generation. We revisit the relationship between partial equilibrium in perfect competition and system optimization of a social planning problem, and we use the latter to show how perfectly competitive electricity prices correspond to marginal water values.

Most industrialized regions of the world have over the last 20 years established wholesale electricity markets that take the form of an auction that matches supply and demand. The exact form of these auction mechanisms varies by jurisdiction, but they typically require offers of energy from suppliers at costs at which they are willing to supply, and clear a market by dispatching these offers in order of increasing cost. Day-ahead markets, such as those implemented in most North American regions, seek to arrange supply well in advance of its demand so that thermal units can be prepared in time. Since the demand cannot be predicted with absolute certainty, these day-ahead markets must be augmented with balancing markets to deal with the variation in load and generator availability in real time.

In this chapter, we study markets with hydroelectric generators that use reservoirs of stored water. Generators with hydroelectric reservoirs face an inventory problem. They would like to optimize the release of water from reservoirs to maximize profits using a stochastic process of prices, but this process is not known and must be deduced by each agent using current and future market conditions and hydrological models of future reservoir inflows. For an agent controlling releases from a hydroelectric reservoir, the marginal cost of supply in the current period involves some modeling of opportunity cost that includes possible high prices in future states of the world with low inflows.

[21]The research described in this chapter has benefited from discussions with Ziming Guan, Tony Downward, Geoffrey Pritchard, and Golbon Zakeri.

In this chapter, we restrict our attention to a setting in which all agents are price-takers who do not act strategically. Under this assumption, it is well known that electricity prices for a single trading period and a single location can be computed as shadow prices from a deterministic convex economic dispatch (ED) model that maximizes total social welfare. In a stochastic setting, where there is a stochastic process of inflows that is known to all agents, and they maximize expected profit using the probability law determining these inflows, a competitive (partial) equilibrium will correspond to the water-release policy that maximizes the expected welfare of a social planner. A stochastic optimization (SO) for maximizing expected social welfare can therefore be used to estimate the marginal cost of electricity supply in future states of the world. These values are estimates of the wholesale electricity prices in these states. This approach is demonstrated in Section 31.2 using a simple model.

In a system with a single storage reservoir, electricity prices can be shown to be equivalent to the expected marginal value of water in the reservoir, in other words the extra social benefit that would be obtained from the optimal deployment of an extra cubic meter of stored water. Observe that this value comes from the optimal marginal use of the water and so the planner must have an optimization problem in mind. Policy iteration algorithms based on an early paper by Stage and Larsson [1719] are often used in practice to compute these marginal values. In Section 31.3, we apply these methods to a simple example.

Most hydroelectricity systems comprise several reservoir storages, and so the social planner's optimization problem becomes a stochastic dynamic programming (DP) problem with a matching number of state variables. When this number exceeds the acceptable state dimension for classical DP (about three or four), the optimization method used most in practice is stochastic dual dynamic programming (SDDP) due to Pereira and Pinto [1464]. In Section 31.4, we outline this method, and we apply it to an example problem from the New Zealand electricity system in Section 31.5. The estimated prices from a case study in 2008 are then compared with their historical counterparts.

31.2 • The Social Optimum

We consider a social planning problem in an electricity system of hydroelectric generating plants fed by reservoirs ($i \in \mathcal{H}$) that are subject to uncertain inflows over some decision horizon of T periods. The system also has generating plants ($j \in \mathcal{T}$) that run on thermal fuel at some known generation cost, and consumer segments $c \in \mathcal{C}$. For simplicity we shall assume that all generating plants sell power into an unconstrained transmission grid so that consumer demand can be aggregated at a single location to which power is sent.

In general, the reservoirs are connected in a set of river chains represented by a network of m nodes (reservoirs and junctions) and l arcs (canals or river reaches). The topology of the network can be represented by the $m \times l$ incidence matrix A, where

$$
a_{ik} = \begin{cases} 1 & \text{if node } i \text{ is the tail of arc } k, \\ -1 & \text{if node } i \text{ is the head of arc } k, \\ 0 & \text{otherwise.} \end{cases}
$$

By adding dummy nodes if necessary, we can ensure that every pair of nodes is joined by at most one arc. If there are several river chains, then the network need not be connected.

We first describe a deterministic model over the planning horizon $t = 1, 2, \ldots, T$. We let $u(t)$ be a vector of flow rates (cubic meters per period) in the arcs in the network in period t. Some arcs correspond to generating stations. Some might correspond to spill around a station. For each arc $k = 1, 2, \ldots, l$, we specify a generation function $G_k(u_k)$ that is zero when the arc does not contain a station, and otherwise is a strictly concave function of flow rate u_k for arcs corresponding to stations, giving the energy output in a period when the flow rate is u_k. Let $x(t)$ denote a vector of reservoir storages in each node at the beginning of period t and $\omega(t)$ be a vector of uncontrolled reservoir inflows (in cubic meters) that have occurred in period t. These define a water balance constraint

$$x(t+1) = x(t) - Au(t) + \omega(t), \; t = 1, 2, \ldots, T.$$

The hydro generators are accompanied by thermal generators $j \in \mathcal{T}$ that produce a vector v of energy that costs $\sum_{j \in \mathcal{T}} C_j(v_j)$ to produce, where C_j is a strictly convex function. Demand of consumer segment $c \in \mathcal{C}$ is represented by d_c, where the strictly concave function $D_c(d_c)$ denotes the welfare accrued by consumer segment c. Note that the derivative D_c' represents the inverse demand curve for consumer c, which is strictly decreasing by assumption. Demand must be satisfied in each time period, so, in a deterministic model,

$$\sum_{k=1}^{l} G_k(u_k(t)) + \sum_{j \in \mathcal{T}} v_j(t) \geq \sum_{c \in \mathcal{C}} d_c(t), \; t = 1, 2, \ldots, T.$$

When inflows to the reservoirs are random variables, the sequence of time intervals $1, 2, \ldots, T$ is replaced by a stochastic process. We represent this by a scenario tree, as shown in Figure 31.1. Here each scenario tree node n spans a period t (which in our models is a week) and corresponds to a realization $\omega(n)$ of reservoir inflows in that period. To account for different inflow sequences, all variables previously indexed by t are now indexed by n. Each node has probability $\phi(n)$, where the probabilities of nodes in a given period sum to one. We denote by $n-$ the unique parent node of node n and by $n+$ its set of children.

The leaf nodes $\mathcal{L}$ of the tree correspond to the end of the decision horizon (period T). For each leaf node $n \in \mathcal{L}$, we specify a strictly convex future-value function $V_n(x(n))$ that measures the total future welfare in state n of having reservoir levels $x(n)$.

We are now in a position to formulate a stochastic social planning (SSP) problem. This is

$$\text{SSP: min } \sum_{n \in \mathcal{N}} \phi(n) \left(\sum_{j \in \mathcal{T}} C_j(v_j(n)) - \sum_{c \in \mathcal{C}} D_c(d_c(n)) \right.$$
$$\left. - \sum_{n \in \mathcal{L}} \phi(n) V(x(n)) \right)$$

$$\text{s.t.} \quad \sum_{k=1}^{l} G_k(u_k(n)) + \sum_{j \in \mathcal{T}} v_j(n) \geq \sum_{c \in \mathcal{C}} d_c(n), \qquad n \in \mathcal{N},$$

$$x(n) = x(n-) - Au(n) + \omega(n), \qquad n \in \mathcal{N},$$

$$d(n) \geq 0, \, v(n) \geq 0, \, x(n) \in \mathcal{X}, \, u(n) \in \mathcal{U}, \, n \in \mathcal{N}.$$

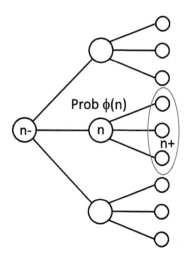

Figure 31.1. *Each node n spans a period and corresponds to a realization $\Omega(n)$ of reservoir inflows in that period.*

Here we assume that $\mathscr{X}$ and $\mathscr{U}$ are compact convex sets and SSP has a nonempty feasible region. We shall assume in what follows that all nonlinear constraints satisfy constraint qualifications to guarantee the existence of Lagrange multipliers. In particular, we have Lagrange multipliers $\phi(n)p(n)$ for each $n \in \mathscr{N}$ so that we can solve SSP by minimizing the Lagrangian

$$\text{LSSP:} \quad \min \quad \sum_{n \in \mathscr{N}} \phi(n)\left(\sum_{c \in \mathscr{C}} p(n)d_c(n) - \sum_{c \in \mathscr{C}} D_c(d_c(n))\right.$$

$$\left. + \sum_{j \in \mathscr{J}}\left(C_j(v_j(n)) - p(n)v_j(n)\right) - V(x(n)) - \sum_{k=1}^{l} p(n)G_k(u_k(n))\right)$$

$$\text{s.t.} \quad x(n) = x(n-) - Au(n) + \omega(n), \qquad n \in \mathscr{N},$$

$$d(n) \geq 0, \, v(n) \geq 0, \, x(n) \in \mathscr{X}, \, u(n) \in \mathscr{U}, \, n \in \mathscr{N}.$$

LSSP separates by agent. The consumer solves

$$\text{CP}(c): \max \quad \sum_{n \in \mathscr{N}} \phi(n)(D_c(d_c(n)) - p(n)d_c(n))$$

$$\text{s.t.} \quad d_c(n) \geq 0.$$

Each thermal generator solves

$$\text{TP}(j): \max \quad \sum_{n \in \mathscr{N}} \phi(n)(p(n)v_j(n) - C_j(v_j(n)))$$

$$\text{s.t.} \quad v_j(n) \geq 0,$$

and the hydro generators together solve

$$\text{HP:} \max \quad \sum_{n \in \mathscr{N}} \phi(n) \sum_{k=1}^{l} p(n)G_k(u_k(n)) + \sum_{n \in \mathscr{L}} \phi(n)V(x(n))$$

$$\text{s.t.} \quad x(n) = x(n-) - Au(n) + \omega(n), \qquad\qquad n \in \mathscr{N},$$

$$x(n) \in \mathscr{X}, \, u(n) \in \mathscr{U}, \qquad\qquad\qquad n \in \mathscr{N}.$$

If all the river chains are operated by a single agent, then HP represents the problem of maximizing this agent's expected revenue and residual value given prices $p(n)$ in each state of the world. If there is more than one hydro agent and HP is not separable by agent, then we require an additional condition to establish a correspondence between competitive equilibrium and an optimal social plan. Let $\mathcal{H}$ denote the set of hydro agents and x_i the vector of water stocks for reservoirs operated by agent i. We require the existence of functions $V_i(x_i)$ so that

$$V(x(n)) = \sum_{i \in \mathcal{H}} V_i(x_i(n)).$$

Even with this condition, to enable a separation into agent problems, in general we require a price $\phi(n)\pi(n)$ for the flow conservation constraint. This gives a Lagrangian for HP, which is

$$\mathcal{L}(x, u, \pi) = \sum_{n \in \mathcal{N}} \phi(n) \sum_{k=1}^{l} p(n) G_k(u_k(n)) + \sum_{n \in \mathcal{L}} \phi(n) \sum_{i \in \mathcal{H}} V_i(x_i(n))$$
$$+ \sum_{n \in \mathcal{N}} \phi(n) \pi(n)^{\top} (x(n-) - Au(n) + \omega(n) - x(n)),$$

that we maximize over $x(n) \in \mathcal{X}$, $u(n) \in \mathcal{U}$. $\mathcal{L}(x, u, \pi)$ can be rearranged to give

$$\mathcal{L}(x, u, \pi) = \sum_{n \in \mathcal{N}} \phi(n) \left(\sum_{k=1}^{l} p(n) G_k(u_k(n)) - \pi(n)^{\top} Au(n) \right)$$
$$+ \sum_{n \in \mathcal{N} \setminus \mathcal{L}} \phi(n) x(n) \left(\frac{\sum_{m \in n+} \phi(m) \pi(m)}{\phi(n)} - \pi(n) \right)$$
$$+ \sum_{n \in \mathcal{L}} \phi(n) \sum_{i \in \mathcal{H}} (V_i(x_i(n)) - \pi_i(n) x_i(n))$$
$$+ \sum_{n \in \mathcal{N}} \phi(n) \pi(n)^{\top} \omega(n).$$

The values of $\pi(n)$ are water prices, often called *marginal water values*. At the end of the planning horizon, if $x_i(n)$ is not at a bound, then we require for a maximum that

$$V_i'(x_i(n)) = \pi_i(n), \quad n \in \mathcal{L}.$$

For flow $u_k(n)$ through station arc k from h to i, we have

$$-\pi(n)^{\top} Au(n) = (\pi_i(n) - \pi_h(n)) u_k(n),$$

so the owner of station k attempts in state of the world n to maximize

$$\sum_{k=1}^{l} p(n) G_k(u_k(n)) + (\pi_i(n) - \pi_h(n)) u_k(n).$$

In other words, the owner maximizes revenue from $u_k(n)$ at price $p(n)$, while paying $\pi_h(n)$ for upstream water and getting paid $\pi_i(n)$ for the water supplied to the downstream reservoir. At the optimal solution u_k^*, we have

$$\pi_h(n) - \pi_i(n) = p(n) G_k'(u_k^*(n)), \tag{31.1}$$

which relates the spot price of energy $p(n)$ to the marginal water value difference and the marginal conversion factor from water to energy.

A reservoir owner in state of the world n makes $x(n)$ as large as possible when the expected marginal water value at the next time period exceeds that at the current period and makes $x(n)$ as small as possible when the expected marginal water value at the next time period is lower than that at the current period. These prices in equilibrium enable different owners on the river to extract the maximum welfare from the water. If there is no possibility of trading in water at market prices, then the market is incomplete, and the competitive equilibrium may not be the same as the optimal social plan. This observation is well known and is studied by Lino et al. [1202].

Henceforth we will assume that all the reservoirs and generating stations on a given river chain are owned and operated by the same generator, and that all the river chains have separate catchments. This means that HP separates into independent river chains, each operated by a single agent. Hydro agent i solves

$$\text{HP}(i):\ \max\ \sum_{n\in\mathcal{N}}\phi(n)\sum_{k=1}^{l(i)}p(n)G_k\left(u_k(n)\right)+\sum_{n\in\mathcal{L}}\phi(n)V_i(x(n))$$

$$\text{s.t.}\quad x(n)=x(n-)-Au(n)+\omega(n),\qquad\qquad n\in\mathcal{N},$$

$$x(n),u(n)\geq 0,$$

where now $x(n)$ $(\omega(n))$ denotes the vector of reservoir storages (inflows) on river chain i and $u(n)\in\mathbb{R}^{l(i)}$ the vector of flows through the $l(i)$ arcs in the river chain.

This defines a perfectly competitive equilibrium defined by the individual optimality conditions and market clearing condition:

$$\text{CE:}\quad u^i(n),x^i(n)\in\arg\max\text{HP}(i),$$

$$v_j(n)\in\arg\max\text{TP}(j),$$

$$d_c(n)\in\arg\max\text{CP}(c),$$

$$0\leq\sum_{i\in\mathcal{H}}\sum_{k=1}^{l(i)}G_k\left(u_k^i(n)\right)+\sum_{j\in\mathcal{T}}v_j(n)-\sum_{c\in\mathcal{C}}d_c(n)\perp p(n)\geq 0.$$

By comparing optimality conditions, we have the following result.

Proposition 31.1. *Suppose all the reservoirs and generating stations on any given river chain are owned and operated by the same generator and*

$$V(x)=\sum_{i\in\mathcal{H}}V_i(x_i).$$

The (unique) welfare-maximizing solution to SSP is the same as the (unique) competitive equilibrium solution to CE.

Each model HP(i) can be solved by computing an optimal water release policy to maximize the expected revenue at energy prices $p(n)$. The optimal policy will generate Lagrange multipliers $\pi(n)$ for the water conservation constraints. These will be related to the energy prices through (31.1). When each hydro agent controls a single reservoir, then it is tempting to suppose that its optimal water release policy can be computed by price decomposition and DP. Unfortunately, this is not always straight-forward because the energy prices $p(n)$ are not guaranteed to be stagewise independent even if the inflows are independent [168]. On the other hand, DP can be applied to the system optimization problem as long as the number of reservoirs (corresponding to states) is not too large. Given a system optimal solution to the social planning problem, it is possible to derive the competitive equilibrium prices and agent policies using the above proposition.

In the next section, we study the computation of competitive equilibrium prices for the simplest possible system, with one hydro agent and one reservoir. We then show how models with several reservoirs might be solved using SDDP [1464, 1676]. We conclude with some results of estimating competitive prices in the New Zealand wholesale electricity market.

31.3 ▪ A Single-Reservoir Model

One of the first models for reservoir water valuation was developed by Stage and Larsson [1719]. It seeks to compute marginal water values $w_i(t)$ corresponding to discrete reservoir levels $i = 0, 1, \ldots, N$ using a form of policy iteration. When the reservoir is full ($i = N$), the marginal value of water is set to zero (here we assume that surplus water can be spilled with no penalty). When the reservoir is empty ($i = 0$), the marginal value of water is set to some high value that we denote as C. Typically this will be a shortage cost, also known as the *value of lost load*. Note that this convention implies that the marginal value at level i is the directional derivative

$$w_i = W_i - W_{i+1},$$

where W_i is the average expected cost of meeting demand in each future period if the current reservoir level is i. If W_i is linearly interpolated, then we obtain a nonsmooth convex function. This implies

$$W_{i-1} - W_i = w_{i-1} \geq w_i = W_i - W_{i+1},$$

with strict inequality occurring as the rule.

The expected marginal value of water at level $i = j$, say, is determined in [1719] by constructing sample paths of reservoir levels that would be produced by an optimal release policy (say to meet a known demand) under random sequences of inflows. Starting at level $i = j$ at time of year t, s out of S such sample paths will reach $i = 0$ before $i = N$. The expected marginal water value $w_j(t)$ is then estimated to be $\frac{s}{S}C$. This process is repeated for all starting levels and times.

Although the Stage and Larsson method is quite general, we will examine it here as applied to a specific problem instance. We assume that reservoir inflows are stagewise independent and have the same distribution at each time of the year. Thus we can dispense with the time index in the formulation and seek values w_i. To simplify the analysis further we will assume that the reservoir has inflow in each period that equals two with probability p and zero with probability $(1-p)$. Demand for energy requires that we release exactly one unit of water in each period. Thus in each time step the reservoir level is a simple random walk, increasing by one unit with probability p and decreasing by one unit with probability $1 - p$. Observe that there is no policy choice since the water usage is prescribed by demand. In this setting, the Stage and Larsson method amounts to iterating toward the solution of a fixed-point problem.

In other words, we obtain the following procedure:

Set $\qquad \nu = 0, \quad w_i^\nu = 0;$
repeat
$\qquad$ for $\quad i = 0$ to N do
$\qquad\qquad$ if $i = 0,$ $\qquad\qquad$ set $w_i^{\nu+1} = \quad C;$
$\qquad\qquad$ if $i = N,$ $\qquad\qquad$ set $w_i^{\nu+1} = \quad 0;$
$\qquad\qquad$ if $0 < i < N,$ $\qquad$ set $w_i^{\nu+1} = \quad (1-p)w_{i-1}^\nu + p w_{i+1}^\nu;$
$\qquad$ end $\quad$ for;
$\qquad\qquad$ set $\quad \nu := \nu + 1$
until $\quad w^{\nu+1} = w^\nu.$

It is easy to see that this procedure converges to the fixed point defined by

$$w_i = (1-p)w_{i-1} + p w_{i+1}, \quad i = 1, 2, \ldots, N-1, \quad w_0 = C, \quad w_N = 0. \qquad (31.2)$$

In fact this simple model can be solved exactly for w because there is a closed-form solution for the probability that any random walk starting at $i = j$ will hit barrier $i = 0$ before it hits barrier $i = N$ (see, e.g., [856]). When $p = 0.5$, this gives $w_j = C\frac{N-j}{N}$. Otherwise, letting $r = \frac{1-p}{p}$, we get

$$w_j = \begin{cases} C\frac{1-r}{1-r^N}\left(r^j + r^{j+1} + \cdots + r^{N-1}\right), & j < N, \\ 0, & j = N. \end{cases}$$

The solution corresponds to the fixed-point calculation

$$\begin{aligned} \eta + W_0 &= C + (1-p)(W_0) + p W_2, \\ \eta + W_i &= (1-p)W_{i-1} + p W_{i+1}, \quad i = 1, 2, \ldots, N, \\ W_{N+1} &= W_N, \end{aligned} \qquad (31.3)$$

for an average cost per period Markov decision process, where η is the long-run average cost per period of meeting demand. Given a solution W to (31.3), it is easy to see that setting $w_i = W_i - W_{i+1}$ solves (31.2).

An alternative model uses discounting. Suppose costs are discounted in each period with rate α, giving a discount factor $\gamma = \frac{1}{1+\alpha}$. This gives a slightly different fixed-point problem:

$$w_i = (1-p)\gamma w_{i-1} + p\gamma w_{i+1}, \quad i = 1, 2, \ldots, N-1, \quad w_0 = C, \quad w_N = 0. \qquad (31.4)$$

The discounted solution corresponds to the fixed-point calculation

$$\begin{aligned} W_0 &= C + \gamma\left((1-p)W_0 + p W_2\right), \\ W_i &= \gamma\left((1-p)W_{i-1} + p W_{i+1}\right), \quad i = 1, 2, \ldots, N, \\ W_{N+1} &= W_N. \end{aligned} \qquad (31.5)$$

Here W_j is the expected discounted cost of meeting demand starting from reservoir level $i = j$. Given a solution W to (31.5), setting $w_i = W_i - W_{i+1}$ solves (31.4).

In both of these models there is an implicit assumption on how shortage costs are incurred. When the reservoir level hits zero, it is assumed that the cost C is incurred immediately. This assumes that none of the inflow occurring over the next period can be used to cover demand. In other words, this is a *here-and-now* or *decision-hazard* problem, which is modeled using (31.3) or (31.5).

In contrast, one might use a *wait-and-see* or *hazard-decision* assumption. Given a reservoir level $i = j$, the random inflow is observed and used with reservoir water to meet demand if possible. Inflow that cannot be stored is spilled. Thus, when the reservoir is empty, with probability $1 - p$ we must meet demand with no inflow, incurring a cost of C, resulting in an empty reservoir, or we meet demand with an inflow of two units, leaving the reservoir with one unit in it. When the reservoir is full and inflow is two, we can use half of it and leave the reservoir full. If the inflow is zero, then the reservoir moves to state $N - 1$. The fixed-point problem (31.3) is now

$$
\begin{aligned}
\eta + S_0 &= (1 - p)(C + S_0) + p S_2, \\
\eta + S_i &= (1 - p)S_{i-1} + p S_{i+1}, \qquad i = 1, 2, \ldots, N, \\
S_{N+1} &= S_N,
\end{aligned}
\tag{31.6}
$$

and (31.5) becomes

$$
\begin{aligned}
S_0 &= (1 - p)(C + \gamma S_0) + p \gamma S_2, \\
S_i &= (1 - p)\gamma S_{i-1} + p \gamma S_{i+1}, \qquad i = 1, 2, \ldots, N, \\
S_{N+1} &= S_N,
\end{aligned}
\tag{31.7}
$$

where in (31.6) S_j is the expected (wait-and-see) discounted cost of meeting demand starting from reservoir level $i = j$, and in (31.7) S_j is the corresponding value in the average cost per period model.

31.3.1 ▪ Thermal Generation

The Stage and Larsson methodology can be extended to a setting where the policy being simulated generates energy using thermal plant (at some cost) as well as using releases from the hydro reservoir. This is the basis for a number of reservoir optimization models in practical use (see, e.g., [327]). It is important to observe that the values w_i derived from such a simulation will only be an expected marginal value for an optimization problem, if the policy being simulated is indeed optimal for that problem. In other words, the expected marginal value of water is the expected extra benefit that accrues from using this water in the most beneficial way. The optimal policy ultimately requires the solution of a stochastic dynamic optimization problem.

To include thermal generation in our example, we alter demand so that it is two units per period and suppose there is a thermal generator that costs $c < C$ to produce one unit of energy. Thus, to meet demand, we can either release two units of water or release one unit of water and run the thermal plant. The thermal generator is assumed to be flexible in that it can be switched on and off with no penalty. It is typical to specify an optimal policy in terms of a marginal water value as follows. If $c < w_i$, then we run the thermal plant. If $c > w_i$, then we release water. If $c = w_i$, then the policy can be any combination of thermal generation and water release.

In this model, the Stage and Larsson methodology (with discounting) might yield the following marginal water value algorithm:

Set $v = 0, \quad w_i^\nu = 0;$

repeat

$\quad\quad$ for $\quad i = 0$ to N do

$\quad\quad\quad\quad$ if $i = 0,$ $\quad\quad\quad\quad$ set $w_i^{\nu+1} = \quad C;$

$\quad\quad\quad\quad$ if $i = N,$ $\quad\quad\quad\quad$ set $w_i^{\nu+1} = \quad 0;$

$\quad\quad\quad\quad$ if $0 < i < N,$ $\quad\quad$ if $w_i^\nu < c, \quad w_i^{\nu+1} = (1-p)\gamma w_{i-2}^\nu + p\gamma w_i^\nu;$

$\quad\quad\quad\quad\quad\quad\quad\quad\quad\quad\quad\quad\quad\quad\quad$ if $w_i^\nu > c, \quad w_i^{\nu+1} = (1-p)\gamma w_{i-1}^\nu + p\gamma w_{i+1}^\nu;$

$\quad\quad\quad$ end $\quad$ for;

$\quad\quad\quad\quad\quad\quad$ set $\quad v := v + 1;$

until $\quad\quad w^{\nu+1} = w^\nu.$

If we relax the integrality constraint so that the thermal plant can generate any amount between zero and one, then the marginal water value will equal c when this plant is partially dispatched. There is a range of fractional water levels between x and $x + 1$, say, for which this will be the optimal action. For almost every instance of this relaxed problem, the marginal value of water will equal c at exactly one integer level i and be different from c for other levels. The correct stationarity conditions (for any choice of γ) that correspond to the minimum cost solution are

$$
\begin{aligned}
w_0 &= C, \\
w_i &= \gamma\big((1-p)w_{i-1} + p w_{i+1}\big), \quad i = 1, \dots, k-1, \\
w_k &= c \in [w_{k+1}, w_{k-1}], \quad\quad\quad\quad i = k, \\
w_i &= \gamma\big((1-p)w_{i-2} + p w_i\big), \quad\quad i = k+1, \dots, N-1, \\
w_N &= 0.
\end{aligned}
\tag{31.8}
$$

The correct value of k can be found by computing an optimal policy. Let W_i denote the expected future discounted cost of meeting demand. The here-and-now fixed-point problem (with discounting) is

$$
\begin{aligned}
W_0 &= C + c + \gamma\big((1-p)W_0 + p W_2\big), \\
W_1 &= c + \gamma\big((1-p)W_0 + p W_2\big), \\
W_i &= \min\{c + \gamma\big((1-p)W_{i-1} + p W_{i+1}\big), \gamma\big((1-p)W_{i-2} + p W_i\big)\}, \quad i = 2, \dots, N, \\
W_{N+1} &= W_N.
\end{aligned}
\tag{31.9}
$$

The optimal solution gives a threshold type of policy, i.e., we release two units of water if

$$
\gamma\big((1-p)W_{i-2} + p W_i\big) < c + \gamma\big((1-p)W_{i-1} + p W_{i+1}\big)
$$

or

$$
c > \gamma\big((1-p)w_{i-2} + p w_i\big).
$$

The right-hand side is the expected discounted marginal water value if we meet demand from hydro only, and water is released if this is less than the cost c of thermal generation.

An optimal policy can be found using linear optimization (LO) and determines a threshold index k below which the thermal plant is always run at capacity and above which it is not. Given this value of k, the optimal policy satisfies

$$
\begin{aligned}
W_0 &= C + c + \gamma\big((1-p)W_0 + p W_2\big), \\
W_1 &= c + \gamma\big((1-p)W_0 + p W_2\big), \\
W_i &= c + \gamma\big((1-p)W_{i-1} + p W_{i+1}\big), \quad i = 2, \dots, k, \\
W_i &= \gamma\big((1-p)W_{i-2} + p W_i\big), \quad\quad i = k+1, \dots, N \\
W_{N+1} &= W_N.
\end{aligned}
$$

By setting $w_i = W_i - W_{i+1}$, we can derive the marginal values w that will hold at optimality. These satisfy

$$w_0 = C,$$
$$w_i = \gamma\left((1-p)w_{i-1} + p\,w_{i+1}\right), \quad i = 1,\dots,k-1,$$
$$w_k = c \in [w_{k+1}, w_{k-1}],$$
$$w_i = \gamma\left((1-p)w_{i-2} + p\,w_i\right), \quad i = k+1,\dots,N-1,$$
$$w_N = 0, \tag{31.10}$$

which are the same equations as (31.8).

The case without discounting is similar. If we form an average cost per period Markov decision problem, then we seek the solution to

$$\begin{aligned}
\max \quad & g \\
& W_0 = -\eta + C + c + (1-p)W_0 + p\,W_2, \\
& W_1 = -\eta + c + (1-p)W_0 + p\,W_2, \\
& W_i \le -\eta + c + (1-p)W_{i-1} + p\,W_{i+1}, \quad i = 2,\dots,N, \\
& W_i \le -\eta + (1-p)W_{i-2} + p\,W_i, \quad i = 2,\dots,N, \\
& W_{N+1} = W_N.
\end{aligned} \tag{31.11}$$

As before, the optimal policy can be found using LO and determines a threshold index k below which the thermal plant is always run at capacity and above which it is not. Given this value of k, we can derive the marginal values w that will hold at optimality. This gives the system

$$w_0 = C,$$
$$w_i = (1-p)w_{i-1} + p\,w_{i+1}, \quad i = 1,\dots,k-1,$$
$$w_k = c,$$
$$w_i = (1-p)w_{i-2} + p\,w_i, \quad i = k+1,\dots,N-1,$$
$$w_N = 0, \tag{31.12}$$

which is the same as (31.8) when $\gamma = 1$. In fact in the average cost per period Markov decision problem for this system, k turns out to be always equal to $N-1$, and

$$w_i = \begin{cases}
C, & i = 0, \\
\frac{ci + C(N-1-i)}{N-1}, & i = 1,\dots,N-2, \\
c, & i = N-1, \\
0, & i = N,
\end{cases} \tag{31.13}$$

when $p = 0.5$ and, recalling $r = \frac{1-p}{p}$,

$$w_i = \begin{cases}
C, & i = 0, \\
C\frac{1-r}{1-r^{N-1}}\left(r^i + r^{i+1} + \cdots + r^{N-2}\right) \\
\quad + c\frac{1-r}{1-r^{N-1}}\left(1 + r + \cdots + r^{i-1}\right), & i = 1,\dots,N-2, \\
c, & i = N-1, \\
0, & i = N,
\end{cases} \tag{31.14}$$

otherwise.

Proposition 31.2. *If $C > c$, then there is a unique monotonic solution to (31.12) defined by choosing $k = N-1$.*

Proof. If $k < N-1$, then

$$w_{k+1} = \big((1-p)w_{k-1} + pw_{k+1}\big),$$

yielding

$$w_{k+1} = w_{k-1},$$

and hence by monotonicity of w_i, we obtain

$$w_{k-1} = w_{k+1} = w_k = c.$$

Solving

$$w_i = (1-p)w_{i-1} + pw_{i+1}$$

for $i = 1, \ldots, k-1$ gives

$$w_i = c, \quad i = 0, \ldots, k+1,$$

contradicting $w_0 = C > c$. Thus $k = N-1$, and the unique solution is given by (31.13) if $p = 0.5$, or (31.14) if $p \neq 0.5$.   □

31.4 • Multiple-Reservoir Models

In most real applications, marginal water values are obtained from the optimal operation of many reservoirs, sometimes linked in a cascaded river system, and almost always linked by some electricity transmission system. When the number of reservoirs is larger than two or three, representing their optimal operation using Markov decision problems becomes more difficult, due to the curse of dimensionality. Approximations of various forms are necessary.

The approximation in widespread use is based on the SDDP algorithm of Pereira and Pinto [1464]. Here the marginal values of water emerge from an approximate stage problem that is modeled as an LO problem. This method can be shown to converge to an optimal policy almost surely under mild assumptions on the sampling process (see [805, 1477]). The subgradients of the optimal value function define the marginal water value.

The class of problems we consider have T stages, denoted by $t = 1, 2, \ldots, T$, in each of which a random right-hand side vector $b_t(\omega_t) \in \mathbb{R}^m$ has a finite number of realizations defined by $\omega_t \in \Omega_t$. We assume that the outcomes ω_t are stagewise independent, and that Ω_1 is a singleton, so the first-stage problem is

$$\begin{aligned} z = \min \quad & c_1^\top x_1 + \mathbb{E}[Q_2(x_1, \omega_2)] \\ \text{s.t.} \quad & A_1 x_1 = b_1, \\ & x_1 \geq 0, \end{aligned} \tag{31.15}$$

where $x_1 \in \mathbb{R}^n$ is the first-stage decision, and $c_1 \in \mathbb{R}^n$ is a cost vector, A_1 is an $m \times n$ matrix, and $b_1 \in \mathbb{R}^m$.

We denote by $Q_2(x_1, \omega_2)$ the second-stage costs associated with decision x_1 and realization $\omega_2 \in \Omega_2$. The problem to be solved in the second and later stages t, given decisions x_{t-1} and realization ω_t, can be written as

$$\begin{aligned} Q_t(x_{t-1}, \omega_t) = \min \quad & c_t^\top x_t + \mathbb{E}[Q_{t+1}(x_t, \omega_{t+1})] \\ \text{s.t.} \quad & A_t x_t = b_t(\omega_t) - E_t x_{t-1} \quad [\pi_t(\omega_t)], \\ & x_t \geq 0, \end{aligned} \tag{31.16}$$

where $x_t \in \mathbb{R}^n$ is the decision in stage t, c_t its cost, and A_t and E_t $m \times n$ matrices. Here $\pi_t(\omega_t)$ denotes the dual variables of the constraints. In the last stage we assume either that $\mathbb{E}[Q_{T+1}(x_T, \omega_{T+1})] = 0$ or that there is a convex polyhedral function that defines the expected future cost after stage T. For all instances of (31.16), we assume relatively complete recourse, whereby (31.16) at stage t has a feasible solution for all values of x_{t-1} that are feasible for the instance of (31.16) at stage $t-1$. Relatively complete recourse can be ensured by introducing artificial variables with penalty terms in the objective.

The SDDP algorithm performs a sequence of major iterations, each consisting of a *forward pass* and a *backward pass* through all the stages, to build an approximately optimal policy. In each forward pass, a set of N scenarios is sampled from the scenario tree and decisions are made for each stage of those N scenarios, starting in the first stage and moving forward up to the last stage. In each stage, the observed values $\bar{x}_t(s)$ of the decision variables x_t, and the costs of each stage in all scenarios s, are saved.

The SDDP algorithm builds a policy that is defined at stage t by a polyhedral outer approximation (OA) of $\mathbb{E}[Q_{t+1}(x_t, \omega_{t+1})]$. This approximation is constructed using cutting planes called Benders cuts, or just *cuts*. In other words, in each tth-stage problem, $\mathbb{E}[Q_{t+1}(x_t, \omega_{t+1})]$ is replaced by the variable θ_{t+1}, which is constrained by a set of linear inequalities

$$\theta_{t+1} - \bar{g}_{t+1,k,s}^\top x_t \geq \bar{h}_{t+1,k,s}, \qquad k = 1, 2, \ldots, K, \quad s = 1, 2, \ldots, N, \qquad (31.17)$$

where K is the number of backward passes that have been completed and $\bar{g}$ and $\bar{h}$ are defined by (31.20) and (31.21) below.

With this approximation, the first-stage problem is

$$\begin{aligned} z = \min \quad & c_1^\top x_1 + \theta_2 \\ \text{s.t.} \quad & A_1 x_1 = b_1, \\ & \theta_2 - \bar{g}_{2,k,s}^\top x_1 \geq \bar{h}_{2,k,s}, \qquad \begin{array}{l} k = 1, 2, \ldots, K, \\ s = 1, 2, \ldots, N, \end{array} \qquad (31.18) \\ & x_1 \geq 0, \end{aligned}$$

and the tth stage problem becomes

$$\begin{aligned} \tilde{Q}_t(x_{t-1}, \omega_t) = \min \quad & c_t^\top x_t + \theta_{t+1} \\ \text{s.t.} \quad & A_t x_t = b_t(\omega_t) - E_t x_{t-1}, \qquad [\pi_t(\omega_t)] \\ & \theta_{t+1} - \bar{g}_{t+1,k,s}^\top x_t \geq \bar{h}_{t+1,k,s}, \qquad \begin{array}{l} k = 1, 2, \ldots, K, \\ s = 1, 2, \ldots, N, \end{array} \qquad (31.19) \\ & x_t \geq 0, \end{aligned}$$

where we interpret the set of cuts as being empty when $K = 0$.

At the end of the forward pass, a convergence criterion is tested, and if it is satisfied then the algorithm is stopped; otherwise it starts the backward pass, which is defined below. In the standard version of SDDP (see [1464]), the convergence test is satisfied when z, the lower bound on the expected cost at the first stage, is statistically close to an estimate of the expected total operation cost obtained by averaging the cost of the policy defined by the cuts when applied to the N sampled scenarios. In this simulation, the total operation cost for each scenario is the sum of the present cost ($c_t^\top x_t$) over all stages t and any end-of-horizon future cost.

If the convergence criterion is not satisfied, then SDDP amends the current policy using a backward pass that adds N cuts to each stage problem, starting at the

penultimate stage and working backward to the first. To compute the coefficients for
the cuts, we solve the next-stage problems for all possible realizations (Ω_{t+1}) in each
stage t and scenario s. The cut for (31.19), the tth (approximate) stage problem in
scenario s, is computed after its solution $\tilde{x}_t^k(s)$ has been obtained in the forward pass
immediately preceding the backward pass k. Solving the $(t+1)$th (approximate) stage
problem for every $\omega_{t+1} \in \Omega_{t+1}$ gives $\tilde{\pi}_{t+1,k,s} = \mathbb{E}[\pi_{t+1}(\omega_{t+1})]$, which defines the cut
gradient

$$\tilde{g}_{t+1,k,s} = -\tilde{\pi}_{t+1,k,s}^\top E_{t+1} \tag{31.20}$$

and its intercept

$$\tilde{b}_{t+1,k,s} = \mathbb{E}[Q_{t+1}(\tilde{x}_t^k(s), \omega_{t+1})] + \tilde{\pi}_{t+1,k,s}^\top E_{t+1}\tilde{x}_t^k(s). \tag{31.21}$$

The SDDP algorithm is initialized by setting $\theta_t = -\infty$, $t = 2,\ldots,T$, $K = 0$, $k = 1$. Thereafter, the algorithm performs the following three steps repeatedly until the
convergence criterion is satisfied:

1. Forward Pass

    For $t = 1$, solve (31.18) and save $\tilde{x}_1^k(s) = x_1$, $s = 1,\ldots,N$, and $\tilde{z}^k = z$;

    For $t = 2,\ldots,T$ and $s = 1,\ldots,N$,

    Solve (31.19) setting $x_{t-1} = \tilde{x}_{t-1}^k(s)$, and save $\tilde{x}_t^k(s)$ and $\tilde{Q}_t(\tilde{x}_{t-1}^k(s), \omega_t)$.

2. Standard Convergence Test (at $100(1-\alpha)\%$ confidence level).

    Calculate the upper bound: $z_u = \frac{1}{N}\sum_{s=1}^{N}\sum_{t=1}^{T} c_t^\top \tilde{x}_t^k(s),$

    $$\sigma_u = \sqrt{\frac{1}{N}\sum_{s=1}^{N}\left(\sum_{t=1}^{T} c_t^\top \tilde{x}_t^k(s)\right)^2 - z_u^2}.$$

    Calculate the lower bound: $z_l = \tilde{z}^k$;

    Stop if

    $$z_l > z_u - \frac{Z_{\frac{\alpha}{2}}}{\sqrt{N}}\sigma_u,$$

    where Z_α is the $(1-\alpha)$ quantile of the standard normal distribution;
    otherwise go to the backward pass.

3. Backward Pass

    For $t = T,\ldots,2$, and $s = 1,\ldots,N$,

    For $\omega_t \in \Omega_t$, solve (31.19) using $\tilde{x}_{t-1}^k(s)$ and save $\tilde{\pi}_{t,k,s} = \mathbb{E}[\pi_t(\omega_t)]$
    and $\tilde{Q}_t(\tilde{x}_{t-1}^k(s), \omega_t)$;

    Calculate the kth cut for s in stage $t-1$ using (31.20) and (31.21).

    Set $K = K+1$, $k = k+1$.

SDDP terminates when the estimated value z_u of the current policy is closer than
$\frac{Z_{\frac{\alpha}{2}}}{\sqrt{N}}\sigma_u$ to a lower bound z_l on the optimal value. The expected value of an optimal

policy for the model (with finitely many inflow outcomes) can then be assumed to lie in the interval $[z_l, z_u + \frac{z_{\frac{\alpha}{2}}}{\sqrt{N}}\sigma_u]$ with probability at least $(1 - \alpha)$.

Energy prices are deduced from the solution to SDDP by simulating a single stage of the (approximately) optimal policy (using cutting planes to represent the Bellman function) and recording the shadow prices on the demand constraint. Marginal water values in each reservoir at level $\bar{x}$ are determined from the subdifferential of $\mathbb{E}[\tilde{Q}_t(x_{t-1}, \omega_t)]$ evaluated at $\bar{x}$. Often this will be a singleton defined by the coefficients of the highest cut at $\bar{x}$.

It is important to observe that the above algorithm operates under a wait-and-see assumption. In other words, we solve (31.19) with full knowledge of the current period's inflow, which can be used to generate electricity to satisfy demand. This means that shortage costs might be avoided even when all reservoir levels are at their minimum (see Section 31.2 above). Since shortages arising in simulations of an optimal policy yield marginal water value estimates, the wait-and-see assumption is likely to underestimate prices.

Based on the single-reservoir analysis, an optimal generation schedule will use a merit order of water release and thermal generation based on marginal water value and marginal cost. With linear dynamics and with linear production functions, this produces a bang-bang type of control policy. It is well known, however, that many such systems admit singular control (i.e., one not uniquely determined by the adjoint solution). For example, the separated continuous LO problems discussed in [81] have this property. Examples in a hydro-scheduling context are given in [1474].

31.5 ▪ Application

We now show how SDDP can be applied to estimate wholesale prices for the New Zealand electricity system under the assumption of perfectly competitive risk-neutral agents. We begin by solving a model in which the inflows are assumed to be stagewise independent. We use the DOASA software described in [1479].

We seek to estimate New Zealand wholesale electricity prices for the week beginning March 10, 2008. The physical characteristics of the system are described in [1479] and [1478] (which together provide a more detailed study of the New Zealand wholesale electricity market over the period 2001–2008). The electricity network shown in Figure 31.2 is approximated by a four-reservoir system by aggregating the storage reservoirs in the Waitaki river system into a single lake (Lake Waitaki). The resulting Waitaki system is shown in Figure 31.3.

Lake Ohau (OHU), which has limited storage capacity, is now treated as a run-of-river resource, and reservoirs Benmore (BEN), Pukaki (PKI), and Tekapo (TEK) are treated as a single lake. Since these occur in a cascade in the river system, the aggregation of water in each must account for their specific energies. Thus Benmore and Pukaki storage and inflows are discounted by their specific energies before being added to the aggregate Lake Waitaki, which has the same specific energy as Tekapo. To allow for the storage ability of Benmore and Pukaki, Lake Waitaki water can be restored to these lower lakes through the arcs connecting Lake Waitaki to its children. These arcs have multipliers that inflate the water volume transferred to correspond to the difference in specific energies of Lake Waitaki and the lower lakes.

The transmission network is approximated by three nodes, one for the South Island and two for the North Island. Limits on power transfer between nodes correspond to reported transmission limits. Energy deficit in any stage is met by load shedding at an

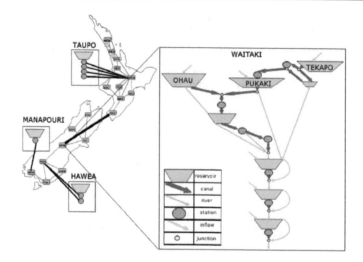

Figure 31.2. *Approximate network representation of New Zealand electricity network showing main hydroelectricity generators. Reprinted with permission from Elsevier* [1479].

increasing shortage cost in three tranches. This is equivalent to having three dummy thermal plants at each node with capacities equal to 5% of load, 5% of load, and 90% of load, for each load sector, and costs as shown in Table 31.1. Load shed above 10% of demand costs a value of lost load of $20,000/MWh, which is roughly equivalent to five hours of load curtailment per year assuming annual amortized peaking-plant costs of $100/kW.

Table 31.1. *Load reduction costs (NZD/MWh) and proportions of industrial load, commercial load, and residential load in each island.*

	Up to 5%	5% to 10%	Above 10%	North Is.	South Is.
Industrial	$1,000	$2,000	$20,000	0.34	0.58
Commercial	$2,000	$4,000	$20,000	0.27	0.15
Residential	$2,000	$4,000	$20,000	0.39	0.27

We choose historical inflows from each week in the years 1997–2006 to give an empirical inflow distribution with 10 outcomes per stage. We assume that these are stagewise independent.

The first model attempts to compute an optimal policy for 52 weeks starting at midnight on March 10, 2008. The computed electricity prices for the week beginning March 11, 2008, are the same at each node. They are $38.45/MWh, $38.76/MWh, and $39.84/MWh, corresponding to off-peak, shoulder, and peak periods. We can simulate the corresponding policy using inflows sampled from the assumed stagewise-independent distribution. This gives storage levels in Lake Waitaki from 100 random sequences, as shown in Figure 31.4. If we simulate the same policy with the 30 most recent historical inflow sequences, then we obtain storage levels in Lake Waitaki as shown in Figure 31.5. The storage levels shown hit both extremes more often than the simulation that assumes stagewise independence.

A heuristic that has been suggested to overcome the optimistic biases from assuming stagewise independence is called *inflow spreading* [1555]. Assuming independence over k weeks results in a lower variance in total inflow over this period than the data

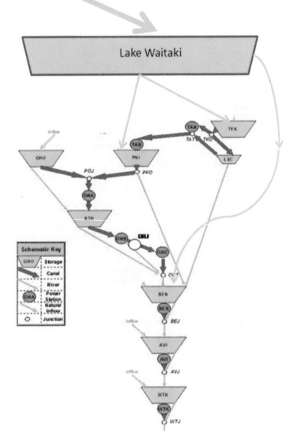

Figure 31.3. *Aggregated model of Waitaki system with storage lakes Pukaki, Tekapo, and Benmore combined to make Lake Waitaki. Their inflows (scaled by specific energies) combine to produce an inflow to Lake Waitaki.*

support. Inflow spreading modifies the inflows using a heuristic that increases the variance of assumed independent inflows so that the variance of k consecutive independent inflows matches the observed variance of this sum.

Consider historical inflow sequence $h(t,y)$ of inflows in week t of year $y = 1, 2, \ldots, N$, and suppose $k = 4$. Let

$$w(t,y) = h(t,y) + h(t+1,y) + h(t+2,y) + h(t+3,y),$$

$$a(t) = \frac{\sum_y h(t,y)}{N}, \quad b(t) = \frac{\sum_y w(t,y)}{N},$$

and

$$d(t,y) = \max\left\{0, a(t) + \frac{(w(t,y) - b(t))}{2}\right\}.$$

The model then uses adjusted inflows

$$\omega(t,y) = \frac{N d(t,y) a(t)}{\sum_y d(t,y)}, \quad t = 1, 2, \ldots, 52, \quad y = 1, 2, \ldots, N.$$

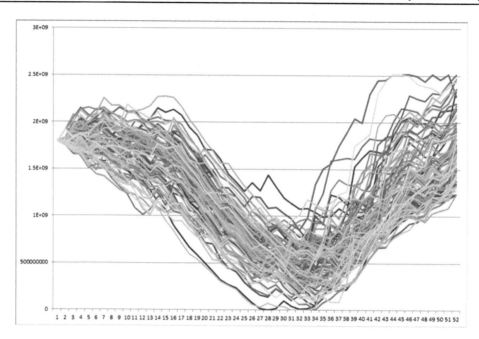

Figure 31.4. *Simulated volumes in Lake Waitaki reservoir for* 100 *randomly generated inflow sequences.*

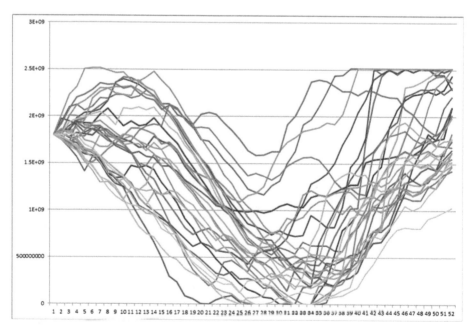

Figure 31.5. *Simulated volumes in Lake Waitaki reservoir for historical inflow sequences* 1979–2008.

An optimal policy for 52 weeks starting at midnight on March 10, 2008, can be computed using adjusted inflows with $k = 4$. We can simulate the corresponding policy using inflows sampled from the assumed stagewise-independent distribution. This gives storage levels in Lake Waitaki from 100 random sequences (with spread inflows)

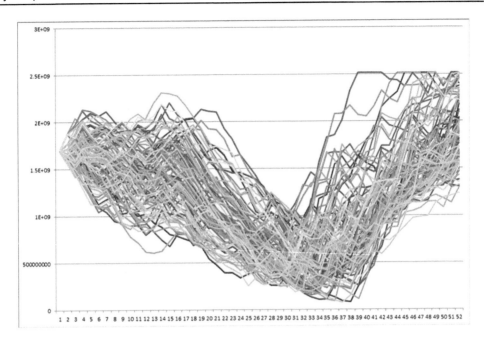

Figure 31.6. *Inflow spreading policy simulated volumes in Lake Waitaki reservoir for* 100 *randomly generated inflow sequences.*

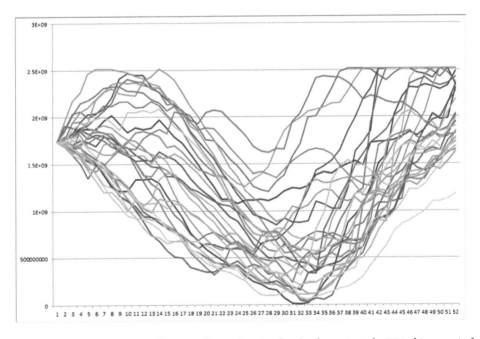

Figure 31.7. *Inflow spreading policy simulated volumes in Lake Waitaki reservoir for* *historical inflow sequences* 1979–2008.

as shown in Figure 31.6. If we simulate the same policy with the 30 most recent historical inflow sequences, then we obtain storage levels in Lake Waitaki as shown in Figure 31.7. The stock levels hit zero in only one year out of 30. The electricity prices for the week beginning March 11, 2008, are computed and are the same at each node. They are $40.10/MWh, $40.81/MWh, and $41.02/MWh, corresponding to off-peak, shoulder, and peak periods. The inflow spreading has resulted in slightly higher prices.

The prices estimated by our model have been computed using SDDP, which makes use of a sampling procedure. This means that there will be some sampling error in their estimation. We first attempt to estimate this error by simulating the policy that we have obtained with perturbed initial reservoir levels. A simulation of the optimal (inflow spreading) policy over 2,500 random inflow sequences yields a sample average optimal cost of $354,861,576. Increasing the initial reservoir volume by 1 million cubic meters and simulating with common random numbers yields a sample average optimal cost of $354,812,679. The difference in costs is $48,897, with a standard error of $2,685. Assuming that the SDDP policy is optimal yields an estimate of marginal water value for Lake Waitaki of $0.0489/$m^3$, with a 95% confidence interval [0.0435, 0.0543]. To convert a cubic meter to an energy value, we must divide by the specific energy of Lake Waitaki, which is 0.001131 MWh/m^3, giving a 95% confidence interval for the perfectly competitive wholesale Benmore price of [$38.46, $48.01].

31.6 ▪ Discussion

One can compare the computed estimates of weekly prices with observed prices in the New Zealand wholesale market in the week beginning March 11, 2008. These range from $60/MWh to $190/MWh. The marginal water values computed by our SDDP algorithm are significantly lower than those reflected in observed electricity prices. There are several possible causes of this.

1. Poor representation of stagewise dependence in inflows: We have represented stagewise dependence in inflows using an inflow spreading heuristic based on the approach currently used by some New Zealand electricity generators. We have also ignored the melting of snowpack, which affects the inflow dynamics in the Waitaki catchment. The true reservoir inflow processes in New Zealand exhibit a more complicated long-range dependence than we have in our model. New inflow models that aim to improve the modeling of this dependence are discussed in [1524].

2. Wait-and-see representation of uncertainty: Our SDDP model assumes that inflows are observed in any week before release decisions are made for that week. As discussed above, this assumption implies that the model will underestimate the probability of a shortage event, which will underestimate marginal water values. On the other hand, local meteorologists make reasonably reliable weather forecasts for the next three to five days, so we conjecture that the wait-and-see model should give reasonably accurate estimates, all else being equal.

3. Assumptions about constant fuel costs not reflecting fuel contract structure: In the New Zealand wholesale market, all large electricity generators (including thermal generators) offer increasing supply functions to the system operator. We have assumed that thermal short-run marginal cost is constant for each plant, so these supply functions in our model are assumed to be constant. Increasing supply functions for thermal plant are often interpreted as evidence of market power exercise, but they have alternative explanations. The owners of these plants typically operate under fuel contracts that affect the cost of their day-to-day fuel consumption. For example, coal is stockpiled, and so if replenishment is expensive

and time-consuming, its marginal value increases as a stock is depleted. On the other hand, gas is typically supplied under take-or-pay contracts over a contract period of several months. This effectively makes above-average consumption in any period more costly as it incurs a risk of exceeding the contract volume and having to pay a penalty.

4. Agents making different assumptions about the probabilities of dry inflow outcomes: We have assumed that all agents have the same probability distribution of inflow outcomes, sampled with equal probability from historical records. Recent climate observations (such as El Nino Southern Oscillation forecasts) might affect agents' subjective assessments of these probabilities.

5. Agents acting strategically in offering generation at higher than competitive prices: We have assumed that all agents act as price-takers. This is not required by the rules governing the operation of the New Zealand wholesale electricity market. Some increase in price might be attributed to agents behaving strategically.

6. Agents not being risk neutral: We have assumed that all agents are risk neutral, and so they act to maximize expected profits. However, generators and purchasers are generally risk averse, so there is a possible discrepancy between a social optimum (and corresponding marginal water values) and a competitive partial equilibrium. This discrepancy is explored in detail in [1476], where competitive agents facing uncertain reservoir inflows are endowed with possibly different coherent time-consistent risk measures. In [1476], it is shown that if enough contracts are made available to agents to trade, and they are not too different in their attitudes to risk, then a social risk measure emerges in equilibrium that can be used to compute a risk-averse social plan that coincides with the partial competitive equilibrium. The risk-averse social plan can be computed using a risk-averse version of SDDP (see [1475]). In this setting, prices will increase as agents trade off their current costs against risk-adjusted future costs, which will make them more conservative than a risk-neutral agent when water is scarce. If the market for risk is incomplete, so thermal plant owners and hydro plant owners have no ability to hedge their different risk positions, then a perfectly competitive equilibrium need not correspond to a social planning solution, even if the planning solution is risk averse. Of course, even if we assume completeness in markets for risk, agents in hydro dominated electricity markets face a plethora of commercial risks apart from uncertain inflows (which are the sole source of uncertainty in our model). It is conceivable that these other uncertainties are regarded as more critical than inflow risk, influencing the price markups we observe in the week beginning March 11, 2008.

Identifying which combination of these factors is driving the difference between modeled and observed prices is difficult. However, for regulators of hydro-dominated electricity markets who monitor inefficiencies arising from imperfect competition or market incompleteness, this is a critically important task. SO models and stochastic equilibrium models are powerful tools to enable this to happen.

Chapter 32

Stochastic Optimization Investment Models with Portfolio and Marginal Risk Constraints

Miguel Lejeune

32.1 ▪ Introduction

Stochastic programming [279, 1522, 1678] models are pervasive in finance because they allow for efficiently and explicitly taking into account the multiple uncertainties affecting investment and credit decisions. The financial optimization area encompasses a variety of problems, such as asset allocation for pension plans and insurance companies, bond and stock selection for portfolio managers, currency hedging for multinational corporations, hedge fund strategies, credit risk models, and asset liability and risk management for large public corporations (see, e.g., [514, 1108, 1980]).

In this chapter, we focus on the construction of optimal portfolios of securities, for which a variety of stochastic programming models have been proposed, including Roy's safety-first risk criterion [1594], Kataoka's model [1043], stochastic dominance [585, 586, 1416, 1584], value-at-risk (VaR) [778], conditional value-at-risk (CVaR) [83, 1106], excess probabilities [1651], and the probabilistic Markowitz model [307, 708]. In particular, we analyze portfolio optimization problems based on the VaR [1353] and CVaR [1574] measures. We first review models including VaR and CVaR constraints restricting the portfolio's overall risk and recall succinctly some of the VaR and CVaR key properties. Next, we discuss how the VaR and CVaR measures can be employed within the emerging risk budgeting and parity approaches [1588], which allocate a risk budget to each individual security. We describe Euler's risk allocation method and explain how it can be used to apportion the portfolio's risk across securities. We show how Euler's method can be implemented with conditional expectation constraints, and we discuss the features and complexity of the marginal VaR and CVaR contribution constraints.

32.2 ▪ Portfolio Optimization Models with VaR and CVaR Portfolio Risk Constraints

Academics and practitioners have long ago recognized the importance of measuring the downside risk of a portfolio. The family of downside risk measures (see [1382] for a history) includes the VaR and CVaR metrics, as well as metrics based on lower-partial moments (e.g., below-target semivariance, below-mean semivariance) [720, 1279], stochastic dominance [586], and the safety-first criterion [1594].

32.2.1 ▪ VaR Models

The emergence of VaR can be traced back to the RiskMetrics framework [1353] that recommends the use of VaR to evaluate the capital adequacy and market risk of commercial banks. VaR is a popular risk management tool and an industry benchmark due in part to central bank regulators' decision to use VaR to calculate the capital requirements of financial institutions. While VaR is homogeneous of degree one and comonotonic additive, VaR is however not a coherent risk measure [99] in general as it does not satisfy the subadditivity axiom, hence not promoting diversification, and is nonconvex. Moreover, it tends to ignore extreme losses beyond itself.

VaR is defined as a threshold loss level $\alpha \in \mathbb{R}$ that can only be exceeded with a low probability level $1 - \beta \in (0, 1)$. Consider an asset universe including n securities, and denote by $\xi \in \mathbb{R}^n$ the vector of stochastic security loss rates. The decision vector is $w \in \mathbb{R}^n$ and represents the proportion of capital invested in each asset. The loss of the portfolio is $f(w, \xi) = w^t \xi$, $\mu \in \mathbb{R}^n$ is the vector of expected security loss, and Σ is the positive definite matrix of variance-covariance.

The probability distribution of ξ has a density $p(\xi)$, and the probability of the loss not exceeding α is given by

$$\Psi(w, \alpha) = \int_{f(w, \xi) \le \alpha} p(\xi) \, d(\xi). \qquad (32.1)$$

The VaR of the portfolio loss at the probability level β is the solution of the following chance-constrained (CC) stochastic problem:

$$\alpha_\beta(w) = \min \{\alpha \in \mathbb{R} : \Psi(w, \alpha) \ge \beta\}. \qquad (32.2)$$

Several portfolio optimization problems with VaR restrictions imposed on the risk of the portfolio have been proposed:

1. Minimization of VaR:

$$\min \quad \alpha \qquad (32.3)$$
$$\text{s.t. } \mathbb{P}(w^t \xi \le \alpha) \ge \beta, \qquad (32.4)$$
$$w \in \mathcal{X}, \qquad (32.5)$$

where $\mathcal{X}$ is a set of deterministic constraints in which one can, for example, prevent shortselling and require investing the entirety of the capital.

2. Maximization of expected portfolio return with an upper bound U on VaR:

$$\max \quad -w^t \mu \qquad (32.6)$$
$$\text{s.t.} \quad \alpha \le U \qquad (32.7)$$
$$(32.4)–(32.5),$$

with the expected return $-w^t \mu$ defined as the negative portfolio loss rate.

3. Mean-VaR model:

$$\max \ -w^t \mu - \lambda \alpha \tag{32.8}$$

$$\text{s.t.} \ (32.4)–(32.5),$$

where λ is a positive scalar denoting the risk aversion of the investor and allowing the determination of the relative importance of expected return and risk.

The so-called safety-first optimization problem [1594] and the probabilistic Markowitz model [307] are two formulations germane to the above VaR models. The safety-risk measure is the very first downside risk metric and consists of maximizing of the probability with which a specified return level R can be attained:

$$\max \ \mathbb{P}(-w^t \xi \geq R) \tag{32.9}$$

$$\text{s.t.} \quad (32.5).$$

The probabilistic Markowitz model [307] involves minimizing the variance under the condition that the return is at least equal to a specified threshold level with a large probability β:

$$\min \quad w^t \Sigma w \tag{32.10}$$

$$\text{s.t.} \ \mathbb{P}(-w^t \mu \geq R) \geq \beta, \tag{32.11}$$

$$(32.5).$$

The optimization models presented above each take the form of a CC problem with random technology matrix [1043]. A key computational challenge resides in the reformulation of the chance constraint in a form that is amenable to an efficient numerical solution. The properties and nature (i.e., reformulation or approximation) of the reformulated problems are highly dependent on the knowledge and/or assumptions about the marginal and joint probability distributions of the security losses. Several reformulations based on various decomposition methods have been proposed in [708] for chance constraints similar to (32.4) and (32.11). Computational tests have shown that the time needed to solve the associated optimization problems can vary by orders of magnitude when the asset universe includes a few hundred or thousand assets.

If the losses of the n assets have a joint normal distribution and assuming $\beta \geq 0.5$, then the chance constraint (32.4) can be equivalently reformulated [1043] as the following second-order cone (SOC) constraint:

$$\alpha \geq w^t \mu + \Phi^{-1}(\beta)\sqrt{w^T \Sigma w}, \tag{32.12}$$

where Φ is the univariate standard normal cumulative probability distribution and $\Phi(\beta)^{-1}$ is its β-quantile.

The normality of the joint distribution of the asset returns is, however, a very strict assumption that is often breached, since most asset return distributions are more leptokurtic, or have a fatter tail, than the normal distribution. In [709], it was shown that the exact value of the β-quantile of the loss distribution can also be obtained when the joint distribution follows a t-distribution. This latter is an elliptically symmetric distribution that permits us to capture the presence of kurtosis and has been shown to model well the return behavior of securities and portfolios under many circumstances (see, e.g., [963, 1281] and the references therein). Under this assumption, an SOC and equivalent representation of the chance constraint (32.4) can also be obtained, as shown in [709].

If the distribution of the portfolio returns is imprecisely known and the quantile is not know exactly, the latter can be approximated by using probability inequalities (i.e., the Cantelli, Chebychev [307], or Camp-Meidell inequalities [1164,1165]) that provide an inner approximation problem of the feasible set of the chance constraint (32.3). The approximation problem takes the form of an SOC problem too (see [307, 1164]).

If the uncertainty of the asset losses is represented by a joint discrete distribution or a finite set K of samples or scenarios, the feasible set of the chance constraint (32.3) can be linearized. We assume that ξ can take $|K|$ different values $\omega^k \in \mathbb{R}^n$ with probability $p^k > 0$ and $\sum_{k \in K} p^k = 1$. The notation ω^k is the vector of joint loss realizations in scenario k. The feasible set of constraint (32.3) can be equivalently represented by the following set of mixed-integer linear inequalities:

$$w^t \omega^k \leq \alpha + \gamma^k M, \; k \in K, \tag{32.13}$$

$$\sum_{k \in K} p^k \gamma^k \leq 1 - \beta, \tag{32.14}$$

where M is a large positive scalar and γ^k is a binary decision variable taking value one if the portfolio loss in scenario k exceeds the VaR α (due to (32.13)) and taking value zero otherwise. The knapsack constraint (32.14) stipulates that the sum of the scenario probabilities in which the loss exceeds the VaR must not exceed the complement $(1-\beta)$ of the reliability level β.

Discussing the concept of p-level efficient points, Prékopa [1523] presents a multiportfolio illustration. This can be viewed as an illustration of the multivariate VaR concept, in which a system of stochastic inequalities of the form $w^t \xi \leq \alpha$ (as in (32.4)) must hold jointly with a prescribed probability level, and each inequality corresponds to a particular portfolio.

32.2.2 ▪ CVaR Models

Rockafellar and Uryasev [1574] have proposed CVaR as an alternative to VaR. In contrast to VaR, CVaR is a coherent risk measure and accounts for the tail of the loss distribution by restricting the expected loss in excess of VaR. Its mathematical expression is given by

$$\psi_\beta(w) = \frac{1}{1-\beta} \int_{f(w,\xi) \leq \alpha_\beta(w)} f(w,\xi) \, p(\xi) \, d(\xi). \tag{32.15}$$

Denoting by $[x]^+$ the positive part of x, it was shown [1574] that the function $F_\beta(w, \alpha) : \mathbb{R}^{n+1} \to \mathbb{R}$ defined by

$$F_\beta(w, \alpha) = \alpha + \frac{1}{1-\beta} \int_\xi [f(w, \xi) - \alpha]^+ \, p(\xi) \, d(\xi) = \alpha + \frac{1}{1-\beta} \mathbb{E}[[f(w, \xi) - \alpha]^+] \tag{32.16}$$

is convex and differentiable, with

$$\psi_\beta(w) = \min_{\alpha \in \mathbb{R}} F_\beta(w, \alpha) \tag{32.17}$$

and

$$\min_{w \in \mathbb{R}^n} \psi_\beta(w) = \min_{(w, \alpha) \in \mathbb{R}^{n+1}} F_\beta(w, \alpha). \tag{32.18}$$

Comparing the VaR and CVaR definitions (32.2) and (32.15) makes it evident that CVaR is a more conservative risk measure since $\alpha_\beta(w) \leq \psi_\beta(w)$. The above properties highlight that (i) CVaR can be computed directly without requiring the a priori calculation of VaR even if CVaR depends on VaR, (ii) VaR can be derived as a by-product of the CVaR calculation, and (iii) the minimum VaR portfolio (assuming its existence) is CVaR efficient at the probability level β [683].

As for VaR, several types of portfolio optimization problems limiting the CVaR of the portfolio loss have been proposed (see, e.g., [83, 1106]):

1. Minimization of CVaR:

$$\min \; \alpha + \frac{1}{1-\beta} \mathbb{E}[[f(w,\xi)-\alpha]^+] \tag{32.19}$$

$$\text{s.t.} \qquad (32.5).$$

2. Maximization of expected portfolio return with an upper bound d on CVaR:

$$\max \qquad -w^t \mu$$

$$\text{s.t.} \;\; \alpha + \frac{1}{1-\beta} \mathbb{E}[[f(w,\xi)-\alpha]^+] \leq V, \tag{32.20}$$

$$(32.4)\text{--}(32.5),$$

where V is an upper bound on the acceptable CVaR level.

3. Mean-CVaR model:

$$\max \; -w^t \mu - \lambda \left(\alpha + \frac{1}{1-\beta} \mathbb{E}[[f(w,\xi)-\alpha]^+] \right) \tag{32.21}$$

$$\text{s.t.} \qquad (32.5).$$

The above models are stochastic optimization (SO) problems with conditional expectation constraints [1521]. The convexity of the CVaR measure provides a compelling computational advantage compared to the nonconvex VaR measure, permitting us to reformulate the above problems as convex optimization problems. The form of the reformulations will again differ depending on the probability distribution of the asset losses.

If the joint distribution of the asset losses is normal, then the CVaR of the portfolio losses is given by [1523]

$$\psi_\beta(w) = w^t \mu - \frac{\varphi(\Phi^{-1}(\beta))}{\beta} \sqrt{w^T \Sigma w}, \tag{32.22}$$

where φ is the univariate standard normal probability density function (pdf). A similar reformulation was proposed in [276] for the class of elliptical distributions.

If the probability distribution of the portfolio losses is not normal and is represented by a set of $|K|$ samples or scenarios, then the CVaR minimization problem can be reformulated as the following linear optimization (LO) problem:

$$\min \; \alpha + \frac{1}{(1-\beta)} \sum_{k \in K} p^k z^k \tag{32.23}$$

$$\text{s.t.} \quad z^k \geq w^t \omega^k - \alpha, \tag{32.24}$$

$$z^k \geq 0, \tag{32.25}$$

$$(32.5),$$

where each $z^k, k \in K$ is an auxiliary decision variable representing the loss amount exceeding the VaR α in each scenario k and $w^t \omega^k$ is the portfolio loss in scenario k. The same linearization approach can also be used for the other two CVaR formulations presented above. The CVaR concept has also been extended to the multivariate case (see, e.g., [519, 1408]).

32.3 ▪ Portfolio Optimization Models with Marginal VaR and CVaR Constraints

32.3.1 ▪ Asset Risk Contribution and Marginal Risk

The subprime crisis has further reduced the risk tolerance of investors, leading to the progressive emergence of a new investment style based on the allocation of a risk budget to each position and named parity or risk budgeting [1588]. The control of the risk associated with a particular position is typically achieved in traditional asset allocation models by adding bounds on the portfolio positions. This approach, however, ignores the effect of correlations among asset returns in risk diversification [1980]. In contrast to this, risk budgets are increasingly used to decompose the total portfolio risk into the risk contribution of each component position. While risk parity assigns the same risk budget to each portfolio position and can be excessively restrictive, risk budgeting allows different risk budgets across positions.

While the attribution of risk to each individual position is a fairly new approach to constructing portfolios of securities, it is common practice in credit risk (see, e.g., [810, 1302]) to calculate the risk associated with a specific loan or obligor and to define the economic capital for this position. Three main types of approaches have been used [1302]. The *standalone* risk approach is based on the CreditMetrics framework [1353] and defines the risk contribution of a loan as the difference between the risk of the entire portfolio and the risk of the portfolio without the given asset. The *incremental* risk contribution of a loan is computed by subtracting the economic capital for the portfolio without the loan from the economic capital corresponding to the entire portfolio. This permits us to capture the amount of capital that would be released if the loan were added to the current portfolio (see, e.g., the notion of incremental VaR [1013]). The *marginal* risk contributions are designed to allocate the diversification benefit among the loans in the portfolios and to measure the amount of the economic capital that should be allocated to each loan viewed as part of a portfolio.

All these approaches have as their objective to calculate the risk contribution of an individual loan to the risk of the entire portfolio. The standalone and incremental risk contribution approaches violate the diversification and linear aggregation axioms defined by Kalkbrener [1020], since the summation of the marginal risks contributed by all loans is generally not equal to the risk of the entire portfolio [1302]. In what follows, we will use the marginal risk contribution for portfolio optimization, since it is additive and possesses the diversification and linear aggregation properties for standard risk measures, such as standard deviation, VaR, and CVaR.

32.3.2 ▪ Marginal VaR and CVaR Constraints and Models

We shall now formulate stochastic constraints that permit the allocation and limitation of VaR and CVaR to assets of a portfolio on the basis of their marginal contribution to the total risk. We first introduce Euler's theorem, which decomposes the risk of a portfolio into a sum of marginal (asset) risk contributions for homogeneous risk functionals of degree one.

Theorem 32.1 (Euler's theorem). *Let $w \in \mathcal{X} \subseteq \mathbb{R}^n$ and ρ be a continuously differentiable function. The function ρ is homogeneous of degree τ if and only if the following equation holds:*

$$\tau \rho(w) = \sum_{i=1}^{n} w_i \frac{\partial \rho(w)}{\partial w_i} . \tag{32.26}$$

For any homogeneous risk function of degree one, Euler's theorem permits us to represent the risk of the portfolio as a sum of sensitivities with respect to each particular position. With this decomposition, each risk contribution can be viewed as a marginal impact on the overall portfolio risk [810].

The positive homogeneity property of the VaR and CVaR measures implies that the risk of a portfolio scales in proportion to the size of the portfolio. The marginal risk contribution of an asset i is given by

$$\rho^r(w_i) = w_i \frac{\partial \rho^r(w_i)}{\partial w_i} , \tag{32.27}$$

where ρ^r is the risk function of type r and r refers here to VaR or CVaR. The Euler-based allocation technique defines a marginal risk contribution using the derivative of the risk measure with respect to the size of that position. The derivative of quantile risk measures may not always exist. In general, under appropriate (smoothness) conditions presented in [1021, 1748, 1749], the derivative of the VaR and CVaR measures can be expressed as a conditional expectation.

Theorem 32.2. *The marginal VaR contribution of any arbitrary asset i is given by*

$$\rho^{VaR}(w_i) = w_i \frac{\partial \alpha_\beta(w)}{\partial w_i} = \mathbb{E}\left[w_i \xi^i \,\middle|\, w^t \xi = \alpha_\beta(w) \right], \quad i = 1,\dots,n . \tag{32.28}$$

Proof. Consider an increase in the asset position i by an amount ϵ. The loss of the resulting new portfolio is $L_\epsilon = X + \epsilon Y$, with $X = w^t \xi$ and $Y = \xi_i$. Let $f(x,y)$ denote the pdf of (X,Y) and $\alpha_\beta(\epsilon)$ be the β-quantile of L_ϵ. For any continuous probability distribution, we have

$$\mathbb{P}\left(X + \epsilon Y \leq \alpha_\beta(\epsilon) \right) = \beta \iff \int \int_{\alpha_\beta(\epsilon)-\epsilon y} f(x,y)\, dx\, dy = \beta .$$

Differentiating with respect to ϵ gives

$$\int \left(\frac{\partial \alpha_\beta(\epsilon)}{\partial \epsilon} - y \right) f(\alpha_\beta(\epsilon) - \epsilon, y)\, dy = 0,$$

which in turn implies

$$\frac{\partial \alpha_\beta(\epsilon)}{\partial \epsilon} = \frac{\int y\, f(\alpha_\beta(\epsilon)) - \epsilon y, y)\, dy}{\int f(\alpha_\beta(\epsilon)) - \epsilon y, y)\, dy} = \mathbb{E}\left[Y \,\middle|\, L_\epsilon = \alpha_\beta(\epsilon) \right].$$

$\square$

Each marginal VaR contribution $\rho^{VaR}(w_i)$ is the conditional expected loss rate due to the position in asset i conditional on a large loss for the entire portfolio.

Similarly, we can demonstrate that the marginal CVaR contribution can be modeled as a conditional expectation:

$$\rho^{CVaR}(w_i) = w_i \frac{\partial \psi_\beta(w)}{\partial w_i} = \mathbb{E}\left[w_i \xi^i \,\middle|\, w^t \xi \geq \alpha_\beta(w)\right], \ i = 1,\ldots,n. \qquad (32.29)$$

The above results show that the marginal VaR and CVaR contributions are conditional expectations of the individual loss random variables, conditioned on rare and large portfolio losses. By contrast to $\mathbb{E}[w_i \xi_i | w^t \xi = \alpha_\beta(w)]$, the marginal CVaR contribution $\mathbb{E}[w_i \xi_i | w^t \xi \geq \alpha_\beta(w)]$ is an elementary conditional expectation because it has a positive probability of occurring.

The possible nonexistence of the derivative of the quantile function and the rarity of these tail events can complicate the computations of the marginal VaR and CVaR contributions. In that respect, kernel methods [1749] and importance sampling techniques [810] have been proposed to approximate the marginal VaR and CVaR contributions.

Portfolio optimization models with marginal and portfolio risk constraints take the form of SO problems including

- up to n conditional expectation constraints and a chance constraint with random technology matrix for the VaR measure, with the following formulation minimizing the portfolio's VaR with constraints on the marginal risk contribution of each asset:

$$\min \quad \alpha$$
$$\text{s.t.} \quad \mathbb{E}\left[w_i \xi^i \,\middle|\, w^t \xi = \alpha_\beta(w)\right] \leq d_i \,, \ i = 1,\ldots,n, \qquad (32.30)$$
$$(32.4)\text{--}(32.5); \text{and}$$

- up to $n+1$ conditional expectations constraints for the CVaR measure:

$$\min \quad \alpha + \frac{1}{1-\beta}\mathbb{E}[[f(w,\xi)-\alpha]^+]$$
$$\text{s.t.} \quad \mathbb{E}\left[w_i \xi_i \,\middle|\, w^t \xi \geq \alpha_\beta(w)\right] \leq e_i \,, \ i = 1,\ldots,n, \qquad (32.31)$$
$$(32.5).$$

While each $d_i \in \mathbb{R}^+$ and $e_i \in \mathbb{R}^+, i = 1,\ldots,n$, is assigned the same value within the risk parity approach, the value taken by the parameters $d_i, e_i, i = 1,\ldots,n$, can differ for risk budgeting approaches. To assign the values of the parameters d_i in (32.30) and e_i in (32.31) and facilitate their interpretation, we can define them as a proportion of the VaR of the portfolio loss

$$d_i = \gamma_i \alpha \,, \ i = 1,\ldots,n,$$

and a proportion of the CVAR of the portfolio loss

$$e_i = \delta_i \left(\alpha + \frac{1}{1-\beta}\mathbb{E}[[f(w,\xi)-\alpha]^+]\right), \ i = 1,\ldots,n.$$

The values of the parameters must be set such that $\sum_{i=1}^{n} \gamma_i \geq 1$ and $\sum_{i=1}^{n} \delta_i \geq 1$ because the marginal VaR and CVaR contributions are additive and $\sum_{i=1}^{n} \rho^{VaR}(w_i) = \alpha_\beta(w)$ and $\sum_{i=1}^{n} \rho^{CVaR}(w_i) = \psi_\beta(w)$. Boudt et al. [328] propose using the percentage marginal CVaR contributions (i.e., the percentage of the CVaR of the portfolio loss). The same authors suggest minimizing the largest marginal CVaR contribution. The corresponding optimization problem can be cast as

$$\min \qquad\qquad z \qquad\qquad\qquad\qquad (32.32)$$

$$\text{s.t.} \quad z \geq \mathbb{E}\left[w_i \xi_i \,\Big|\, w^t \xi = \psi_\beta(w)\right], \; i = 1, \dots, n, \qquad (32.33)$$

$$(32.5).$$

Although risk budgeting is widely used by finance practitioners, not many studies about the behavior of portfolios constructed with a risk budgeting approach have appeared [1588]. Marginal risk measures have been predominantly employed only in ex-post analysis of the risk contribution of an individual asset (see, e.g., [328, 1980]). The literature about ex-ante use on marginal risk measures is scant. Few portfolio optimization models in the literature incorporate direct marginal risk control into the construction of an optimal portfolio. It is conjectured [1980] that this might be due to the difficulty of solving the associated optimization problems, which are often nonconvex. The few available results have essentially been derived for the variance risk measure (see, e.g., [1980]), while downside risk measures have generally been overlooked.

Part IX

Inventory and Supply Chain Optimization

Chapter 33

Inventory and Supply Chain Optimization

Lawrence V. Snyder

33.1 ▪ Introduction

Inventory and supply chain optimization can be viewed from two perspectives. On the one hand, they are applications of a wide range of optimization techniques, including many that are discussed elsewhere in this book, such as linear, nonlinear, mixed-integer, robust, and stochastic optimization. On the other hand, they constitute, in and of themselves, a rich set of optimization tools that are widely applied, not only to supply chains, but also to other sectors, such as healthcare, energy, and humanitarian relief. This part of the book takes the latter perspective, discussing the theoretical foundations of inventory and supply chain optimization in this chapter, and their applications to other sectors in the chapters that follow.

In this chapter, we discuss models and algorithms for inventory and supply chain optimization, focusing on three classes of problems: inventory optimization, facility location, and decentralized supply chain optimization. For further reading on inventory and supply chain optimization, there are many books on the subject, e.g., [124, 463, 798, 1505, 1695–1697, 1711, 1984].

33.2 ▪ Inventory Optimization

In this section we discuss models and algorithms for inventory optimization. We discuss deterministic models (the classical economic order quantity (EOQ) and dynamic economic lot-sizing problems and their variants) in Section 33.2.1 and then turn our attention to stochastic models (the newsvendor problem, dynamic programming (DP) approaches to inventory optimization, and multiechelon models) in Section 33.2.2.

The chapters that follow in this part of the book discuss how classical inventory models have been extended and applied to energy storage (Section 34.2), humanitarian logistics (Section 36.3), and healthcare (Sections 35.2 and 35.3). A closely related class of models—for capacity expansion or planning—is also applied in these sectors; see Sections 34.1.1 and 35.1.

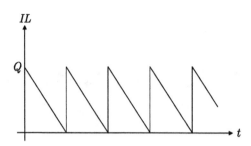

Figure 33.1. *EOQ inventory level process.*

33.2.1 ▪ Deterministic Inventory Optimization

33.2.1.1 ▪ The EOQ Problem

The EOQ problem is arguably the most famous inventory optimization problem. It dates to Harris in 1913 [913], whose coverage was less mathematical than the approach commonly used today, and whose original work was unfortunately overlooked (and misattributed) for many decades before it began to be widely cited in the 1980s. (See the histories of the problem by [674–676], and the book [462] and journal issue [405] in honor of its centennial in 2013.)

The EOQ problem aims to determine the "economic" (i.e., optimal) order quantity that minimizes the average cost per unit time. It is a continuous-review model, i.e., the inventory process is assumed to be monitored continuously, and replenishment orders may be placed at any time. The demand is assumed to be constant and continuous, occurring at a rate of λ items per unit time, and stockouts are not allowed; that is, the inventory must be sufficient at all times to satisfy the demand from stock. We will assume the lead time is zero—shipments are received as soon as they are ordered—though it is easy to relax this assumption to allow nonzero but deterministic lead times. On-hand inventory incurs a *holding cost* of h per item per unit time, and each order placed incurs a *fixed cost* (also known as a *setup cost*) of K, which is independent of the order quantity. One may also include a *purchase cost* c per item, but the total purchase cost per unit time is a constant, independent of the decision variable, so it is often omitted from the optimization problem, as we do here. The central trade-off in the EOQ problem is between fixed and holding costs: If we place many small orders, we incur large fixed costs but small holding costs; we incur the reverse if we place few large orders; the goal is to find the happy medium.

Theorem 33.1. *The optimal ordering strategy for the EOQ problem has the following properties:*

1. Zero-inventory ordering (ZIO): *No order is placed unless the inventory level equals zero.*

2. Stationary intervals: *Orders are equally spread out over time (and therefore all order quantities are equal).*

Therefore, the EOQ problem has a single decision variable, Q, representing the order quantity. A plot of the inventory level over time is given in Figure 33.1. Note that the diagonal line segments have slope $-\lambda$.

If the order quantity is Q, then there are λ/Q orders placed per unit time, so the total fixed cost per unit time is $K\lambda/Q$. The average inventory level is $Q/2$, so the total holding cost per unit time is $hQ/2$. Therefore, the EOQ problem can be formulated as follows:

$$\min \quad C(Q) = \frac{K\lambda}{Q} + \frac{hQ}{2} \tag{33.1a}$$

$$\text{s.t.} \quad Q > 0. \tag{33.1b}$$

Since the objective function $C(Q)$ is convex, the optimal Q can be obtained from the first-order condition $C'(Q) = 0$, which yields

$$Q^* = \sqrt{\frac{2K\lambda}{h}}. \tag{33.2}$$

The availability of a closed-form solution for the EOQ problem allows us to prove properties of optimal solutions that would not be evident if the model had to be solved numerically. For example, it follows immediately from the first-order condition that

$$\frac{K\lambda}{Q^*} = \frac{hQ^*}{2}, \tag{33.3}$$

i.e., the fixed and holding cost components of the objective function are equal at optimality. One can also show that

$$C(Q^*) = \sqrt{2K\lambda h} = hQ^* \tag{33.4}$$

and use (33.2) and (33.4) to examine how the optimal solution changes in response to changes in the parameters. (For example, Q^* and $C(Q^*)$ are increasing concave functions of K.) Equation (33.4) also leads to the *sensitivity analysis* result that for all $Q > 0$,

$$\frac{C(Q)}{C(Q^*)} = \frac{1}{2}\left(\frac{Q}{Q^*} + \frac{Q^*}{Q}\right). \tag{33.5}$$

The closed-form expression for $C(Q^*)$ also allows the EOQ problem to be embedded compactly into other optimization models. For example, [556, 1683] incorporate it into the objective function of a facility location problem to account for inventory costs when making location decisions.

Many extensions of the EOQ problem relax assumptions or generalize the problem in other ways. For example, it is straightforward to allow deterministic, constant, nonzero lead times L: we simply place each order L time units before the inventory level will reach zero. In another variant, planned backorders (stockouts) are allowed, at a cost, a problem known as the *EOQ with backorders* (EOQB) problem, for which the optimal solution is given by

$$Q^* = \sqrt{\frac{2K\lambda(h+p)}{hp}} \tag{33.6}$$

in place of (33.2). This solution can be derived easily from the fact that the EOQB is mathematically equivalent to (33.1) with h replaced by a modified constant. The same can be said for a variant in which the lead times are nonconstant (but deterministic) and proportional to the number of products ordered, i.e., treating the production rate as finite (whereas the EOQ model treats this rate as infinite), known as the *economic production quantity* (EPQ) problem. More complicated ordering cost structures such as quantity discounts can be incorporated; this disrupts the tractability of the original model somewhat, but the problem can still be solved easily using customized algorithms with the EOQ problem as a subproblem.

Other variants are considerably more difficult, especially those involving multiple products linked by a cost or capacity constraint. One such problem is the *joint replenishment problem* (JRP), in which each product in an order incurs a "minor" fixed cost and the order as a whole incurs a "major" fixed cost. Another related problem is the *economic lot scheduling problem* (ELSP), in which multiple products must be produced, sequentially, on a single machine or other capacitated resource. Like the EPQ problem, the production rate is finite. Both problems are NP-hard [96, 780], and for both, heuristics generally use the EOQ or EPQ problem as a subproblem.

33.2.1.2 ▪ The Dynamic Economic Lot-Sizing Problem

Now consider a periodic-review setting. The inventory process is monitored only at discrete, regularly spaced points in time, and replenishment orders may only be placed at those moments. The demand is deterministic but nonstationary (time-varying), and the planning horizon is finite. As in the EOQ problem, stockouts are prohibited. This problem is known as the *dynamic economic lot-sizing* (DEL) problem (or sometimes the *Wagner–Whitin problem* after its original authors [1861]).

There are T periods in the planning horizon, and the demand in period t is denoted by d_t. The system incurs a fixed cost of K per order and a holding cost of h per item per period. As in the EOQ problem, we will assume that the lead time and purchase cost are both zero, though these assumptions are easy to relax. The goal of the DEL problem is to determine which periods to place orders in, and how much to order in each period to minimize the total cost over the horizon.

The DEL problem can be formulated as a mixed-integer optimization problem, but it is usually instead formulated and solved using DP. The DP formulation relies on the fact that the DEL problem, like the EOQ problem, has the ZIO property [1861], and so in period t, we will order exactly the demand for periods t through s for some $s \geq t$. This allows us to ignore the "how much?" decision and focus on the "when?" decision.

Let θ_t be the cost-to-go function, i.e., the optimal cost in periods $t, t+1, \ldots, T$ if we place an order in period t. Create a dummy period $T+1$ and let $\theta_{T+1} \equiv 0$. Then the DP formulation is given by

$$\theta_t = \min_{t < s \leq T+1} \left\{ K + h \sum_{i=t}^{s-1} (i-t)d_i + \theta_s \right\}. \tag{33.7}$$

If s is the minimizer in (33.7), then the next period in which we order will be s, assuming we also order in t. If $s = T+1$, then we do not order again before the end of the horizon. The cost to order in period t for periods $t, t+1, \ldots, s-1$ is given inside the braces, noting that the items for period i will be held for $i - t$ periods. The DP problem (33.7) can be solved using standard DP methodologies, and the optimal solution can be obtained by backtracking. This algorithm is known as the *Wagner–Whitin algorithm* [1861]. It has complexity $O(T^2)$. Faster, $O(T)$, algorithms have also been developed [693, 1860]. Heuristics for this problem are also common in practice (despite the availability of efficient exact algorithms); these heuristics and their resulting solutions tend to be more intuitive for managers (see, e.g., [1695]).

The DEL problem can also be formulated as a shortest-path problem (given the equivalence between such problems and DP problems) on a network with $T+1$ nodes, each of which represents a period (or the dummy period). Each node t has a forward arc to each node $s > t$, with cost given by the value inside the braces in (33.7), excluding the θ_s term. Selecting this arc represents ordering in period t and then not again until period s.

The DEL problem as discussed above can be modified easily to handle nonstationary costs, nonzero lead times, and positive initial inventories. On the other hand, concave order costs increase the computational complexity (while still maintaining polynomial solvability) [1948], whereas ordering capacities make the problem NP-hard [733].

33.2.2 ▪ Stochastic Inventory Optimization

33.2.2.1 ▪ The Newsvendor Problem

The classical newsvendor problem dates back to Edgeworth in 1888 [652], with mathematical foundations developed later by [98, 1356, 1886]; see [1506] for a history. The problem considers a single firm selling a single product with stochastic demand. The firm purchases inventory before the start of the selling season, and no additional orders may be placed. Unmet demands are lost, and unsold inventory must be scrapped, possibly earning a revenue or incurring a cost. Because excess inventory cannot be carried over until the next selling season, nor can unmet demands, the product is considered to be *perishable*, a category that for inventory theorists includes any product that has a date after which it cannot be sold—for example, milk, vegetables, and flowers, but also airplane seats, high-tech products, and newspapers. In fact, the story that gives the problem its name imagines a newsvendor who purchases newspapers from the publisher each morning, sells them throughout the day, and returns unsold newspapers to the publisher for a partial reimbursement (i.e., salvages them).

The newsvendor problem can be formulated in a number of ways. Here we consider a general formulation that can accommodate a number of cost/revenue structures. Let h be the holding cost and p be the stockout cost per unit of unmet demand. These costs, also known as the overage cost and underage cost (respectively), represent the losses incurred for having one too many or one too few (respectively) units on hand for the demand during the selling period. For example:

- Suppose the firm buys products for c per unit, sells them for r per unit, and salvages unsold items for v per unit. Then $h = c - v$ and $p = r - c$.

- Suppose the firm buys products for c per unit, earns a profit of π per unit sold, incurs a *loss-of-goodwill* cost (representing customers' dissatisfaction) of g per unit of unmet demand, and pays d per unit to dispose of unused inventory. Then $h = c + d$ and $p = \pi + g$.

The demand during the selling season is a random variable X with probability distribution function (pdf) $f(\cdot)$ and cumulative distribution function (cdf) $F(\cdot)$. We assume $F(0) = 0$, i.e., the demand is always nonnegative. Let Q be the quantity ordered before the start of the selling season. Then the expected cost can be expressed as a function of the decision variable Q as follows:

$$C(Q) = h \int_0^Q (Q - x) f(x) dx + p \int_Q^\infty (x - Q) f(x) dx. \qquad (33.8)$$

The optimization problem to be solved is

$$\min \quad C(Q) \qquad\qquad (33.9a)$$
$$\text{s.t.} \quad Q > 0. \qquad\qquad (33.9b)$$

The objective function $C(Q)$ is convex, and therefore the optimal Q can be obtained from the first-order condition. $C'(Q)$ can be evaluated using Leibniz's rule, or by rewriting (33.8) in terms of the loss function $n(x)$ corresponding to the demand distribution, and its complement $\tilde{n}(x)$. In particular,

$$C(Q) = h\tilde{n}(Q) + pn(Q), \qquad\qquad (33.10)$$

where

$$n(x) = E[(X - x)^+], \qquad\qquad (33.11)$$
$$\tilde{n}(x) = E[(X - x)^-]. \qquad\qquad (33.12)$$

(We use the notation $a^+ = \max\{0, a\}$ and $a^- = \max\{0, -a\}$.) These functions have derivatives

$$n'(x) = F(x) - 1, \qquad\qquad (33.13)$$
$$\tilde{n}'(x) = F(x). \qquad\qquad (33.14)$$

Then the first-order condition is given by

$$hF(Q) + p(F(Q) - 1) = 0$$

and is solved by

$$Q^* = F^{-1}\left(\frac{p}{h+p}\right). \qquad\qquad (33.15)$$

The quantity $\alpha \equiv p/(h + p)$ in (33.15) is known as the *critical fractile* (or *critical ratio*). $F(Q)$ is the probability that the demand does not exceed the supply, i.e., the probability that there are no stockouts, also known as the *type-1 service level*. Therefore, (33.15) says that, at optimality, the type-1 service level equals α.

If the demand is normally distributed with mean μ and variance σ^2, then (33.15) becomes

$$Q^* = \mu + z_\alpha \sigma, \qquad\qquad (33.16)$$

where z_α is the αth fractile of the standard normal distribution.

If, instead, the demand has a discrete distribution (e.g., Poisson), then the cost function (33.8) is replaced by

$$C(Q) = h\sum_{x=0}^{Q}(Q-x)f(x) + p\sum_{x=Q}^{\infty}(x-Q)f(x), \qquad\qquad (33.17)$$

and $Q^* = n$, where n is the smallest integer that satisfies

$$F(n) \geq \frac{p}{h+p}. \qquad\qquad (33.18)$$

(Note the similarity to (33.15).)

33.2.2.2 ▪ The Finite-Horizon Problem

If the product is not perishable, then excess inventory can be carried from period to period at a holding cost of h per unit per period. We assume unmet demands are *backordered* at a cost of p per unit per period, which includes loss of goodwill as well as any administrative cost to manage the backorders. (In an alternative assumption, known as *lost sales*, unmet demands are lost, and p represents the lost profit as well as loss-of-goodwill cost.) We assume that each order incurs a fixed cost of K and a purchase cost of c. The demand per period is a random variable with pdf $f(\cdot)$ and cdf $F(\cdot)$. The costs and demand distribution are assumed here to be stationary, but this assumption can be relaxed easily. It is also straightforward to incorporate discounting, i.e., the time value of money, which we ignore here.

It is common to interpret backorders as negative inventory, and to define the *inventory level* as the on-hand inventory minus the backorders. When a replenishment order arrives, the new inventory level equals the old inventory level plus the order quantity, whether the old inventory level was positive or negative.

Assume the horizon is finite, with T time periods. (We consider the infinite-horizon case in the next section.) We formulate and solve this problem using DP. The cost-to-go function is given by $\theta_t(x)$, representing the optimal expected cost in periods $t, t+1, \ldots, T$, assuming we begin period t with an inventory level of x. Rather than treating the order quantity Q as the decision variable, it is convenient to work with the *order-up-to level* y, defined as $y = x + Q$. The DP formulation is then given by

$$\theta_t(x) = \min_{y \geq x} \left\{ c(y-x) + K\delta(y-x) + C(y) + E[\theta_{t+1}(y-D)] \right\}, \qquad (33.19)$$

where $C(\cdot)$ is as defined in (33.8) or (33.10) and $\delta(z) = 1$ if $z > 0$, and 0 otherwise. For a given order-up-to level y, the first two terms inside the braces calculate the order costs, the third term calculates the expected holding and stockout costs that will be incurred in period t, and the fourth term captures the expected future cost, accounting for the fact that we will begin period $t+1$ with an inventory level of $y - D$, where D is the random demand in period t. To initialize the recursion, we define a function $\theta_{T+1}(x)$, known as the *terminal value function*, which describes the costs and/or revenues incurred at the end of the horizon as a function of the ending inventory level. The terminal value function can take many forms, and it is often assumed to be convex.

This formulation assumes that both the demand distribution and the feasible range of x values (inventory levels) have been truncated and discretized; in general, wider truncation ranges and narrower discretization intervals yield more accurate solutions but slower run times.

The DP formulation (33.19) is useful as a computational tool, but the structure of the problem itself can also be used to prove structural properties of its optimal solutions. For example, Figure 33.2 plots $y_t(x)$ (the y that attains the minimum in (33.19)) as a function of x for fixed t for problems with $K = 0$ and $K > 0$ in parts (a) and (b), respectively. Figure 33.2(a) suggests that the optimal order-up-to level follows a pattern: if $x < S_t$, then order up to S_t, and if $x \geq S_t$, then do nothing, where S_t is the breakpoint of the function $y_t(x)$. Such an ordering strategy is called a *base-stock policy*. Similarly, Figure 33.2(b) suggests that if $x < s_t$, we should order up to S_t, and if $x \geq s_t$, we should do nothing, for appropriate constants s_t and S_t; this is an (s, S) *policy*. These inventory policies are attractive because they simplify the decision-making rules considerably: rather than requiring a list of the optimal actions $y_t(x)$ for all t and x, we simply need the policy parameter S_t (in the case of a base-stock policy) or the

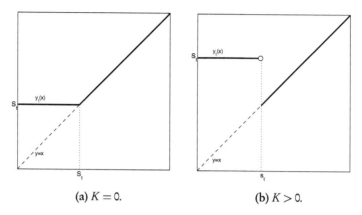

(a) $K = 0$.   (b) $K > 0$.

Figure 33.2. $y_t(x)$ *as a function of* x *for fixed* t.

parameters s_t and S_t (in the case of an (s, S) policy) for each t. Indeed, these policies are always optimal for this problem, as the next theorem establishes.

Theorem 33.2. *If* $\theta_{T+1}(x)$ *is convex, then in each period of the finite-horizon problem,*

(a) *a base-stock policy is optimal if* $K = 0$;

(b) *an* (s, S) *policy is optimal if* $K \geq 0$.

(See, e.g., [1505, 1711, 1984] for proofs.) To determine the optimal policy parameters s_t and/or S_t, we must still solve the DP problem (33.19), except in certain special cases. Note that a base-stock policy is a special case of an (s, S) policy in which $s = S$, which justifies the $\geq$ in Theorem 33.2(b).

33.2.2.3 ▪ Infinite-Horizon Problems

We now turn our attention to infinite-horizon stochastic inventory problems. We first consider an infinite-horizon version of the periodic-review problem in the previous section and then discuss its continuous-review counterpart. In continuous time, the relevant policy is an (r, Q) policy: whenever the inventory level reaches the *reorder point* (denoted r), we place an order of size Q.

Theorem 33.2 continues to hold if $T = \infty$. Assume first that $K = 0$; then a base-stock policy is still optimal in each period of the infinite horizon, and if the costs and demand parameters are stationary, then the optimal base-stock level is the same in every period and is given by (33.15). If, instead, $K \geq 0$, then an (s, S) policy is still optimal in every period, but the optimal parameters cannot be found in closed form. Zheng and Federgruen [1973] propose a simple algorithm, similar to Algorithm 33.1 below, to find exact optimal values for s^* and S^*. Several approximations are also in common use. Some involve computing parameters for an (r, Q) policy (exactly or heuristically) and then converting that policy into an (s, S) policy by setting $s = r$ and $S = r + Q$, since both s and r represent reorder points, while S represents the order-up-to level after an order of size Q is placed. Another approximation, known as the *power approximation* [656], sets s and S using closed-form expressions that were developed by postulating a functional form and fitting regression models to the solutions of many problem instances.

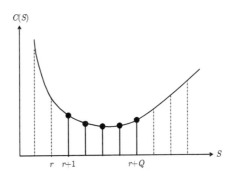

$C(S)$

r $r+1$ $r+Q$ S

Figure 33.3. *Choosing r for fixed Q.*

Now consider a continuous-review version of the problem: demands occur with rate λ (items per unit time), and the inventory level is monitored continuously. Replenishment orders may be placed at any time and arrive after a fixed lead time of L time units. Holding and stockout costs accrue at a rate of h and p (respectively) per item per unit time, and a fixed cost of K is incurred for each order. The objective is to minimize $C(r,Q)$, the long-run expected cost per unit time. An (r,Q) policy is known to be optimal for this problem [1984].

Suppose first that the demands arrive according to a Poisson process or other discrete process. It is convenient to focus on the *lead-time demand* (the demand occurring during a lead time) as the random variable, rather than the demand per unit time. Let D represent this random variable, and let $f(\cdot)$ and $F(\cdot)$ be its probability mass function (pmf) and cdf, respectively. (If demands arrive according to a Poisson process, then $D \sim \text{Pois}(\lambda L)$.) One can show that the expected cost per unit time is given by

$$C(r,Q) = \frac{K\lambda + \sum_{S=r+1}^{r+Q} C(S)}{Q}, \tag{33.20}$$

where $C(S)$ is as defined in (33.17). $C(r,Q)$ is jointly convex in r and Q. For fixed Q, we need to choose r to minimize $\sum_{S=r+1}^{r+Q} C(S)$ (see Figure 33.3); let $r(Q)$ be this r. Since $C(S)$ is convex, one can argue (e.g., [1711, 1984]) that $r(Q+1)$ equals either $r(Q)$ or $r(Q) - 1$. Moreover, for $Q = 1$, $r(Q) = S$, where S is the smallest integer x that satisfies (33.18). Therefore, we can optimize recursively using Algorithm 33.1, due to [695].

ALGORITHM 33.1. **Optimize** $C(r, Q)$.

1: $Q \leftarrow 1; r(Q) \leftarrow S - 1$, where S is the smallest integer x satisfying (33.18)
2: **if** $C(r) < C(r+Q+1)$ **then**
3:    $r(Q+1) \leftarrow r(Q) - 1$
4: **else**
5:    $r(Q+1) \leftarrow r(Q)$
6: **end if**
7: **if** $C(r(Q+1), Q+1) > C(r(Q), Q)$ **then** STOP
8: **else**
9:    $Q \leftarrow Q+1$; goto 2
10: **end if**

Suppose now that the demand follows a continuous distribution (e.g., the demand in any time interval of length Δt is distributed as $N(\lambda \Delta t, \sigma^2 \Delta t)$). The objective function is identical to (33.20) with the sum replaced by an integral whose limits are r and $r + Q$. It is common to solve this problem using heuristics. The first-order conditions imply that for given Q, the optimal r satisfies

$$C(r) = C(r + Q), \tag{33.21}$$

where $C(\cdot)$ is as defined in (33.8) or (33.10) [1972]. One common heuristic is therefore to set Q equal to the optimal order quantity from the EOQ problem with backorders (33.6) and then set r to satisfy (33.21). This heuristic had been known empirically to perform well, an observation that is made rigorous by Zheng [1972], who proves a fixed worst-case bound of $\frac{1}{8} = 0.125$. Axsäter [123] improves the error bound to $(\sqrt{5}-2)/2 \approx 0.118$.

Another heuristic for the continuous-demand problem formulates an approximate expected cost as

$$C(r, Q) = h\left(r - \lambda L + \frac{Q}{2}\right) + \frac{K\lambda}{Q} + \frac{p\lambda n(r)}{Q}, \tag{33.22}$$

where $n(\cdot)$ is the loss function corresponding to the demand distribution. This approximation is the result of two simplifying assumptions: (1) the holding cost is incurred at a rate of h times the inventory level, even if the inventory level is negative, and (2) the stockout cost is incurred once per stockout, not per unit time [1711]. The first-order conditions yield two equations in two unknowns (r and Q) that are usually solved using an iterative approach.

33.2.2.4 ▪ Supply Uncertainty

An important class of inventory problems considers supply uncertainty (as opposed to demand uncertainty), typically in the form of *disruptions* (the supply is completely or mostly unavailable for certain periods of time), *yield uncertainty* (the quantity delivered differs randomly from the quantity ordered), *capacity uncertainty* (the capacity available to produce the item is random), or *lead-time uncertainty* (the delivery lead time is random). Often these models are formulated as extensions of the classical inventory models discussed above, but their tractability is often affected by the new source of uncertainty. For reviews, see [108, 1371, 1709, 1925]. For energy and healthcare applications, see Sections 34.1.2 and 35.2, respectively.

For example, suppose that, in the newsvendor problem (Section 33.2.2.1), the demand is deterministic and equal to d, but the yield is uncertain; in particular, if we order Q units, then we actually receive $Q + Y$, where Y is a random variable with cdf $F(\cdot)$ that does not depend on Q. One can show [1711] that

$$Q^* = d - F^{-1}\left(\frac{h}{h + p}\right). \tag{33.23}$$

Note the similarity between (33.23) and (33.15); here, the larger the likely yield, the smaller the order quantity. Now suppose instead that the supply is subject to complete disruptions, and let $F(n)$ be the steady-state probability that in the current period, we are in a disruption that has lasted n periods or fewer. Then one can show [1711] that $Q^* = n(d + 1)$, where n is the smallest integer that satisfies (33.18).

33.2.3 ▪ Multiechelon Inventory Optimization

The problems in Sections 33.2.1 and 33.2.2 consider only a single inventory site. Today's supply chains often consist of tens, hundreds, or even thousands of interconnected *stages* (which can represent locations, products, activities, etc.), often grouped into tiers or *echelons*. This motivates the development of *multiechelon inventory optimization* models in which inventory decisions are optimized globally across all stages. These problems are almost always nonlinear, and for many, it is difficult or impossible even to formulate an analytical expression for the expected cost. Therefore, research on multiechelon inventory optimization has focused mostly on simple network structures such as serial systems, in which each stage has a single predecessor and successor (except for the source and sink stages).

Clark and Scarf [480] lay the groundwork for the study of multiechelon inventory optimization, introducing the notion of an *echelon base-stock policy*, in which stage j places orders so that the sum of the inventory at j and its downstream stages equals the *echelon base-stock level*. They prove that an echelon base-stock policy is optimal at each stage of a serial system with no fixed costs in a finite horizon. (The result is also true for infinite-horizon problems [696].) They also introduce an exact algorithm for finding the optimal echelon base-stock levels by recursively solving single-variable, convex optimization problems, one for each stage. Their algorithm is refined in [441], resulting in the following theorem (see also [1984]).

Theorem 33.3. *Consider a serial system without fixed costs whose stages are indexed so that stage N is farthest upstream and stage 1 is farthest downstream. Let $\underline{C}_0(x) = (p + h_1')x^-$. Then the optimal echelon base-stock vector, $(S_1^*, \ldots, S_N^*)$, is the solution to the following recursive system of equations:*

$$\hat{C}_j(x) = h_j x + \underline{C}_{j-1}(x), \tag{33.24a}$$

$$C_j(y) = E\left[\hat{C}_j(y - D_j)\right], \tag{33.24b}$$

$$S_j^* = \operatorname{argmin}\{C_j(y)\}, \tag{33.24c}$$

$$\underline{C}_j(x) = C_j(\min\{s_j^*, x\}). \tag{33.24d}$$

In Theorem 33.3, h_j' is the *local holding cost* at stage j (the type of holding cost we have considered thus far), whereas h_j is its *echelon holding cost*, defined as $h_j = h_j' - h_{j+1}'$ and $h_{N+1}' \equiv 0$. Note that $C_j(y)$ is convex for each j. Zipkin [1984] calls (33.24) the "fundamental equation of supply-chain theory."

Motivated by the fact that each stage in the serial system functions like a newsvendor (Section 33.2.2.1), Shang and Song [1674] introduce a heuristic in which the base-stock level at each stage is set equal to the average of the solutions to two newsvendor problems. This newsvendor heuristic is considerably easier to implement than (33.24), yields closed-form solutions that lend insight into the structure of the problem, and performs quite well, with average optimality errors typically less than 1%.

Another approach to optimizing multiechelon inventory systems, known as the *guaranteed-service model*, assumes that the demand is bounded above and requires each stage to satisfy orders from its downstream customers within a fixed lead time, called the *committed service time* (CST). This allows an explicit expression for the expected cost and a formulation that lends itself to classical optimization approaches. Let A be the set of arcs in the network (representing supplier-customer relationships). Let S_j be the CST promised by stage j to its customers[22] and SI_j be the longest CST promised to j by its suppliers, and let T_j be the processing time (lead time) at stage j. Let D be the set of stages that face external demand, i.e., stages j for which there is no $(j,i) \in A$; we assume the CST is bounded above by a constant $\bar{S}_j$ for all $j \in D$. Assume the demand in τ time periods is bounded above by $\mu\tau + k\sigma\sqrt{\tau}$, where μ and σ are the single-period demand mean and standard deviation. Let h_j be the holding cost at stage j. Then the problem to solve is [848]

$$\min \quad \sum_{j=1}^{N} h_j k\sigma \sqrt{SI_j + T_j - S_j} \qquad (33.25a)$$

$$\text{s.t.} \quad SI_j - SI_j \leq T_j \qquad \forall j = 1,\ldots,N, \qquad (33.25b)$$

$$\qquad SI_j - S_i \geq 0 \qquad \forall (i,j) \in A, \qquad (33.25c)$$

$$\qquad S_j \leq \bar{S}_j \qquad \forall j \in D, \qquad (33.25d)$$

$$\qquad S_j, SI_j \geq 0 \qquad \forall j = 1,\ldots,N. \qquad (33.25e)$$

For serial systems, Graves [847] proposes solving (33.25) using DP; see [1711] for details. If the network contains no undirected cycles, i.e., it is a tree, the problem can be solved with a DP algorithm that makes use of two separate cost-to-go functions [848, 849]. A more general method that applies to any network with no directed cycles (otherwise a stage may be a raw material for itself), including networks containing undirected cycles, is proposed by [1252]. Their method makes use of (33.25) directly, adding redundant constraints that trigger the commercial solver CPLEX to generate flow cover cuts automatically.

33.3 • Facility Location

Facility location problems choose sites for facilities (factories, warehouses, retailers, etc.) to strike a balance between the cost of locating facilities and their proximity to the customers they serve. Facility location problems are among the most well studied classical applied optimization areas. They have been implemented in a variety of applied settings, in both the public sector (power plants, fire stations, blood banks) and the private sector (fast food restaurants, retail stores, warehouses), as well as for some surprising types of "facilities," such as political party platforms, satellite orbits, and bank accounts [526]. They also arise as subproblems in optimization algorithms for other problems, such as vehicle routing. Most facility location problems are simple to formulate but are NP-hard, making them a popular testbed for demonstrating nearly every type of discrete optimization approach, exact or heuristic. We provide only a brief overview of facility location problems. For surveys, see [338, 554, 897, 1075, 1560, 1561, 1704, 1708], or for textbooks, see [555, 627, 628, 658, 659, 901, 970, 1327]. Facility location problems are also related to network design problems, which consider yes/no decisions on nodes and/or arcs in a general network [140, 1251].

The subsequent chapters in this part of the book discuss applications of facility location models in energy (Sections 34.3 and 34.4) and humanitarian logistics (Section 36.2).

Continuous location models allow facilities to be placed anywhere in a continuous region. The single-facility version of the problem is known as the *Weber problem* [629];

---

[22]We have opted to use the notation common in the literature for this model, in which S represents a service time, not to be confused with S in the previous model, which represents a base-stock level.

it aims to choose the location (x, y) of the facility that minimizes the total weighted distance between the facility and its customers:

$$\min \quad \sum_{i \in I} h_i c_i(x, y) \tag{33.26a}$$

$$\text{s.t.} \quad (x, y) \in \Omega. \tag{33.26b}$$

In (33.26), I is the set of customers, h_i is the demand (or other weight) of customer i, $c_i(x, y)$ is a function specifying the Euclidean (or other) distance between location (x, y) and customer i, and Ω is the allowable region for the facility. The problem is usually solved using the Weiszfeld procedure [1881] or another iterative method, or using standard nonlinear optimization (NLO) techniques. Multifacility problems are more difficult to solve; because of this, and because facilities usually must be located within a much more restrictive set of potential sites, the Weber problem and other continuous location problems have not been applied widely in a supply chain context.

More commonly, supply chain optimizers solve *network location models* (which allow facilities to be located at the nodes or along the edges of a network) or *discrete location models* (which restrict locations to the nodes). For many problems, the two are equivalent due to the *Hakimi property*, which says that an optimal solution exists for which all facilities are located at nodes. We focus on the discrete case in this chapter.

The *uncapacitated fixed-charge location problem* (UFLP) (which also goes by a variety of other names) chooses facility locations out of a discrete set of potential sites to minimize the sum of the fixed costs to locate the facilities and the transportation costs to serve the customers. Let I be the set of customers and J be the set of potential facility sites; let h_i be the demand of customer i and c_{ij} be the transportation cost between i and j. (The costs c_{ij} may be arbitrary or may be functions of the distance between i and j, which in turn may be computed using an ℓ_p-norm, great-circle distances, road network (shortest-path) distances, or any other method.) Define decision variables x_j, which equals one if we locate a facility at site j (and zero otherwise), and y_{ij}, which equals one if customer i is assigned to facility j (and zero otherwise). Then the UFLP can be formulated as follows:

$$\min \quad \sum_{j \in J} f_j x_j + \sum_{i \in I} \sum_{j \in J} h_i c_{ij} y_{ij} \tag{33.27a}$$

$$\text{s.t.} \quad \sum_{j \in J} y_{ij} = 1 \qquad \forall i \in I, \tag{33.27b}$$

$$y_{ij} \leq x_j \qquad \forall i \in I, \forall j \in J, \tag{33.27c}$$

$$x_j \in \{0, 1\} \qquad \forall j \in J, \tag{33.27d}$$

$$y_{ij} \geq 0 \qquad \forall i \in I, \forall j \in J. \tag{33.27e}$$

The objective function (33.27a) calculates the total fixed plus transportation cost. Constraints (33.27b) require each customer to be assigned to exactly one facility. Constraints (33.27c) prohibit a customer from being assigned to a facility that has not been opened. Constraints (33.27d) require the x variables to be binary and constraints (33.27e) require the y variables to be nonnegative, though there always exists an optimal solution in which they are binary.

The UFLP is NP-hard but can be solved quickly using off-the-shelf branch-and-bound approaches because the linear programming (LP) relaxation tends to be very tight—often the solution to the LP relaxation even has binary variables. Many other exact optimization methods have been proposed for the UFLP and its variants, including branch-and-cut, LP decomposition, and the dual-ascent procedure DUALOC [673]. Heuristics include greedy methods, neighborhood search and other improvement techniques, and a wide range of metaheuristics.

One of the most common approaches for solving the UFLP is Lagrangian relaxation [722,723], which can be interpreted as an exact algorithm or a heuristic depending on whether the Lagrangian procedure is embedded into branch-and-bound to close the optimality gap. In the most common version, we relax constraints (33.27b), letting λ be the Lagrange multipliers, to obtain the following subproblem:

$$\min \quad \sum_{j\in J} f_j x_j + \sum_{i\in I}\sum_{j\in J} h_i c_{ij} y_{ij} + \sum_{i\in I} \lambda_i \left(1 - \sum_{j\in J} y_{ij}\right)$$

$$= \sum_{j\in J} f_j x_j + \sum_{i\in I}\sum_{j\in J} (h_i c_{ij} - \lambda_i) y_{ij} + \sum_{i\in I} \lambda_i \qquad (33.28)$$

s.t.   (33.27c)–(33.27e)

This subproblem decomposes by facility. If facility j is open in the optimal solution to the subproblem, then customer i will be assigned to it if and only if $h_i c_{ij} - \lambda_i$. Therefore, the net change in the objective function (33.28) if we open facility j is $f_j + \beta_j$, where

$$\beta_j = \sum_{i\in I} \min\{0, h_i c_{ij} - \lambda_i\}. \qquad (33.29)$$

It is optimal to set $x_j = 1$ if $f_j + \beta_j < 0$ and to set $y_{ij} = 1$ if $x_j = 1$ and $h_i c_{ij} - \lambda_i < 0$. Then (33.28) provides a lower bound on the optimal objective function value of (33.27). A solution to the subproblem can be converted to a feasible solution (thereby obtaining an upper bound) by assigning each customer to its nearest open facility. The Lagrange multipliers are then updated. Algorithm 33.2 summarizes the method in pseudocode, assuming the multipliers are updated using subgradient optimization. (Other approaches, such as bundle methods, have also been used successfully.) In the algorithm, α^t is a constant that is typically reduced after a certain number of consecutive nonimproving iterations.

ALGORITHM 33.2. Lagrangian Relaxation Algorithm for UFLP.

```
1: t ← 1; λ ← initial values                                      ▷ Initialization
2: for all j ∈ J do                                               ▷ Lower Bound
3:     β_j ← Σ_{i∈I} min{0, h_i c_{ij} − λ_i}
4:     if f_j + β_j < 0 then
5:         x_j ← 1; y_{ij} ← 1 ∀i ∈ I s.t. h_i c_{ij} − λ_i < 0
6:     else
7:         x_j ← 0; y_{ij} ← 0 ∀i ∈ I
8:     end if
9: end for
10: ℒ^t ← Σ_{j∈J}(f_j + β_j)x_j + Σ_{i∈I} λ_i
11: for all i ∈ I do                                              ▷ Upper Bound
12:     j* ← argmin_{j∈J}{c_{ij} : x_j = 1}
13:     ȳ_{ij*} ← 1
14: end for
15: 𝒰^t ← Σ_{j∈J} f_j x_j + Σ_{i∈I} Σ_{j∈J} h_i c_{ij} ȳ_{ij}
```

16: **if** $\mathcal{U}^t = \min_\tau \{\mathcal{U}^\tau\}$ **then**
17:     $x^* \leftarrow x; \, y^* \leftarrow \bar{y}; \, \mathcal{U}^* \leftarrow \mathcal{U}^t$
18: **end if**
19: $\Delta^t \leftarrow \alpha^t(\mathcal{U}^* - \mathcal{L}^t)/\sum_{i \in I}\left(1 - \sum_{j \in J} y_{ij}\right)^2$ ▷ Multiplier Update
20: **for all** $i \in I$ **do**
21:     $\lambda_i^{t+1} \leftarrow \lambda_i^t + \Delta^t\left(1 - \sum_{j \in J} y_{ij}\right)$
22: **end for**
23: **if** termination criteria reached **then**   ▷ Check for Termination
24:     **return** (x^*, y^*)
25: **else**
26:     $t \leftarrow t + 1$; go to 2
27: **end if**

A close cousin of the UFLP, the *P-median problem*, omits the fixed cost from the objective function and instead adds a constraint restricting the number of facilities located to be no more than P. The Lagrangian relaxation algorithm for the UFLP can be modified easily for the P-median problem: the optimal solution to the subproblem sets $x_j = 1$ for the facilities j with the P smallest (most negative) values of β_j, where β_j is as defined in (33.29). Most of the other algorithms discussed above for the UFLP have also been adapted for the P-median problem; see the annotated bibliography [1558].

Another important class of facility location models is *covering models*, in which customers only count as "covered" if they are within a fixed radius of an open facility [1707]. The *set covering location problem* [896] finds the minimum number of facilities required to cover every customer (and is a generalization of the graph-theoretic vertex cover problem), while the *maximal covering location problem* [470] covers the maximum population possible with a fixed number P of facilities. These problems often arise in public-sector facility location problems, e.g., when locating ambulances, fire stations, or libraries.

33.4 ▪ Decentralized Supply Chain Optimization

The models discussed thus far in this chapter assume that a single decision maker optimizes all aspects of the system that are captured by the model. However, modern supply chains are decentralized, with each supply chain partner wishing to optimize its own best interests. This local optimization usually results in the supply chain as a whole failing to achieve the global optimum. In this section, we discuss a game-theoretic analysis of *contracts* that can be used to "coordinate" the supply chain. See [385] for more detailed coverage. See Section 35.4 for a discussion of the application of these models in healthcare.

Consider a retailer who faces stochastic demand in a single period, functioning like the newsvendor in Section 33.2.2.1. Now, however, we consider the retailer's supplier as an active player in the game. Let D be the random demand during the period, with pdf $f(\cdot)$ and cdf $F(\cdot)$. The retailer earns a revenue of r per unit sold and incurs a cost of c_r per unit ordered from the supplier. (This cost is lost to the system—it is not paid to the supplier.) The supplier incurs a cost of c_s per unit ordered by the retailer. We assume $c_r + c_s < r$. Unsatisfied demands are lost and incur stockout costs of p_r and p_s at the retailer and supplier, respectively. Unsold inventory can be salvaged for a revenue of v per unit, with $v < c_r$. Let $c \equiv c_r + c_s$ and $p \equiv p_r + p_s$.

Suppose first that the retailer pays the supplier w per unit ordered, and that the supplier can choose w. This arrangement is called the *wholesale price contract*, and w is called the *wholesale price*. The supplier and retailer each have one decision variable, the wholesale price w and the order quantity Q, respectively. We assume the supplier chooses w first, and the retailer responds by choosing Q. The supplier wishes to maximize its profit, denoted $\pi_s(Q,w)$, and the retailer wishes to maximize its profit, denoted $\pi_r(Q,w)$. Let $S(Q)$ denote the expected sales as a function of Q; then

$$S(Q) = E[\min\{Q,D\}] = Q - \bar{n}(Q), \tag{33.30}$$
$$S'(Q) = 1 - F(Q), \tag{33.31}$$

where $\bar{n}(\cdot)$ is the complement of the loss function, defined in (33.12). One can show [385, 1711] that

$$\pi_r(Q,w) = (r - v + p_r)S(Q) - (w + c_r - v)Q - p_r\mu, \tag{33.32}$$
$$\pi_s(Q,w) = p_s S(Q) + (w - c_s)Q - p_s\mu. \tag{33.33}$$

The supplier and retailer are engaged in a Stackelberg game, which gives rise to a bilevel optimization problem for the supplier:

$$\max_w \quad \pi_s(Q(w), w) \tag{33.34a}$$

$$\text{s.t.} \quad Q(w) \in \operatorname*{argmax}_Q\{\pi_r(Q,w)\}. \tag{33.34b}$$

The supplier cannot choose Q directly, but it can induce the retailer to choose whatever Q it wishes through its choice of w. Both π_r and π_s are concave, so the retailer's and supplier's optimal order quantities, Q_r^* and Q_s^*, satisfy the first-order conditions

$$\left.\frac{\partial \pi_r(Q,w)}{\partial Q}\right|_{Q=Q_r^*} = (r - v + p_r)S'(Q_r^*) - (w + c_r - v) = 0, \tag{33.35}$$

$$\left.\frac{\partial \pi_s(Q,w)}{\partial Q}\right|_{Q=Q_s^*} = p_s S'(Q_s^*) + (w - c_s) = 0. \tag{33.36}$$

The key question is whether there exists a w such that Q_r^* and Q_s^* are equal to each other and equal to the order quantity Q^0 that maximizes the profit of the supply chain as a whole, denoted $\Pi(Q,w) \equiv \pi_r(Q,w) + \pi_s(Q,w)$. If so, the supply chain is said to be *coordinated*; if not, then the two players fail to achieve the globally optimal profit, i.e., they leave money on the table. The next theorem demonstrates that such a w exists, but it ensures that the supplier earns a negative profit, and therefore the supply chain is not coordinated.

Theorem 33.4. $Q_r^* = Q_s^* = Q^0$ *if and only if*

$$w = c_s - \frac{c - v}{r - v + p}p_s. \tag{33.37}$$

The proof follows from (33.35)–(33.36); see [385, 1711]. Since $v < c_r \le c < r$, (33.37) implies that $w < c_s$; that is, the supplier's revenue w per unit is less than its cost, and thus it earns negative profit.

A wide range of alternative contracts have been proposed that *do* coordinate the supply chain. We discuss the *buyback contract* [1457]; others are the revenue sharing contract, the quantity flexibility contract, the quantity discount contract, etc. In the buyback contract, the supplier charges a wholesale price of w (as in the wholesale price contract), but it also buys back unsold inventory from the retailer for b per unit. We assume that $0 \le b \le r - v + p_r$, so that the retailer cannot earn more by selling an item back to the supplier (for a net revenue of $b + v - p_r$) than by selling it to the customer (for a revenue of r).

One can show [385, 1711] that under the buyback contract,

$$\pi_r(Q, w, b) = (r - v + p_r - b)S(Q) - (c_r - v + w - b)Q - p_r\mu, \qquad (33.38)$$
$$\pi_s(Q, w, b) = (p_s + b)S(Q) - (c_s - w + b)Q - p_s\mu. \qquad (33.39)$$

For fixed b, let the wholesale price be given by

$$w(b) = b + c_s - (c - v)\frac{b + p_s}{r - v + p}. \qquad (33.40)$$

Then the supply chain is coordinated.

Theorem 33.5. *Under the buyback contract, for any b such that $0 \le b \le r - v + p_r$, if $w(b)$ is set according to (33.40), then $Q_r^* = Q_s^* = Q^0$.*

The proof follows from the first-order conditions; an alternative proof expresses each profit function as a positive constant times $\Pi(\cdot)$ plus a constant (and thus all three functions have the same optimizer). It can be shown that there always exists a $(b, w(b))$ pair such that both players earn positive profit; in fact, both earn more than they would without the contract.

Chapter 34

Supply Chain Design and Optimization with Applications in the Energy Industry

Owen Q. Wu and Yanfeng Ouyang

In this chapter, we discuss a few supply chain design concepts and optimization models that have been applied in the energy sector. It is not our intention to provide a comprehensive review of the vast literature. Our goal is to provide the reader with pointers to some interesting and challenging problems, thereby triggering thoughts on the synergies between supply chain optimization and energy sustainability. To this end, we select four supply chain research areas that have seen substantial synergies with energy research. These four areas are strategic sourcing, inventory management, supply chain competition, and network design. This chapter discusses these areas and related applications in the energy industry.

34.1 ▪ Strategic Sourcing and Power System Management

Strategic sourcing in supply chain management involves understanding supply characteristics and making decisions such as supplier selection, procurement quantities, and managing supply uncertainties. When suppliers are reliable (i.e., no supply uncertainty), the sourcing strategy hinges on the trade-off between the efficiency and responsiveness of the suppliers. When some suppliers are unreliable but offer low-cost supply, one must strike a balance between the cost advantage of unreliable suppliers and the cost of mitigating supply uncertainties. Both of these trade-offs manifest themselves in power system management, which we discuss in this section.

34.1.1 ▪ Efficient and Responsive Sourcing in Power System Capacity Planning

Suppliers with short lead times allow a supply chain to quickly respond to demand fluctuations, but the speed typically means extra cost to the supply chain. To manage the trade-off between efficiency and responsiveness, a supply chain can choose to have a mixture of efficient and responsive suppliers. A well-known example is the "dual-response" manufacturing in the supply chain for Hewlett Packard inkjet printers [270]. One supplier has low production cost but long lead time; the other has short lead time but high production cost. Using both suppliers allows Hewlett Packard to serve a large portion of its demand efficiently while meeting short-term demand fluctuations responsively. Other examples of using hybrid modes of production can be seen in the fashion clothing industry [721].

Capacity planning for electric power systems also involves the trade-off between efficiency and responsiveness, but with different features. Imagine yourself making capacity investment decisions in an electric utility company. How would you plan a portfolio of power generation technologies to meet uncertain electric demand over the next 20 years? You can choose from a variety of technologies with very different cost structures and construction lead times. It may take more than 10 years to undergo the approval and construction processes of a nuclear power plant, whereas gas-fired generators can be installed within two years. The responsiveness in this context pertains not to the production lead time but to the capacity construction lead time. The total cost comprises the capital, operating, and outage costs.

The above utility capacity planning problem was first studied by Gardner and Rogers [786], who extended the traditional planning methods by taking differences in technology lead times into account. They considered two groups of technologies differentiated by construction lead times. The capacity investment of long lead time technologies must be decided prior to the resolution of uncertain demand, whereas the decisions for short lead time technologies need not be made until demand realizes. The problem is formulated as a two-stage stochastic program with recourse. The solution is termed as an "act, learn, then act" solution and compared with the solutions from traditional planning methods that ignore the difference in technology lead times. One traditional approach is "act, then learn," in which the capacity mix is decided under a given demand forecast; no recourse is considered. Another traditional approach is "learn, then act," in which a capacity mix is found for each given demand realization, and then the solutions are combined, in an ad hoc fashion, to arrive at an implementable solution.

The analysis in [786] reveals that the traditional planning methods may be seriously flawed. There are circumstances where some short lead time technologies are screened out by the traditional planning methods but enter the optimal solution; there are also circumstances where some long lead time technologies are used in the traditional solutions but dropped in the optimal solution. The optimal solution tends to utilize the responsiveness provided by the short lead time technologies and thus forgoes some cost advantage of the long lead time technologies. The paper informs the system planners that they need to examine the extent to which technology lead times can be traded off against capital and/or operating costs.

Beyond uncertain demand, the utility capacity planning problem is often complicated by many sources of uncertainty. The Fukushima tragedy has spurred reevaluation of nuclear power technology and resulted in regulatory changes in many countries. The tightened Environmental Protection Agency (EPA) regulations on emissions have been pushing many coal-fired power generators toward retirement. The shale gas boom has made natural gas power generation technologies more economical, amidst regulatory and geopolitical uncertainties. Increasing uncertainties require utility planners to build more flexibility into the power systems planning process. Recognizing the value of flexibility also encourages the development of technologies with shorter construction lead time.

34.1.2 ▪ Random Capacity and Volume Flexibility in Power System Operations

In a typical power system, resources are coordinated by unit commitment (UC) and economic dispatch (ED) programs. The UC program is run every day to determine which generators (i.e., units) are committed to power generation for each hour of the next day, and the ED program is run in real time to determine the output levels of the committed generators. These programs involve sophisticated system modeling and optimization techniques and thus present great opportunities for applying operations research and analytics.[23] Although these programs reflect high granularity of the reality, they do not directly serve the purpose of designing energy policies.

For policy design, models need (at least initially) to be simpler than reality but complicated enough to capture the essential trade-offs in reality. Such models will allow various stakeholders to understand the mechanisms by which certain policies affect key trade-offs and system performance. Large system models can then be used to simulate the system performance and estimate the impact of certain policies.

In supply chain research, there has been a significant amount of work devoted to managing supply uncertainties, as discussed in Section 33.2.2. We refer the reader to [694, 1709, 1775, 1925] and the references therein. Below, we provide a perspective of thinking about power generation systems that is useful for policy research. This perspective will lead to models that share some features with the supply chain literature, yet present unique characteristics.

Power generators can be categorized based on capacity certainty and volume flexibility:

- random capacity, very low marginal cost;
- certain capacity, volume inflexible, low marginal cost; and
- certain capacity, volume flexible, high marginal cost.

Type 1 capacity refers to intermittent generation from renewable sources, such as wind and solar power. Their marginal cost of production is nearly zero, but they have inherent uncertainties. Type 2 capacity includes nuclear power generators, which have low marginal cost and are typically designed to run at a constant power output level. Type 3 capacity consists of generators with varying degrees of flexibility. They are more flexible than type 2 but also more costly to run. Coal-fired generators have a higher marginal cost than nuclear power generators, but they can adjust their output at a certain rate (known as the ramp rate). A higher ramp rate means a shorter lead time for changing the output level. The most flexible generators are oil- and natural gas–fired combustion turbines, which can meet demand fluctuations from minute to minute, but these generators have high operating cost and thus are known as peaking generators. There are also gas-fired combined-cycle generators whose flexibility is in between coal-fired generators and peaking generators. In terms of marginal cost, combined-cycle generators have become competitive to coal-fired generators due to the lower price of natural gas in recent years.

With the above taxonomy, it is possible to look at power system operations from the supply chain optimization angle. The combination of type 2 and 3 resources is

---

[23]For example, the Midcontinent Independent System Operator (formerly named Midwest ISO) won the 2011 INFORMS Edelman Award [977] for using operations research to improve the reliability and efficiencies of the region's power plants and transmission assets.

similar to the dual-response manufacturing discussed previously, with type 2 capacity serving the baseload and type 3 capacity meeting demand fluctuations. There are two key differences. First, the trade-off between efficiency and responsiveness in power system operations occurs in a much shorter time frame, and power generation and consumption must be constantly balanced. Second, the cost structures of power generators have their unique features, which we elaborate below.

Wu and Kapuscinski [1908] model two subgroups within type 3 generators: fully flexible generators (peaking generators) and intermediate generators. Fully flexible generators can adjust their output almost instantaneously, whereas intermediate generators have limited flexibility reflected by the four cost components illustrated in Figure 34.1: (i) *Cycling cost.* Cycling an intermediate generator increases the wear and tear cost and requires extra fuel during the startup process. The dispatchable intermediate capacity (solid curve) represents the intermediate capacity that is started and can be dispatched to produce energy. (ii) *Part-load penalty.* Intermediate generators are most efficient when producing at a full load (i.e., all dispatchable capacity is utilized). Operating at any lower load increases the average production cost; this extra cost is the part-load penalty. (iii) *Min-gen penalty.* In normal operating conditions, the part load should stay above a minimum generation level (e.g., 50% of the dispatchable capacity); otherwise a min-gen penalty will be incurred. (iv) *Peaking premium.* The dispatchable intermediate capacity cannot be adjusted instantaneously, and thus peaking generators may be needed even if the load on flexible resources is below the total intermediate capacity, which occurs in the areas labeled as (iv) in Figure 34.1.

The system operator aims to minimize the total operating cost, which entails continuously balancing the above cost components, whether or not intermittent generation is present. The growth of intermittent generation resources (type 1) poses increasing management challenges. If we meet 20% of energy demand from renewable sources (mandated by the renewable portfolio standards in many states), the actual percentage of demand met from renewable sources can vary wildly from 0% to 100%, depending on the weather. These fluctuations introduce additional variability into power systems, which complicates the trade-off among the aforementioned cost components.

Wu and Kapuscinski [1908] model the above cost components and study the policies for using intermittent renewable energy. When intermittent generation was introduced into most countries and regions, it was given priority to be used; this policy is referred to as the *priority dispatch* policy. Implementing such a policy requires little

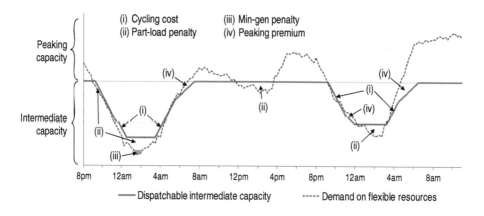

Figure 34.1. *Costs of balancing electrical systems: An example.*

change to the system optimization programs because intermittent generation is simply subtracted from the demand before the programs are run. With the rapid growth in renewable energy penetration, the intermittency began to challenge the systems' ability to balance supply with demand. Curtailment thus became necessary when excessive energy from intermittent resources threatened system reliability. In some circumstances, although curtailment is not absolutely necessary, it provides the system operator with an additional lever to manage variability, thereby reducing system operating costs. Such curtailment is allowed under the *economic curtailment* policy, but not under the priority dispatch policy.

In [1908], the authors compare the two policies and identify the sources of the operational benefits of the economic curtailment policy. Among the four cost components discussed above, economic curtailment policy significantly reduces cycling cost and peaking premium. Curtailing intermittent generation during low-demand periods helps reduce the need for cycling intermediate generators (i.e., reduces the depth of the valleys in the dispatchable intermediate capacity in Figure 34.1). Curtailment also allows more intermediate generators to start up earlier in the morning (i.e., shifts the increasing part of the dispatchable intermediate capacity in Figure 34.1 toward the left), reducing the peaking premium that would otherwise be incurred to meet the rising morning demand. In addition to these operational benefits, economic curtailment also increases the utilization of cheaper inflexible generators.

It is worth noting that the model in [1908] is a stochastic dynamic programming model. The value of economic curtailment is higher under the deterministic optimization programs used prevalently in practice. This is because curtailment serves as a recourse for the decisions generated by deterministic optimization programs, but this recourse is not as valuable under stochastic dynamic programs because the decisions are already adjusted in response to the weather and demand fluctuations.

The recent work by Al-Gwaiz et al. [40] is another example of utilizing the taxonomy introduced earlier to study energy policies. This work focuses on modeling and analyzing the power market competition, which features supply function competition (i.e., each firm submits a supply function that specifies the amount of power it is willing to produce at each price). Different from the classical supply function equilibrium literature, which studies the competition involving only generators of type 3, the authors study the supply function competition among all inflexible and flexible generators. Furthermore, the authors introduce intermittent generation into the model and analyze how it affects the competitive behavior of the other generators. This research opens a promising avenue for analyzing how random capacity and volume flexibility impact power market competition.

34.2 ▪ Inventory Management for Energy Storage Facilities

Energy storage is to grids as inventory is to manufacturing firms. Energy storage is used to buffer against predictable variability (e.g., diurnal demand cycles) and unpredictable variability (supply or demand shocks) to smooth conventional resources' power output. Smoothing production reduces cost because the power generation cost function is highly convex: the marginal cost of nuclear power is below $5 per MWh, whereas that of a peaking unit can be $80 per MWh. The classic inventory optimization theory discussed in Section 33.2 focuses on minimizing inventory-related costs under linear production/purchasing cost. Convex production cost has also been considered in the literature, pioneered by Modigliani and Hohn [1343], who examine the optimal production schedule for meeting demand over a planning horizon. However, energy storage operations involve different cost structures and thus present opportunities to develop inventory theory for energy storage applications.

Electricity per se cannot be stored; to be stored, electricity must be converted into other forms of energy, such as potential or chemical energy. This conversion process involves energy loss, known as the *conversion loss*. The other closely related measure is *storage efficiency*, which is equal to 1– conversion loss rate. For example, the storage efficiency of a lithium-ion battery ranges from 80 to 90%. The stored energy does slowly decrease over time (similar to inventory holding cost), but this type of energy loss is often negligible compared to the conversion loss, because energy storage typically operates on daily cycles or more frequently.

The cost model represented in Figure 34.1 has been extended in [1908] to include costs of storage operations. It is interesting to study how storage operations impact emissions. First, storage allows more clean intermittent energy to be used (instead of being curtailed) and thus reduces emissions. Second, storage reduces the peaking cost while increasing the use of intermediate capacity, which leads to more or less emissions depending on types of fuels. Third, energy conversion losses during storage operations increase emissions. The net effect of storage on emissions depends on the relative strengths of these three factors and is detailed in [1908].

Secomandi [1656] develops a model for natural gas storage facilities, which can also be applied to energy storage for power systems, because the model incorporates injection and withdrawal loss factors (mathematically equivalent to conversion loss) and holding cost. The author also considers a constraint on the rate at which energy can be injected and withdrawn—important for both natural gas storage and energy storage. The problem is formulated as a stochastic dynamic program, and structural properties of the optimal policy are derived. The optimal policy is characterized by two stage- and price-dependent base-stock targets: if inventory falls between the two targets, it is optimal not to do anything; otherwise the firm should inject or withdraw to bring the inventory as close as possible to the closer target.

Wu et al. [1909] focus on understanding the types of real options in energy storage operations and how one should trade off among these options. The authors analyze a heuristic policy commonly used in practice (the rolling intrinsic policy, which solves a deterministic problem every period using up-to-date price information) and point out that this heuristic policy does not attempt to capture the options' extrinsic values that arise from the stochastic evolution of the prices. The authors then design a new heuristic policy, in which the prices are adjusted to approximate the extrinsic values before applying the traditional policy. This simple idea turns out to be very effective: in a three-period setting, the new policy is optimal, and in multiperiod settings, numerical results for natural gas storage show that the new policy recovers a significant portion of the value loss of the traditional policy.

It is important to note that many electricity markets include not only an energy market but also an operating reserve market (also known as an ancillary services market). Operating reserve is the reserved capacity that allows the system operator to manage supply-demand imbalances caused by normal fluctuations or unexpected disruptions. Energy storage can serve as an operating reserve, and thus the storage value needs to incorporate the values derived from both energy and operating reserve markets. Drury et al. [633] quantify the value of compressed-air energy storage (CAES) derived from both markets. They find that the value from the energy market alone (i.e., the energy arbitrage value) cannot support CAES investment in most locations, but the addition of the revenues from providing operating reserves can support CAES investment in several locations.

A promising research avenue is to construct rigorous models for valuing energy storage participating in both the energy and operating reserve markets. The allocation of storage capacity to each market is nontrivial. As discovered in [633], the optimal allocation of storage capacity to provide operating reserves and energy arbitrage has seasonal trends and can shift significantly based on market conditions. Energy storage capacity needs to be dynamically allocated to maximize its market value.

Storage location choice is another important research direction. Denholm and Sioshansi [581] consider the trade-off between colocating storage with a wind farm and locating storage closer to the load. When storage is colocated with a remote wind farm, the main advantage is the downsized transmission line and increased utilization of the transmission line. However, being remote to the load, storage is not as valuable as if it were closer to the load. The paper investigates whether the reduced transmission costs exceed the costs associated with locating energy storage away from the load.

34.3 ▪ Competitive Feedstock Procurement for Biofuel Production

The biofuel supply chain resembles any other multiechelon chain in that it involves a number of stages for biomass harvesting, storage, processing, and transportation, and those for biofuel manufacturing, transportation, and blending. The design problem could be considered an extension of the ones discussed in Section 33.3. A unique feature of the biofuel supply chain, however, is that the increasing demand for bioenergy crops leads to intensive competition for agricultural land—an already scarce resource worldwide—among uses for energy production, food production, and environmental conservation [1770]. While traditional inventory management theories normally consider resource competition among similar vendors, the competition feature of biofuel supply chains leads to two direct consequences. First, ill-planned biofuel industry growth may result in suboptimal land use, significantly reducing food supply; in turn, this will lead to higher food prices, higher greenhouse-gas emissions, and reduced biodiversity. This probably explains why U.S. corn prices have increased dramatically since 2006 to record high in recent years [1543]. Second, desirable economic returns from biofuel production have renewed farmers' interest in reclaiming idle marginal lands as substitutes for regular farmland. Marginal land has long served as a source of environmental conservation (e.g., CO_2 sequestration, habitat preservation, soil productivity restoration); however, since 2007, two million hectares of conserved land in the U.S. has been reclaimed, causing significant environmental hazards, such as soil erosion and pollution from fertilizer runoff. These issues directly involve intriguing organizational, operational, and infrastructure interdependencies among multiple industry sectors (e.g., energy, environment, agriculture) that are difficult for any single industry stakeholder to handle. Such issues often require holistic government intervention and policy regulations, which could be designed using game-theoretic modeling techniques such as those discussed in Section 33.4.

For example, the biofuel production goals (as specified by the U.S. government) have raised a number of pressing questions: Are strategic changes in agricultural land use and feedstock production (e.g., mix of feedstocks) required? How will government regulations and climate control policies affect industry development? What is the optimal size and locational distribution of biofuel refinery plants, how should the feedstock supply contracts be priced, and to what extent is there a divergence between privately profitable and socially optimal designs? In particular, the government faces a difficult food-energy-environment trilemma: how to stimulate the growth of the bio-fuel industry while, at the same time, protecting food security and environmental sus-tainability.

Addressing these challenges requires a comprehensive analysis that holistically ad-dresses the biofuel industry, the food sector, the environmental sector, and the involved farmland markets. Integrating multiple layers of decisions into one overarching model-ing framework is challenging because such decisions are often planned and managed by different stakeholders, who often have independent, if not conflicting, objectives—this generally results in extremely complicated dynamic interactions and requires novel so-lution methods. These types of problems seem to be related to the earlier research on spatial location equilibrium, as first proposed by [1771], where a firm determines the location and production level of its facilities, knowing that these decisions will have direct impacts on the sales prices of products in spatially distributed markets. The con-cept was later extended to a plethora of supply chain network equilibrium models, originating in [1369], to address Nash or Stackelberg types of competitions among decision makers in multiechelon supply chain networks.

In the biofuel supply chain setting, the emerging industry (e.g., biofuel sector) pene-trates into an existing business (e.g., food sector) and competes for feedstock/farmland supply through existing or new spatially distributed sources (or markets). The emerg-ing industry seeks the best strategic design configuration (e.g., refinery location and capacity, supply pricing and procurement, and transportation logistics) to maximize its own profit. Meanwhile, the existing business sector reacts to the new business by re-arranging its supply chain operations (e.g., adjusting production level and alternating supply allocation), and each party looks for ways to maximize its benefit under the changing business world. The introduction of the emerging industry often involves spatial equilibrium of commodity flow, market demand, and resource supply.

In an exploratory effort, Bai et al. [134] propose a bilevel leader-follower game model that incorporates farmers' decisions on land use and market choice into the bio-fuel manufacturers' supply chain design problem. The model determines the optimal number and locations of biorefineries, the required prices for these refineries to com-pete for feedstock resources, and farmers' land use choices between food and energy. The model is solved by transforming the mixed-integer bilevel problem into a mixed-integer quadratic program based on KKT conditions. Noncooperative and cooperative games are studied respectively to address possible business partnership scenarios (e.g., via long-term leases) between feedstock suppliers and biofuel manufacturers. Using corn as an example of feedstock crops, spatial market equilibrium is utilized to model the relationship between corn supply and demand and the associated price variations in local grain markets. It is found that biofuel supply chain design does have a direct impact on land use choices for farms in the area. Compared with the noncooperative game scenario, cooperation among the industry and the farmers tends to save trans-portation cost and generate higher profit for the whole supply chain.

In a follow-up study [133], the same authors extend the framework by introduc-ing government regulations on farmland use and an associated marginal land market into the Stackelberg game. This model better represents the problem realism with more land use options, including the possibility of marginal land reclamation and en-ergy/food market equilibria, thus providing more comprehensive economic insights. Noting that farmers are generally independent stakeholders, a land use allowance con-cept and a cap-and-trade mechanism are introduced to provide indirect economic in-centives for the farmers to comply with government restrictions. These two models are proved to achieve equivalent land use patterns at optimality, and the proposed land use constraints are shown to be effective in balancing the amount of farmland used for food and energy production. In some cases, the proposed cap-and-trade mechanism could result in less profit for the leading biofuel manufacturer but higher social welfare for the entire system (including food, fuel, and land markets).

Wang et al. [1872] further incorporate the blenders into the scope of the biofuel supply chain. The biofuel consumption mandate is enforced via the Renewable Identification Number (RIN) system, a tracking mechanism that monitors obligated parties' compliance. The biofuel manufacturers obtain an RIN for each batch of biofuel production from the EPA; RINs are then transferred to blenders (e.g., energy companies) during biofuel consumption, and they can be traded among blenders; finally, the blenders are mandated to hand in specified number of RINs to the EPA at the end of each year, or else penalties will be imposed. In this work, competition among food and biofuel industry players (including among multiple biofuel manufacturers) is addressed via Nash equilibrium models and bilevel Stackelberg leader-follower models. Based on these models, the advantages and shortcomings of the current biofuel production mandate are analyzed.

These biofuel supply chain studies generally formulate the problems into discrete mathematical programs with equilibrium constraints (MPECs), which are generally nonlinear, nonconvex, and hence quite hard to solve. Solution methods are generally based on relaxation, decomposition, and transformation. Finding efficient solution approaches for such problems remains a challenge. As a side note, some approximation schemes for large-scale discrete decisions (e.g., facility location) into differentiable continuous counterparts (e.g., facility density) have been proposed to reduce the complexity of the problems [536].

34.4 ▪ Supply Network Design under Transportation Congestion and Infrastructure Deterioration

The changes in the energy industry have created unique challenges for many critical lifeline infrastructure systems far beyond those in the energy sector. Expanding ethanol production, for example, will not only lead to the expansion of biorefinery systems but also strain existing supporting infrastructures that are already aging and degrading (see [1400] for a review). In particular, the already congested local and regional transportation networks are experiencing increasing freight demands for supplying feedstocks to refineries and delivering ethanol to consumers. Due to the low energy density of feedstock biomass, transportation of the bulky feedstock (and ethanol) incurs one of the major operational costs in biofuel supply chain systems. Trucking remains the dominant mode of transportation because alternative modes would either require heavy investment or be unsuitable for the emerging biofuel industry—for example, the current pipeline infrastructure cannot be used for ethanol transportation due to erosion concerns. Most bioenergy production facilities are designed with a very large production capacity to achieve economies of scale. As such, a large number of trucks must be added to the highway network to ship sufficient low-energy-density biomass to satisfy the enormous ethanol production requirement.

Earlier work on the bioenergy supply chain [1031] formulates a standard discrete facility location model to optimize the biofuel supply chain, where the point-to-point costs from transporting biomass, ethanol, and by-products are assumed to be exogenously given. Establishment of industry facilities, however, often induces heavy vehicle traffic that exacerbates congestion and infrastructure (e.g., bridge, pavement) deterioration in the neighboring highway network. This has been the case for the booming energy industry, especially when new production facilities are built near neighborhoods that were not originally built for heavy traffic. For instance, Iowa's growing renewable energy industries have had significant impacts on the quality of its transportation infrastructure, such that pavement repairs and maintenance costs in multiple Iowa rural counties increased significantly during and after the construction of biofuel production plants [808]. Such unintended consequences of energy production facility development increase the social cost to the general public (e.g., due to traffic delay and highway maintenance), and in turn have a negative impact on the efficiency of the freight shipments associated with these facilities.

Planning of biorefinery locations and biofuel supply chains, therefore, should be made cautiously to establish a sustainable bioenergy economy in which the investment in refinery construction and operations, the cost for biomass and ethanol transportation, and the related socioeconomic impact are minimized. Bai et al. [132] develop a model to plan biofuel refinery locations where the total system cost for refinery investment, feedstock and product transportation, and public travel is minimized. Shipment routing of both feedstock and product in the biofuel supply chain and the resulting traffic congestion impact are incorporated into the model to decide optimal locations of biofuel refineries. A Lagrangian relaxation–based heuristic algorithm is introduced to obtain near-optimal feasible solutions efficiently. It is found through computational case studies that ignoring congestion in biofuel supply chain design could lead to much higher transportation costs for not only the biomass shipments but also the public. Hajibabai and Ouyang [894] further extend the model to allow for possible highway/railroad capacity expansion at chokepoints around the network. It is found that significant cost reductions can be achieved by simultaneously improving the capacity of the transportation network and expanding the biofuel supply chain.

Hajibabai et al. [893] present an integrated facility location model that simultaneously considers traffic routing under congestion and pavement rehabilitation under deterioration. The objective is to minimize the total cost due to facility investment, transportation cost including traffic delay, and pavement life-cycle costs. Building upon analytical results on optimal pavement rehabilitation, the problem is formulated into a bilevel mixed-integer nonlinear programming (MINLP) problem, with facility location, freight shipment routing, and pavement rehabilitation decisions in the upper level and traffic equilibrium in the lower level. This problem is then reformulated into an equivalent single-level problem based on the KKT conditions and piecewise-linear approximation of traffic delay functions. Computational analysis shows that the proposed model can improve supply chain sustainability and minimize its negative societal impacts from congestion and pavement damage. In particular, significant reductions in pavement-related costs (e.g., agency cost and users' vehicle operating cost) as well as overall systemwide cost are observed, indicating that the joint optimization of the biofuel supply chain and the supporting transportation infrastructure not only results in a potential for Pareto improvement but also provides incentives for policy making and mechanism design through benefit/cost reallocation.

The supporting infrastructure is not just impacted by biofuel supply chains; similar problems are seen in a wide range of other energy industries. For example, in Pennsylvania and South Dakota, the heavy truck traffic induced by the emerging natural gas industry (e.g., for transporting water and supplies in support of the hydraulic fracturing process) has caused not only congestion to the residents in nearby towns but also severe damage to state and local roads, resulting in hundreds of millions of dollars spent on pavement repair and replacement. More generally, the development and transmission of energy can produce an array of effects at the community level, not only due to road network congestion and pavement deterioration but also including overburdened municipal services, reduced water availability for conventional uses, economic volatility, disruption of social and cultural patterns, and the stigma associated with environmental health risk and industrialization. A holistic coupled modeling approach, with embedded physical and social processes, is needed to design and analyze the energy supply networks.

34.5 ▪ Concluding Remarks

Supply chain design and optimization aim at matching supply with demand at minimum total cost, which is exactly the goal of the energy industry. With this common goal, it is not surprising that synergies exist between the two research fields. The purpose of this chapter is to highlight some of the existing synergies and provide the reader with some starting points for further reading. We hope the discussion in this chapter will foster more synergies between the two important fields in the future.

Chapter 35

Supply Chain Optimization in Healthcare

Hui Zhao

Among the large amount of research on healthcare, relatively limited literature has been devoted to the healthcare supply chain domain. In this chapter, we particularly discuss issues with supply chains of healthcare products (e.g., drugs) and supplies, a less-studied area. We cover supply chain issues related to pharmaceutical capacity planning, production planning, inventory management, and supply chain contracts, each in a different section. In each section, we start with the unique challenges in the specific setting, followed by sampling papers targeted at these problems for basic ideas of solution and main results. Due to the limited space and the desire to expose to the readers the richness in this area, our coverage for each topic is far from sufficient but is a good starting point for interested readers to explore further.

35.1 ▪ Pharmaceutical Capacity Planning and Outsourcing

Capacity planning for new drugs is one of the most challenging tasks facing pharmaceutical manufacturers because of several conflicting factors: (1) long lead time to construct capacity (having to build in advance) but high uncertainty in product launch due to uncertain clinical trial and Food and Drug Administration (FDA) approval results, and (2) high demand uncertainty but high service level requirement immediately after drug approval (patients tend to choose different drugs in case of short supply).

As part of the solution to these dilemmas, many companies have moved from sole in-house production to a mixture of in-house and outsourcing. However, there has been limited guidance in the timing of the in-house investment and how to balance in-house and outsourcing based on the unique features and problem characteristics (e.g., drug demand variability) in the pharmaceutical industry. Current practices treat such problems as one-time decisions, ignoring the fact that drug patent periods and the many uncertainties have important implications for the dynamics in such decisions. Okajima et al. [1419] study the dual capacity sourcing problem considering finite patent protection horizon and important uncertainties frequently observed in practice: product launching risks and temporally dependent demand costs. Questions tackled include the following: Should a manufacturer dynamically adjust the outsourcing decisions at different stages of the planning horizon, and if so, how? How should the manufacturer plan its in-house capacity investment accordingly? How do these decisions change with system characteristics such as the demand variability of the drug?

A finite-horizon stochastic dynamic programming model is built to tackle this problem. In each period, a manufacturer can invest in two types of capacity: in-house capacity that has a longer (two-period) acquisition lead time but can be used throughout the remaining periods, and outsourcing capacity, reserved from a contract manufacturing organization (CMO) as options, that has a shorter (one-period) reservation lead time but can only be used in the subsequent period. Results show that, indeed, the optimal decisions are dynamically changing in the planning horizon. The optimal total (in-house plus outsourcing) capacity is characterized by state-dependent base-capacity levels. However, the structure of the optimal in-house capacity policy is intractable. Two well-behaved approximation approaches are developed with extensive numerical analysis. Based on the analysis, guiding principles for effective dual capacity sourcing policies are suggested.

Okajima et al. [1420] extend the dual-capacity sourcing model to multiple drugs because many manufacturers consider building flexible capacity for risk pooling, in addition to using outsourcing to reduce risk. The paper investigates whether outsourcing and flexible capacity can be used as substitutes or complements and how the decisions change along the planning horizon based on a two-stage capacity planning model: in-house flexible capacity is only built up through three periods after drug approval (referred to as the capacity expansion stage). Demand in the expansion stage can be met by either in-house flexible capacity or outsourcing. In the rest of the periods (referred to as the capacity allocation stage), demand can be met by allocation of the flexible capacity between the two drugs (capacity switching) as well as outsourcing. Results show that outsourcing and flexible capacity are complementary as long as both drugs have uncertain demands. Interestingly, the optimal policy in the expansion stage is to prioritize in-house capacity investment on *one* of the drugs (betting one) in period 1 (even if both drugs have similar risks) and to use capacity switching to meet the demand of the other, if needed, after product-launching risks for both drugs have been resolved. Any capacity requirement that cannot be met from switching the in-house flexible capacity will be fulfilled by outsourcing.

35.2 ▪ Vaccine Production Planning

One important factor differentiating production planning in the healthcare setting is the consideration of social welfare. Although for-profit parties, e.g., manufacturers, still aim at profitability, government and other health organizations take social welfare as at least part of their objectives. One of the challenges is to align these different objectives. In this section, we will use the influenza (flu) vaccine supply chain as an example to demonstrate this aspect of the research. Later, in the section on supply chain contracting, we will touch on social welfare again in the drug shortage problem.

The influenza vaccine supply chain resembles the newsvendor problem discussed in Section 33.2.2.1 (e.g., a generally one-season product) but possesses several differentiating characteristics. The most important are probably the uncertainty in production yield and the nonlinear value (health benefits) of vaccination due to infection dynamics. There has been extensive literature on the flu vaccine value chain. As mentioned, we will focus on aligning manufacturer choices of production volume for profitability and the governmental choices that balance costs and social welfare. We sample two of these works, Chick et al. [456] and Arifoglu et al. [94]. At the end of this section we will also briefly discuss other challenges with the influenza supply chain.

First, there are a couple of important components to be included in the vaccine planning models:

- The number of individuals infected in the flu season is denoted by $T(f)$, a nonincreasing function of f, where f is the fraction vaccinated. This component is important for infection dynamics and social welfare. Both works assume specific properties for $T(f)$ for tractability. For example, Chick et al. [456] looked at both piecewise-linear $T(f)$ and convex $T(f)$, while Arifoglu et al. [94] adopted a function of the expected value of $T(f)$ that is decreasing and convex.

- There is uncertainty in production yield. Most works assume that the obtained number of vaccine doses, Q_r, is a stochastic proportion of the planned production quantity, Q. That is, $Q_r = UQ$, where yield U is a random variable with mean μ and standard deviation σ, following a continuous and differentiable probability distribution. (See Section 33.2.2.4.)

Chick et al. [456] consider a supply chain with a single manufacturer (whose decision is on production quantity Q) and a single buyer, the government (whose decision is on the vaccination fraction, f, which bounds up units bought). Showing that production risks (due to random yield) taken by the manufacturer lead to an insufficient supply of vaccine, the authors pursue coordinating contracts. They demonstrate that neither the wholesale price contract nor the buyback contract can coordinate because they are executed on the production *output* and are silent about the risks related to random yield, e.g., they do not compensate the manufacturer when the production volume (Q) is high and the yield (UQ) is low. Further, a cost-sharing contract in which the government pays proportionally to the *production volume Q* can coordinate when $T(f)$ is piecewise-linear nonincreasing, but not when $T(f)$ is convex, since government does not order enough in this case (i.e., f is too small). Hence, a coordinating contract should provide an incentive for the government to vaccinate a higher fraction of the population (higher f) and provide a manufacturer with an incentive to produce enough. A variant of the cost-sharing contract can achieve the goal. Specifically, government receives a quantity discount and shares a portion of the manufacturer's production cost. The more people the government plans to vaccinate, i.e., the higher f is, the greater discount it gets for the units bought and the higher the portion the government pays of the production cost. This contract, however, requires true information sharing between the manufacturer and the government about the production quantity and the production yield.

Arifoglu et al. [94] incorporate the demand side, in particular, consumers' strategic behavior, into the vaccine supply chain problem. They consider a profit-maximizing manufacturer selling directly to rational utility-maximizing individuals who make their own vaccination decisions. The manufacturer first chooses the production quantity. Then, after observing the obtained number of doses (random yield realized), individuals, each with a different disutility if infected (δ), decide independently whether they will seek vaccination.

- When not seeking vaccination, an individual has a probability of $p(f)$ to be infected, where $p(f)$ is decreasing in the vaccination fraction f. If infected, the individual has a utility of $\bar{V} - \delta$, while if not infected, he enjoys a utility $\bar{V}$.

- When seeking vaccination, the individual will incur a disutility of θ for searching. With probability ϕ, she will be vaccinated, incurring a net utility of $\bar{V} - r - \theta$, where r includes disutility from the vaccine price and side effects from the vaccine. With probability $1 - \phi$, she will not be vaccinated and will incur a net utility equal to $\bar{V} - \delta p(f) - \theta$.

Putting all the above information together, an individual with infection disutility δ will search for the vaccine if the expected net utility of searching is greater than that of not searching, i.e., $\theta < \phi[\delta p(f) - r]$. Thus, knowing the distribution of δ, we can calculate the fraction of the population seeking vaccination, i.e., the demand for vaccination. When incorporating the uncertain yield for supply and allocation priority (i.e., vaccines are allocated first to individuals with infection disutility exceeding a threshold), we can find the fraction of vaccination.

Consider that consumers' decisions reveal the demand-side inefficiencies, when there is no concern about sufficient supply (e.g., previous economic studies), the equilibrium demand is always less than the socially optimal demand because self-interested individuals do not internalize the social benefit of protecting others via reduced infectiousness (positive externality). In contrast, when there is limited supply due to yield uncertainty and manufacturer's incentives, equilibrium demand can be greater than the socially optimal demand because of a second (negative) externality: self-interested individuals ignore the fact that vaccinating people with high infection costs is more beneficial for society when supply is limited. The negative externality effect is exacerbated if there is less-efficient allocation of the vaccine during shortage.

Finally, the above analysis motivates studies on the government's demand-side intervention (e.g., compulsory vaccination, tax/subsidy mechanism) to change consumer behavior to mitigate demand-side inefficiency (e.g., [1258] developed a demand-side subsidy/tax program that can induce socially optimal vaccine coverage in an oligopoly market with identical vaccine producers and consumer population). It also motivates supply-side intervention to provide incentives to manufacturers for system-optimal production quantity (e.g., [456]).

As we mentioned, the above discussion mainly centers around the consideration of social welfare. Another unique challenge of influenza production planning lies in the strain selection problem. Since the influenza virus evolves over time, the selection of strains for the vaccine is crucial in the manufacturing process. Wu et al. [1907] formulate the annual vaccine strain selection problem as a stochastic dynamic program. Kornish and Keeney [1098] study the annual influenza vaccine composition problem: deciding between strains of the virus to include with a deadline. Cho [458] proposes a dynamic selection policy subject to random production yields. Ozaltin et al. [1432] determine the optimal vaccine composition and its timing through a multistage stochastic mixed-integer program.

There is also recent research focused on the vaccine industry and vaccine issues involving multiple countries. Deo and Corbett [587] discuss how yield uncertainty can help to explain the high concentration in the influenza vaccine industry. Sun et al. [1741] and Wang et al. [1871] examine how two countries would allocate resources at the onset of an epidemic when they seek to protect their own populations by minimizing the total number of infectives over the entire time horizon. More recently, Mamani et al. [1259] have extended the work of [456] by considering the contract design problem, which involves multiple purchasers and the possibility of cross-national disease transmission.

35.3 ▪ Inventory Management at Healthcare Facilities

Inventory management can be difficult in healthcare settings due to two important factors: perishable goods (e.g., blood, drugs) and special phenomena in the health setting. In this section, we will discuss inventory management for perishable items in the context of blood management. In the next section, we discuss inventory issues related to special phenomena, one for brand-name drugs under steady price increase, and the other for generic drugs under the recent drug shortage.

Inventory management for perishable goods is complex because, compared to standard inventory management, where generally only the inventory position information is needed, the optimal ordering policy for perishables requires information about the amount of inventory for every age. As a result, most of the work in this area focuses on optimal or near-optimal ordering policies to minimize operating costs under a single demand stream. Nahmias and Pierskalla [1373] were the first to examine the optimal order policy for a two-period lifetime problem. They determined properties of the stationary, state-dependent optimal policy. The two-period lifetime problem was later extended to m-period lifetimes by [766] and [1370], but the structure of the optimal policy was not found. Hence, later works focus on exploring heuristic policies/solutions to find the near-optimal policies. Readers are referred to [1037] and [1372] for excellent reviews of perishable goods inventory models and [1482] for research focused on blood management. More recently, [449] studied coordinating inventory control and pricing strategies for perishable products. In this section, we will sample two issues tackled recently in managing blood inventory: regular and expedited ordering, and substitution due to different classes of demand for products with different ages.

Zhou et al. [1975], motivated by inventory management of platelets in hospitals (which have a three-day lifespan excluding time for transportation, testing, and arrangement), extend the perishable problems from two periods to three periods while incorporating dual models of replenishment — fresh regular orders of Q_i (which can last for three periods (days)) placed at the beginning of each cycle i and an expedited replenishment (for one-period-old product), which can be placed following an order-up-to level (s_i) between regular orders. The authors prove the existence of a unique optimal solution and derive the necessary and sufficient conditions for the optimal policy. An algorithm is designed to approximate the optimal solution for the multicycle problem.

While most papers focus on a single demand stream, [582] looks at the problem where there is substitution among multiple types of demand. The blood product industries tend to substitute freely between products of different ages as long as allowed. This paper, however, looks at all four scenarios: (1) no substitution, (2) downward substitution (excess demand for old items can be satisfied by new but not the reverse), (3) upward substitution (demand for new can be satisfied by old but not the reverse), and (4) full substitution (both directions of substitution). They consider a single product with two periods of lifetime under two simple replenishment policies: base-stock policies based on either the total amount of inventory of all ages or the amount of new items in a periodic, infinite-horizon setting. One nice feature of the paper is that a sample path method is used, so no demand distribution is assumed. Readers are also referred to [478], which looks at the substitution aspect of platelets inventory management, and [891, 892] for optimizing platelet production.

35.4 ▪ Supply Chain Contracting in the Healthcare Industry

Contracting (see Section 33.4) is very important in healthcare supply chains because of the many parties involved. Contracting in the healthcare realm can be challenging due to unique situations/partnerships and the consideration of regulation/government (agencies) most of the time. In this section, we sample three types of contracts in the healthcare setting, all related to unique situations the industry is facing.

35.4.1 ▪ Licensing Contracts in Collaborative R&D for New Drugs

Due to the extremely high risks and soaring costs for developing new drugs, in recent years, many pharmaceutical companies have turned to collaborative R&D. Instead of one company covering a long value chain from testing promising molecules to bringing the drug to market, many large companies seek licensing arrangements with smaller, e.g., biotechnology, companies. Such smaller companies, usually bound by financial capabilities, typically develop new drugs up to proof of principle and then look to out-license the project to larger pharmaceutical partners who have the expertise and funding to take the potential drugs through the strict FDA approval process and eventually to the market. Such arrangements, beneficial to both sides, are usually governed by licensing contracts. A licensing contract may include an upfront fee (m_0); a milestone payment upon completion of specific stages of the product development (m_1), e.g., after successful completion of the project; and a royalty percentage of the sales (r), all paid by the licensee (e.g., the small company) to the licensor (the larger pharma company). The licensee, if successful in licensing the project, exerts an effort, x, to bring the product to market.

A few questions are interesting to study. A key one is the structure of the optimal contract, i.e., whether the contract should have all (a three-part tariff) or only parts of the components m_0, m_1, and r, and what values each of these components should have. Crama et al. [524], from which we draw most of our discussion, study the optimal contract structure and the corresponding fees from the licensor's perspective (i.e., the licensor has higher bargaining power). The optimal structure from the licensee's perspective is still an open question.

Depending on the success rate of the project in the future (probability of technical success, PTS), a project can be a high type (high probability of success) or a low type. Different estimates of PTS will result in different valuations of the projects, hence different licensing fees. One distinctive challenge of this problem is that the licensor (he) and the licensee (she) may have different estimates of PTS. Hence, when the licensee's PTS estimate and her project valuation are her private information, unknown to the licensor, the problem involves *hidden information* or *adverse selection*. Thus, the licensor has to design incentive compatible contracts such that a high-type licensee will not pretend to be a low-type. In addition, because the licensee's effort x is unknown at the contracting stage, the model also involves *hidden action* or *moral hazard* (MH).

This problem draws tools from game theory, principal-agent problems, and contract design. It is also concerned with risk preferences (i.e., risk neutral versus risk averse), an important factor in this context. Specifically, the licensor is either risk neutral or risk averse, while the licensee is risk neutral. With a cost c, the licensor proposes a project to the licensee with a contract $T = \{m_0, m_1, r\}$. While the licensor's estimate of PTS is p^o, the licensee's estimate of PTS, $p^e \in [\underline{p}^e, \bar{p}^e] \subset [0,1]$, is her private information, but the licensor knows its probability density function (pdf) $f(p^e)$ and cumulative distribution function (cdf) $F(p^e)$. Once contracted, the licensee makes an effort, x, which is directly related to the final payoff of the project, $s(x)$.

If the licensee pretends to be a q-type given her true type is p^e, she receives a value

$$V^e(q, p^e) = -c - m_0(q) - x^*(r(q), p^e) + p^e[(1 - r(q))s(x^*(r(q), p^e)) - m_1(q)],$$

where $x^*(r(q), p^e) = arg\,max_x\{-c - m_0(q) - x + p^e[(1 - r(q))s(x) - m_1(q)]\}$ is her optimal effort level if a p^e-type licensee pretends to be q-type. As for the licensor, with probability $1 - p^o$, he receives the contract signing fee m_0 only, and with probability p^o, he receives a total of $m_0 + m_1 + r\,s(x)$. Thus, assuming $u^o(z)$ is the licensor's Von Neumann–Morgenstern utility function with $u_z^o \geq 0$ and $u_{zz}^o \leq 0$, the licensor solves an optimization problem to propose the contract, maximizing his total expected utility over the licensee's types:

$$max_{m_0(), m_1(), r(), x^*} \int_{\underline{p}^e}^{\bar{p}^e} [p^o u^o(m_0(p^e) + m_1(p^e) + r(p^e)s(x^*(r(p^e), p^e)))$$
$$+ (1 - p^o)u^o(m_0 p^e)]dF(p^e),$$

subject to a few constraints: (1) the incentive compatibility (IC) constraints (based on the revelation principle), which ensure the licensee reveals her true type, that is, $q^*(T, p^e) = p^e$, where $T = \{m_0, m_1, r\}$, and (2) the individual rationality (IR) constraints, which ensure the licensee obtains at least her bottom-line payoff by accepting the contract. Readers are directed to [524] for the solution and results of this problem and the discussion of the advantage of the three-part tariff over a two-part tariff.

Xu and Zhao [1918] also develop a model to study drug R&D but focus on drug approval regulation in terms of trading off between evidence generation and faster access to new drugs. In particular, the authors look at the efficiency of the current accelerated approval process, analyzing why such a well-intentioned regulation process did not achieve its original purpos and proposing mechanisms to improve the current situation.

35.4.2 ▪ Fee-for-Service Contracts in Managing Brand-Name Drug Supply Chains

Compared to other consumer product supply chains, brand-name drug distribution supply chains possess the following unique features: (1) The distributor earns the majority of its margins from upstream (i.e., manufacturer) [1652] instead of from both upstream and downstream partners, as in most other supply chains. (2) The *brand-name* drug price continues to increase each year in the range of 6 to 15%. These large and almost predictable price increases have a profound impact on the business strategies of the manufacturer and the distributor in the industry.

Before 2004, investment buying (IB), speculation of inventory in anticipation of drug price increases, was the way distributors made most of their profits! IB brought many disadvantages, the most prominent ones for production/inventory planning for the manufacturer being nontransparency of inventory in the supply chain and demand volatility due to the demand spike from the distributor before the price increase and zero demand (for an uncertain number of periods) afterward. Later, IB was replaced by fee-for-service (FFS) contracts due to the Securities and Exchange Commission's (SEC's) investigation, which dramatically changed the way that pharmaceutical distribution supply chains are managed. Under FFS contracts, the manufacturer limits the amount of inventory distributors can carry at any time (by imposing an inventory cap, $\bar{y}$) and requires inventory information sharing from the distributor while compensating it with a per-unit fee, u (possibly the only supply chain where the manufacturer does not like the distributor to hold more!). In spite of its widespread popularity, the FFS model has never been rigorously analyzed, and its impact, compared to the IB model, on the manufacturer, the distributor, and the supply chain, a question asked by the industry, has never been carefully studied. Zhao et al. [1965] address these questions by formulating the multiperiod stochastic production/inventory problems faced by the manufacturer and the distributor under the FFS and IB models, deriving their optimal policies, and computing the policy parameters. The difficult part of these models is how to capture the players' reactions to a price increase, δ at the beginning of period n.

The distributor's problem is quite straightforward, and its optimal policy is proved to be a nonstationary base-stock policy, where the distributor should maintain its "newsvendor" base-stock level in all periods except in the period right before the price increase (period $n-1$) in which it will do IB as much as allowed by the inventory cap. This means the distributor will start period $n-1$ with $\bar{y}$ under FFS and with y_{n-1}^{d*} under IB since $\bar{y} = \infty$ in this case, where y_{n-1}^{d*} is the distributor's optimal order quantity in period $n-1$ without the inventory cap.

The manufacturer's problem, however, is much more complex because estimating downstream demand from the distributor (which is necessary for planning due to the production lead time) is difficult for two reasons: (1) the manufacturer may see one or more period(s) without demand from the distributor after the price increase because of IB before the price increase (referred to as transitional periods), and (2) the number of transitional periods and the first demand seen by the manufacturer after the transitional periods are also uncertain.

To tackle this problem, under FFS, since the distributor shares its on-hand inventory level at the beginning of each period (which means the manufacturer can find out the distributor's downstream demand for the period past), the manufacturer, knowing that the distributor starts period $n-1$ with $\bar{y}$, can use the total downstream demand since period $n-1$ to predict the probability of seeing demand from the distributor in the next period. Define the state during the transitional periods as $i = 0, 1, \ldots, \bar{y}-y^*-1$, representing the total downstream demand realized at the distributor since period $n-1$. Now define p_i, $0 \leq i \leq \bar{y}-y^*-1$ (i.e., p_i is defined only for the transitional periods), as the probability that the manufacturer sees the distributor resume ordering after period $n-1$ in state i. Hence, $p_i = Prob\{i+D > \bar{y}-y^* | i < \bar{y}-y^*\}$, where D is the random demand seen by the distributor in one period. We assume, if the distributor resumes ordering in state i, that the ordering quantity, η_i, equal to $i+D-(\bar{y}-y^*) = D-(\bar{y}-y^*-i)$ in this case, follows a distribution with a cdf of $\Theta_i(\eta_i)$. It is easy to verify that $p_i \leq p_{i+1}$ and $\Theta_i(\cdot) \leq_{st} \Theta_{i+1}(\cdot)$ for $i < \bar{y}-y^*-1$.

In the IB model, however, since the distributor does not share its on-hand inventory information, during the transitional periods, the manufacturer can only estimate the demand from the distributor based on how many periods have elapsed since the distributor's order in period $n-1$ (recall that the distributor starts period $n-1$ with y_{n-1}^{d*}). Hence, during the transitional periods, state i is redefined as the number of periods that have elapsed since period $n-1$, $0 < i \leq N-n+1$, $i \in I^+$ (positive integers). We define p_i, $0 < i \leq N-n+1$, as the probability of the manufacturer seeing the distributor resuming ordering in state i:

$$p_i = Prob\left(\sum_{k=n-2}^{i+n-1} D_k > y_{n-1}^{d*} - y^* \,\middle|\, \sum_{k=n-1}^{i+n-3} D_k < y_{n-1}^{d*} - y^* \right).$$

We also say the distributor's demand to the manufacturer (η_i), if realized in state i, follows a distribution with a cdf of $\Psi_i(\eta_i)$. If the end-item retailer demand has an increasing failure rate (IFR) distribution (e.g., normal, uniform, Erlang), then $p_i \leq p_{i+1}$ and $\Psi_i(\cdot) \leq_{st} \Psi_{i+1}(\cdot)$.

With the above, in both the FFS and the IB case, a nonstationary produce-up-to policy is shown to be optimal for the manufacturer with nested produce-up-to levels for the transitional periods:

$$FFS: y^0 \leq y^1 \leq y^2 \leq \cdots \leq y^{\tilde{y}-y^*-1} \leq y^{0^+} = \Phi^{-1}\left(\frac{p}{p+h}\right)$$

and

$$IB: y^1 \leq y^2 \leq \cdots \leq y^{N-n+1} \leq y^{0^+} = \Phi^{-1}\left(\frac{p}{p+h}\right).$$

Based on the optimal policy, the authors show that FFS contracts can improve the total supply chain profit compared to IB, so there exists a range of the per-unit fees (u) that leads to Pareto improvement. A heuristic is developed to calculate the near-to-optimal Pareto fee ranges.

It is worth mentioning that while FFS contracts remain the main drug purchase contracts for brand-name drugs, the pharmaceutical industry is experiencing an increased number of specialty drugs among all the brand-name drugs developed. Mani and Zhao [1264] investigate reasons for such a trend and its impact on manufacturers' financial performance, while [1917] explores the changes in distribution channel choices under this trend. Both works use extensive industry data and empirical methods. More optimization models will probably be developed after the empirical works better expose the issues involved.

35.4.3 ▪ Mitigating Generic Drug Shortages with Fail-to-Supply Contracts

Drug shortage is probably one of the most challenging problems currently facing the pharmaceutical industry, the healthcare industry as a whole, and the U.S. government. Drug shortages continued to increase, with the number of drugs in shortage in the United States tripling from 2005 to 2010 [691]. The problem has affected not only social welfare but also hospital operations and government spending because many of these drugs are reimbursed from programs such as Medicare. Although the problem has drawn tremendous attention, few proposed solutions are based on rigorous research and even fewer are taken from the supply chain's perspective, which is believed to be one of the keys to the problem.

Current drug shortages concentrate on generic sterile injectable drugs, which often have low profit margin (old generics) and require relatively complex manufacturing processes. These characteristics make these drugs unappealing to manufacturers, resulting in barely sufficient capacity/high capacity utilization and hence high vulnerability to supply/manufacturing disruptions. At the same time, holding sterile injectable drugs is costly [812], and failure to satisfy customer orders incurs a low cost due to very weak fail-to-supply (FTS) contracts, e.g., in many cases, buyers are only compensated when they can find alternative sources of supply, which is unlikely to happen during shortage.

Many who believe low profit margin is the cause of drug shortage advocate for price increases. However, from the supply chain's perspective, the combination of low price margins and weak FTS contracts seems to be more detrimental. Given that price and FTS penalty affect many parties on the supply chain, any study from the supply chain perspective must take into consideration not only shortage reduction but also the manufacturer's profitability, the government's balance between spending and social welfare, and healthcare providers' cost, to propose sustainable solutions that are acceptable to all.

Jia and Zhao [1000] develop a model of the pharmaceutical supply chain pertaining to drug shortage, based on which they propose Pareto-improving FTS contracts that mitigate drug shortage, ensure drug manufacturers' and GPOs' (group purchasing organizations') profit, and cut/maintain government spending (considering its trade-off between Medicare drug reimbursement cost and loss of social welfare due to drug shortage). The model includes three levels of analysis pertaining to shortage: (1) the balance between price and FTS in the drug purchase contract between drug manufacturers and GPOs, (2) capacity adjustment by drug manufacturers based on the contract, and (3) production/inventory decisions at drug manufacturers based on the contract and the capacity decision. The model captures key characteristics of the current drug shortage problem, including capacity utilization–related disruptions, uncertain recovery time from disruptions, and a combination of backorder and lost sales. In addition to deriving optimal policies and proposing Pareto-improving contracts, the model contributes to the literature by incorporating capacity-based disruption, which is not covered in previous literature but is essential to the drug shortage problem. Due to the lack of data on the relationship between disruption and capacity, the authors used available shortage data to approximate such relationships and used the data in its optimization model for the manufacturer's decisions and for capturing drug shortage measures. Results from the model show that only increasing prices (as advocated by many) cannot improve drug shortage. Increased prices must be combined with a strengthened FTS penalty to cause a potentially significant reduction in drug shortages. Numerical studies provide the range of effective price-FTS penalty pair values for a wide range of scenarios, as well as their impact on government spending.

Chapter 36

Humanitarian Applications of Supply Chain Optimization

Melih Çelik, Özlem Ergun, Pınar Keskinocak, Mallory Soldner, and Julie Swann

36.1 ▪ Introduction

The global impact of disaster-related and long-term humanitarian issues has been steadily increasing. Over the last 10 years, an annual average of 106,000 people have been killed in natural disasters, resulting in an annual economic loss of approximately $200 billion [90]. With the growing population and increased urbanization, these effects are expected to increase in the future. Ongoing issues such as hunger; poverty; child mortality; maternal health; spread of HIV/AIDS, malaria, and other infectious diseases; and environmental sustainability contribute to human suffering and economic losses. For example, as of 2013, 25% of under-five-year-old children are malnourished throughout the world [1809], and more than 1.5 million children per year in sub-Saharan Africa die due to neonatal infections [1810]. Disasters can be classified based on the cause (natural, such as a hurricane, versus human-made, such as a terrorist attack), or the onset (sudden-onset, such as an earthquake, versus slow-onset, such as a drought), or predictability (e.g., timing and location of hurricanes may be easier to predict than those of earthquakes). Long-term development and healthcare-related issues differ from disasters in that their causes often cannot be traced back to a specific catastrophic event [418].

Humanitarian applications of supply chain optimization aim to address these issues and reduce their adverse effects by utilizing analytics. In recent years there has been an increasing interest in the literature in humanitarian applications of supply chain optimization (see [62, 92] and [418] for reviews). Despite the similarities of the decisions to those in traditional supply chain models, such as facility location, inventory planning, vehicle routing, network flows, and capacity expansion, humanitarian supply chains pose additional challenges due to their unique characteristics. These include the existence of multiple stakeholders (e.g., donors, governments, NGOs, beneficiaries) with multiple and often conflicting objectives; high levels of uncertainty in demand, supply, and information (for reasons such as lack of visibility into the supply chain, uncertainty on the effects of the disaster, and lack of information infrastructure); and fairness considerations in decisions in addition to efficiency and effectiveness. These characteristics significantly impact the models and the decisions. For example, conflicting objectives of the stakeholders lead to the formulation of multiobjective optimization models, and inclusion of fairness considerations requires novel approaches in defining the objective function(s) and/or the constraints.

In the remainder of this chapter, we discuss three humanitarian applications of supply chain optimization. In Section 36.2, we describe a facility location model addressing network expansion of a donated breast milk supply chain in South Africa. In Section 36.3, we present an inventory model for allocating scarce perishable commodities in a developing country. Section 36.4 considers humanitarian aid transport networks. For each of the three applications, we provide an overview of the problem domain and its humanitarian applications in the literature, describe one specific application in detail, and point to potential research directions regarding humanitarian applications of this domain.

36.2 ▪ Facility Location: Expansion of Donated Breast Milk Bank Supply Chain in South Africa

Facility location refers to a set of problems in which the aim is to decide on the locations of a set of facilities with objectives or constraints on service level and cost; see Section 33.3. Facility location has received significant attention in the humanitarian logistics literature [418], especially in disaster mitigation and preparedness. An important problem in the mitigation stage is the reliable facility location problem (e.g., [233, 1417, 1710]), where the aim is to locate facilities so that potential disaster damage on these facilities is minimized and postdisaster service can be maintained. In the preparedness stage, the adverse effects of the disaster can be reduced by the location of early warning systems (e.g., [237, 1916]), which are activated when the disaster is imminent or after it strikes, or protection systems (e.g., [402, 1639]) that strengthen various parts of the network. A common objective in these problems is to maximize the "coverage" of the network, i.e., the number of beneficiaries or network locations receiving service from the facilities. In the disaster preparedness stage, facility location decisions are generally coupled with supply prepositioning (e.g., [624] and [1316]), and the problems are solved using a two-phase stochastic programming approach, with location and prepositioning decisions before the disaster hits, and transportation decisions in the aftermath. Examples of facility location problems in the context of long-term humanitarian issues can be found in [153, 399, 645, 854, 1713].

In humanitarian supply chains, as in the case of public-sector applications, providing equitable service to the beneficiaries is important. In the facility location literature, equity has been modeled in a number of ways. Extensive reviews of these models are provided in [657] and [1284]. In addition to maximizing the minimum coverage over all nodes, lexicographic approaches sequentially aim to maximize the second-worst coverage, third-worst coverage, and so on [1243, 1414]. Other objective functions to model equity include maximization of the minimum dispersion of facilities over the communities [1008], minimizing the Gini coefficient (an inequality measure that is based on the area between the Lorenz curve, which plots the percent demand-weighted distance traveled versus the percent population, and the line of equality) [626], minimizing the maximum perceived risk in treatment of hazardous waste [799], minimizing customer disutility [1009], and minimizing the difference between (i) maximum and minimum distance of beneficiaries to locations [1526] and (ii) maximum and minimum number of beneficiaries assigned to facilities [1276]. Throughout the remainder of this section, we present a modeling approach that aims to capture the effects of various equity objectives in facility location, within the context of donated breast milk distribution. Earlier work on this problem is described in [399].

Globally, around 3% of all children die within the first four weeks after being born. A vast majority of these deaths are caused by neonatal infections [991], which can be prevented by proper breastfeeding [1810]. However, in many developing parts of the world, factors such as maternal death during/after birth, maternal disease (e.g., HIV, TB-meningitis), and lack of rooming-in facilities in public healthcare make proper breastfeeding impossible. The most effective solution to this problem is to provide infants in need with donated breast milk.

The South African Breastmilk Reserve (SABR) strives to provide equitable distribution of donated breast milk to in-need neonatals in South Africa. It operates the supply chain for milk delivery by providing setup, management, and coordination activities, while operational decisions are made by individual facilities. Donations can be made by (i) external donors, who express their milk at home and take this unpasteurized donor breast milk (UDBM) to *corners*, which act as temporary storage hubs, and (ii) lactating mothers in maternity wards of hospitals. In both cases, UDBM is transported to in-hospital *milk banks*, where it is pasteurized and stored. Pasteurized donor breast milk (PDBM) is dispensed to neonatal intensive care units (NICUs) upon demand.

SABR currently operates a supply chain with 14 milk banks and 15 corners, serving more than 40 NICUs, mostly in the densely populated Gauteng region. In recent years, increasing demand for the services of SABR has resulted in plans to expand and evolve its supply chain network. Particular decisions include (i) which corners and milk banks to open throughout the country; (ii) how to assign corners to milk banks and milk banks to NICUs; (iii) the flow of milk between corners, milk banks, and NICUs; and (iv) whether or not to switch from current volunteer-based transportation to courier-based transportation at the expense of extra transportation cost, but with the advantage of increased reliability of deliveries. Due to the geographical and demographic diversity of South Africa, it is challenging to estimate supply and demand and to establish an equitable distribution mechanism.

The supply chain expansion problem of SABR has been modeled using a mixed-integer program in [399], with the following sets:

C	Corners,	M	Milk banks,
N	NICUs,	R	Regions,
N_r	Set of NICUs in region $r \in R$.		

To analyze the trade-off in reliability and cost between volunteer and courier transportation, it is assumed that (i) $(1-v)$ fraction of the planned flow is actually delivered for each link under volunteer transportation, and (ii) transportation cost is multiplied by a factor of $(1+m)$ when couriers are utilized. The following parameters are used in the model:

v	Percent reliability lost on each shipment due to volunteer transportation,
m	Extra markup fraction per kilometer for courier transportation,
d_{ij}^1, d_{jk}^2	Distance between corner i and milk bank j, and milk bank j and NICU k,
C_i^1, C_j^2	Capacities (in units) of corner i and milk bank j,
D_k	Demand (in bottles) for milk at NICU k,
B	Budget for facility opening and transportation costs.

The following variables address the decisions (i), (ii), (iii), and (iv) discussed above:

x_{ij}^1 Planned flow of UDBM between corner i and milk bank j,

x_{jk}^2 Planned flow of PDBM between milk bank j and NICU k,

y_i^1, y_j^2 Binary variables for opening corner i and milk bank j,

z_{ij}^1, z_{jk}^2 Binary variables indicating assignment of corner i to milk bank j, and milk bank j to NICU k.

To analyze the effects of incorporating equity objectives into the model, three alternative objective functions are proposed, varying the levels at which equity is enforced. As a baseline, the first objective focuses solely on efficiency of coverage and maximizes the total satisfied demand:

$$z_1 = \max \sum_{j \in M} \sum_{k \in N} (1-v) x_{jk}^2. \tag{36.1}$$

At the other extreme, the second objective focuses on equity of distribution and maximizes the minimum fraction of demand satisfied over all NICUs:

$$z_2 = \max \ l \tag{36.2}$$

$$\text{s.t. } l \leq \frac{\sum_{j \in M} (1-v) x_{jk}^2}{D_k} \quad \forall k \in N. \tag{36.3}$$

The third objective is a compromise between the first two. In this case, the country is divided into regions based on demographic and geographical similarities. For each region, a minimum coverage level is determined by finding the NICU with the minimum fraction of demand covered. The objective is to maximize the sum of these minimum coverage levels over all regions:

$$z_3 = \max \sum_{r \in R} l_r \tag{36.4}$$

$$\text{s.t. } l_r \leq \frac{\sum_{j \in M} (1-v) x_{jk}^2}{D_k} \quad \forall k \in N_r. \tag{36.5}$$

Under each objective, the model uses several constraints:

- They limit the total of fixed costs and transportation costs by the budget. Since a single vehicle is sufficient to transport the UDBM or PDBM, the cost of transportation depends only on a fixed charge for using a link.
- They ensure that UDBM or PDBM can be sent from a corner or milk bank only if the facility is open, and that the outgoing flow cannot exceed the capacity of the facility.
- They force link assignment variables to take a value of one if there is flow on the link.
- They balance the flow through each milk bank.
- They ensure that PDBM flow into an NICU cannot exceed its demand, as otherwise unused units of PDBM would inaccurately contribute to the coverage objectives of z_1, z_2, and z_3.

In addition to the modeling and analysis of equity, two important contributions are made in [399]. First, a decision support tool is developed for determining supply and demand under various scenarios, depending on factors such as population data, income level, HIV prevalence, and education. Second, while maximizing coverage (i.e., z_1) on a network with 201 candidate corners and 136 candidate milk banks (with 136 NICUs) can be optimally solved within a few minutes, finding an optimal equitable solution (i.e., one that minimizes z_2) is not possible using a commercial solver such as CPLEX, even after long running times, underlining the modeling challenge posed by equity objectives. To overcome this issue, a variable fixing heuristic is proposed. First, the problem is solved by relaxing the integrality of assignment (z_{ij}^1 and z_{jk}^2) variables.

Then, variables with value less than a threshold are fixed to zero, and those over the threshold are fixed to one. The model is then solved with fixed values. Over the instances solved, the average optimality gap of the solutions resulting from this heuristic is less than 10%.

Computational experiments under four different supply-demand scenarios yield the following insights:

- Due to increased transportation costs, the cost of equity is higher when the supply is limited.
- Under z_2, the fraction of total demand satisfied is close to the minimum fraction of demand satisfied over all NICUs, implying that the variance of the fraction of demand over the NICUs is not high.
- Under z_1, there are diminishing returns for additional budget in terms of coverage. For z_2 and z_3, the marginal returns from higher budgets increase up to a certain level, after which there are diminishing returns.
- Courier transportation significantly outperforms volunteer-based transportation, as increased reliability more than makes up for the extra cost.

Lessons learned from the work in [399] have been developed into a case study and a teaching game. Future research directions involve analyzing the robustness of the results to deviations on the instance parameters (particularly the trade-off between volunteer and courier transportation), as well as the incorporation of lexicographic equity approaches, particularly for cases where supply is too tight to provide a minimum coverage level of more than zero.

Despite the abundance of work on facility location spanning disaster mitigation and preparedness stages as well as long-term humanitarian issues, one direction that deserves attention is the location of disaster response facilities, such as shelters or points of distribution. Fairness and equity in resource allocation are significantly affected by the location of facilities, which points to the need for extending the existing research in equity in facility location. This is particularly applicable for problems in public health, where access to care is significantly variable over the network. Equity is also important for other humanitarian applications; therefore incorporating these considerations for other humanitarian problems is another important direction for further research [177, 964].

36.3 ▪ Inventory Modeling: Perishable Commodity Allocation with Multiple Demand Types

Unlike facility location models, very few studies in the humanitarian logistics literature focus directly on inventory modeling [92] (see Section 33.2). Existing inventory models in the humanitarian logistics literature mainly consider the preparedness and response stages in an integrated manner by considering prepositioning decisions in the preparedness stage, whereas the response stage involves inventory relocation, reordering, and transportation decisions. For example, in [935], the storage levels at each location and delivery schedules for each vehicle are determined given a set of alternative disaster scenarios, with the objective of minimizing costs, travel times, and unsatisfied demand. In [1620], a multiperiod model is developed for food storage and distribution. In [1216], the focus is on demand surges for emergency items before hurricanes, and a stochastic model is developed with Bayesian information updates. This work is extended in [565] to include equity of service and congestion in the network. In [179], regular and emergency supplies are considered separately, and two different reorder levels and order quantities are derived. The method is applied to a crisis response case in Sudan [178]. Decisions in the response stage are addressed in [1593] by focusing on inventory relocation.

Parallel to traditional supply chains, in many health and humanitarian supply chains, beneficiaries in need of the commodities might need them at different levels of urgency. For example, patients receiving donated blood units have varying levels of need, as in the case of emergency patients with class IV hemorrhage versus those undergoing routine elective surgeries. When supply of the commodity is scarce, this differentiation of demand classes, along with the uncertainty of demand, poses an important challenge for the allocation of the commodity to distribution locations, which in turn apply a certain *usage policy* (e.g., prioritizing the higher classes) to serve the beneficiaries.

Throughout the remainder of this section, we consider the problem of allocating a limited supply of a perishable relief commodity from an allocation center to multiple distribution locations, each serving multiple classes of beneficiaries (details given in [419]). The perishability of the commodity, along with the lack of appropriate infrastructure, prevents redistribution of inventory once allocation is made. Consequently, the problem is modeled for a single period, at the end of which the satisfied demand accrues class-dependent benefits. Since supply of the commodity is scarce, any allocated commodity that is unused at the end of the period incurs a certain benefit loss. Given the historical distribution of demand from each class and usage policies at the distribution centers, the objective is to determine the allocation quantities that maximize the expected net benefit. The research is motivated by the allocation of donated blood units in Namibia and Zambia. A key factor that we consider in modeling is that the commodity is perishable, as might be true for disaster relief and healthcare commodities such as donated organs, food supplies, or blood products. We also study a system with decentralized decision makers who are not controlled by a centralized entity, as is true both in humanitarian applications and in many public health systems. Finally, we focus on finding policies that are easy to implement.

The allocation problem can be classified as an extension of the newsvendor problem (Section 33.2.2.1), with multiple levels in the supply chain, multiple distribution locations, and multiple customer types. The newsvendor problem has been extensively studied in the literature (see [1530] for a recent review). The cases with multiple customer types (e.g., [1657]) and multiple distribution locations (e.g., [220, 234, 235]) have been well studied. On the other hand, to the best of our knowledge, these two aspects have not been simultaneously considered. In addition to simultaneous consideration of these aspects, we also analyze of the effects of usage policies and centralization on the structure of optimal allocations.

Also relevant to this context is the distribution of blood in a developing country. Single-period blood allocation models are studied in [692] and [1518], whereas a recent work on multiperiod blood distribution is provided in [1529]. For an extensive review on perishable commodity inventory management and blood supply chains, the interested reader is referred to [1037] and [186], respectively.

The problem is modeled as follows. A single *allocation center* with a capacity of C units of a commodity needs to allocate this commodity to a number of *distribution locations*, indexed by the set I. Each distribution location faces demand from multiple demand classes, indexed by J. A unit of demand satisfied from class j accrues w_j units of benefit. Without loss of generality, we assume $w_1 \geq w_2 \geq \cdots \geq w_{|J|}$. For each unit that is allocated to a distribution location, but is not used to satisfy demand until the end of the period, there is a benefit loss of w_0. This benefit loss can be considered as the opportunity cost of the units not being assigned to another distribution location where they could have been used.

The demand for each class j in each location i is a random variable D_j^i that follows a distribution with function $f_j^i(.)$, for which the cumulative distribution function (cdf) is denoted by $F_j^i(.)$. It is assumed that the allocation center knows these distributions based on historical demand or forecasts. Once demand starts arriving at a location, it is satisfied based on a *usage policy*. Two particular usage policies are considered: (i) *prioritization* of higher classes over the lower ones, and (ii) *first-come, first-served (FCFS)* usage. For each usage policy k, $E_k(z_i|x_i)$ denotes the expected benefit when x_i units are allocated to distribution location i. Based on these, the optimization problem is given by

$$\pi_k = \max \sum_{i=1}^{|I|} E_k(z_i|x_i)$$

$$\text{s.t.} \sum_{i=1}^{|I|} x_i \leq C \qquad\qquad \forall i \in I,$$

$$x_i \geq 0 \qquad\qquad \forall i \in I.$$

Under prioritized usage, demand from a lower class can be satisfied only if there are leftovers after all the demand from higher classes is satisfied. Within the context of dynamic arrivals, this implies that lower-class demand may need to wait until the demand of higher classes is completely satisfied. Prioritization can also be applied by serving the demand classes sequentially, each within a certain subperiod. When demand is (without loss of generality) a continuous random variable, let the expected benefit in a given distribution location be $E_P(z|x)$. It can be shown that $E_P(z|x)$ is concave, and for an uncapacitated system, the optimal quantity is given by the stationary point of the expectation function [419]. The equation for finding the stationary point has an intuitive structure in that it gives the *marginal benefit* of allocating one more unit to the distribution location (denoted by $MB_P^i(x)$) by multiplying the unit benefit for each class with the probability that the next unit of demand is from that class and subtracting the expected benefit loss when there is no additional unit of demand. The marginal benefit function is concave for continuous demand, and the differences between consecutive marginal benefits are nonincreasing for discrete demand. When demand is a discrete random variable, the optimal allocation is the smallest possible value of x for which the left-hand side of the equation is less than or equal to the right-hand side.

In the case of a capacitated allocation center and multiple distribution locations, the intuition behind the marginal benefit equation yields an algorithm that finds the optimal allocation. Assume, without loss of generality, that demand is discrete. First, $MB_P^i(0)$ are calculated for all locations. If all are negative, no allocation is made.

Otherwise, the first unit is allocated to the location with the highest marginal bene-fit. Proceeding in this way, until either capacity is exhausted or all marginal benefits are negative, in each step, the location with the highest marginal benefit is found, its allocation is incremented by one, and its marginal benefit is updated.

Under FCFS usage, the order of the arrivals becomes important; therefore it is assumed that arrivals of demand from each class j follow a Poisson process with rate λ_j to make use of the uniformity of arrivals. The expected net benefit function is denoted by $E_F(z|x)$ and has the same intuitive properties as in the case of prioritization.

Therefore, a greedy algorithm finds the optimal allocation for the case with a ca-pacitated allocation center and multiple distribution locations. The algorithms differ only by the calculation of the marginal benefit function.

When there are two demand classes, each following a Poisson process, and the sys-tem is uncapacitated, the optimal allocation under prioritization for a given distribu-tion location never exceeds that under FCFS usage when the expected rate for each class is sufficiently high.

Under the same settings, the worst-case ratio of the expected net benefit for optimal allocation under FCFS use (π_F) relative to that under prioritized use (π_P) is given in the following theorem.

Theorem 36.1. *In the case where* $\frac{\lambda_2}{\lambda_1} = K$ *for some constant* $K \geq 0$, $\frac{w_1}{w_2} = L$ *for some constant* $L \geq 1$, *and* $\min\{\lambda_1, \lambda_2\} \geq T = \frac{w_0}{w_2}$ *for some constant* $T \geq 0$,

$$\frac{\pi_F}{\pi_P} \geq \frac{K+L}{L(K+1)}, \tag{36.6}$$

and the bound is asymptotically tight.

By Theorem 36.1, two main factors determine the ratio between the performances of allocations under the two usage policies: (i) the ratio of the unit benefits of the two classes, the increase of which increases the difference in performances, and (ii) the ratio of demand rates for each class. With everything else constant, if the lower-class demand rate is increased, the performance of FCFS decreases more because first-class demand will be left farther behind, whereas prioritization has the advantage of satisfying the first-class demand before the second class and hence is unaffected by the increasing rate.

The next two results are related to centralization of the distribution locations, that is, a single distribution center that faces and satisfies the aggregate demand for each class in the entire system. The results underline the fact that prioritizing demand makes bet-ter use of the potential increase in the benefit by centralizing the locations, extending the results in [671] and [1732].

Theorem 36.2. *For an uncapacitated setting, under FCFS use, centralizing the distribu-tion locations performs at least as well as the decentralized system. However, under a ca-pacitated system, centralizing the demand may worsen the expected net benefit.*

Theorem 36.3. *Under prioritized use, regardless of the capacity, centralizing the distribu-tion locations always performs at least as well as the decentralized system.*

Computational experiments based on instances from the 2009 U.S. H1N1 vaccina-tion campaign yield the following insights [419]:

- As demand rates increase relative to population size, the performance difference between prioritization and FCFS use increases (at a slower rate when demand is centralized).
- Both centralizing the demand and prioritization play an important role in increasing equity of distribution by decreasing the variance of expected benefit per person over the network.
- Heuristic distributions using the urgent or total demand rate perform well when the unit benefit loss due to unsatisfied demand is low.

Future research directions regarding this work include analysis of real-time allocations and reallocations considering commodity age and deriving necessary and sufficient conditions for a decentralized system to outperform a centralized one, particularly when demand rates are identical over all distribution locations, as in the case of pooled queues [1817, 1818].

While results and insights from traditional supply chains can be applied to inventory optimization in humanitarian supply chains, several issues are specific to humanitarian applications and require specific attention when planning for inventories. For example, due to uncertainty of funding, it may not always be possible to implement an inventory policy, even if it can be identified. As in the case of facility location, inventory planning for fair/equitable demand satisfaction is an important research direction. With the existence of multiple humanitarian aid organizations, collaboration and coordination of inventory management is an important challenge. Addressing the decisions of coordinating what, how much, and where to preposition requires novel approaches compared to those in traditional supply chains.

36.4 ▪ Managing Bottlenecks in Humanitarian Transport Networks

Delays in transportation in humanitarian supply chains prevent life-saving aid from reaching beneficiaries when needed. Characterizing and reducing these delays is the focus of this section. Motivated by the lack of systematically available real-time data in humanitarian transport networks on costs, capacity, and alternative options, this research develops a modeling paradigm that will yield relevant tactical and strategic insights. This approach contrasts with models requiring extensive data inputs, which are often assumed in traditional transport network modeling, e.g., time-expanded multicommodity flow models, and even existing humanitarian models [29, 63, 86, 231, 993, 1689]. Further, in public impact applications, transportation delays are often driven by reliability issues (e.g., at congested ports), and these delays can change greatly depending on the quantities arriving. We describe a queuing model with breakdowns to model delays in port and transportation corridors (the overland travel from discharge ports to delivery points). Using the model, we gain insight into where delays or the costs of delays are most detrimental to system performance (i.e., the network's "bottleneck") in port and transportation corridors. We then include our delay modeling in a convex cost network flow model that determines optimal routing when several port options are available.

36.4.1 ▪ Modeling Port and Corridor Delays

At discharge ports, delays can occur at sea when the arriving vessels exceed the port capacity (berth space, bagging machines, etc.) or once cargo is on land for unloaded tonnage awaiting offtake into the corridor [703, 1705, 1805]. Offtake delays into the corridor can occur due to capacity restrictions (e.g., the trucking tonnage contracted by a humanitarian organization for the month) and also due to disruptions (e.g., security issues, limited truck access to the port, and even labor strikes) [100, 459, 1806]. We model these delays through a two-station, tandem queuing model where the first station models delays at the port and the second station models offtake delays into the corridor, with the inclusion of stochastic breakdowns. Since the next subsection deals with routing flow among multiple ports, the notation is specified in terms of port-corridor pair i, where the port station i feeds into the corridor station i.

Instead of focusing on what should be done in managing daily operations or on exceptional time periods where the port sees a rapid scale-up of activity or a lengthy closure, the chosen model captures a high-level, strategic analysis of where and how lengthy the delays in the network will be. The resulting long-term characterization of expected delays allows humanitarian logisticians to better understand and plan for systemic delays and incorporate them into the decision on routing aid through the ports. Such long-term analysis of ongoing operations is quite relevant for humanitarian organizations. For example, a majority of the food aid transported overseas for the UN World Food Programme in 2011 went toward nonemergency projects as part of normal port operations [1807]. Limited models exist for incorporating congestion delays into humanitarian transport models; [1955] addresses bottlenecks with respect to location modeling, while [1170] and [1206] address port disruptions in supply chain decisions in the nonhumanitarian context. More generally, congestion in transportation networks has been addressed in [282, 406, 537]. The following nonnegative parameters are used in our model:

λ_i Vessel arrival rate to port-corridor pair i, according to a Poisson process,

μ_{pi} Exponential processing rate at port i (capturing berths, bagging machines, and other port factors),

μ_{ci} Exponential processing rate at corridor i (capturing offtake trucking capacity and other corridor factors),

$f_i > 0$ Mean time to failure in corridor i, according to a Poisson failure process with rate $\frac{1}{f_i}$,

r_i Mean time to recovery in corridor i after a failure,

v_i Variance for recovery time in corridor i.

The first station in the tandem queuing system models port delays as an M/M/1 queue. The exponential service rate suits the goal of providing closed-form expected case analysis while also incorporating uncertainty and variability, with relatively few data inputs, though other distributions could be used in modeling delays in a specific port and corridor. Similarly, a single server at the port is assumed since the objective is not to capture the specific movements in and out of the berths and through the bagging machine stations, but rather the overall delay time spent at the port. If the station is stable (i.e., if $\rho_{pi} = \frac{\lambda_i}{\mu_{pi}} < 1$), then the expected time at the port in queue and in service per arrival, W_{pi}, is given by

$$W_{pi} = \frac{1}{\mu_{pi} - \lambda_i}. \tag{36.7}$$

Departures from the port i station are assumed to proceed directly into the queue at the corridor i station. For the corridor, the addition of stochastic failures to the offtake server causes the station to be an M/G/1 queue. Stochastic failures are assumed to be preemptive (meaning that a disruption can occur in the middle of service and service will be preempted until the server is working again) due to the uncertain environment in which humanitarian operations occur. The single-server model assumption allows the halting of all offtake into the corridor, which may occur in practice for disruptions such as security issues or strikes. We model these "corridor breakdowns" according to the preemptive failure modeling framework described in [953], which uses f_i, r_i, and v_i. The long-run proportion of time that the corridor is available for transport, represented as A_i, is then $A_i = \frac{f_i}{f_i + r_i}$. A_i can be used to verify that the parameters used for a specific port-corridor network match the real-life performance. If the station is stable (i.e., if $\rho_{ci} = \frac{(f_i + r_i)\lambda_i}{f_i \mu_{ci}} = \frac{\lambda_i}{A_i \mu_{ci}} < 1$), then we derive the expected time at the corridor in queue and in service per arrival, W_{ci}, as

$$W_{ci} = \frac{2(f_i + r_i) + (r_i^2 + v_i)\lambda_i}{2f_i \mu_{ci} - 2(f_i + r_i)\lambda_i}. \tag{36.8}$$

The following insights, described more fully in [1714], result from our delay modeling framework:

- Delays do not scale linearly with the arrival rate. Instead, each additional arriving vessel has an increasing marginal delay cost (W_{pi}, $\lambda_i W_{pi}$, W_{ci}, and $\lambda_i W_{ci}$ are all increasing and convex with respect to λ_i, subject to stability).
- Considering the impact of breakdowns is key, especially in the uncertain environment of many humanitarian operations. Even if $\mu_{p1} < \mu_{p2}$ and $\mu_{c1} < \mu_{c2}$, expected delays for port-corridor 1 can be longer than for port-corridor 2 if $A_1 > A_2$.
- Variance can have a confounding influence. For example, consider two corridors with equal processing rates and availability, corridor 1 with shorter failures more often and corridor 2 with longer failures less often, and further assume the two corridors have equal coefficients of variance for recovery. While intuition might suggest that $W_{c1} = W_{c2}$, corridor 2, with longer failures less often, will have a longer average delays due to the impact of variance.
- Potential returns on investment (reduction in delay over the cost of network improvement) can be computed for a given port and corridor i using our modeling framework. For example, contracting more offtake trucking capacity (increasing μ_{ci}) might result in bigger delay reductions for less money than purchasing extra bagging machines (increasing μ_{pi}).

The total expected delay per arrival to port-corridor i is $W_i = W_{pi} + W_{ci} = \frac{1}{\mu_{pi} - \lambda_i} + \frac{2(f_i + r_i) + (r_i^2 + v_i)\lambda_i}{2f_i \mu_{ci} - 2(f_i + r_i)\lambda_i}$. This closed-form expression can be a meaningful addition to planning lead times that might otherwise only include best-case processing times without factoring in the impact of congestion. We next discuss incorporating delay modeling into routing decisions among ports.

36.4.2 ▪ Incorporating Delay Modeling into Delivery Routing

Often, routing options through multiple discharge ports are available for a humanitarian operation. For example, in Mali, the main humanitarian hubs can be reached through several ports in Western Africa (Dakar, Abidjan, Tema, Lome, and Cotonou).

Likewise, since multiple operations (e.g., ones in Mali, Niger, and other countries in the Sahel) share capacity at the ports, balancing the bottlenecks and the impact of congestion of the collective routing decisions is desired. Next, we incorporate port and offtake congestion delays into a routing model. Since total delay time at the port and offtake in the corridor can be shown to be convex with respect to the arrival rate [1714], the following convex cost network flow model can be used:

$$\min \sum_{(i,j)\in A} C_{ij}(\lambda_{ij}) \tag{36.9}$$

$$\text{s.t.} \sum_{j:(i,j)\in A} \lambda_{ij} - \sum_{j:(j,i)\in A} \lambda_{ji} = b(i), \forall i \in N, \tag{36.10}$$

$$0 \le \lambda_{i,j} \le u_{ij} \forall (i,j) \in A. \tag{36.11}$$

In the objective function (36.9), each arc has a convex cost/delay function with respect to the flow decision variable (e.g., Section 36.4.1's convex port and corridor delay functions). Flow balance is ensured by constraint (36.10), and constraint (36.11) prevents arc capacity from being exceeded.

One possible configuration of a convex cost flow model incorporating congestion delays appears in Figure 36.1, where N discharge ports are available to route a total of λ flow (e.g,. the monthly demand) through and onward to M delivery points on delivery arcs $A_D = \{1,\ldots,N\} \times \{1,\ldots,M\}$. Each port-corridor $i \in \{1,\ldots,N\}$ is characterized by its service and failure parameters, μ_{pi}, μ_{ci}, f_i, r_i, and v_i, and each delivery point $j \in \{1,\ldots,M\}$ has demand $b(j)$. We assume that the supply is equal to the demand ($\lambda = \sum_{j\in\{1,\ldots,M\}} b(j)$). A mathematical program with objective (36.12), which has convex costs on each arc, can be solved to find the minimum cost flow through the network, where scalars α_i, β_{pi} and β_{ci}, and γ_{ij} quantify port fees, port and corridor delay costs, and delivery costs, respectively. Details of the model and results appear in [1714]:

$$\min \sum_{i\in\{1,\ldots,N\}} \lambda_i(\alpha_i + \beta_{pi} W_{pi} + \beta_{ci} W_{ci}) + \sum_{(i,j)\in A_D} \gamma_{ij} \lambda_{ij}. \tag{36.12}$$

In the humanitarian context, delays, congestion, and disruptions are especially important to incorporate into transportation models. While a traditional linear-cost network flow model might choose to route a much higher percentage of flow through the cheapest port (causing longer delays due to congestion as the capacity is approached, plus bigger backups when breakdowns occur), our model can weigh the trade-offs in

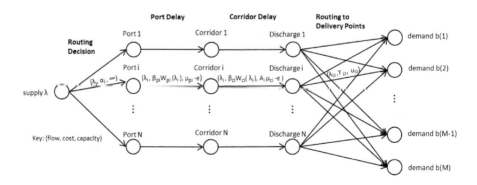

Figure 36.1. *Humanitarian supply chain convex cost routing model with deliveries.*

congestion and breakdowns to spread out the flow among several ports and reduce the impact of bottlenecks in the network.

Further research directions for this work involve how to allocate a limited budget for improvements (by increasing service rates or decreasing the duration, frequency, or variance of disruptions to service) and modeling this set of decisions in conjunction with the routing decisions. More generally in the humanitarian context, though many results and insights from traditional supply chain transportation models apply, two context-specific areas deserving special attention and novel approaches are (i) incorporating systemic congestion delays and breakdowns and (ii) creating models that do not require extensive inputs. Further, modeling ongoing operations in a nonemergency setting is an important research area. The specific challenges arising in humanitarian applications require novel approaches compared to those in traditional supply chains.

36.5 ▪ Discussion and Future Perspectives

Humanitarian supply chain optimization has been a widely studied area in recent years, due both to its theoretical contributions to the supply chain optimization literature and to its practical impacts in humanitarian applications. In this chapter, we discussed three such applications: (i) a facility location problem in a donated breast milk supply chain in South Africa, (ii) a newsvendor problem for allocating a perishable commodity to multiple demand locations with stochastic multiclass demand, and (iii) a queueing and network flow model for aid delivery.

Despite the vast literature on humanitarian supply chain optimization, several research directions need further attention. Regarding different stages of the disaster life cycle, decisions in the recovery stage (e.g., rebuilding of roads and critical facilities, mass care for displaced populations, restoration of lifeline services) have not received much attention. Compared to disasters, long-term development issues have been largely overlooked, possibly because their effects are not easily quantifiable. Modeling objective functions is another important challenge. While the literature generally uses multiobjective models with conventional objectives from the supply chain optimization literature, there is some recent work in using "humanitarian-specific" objective functions for these cases (e.g., [949]). Nevertheless, there is considerable need for future work in this area. Similarly, incorporating fairness into the objective function or constraints (e.g., [964]) presents potential research challenges. Lastly, the technical complexity of the models prevents usage in the field for disasters and long-term development issues. This can be overcome by developing user-friendly decision support tools (e.g., [1220]) that provide a user interface to decrease the technical complexity.

A growing number of studies in the area of global healthcare logistics focus on the uncertainty of various aspects of the system, including the uncertainty of funding from multilateral organizations for inventory and the availability of commodities (e.g., [1379, 1548]) as well as the uncertainty of supplies and how this affects the treatment of new and existing patients [588]. A recent stream of literature considers the distribution of drugs from a central warehouse to multiple clinics under limited supply [784, 1168, 1456, 1855], a problem particularly relevant to that discussed in Section 36.3.

Part X

Methodologies and Applications in Black-Box and Derivative-Free Optimization

Chapter 37

Methodologies and Software for Derivative-Free Optimization

Ana Luísa Custódio, Katya Scheinberg, and Luis Nunes Vicente[24]

37.1 ▪ Introduction

Derivative-free optimization (DFO) methods [502] are typically considered for the minimization/maximization of functions for which the corresponding derivatives neither are available for use nor can be directly approximated by numerical techniques. Constraints may be part of the problem definition, but, similar to the objective function, it is possible that their derivatives are not available. Problems of this type are common in engineering optimization, where the value of the functions is often computed by simulation and may be subject to statistical noise or other forms of inaccuracy. In fact, expensive function evaluations would prevent approximation of derivatives, and, even when computed, noise would make such approximations less reliable. In the past couple of decades, intense research has resulted in robust and efficient DFO methods, accompanied by convergence theory and numerical implementations.

The purpose of this chapter is to provide an overview of the main classes of state-of-the-art DFO methods, with a focus on the underlying ideas and on the respective classes of problems to which these methods are applicable. Only short descriptions of the methods and algorithms will be given, highlighting the motivational aspects that led to their rigorous properties. We provide references to detailed algorithmic descriptions, theoretical results, and available software packages.

This chapter is structured around different problem features, rather than around classes of DFO methods, as was the case in [502]. Such a structure is more accessible to users of DFO as it directs the reader to the appropriate DFO algorithm suited for the given problem.

[24]This work has been partially supported by NSF grants DMS 10-16571 and DMS 13-19356, AFOSR grant FA9550-11-1-0239, and DARPA grant FA 9550-12-1-0406 negotiated by AFOSR and by FCT under grants UID/MAT/00324/2013, UID/MAT/00297/2013, and P2020 SAICTPAC/0011/2015.

Little notation or terminology needs to be introduced as the contents are given at a general level. However, we point out that by global convergence we mean convergence to some form of stationarity regardless of the starting point. The vector norms will be ℓ_2-norms. The symbol $\mathscr{C}^k$ denotes the space of real n-dimensional functions whose derivatives are continuous up to the order k. The notation $\mathcal{O}(A)$ will mean scalar times A, where the scalar does not depend on the iteration counter of the method under analysis (thus depending only on the problem or on algorithmic constants). The dependence of A on the dimension n of the problem will be made explicit whenever appropriate. The chapter is organized as follows. Section 37.2 covers unconstrained optimization. Bound and linearly constrained problems are addressed in Section 37.3. Section 37.4 is devoted to other types of problem constraints. Extensions to global optimization, multiobjective optimization, mixed-integer problems, and some additional practical issues are briefly surveyed in Section 37.5.

37.2 • Unconstrained Optimization

37.2.1 • Smooth Functions

In this section we consider the unconstrained minimization of an objective function $f : \mathbb{R}^n \rightarrow \mathbb{R}$ that is at least once continuously differentiable and bounded from below (for which gradients neither are available for use nor can be accurately approximated).

37.2.1.1 • Sampling and Modeling

Let us discuss model-based trust-region methods for DFO. At each iteration of a trust-region method [496], one typically considers minimizing a model $m_k(x_k + s) = f(x_k) + s^\top g_k + \frac{1}{2} s^\top H_k s$ in a region around the current iterate x_k to obtain a trial point $x_k + s_k$. The region is frequently defined as a ball of type $B(x_k; \Delta_k) = \{x_k + s \in \mathbb{R}^n : \|s\| \leq \Delta_k\}$, where Δ_k denotes the trust-region radius. The model m_k serves as a local approximation of the function, in particular of its curvature. The vector g_k can be set to $\nabla f(x_k)$ in the presence of first-order derivatives (similarly for H_k), but DFO trust-region methods are based on models built from sampling and some form of interpolation [503, 1509, 1895].

How well the model approximates the function is reflected by the ratio $\rho_k = [f(x_k) - f(x_k + s_k)] / [m_k(x_k) - m_k(x_k + s_k)]$. The algorithm proceeds by accepting the trial point $x_k + s_k$ when $\rho_k \geq \eta_0$ for some $\eta_0 > 0$. If $\rho_k < \eta_1$, with $\eta_1 \geq \eta_0$, then the quality of the model may be improved if not deemed sufficiently good, or, if the quality of the model is believed to be good, the trust-region radius is reduced since the step is then deemed to be too large. If x_k is nonstationary and m_k has good quality, the algorithm succeeds in accepting a trial point $x_k + s_k$ as a new iterate (at which the function value is improved) in a finite number of reductions of the trust-region radius Δ_k (see [502, Lemmas 10.6 and 10.17]).

In first-order approaches, the quality of a model is measured by its ability to provide accuracy similar to a first-order Taylor expansion:

$$\begin{aligned} |f(y) - m_k(y)| &\leq \varkappa_f \Delta^2, \\ \|\nabla f(y) - \nabla m_k(y)\| &\leq \varkappa_g \Delta \qquad \forall y \in B(x_k; \Delta), \end{aligned}$$

where $\varkappa_f$ and $\varkappa_g$ are positive constants. Models that are $\mathscr{C}^1$ (with a Lipschitz-continuous gradient) and satisfy the above bounds are called fully linear [499]. It was shown in [497] that a subsequence of the iterates generated by a model-based trust-region method drives the gradient to zero, under the condition that fully linear models are available when necessary. This result was further improved in [501] for the whole sequence of iterates, including the case where $\eta_0 = 0$, which means that any decrease in the function value is sufficient to accept a new point.

If convergence to second-order stationarity points is desired, then fully quadratic models [499] need to be considered. In this case the models should be $\mathscr{C}^2$ (with a Lipschitz-continuous Hessian) and satisfy

$$|f(y) - m(y)| \leq \varkappa_f \Delta^3,$$
$$\|\nabla f(y) - \nabla m(y)\| \leq \varkappa_g \Delta^2,$$
$$\|\nabla^2 f(y) - \nabla^2 m(y)\| \leq \varkappa_h \Delta \; \forall y \in B(x; \Delta).$$

Convergence to second-order stationary points is established in [501].

Building a (fully linear or fully quadratic) model based on a sample set raises questions related to the choice of the basis functions used in the model definition and to the geometry of the sample set. The use of polynomial models is quite attractive due to its simplicity, and in [499, 500] a first systematic approach to the subject of sampling geometry when using this class of functions was proposed (introducing the notion of Λ-poised sets, which is related to Lagrange polynomials and ensures fully linear or fully quadratic models). The strict need to control geometry or consider model improvement steps was questioned in [689], where good numerical results were reported for an interpolation-based trust-region method (using complete quadratic models) that ignores the geometry of the sample sets. In [1644], an example was given showing that geometry cannot be totally ignored and that some form of model improvement is necessary, at least when the size of the model gradient becomes small (a procedure known as the criticality step, which then ensures that the trust-region radius converges to zero). In [1644], an interpolation-based trust-region method was proposed that resorts to geometry-improving steps only when the model gradient is small. Global convergence for this method is the result of a self-correction property inherent in the combination of trust regions and polynomial interpolation models.

Quadratic functions are particularly well suited to capturing curvature [502]. In the context of expensive function evaluations, constructing a complete quadratic model, which requires $(n + 1)(n + 2)/2$ function evaluations, could be unaffordable. A typical approach is to consider minimum Frobenius norm models, which are commonly built when at least $n + 1$ sampling points are available for use, allowing one to at least compute a fully linear model. Some variants minimize the Frobenius norm of the model Hessian [498], since the norm of the model Hessian is connected with the accuracy of the model. Other approaches, inspired by quasi-Newton methods, use a least updating minimum Frobenius norm strategy, minimizing the difference between the current and the previous model Hessian [1512]. Minimizing the ℓ_1-norm of the model Hessian has also been proposed to build accurate models from relatively small sample sets [157]. Inspired by the sparse solution recovery theory developed in compressed sensing, the underlying idea is to take advantage of the sparsity of the Hessian in cases where the sparsity structure is not known in advance. Algorithms to compute fully linear and fully quadratic models, in the context of polynomial interpolation or regression, can be found in [499, 500] (see also [502]).

An alternative to polynomial bases is radial basis functions (RBFs) [353, 1510]. An RBF is defined by the composition of a univariate function and a function measuring the distance to a sample point. Thus, it is constant on a sphere and has a structure different from polynomials (more nonlinear; potentially more nonconvex). Models based on RBFs typically involve a linear polynomial tail and can be made fully linear. The use of RBFs in model-based trust-region methods was analyzed in [1892].

Currently, several solvers implementing interpolation-based trust-region methods are available to the community. Quadratic polynomial models are at the heart of DFO [595] and NEWUOA [1513] computational codes. In the first case, when the size of the sampling set is not large enough to build a complete quadratic interpolation model, minimum Frobenius norm models are computed. In contrast, NEWUOA [1513] uses the least updating minimum Frobenius norm strategy described above. Good numerical results on unconstrained problems were also reported for the BC-DFO code [843], an interpolation-based trust-region method developed for bound constrained optimization (see Section 37.3 below). Models based on RBFs are implemented in ORBIT [1891].

37.2.1.2 ▪ Sampling Using Simplex Sets

In turn, direct-search methods use function values from sampling only to make algorithmic decisions, without explicit or implicit modeling of the function. However, the geometry of the sample sets continues to play a crucial role in the algorithmic design and convergence properties.

One possibility is to sample at the vertices of a simplex set, of which there are $n+1$ in number, exactly as many points as required to build a fully linear model. The goal of each iteration in the well-known Nelder–Mead algorithm [1386] is to improve the worst vertex of a simplex, and toward this purpose a number of operations are performed (reflection, expansion, outside contraction, inside contraction, and shrink). The various simplex operations allow the method to follow the curvature of the function, which explains its good performance in many problems.

However, since all simplex operations (except for shrinking) can cause the simplex geometry to deteriorate (an evident example is when an expansion occurs), it is difficult to establish convergence for the original algorithm. In fact, a $\mathscr{C}^2$ strictly convex function has been constructed in [1307] for $n = 2$ for which the algorithm [1386] fails to converge to the minimizer (by generating an infinite sequence of inside contractions). Convergence can be established for $n = 1$ (see [1116] or Exercise 7 of Chapter 8 of [502]) and for $n = 2$ for functions where the Hessian is always positive definite and when no simplex expansions are allowed [1115]. Modified variants have been proposed, yielding global convergence in $\mathbb{R}^n$, by including strategies like monitoring the simplex geometry and then possibly attempting a poll-type step (see below) together with using a sufficient decrease condition for accepting new points [1789] (see the survey in [502]).

Numerical implementations of variants of the Nelder–Mead method can be found in [1387] or in the Matrix Computation Toolbox [1297] (see the function NMSMAX).

37.2.1.3 ▪ Sampling Using Positive Spanning Sets

Direct-search methods can also be of directional type, where the function is evaluated along directions in positive spanning sets (PSSs) [563]. (A PSS is a set of vectors that spans $\mathbb{R}^n$ with nonnegative coefficients.)

Typically, these methods evaluate the objective function at points of the form $x_k + \alpha_k d$, $d \in D_k$, where x_k represents the current iterate, α_k the current step size parameter, and D_k a PSS. This procedure (called polling) is attempted with the goal of decreasing the current best function value. When only simple decrease is required, polling is successful if $f(x_k + \alpha_k d) < f(x_k)$ for some $d \in D_k$. Similarly to trust-region methods, several authors propose the use of sufficient decrease strategies [1089, 1234], where success requires $f(x_k + \alpha_k d) < f(x_k) - \rho(\alpha_k)$ for some $d \in D_k$, and where $\rho(\cdot)$ represents a forcing function (namely a nonnegative, nondecreasing function satisfying $\rho(t)/t \to 0$ when $t \to 0$). When no improvement is found, α_k is decreased. When polling is successful, α_k is kept constant or increased.

A property of a PSS essential to minimizing a smooth function is that at least one of its vectors is a descent direction, regardless of where the negative gradient is [511, 1089]. Thus, unless the current iterate is already a first-order stationary point, the algorithm will succeed in finding a better point in a finite number of reductions of the step size. As in model-based trust-region methods, where the trust-region radius is guaranteed to converge to zero, in direct search a subsequence of step sizes will also converge to zero. In fact, imposing sufficient decrease promotes unsuccessful iterations with consequent reductions of the step size, and using the boundedness from below of the function, one can easily ensure convergence to zero for a subsequence of step sizes [1089]. When only simple decrease is required, one has to implicitly keep a distance of the order of the step size among all iterates, and the typical way to achieve this is by generating PSSs such that all trial points lie in underlying integer lattices [112, 584, 1778].

Using simple decrease and a finite number of PSSs throughout the iterations, it is proved in [1778] that the gradient is driven to zero for a subsequence of the iterates. Such an algorithmic framework is improved, generalized, and analyzed in [112] and coined as *generalized pattern search* (see also [3]). It is shown in [1089] that an infinite number of PSSs can be used when sufficient decrease is imposed (an approach known as generating set search), as long as they are uniformly nondegenerate (meaning that their *cosine measure* [1089] is bounded away from zero).

Polling can be opportunistic (when moving to the first point $x_k + \alpha_k d$ yielding the desired decrease) or complete (when the best of the points $x_k + \alpha_k d$ is taken in D_k and then compared with x_k). Complete polling leads to the convergence of the whole sequence of gradients to zero [1089] (under the additional condition that the step size converges to zero, which occurs naturally when imposing sufficient decrease or if the step size is never increased). When polling is not complete, for instance when the first poll direction leading to descent is taken, the order of the poll directions has some influence on the numerical performance of the method (see [114, 533]).

Nowadays, several implementations of direct-search methods of directional type are available, such as DFL [594], HOPSPACK [954], NOMAD [1404], and SID-PSM [1693]. Even if most of these solvers offer additional features, polling is common to all of them.

37.2.2 ▪ Nonsmooth Functions

In the presence of nonsmoothness, the cone of descent directions can be arbitrarily narrow (see the example provided in [1089, page 441]). Thus, the use of a finite number of PSSs may not guarantee the existence of a descent direction among the poll vectors and can cause stagnation of the optimization process. This fact was the main motivation for considering more general sets of directions [510] (see also [44], where the motivation arose from a practical context).

To rigorously avoid stagnation and guarantee some form of convergence (as defined below), poll vectors must therefore be asymptotically dense in the unit sphere. When simple decrease is used, all the generated trial points are required to belong to integer lattices, and mesh adaptive direct search (MADS) [114] offers a framework to do so while using infinitely many directions (and taking them from PSSs if desired). If sufficient decrease is imposed, then the computation of new points is free of rules, and the set of poll directions can be simply randomly generated in the unit sphere [1846] (an approach here denoted by RdDS).

In the absence of smoothness, convergence can be established by proving the nonnegativity of some form of generalized directional derivatives at a limit point of the sequence of iterates and along all normalized directions. To do so, the authors in [112, 114] propose the use of Clarke [482] analysis for locally Lipschitz-continuous functions. As a consequence of using asymptotically dense sets of directions, a hierarchy of convergence results is derived in [114], depending on the level of nonsmoothness present in the function. More recently, using Rockafellar generalized directional derivatives [1573], the convergence results were extended to discontinuous functions [1846]. Second-order results can be found in [4].

Simplex gradients [1048] have been suggested as a possibility to define directions of potential descent. A simplex gradient can be regarded as the gradient of a particular linear interpolation model, requiring the evaluation of the function in a simplex (its quality as an approximation of the gradient in the continuously differentiable case is analyzed in [502, 1048]). Simplex gradients are also a possibility for approximating a direction in the Clarke subdifferential [482], defined for Lipschitz-continuous functions as the set $\partial f(x) = \{\zeta \in \mathbb{R}^n : f^\circ(x;d) \geq \zeta^\top d \text{ for all } d \in \mathbb{R}^n\}$, where $f^\circ(x;d)$ represents the Clarke generalized directional derivative at x along d (its quality as an approximation to such generalized gradients is analyzed in [529]).

In practice, nonsmooth functions are frequently nonsmooth compositions of smooth functions. Lower-$\mathscr{C}^k$ functions [1577], for instance, are characterized by being locally given as a maximum of $\mathscr{C}^k$ functions. Convex functions are lower-$\mathscr{C}^2$ [1577]. Trivially, $f = \max\{f_1,\ldots,f_m\}$ is a lower-$\mathscr{C}^k$ function provided that each f_i is $\mathscr{C}^k$. In [909] (see references therein), minmax problems of this form are addressed, when the f_i's are $\mathscr{C}^1$ functions, by considering simplex gradients as approximations to generalized gradients in a line-search approach. The general lower-$\mathscr{C}^2$ case is considered in [173], adapting ideas from convex nonsmooth optimization.

Another possibility for optimizing a nonsmooth function without derivatives is to approximate it by a family of smoothing functions (see [789, 1070, 1395]). The smoothing functions typically depend on a parameter, which must then be driven asymptotically and may require prior knowledge of the nonsmooth structure of the function.

Regarding numerical implementations, NOMAD [1138, 1404] is a reference for nonsmooth unconstrained DFO using direct search. In this solver, two different instances are available to build the asymptotically dense sets of directions in the unit sphere fulfilling the integer lattice requirements, namely the probabilistic LTMADS [114] and the deterministic ORTHOMADS [6].

37.2.3 ▪ Noisy Functions

Simplex gradients are also used as search directions to optimize noisy functions. In implicit filtering [322], a (not too refined) line search is performed along a negative simplex gradient. A quasi-Newton scheme is then used for curvature approximation. Such ingredients adapt the method in [322] to noisy problems in the hope that it can escape from spurious minimizers. A detailed description of the algorithm and corresponding convergence results can be found in [1050]. A numerical implementation, called IFFCO, is available at [976].

In the presence of noise it is natural to consider least-squares regression techniques (see Chapter 4 in [502]) and use them in trust-region methods. However, when the level of noise is large, this type of model may overfit the available data. In [1032], assuming the knowledge of an upper bound for the level of noise present in function evaluations, relaxing the interpolation conditions using the corresponding bound is suggested. In [273], incorporating the knowledge about the noise level from each function evaluation in a weighted regression is suggested instead. When the level of noise is sufficiently small relative to the trust-region radius, trust-region methods based on weighted regression models retain global convergence to stationary points [273].

If the noise present in the function evaluation has a stochastic nature, then a simple possible approach is to replicate function evaluations performed at each point, conferring accuracy to the estimation of the real corresponding function value. This procedure is followed to adapt simplex-type methods [80] and interpolation-based trust-region methods [580] to noisy optimization. Recently, in the context of direct search using PSSs [448], replication techniques were also applied to smooth and nonsmooth functions computed by Monte Carlo simulation. Statistics-based approaches, namely by using hypothesis tests, are also suggested to provide confidence to the decision of accepting a new point when using direct search in the presence of stochastic noise [1718, 1786].

37.2.4 ▪ Worst-Case Complexity and Global Rates

The analysis of global convergence of algorithms can be complemented or refined by deriving worst-case complexity (WCC) bounds for the number of iterations or function evaluations, information that may be valuable in many practical instances. Derivative-free or zero-order methods have also been recently analyzed to establish their WCC bounds. As in gradient-based methods (see [411, 842, 1394]), it is shown in [1845] that there is a WCC bound of $\mathcal{O}(\epsilon^{-2})$ for the number of iterations of direct-search methods (using PSSs and imposing sufficient decrease) when applied to a smooth, possibly nonconvex function. This type of bound translates into a sublinear global rate of $1/\sqrt{k}$ for the decay of the norm of the gradient. Note that these rates are called global because they are obtained independent of the starting point. In DFO, it also becomes important to measure the effort in terms of the number of function evaluations: the corresponding WCC bound for direct search is $\mathcal{O}(n^2\epsilon^{-2})$. DFO trust-region methods achieve similar bounds and rates [788]. The authors in [412] derive a better WCC bound of $\mathcal{O}(n^2\epsilon^{-3/2})$ for their adaptive cubic overestimation algorithm, but they use finite differences to approximate derivatives.

In the nonsmooth case, using smoothing techniques, a WCC bound of $\mathcal{O}((-\log(\epsilon))\epsilon^{-3})$ iterations (and $\mathcal{O}(n^3(-\log(\epsilon))\epsilon^{-3})$ function evaluations) is established for the zero-order methods in [788, 789, 1395], where the threshold ϵ refers now to the gradient of a smoothed version of the original function and the size of the smoothing parameter. Composite DFO trust-region methods [788] can achieve $\mathcal{O}(\epsilon^{-2})$ when the nonsmooth part of the composite function is known.

In [1394, Section 2.1.5] it is also shown that the gradient method achieves an improved WCC bound of $\mathcal{O}(\epsilon^{-1})$ if the function is convex and the solution set is nonempty. Correspondingly, the global decay rate for the gradient is improved to $1/k$. Due to convexity, the rate $1/k$ also holds for the error in function values. For DFO, direct search [604] attains the $\mathcal{O}(\epsilon^{-1})$ bound ($\mathcal{O}(n^2\epsilon^{-1})$ for function evaluations) and a global rate of $1/k$ in the convex (smooth) case. As in the gradient method, direct search achieves an r-linear rate of convergence in the strongly convex case [604]. The analysis can be substantially simplified when direct search does not allow an increase in the step size (see [1093]).

The factor of n^2 has been proved to be approximately optimal, in a certain sense, in the WCC bounds for the number of function evaluations attained by direct search (see [605]).

37.2.5 ▪ Models and Descent of Probabilistic Type

The development of probabilistic models in [157] for DFO and the benefits of randomization for deterministic first-order optimization led to the consideration of trust-region methods where the accuracy of the models is given with some positive probability [158]. It has been shown that provided the models are fully linear with a certain probability, conditioned to the prior iteration history, the gradient of the objective function converges to zero with probability one. In this trust-region framework, if $\rho_k \geq \eta_0 > 0$ and the trust-region radius is sufficiently small relative to the size of the model gradient g_k, then the step is taken and the trust-region radius is possibly increased. Otherwise the step is rejected and the trust-region radius is decreased. It is shown in [158] that global convergence to second-order stationary points is also almost surely attainable.

Not surprisingly, one can define descent probabilistically in a similar way as for fully linear models. A set of directions has *probabilistic descent* if at least one of them makes an acute angle with the negative gradient with a certain probability. Direct search based on probabilistic descent has been proved globally convergent with probability one [841]. Polling based on a reduced number of randomly generated directions (which can go down to two) satisfies the theoretical requirements [841] and can provide numerical results that compare favorably to the traditional use of PSSs.

It is proved in [841] that both probabilistic approaches (for trust regions and direct search) enjoy, with overwhelmingly high probability, a gradient decay rate of $1/\sqrt{k}$ or, equivalently, that the number of iterations taken to reach a gradient of size ϵ is $\mathcal{O}(\epsilon^{-2})$. Interestingly, the WCC bound in terms of function evaluations for direct search based on probabilistic descent is reduced to $\mathcal{O}(nm\epsilon^{-2})$, where m is the number of random poll directions [841].

Recently, in [445], a trust-region model-based algorithm for solving unconstrained stochastic optimization problems was proposed and analyzed that uses random models obtained from stochastic observations of the objective function or its gradient. Almost sure global convergence was established.

37.3 ▪ Bound and Linearly Constrained Optimization

We now turn our attention to linearly constrained optimization problems in which $f(x)$ is minimized subject to $b \leq Ax \leq c$, where A is an $m \times n$ matrix and b and c are m-dimensional vectors. The inequalities are understood componentwise. In particular, if A is the identity matrix, then we have a bound constrained optimization problem. Again, we consider the derivative-free context, where it is not possible to evaluate derivatives of f.

37.3.1 ▪ Sampling Along Directions

In a feasible method, where all iterates satisfy the constraints, the geometry of the boundary near the current iterate should be taken into account when computing search directions (to allow for sufficiently long feasible displacements). In direct search, this can be accomplished by computing sets of positive generators for tangent cones of nearby points and then using them for polling. (A set of positive generators of a convex cone is a set of vectors that spans the cone with nonnegative coefficients.) If there are only bounds on the variables, such a scheme is ensured simply by considering all the coordinate directions [1171]. For general nondegenerate linear constraints, there are schemes to compute such positive generators [1172] (for the degenerate case, see [7]). If the objective function is continuously differentiable, the resulting direct-search methods are globally convergent to first-order stationary points [1172] (see also [1089]), in other words, to points where the gradient is in the polar of the tangent cone, implying that the directional derivative is nonnegative for all directions in the tangent cone. Implementations are given in HOPSPACK [954] and PSwarm [1525].

If the objective function is nonsmooth, one has to use polling directions that are asymptotically dense in the unit sphere (for which there are two main techniques, either MADS [114] or RdDS [1846]). We have seen that in unconstrained optimization, global convergence is attained by proving that the Clarke generalized derivative is nonnegative at a limit point for all directions in $\mathbb{R}^n$, which, in the presence of bounds/linear constraints, trivially includes all the directions of the tangent cone at the limit point. One can also think of hybrid strategies combining positive generators and dense generation (see the algorithm CS-DFN [688] for bound constrained optimization where the coordinate directions are enriched by densely generated ones when judged efficient).

37.3.2 ▪ Sampling and Modeling

Active-set-type approaches have also been considered in the context of trust-region methods for derivative-free bound constrained optimization. One difficulty is that the set of interpolation points may get aligned at one or more active bounds and deteriorate the quality of the interpolation set. In [843] an active-set strategy is considered by pursuing minimization in the subspace of the free (nonactive) variables, circumventing this difficulty and saving function evaluations from optimization in lower-dimensional subspaces. The code is called BC-DFO [843].

In other strategies, all the constraints are included in the trust-region subproblem. This type of trust-region method is implemented in the codes BOBYQA [1514] (a generalization of NEWUOA [1513] for bound constrained optimization) and DFO [595] (which also considers feasible regions defined by continuously differentiable functions for which gradients can be computed). Recently, extensions to linearly constrained problems have been provided in the codes LINCOA [1515] and LCOBYQA [872].

37.4 ▪ Nonlinearly Constrained Optimization

Consider now the more general constrained problem

$$\begin{aligned} \min \quad & f(x) \\ \text{s.t.} \quad & x \in \Omega = \Omega_r \cap \Omega_{nr}. \end{aligned} \tag{37.1}$$

The feasible region of this problem is defined by relaxable and/or unrelaxable constraints. The nonrelaxable constraints correspond to $\Omega_{nr} \subseteq \mathbb{R}^n$. Such constraints have to be satisfied at all iterations in an algorithmic framework for which the objective function is evaluated. Often they are bounds or linear constraints, as considered above, but they can also include hidden constraints (constraints that are not part of the problem specification/formulation that manifest themselves as some indication that the objective function could not be evaluated). In contrast, relaxable constraints, corresponding to $\Omega_r \subseteq \mathbb{R}^n$, need only be satisfied approximately or asymptotically and are often defined by algebraic inequality constraints.

Most of the globally convergent derivative-free approaches for handling nonlinearly constrained problems are of direct-search or line-search type, and we summarize these next.

37.4.1 ▪ Unrelaxable Constraints

Feasible methods appear to be the only option when all the constraints are unrelaxable ($\Omega_r = \mathbb{R}^n$). They typically generate a sequence of feasible points, thus allowing the iterative process to be terminated prematurely with a guarantee of feasibility for the best point tested so far. This is an important feature in engineering design problems because engineers typically do not want to spend a large amount of computing time without obtaining a useful (i.e., feasible) solution. One way of designing feasible methods is to use the barrier function (coined *extreme barrier* in [114])

$$f_{\Omega_{nr}}(x) = \begin{cases} f(x) & \text{if } x \in \Omega_{nr}, \\ +\infty & \text{otherwise.} \end{cases}$$

It is not necessary to evaluate f at infeasible points where the value of the extreme barrier function can be set directly to $+\infty$. Hidden constraints are fundamentally different because it is not known a priori if the point is feasible. Direct-search methods take action solely based on function value comparisons and are thus appropriate to use with an extreme barrier function. In the context of direct-search methods of directional type for nonsmooth functions, we have seen that there are two known ways of designing globally convergent algorithms (MADS [114] and RdDS [1846]). In each case, one must use sets of directions whose union (after normalization if needed) is asymptotically dense in the unit sphere of $\mathbb{R}^n$. The resulting approaches are then globally convergent to points where the Clarke directional derivative is nonnegative along all directions in the (now unknown) tangent cone. An alternative to extreme barrier when designing feasible methods is to project onto the feasible set, although this might require knowledge of the derivatives of the constraints and be expensive or impractical in many instances (see [1235] for an example of such an approach).

37.4.2 ▪ Relaxable Constraints

In the case where there are no unrelaxable constraints (rather than those of type $b \le Ax \le c$), one can use a penalty term by adding to the objective function a measure of constraint violation multiplied by a penalty parameter, thus allowing starting points that are infeasible with respect to the relaxable constraints. In this vein, an approach based on an augmented Lagrangian method is suggested (see [1173]), considering the solution of a sequence of subproblems where the augmented Lagrangian function takes into account only the nonlinear constraints and is minimized subject to the remaining ones (of the type $b \le Ax \le c$). Each problem can then be approximately solved using an appropriate DFO method, such as a (directional) direct-search method. This application of augmented Lagrangian methods yields global convergence results to first-order stationary points of the same type as those obtained in the presence of derivatives. In [599], a more general augmented Lagrangian setting is studied where the problem constraints imposed in the subproblems are not necessarily of linear type.

In turn, algorithms for inequality constrained problems, based on smooth and nonsmooth penalty functions, are developed and analyzed in [688, 1209, 1212], imposing sufficient decrease and handling bound/linear constraints separately, proving that a subset of the set of limit points of the sequence of iterates satisfies the first-order necessary conditions of the original problem. Numerical implementations can be found in the DFL library [594].

Filter methods from derivative-based optimization [727] have also been used in the context of relaxable constraints in DFO. In a simplified version, these methods treat a constrained problem as a biobjective unconstrained one, considering two goals - minimizing both the objective function and a measure of the constraints violation, while giving priority to the latter. Typically a restoration procedure is considered to compute nearly feasible points. A first step in this direction in DFO is suggested in [113] for direct-search methods using a finite number of PSSs. The filter approach in [583] (where an envelope around the filter is used as a measure of sufficient decrease) guarantees global convergence to a first-order stationary point. Inexact restoration methods from derivative-based optimization [1288] have also been applied to DFO, again with algorithms alternating between restoration and minimization steps. In [1289], an algorithm is proposed for problems with "thin" constraints, based on relaxing feasibility and performing a subproblem restoration procedure. Inexact restoration has been applied in [352] to optimization problems where derivatives of the constraints are available for use, thus allowing derivative-based methods in the restoration phase.

37.4.3 ▪ Relaxable and Unrelaxable Constraints

The first general approach to considering both relaxable and unrelaxable constraints is called progressive barrier [115]. It allows the handling of both types of constraints by combining MADS for unrelaxable constraints with nondominance filter-type concepts for the relaxable constraints (see the consequent developments in [117]). An alternative to progressive barrier is proposed in [844], handling the relaxable constraints using a merit function instead of a filter, and using RdDS for the unrelaxable ones. The merit function and the corresponding penalty parameter are only used to evaluate an already computed step. An interesting feature of these two approaches is that constraints can be considered relaxable until they become feasible, whereupon they can be transferred to the set of unrelaxable constraints. Both of them exhibit global convergence properties.

37.4.4 ▪ Model-Based Trust-Region Methods

On the model-based trust-region side of optimization without derivatives, nonlinear constraints have been considered mostly in implementations and in a relaxable mode.

Two longstanding software approaches are COBYLA [1511] (where all the functions are modeled linearly by interpolation; see also [340]), and DFO [498] (where all the functions are modeled quadratically by interpolation).

Another development avenue has been along composite step–based sequential quadratic programming (SQP) [496, Section 15.4]. Here the objective function is modeled by quadratic functions and the constraints by linear functions. The first approach is proposed in [489] and [229], using, respectively, filters and merit functions for step evaluation.

More recently, a trust-funnel method (where the iterates can be thought as flowing toward a critical point through a funnel centered on the feasible set; see [834]) was proposed in [1621] for the particular equality constrained case. Another approach (with implementation code NOWPAC) is proposed in [121] for equalities and inequalities and inexact function evaluations.

37.5 ▪ General Extensions

In real-world applications, it is often the case that the user can supply a starting point for the optimization process and that some (local) improvement over the provided initialization may be the goal. Nevertheless, there are situations where global minimizers are requested and/or good starting points are unknown. Extensions of DFO to global optimization aim at such additional requirements. One possibility is to partition the feasible region into subdomains, which are locally explored by a DFO procedure in an attempt to identify the most promising ones. DIRECT [1012] and MCS [971] follow this approach, the latter being enhanced by local optimization based on quadratic polynomial interpolation (see the corresponding codes in [710] and [1308]). An alternative is to multistart different instances of a DFO algorithm from distinct feasible points. Recently, in the context of direct search, it was proposed to merge the different starting instances when sufficiently close to each other [530] (see the corresponding code GLODS [814]). Heuristics have been tailored to global optimization without derivatives, and an example providing interesting numerical results uses evolutionary strategies like CMA-ES [903] (for which a modified version is globally convergent to stationary points [601]).

DFO algorithms can be equipped with a search step to improve their local or global performance (such steps are called *magical* in [496]). The paper [312] proposes a search-poll framework for direct search, where a search step is attempted before the poll step. A similar idea can be applied to model-based trust-region algorithms [845]. The search step is optional and does not interfere with the global convergence properties of the underlying methods. Surrogate models (see [1215, Section 3.2] and [502, Section 12]) can be built and optimized in a search step, as in [532] for quadratics or in [70] for RBFs. Another possibility for its use is the application of global optimization heuristics [110, 1833]. See the various solvers [1404, 1525, 1693].

Parallelizing DFO methods is desirable in the presence of expensive function evaluations. The poll step of direct search offers a natural parallelization by distributing the poll directions among processors [961]. Asynchronous versions of this procedure [855] are relevant in the presence of considerably different function evaluation times. Several codes [954, 1404, 1525] offer parallel modes. Subspace decomposition in DFO is also attractive for parallelization and surrogate building [116, 846].

The extension of DFO methods to problems involving integer or categorical variables has also been considered. The methodologies alternate between a local search in the continuous space and some finite exploration of discrete sets for the integer variables. Such discrete sets or structures could be fixed in advance [5] or be adaptively defined [1211]. Implementations are available in NOMAD [1404] and DFL library [594], respectively.

Multiobjective optimization has also been the subject of DFO. A common approach to compute Pareto fronts consists of aggregating all the functions into a single parameterized one, which has been done in DFO (see [120] and references therein). In [531], the concept of Pareto dominance was used to generalize direct search to multiobjective DFO without aggregation. Implementations are available in the codes NOMAD [1404] and DMS [602], respectively.

Chapter 38

Implicit Filtering and Hidden Constraints

C. T. Kelley[25]

38.1 ▪ Introduction

A hidden constraint, a yes-no constraint [409], or a virtual constraint [498] in an optimization problem is one without an explicit representation as an equality or inequality. In many cases, one detects constraint violation if the objective function fails to return a value. At other times, the constraint is tested by a separate computation but only returns a flag for failure or success. Our view is that if one cannot quantify the degree of infeasibility, then the constraints are hidden from the optimization.

Hidden constraints arise, for example, if computing the objective function depends on an iteration that fails to converge. Hidden constraints may also depend on the success or failure of an internal simulation (see [557, 558] for an example of this). Hidden constraints are related to yes-no constraints in that they are either satisfied or not. Yes-no constraints, however, need not be hidden.

Such problems confound methods based on smoothness assumptions, even if the methods do not compute gradients directly. Sampling based on the *search-poll* paradigm is better equipped to handle hidden constraints. The search-poll idea divides the optimization into a search phase, which can be anything one likes, and a poll step, which (1) is needed to recover when the search phase is unable to make progress and (2) enables one to prove convergence results.

In this chapter we discuss one way to handle hidden constraints in the context of stencil-based sampling algorithms and describe how that approach is realized in the MATLAB code `imfil.m` [1050], an implementation of the implicit filtering algorithm.

[25]This work has been partially supported by the Consortium for Advanced Simulation of Light Water Reactors (www.casl.gov), an Energy Innovation Hub (http://www.energy.gov/hubs) for Modeling and Simulation of Nuclear Reactors under U.S. Department of Energy Contract No. DE-AC05-00OR22725, National Science Foundation grants CDI-0941253, DMS-1406349, and Army Research Office grant W911NF-11-1-0367.

Our approach is motivated by several examples. Implicit filtering was invented to solve a problem in yield optimization [804, 1725] for semiconductors. The problem is to design a manufacturing process that maximizes the fraction of devices that fall within acceptable performance bounds. This is a very different problem from designing the process to optimize the average device performance. The yield optimization problem must include a model of the manufacturing process and its random variations. We detected a violation of the hidden constraint when the internal device simulator reported a power-added efficiency of more than 100%, which is clearly nonphysical.

In an application to the design of automotive valve trains [460, 557, 558], the optimization could (and did) guide the simulator to parameter values at which the dynamics were unstable or stiff. The simulator, which used a variable-step explicit Runge–Kutta integrator [101, 1673] for the dynamics, would fail with such parameter values. The integrator reported the failure with a message that the time step had become too small to proceed.

In [410] we report on a hybrid algorithm with implicit filtering and the DIRECT method [1012] for noisy problems in gas transmission pipeline design. The hidden constraints in this application were consequences of an internal solver for a mixed-integer optimization problem that determined if compressor stations should be shut down. In one case, only 3% of the points that satisfied the bound constraints also satisfied the hidden constraints.

The objective function in a water resource policy application in [433, 1050, 1069] is the cost of water rights for a region in Texas. The costs are computed by using a regional rainfall model to determine the mix of options, leases, and purchases of water rights to meet the supply target. The regional rainfall model randomly samples from historical records and uses the results to determine a year's rainfall. The rainfall estimate is used to compute cost and reliability for the various policy scenarios. The objective was to design a policy for a municipal water supply that both minimized cost and controlled risk. The hidden constraints were bounds on conditional value-at-risk (CVaR) and expected reliability, quantities that could only be computed after the model had run and were not given by explicit formulas. This problem was made more difficult by the randomness in the model itself.

The accuracy in the cost model improved as one averaged over more model runs. The image we show in the left half of Figure 38.1 is a realization of the objective function for the water resource policy application using 500 calls to the rainfall model. In [433] we used 10,000 calls to the rainfall model for each set of parameter values. In [1050] we found that 500 calls gave equally good results, and we use 500 calls in the experiments in Section 38.5.

The two images in Figure 38.1 were taken from the water resources application [1050] (left) and the gas pipeline application [410] (right). The landscapes are plots of objective function values over the bound constraints for two of the variables in the problems. The objective function returns "not a number" (NaN) when a hidden constraint is violated, which leads to "holes" in the landscape. The randomness in the water resource policy application leads to the missing values in the center of the landscape. The two hidden constraints are responsible for cutting off the corner and the leftmost part of the landscape.

Implicit filtering accommodates hidden constraints naturally. We describe the implicit filtering method in Section 38.2. We describe our precise formulation of hidden constraints in Section 38.3 and state some convergence results from [114, 712, 1050]. In Section 38.4 we show how the implicit filtering code `imfil.m` from [1050] can be used to implement some of the ideas from the theory. The convergence results require

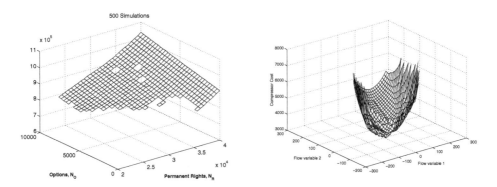

Figure 38.1. *Optimization landscapes.*

one to expand the set of search directions in sampling methods, and we report on some experiments with that larger direction set in Section 38.5.

38.2 ▪ Implicit Filtering

Implicit filtering solves

$$\min_{\Omega} f(x), \tag{38.1}$$

where

$$\Omega = \{x \mid L \le x \le U\} \subset R^N. \tag{38.2}$$

In (38.2), $L, U \in R^N$ and the inequalities are componentwise.

Implicit filtering augments coordinate search with a finite difference approximation of the gradient of f and a quasi-Newton model Hessian. We refer the reader to [1050] for the details of the quasi-Newton part of the algorithm.

Given a current point x_c and a scale h_c, coordinate search proceeds by evaluating f on the stencil

$$S(x_c, h_c) = x_c \cup \{\{x_c \pm h_c e_j\}_{j=1}^N \cap \Omega\},$$

where $\{e_j\}_{j=1}^N$ are the coordinate directions. One could use a different direction set [114, 1050, 1089], and, as we will see in Section 38.3, must use a larger one if there are hidden constraints. If one finds a better point, one accepts that as the new point x_+. Otherwise,

$$f(x_c) \le \min_{z \in S(x_c, h_c)} f(z), \tag{38.3}$$

a condition we will call *stencil failure*. In that case one sets the new point $x_+ = x_c$ and reduces h_c, usually by a factor of two, to obtain a new scale h_+.

The convergence theorem for coordinate search says that a subsequence of the coordinate search iteration asymptotically satisfies the necessary condition

$$x = \mathcal{P}(x - \nabla f(x)),$$

where $\mathcal{P}$ is the projection onto Ω,

$$\mathcal{P}(x) = \max(L, \min(x, U)), \tag{38.4}$$

and the maxima and minima are understood to be componentwise.

In the case where f is defined and differentiable on all of Ω, the result [1050] is given in the following theorem.

Theorem 38.1. *Let f be Lipschitz continuously differentiable in Ω. Let $\{x_n\}$ be the sequence of coordinate search iterations for problem (38.1). Then*

$$\liminf_{n \to \infty} \|x_n - \mathcal{P}(x_n - \nabla f(x_n))\| = 0. \tag{38.5}$$

The proof of this result follows the pattern-search ideas from [1050, 1089, 1777] and uses the fact that the iterations for each fixed h lie on a grid in Ω that has finitely many points. Methods like implicit filtering do not remain on the grid, so one must assume that (38.3) holds for infinitely many n. With that additional assumption, (38.5) holds [1050]. The formal convergence theorem is as follows.

Theorem 38.2. *Let f be Lipschitz continuously differentiable in Ω. Let $\{x_n\}$ be the sequence of implicit filtering iterations for problem (38.1). Assume that (38.3) holds infinitely often. Then (38.5) holds.*

One may take more general direction sets, but these sets must include the positive and negative coordinate directions if the sequence is to satisfy (38.5) [804, 1050, 1171]. As we shall see in the remainder of this chapter, one must enlarge the direction set to handle hidden constraints.

38.3 • Hidden Constraints and Convergence Results

The feasible set for problems with hidden constraints is a set $\mathcal{D} \subset \Omega$ defined by an explicit set of inequalities. One can only test for feasibility by attempting to evaluate f.

Algorithms that handle hidden constraints must accept a flag from the objective function, and objective functions must be designed to return a flag that indicates failure. One very poor way to do this in the context of implicit filtering is for the objective to return a very large value. The problem with that approach is that the approximate gradient will bias the optimization away from the region near the hidden constraint boundary, which is often where the optimal point will be.

38.3.1 • Geometry of $\mathcal{D}$ and Necessary Conditions

Convergence theorems for such problems must make assumptions on the geometry of $\mathcal{D}$ [114, 448, 712, 1050, 1404]. We will express these assumptions using ideas from nonsmooth optimization [114, 115, 482, 1050]. Define the Clarke cone or the tangent cone to $\mathcal{D}$ at $x \in \mathcal{D}$ by

$$T_{\mathcal{D}}^{CL}(x) = cl\{v \mid x + tv \in \mathcal{D} \text{ for all sufficiently small } t > 0\}. \tag{38.6}$$

We will assume that $\mathcal{D}$ is regular in the sense that $T_{\mathcal{D}}^{CL}(x)$ is the closure of its nonempty interior for all $x \in \mathcal{D}$. If f is differentiable everywhere, then the first-order necessary conditions for optimality at a point $x \in \mathcal{D}$ are

$$\partial f(x)/\partial v \geq 0 \text{ for all } v \in T_{\mathcal{D}}^{CL}(x). \tag{38.7}$$

If $\mathcal{D}$ is given by smooth inequality constraints, (38.7) is equivalent to the usual necessary conditions for constrained problems [482, 724].

38.3.2 ▪ Rich Sets of Search Directions

If one wants to apply implicit filtering or any other algorithm that samples points on a pattern, constraints present a problem. Figure 38.2, taken from [1050], illustrates the problem. In the figure, $\mathcal{D}$ is the lower triangle of the unit square

$$\Omega = \{x = (\xi, \eta)^T \mid 0 \le \xi, \eta \le 1\}.$$

Consider the objective function $f(x) = 1 - \eta$. The feasible points on the stencil are x_{left} and x_{down}, both of which do not improve $f(x)$. x_{up} is a better point, but it is not in $\mathcal{D}$. Hence the stencil fails and we would reduce h. This reduction in h would never terminate and the optimization would converge to x_0, which is not optimal. The failure is a result of the fixed direction set's inability to capture the geometry of $\mathcal{D}$. If one added the new point x_{new}, which captures the decrease of f in $\mathcal{D}$, to the stencil, then $f(x_{new}) < f(x_0)$ and the optimization would continue and converge to the correct result of $x^* = (1, 1)^T$.

In the case of linear constraints [1050, 1172], for example, one can determine in advance the tangent directions of the constraints and add additional directions when the iteration approaches a constraint boundary. For hidden constraints, a fixed set of directions will not suffice. In this case, one must vary the direction set as the iteration progresses so every direction is an accumulation point of the sequence of directions [114, 1050].

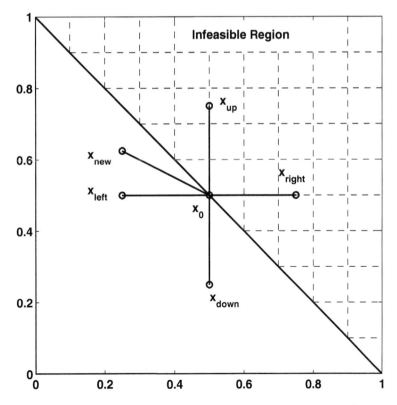

Figure 38.2. *Coordinate search misses constraint boundary* [1050].

So, if x_n and h_n are the nth point and scale in the iteration, the stencil is

$$S(x_n, h_n) = S(x_n, h_n, V_n) = \{z \mid z = x_n + h_n v_i, \; v_i \in V_n \; 1 \le i \le K_n\}, \qquad (38.8)$$

and the nth direction set is

$$V_n = \{v_i^n\}_{i=1}^{K_n}. \qquad (38.9)$$

Now let $\mathcal{V} = \{V_n\}$ be the sequence of direction sets. We say that $\mathcal{V}$ is *rich* if for any unit vector $v \in R^N$ and any subsequence $\mathcal{W} = \{W_{n_j}\}$ of $\mathcal{V}$,

$$\liminf_{j \to \infty} \; \min_{w \in W_{n_j}} \|w - v\| = 0. \qquad (38.10)$$

One way to obtain a rich sequence of directions sets (with probability one, of course) is to add one or more random directions to a fixed stencil. This is the approach we will take in the examples in Section 38.5. Various formulations of the MADS (mesh adaptive direct search) algorithm [6, 114, 650] use both random and deterministic approaches.

38.3.3 ▪ Approximating the Gradient

In this section we will assume that f is smooth and $\mathcal{D}$ is regular. Implicit filtering builds a difference approximation of the gradient at each point from the evaluations of f at the points of the stencil. If f is everywhere defined in Ω, then the components of the difference gradient are one-sided or centered differences, depending on whether both endpoints of the difference are in Ω or not. The scaling and the requirement that $h \le 1/2$ imply that at least one of $x \pm h e_j \in \Omega$ for each $j = 1, \dots, N$.

If there are hidden constraints and we use rich sequences of direction sets, the points used for the difference approximation may be more than we need in some directions. We may also fail to have enough data to approximate some components of the gradient at all. The solution is to compute the difference gradient as the solution of a linear least-squares problem.

Suppose $x \in \Omega$ and

$$x + h v_i \in \mathcal{D}$$

for $1 \le i \le K$. One way to extract an approximate gradient from the values of f at the feasible points is to take the minimum norm solution of the linear least-squares problem

$$\min_{y \in R^N} \|h V^T y - \delta(f, x, V, h)\|, \qquad (38.11)$$

where V is an $N \times K$ matrix with columns $\{v_i\}_{i=1}^K$ and

$$\delta(f, x, V, h) = \begin{pmatrix} f(x + h v_1) - f(x) \\ f(x + h v_2) \quad f(x) \\ \vdots \\ f(x + h v_K) - f(x) \end{pmatrix}.$$

The solution y of (38.11) is called the stencil gradient in [1050]. In [322] we called this the "simplex gradient" because the stencil in that paper was the set of vertices of a simplex. The stencil in this chapter is much more general. This is the approximation we use in the implicit filtering code imfil.m.

The convergence theory does not depend on the details of implicit filtering. The only requirements are that $\mathcal{D}$ be regular; that the sequence of stencils be rich; and that stencil failure,

$$f(x_n) \leq f(x_n + h_n v_i^n) \text{ for all } i \text{ such that } x_n + h_n v_i^n \in \mathcal{D}, \qquad (38.12)$$

occur infinitely often. While the option of taking the quasi-Newton step greatly improves the practical performance of the algorithm, it is irrelevant to the theory at this level of generality. We point out that one can exploit the quasi-Newton method to obtain convergence rates for some smooth problems [461, 1050] that are defined everywhere in Ω.

We will state the convergence theory in general at the end of the next section.

38.3.4 ▪ Lipschitz-Continuous f

Once we have a regular domain of definition $\mathcal{D}$ for f, generalizing our convergence results to nonsmooth but Lipschitz-continuous f requires appropriately formulating the necessary conditions, but the analysis [712, 1050] is very close to the smooth case. If f is Lipschitz continuous in $\mathcal{D}$, define the generalized directional derivative [482] of f at $x \in \mathcal{D}$ in the direction v as

$$f^\circ(x; v) = \limsup_{\substack{y \to x,\ y \in \mathcal{D} \\ t \downarrow 0,\ y + tv \in \mathcal{D}}} \frac{f(y + tv) - f(y)}{t}. \qquad (38.13)$$

The first-order necessary condition in this case is

$$f^\circ(x^*; v) \geq 0 \text{ for all } v \in T_{\mathcal{D}}^{CL}(x^*). \qquad (38.14)$$

The gradient approximation in implicit filtering may fail to find a better point but will do no harm if stencil failure (38.12) happens infinitely often. We can now state the convergence result.

Theorem 38.3. *Let $\mathcal{V}$ be a rich sequence of stencils. Let f be Lipschitz continuous and defined on $\mathcal{D} \subset \Omega$, where $\mathcal{D}$ is regular. Let $\{x_{n_j}\} \subset \mathcal{D}$ be a subsequence of the iteration such that stencil failure occurs for all but finitely many n_j. Then any limit point of the subsequence satisfies (38.14).*

The theorem applies not only to implicit filtering but also to any search algorithm that uses a rich sequence of stencils. One such deterministic algorithm is the DIRECT [712, 1012] method. If one creates a rich sequence by adding random directions, then the result holds with probability one.

We have ignored the possibility of errors in evaluating f and errors in detecting hidden constraints. If the errors in f can be bounded in terms of the scale h, then the theory can readily be extended [1050]. A more interesting case is if the errors in f are random with bounds on the variance. We have recently extended Theorem 38.3 to that case as well [448].

The result in [448] extends Theorem 38.3 by allowing for randomness in the function evaluation. Suppose function evaluation returns $\hat{f}$ rather than f, where $\hat{f}$ is the outcome of a Monte Carlo simulation with N trials. If $\hat{f}$ is centered at f with variance

$O(1/\sqrt{N})$ and the probability that $\hat{f}$ either returns a value with $x \in \mathcal{D}$ or fails to return a value when $x \in \mathcal{D}$ is also $O(1/\sqrt{N})$, then one can prove asymptotic convergence provided that N increases as h decreases and $(h_n \sqrt{N_n}) = o(1)$ as $n \to \infty$. Under these assumptions and the assumptions of Theorem 38.3, then the conclusions of Theorem 38.3 hold with probability one.

38.3.5 ▪ Limitations of the Theory

Convergence theorems are asymptotic results. Methods like implicit filtering are typically used for problems with noisy objective functions that may be expensive to evaluate. The goal of the optimization is usually not to compute a high-precision result but rather to improve a suboptimal point that one has in hand. The iteration for methods like implicit filtering is often managed by giving the optimization algorithm a budget of calls to f and terminating the optimization shortly after that budget has been exceeded.

Because of this, asymptotic convergence results provide guidance that is often very useful but does not precisely predict the performance of the iteration. For example, in the theory in this section, we see that adding random vectors to the stencil at each iteration will make the sequence of direction sets rich. While satisfying, the theory gives no insight into how many random directions one would need or even if adding such directions would improve an actual computation that may run for only a few iterations.

We can extract two general ideas from the discussion above. One is that adding random directions can improve the optimization performance. The other is that if there is randomness in evaluating f of the type covered by [448], then one could use fewer trials (lower N) in the early phase of the iteration with no loss of quality in the optimization iteration. We will look into these ideas in Section 38.5.

38.4 ▪ `imfil.m` and Hidden Constraints

In this section we will discuss how to use `imfil.m` to handle a problem with hidden constraints. The first step is to configure the MATLAB code for the objective function to return NaN when the function is undefined. This may require adding to an existing code or setting traps for errors. We will assume that this has been done.

We refer the reader to [1050] for most of the details on the use of `imfil.m`. We will talk only about the things that must be done to deal with hidden constraints.

Implicit filtering is prepared to receive NaNs and remove infeasible points both from the set of search directions and from the computation of the difference gradient. The implementation in `imfil.m` maintains a `complete_history` structure that includes a record of the failed points. One can use that structure to gauge the difficulty of the problem or estimate the set $\mathcal{D}$ by looking at how often and where the function fails.

One defines the required `options` argument to `imfil.m` with the `imfil_optset.m` code. This enables one to add random directions to the default stencil, for example. The command to add k random directions, uniformly distributed on the unit sphere, to the default stencil (the positive and negative coordinate directions) at each iteration is

```
options = imfil_optset('random_stencil',k,options_in)
```

Here `options_in` is the options structure one wishes to modify. If one is setting up options for the first time in a computation, that argument should be omitted. `options` will be the argument sent to `imfil.m` as part of the call.

While adding one or more random directions will make the sequence of direction sets rich, one might ask how necessary it is in practice. We will investigate that question with an example in Section 38.5.

38.5 ▪ An Example with Random Directions

In this section we consider the problem [433, 1050] that generated the landscape on the left side of Figure 38.1. We use the MATLAB codes that accompany [1050]. These codes can be downloaded from the SIAM webpage for this reference (click on "Latest version of software"):

http://bookstore.siam.org/SE23

This problem has six unknowns. The initial data and algorithmic parameters are described in Chapter 10 of [1050]. The only changes in the default parameters are the number of random vectors in the direction set and an increase in the budget of function evaluations to 800. The optimization in implicit filtering, as implemented by imfil.m, terminates when the budget of function evaluations is exceeded or when the sequence of scales $\{h_k\}$ has been exhausted. In this computation the scales are $h_k = 2^{-k}$ for $k = 1, \ldots, 10$.

We will explore how adding random vectors to the stencil (which we did not do in [433, 1050]) affects the performance of the algorithm. We will consider four cases, of $0, 1, 8$, and 16 random vectors.

In this example, the number N refers to the calls f makes to a water resource model. Each call to the model randomly samples historical rainfall data. Constraints on the reliability of the water supply and CVaR are tested after the evaluation of f is complete. We will treat these as hidden constraints. We refer to [433] for the details.

We will use $N = 500$ model runs to evaluate the cost function. In Figure 38.3 we plot the objective function value against the number of calls to the function. The problem also has linear constraints, which we also treat as hidden, but add the tangent directions to the direction set if the stencil at the current point crosses a constraint boundary. If the linear constraints are violated, we flag the function evaluation as a failure but do not count that against the cost of the optimization. We account for the randomness in the function evaluation by performing the optimization 16 times for each case.

The results show that adding random vectors certainly does no harm. Even for only $k = 1$ random direction, the optimization does better in the sense that more of the iteration histories converge to lower function values. That effect is more pronounced for $k = 16$, but $k = 8$ is little better than $k = 1$. Note that the number of function calls increases as k increases, doubling as k goes from 0 to 16. Notice that as k increases, the optimization histories seem to focus on two local minima, telling us something about the optimization landscape.

For the next set of experiments, we will use $N = 25$ model runs for $h = 1/2$, $N = 125$ for $h = 1/4$, and $N = 500$ thereafter. This reflects the fast increase in N relative to $1/h^2$ from the theory in [448] for the first two iterations, while limiting N to keep the optimization inexpensive. In [448] we allow N to increase to nearly 5,000,000 to examine the asymptotic convergence of the optimization. That is neither necessary nor wise in practice, of course.

While there is no proof that the function from [433] satisfies the hypotheses of the results in [448], the general ideas from that paper support using fewer model runs early in the optimization.

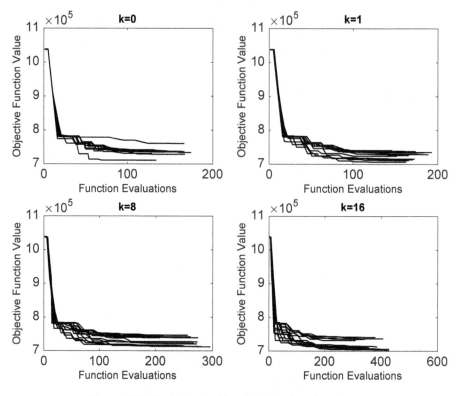

Figure 38.3. *Optimization histories:* k *random directions.*

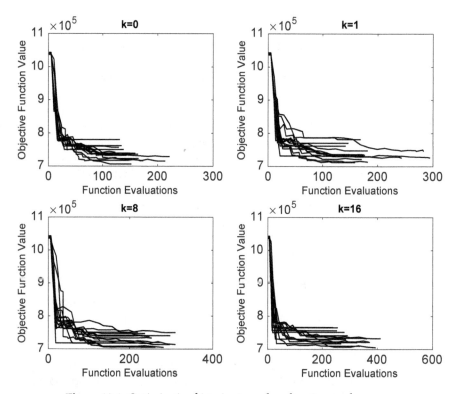

Figure 38.4. *Optimization histories:* k *random directions, scale aware.*

For the next experiments we must use the `scale_aware` option in `imfil.m`. This option means that `imfil.m` will pass the scale h to the function, which must be prepared to respond. In this case the scale-aware function will vary N as a function of h. We plot the results in Figure 38.4.

The low-accuracy function evaluations (small N) early in the iteration result in lower cost but a wider spread in the optimization histories. Leaving N fixed at a larger value may be justified by the more consistent results one sees in Figure 38.3.

38.6 ▪ Conclusions

Stencil-based sampling methods can handle hidden constraints if the direction set for the poll step is rich. Asymptotic convergence results provide useful guidance but do not completely describe the performance of algorithms in practice. One reason for this deviation of practice from theory is that one can usually take only a limited number of iterations that use function values of modest accuracy.

Chapter 39

Volumetric Alignment of Protein-Binding Cavities

Ruobing Chen, Brian Y. Chen, and Katya Scheinberg[26]

39.1 ▪ Introduction

Proteins perform chemical functions by binding to other molecules. Many perform the same function and specialize by preferentially binding certain partners. Together, the binding preferences, or *specificities*, of many proteins organize the cell into systems of coordinated molecules where they might otherwise appear to be chaotic. The desire to understand and engineer these systems motivates many investigators to study why proteins prefer to bind some partners, and how they can reengineer them to bind others. Discovering why mutations in cancer proteins cause them to prefer not to bind chemotherapy drugs and thereby resist treatment [952] or why regulator proteins can selectively bind many partner molecules [623] are just a few examples of the broad space of investigations that seek to discover the molecular mechanisms that control specificity. To understand how specificity is achieved, structural biologists examine the molecular shape, charge, and other biophysical properties of proteins to identify which parts of the protein influence specificity, and how they do so.

One way to examine these properties is to visualize three-dimensional superpositions of two or more proteins. Superpositions can reveal where the proteins are similar, and where they are different. Similarities can point to regions that stabilize similar molecular fragments, while differences in specificity might arise from regions where binding sites vary. Making observations like these depends on accurate superpositions of protein structures. An accurate superposition should align similar elements of shape or charge as much as possible, to avoid mischaracterizing them as differences that accommodate different binding partners. An ideal superposition should also accentuate actual structural and electrostatic differences and not let them be obscured by incidental similarities. Current techniques for generating superpositions are not equipped to detect all such similarities and differences.

[26]This work has been partially supported by NSF grant CCF-1320137 and AFOSR grant FA9550-11-1-0239.

Orienting molecular structures to achieve these comparisons is typically achieved by superposing corresponding atoms in two or more protein structures [274, 436, 950, 1690, 1911, 1924, 1927]. This kind of superposition ensures that many atoms overlap, enabling similar proteins to be well aligned, but it also has two major shortcomings. First, the required correspondences between atoms cannot be fully constructed between sidechain atoms, because sidechains have different lengths. This underlying variability forces atom-based superpositions to simplify amino acid geometry into backbone-only [436, 950, 1690, 1924, 1927], surrendering detail. Second, while electrostatic potentials can be represented at the molecular surface [1067] or labeled on specific atoms, the electrostatic field is not represented at longer ranges that are remote from the protein. Superpositions are thus unable to incorporate the general shape of the electrostatic field into the alignment. The work described below explores an alternative approach to comparative superposition that mitigates these issues.

39.1.1 ▪ Volumetric Analysis of Surface Properties Software

In earlier work, we demonstrated [435] that a different representation of molecular shape, based on solid representations rather than atoms, can be used to generate meaningful superpositions of small-molecule (ligand) binding cavities. This was achieved by optimizing the overlapping volume of two cavities, using the derivative-free optimization (DFO) model-based trust-region method (DFO-TR) and the volumetric analysis of surface properties (VASP) method [437]. VASP is used as a black-box function, and it evaluates the volume of overlap between two cavities, based on an input superposition. Used together, DFO-VASP examines hundreds of possible superpositions in a systematic way in the search for an individual superposition with the greatest overlapping volume.

In VASP, the overlapping volume is evaluated using the marching cube algorithms [1223], a technique for generating a polyhedral surface for a closed three-dimensional volume. In the abstract, this process identifies the overlapping region of two areas A and B by first decomposing space into a fine cubic lattice. A cubic lattice can be a set of cubes, segments of cubes, or points that form the corners of cubes. The marching cube algorithm operates by identifying the corner points that are inside both A and B. We refer to these as interior points. The cube segment between any interior point and a noninterior point must exit the overlapping region. For all such cube segments, the marching cube algorithm identifies the intersection points between the cube segment and the boundary of the intersecting region. Finally, the set of all intersection points is combined to create a polyhedral mesh that approximates the intersecting region. The volume of the intersecting region can be calculated using the surveyor's formula [1642].

The nature of this approximation affects the accuracy of DFO-VASP: intersections computed on lattices with finely sized cubes are more precise approximations of the surface, while lattices with coarser cubes have greater, though bounded, inaccuracies. As one surface is rotated and translated in the search for increasingly greater overlapping volumes, the surface intersects the lattice, which remains axis aligned, in different ways, generating noise in the approximation. This noise can be substantial because of the complex shape of molecular surfaces and cavities based on the molecular surface. To capture the complexity of the molecular surface with higher accuracy requires higher resolution of the lattice used in DFO-VASP. Higher resolutions are essential when smaller cavities are aligned, but they also result in a larger computational burden, as we will illustrate below.

39.1.2 ▪ Alignments of Electrostatic Data

In addition to molecular shape, other electric fields also influence function. To consider this second range of data, DFO-VASP can also be used to superpose electrostatic isopotentials. Electrostatic isopotentials represent a spatial region where positive electrostatic potentials are greater than a given threshold, or negative electrostatic potentials are smaller than a given threshold. Because electrostatic isopotentials are necessarily closed regions, the superposition of two isopotentials can be achieved by the same general approach as the superposition of ligand-binding cavities. Electrostatic potentials used here represent entire proteins rather than regional binding sites, and electrostatic potentials can be generated at different thresholds for different comparison purposes.

Electrostatic isopotentials, especially of whole proteins, can be dramatically larger than ligand-binding cavities. Differences in size require different resolution thresholds to be considered to maintain efficiency. While coarser resolutions exhibit greater absolute inaccuracy, relative to isopotential volume, inaccuracy from noisy comparison is no larger than for ligand-binding cavities. For this reason it is essential for DFO-VASP to adjust the range of resolutions considered in the superposition problem when considering isopotentials. In Section 39.3.2 we will illustrate the range of resolutions that we found efficient for aligning isopotentials. Considering the superposition of electrostatic isopotentials enables us to critically examine how DFO can be used to generate efficient superpositions in spite of noisy and very diverse data.

39.2 ▪ Description of the Basic DFO Method

We consider the problem of maximizing the overlapping volume in the protein alignment as the unconstrained optimization problem

$$\min_{x \in R^n} f(x), \tag{39.1}$$

where $f(x)$ is the negative volume of intersection of two or more protein structures (or their parts), which is approximately computed by the VASP software given their relative positions, i.e., rotation and translation. Hence, in this case x defines the relative position, and the number of parameters for optimization can range from seven (three specifying the rotation axis, one specifying the rotation angle, and three specifying the translation vector) to multiples of seven, depending on the number of structures one chooses to align. Notice that, to normalize the rotation axis, we need to add an equality constraint, which in the case of two-protein alignment can be expressed as $\|x_a\|_2 = 1$, where $x_a \in R^3$ is a vector with the entries being the first three entries of $x \in R^7$. However, to avoid solving problems with nonlinear constraints, we simply add the constraint as a penalty term, $\lambda(\|x_a\| - 1)^2$, to the objective function. Since this constraint only serves to eliminate multiple and badly scaled solutions, it does not have to hold exactly. By choosing a small and constant value for the penalty parameter λ we produce a stable unconstrained formulation for our problem.

The model-based DFO algorithm that we use is based on the trust-region framework described in [502]. This framework relies on constructing fully linear interpolation models of the objective function. The definition and details of fully linear models can be found in [502] and in Chapter 37. This trust-region algorithmic framework can be roughly described as follows.

Step 0: Initialization. Choose an initial point x_0, trust-region radius Δ_0, and initial interpolation set $Y_0 \subset \mathcal{B}(x_0, \Delta_0)$, which in turn defines an interpolation model m_0 around x_0. Choose $\eta > 0$ and $\gamma > 1, 1 > \theta > 0$. Set $k = 0$.

Step 1: Criticality step. If $\|\nabla m_k(x_k)\| < \theta\Delta_k$, reduce Δ_k and recompute a fully linear model in $\mathcal{B}(x_k, \Delta_k)$. Repeat until $\|\nabla m_k(x_k)\| \geq \theta\Delta_k$.

Step 2: Compute a trial point. Let $m_k(x)$ be the model built around an iterate x_k that is assumed to represent f sufficiently well in a trust region $\mathcal{B}(x_k, \Delta_k)$. Compute x_k^+ such that

$$m_k(x_k^+) = \min_{x \in \mathbb{B}_k} m_k(x),$$

where $m_k(x_k^+)$ is "sufficiently small" compared to $m_k(x_k)$.

Step 3: Evaluate the trial point. Compute $f(x_k^+)$ and $\rho_k = \frac{f(x_k) - f(x_k^+)}{m_k(x_k) - m_k(x_k^+)}$.

Step 4: Define the next iteration.

    Step 4a: Successful iteration. If $\rho_k \geq \eta$, define $x_{k+1} = x_k^+$ and choose $\Delta_{k+1} \geq \Delta_k$. Obtain Y_{k+1} by including $\{x_k^+\}$ and dropping one of the existing interpolation points if necessary.

    Step 4b: Unsuccessful iteration. If $\rho_k < \eta$, then define $x_{k+1} = x_k$ and set $\Delta_{k+1} = \gamma^{-1}\Delta_k$ if m_k is fully linear. Update Y_{k+1} to include x_{k+1} and possibly exclude some faraway points.

Step 5: Update the model. If the model m_k is not fully linear, then improve Y_k to get Y_{k+1}. $k = k + 1$. Go to step 1.

This algorithmic framework has been shown to converge to a local first-order optimal solution in the absence of noise [502]. Second-order optimal solutions can be obtained via the use of fully quadratic models, but in our applications this was deemed unnecessary. The numerical implementation of this algorithm terminates its execution when the step size parameter falls below a given threshold. For theoretical guarantees, a different, more computationally costly stopping criterion needs to be employed, such as a criticality step (see [502]), but in practice, a simple threshold strategy is used. See [496] for a detailed description of trust-region algorithms.

As one can observe, the essential mechanism of the above algorithm lies in checking the function reduction in step 3 by examining the ratio ρ_k. However, when the underlying function f is computed with noise, the ratio becomes $\rho_k' = \frac{f(x_k)+\epsilon_k - f(x_k^+)-\epsilon_k^+}{m_k(x_k)-m_k(x_k^+)}$, where ϵ_k and ϵ_k^+ are unknown noise components. It is easy to see that if the noise level is comparable to $m_k(x_k) - m_k(x_k^+)$, then the information about the achieved reduction in f provided by the noisy estimate ρ_k' is possibly corrupted. Hence, false steps can be taken by the algorithm; for example, a trial point x^+ may get accepted as the new iterate when $f(x_k^+) > f(x_k)$ or the trust-region radius may get reduced and the step may get rejected when $f(x_k^+) < f(x_k)$. Therefore, a special modification of this algorithm is necessary for noisy VASP evaluations.

39.3 ▪ Noise-Handling Strategies for VASP

39.3.1 ▪ Noise Analysis and Reduction

The presence of relative noise in the function values introduces a great deal of difficulty in optimization [1352]. Fortunately, in the case of VASP volume evaluations,

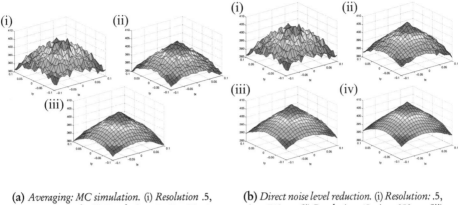

(a) *Averaging: MC simulation. (i) Resolution .5, time: 710 seconds. (ii) Resolution .5, .53, .57, .6, Time: 1510 seconds. (iii) Resolution .5, .51, .52, ..., .6, Time: 3250 seconds.*

(b) *Direct noise level reduction. (i) Resolution: .5, time: 710 sec. (ii) Resolution .45, time: 850 sec. (iii) Resolution .35, time: 1510 sec. (iv) Resolution .3, time: 2300 sec.*

Figure 39.1. *Noise reduction* [444].

the level of noise can be reduced by two strategies, i.e., by *averaging* and *direct noise level reduction*. Both of these approaches control the noise level by utilizing "resolution," an input parameter for VASP. The resolution represents the lattice cube size that VASP uses to discretize the shapes of protein structures. Smaller resolution means a finer lattice is used to approximate the shapes, which in turn means high accuracy of the estimates, but also a larger number of corner points that need to be examined and greater computational time. There is thus a trade-off between noise level and computational cost. We seek to exploit this trade-off to reduce the overall computational effort.

The first approach to noise reduction is simple *averaging* of sample values. It exploits the same idea and the standard Monte Carlo method, taking advantage of the fact that larger lattice sizes (i.e., resolution) correspond to fast volume computations and that small modifications in the resolution values results in nearly random noise in VASP evaluations. In other words, small changes in resolution result in a random change in the noise component, while the true volume remains the same. Hence, by computing multiple function values with different resolutions and averaging these values, we can reduce the noise level and get a better average estimate than each individual estimate. Figure 39.1(a) illustrates how the averaged function surfaces get smoother by averaging surfaces computed with 1, 4, and 11 distinct resolution values and lists the corresponding runtimes.

It turns out that while *averaging* reduces the noise level, it cannot make this level arbitrarily small, because the noise is not truly random. Moreover, the additional computational cost is substantial in the sequential environment. We now describe the second approach, *direct noise level reduction*, which simply changes resolution to achieve a certain level of accuracy. Figure 39.1(b) presents the resulting smoother function surfaces by reducing resolution from 0.5, 0.45, 0.35, to 0.3, and the corresponding runtimes. By comparing Figure 39.1(a) and Figure 39.1(b), one can observe that *direct noise level reduction* produces smoother objective functions than the *averaging* approach with less computational cost. This result justifies using *direct noise level reduction* as our smoothing strategy of choice.

39.3.2 ▪ Dynamic Adjustment of Accuracy

While the fast, low-accuracy function evaluations may be sufficient at the early stages of the algorithm, eventually the trust-region radius (and hence the step size) becomes small, and so does the predicted reduction achieved by a trial step. Once the value of this reduction is comparable to the noise level, this step acceptance criterion is no longer reliable. In that case, noise level reduction becomes imperative to ensure progress. Because higher-accuracy evaluations take more time, we try to resort to them only when necessary. Hence, it is advantageous to increase the accuracy dynamically as the algorithm progresses. Here we make use of the specific mechanism of our DFO algorithm. Because we need to compute maximum volume alignment of many pairs of proteins, for all of whom the accuracy versus time trade-offs are nearly the same, we precompute several estimates of the noise level for different resolution values and apply them in the dynamic strategy of adjusting the resolution parameter. Specifically, at iteration k, given current relative noise level δ_l and a constant $\theta > 1$, if $m_k(x_k) - m_k(x_k^+) < \theta f(x_k) \cdot \delta_l$, we reduce the noise to the next level δ_{l+1} and compute a new model in $B(x_k, \Delta_k)$.

To obtain the noise level estimates we assume that for a fixed resolution value, the noise level does not depend on the value of x (which is not true in the case of VASP, strictly speaking, but appears to produce reasonable results, since the noise level is more affected by the resolution value than by the change in rotation and translation). Given resolution (i.e., lattice cube size) r_v, for any x, the relative noise is defined as

$$\delta_{r_v} = \frac{f^*(x) - f_{r_v}(x)}{f^*(x)},$$

where $f^*(x)$ represents the noise-free true function value and $f_{r_v}(x)$ is the computed function value by VASP with r_v as the resolution value. Letting r_v^* be the smallest resolution that is practically computable, the noise level can then be estimated as

$$\tilde{\delta}_{r_v} = \frac{f_{r_v^*}(x) - f_{r_v}(x)}{f_{r_v^*}(x)}.$$

Figure 39.2 shows the relative noise estimation and related computational burden with respect to r_v. We observe that as r_v decreases (approaching 0.08), the runtime increases superlinearly. This is natural, as the number of corner points that VASP needs to examine grows superlinearly as r_v decreases. We also note that the reduction in noise level decreases superlinearly to zero, while the noise level itself does not decrease to zero. Nevertheless, in our experiments, the smaller noise levels were sufficient to achieve solutions of acceptable accuracy. Lower values of r_v result in very costly function evaluations, but these levels were not necessary to obtain practical solutions for examining protein-binding specificity. Similar trade-offs were observed between ligand-binding cavities and whole-protein electrostatic isopotentials, demonstrating that the dynamic adjustment of accuracy is effective at absolute sizes and resolution levels.

Based on these observations, we selected several noise levels based on the exchange between relative noise and runtime. We chose resolution values that sufficiently improve computation accuracy but avoid unnecessary calculations.

39.3.3 ▪ Warm Start Versus Random Start

The DFO framework in [502] converges to a local stationary point. In practice this method tends to find "good" local maxima; however, no guarantee of a global solution

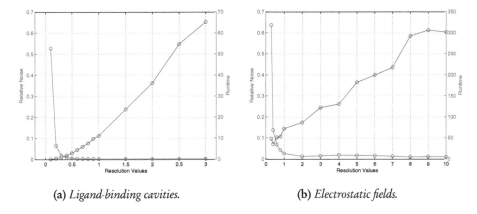

(a) *Ligand-binding cavities.*   (b) *Electrostatic fields.*

Figure 39.2. *Trade-off between the relative noise and runtime in computing the volumes of ligand-binding cavities and electrostatic fields. Part* (b) © *2015 IEEE. Reprinted with permission from* [880].

can be provided. Hence, different starting points may produce different final results if the optimization problem has multiple optima. Since the atomic and maximum volume superpositions may be closely related, it is natural to use the atomic superposition as a starting point (superposition) for volume optimization. We refer to alignments obtained by this strategy as *warm-started* alignments. However, as discussed earlier, superpositions of corresponding atoms do not necessarily yield maximal overlapping volume between ligand-binding cavities, and the lack of similarities in backbone structure may imply that the atomic superposition is a biased starting point. Therefore, as an alternative, we initiate our algorithm using randomly generated alignments and compare the results against warm-started alignments.

In the random-start approach, starting points are chosen by the Latin hypercube sampling (LHS) technique [1306], which has been successfully used in global DFO. It is a statistical method of generating a distribution of starting values of parameters from a multidimensional distribution. It selects m different values from each of n variables $X_1, \ldots, X_k$ so each sample is the only one in each axis-aligned hyperplane containing it. This LHS scheme ensures that the ensemble of random starting points represents the variability in the domain. In our experiments, we used 10 starting points (for the problem with seven variables). The range of each variable is divided into 10 equally probable intervals. After independently initiating DFO from these 10 starting points, the solution with the largest objective function value is returned.

39.4 ▪ Computational Experiments

39.4.1 ▪ Data Set Construction

39.4.1.1 ▪ Protein Families

To evaluate the effectiveness of DFO-VASP in detecting binding preferences of proteins, we tested it on two classes of proteins, the serine protease superfamily (set 1) and the enolase superfamily (set 2). Based on a wealth of biochemical research, each superfamily is known to contain three subfamilies with distinct binding preferences that are achieved by well-known differences in binding site shape. We treat this classification as ground truth and use it to evaluate whether or not DFO-VASP can reproduce these experimentally determined classifications.

We also chose this data set because each subfamily exhibits at least two sequentially nonredundant representatives. This requirement ensures that differences between proteins with different binding preferences can be contrasted with the much smaller differences between proteins with similar binding preferences. With multiple members in each subfamily, this kind of validation would not be possible. The proteins used can be found in the protein data bank (PDB) [232] and are listed by their PDB identification code in Table 39.1.

Table 39.1. *PDB codes of structures used.*

Set 1: Serine Protease subfamilies:			Set 2: Enolase subfamilies:		
Trypsins:	Chymo-trypsins:	Elastases:	Enolases:	Mandelate Racemase:	Muconate Lactonizing Enzyme:
2f91, 1fn8, 2eek, 1h4w, 1bzx, 1aq7 1ane, 1aks, 1trn, 1a0j	1eq9, 8gch	1elt, 1b0e	1e9i, 1iyx, 1pdy, 2pa6, 3otr, 1te6	1mdr, 2ox4	2pgw, 2zad

We also compared electrostatic isopotentials from structures in set 1. The details are described in [880]. To compare DFO-VASP to an existing atom-based superposition method, we used Ska [1924], an algorithm for whole-protein structure alignment. We superposed all pairs of set 1 structures and all pairs of set 2 structures, generating an alternative superposition of all cavities.

39.4.2 ▪ Experimental Results

39.4.2.1 ▪ Validating DFO-VASP Superpositions

DFO-VASP searches for the superposition of two cavities that maximizes their overlapping volume, but this strategy does not inherently guarantee that superposing cavities from similar proteins will result in a biochemically relevant superposition. To evaluate if it does, we generated superpositions of all pairs of set 1 cavities and all pairs of set 2 cavities. Visually examining all 91 pairs of superposed set 1 cavities, we observed that in all 91 cases, superposed cavities were logically oriented: entrances to each cavity were oriented in exactly the same direction, and conserved cavity shapes were strongly superposed. An example of a superposition like this is Figure 39.3(b). Thirty-three of the 45 pairs of superposed set 2 cavities were also superposed in logical orientations, with cavity entrances oriented in nearly identical directions (e.g., Figure 39.3(a)). From the remaining 12, six pairs of set 2 cavities were superposed with entrances at an angle of

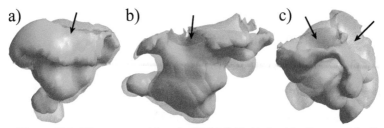

Figure 39.3. *Three superpositions by DFO-VASP. (a) Cavities from 1e9i (teal) and 1te6 (yellow, transparent). (b) Cavities from 1ane (teal) and 1a0j (yellow, transparent). (c) Cavities from 1e9i (teal) and 2pa6 (yellow, transparent). Black arrows indicate the entrance and direction of the cavity.* © 2012 IEEE. Reprinted with permission from [435].

approximately 45°—an angle for which ligand access to both cavities would have been difficult—and six more cavities were superposed at an angle of approximately 90°, for which ligand access to both cavities is impossible. Figure 39.3(c) is an example of this kind of erroneous superposition. In total, 124 out of the 136 superpositions produced cavities superposed in biochemically consistent orientations.

All superpositions observed here, however, differed in some respects from backbone superposition. Cavities in set 1 have different lengths, causing DFO-VASP to "center" smaller cavities along longer cavities. The entrance to these cavities is defined in part by backbone shape, and, as a result, backbone superpositions generally superposed the cavity entrances more closely than the whole volume. Set 1 cavities generally had similar depth, and this effect did not occur.

39.4.2.2 ▪ Comparison to Backbone Superpositions

To further evaluate DFO-VASP, we compared superpositions of cavity pairs from both data sets. Optimal superpositions were computed using random starting positions, as described in Section 39.3.3, and also by warm-starting with backbone superposition. The volumes of intersection generated by these two methods were compared to the volume generated by backbone superposition.

Random-started superpositions and warm-started superpositions both exhibited greater volumes of superposition than backbone superpositions. This is apparent in Figure 39.4, which indicates that volumes of intersection for DFO-based superpositions are greater than with backbone superpositions of the same pairs of cavities, because the differences are always greater than zero. One can also observe that warm-started superpositions performed more dependably than random-started superpositions.

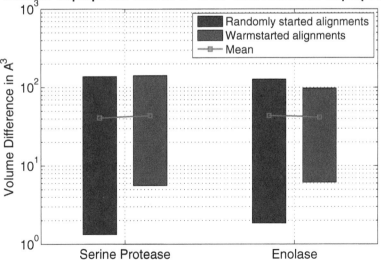

Figure 39.4. *Comparison of alignment volumes. The difference in intersection volume between random-started superpositions and backbone superpositions is shown in blue. The difference in intersection volume between warm-started superpositions and backbone superpositions is shown in red. Average differences are indicated by red lines connecting red squares.*

39.5 ▪ Discussion

We have observed that DFO of the intersection volume of superposed binding cavities can be used to achieve biologically meaningful superpositions of ligand-binding cavities. Using VASP as a black-box function, with variable resolution, achieves trade-offs in runtime, precision, and noise that yield unique optimization challenges. An analysis of noise in this calculation reveals that a dynamic approach to setting the resolution parameter points to reasonable trade-offs between runtime and relative noise that are applicable for both binding cavities and electrostatic isopotentials. When comparing random-started and warm-started superpositions, we observed that final intersection volumes from random-started superpositions were more randomly distributed than warm-started superpositions, but both approaches to superposition yielded greater superposition volumes than backbone superposition. These results demonstrate that DFO-VASP is capable of generating superpositions independent of other protein structure data, creating a unique approach with significant potential applications in protein structure alignment.

Chapter 40

POUNDERS in TAO: Solving Derivative-Free Nonlinear Least-Squares Problems with POUNDERS

Stefan M. Wild[27]

40.1 ▪ Introduction

Much of the research and software tools in derivative-free optimization (DFO) focus on *black-box optimization problems*. These are problems where the objective is effectively a black-box function, such as the scalar-valued output of an executable-only code, of the inputs. In practice, however, one often has additional knowledge (e.g., sparsity structures, partial separability, nonlinearity, convexity, form of nonsmoothness) about the problem, while still not having access to complete derivative information. This knowledge, which we characterize as defining a *grey-box optimization problem*, can be exploited to reduce the solve time and/or obtain more accurate solutions. On the other hand, when designing software for DFO, one must balance the customization needed to exploit this type of information with ease of use. Achieving this balance is especially challenging in DFO because of the size and diversity of the user pool, which includes many application scientists and engineers. For some of these users, DFO is a point of entry to optimization and related techniques such as algorithmic differentiation, nonlinear optimization (NLO), and other structure-exploiting algorithms.

Here we address nonlinear least squares, a particular form of structural information that occurs frequently in DFO and that requires minimal input from a user. Formally, we seek local solutions to

$$\min\left\{f(x) = \frac{1}{2}\|F(x)\|_2^2 = \frac{1}{2}\sum_{i=1}^{p} F_i(x)^2 : x \in \Omega \subseteq \mathbb{R}^n\right\}, \qquad (40.1)$$

[27]The submitted manuscript has been created by UChicago Argonne, LLC, Operator of Argonne National Laboratory ("Argonne"). Argonne, a U.S. Department of Energy Office of Science laboratory, is operated under Contract No. DE-AC02-06CH11357. The U.S. Government retains for itself, and others acting on its behalf, a paid-up, nonexclusive, irrevocable worldwide license in said article to reproduce, prepare derivative works, distribute copies to the public, and perform publicly and display publicly, by or on behalf of the Government.

where $F : \mathbb{R}^n \to \mathbb{R}^p$ defines the system of nonlinear equations/residuals, the Jacobian $\nabla_x F(x)$ is unavailable, $\Omega = \{x \in \mathbb{R}^n : l_i \le x_i \le u_i, i = 1,\dots,n\}$, and the bounds $l_i < u_i$ are not necessarily finite.

These problems arise frequently in settings such as nonintrusive model calibration, where one has data d_i at design sites θ^i and wishes to estimate the parameters x of a nonlinear model $S(\theta; x)$ that best fit the data. A maximum likelihood approach directly yields (40.1), with $F_i(x) = S(\theta^i; x) - d_i$, while many Bayesian approaches can be captured through weighted residuals $F_i(x) = \frac{1}{w_i}(S(\theta^i; x) - d_i)$.

Other examples of algebraic structures that have been exploited in DFO are partial separability [491], nonsmoothness of constraints [1033], and bilevel problems [504]. The solver we describe here is called POUNDERS (Practical Optimization Using No Derivatives for sums of Squares). Other derivative-free approaches to least-squares problems are Implicit Filtering [1049, 1050] and DFLS (derivative-free least squares [1953, 1954]), both of which are described later, and LMDIF [1350], which is a finite difference–based implementation of the Levenberg–Marquardt method. As with these three methods, the usual caveats apply with POUNDERS. In particular, we seek local solutions to a potentially multimodal problem (40.1), and convergence theory (which is not our focus here) is generally limited to problems with sufficiently smooth residuals (despite the derivatives of these residuals being unavailable).

40.2 ▪ Smooth Residual Models

The main device that POUNDERS uses to exploit the known structure of (40.1) is a collection of smooth surrogate models, one for each residual F_i. We briefly review the general use of models in DFO and introduce our approach for exploiting the known structure.

40.2.1 ▪ Quadratic Interpolation Models

Many forms of models have been employed for model-based optimization, from classical polynomials [502] to radial basis functions [1893] to sparse polynomials [157]. Here we focus on quadratic models

$$q_k(x) = c + (x - x^k)^\top g + \frac{1}{2}(x - x^k)^\top H(x - x^k), \qquad (40.2)$$

where we have intentionally centered these models around a point $x^k \in \Omega$. The model q_k is defined by the $\frac{(n+1)(n+2)}{2}$ parameters $c \in \mathbb{R}$, $g \in \mathbb{R}^n$, $H = H^\top \in \mathbb{R}^{n \times n}$, where we have dropped the explicit dependence of these parameters on x^k.

Given a set $\mathcal{Y} = \{y^1, \dots, y^{|\mathcal{Y}|}\} \subseteq \Omega$ and corresponding function values $\{f(y^i) : y^i \in \mathcal{Y}\}$, one can demand that the quadratic model interpolate f on $\mathcal{Y}$ by determining parameters (c, g, H) such that

$$q_k(y^i) = f(y^i), \quad i = 1, \dots, |\mathcal{Y}|. \qquad (40.3)$$

If the point x^k belongs to the interpolation set $\mathcal{Y}$, one can easily show that $c = f(x^k)$.

Given a basis ϕ for quadratic functions on $\mathbb{R}^n$ (for example, the monomial basis $\{1, x_1, \dots, x_n, x_1^2, x_1 x_2, \dots, x_n^2\}$), satisfying the interpolation conditions (40.3) is equivalent to satisfying the linear system

$$\Phi(\mathcal{Y})z = \begin{bmatrix} \phi(y^1) \\ \vdots \\ \phi(y^{|\mathcal{Y}|}) \end{bmatrix} z = \begin{bmatrix} f(y^1) \\ \vdots \\ f(y^{|\mathcal{Y}|}) \end{bmatrix} = f(\mathcal{Y}), \qquad (40.4)$$

where the solution z defines the model parameters (c, g, H). Whether this system has a unique solution for arbitrary function values $f(\mathcal{Y})$ depends solely on the interpolation set $\mathcal{Y}$; in particular we note that, unlike when doing univariate interpolation, having $\frac{(n+1)(n+2)}{2}$ distinct points in $\mathcal{Y}$ is not a sufficient condition for a unique solution to (40.4) when $n \geq 2$. The conditioning of this system clearly depends on the basis employed, and most methods take this into consideration when selecting particular forms of $\mathcal{Y}$ and/or ϕ. Measures of the quality of the sample set $\mathcal{Y}$ and the approximation properties of the resulting model are discussed in [502]. Here a primary concern is the ability of such models to approximate a function in a neighborhood of x^k; doing so requires that the sample points be within a certain proximity of x^k.

Interpolation is not the only form of model that one could consider. Other forms just correspond to different ways of "solving" (40.4). For example, for overdetermined $(|\mathcal{Y}| > \frac{(n+1)(n+2)}{2})$ sample sets one could obtain a regression-based quadratic by minimizing $\|\Phi(\mathcal{Y})z - f(\mathcal{Y})\|$, and for underdetermined sample sets one could find the interpolating quadratic for which $\|z\|$ is minimized.

For expensive problems, we generally find that we have fewer than $\frac{(n+1)(n+2)}{2}$ nearby points. A popular way of resolving the extra degrees of freedom is the approach of Powell [1512], which finds the interpolating quadratic whose Hessian is closest to the prior model's Hessian, H^{k-1}:

$$\min_{c,g,H=H^\top} \left\{ \|H - H^{k-1}\|_F^2 : q_k(y^i) = f(y^i), i = 1, \dots, |\mathcal{Y}| \right\}. \tag{40.5}$$

This corresponds to minimizing a seminorm of z for a particular choice of basis and again places certain demands on the interpolation set $\mathcal{Y}$. We refer the interested reader to [1890] for details of the solution procedure used in POUNDERS for solving (40.5).

We note that a common measure of approximation quality for interpolation-based models in DFO is based on Taylor-like conditions [502]. For example, given a continuously differentiable function f, a model m is said to be a fully linear approximation on $\mathcal{B}(x^k, \Delta) = \{x \in \Omega : \|x - x^k\| \leq \Delta\}$ of f if

$$|f(x) - m(x)| \leq \nu_1 \Delta^2 \quad \text{and} \quad |\nabla f(x) - \nabla m(x)| \leq \nu_2 \Delta \qquad \forall x \in \mathcal{B}(x^k, \Delta) \tag{40.6}$$

for some positive constants ν_1, ν_2 (independent of x and Δ).

40.2.2 ▪ Modeling Residuals

Given that we have a vector mapping F, in POUNDERS we form a quadratic model

$$q_k^{(i)}(x) = c^{(i)} + (x - x^k)^\top g^{(i)} + \frac{1}{2}(x - x^k)^\top H^{(i)}(x - x^k) \tag{40.7}$$

of each residual $F_i(x)$ for $i = 1, \dots, p$. At first glance, it may appear that we have substantially increased the linear algebraic overhead of determining the $\frac{(n+1)(n+2)}{2}$ coefficients for these p models. If we demand that the models employ a common interpolation set $\mathcal{Y}$, however, the system (40.4) becomes

$$\Phi(\mathcal{Y})Z = \begin{bmatrix} \phi(y^1) \\ \vdots \\ \phi(y^{|\mathcal{Y}|}) \end{bmatrix} \begin{bmatrix} z^1 & \cdots & z^p \end{bmatrix} = \begin{bmatrix} F_1(y^1) & \cdots & F_p(y^1) \\ \vdots & & \vdots \\ F_1(y^{|\mathcal{Y}|}) & \cdots & F_p(y^{|\mathcal{Y}|}) \end{bmatrix} = F(\mathcal{Y})^\top. \tag{40.8}$$

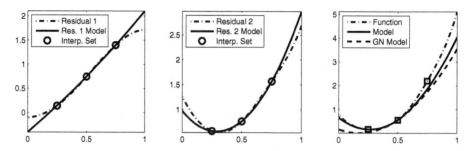

Figure 40.1. *Illustration of quadratic interpolation models on an* $n = 1$-*dimensional problem with two residuals (left two plots). The master model no longer interpolates the objective on the interpolation set; the Gauss–Newton (GN) model neglects the model Hessian terms (rightmost plot).*

Thus, the coefficients for this collection of models are determined from a single linear system with multiple right-hand sides. In practice, the basis matrix $\Phi(\mathcal{Y})$ and derived matrices associated with the approach in (40.5) are dense and therefore solved with a direct method. Since the main expense will be forming the inverse of this matrix (or a factorization of this inverse), the cost of obtaining Z grows slowly in p.

Furthermore, since the conditioning of (40.5) depends solely on $\mathcal{Y}$ (and the basis ϕ), approximation properties satisfied by any one of the models will be shared by the collection of models, provided that the residuals $\{F_i\}$ satisfy common regularity conditions. For example, if $q^{(1)}$ satisfies (40.6), then every $q^{(i)}$ determined in a similar manner—whether from (40.8) or the analogue of (40.5)—is a fully linear approximation on $\mathcal{B}(x^k, \Delta)$ of the corresponding F_i, provided that all the residuals belong to the same class of functions (i.e., with regard to smoothness and boundedness).

The left two plots in Figure 40.1 illustrate quadratic interpolation models using a common interpolation set $\mathcal{Y}$ on a one-dimensional problem. The rightmost plot illustrates a coupling of these models, described next.

40.2.3 ▪ Master Model for Nonlinear Least Squares

Given a quadratic model for each residual, there are several ways to construct a model for the objective f in (40.1). For example, direct substitution using $q^{(i)} \approx F_i$ in (40.1) yields the fourth-order polynomial $\pi(x) = \frac{1}{2}\sum_{i=1}^{p} q^{(i)}(x)^2$.

Provided that the residuals are twice differentiable, we have that the first- and second-order derivatives of the objective are

$$\nabla f(x) = \sum_{i=1}^{p} \nabla F_i(x) F_i(x) \quad \text{and} \quad \nabla^2 f(x) = \sum_{i=1}^{p} \nabla F_i(x) \nabla F_i(x)^\top + \sum_{i=1}^{p} F_i(x) \nabla^2 F_i(x),$$

respectively. The implicit filtering method [1049, 1050] uses a Gauss–Newton model, whereby the second term in the Hessian is neglected and the gradient of a linear model is used in place of each (unavailable) ∇F_i.

In POUNDERS we employ the full second-order information for the Hessian of f and define the master model

$$m_k(x^k + \delta) = f(x^k) + \delta^\top \sum_{i=1}^{p} F_i(x^k) g(i) + \frac{1}{2}\delta^\top \sum_{i=1}^{p} \left(g^{(i)}(g^{(i)})^\top + F_i(x^k) H^{(i)} \right) \delta,$$

$$(40.9)$$

where the first term assumes that x^k belongs to the interpolation set $\mathcal{Y}^k$ and hence

$$\frac{1}{2}\sum_{i=1}^{p}\left(q^{(i)}(x^k)\right)^2 = \frac{1}{2}\sum_{i=1}^{p}\left(F_i(x^k)\right)^2.$$

We refer to this model as the "full-Newton" model.

An approach that falls somewhere between the Gauss–Newton and full-Newton models is proposed in [1954], whereby the Gauss–Newton model Hessian is adaptively regularized. A key strength of this approach in the unconstrained case is that this regularization yields fast local convergence for a class of zero-residual problems [1953].

An interesting observation about both the master model in (40.9) and the regularized model used in [1954] is that these models no longer interpolate the nonlinear least-squares objective f on the set $\mathcal{Y}$. For a univariate example, the rightmost plot in Figure 40.1 shows that the master model in (40.9) interpolates f only at the designated center point (in this case $x = 0.5$) but not at the remaining points in $\mathcal{Y}$. This plot also illustrates potential differences between the model in (40.9) and a Gauss–Newton model using an interpolation-based Jacobian estimate.

40.3 ▪ The Model-Based Algorithm Underlying POUNDERS

We now discuss the basic form of the algorithm underlying POUNDERS. We use a trust-region framework, wherein the master model m_k in (40.9) is used as a quadratic surrogate for the objective f in a local neighborhood of the current iterate x^k.

ALGORITHM 40.1. **Iteration k of Model-Based Algorithm Underlying POUNDERS.**

Given $\mathcal{H}^k, x^k \in \Omega, \Delta_k > 0$ and constants $\eta > 0, \varepsilon > 0, \Delta_{\max} \le \min_i \{u_i - l_i\}$:

1. Define $\mathcal{Y}^k$ based on $\mathcal{H}^k$, form m_k, and determine if m_k is *valid*.

2. If $\|P(\nabla m_k(x^k), x^k, l, u)\| \le \varepsilon$, check for termination; otherwise proceed to step 3.

3. Solve the trust-region subproblem (40.13) to obtain x^+.

4. If $\|x^k - x^+\| \ge \frac{1}{100}\Delta_k$ or m_k is *valid*, proceed to step 4a; otherwise, set $x^{k+1} = x^k$, $\Delta_{k+1} = \Delta_k$, and $\rho_k = -1$, and go to step 5.

   4a. Evaluate f at x^+ and compute $\rho_k = \frac{f(x^k) - f(x^+)}{m_k(x^k) - m_k(x^+)}$.
   4b. Update the trust region via

   $$x^{k+1} = \begin{cases} x^+ & \text{if } \rho_k \ge \eta, \\ x^+ & \text{if } \eta > \rho_k > 0 \text{ and } m_k \text{ is } valid, \\ x^k & \text{otherwise;} \end{cases} \quad (40.10)$$

   $$\Delta^{k+1} = \begin{cases} \min\{2\Delta_k, \Delta_{\max}\} & \text{if } \rho_k \ge \eta \text{ and } \|x^k - x^+\| > \frac{3}{4}\Delta_k, \\ \frac{\Delta_k}{2} & \text{elseif } m_k \text{ is } valid, \\ \Delta_k & \text{otherwise.} \end{cases} \quad (40.11)$$

5. If m_k is not *valid* and $\rho_k < \eta$, evaluate f at model-improving point and iterate; otherwise iterate.

The steps specified in Algorithm 40.1 are repeated until a specified budget of function evaluations has been exhausted or the criticality test (step 2 and detailed below) has been satisfied.

We collect in $\mathcal{H}^k$ the history of points for which the residual values are available. In each iteration, an interpolation set $\mathcal{Y}^k \subseteq \mathcal{H}^k$ for the submodels (40.7) is constructed, from this history, based on the current iterate x^k and trust-region radius Δ_k. Similarly, we say that the model m_k is *valid* if it satisfies certain approximation guarantees (such as (40.6)) based on (x^k, Δ_k). Our technique for selecting points from $\mathcal{H}^k$ also determines the validity of the master model m_k and is discussed in [1893]. Here we note only that the interpolation set is constructed in a way that ensures that any "model-improving points" evaluated (in step 5) since the last iterate change are included in $\mathcal{Y}^k$. This ensures that in no more than n consecutive iterations will model-improving points be evaluated before the resulting master model is deemed *valid*.

The criticality test is applied depending on a measure of the projected gradient step, with the ith component being defined by

$$[P(g, x, l, u)]_i = \begin{cases} 0 & \text{if } x_i = l_i \text{ and } g_i \geq 0, \\ 0 & \text{if } x_i = u_i \text{ and } g_i \leq 0, \\ g_i & \text{otherwise.} \end{cases} \tag{40.12}$$

The tolerance $\varepsilon > 0$ is specified by a user as input. If $\|P(\nabla m_k(x^k), x^k, l, u)\| \leq \varepsilon$ and m_k is not *valid*, then the trust region is maintained, $(x^{k+1}, \Delta_{k+1}) = (x^k, \Delta_k)$; we set $\rho_k = -1$; and we go to step 5 to evaluate a model-improving point. On the other hand, if $\|P(\nabla m_k(x^k), x^k, l, u)\| \leq \varepsilon$ and m_k is *valid*, then we must ensure that the trust region (to which approximation quality is deeply tied; see (40.6)) is sufficiently small. In this case, either $\Delta^k \leq \varepsilon$ and we terminate, or $\Delta^k > \varepsilon$ and we set $(x^{k+1}, \Delta_{k+1}) = (x^k, \varepsilon)$ and iterate (proceeding to step 1).

In each iteration where the criticality test is not invoked, a candidate point $x^+ \in \Omega$ is obtained by solving the trust-region subproblem

$$\min \left\{ m_k(x) : x \in \mathcal{B}(x^k, \Delta_k) \right\}, \tag{40.13}$$

where we recall that the definition of the trust region $\mathcal{B}(x^k, \Delta_k)$ includes any bound constraints, thus ensuring that $x^+ \in \Omega$. We are purposely ambiguous about the norm $\|\cdot\|$ defining the trust region because, as discussed in the next section, in POUNDERS this norm (e.g., ℓ_2, ℓ_∞) depends on whether bound constraints are present and which subproblem solver is employed.

In typical trust-region algorithms, the candidate point x^+ is then evaluated. We avoid performing this evaluation, however, if both the resulting step is small and the current model is not deemed *valid*. In this case we instead perform the evaluation at a model-improving point. In all other cases, the candidate point is evaluated, and the usual ratio of actual decrease to predicted decrease (ρ_k) is computed. Using (40.10), the iterate is updated if this ratio is sufficiently large or if the model was *valid* and a strict decrease in the function value was obtained. Using (40.11), the trust-region radius is increased only if both the ratio and the step length are sufficiently large. The radius is decreased only if the model is *valid*.

We note that if m_k is not *valid* and $\rho_k < \eta$, then the trust region (and hence the implied definition of validity) remain unchanged. In this case we evaluate F at a model-improving point, which in POUNDERS is actually generated at the same time as the model is determined not to be *valid* (in step 1). A key difference between the procedure employed in [1893] and that used in POUNDERS, however, occurs when finite bounds are imposed. In the development of POUNDERS, a concerted effort was made to ensure that the model-improving points respect these bounds. This decision was made because these bounds are unrelaxable (see [1139]) for many problems in practice; violating a bound could mean crashing the corresponding simulation evaluation (e.g., when a negative hydraulic conductivity is passed to a subsurface flow simulator). Because of our requirement that all points evaluated by the algorithm remain in Ω, we must bound Δ_{max}, the maximum trust-region radius. If the trust-region radius were allowed to grow significantly, one could not ensure that a model m_k is *valid* solely by using model-improving points that respect the bound constraints.

40.4 ▪ POUNDERS in the Toolkit for Advanced Optimization

We now describe further details of a specific implementation of POUNDERS.

40.4.1 ▪ The Toolkit for Advanced Optimization

The Toolkit for Advanced Optimization (TAO, [1362]) is a software package designed for solving optimization problems on high-performance architectures. TAO has an open-source license and is available at http://www.mcs.anl.gov/tao/. The Portable, Extensible Toolkit for Scientific Computation (PETSc, [152]) provides the core scalable data structures and linear algebra routines that enable the parallel scalability of TAO. Consequently, TAO is used to solve problems on machines ranging from single-core laptops to massively parallel leadership-class supercomputers.

In addition to POUNDERS, the current version of TAO includes solvers for unconstrained optimization (e.g., limited-memory, variable metric quasi-Newton; Newton line search; and Newton trust-region methods), bound-constrained optimization (e.g., TRON [1197], interior-point Newton method), PDE-constrained optimization, and complementarity problems. The use of parallel data structures and linear algebra routines makes the solvers in TAO especially amenable to solving large-scale problems.

For our purposes, however, the key benefit of these parallel capabilities in TAO is not for linear algebraic operations but for objective function evaluation. In particular, the TAO separable objective functionality allows one to evaluate the residual F at a single point x using parallel resources. For example, if each residual component F_i can benefit from shared-memory parallelism to scale up to c cores, the separable objective capabilities allow for internode parallelism of the p residuals so that the wall clock time for an objective evaluation can exhibit a potential speedup of cp. Examples of this functionality are included with TAO.

40.4.2 ▪ POUNDERS Inputs

POUNDERS is available in TAO as the solver tao_pounders. The solver requires a minimal number of inputs:

- $x^0 \in \Omega$, an initial starting point;
- $\Delta_0 > 0$, the initial trust-region radius;
- a routine for evaluating the residual vector F at any given $x \in \Omega$;
- convergence criteria (e.g., a maximum number of function evaluations, a model gradient tolerance $\varepsilon > 0$, a desired function value).

The values for the remaining internal constants in Algorithm 40.1 that are used by POUNDERS are $\eta = \frac{1}{10}$ and $\Delta_{max} = \min\left\{\frac{1}{2}\min_i\{u_i - l_i\}, 1000\Delta_0\right\}$.

Several other options are available:

- how the trust-region subproblem is solved, including different solver types (e.g., an interior-point Newton method, the GQT (named after Goldfeld, Quandt, and Trotter) routine [1351]), different subproblem tolerances, and different norms for the trust-region radius, where by default, an infinity-norm trust region is used, and the resulting bound-constrained quadratic program is solved by the TAO solver TRON.
- a maximum number of interpolation points $|\mathcal{Y}|$ (in $\{n+2, \ldots, \frac{(n+1)(n+2)}{2}\}$), which, by default, is set to $2n + 1$;
- finite lower and/or upper bounds, where, by default, the problem is assumed to be unconstrained; and
- a set of points $\mathcal{H}^0$ (and the corresponding residuals) at which the residual vector F has been evaluated prior to calling POUNDERS, where, by default, this set is assumed to be empty.

Since the trust-region norms employed in POUNDERS treat each variable identically, the scaling of variables is an important consideration. Other solvers in TAO employ gradient information to scale each of the variables. Since POUNDERS does not have access to actual derivative information, the user must ensure that the problem is well-scaled. If the bounds are selected so that the residual variation across these bounds is similar for each variable, then we advocate shifting and scaling the problem so that the bounds correspond to the unit hypercube $[0, 1]^n$. For unconstrained variables, one would similarly scale each variable so that unit changes in each variable result in similar order-of-magnitude changes.

40.5 ▪ Calibrating Energy Density Functionals

We now illustrate the application of POUNDERS on problems arising in the Universal Nuclear Energy Density Functional (UNEDF) low-energy physics project [294].

A grand-challenge problem in low-energy nuclear physics is to determine an energy density functional (EDF) that describes properties of atomic nuclei across the nuclear landscape (see, e.g., Figure 40.2). One of the focuses of the UNEDF project was developing EDFs based on density functional theory. Such EDF approaches depend on phenomenological constants, for example, in the form of low-energy coupling constants. Values for these constants are typically optimized based on fits to available experimental data and pseudodata (derived from experiment or ab initio approaches). Selection of these fit observables depends on which coupling constants must be determined and on the desired properties for the resulting functional. A starting point for the optimization, which can play a critical role in the solution quality obtained by a local optimization solver, is often readily available in these calibration problems, for example, from the values of a previous-generation EDF optimization or from "natural values" for the coupling constants.

The number of optimization variables is typically on the order of a dozen; see Table 40.1. Although some coupling constants have natural ranges, the majority are effectively free: the physics-based view is that they are "constrained by the fit observables." In practice, however, the simulation codes that evaluate a particular observable are not expected to produce meaningful output for arbitrary values of coupling constants.

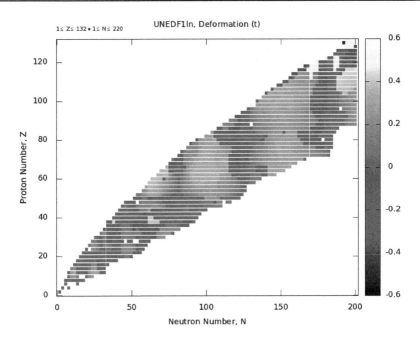

Figure 40.2. *The nuclear landscape as shown by a table of nuclides (with each column representing an element and its isotopes). The shading for each nucleus shows the total deformation as computed by an EDF code using the UNEDF$_1$ functional optimized by POUNDERS [1102].*

Table 40.1. *Problem characteristics.*

Problem	UNEDF$_0$	UNEDF$_1$	UNEDF$_2$	NNLO$_{opt}$	BPW$_{opt}$
Reference	[1100]	[1102]	[1101]	[660]	[241]
# Variables, n	12	12	14	14	17
# Residuals, p	108	115	130	2173	2049
# Nuclei Calc.	72	79	98	2173	2049

In the worst case, the code can even fail (e.g., because the underlying self-consistent equation solver or eigensolver fails to converge). Thus, for computational reasons, bound constraints are often specified by application users as a way to restrict the domain in which the optimization solver is allowed to operate. This is one of the reasons POUNDERS respects bound constraints, not only for the trust-region subproblem (40.13) but also for model-improving points. In the case of the EDF calibration problems, the majority of the bounds are specified to ensure convergence of the underlying simulation routines, and thus these bounds are expected to be inactive at a solution. For example, for the UNEDF$_2$ solution [1101], only two of the 14 variables attained one of their bounds.

Observables used in these optimizations have included a wide range of nuclear properties. For example, in the UNEDF$_2$ study [1101], the $p = 108$ residuals involved 47 deformed binding energies, 29 spherical binding energies, 28 proton point radii, 13 OES values, four fission isomer excitation energies, and nine single-particle level splittings; in the study [241] (which we refer to as BPW$_{opt}$), only binding energies were considered, but pseudodata from over 2,000 nuclei were used. In all problems, the residual vector passed to POUNDERS corresponds to the scaled difference between a simulated observable of a particular nucleus and its corresponding experimental data or pseudodata value. The scaling weights are typically based on the uncertainty in

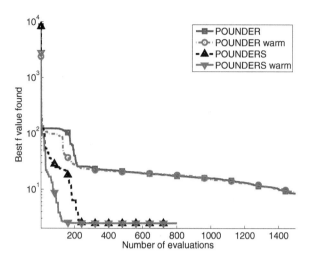

Figure 40.3. *Best function value found (log scale) for the energy density calibration problem in* [241]. *This figure shows the benefit of exploiting the least-squares structure (labeled "POUNDERS") over not doing so (labeled "POUNDER"); this benefit persists if both methods are warm-started with evaluations for an initial space-filling design.*

the data and simulation, and the effects of these weights are typically analyzed at the solution to the optimization; see [1100].

Computationally, the overwhelming expense in running POUNDERS on such problems can be attributed to the time required to evaluate the residual vector. The CPU time required to perform a single nucleus simulation (which results in multiple observable outputs for some problems; see Table 40.1) at a particular $x \in \Omega$ value ranged from 10 seconds in [241] to 12 minutes in [1100].

Taking advantage of the separable function capabilities in TAO, one can reduce the wall clock time needed to evaluate the residual vector by simulating each of the nuclei concurrently. If the simulator can itself exploit parallelism, then each nucleus calculation can be sped up further. For example, in the UNEDF$_1$ study, each of the 79 nuclei calculations employed a single node consisting of eight cores, with the master TAO driver operating on another node. Using these 632 cores, the optimization evaluated 218 points in 5.67 hours of wall clock time. The same run performed on a single core would have consumed roughly 70 days (the speedup being less than ideal because of load imbalance and the eight cores not being perfectly utilized). For the UNEDF$_2$ run, the computational footprint grew to nearly 1,600 cores (with 16 cores for each of 98 nuclei calculations).

However, benefiting from larger computational resources is not the only factor accelerating the solution of expensive EDF calibration problems. Algorithmic improvements, which result in reductions in the number of points at which the simulators must be run in the course of the optimization, are responsible for a multiplicative scaling of this speedup. For the BPW$_{opt}$ problem, Figure 40.3 shows the objective function values obtained with two different model-based trust-region algorithms: POUNDERS exploits the availability of the residual vector F, while the algorithm labeled "POUNDER" treats the scalar-valued aggregate f as a black box. Comparable objective values are obtained in a factor of 10 fewer evaluations when the structure is exploited. An even larger factor was seen when comparing POUNDERS with TAO's Nelder–Mead code on the UNEDF$_0$ problem [1100].

Furthermore, POUNDERS exploits residual evaluations done externally, for example, as a result of a variable scaling study, an efficient initial sampling, or a globalization strategy. For the BPW_{opt} problem, Figure 40.3 illustrates the benefits of using this information to warm-start the initial submodels used by POUNDERS.

40.5.1 ▪ Discussion

The POUNDERS solver has been used to make a number of advances in EDFs by solving computationally expensive nonlinear least-squares problems in the absence of Jacobian information. We attribute the algorithmic benefits of the approach to the model-based framework, which has proved effective in the general black-box case, and to taking advantage of the additional information (in this case, the residual vector) provided in grey-box problems. Because of their core similarities, we expect that the DFLS algorithm would perform similarly to POUNDERS on unconstrained problems; we also observe that DFLS compares favorably to other derivative-free methods on a large set of mathematical test problems [1953]. Lowering the barrier to running many simulations concurrently has also been a key strength of POUNDERS.

Bibliography

[1] J. ABELLO, S. BUTENKO, P. PARDALOS, AND M. RESENDE, *Finding independent sets in a graph using continuous multivariable polynomial formulations*, Journal of Global Optimization, 21 (2001), pp. 111–137. (Cited on p. 165)

[2] K. ABHISHEK, S. LEYFFER, AND J. T. LINDEROTH, *FilMINT: An outer-approximation-based solver for convex mixed integer nonlinear programs*, INFORMS Journal on Computing, 22 (2010), pp. 555–567. (Cited on pp. 279, 289)

[3] M. A. ABRAMSON, *Second-order behavior of pattern search*, SIAM Journal on Optimization, 16 (2005), pp. 515–530. (Cited on p. 499)

[4] M. A. ABRAMSON AND C. AUDET, *Convergence of mesh adaptive direct search to second-order stationary points*, SIAM Journal on Optimization, 17 (2006), pp. 606–619. (Cited on p. 500)

[5] M. A. ABRAMSON, C. AUDET, J. W. CHRISSIS, AND J. G. WALSTON, *Mesh adaptive direct search algorithms for mixed variable optimization*, Optimization Letters, 3 (2009), pp. 35–47. (Cited on p. 506)

[6] M. A. ABRAMSON, C. AUDET, J. E. DENNIS, JR., AND S. LE DIGABEL, *OrthoMADS: A deterministic MADS instance with orthogonal directions*, SIAM Journal on Optimization, 20 (2009), pp. 948–966. (Cited on pp. 500, 512)

[7] M. A. ABRAMSON, O. A. BREZHNEVA, J. E. DENNIS, JR., AND R. L. PINGEL, *Pattern search in the presence of degenerate linear constraints*, Optimization Methods and Software, 23 (2008), pp. 297–319. (Cited on p. 503)

[8] L. ACHENIE AND L. T. BIEGLER, *Algorithmic synthesis of chemical reactor networks using mathematical programming*, Industrial and Engineering Chemistry Fundamentals, 25 (1986), pp. 621–627. (Cited on p. 319)

[9] ———, *A superstructure based approach to chemical reactor network synthesis*, Computers and Chemical Engineering, 14 (1990), pp. 23–40. (Cited on p. 319)

[10] L. ACHENIE, V. VENKATASUBRAMANIAN, AND R. GANI, *Computer Aided Molecular Design: Theory and Practice*, Elsevier, 2002. (Cited on pp. 77, 329)

[11] T. ACHTERBERG, *SCIP: Solving constraint integer programs*, Mathematical Programming Computation, 1 (2009), pp. 1–41. (Cited on pp. 196, 290, 304, 313)

[12] T. ACHTERBERG AND T. BERTHOLD, *Improving the feasibility pump*, Discrete Optimization, 4 (2007), pp. 77–86. (Cited on p. 57)

[13] T. ACHTERBERG, T. KOCH, AND A. MARTIN, *Branching rules revisited*, Operations Research Letters, 33 (2005), pp. 42–54. (Cited on p. 285)

[14] W. ACHTZIGER, *Truss topology optimization including bar properties different for tension and compression*, Structural Optimization, 12 (1996), pp. 63–73. (Cited on pp. 17, 23)

[15] ———, *Multiple-load truss topology and sizing optimization: Some properties of minimax compliance*, Journal of Optimization Theory and Applications, 98 (1998), pp. 255–280. (Cited on p. 20)

[16] W. ACHTZIGER, M. BENDSØE, A. BEN-TAL, AND J. ZOWE, *Equivalent displacement based formulations for maximum strength truss topology design*, IMPACT of Computing in Science and Engineering, 4 (1992), pp. 315–345. (Cited on pp. 14, 19, 20, 21, 22, 137, 139)

[17] W. ACHTZIGER AND C. KANZOW, *Mathematical programs with vanishing constraints: Optimality conditions and constraint qualifications*, Mathematical Programming, 114 (2008), pp. 69–99. (Cited on p. 25)

[18] W. ACHTZIGER AND M. KOČVARA, *Structural topology optimization with eigenvalues*, SIAM Journal on Optimization, 18 (2007), pp. 1129–1164. (Cited on pp. 25, 141)

[19] W. ACHTZIGER AND M. STOLPE, *Global optimization of truss topology with discrete bar areas—Part II: Implementation and numerical results*, Computational Optimization and Applications, 44 (2009), pp. 315–341. (Cited on p. 145)

[20] J. ADAMS, E. BALAS, AND D. ZAWACK, *The shifting bottleneck procedure for job shop scheduling*, Management Science, 34 (1988), pp. 391–401. (Cited on p. 57)

[21] J. W. ADAMS, *FIR digital filters with least-squares stopbands subject to peak-gain constraints*, IEEE Transactions on Circuits and Systems, 39 (1991), pp. 376–388. (Cited on pp. 28, 29)

[22] J. W. ADAMS AND J. L. SULLIVAN, *Peak-constrained least-squares optimization*, IEEE Transactions on Signal Processing, 46 (1998), pp. 306–321. (Cited on p. 28)

[23] W. P. ADAMS AND H. D. SHERALI, *A tight linearization and an algorithm for zero-one quadratic programming problems*, Management Science, 32 (1986), pp. 1274–1290. (Cited on p. 282)

[24] ———, *A hierarchy of relaxations leading to the convex hull representation for general discrete optimization problems*, Annals of Operations Research, 140 (2005), pp. 21–47. (Cited on p. 282)

[25] N. ADHYA, M. TAWARMALANI, AND N. V. SAHINIDIS, *A Lagrangian approach to the pooling problem*, Industrial and Engineering Chemistry Research, 38 (1999), pp. 1956–1972. (Cited on pp. 213, 215, 216)

[26] C. S. ADJIMAN, I. P. ANDROULAKIS, AND C. A. FLOUDAS, *A global optimization method, αBB, for general twice-differentiable constrained NLPs—II Implementation and computational results*, Computers and Chemical Engineering, 22 (1998), pp. 1159–1179. (Cited on p. 283)

[27] ———, *Global optimization of mixed-integer nonlinear problems*, AIChE Journal, 46 (2000), pp. 1769–1797. (Cited on p. 283)

[28] C. S. ADJIMAN, S. DALLWIG, C. A. FLOUDAS, AND A. NEUMAIER, *A global optimization method, αBB, for general twice-differentiable constrained NLPs—I. Theoretical advances*, Computers and Chemical Engineering, 22 (1998), pp. 1137–1158. (Cited on pp. 169, 173, 283)

[29] A. AFSHAR AND A. HAGHANI, *Modeling integrated supply chain logistics in real-time large-scale disaster relief operations*, Socio-Economic Planning Sciences, 46 (2012), pp. 327–338. (Cited on p. 487)

[30] A. AGGARWAL AND C. A. FLOUDAS, *A decomposition strategy for optimum search in the pooling problem*, ORSA Journal of Computing, 2 (1990), pp. 225–235. (Cited on p. 212)

[31] R. AGRAWAL, *A method to draw fully thermally coupled distillation column configurations for multicomponent distillation*, Chemical Engineering Research and Design, 78 (2000), pp. 454–464. (Cited on p. 323)

[32] ———, *Synthesis of multicomponent distillation column configurations*, AIChE Journal, 49 (2003), pp. 379–401. (Cited on p. 323)

[33] R. AGRAWAL AND Z. T. FIDKOWSKI, *More operable arrangements of fully thermally coupled distillation columns*, AIChE Journal, 44 (1998), pp. 2565–2568. (Cited on p. 323)

[34] A. A. AHMADI, A. OLSHEVSKY, P. A. PARRILO, AND J. N. TSITSIKLIS, *NP-hardness of deciding convexity of quartic polynomials and related problems*, Mathematical Programming, 137 (2013), pp. 453–476. (Cited on p. 166)

[35] S. AHMED, *Convexity and decomposition of mean-risk stochastic programs*, Mathematical Programming, 106 (2006), pp. 433–446. (Cited on p. 391)

[36] ———, *Two-stage stochastic integer programming: A brief introduction*, in [485], 2011. (Cited on p. 391)

[37] S. AHMED AND N. V. SAHINIDIS, *Robust process planning under uncertainty*, Industrial and Engineering Chemistry Research, 37 (1998), pp. 1883–1892. (Cited on p. 393)

[38] E. AHMETOVIĆ AND I. E. GROSSMANN, *Global superstructure optimization for the design of integrated process water networks*, AIChE Journal, 57 (2011), pp. 434–457. (Cited on pp. 77, 326, 327)

[39] E. AHMETOVIĆ AND Z. KRAVANJA, *Simultaneous synthesis of process water and heat exchanger networks*, Energy, 57 (2013), pp. 236–250. (Cited on p. 327)

[40] M. AL-GWAIZ, X. CHAO, AND O. Q. WU, *Understanding how generation flexibility and renewable energy affect power market competition*, Manufacturing and Service Operations Management, published online November 16, 2016. (Cited on p. 461)

[41] F. A. AL-KHAYYAL AND J. E. FALK, *Jointly constrained biconvex programming*, Mathematics of Operations Research, 8 (1983), pp. 273–286. (Cited on p. 282)

[42] A. M. ALATTAS, I. E. GROSSMANN, AND I. PALOU-RIVERA, *Integration of nonlinear crude distillation unit models in refinery planning optimization*, Industrial and Engineering Chemistry Research, 50 (2011), pp. 6860–6870. (Cited on pp. 40, 41)

[43] ———, *Refinery production planning: Multiperiod MINLP with nonlinear CDU model*, Industrial and Engineering Chemistry Research, 51 (2012), pp. 12852–12861. (Cited on p. 327)

[44] P. ALBERTO, F. NOGUEIRA, H. ROCHA, AND L. N. VICENTE, *Pattern search methods for user-provided points: Application to molecular geometry problems*, SIAM Journal on Optimization, 14 (2004), pp. 1216–1236. (Cited on p. 499)

[45] D. L. ALDERSON, G. G. BROWN, W. M. CARLYLE, AND R. K. WOOD, *Solving defender-attacker-defender models for infrastructure defense*, in Operations Research, Computing, and Homeland Defense, R. K. Wood and R. F. Dell, eds., INFORMS, Hanover, MD, 2011, pp. 28–49. doi 10.1287/ics.2011.0047. (Cited on p. 341)

[46] D. M. ALEMAN, A. KUMAR, R. K. AHUJA, H. E. ROMEIJN, AND J. F. DEMPSEY, *Neighborhood search approaches to beam orientation optimization in intensity modulated radiation therapy treatment planning*, Journal of Global Optimization, 42 (2008), pp. 587–607. (Cited on p. 95)

[47] M. ALFAKI AND D. HAUGLAND, *Comparison of discrete and continuous models for the pooling problem*, in 11th Workshop on Algorithmic Approaches for Transportation Modelling, Optimization, and Systems, 2011, pp. 112–121. (Cited on pp. 214, 217)

[48] ——, *A multi-commodity flow formulation for the generalized pooling problem*, Journal of Global Optimization, 56 (2013), pp. 917–937. (Cited on p. 211)

[49] ——, *Strong formulations for the pooling problem*, Journal of Global Optimization, 56 (2013), pp. 897–916. (Cited on pp. 207, 211, 215, 216, 217)

[50] F. ALIZADEH, *Combinatorial Optimization with Interior Point Methods and Semidefinite Matrices*, PhD thesis, University of Minnesota, Minneapolis, MN, 1991. (Cited on p. 11)

[51] F. ALIZADEH AND D. GOLDFARB, *Second-order cone programming*, Mathematical Programming, 95 (2003), pp. 3–51. (Cited on pp. 111, 116, 154, 159, 274)

[52] B. ALKIRE AND L. VANDENBERGHE, *Convex optimization problems involving finite autocorrelation sequences*, Mathematical Programming Series A, 93 (2002), pp. 331–359. (Cited on pp. 29, 30)

[53] G. ALLAIRE AND C. CASTRO, *A new approach for the optimal distribution of assemblies in a nuclear reactor*, Numerische Mathematik, 89 (2001), pp. 1–29. (Cited on p. 292)

[54] N. ALLAUDEEN, J. L. SCHNIPPER, E. J. ORAV, R. M. WACHTER, AND A. R. VIDYARTHI, *Inability of providers to predict unplanned readmissions*, Journal of General Internal Medicine, 26 (2011), pp. 771–776. (Cited on p. 96)

[55] N. ALLAUDEEN, A. VIDYARTHI, J. MASELLI, AND A. AUERBACH, *Redefining readmission risk factors for general medicine patients*, Journal of Hospital Medicine, 6 (2011), pp. 54–60. (Cited on p. 97)

[56] R. ALMGREN, C. THUM, E. HAUPTMANN, AND H. LI, *Equity market impact*, Risk, 18 (2005), pp. 57–62. (Cited on p. 153)

[57] H. ALMUTAIRI AND S. ELHEDHLI, *A new Lagrangean approach to the pooling problem*, Journal of Global Optimization, 45 (2009), pp. 237–257. (Cited on pp. 213, 216)

[58] A. ALONSO-AYUSO, L. F. ESCUDERO, AND F. J. MARTÍN-CAMPO, *Collision avoidance in air traffic management: A mixed-integer linear optimization approach*, IEEE Transactions on Intelligent Transportation Systems, 12 (2011), pp. 47–57. (Cited on pp. 295, 300)

[59] ——, *A mixed 0-1 nonlinear optimization model and algorithmic approach for the collision avoidance in ATM: Velocity changes through a time horizon*, Computers and Operations Research, 39 (2012), pp. 3136–3146. (Cited on pp. 295, 300)

[60] A. ALONSO-AYUSO, L. F. ESCUDERO, F. J. MARTÍN-CAMPO, AND N. MLADENOVIĆ, *A VNS metaheuristic for solving the aircraft conflict detection and resolution problem by performing turn changes*, Journal of Global Optimization, 63 (2014), pp. 583–596. (Cited on pp. 295, 300, 301)

[61] O. ALSAC AND B. STOTT, *Optimal load flow with steady-state security*, IEEE Transactions on Power Apparatus and Systems, 3 (1974), pp. 745–751. (Cited on p. 191)

[62] N. ALTAY AND W. G. GREEN III, *OR/MS research in disaster operations management*, European Journal of Operational Research, 175 (2006), pp. 475–493. (Cited on p. 479)

[63] R. ALVARENGA, Ö. ERGUN, J. LI, F. MATA, N. SHEKHANI, D. SLATON, J. STONE, A. VASUDEVAN, AND E. YANG, *World Food Programme East African Corridor Optimization*, technical report, H. Milton Stewart School of Industrial and Systems Engineering, Georgia Institute of Technology, Atlanta, GA, 2010. Senior Design Final Report. (Cited on p. 487)

[64] D. ALVRAS AND M. W. PADBERG, *Linear Optimization and Extensions: Problems and Solutions*, Springer-Verlag, Berlin, Germany, 2001. (Cited on p. 72)

[65] B. ALZALG, *The algebraic structure of the arbitrary-order cone*, Journal of Optimization Theory and Applications, 169 (2016), pp. 32–49. (Cited on p. 112)

[66] L. AMBROSIO, *Transport equation and Cauchy problem for non-smooth vector fields*, in Calculus of Variations and Nonlinear Partial Differential Equations, B. Dacorogna and P. Marcellini, eds., vol. 1927 of Lecture Notes in Mathematics, Springer-Verlag, Berlin, Heidelberg, 2008. (Cited on pp. 127, 132)

[67] AMERICAN CANCER SOCIETY, http://www.cancer.org/research/cancerfacts-statistics/all-cancer-facts-figures/cancer-facts-figures-2017.html, 2017. (Cited on p. 99)

[68] AMERICAN MEDICAL ASSOCIATION, *Physician Characteristics and Distribution in the U.S.*, American Medical Association, Atlanta, Georgia, 2010. (Cited on p. 345)

[69] F. AMOS, M. RÖNNQVIST, AND G. GILL, *Modelling the pooling problem at the New Zealand Refining Company*, The Journal of the Operational Research Society, 48 (1997), pp. 767–778. (Cited on p. 207)

[70] L. T. H. AN, A. I. F. VAZ, AND L. N. VICENTE, *Optimizing radial basis functions by d.c. programming and its use in direct search for global derivative-free optimization*, TOP, 20 (2012), pp. 190–214. (Cited on p. 506)

[71] E. ANDERSEN AND K. D. ANDERSEN, *The MOSEK interior point optimizer for linear programming: An implementation of the homogeneous algorithm*, in High Performance Optimization, H. Frenk, K. Roos, T. Terlaky, and S. Zhang, eds., Kluwer Academic, 2000, pp. 197–232. (Cited on p. 117)

[72] E. D. ANDERSEN, J. GONDZIO, CS. MÉSZÁROS, AND X. XU, *Implementation of interior point methods for large scale linear programming*, in Interior Point Methods of Mathematical Programming, T. Terlaky, ed., Kluwer A. P. C., Dordrecht, Netherlands, 1996, pp. 189–252. (Cited on pp. 8, 10)

[73] E. D. ANDERSEN, B. JENSEN, R. SANDVIK, AND U. WORSØE, *The Improvements in MOSEK Version 5*, technical report 1-2007, MOSEK ApS, Fruebjergvej 3 Box 16, 2100 Copenhagen, Denmark, 2007. (Cited on p. 10)

[74] E. D. ANDERSEN, C. ROOS, AND T. TERLAKY, *Notes on duality in second order and p-order cone optimization*, Optimization, 51 (2002), pp. 627–643. (Cited on p. 113)

[75] ———, *On implementing a primal-dual interior-point method for conic quadratic optimization*, Mathematical Programming, 95 (2003), pp. 249–277. (Cited on pp. 8, 10, 111)

[76] E. D. ANDERSEN AND Y. YE, *Combining interior-point and pivoting algorithms for linear programming*, Management Science, 42 (1996), pp. 1719–1731. (Cited on pp. 9, 10)

[77] ——, *A computational study of the homogeneous algorithm for large-scale convex optimization*, Computational Optimization and Applications, 10 (1998), pp. 243–280. (Cited on pp. 8, 10)

[78] ——, *On a homogeneous algorithm for the monotone complementarity problem*, Mathematical Programming, Series A, 84 (1999), pp. 375–399. (Cited on p. 8)

[79] C. L. ANDERSON, N. BURKE, AND M. DAVISON, *Optimal management of wind energy with storage: Structural implications for policy and market design*, Journal of Energy Engineering, 141 (2014), p. B4014002. (Cited on p. 373)

[80] E. J. ANDERSON AND M. C. FERRIS, *A direct search algorithm for optimization with noisy function evaluations*, SIAM Journal on Optimization, 11 (2001), pp. 837–857. (Cited on p. 501)

[81] E. J. ANDERSON AND P. NASH, *Linear Programming in Infinite-Dimensional Vector Spaces*, John Wiley and Sons, 1987. (Cited on p. 419)

[82] W. F. ANDERSON, *Prospects for human gene therapy*, Science, 226 (1984), pp. 401–409. (Cited on pp. 175, 182)

[83] F. ANDERSSON, H. MAUSSER, D. ROSEN, AND S. URYASEV, *Credit risk optimization with conditional value-at-risk criterion*, Mathematical Programming, 89 (2001), pp. 273–291. (Cited on pp. 427, 431)

[84] M. J. ANDRECOVICH AND A. W. WESTERBERG, *An MILP formulation for heat-integrated distillation sequence synthesis*, AIChE Journal, 31 (1985), pp. 1461–1474. (Cited on p. 323)

[85] I. P. ANDROULAKIS, C. D. MARANAS, AND C. A. FLOUDAS, *αBB: A global optimization method for general constrained nonconvex problems*, Journal of Global Optimization, 7 (1995), pp. 337–363. (Cited on pp. 169, 173, 283)

[86] V. DE ANGELIS, M. MECOLI, C. NIKOI, AND G. STORCHI, *Multiperiod integrated routing and scheduling of World Food Programme cargo planes in Angola*, Computers and Operations Research, 34 (2007), pp. 1601–1615. (Cited on p. 487)

[87] M. F. ANJOS AND J. B. LASSERRE, *Handbook on Semidefinite, Conic and Polynomial Optimization*, International Series in Operations Research & Management Science, Springer-Verlag, 2011. (Cited on pp. 107, 118, 120, 608, 651, 660)

[88] K. M. ANSTREICHER, *Semidefinite programming versus the reformulation-linearization technique for nonconvex quadratically constrained quadratic programming*, Journal of Global Optimization, 43 (2009), pp. 471–484. (Cited on p. 282)

[89] A. C. ANTOULAS, D. C. SORENSEN, AND S. GUGERCIN, *A survey of model reduction methods for large-scale systems*, Contemporary Mathematics, 280 (2001), pp. 193–220. (Cited on p. 318)

[90] AON BENFIELD, *Annual Global Climate and Catastrophe Report.* http://thought leadership.aonbenfield.com/Documents/20140113_ab_if_annual_climate_catastrophe_report.pdf, 2014. (Cited on p. 479)

[91] D. L. APPLEGATE, R. E. BIXBY, V. CHVÁTAL, AND W. J. COOK, *The Traveling Salesman Problem*, Princeton University Press, 2006. (Cited on p. 53)

[92] A. APTE, *Humanitarian logistics: A new field of research and action*, Foundation and Trends in Technology, Information, and Operations Management, 3 (2009), pp. 1–100. (Cited on pp. 479, 483)

[93] C. ARAZ, H. SELIM, AND I. OZKARAHAN, *A fuzzy multi-objective covering-based vehicle location model for emergency services*, Computers and Operations Research, 34 (2007), pp. 705–726. (Cited on p. 95)

[94] K. ARIFOGLU, S. DEO, AND S. IRAVANI, *Consumption externality and yield uncertainty in the influenza vaccine supply chain: Interventions in demand and supply sides*, Management Science, 58 (2012), pp. 1072–1091. (Cited on p. 471)

[95] R. ARINGHIERI AND F. MALUCELLI, *Optimal operations management and network planning of a district heating system with a combined heat and power plant*, Annals of Operations Research, 120 (2003), pp. 173–199. (Cited on p. 304)

[96] E. ARKIN, D. JONEJA, AND R. ROUNDY, *Computational complexity of uncapacitated multi-echelon production planning problems*, Operations Research Letters, 8 (1989), pp. 61–66. (Cited on p. 442)

[97] M. ARONSSON, M. BOHLIN, AND P. KREUGER, *MILP formulations of cumulative constraints for railway scheduling—A comparative study*, in Proceedings of 9th Workshop on Algorithmic Approaches for Transportation Modeling (ATMOS), 2009. (Cited on p. 69)

[98] K. J. ARROW, T. HARRIS, AND J. MARSCHAK, *Optimal inventory policy*, Econometrica, 19 (1951), pp. 250–272. (Cited on p. 443)

[99] P. ARTZNER, F. DELBAEN, J.-M. EBER, AND D. HEATH, *Coherent measures of risk*, Mathematical Finance, 9 (1999), pp. 203–228. (Cited on pp. 350, 394, 428)

[100] J. F. ARVIS, G. RABALLAND, AND J. F. MARTEAU, *The Cost of Being Landlocked: Logistics Costs and Supply Chain Reliability*, Working Paper 4258, World Bank Policy Research, 2007. (Cited on p. 488)

[101] U. M. ASCHER AND L. R. PETZOLD, *Computer Methods for Ordinary Differential Equations and Differential Algebraic Equations*, SIAM, Philadelphia, 1998. (Cited on p. 508)

[102] M. F. ASHBY AND D. R. H. JONES, *Engineering Materials 1: An Introduction to Properties, Applications and Design*, 3rd edition, Butterworth-Heinemann Ltd, 2005. (Cited on p. 16)

[103] ASHRAE, *ANSI/ASHRAE Standard 55-2004: Thermal Environment Conditions for Human Occupancy*, American Society of Heating, Refrigerating and Air-Conditioning Engineers, 2004. (Cited on p. 263)

[104] ASPEN TECHNOLOGY, INC., *Aspen Plus User's Guide*, 2002. (Cited on p. 242)

[105] ASTRO, http://www.rtanswers.org/statistics/aboutradiationtherapy/, 2017. (Cited on p. 99)

[106] A. ATAMTÜRK AND V. NARAYANAN, *Conic mixed integer rounding cuts*, Mathematical Programming, 122 (2010), pp. 1–20. (Cited on p. 286)

[107] A. ATAMTÜRK AND M. ZHANG, *Two-stage robust network flow and design under demand uncertainty*, Operations Research, 55 (2007), pp. 662–673. (Cited on p. 333)

[108] Z. ATAN AND L. V. SNYDER, *Inventory strategies to manage supply disruptions*, in Supply Chain Disruptions: Theory and Practice of Managing Risk, H. Gurnani, A. Mehrotra, and S. Ray, eds., Springer-Verlag, London, 2012, ch. 5, pp. 115–139. (Cited on p. 448)

[109] J. P. AUBIN AND I. EKELAND, *Estimates of the duality gap in nonconvex optimization*, Mathematics of Operations Research, 1 (1976), pp. 225–245. (Cited on p. 169)

[110] C. AUDET, V. BÉCHARD, AND S. LE DIGABEL, *Nonsmooth optimization through mesh adaptive direct search and variable neighborhood search*, Journal of Global Optimization, 41 (2008), pp. 299–318. (Cited on p. 506)

[111] C. AUDET, J. BRIMBERG, P. HANSEN, S. LE DIGABEL, AND N. MLADENOVIC, *Pooling problem: Alternate formulations and solution methods*, Management Science, 50 (2004), pp. 761–776. (Cited on pp. 211, 212, 216)

[112] C. AUDET AND J. E. DENNIS, JR., *Analysis of generalized pattern searches*, SIAM Journal on Optimization, 13 (2002), pp. 889–903. (Cited on pp. 499, 500)

[113] ———, *A pattern search filter method for nonlinear programming without derivatives*, SIAM Journal on Optimization, 14 (2004), pp. 980–1010. (Cited on p. 505)

[114] ———, *Mesh adaptive direct search algorithms for constrained optimization*, SIAM Journal on Optimization, 17 (2006), pp. 188–217. (Cited on pp. 499, 500, 503, 504, 508, 509, 510, 511, 512)

[115] ———, *A progressive barrier for derivative-free nonlinear programming*, SIAM Journal on Optimization, 20 (2009), pp. 445–472. (Cited on pp. 505, 510)

[116] C. AUDET, J. E. DENNIS, JR., AND S. LE DIGABEL, *Parallel space decomposition of the mesh adaptive direct search algorithm*, SIAM Journal on Optimization, 19 (2008), pp. 1150–1170. (Cited on p. 506)

[117] ———, *Globalization strategies for mesh adaptive direct search*, Computational Optimization and Applications, 46 (2010), pp. 193–215. (Cited on p. 505)

[118] C. AUDET, P. HANSEN, B. JAUMARD, AND G. SAVARD, *A branch and cut algorithm for nonconvex quadratically constrained programming*, Mathematical Programming Series A, 87 (2000), pp. 131–152. (Cited on p. 212)

[119] C. AUDET, P. HANSEN, AND G. SAVARD, eds., *Essays and Surveys in Global Optimization*, Springer, 2005. (Cited on p. 173)

[120] C. AUDET, G. SAVARD, AND W. ZGHAL, *A mesh adaptive direct search algorithm for multiobjective optimization*, European Journal of Operational Research, 204 (2010), pp. 545–556. (Cited on p. 506)

[121] F. AUGUSTIN AND Y. M. MARZOU, *NOWPAC: A Provably Convergent Nonlinear Optimizer with Path-Augmented Constraints for Noisy Regimes*, arXiv:1403.1931v1, 2014. (Cited on p. 506)

[122] M. P. AVRAAM, N. SHAH, AND C. C. PANTELIDES, *Modelling and optimisation of general hybrid systems in the continuous time domain*, Computers and Chemical Engineering, 22 (1998), pp. S221–S228. (Cited on p. 328)

[123] S. AXSÄTER, *Using the deterministic EOQ formula in stochastic inventory control*, Management Science, 42 (1996), pp. 830–834. (Cited on p. 448)

[124] ———, *Inventory Control*, 2nd edition, Springer, 2006. (Cited on p. 439)

[125] M. BAGAJEWICZ, *A review of recent design procedures for water networks in refineries and process plants*, Computers and Chemical Engineering, 24 (2000), pp. 2093–2113. (Cited on p. 326)

[126] M. BAGAJEWICZ AND D. C. FARIA, *On the appropriate architecture of the water/wastewater allocation problem in process plants*, Computer Aided Chemical Engineering, 26 (2009), pp. 1–20. (Cited on p. 326)

[127] M. BAGAJEWICZ AND V. MANOUSIOUTHAKIS, *Mass/heat-exchange network representation of distillation networks*, AIChE Journal, 38 (1992), pp. 1769–1800. (Cited on p. 327)

[128] M. BAGAJEWICZ, R. PHAM, AND V. MANOUSIOUTHAKIS, *On the state space approach to mass/heat exchanger network design*, Chemical Engineering Science, 53 (1998), pp. 2595–2621. (Cited on p. 327)

[129] M. BAGAJEWICZ, M. RIVAS, AND M. J. SAVELSKI, *A robust method to obtain optimal and sub-optimal design and retrofit solutions of water utilization systems with multiple contaminants in process plants*, Computers and Chemical Engineering, 24 (2000), pp. 1461–1466. (Cited on p. 326)

[130] M. BAGAJEWICZ AND M. SAVELSKI, *On the use of linear models for the design of water utilization systems in process plants with a single contaminant*, Chemical Engineering Research and Design, 79 (2001), pp. 600–610. (Cited on p. 326)

[131] X. BAI, K. SCHEINBERG, AND R. TÜTÜNCÜ, *Least-squares approach to risk parity in portfolio selection*, Quantitative Finance, 16 (2016), pp. 357–376. (Cited on p. 159)

[132] Y. BAI, T. HWANG, S. KANG, AND Y. OUYANG, *Biofuel refinery location and supply chain planning under traffic congestion*, Transportation Research Part B, 45 (2011), pp. 162–175. (Cited on p. 466)

[133] Y. BAI, Y. OUYANG, AND J. S. PANG, *Enhanced models and improved solution for competitive biofuel supply chain design under land use constraints*, European Journal of Operational Research, 249(1) (2016), pp. 281–297. (Cited on p. 464)

[134] ———, *Biofuel supply chain design under competitive agricultural land use and feedstock market equilibrium*, Energy Economics, 34 (2012), pp. 1623–1633. (Cited on p. 464)

[135] R. BAKER AND C. L. E. SWARTZ, *Interior point solution of multilevel quadratic programming problems in constrained model predictive control applications*, Industrial and Engineering Chemistry Research, 47 (2008), pp. 81–91. (Cited on p. 43)

[136] T. E. BAKER AND L. S. LASDON, *Successive linear programming at Exxon*, Management Science, 31 (1985), pp. 264–274. (Cited on pp. 39, 40, 212)

[137] S. BALAKRISHNA AND L. T. BIEGLER, *Constructive targeting approaches for the synthesis of chemical reactor networks*, Industrial and Engineering Chemistry Research, 31 (1992), pp. 300–312. (Cited on p. 320)

[138] ———, *Targeting strategies for the synthesis and energy integration of nonisothermal reactor networks*, Industrial and Engineering Chemistry Research, 31 (1992), pp. 2152–2164. (Cited on p. 320)

[139] ———, *A unified approach for the simultaneous synthesis of reaction, energy, and separation systems*, Industrial and Engineering Chemistry Research, 32 (1993), pp. 1372–1382. (Cited on p. 320)

[140] A. BALAKRISHNAN, T. L. MAGNANTI, AND R. T. WONG, *A dual-ascent procedure for large-scale uncapacitated network design*, Operations Research, 37 (1989), pp. 716–740. (Cited on p. 450)

[141] E. BALAS, *An additive algorithm for solving linear programs with 0-1 variables*, Operations Research, 13 (1965), pp. 517–546. (Cited on p. 53)

[142] ———, *Intersection cuts—A new type of cutting planes for integer programming*, Operations Research, 19 (1971), pp. 19–39. (Cited on p. 54)

[143] ———, *Disjunctive programming*, Annals of Discrete Mathematics, 5 (1979), pp. 3–51. (Cited on pp. 54, 56, 69, 286)

[144] ———, *Disjunctive programming: Properties of the convex hull of feasible points*, Discrete Applied Mathematics, 89 (1998), pp. 1–44. (Cited on pp. 54, 56)

[145] E. BALAS AND M. C. CARRERA, *A dynamic subgradient-based branch and bound procedure for set covering*, Operations Research, 44 (1996), pp. 875–890. (Cited on p. 57)

[146] E. BALAS, S. CERIA, AND G. CORNUÉJOLS, *A lift-and-project cutting plane algorithm for mixed 0-1 programs*, Mathematical Programming, 58 (1993), pp. 295–324. (Cited on pp. 55, 56, 286, 287)

[147] ———, *Mixed 0-1 programming by lift-and-project in a branch-and-cut framework*, Management Science, 42 (1996), pp. 1229–1246. (Cited on p. 56)

[148] E. BALAS, S. CERIA, G. CORNUÉJOLS, AND N. NATRAJ, *Gomory cuts revisited*, Operations Research Letters, 19 (1996), pp. 1–9. (Cited on p. 56)

[149] E. BALAS AND M. PERREGAARD, *A precise correspondence between lift-and-project cuts, simple disjunctive cuts, and mixed integer Gomory cuts for 0-1 programming*, Mathematical Programming, Series B, 94 (2003), pp. 221–245. (Cited on p. 56)

[150] B. BALASUNDARAM AND S. BUTENKO, *Constructing test functions for global optimization using continuous formulations of graph problems*, Optimization Methods and Software, 20 (2005), pp. 439–452. (Cited on p. 165)

[151] ———, *On a polynomial fractional formulation for independence number of a graph*, Journal of Global Optimization, 35 (2006), pp. 405–421. (Cited on p. 165)

[152] S. BALAY et al., *PETSc Users-Manual*, Tech. Report ANL-95/11 -Revision 3.7, Argonne National Laboratory, 2016. (Cited on p. 535)

[153] B. BALCIK AND B. BEAMON, *Facility location in humanitarian relief*, International Journal of Logistics: Research and Applications, 11 (2008), pp. 101–121. (Cited on p. 480)

[154] R. BALDICK, *The generalized unit commitment problem*, IEEE Transactions on Power Systems, 10 (1995), pp. 465–473. (Cited on p. 360)

[155] R. C. BALIBAN, J. A. ELIA, AND C. A. FLOUDAS, *Toward novel hybrid biomass, coal, and natural gas processes for satisfying current transportation fuel demands, 1: Process alternatives, gasification modeling, process simulation, and economic analysis*, Industrial and Engineering Chemistry Research, 49 (2010), pp. 7343–7370. (Cited on p. 318)

[156] M. L. BALINSKI AND A. W. TUCKER, *Duality theory of linear programs: A constructive approach with applications*, SIAM Review, 11 (1969), pp. 347–377. (Cited on p. 9)

[157] A. S. BANDEIRA, K. SCHEINBERG, AND L. N. VICENTE, *Computation of sparse low degree interpolating polynomials and their application to derivative-free optimization*, Mathematical Programming, 134 (2012), pp. 223–257. (Cited on pp. 497, 502, 530)

[158] ———, *Convergence of trust-region methods based on probabilistic models*, SIAM Journal on Optimization, 24 (2014), pp. 1238-1264. (Cited on p. 502)

[159] X. BAO AND N. V. SAHINIDIS, *Finite algorithms for global minimization of separable concave programs*, in [1445], pp. 17–30. (Cited on p. 170)

[160] X. BAO, N. V. SAHINIDIS, AND M. TAWARMALANI, *Multiterm polyhedral relaxations for nonconvex, quadratically constrained quadratic programs*, Optimization Methods and Software, 24 (2009), pp. 485–504. (Cited on pp. 282, 289)

[161] ———, *Semidefinite relaxations for quadratically constrained quadratic programming: A review and comparisons*, Mathematical Programming, 129 (2011), pp. 129–157. (Cited on p. 282)

[162] R. BARANIUK, *Compressive sensing*, IEEE Signal Processing Magazine, 24 (2007), pp. 118–121. (Cited on pp. 31, 33)

[163] J. F. BARD AND H. W. PURNOMO, *Preference scheduling for nurses using column generation*, European Journal of Operational Research, 164 (2005), pp. 510–534. (Cited on p. 96)

[164] R. J. BARTHELMIE, K. HANSEN, S. T. FRANDSEN, O. RATHMANN, J. G. SCHEPERS, W. SCHLEZ, J. PHILLIPS, K. RADOS, A. ZERVOS, E. S. POLITIS, AND P. K. CHAVIAROPOULOS, *Modelling and measuring flow and wind turbine wakes in large wind farms offshore*, Wind Energy, 12 (2009), pp. 431–444. (Cited on p. 368)

[165] P. I. BARTON AND C. K. LEE, *Modeling, simulation, sensitivity analysis, and optimization of hybrid systems*, ACM Transactions on Modeling and Computer Simulation (TOMACS), 12 (2002), pp. 256–289. (Cited on p. 328)

[166] ———, *Design of process operations using hybrid dynamic optimization*, Computers and Chemical Engineering, 28 (2004), pp. 955–969. (Cited on p. 328)

[167] M. BARTTFELD, P. A. AGUIRRE, AND I. E. GROSSMANN, *Alternative representations and formulations for the economic optimization of multicomponent distillation columns*, Computers and Chemical Engineering, 27 (2003), pp. 363–383. (Cited on p. 321)

[168] K. BARTY, P. CARPENTIER, G. COHEN, AND P. GIRARDEAU, *Price Decomposition in Large-Scale Stochastic Optimal Control*, technical report, 2010. arXiv:1012.2092. (Cited on p. 411)

[169] A. BARVINOK, *A Course in Convexity*, American Mathematical Society, Providence, RI, 2002. (Cited on p. 125)

[170] J. H. BASS, M. REBBECK, L. LANDBERG, AND A. HUNTER, *An Improved Measure-Correlate-Predict Algorithm for the Prediction of the Long Term Wind Climate in Regions of Complex Environment*, technical report May 1998, Renewable Energy Systems Limited (UK), Risø National Laboratory (Denmark), Ecotècnia (Spain), University Of Sunderland (UK), 2000. (Cited on pp. 368, 370)

[171] M. H. BAUER AND J. STICHLMAIR, *Design and economic optimization of azeotropic distillation processes using mixed-integer nonlinear programming*, Computers and Chemical Engineering, 22 (1998), pp. 1271–1286. (Cited on pp. 318, 321)

[172] B. T. BAUMRUCKER, J. G. RENFRO, AND L. T. BIEGLER, *MPEC problem formulations and solution strategies with chemical engineering applications*, Computers and Chemical Engineering, 32 (2008), pp. 2903–2913. (Cited on p. 243)

[173] H. H. BAUSCHKE, W. L. HARE, AND W. M. MOURSI, *A derivative-free comirror algorithm for convex optimization*, Optimization Methods and Software, 30 (2015), pp. 706–726. (Cited on p. 500)

[174] E. BEALE, *Cycling in the dual simplex method*, Naval Research Logistics Quarterly, 2(4) (1955), pp. 269–276. (Cited on p. 6)

[175] ———, *On minimizing a convex function subject to linear inequalities*, Journal of the Royal Statistical Society. Series B (Methodological), 17 (1955), pp. 173–184. (Cited on p. 390)

[176] E. BEALE AND J. TOMLIN, *Special facilities in a general mathematical programming system for nonconvex problems using ordered sets of variables.*, in Proceedings of Fifth International Conference on Operational Research, J. Lawrence, ed., Tavistock Publications, London, 1970, pp. 447–454. (Cited on p. 284)

[177] B. BEAMON AND B. BALCIK, *Performance evaluation in humanitarian supply chains*, International Journal of Public Sector Management, 21 (2008), pp. 4–25. (Cited on p. 483)

[178] B. BEAMON AND S. A. KOTLEBA, *Inventory management support systems for emergency humanitarian relief operations in South Sudan*, The International Journal of Logistics Management, 17 (2006), pp. 187–212. (Cited on p. 484)

[179] ———, *Inventory modelling for complex emergencies in humanitarian relief operations*, International Journal of Logistics: Research and Applications, 9 (2006), pp. 1–18. (Cited on p. 484)

[180] N. BEAUMONT, *An algorithm for disjunctive programs*, European Journal of Operational Research, 48 (1990), pp. 362–371. (Cited on p. 316)

[181] A. BECK AND A. BEN-TAL, *Duality in robust optimization: primal worst equals dual best*, Operations Research Letters, 37 (2009), pp. 1–6. (Cited on p. 337)

[182] A. BECK AND M. TEBOULLE, *Global optimality conditions for quadratic optimization problems with binary constraints*, SIAM Journal on Optimization, 11 (2000), pp. 179–188. (Cited on p. 168)

[183] N. I. BEDENIK, B. PAHOR, AND Z. KRAVANJA, *An integrated strategy for the hierarchical multilevel MINLP synthesis of overall process flowsheets using the combined synthesis/analysis approach*, Computers and Chemical Engineering, 28 (2004), pp. 693–706. (Cited on p. 318)

[184] G. BEDNARZ, D. MICHALSKI, C. HOUSER, M. S. HUQ, Y. XIAO, P. R. ANNE, AND J. M. GALVIN, *The use of mixed-integer programming for inverse treatment planning with pre-defined field segments*, Physics in Medicine and Biology, 47 (2002), p. 2235. (Cited on p. 94)

[185] L. L. BEGHINI, A. BEGHINI, N. KATZ, W. F. BAKER, AND G. H. PAULINO, *Connecting architecture and engineering through structural topology optimization*, Engineering Structures, 59 (2014), pp. 716–726. (Cited on p. 13)

[186] J. BELIËN AND H. FORCÉ, *Supply chain management of blood products: A literature review*, European Journal of Operational Research, 217 (2012), pp. 1–16. (Cited on p. 485)

[187] M. L. BELLOWS, H. K. FUNG, M. S. TAYLOR, C. A. FLOUDAS, A. LÓPEZ DE VICTORIA, AND D. MORIKIS, *New compstatin variants through two de novo protein design frameworks*, Biophysical Journal, 98 (2010), pp. 2337–2346. (Cited on pp. 176, 182)

[188] M. L. BELLOWS, M. S. TAYLOR, P. A. COLE, L. SHEN, R. F. SILICIANO, H. K. FUNG, AND C. A. FLOUDAS, *Discovery of entry inhibitors for HIV-1 via a new de novo protein design framework*, Biophysical Journal, 99 (2010), pp. 3445–3453. (Cited on pp. 176, 182)

[189] P. BELOTTI, *Couenne: A User's Manual*, https://projects.coin-or.org/Couenne/browser/trunk/Couenne/doc/couenne-user-manual.pdf?format=raw. Lehigh University. (Cited on pp. 304, 313)

[190] ——, *Disjunctive cuts for nonconvex MINLP*, in [1160], pp. 117–144. (Cited on p. 287)

[191] P. BELOTTI, S. CAFIERI, J. LEE, AND L. LIBERTI, *Feasibility-based bounds tightening via fixed points*, in Combinatorial Optimization and Applications, Springer, 2010, pp. 65–76. (Cited on p. 281)

[192] P. BELOTTI, C. KIRCHES, S. LEYFFER, J. LINDEROTH, J. LUEDTKE, AND A. MAHAJAN, *Mixed-integer nonlinear optimization*, Acta Numerica, 22 (2013), pp. 1–131. (Cited on pp. 91, 274, 285, 315)

[193] P. BELOTTI, J. LEE, L. LIBERTI, F. MARGOT, AND A. WÄCHTER, *Branching and bounds tightening techniques for non-convex MINLP*, Optimization Methods and Software, 24 (2009), pp. 597–634. (Cited on pp. 196, 281, 285, 289)

[194] A. BEMPORAD AND M. MORARI, *Control of systems integrating logic, dynamics, and constraints*, Automatica, 35 (1999), pp. 407–427. (Cited on pp. 77, 328)

[195] O. BEN-AYED AND C. E. BLAIR, *Computational difficulties of bilevel linear programming*, Operations Research, 38 (1990), pp. 556–560. (Cited on p. 362)

[196] A. BEN-TAL, D. BERTSIMAS, AND D. B. BROWN, *A soft robust model for optimization under ambiguity*, Operations Research, 58 (2010), pp. 1220–1234. (Cited on p. 333)

[197] A. BEN-TAL, S. BOYD, AND A. NEMIROVSKI, *Extending scope of robust optimization*, Mathematical Programming, Series B, 107 (2006), pp. 63–89. (Cited on pp. 340, 342)

[198] A. BEN-TAL, D. DEN HERTOG, A. D. WAEGENAERE, B. MELENBERG, AND G. RENNEN, *Robust solutions of optimization problems affected by uncertain probabilities*, Management Science, 59 (2013), pp. 341–357. (Cited on p. 375)

[199] A. BEN-TAL, D. DEN HERTOG, AND J.-P. VIAL, *Deriving robust counterparts of nonlinear uncertain inequalities*, Mathematical Programming, 149 (2015), pp. 265–299. (Cited on pp. 338, 339, 341)

[200] A. BEN-TAL, B. D. CHUNG, S. R. MANDALA, AND T. YAO, *Robust optimization for emergency logistics planning: Risk mitigation in humanitarian relief supply chains*, Transportation Research Part B: Methodological, 45 (2011), pp. 1177–1189. (Cited on p. 333)

[201] A. BEN-TAL, G. EIGER, AND V. GERSHOVITZ, *Global optimization by reducing the duality gap*, Mathematical Programming, 63 (1994), pp. 193–212. (Cited on pp. 169, 210, 215, 216, 283)

[202] A. BEN-TAL, L. EL GHAOUI, AND A. NEMIROVSKI, *Robust optimization methodology and applications*, Mathematical Programming, 92 (2002), pp. 453–480. (Cited on p. 66)

[203] ——, *Robust Optimization*, Princeton Series in Applied Mathematics, Princeton University Press, 2009. (Cited on pp. 149, 156, 358)

[204] A. BEN-TAL, B. GOLANY, A. NEMIROVSKI, AND J.-P. VIAL, *Retailer-supplier flexible commitments contracts: A robust optimization approach*, Manufacturing and Service Operations Management, 7 (2005), pp. 248–271. (Cited on p. 333)

[205] A. BEN-TAL, A. GORYASHKO, E. GUSLITZER, AND A. NEMIROVSKI, *Adjustable robust solutions of uncertain linear programs*, Mathematical Programming, Series A, 99 (2004), pp. 351–376. (Cited on pp. 339, 340, 341, 342)

[206] A. BEN-TAL, M. KOČVARA, A. NEMIROVSKI, AND J. ZOWE, *Free material design via semidefinite programming: The multiload case with contact conditions*, SIAM Review, 42 (2000), pp. 695–715. (Cited on p. 144)

[207] A. BEN-TAL AND A. NEMIROVSKII, *Potential reduction polynomial time method for truss topology design*, SIAM Journal on Optimization, 4 (1994), pp. 596–612. (Cited on p. 25)

[208] ———, *Robust truss topology design via semidefinite programming*, SIAM Journal on Optimization, 7 (1997), pp. 991–1016. (Cited on pp. 25, 333)

[209] ———, *Robust solutions of uncertain linear programs*, Operations Research Letters, 25 (1999), pp. 1–13. (Cited on pp. 335, 336, 349, 352, 394)

[210] ———, *Robust solutions of linear programming problems contaminated with uncertain data*, Mathematical Programming, Series A, 88 (2000), p. 411. (Cited on p. 394)

[211] ———, *Lectures on Modern Convex Optimization: Analysis, Algorithms, and Engineering Applications*, MPS-SIAM Series on Optimization, SIAM, Philadelphia, 2001. (Cited on pp. 11, 141)

[212] ———, *Robust optimization—methodology and applications*, Mathematical Programming, Series B, 92 (2002), pp. 453–480. (Cited on pp. 333, 334, 335)

[213] ———, *Selected topics in robust convex optimization*, Mathematical Programming, 112 (2008), pp. 125–158. (Cited on p. 334)

[214] A. BEN-TAL, A. NEMIROVSKI, AND J. ZOWE, *Interior Point Polynomial Time Method for Truss Topology Design*, research report 3/92, TECHNION, Haifa, 1992. (Cited on p. 137)

[215] J. BENBASSAT AND M. TARAGIN, *Hospital readmissions as a measure of quality of health care: Advantages and limitations*, Archives of Internal Medicine, 160 (2000), pp. 1074–1081. (Cited on p. 96)

[216] A. BEN-TAL AND M. P. BENDSØE, *A new method for optimal truss topology design*, SIAM Journal on Optimization, 3 (1993), pp. 322–358. (Cited on p. 20)

[217] M. P. BENDSØE, A. BEN-TAL, AND J. ZOWE, *Optimization methods for truss geometry and topology design*, Structural Optimization, 7 (1994), pp. 141–159. (Cited on p. 22)

[218] M. P. BENDSØE, G. I. N. ROZVANY, AND U. KIRSCH, *Layout optimization of structures*, Applied Mechanics Reviews, 48 (1995), pp. 41–119. (Cited on p. 22)

[219] M. P. BENDSØE AND O. SIGMUND, *Topology Optimization: Theory, Methods and Applications*, Springer, 2003. (Cited on pp. 13, 19, 20)

[220] S. BENJAAFAR, Y. LI, D. XU, AND S. ELHEDHLI, *Demand allocation in systems with multiple inventory locations and multiple demand sources*, Manufacturing and Service Operations Management, 10 (2008), pp. 43–60. (Cited on p. 484)

[221] K. P. BENNETT AND O. L. MANGASARIAN, *Robust linear programming discrimination of two linearly inseparable sets*, Optimization Methods and Software, 1 (1992), pp. 23–34. (Cited on p. 30)

[222] H. P. BENSON, *Concave minimization: Theory, applications, and algorithms*, in [957], pp. 43–148. (Cited on p. 171)

[223] ———, *Deterministic algorithms for constrained concave minimization: A unified critical survey*, Naval Research Logistics, 43 (1996), pp. 765–795. (Cited on p. 171)

[224] H. Y. BENSON, *Mixed integer nonlinear programming using interior-point methods*, Optimization Methods and Software, 26 (2011), pp. 911–931. (Cited on p. 289)

[225] H. Y. BENSON AND U. SAGLAM, *Mixed-integer second-order cone programming: A survey*, in Tutorials in Operations Research, INFORMS, 2013, pp. 13–36. (Cited on p. 274)

[226] S. J. BENSON AND Y. YE, *Algorithm 875: DSDP5–software for semidefinite programming*, ACM Transactions on Mathematical Software (TOMS), 34 (2008), Article 16. (Cited on p. 118)

[227] S. J. BENSON, Y. YE, AND X. ZHANG, *Solving large-scale sparse semidefinite programs for combinatorial optimization*, SIAM Journal on Optimization, 10 (2000), pp. 443–461. (Cited on p. 118)

[228] M. L. BERGAMINI, P. AGUIRRE, AND I. GROSSMANN, *Logic-based outer approximation for globally optimal synthesis of process networks*, Computers and Chemical Engineering, 29 (2005), pp. 1914–1933. (Cited on p. 284)

[229] F. V. BERGHEN, *CONDOR: A Constrained, Non-Linear, Derivative-Free Parallel Optimizer for Continuous, High Computing Load, Noisy Objective Functions*, PhD thesis, Université Libre de Bruxelles, 2004. (Cited on p. 505)

[230] A. B. BERKELAAR, B. JANSEN, C. ROOS, AND T. TERLAKY, *Basis- and tripartition identification for quadratic programming and linear complementarity problems. From an interior point solution to an optimal basis and vice versa*, Mathematical Programming, 86 (1999), pp. 261–282. (Cited on pp. 4, 8, 9)

[231] D. BERKOUNE, J. RENAUD, M. REKIK, AND A. RUIZ, *Transportation in disaster response operations*, Socio-Economic Planning Sciences, 46 (2012), pp. 23–32. (Cited on p. 487)

[232] H. M. BERMAN, J. WESTBROOK, Z. FENG, G. GILLILAND, T. BHAT, H. WEISSIG, I. N. SHINDYALOV, AND P. E. BOURNE, *The protein data bank*, Nucleic Acids Research, 28 (2000), pp. 235–242. (Cited on p. 526)

[233] O. BERMAN, D. KRASS, AND M. B. C. MENEZES, *Facility reliability issues in network p-median problems: Strategic centralization and co-location effects*, Operations Research, 55 (2007), pp. 332–350. (Cited on p. 480)

[234] F. BERNSTEIN AND A. FEDERGRUEN, *Pricing and replenishment strategies in a distribution system with competing retailers*, Operations Research, 51 (2003), pp. 409–426. (Cited on p. 484)

[235] ———, *Decentralized supply chains with competing retailers under demand uncertainty*, Management Science, 51 (2005), pp. 18–29. (Cited on p. 484)

[236] R. BERRETTA, A. MENDES, AND P. MOSCATO, *Integer programming models and algorithms for molecular classification of cancer from microarray data*, in Proceedings of the Twenty-eighth Australasian conference on Computer Science—Volume 38, Australian Computer Society, Inc, 2005. (Cited on p. 93)

[237] J. BERRY, W. E. HART, C. A. PHILLIPS, J. G. UBER, AND J. P. WATSON, *Sensor placement in municipal water networks with temporal integer programming models*, Journal of Water Resources Planning and Management, 132 (2006), pp. 218–224. (Cited on p. 480)

[238] L. BERTACCO, M. FISCHETTI, AND A. LODI, *A feasibility pump heuristic for general mixed-integer programs*, Discrete Optimization, 4 (2007), pp. 63–76. (Cited on p. 57)

[239] T. BERTHOLD, *RENS*, Mathematical Programming Computation, 6 (2014), pp. 33–54. (Cited on p. 286)

[240] T. BERTHOLD AND A. M. GLEIXNER, *Undercover: A primal MINLP heuristic exploring a largest sub-MIP*, Mathematical Programming, 144 (2014), pp. 315–346. (Cited on p. 286)

[241] M. BERTOLLI, T. PAPENBROCK, AND S. M. WILD, *Occupation number-based energy functional for nuclear masses*, Physical Review C, 85 (2012), p. 014322. (Cited on pp. xxii, 537, 538)

[242] D. P. BERTSEKAS, *Nonlinear Programming*, 2nd edition Athena Scientific, September 1999. (Cited on p. 222)

[243] D. P. BERTSEKAS AND R. GALLAGER, *Data Networks*, 2nd edition, Prentice-Hall, Upper Saddle River, NJ, USA, 1992. (Cited on p. 292)

[244] D. BERTSIMAS AND D. B. BROWN, *Constructing uncertainty sets for robust linear optimization*, Operations Research, 57(6) (2009), pp. 1483–1495. (Cited on p. 344)

[245] D. BERTSIMAS, D. B. BROWN, AND C. CARAMANIS, *Theory and applications of robust optimization*, SIAM Review, 53 (2011), pp. 464–501. (Cited on pp. 91, 334, 335, 340, 370)

[246] D. BERTSIMAS AND C. CARAMANIS, *Finite adaptability in multistage linear optimization*, IEEE Transactions on Automatic Control, 55(12) (2010), pp. 2751–2766. (Cited on pp. 333, 340, 342, 343)

[247] D. BERTSIMAS AND I. DUNNING, *Multistage robust mixed-integer optimization with adaptive partitions*, Operations Research, 64 (2016), pp. 980–998. (Cited on p. 343)

[248] D. BERTSIMAS AND V. GOYAL, *On the power and limitations of affine policies in two-stage adaptive optimization*, Mathematical Programming, Series A, 134 (2012), pp. 491–531. (Cited on pp. 333, 340, 341, 342)

[249] D. BERTSIMAS, V. GOYAL, AND X. A. SUN, *A geometric characterization of the power of finite adaptability in multistage stochastic and adaptive optimization*, Mathematics of Operations Research, 36(1) (2011), pp. 24–54. (Cited on pp. 333, 340, 341, 343)

[250] D. BERTSIMAS, V. GUPTA, AND N. KALLUS, *Data-Driven Robust Optimization*, submitted to Operations Research, 2013. (Cited on p. 344)

[251] D. BERTSIMAS, D. A. IANCU, AND P. A. PARRILO, *Optimality of affine policies in multistage robust optimization*, Mathematics of Operations Research, 35(2) (2010), pp. 363–394. (Cited on pp. 333, 342)

[252] ——, *A hierarchy of near-optimal policies for multistage adaptive optimization*, IEEE Transactions on Automatic Control, 56(12) (2011), pp. 2809–2824. (Cited on pp. 340, 342)

[253] D. BERTSIMAS, E. LITVINOV, X. A. SUN, J. ZHAO, AND T. ZHENG, *Adaptive robust optimization for the security constrained unit commitment problem*, IEEE Transactions on Power Systems, 28 (2013), pp. 52–63. (Cited on pp. xxiv, 333, 358, 360, 362, 363)

[254] D. BERTSIMAS, O. NOHADANI, AND K. M. TEO, *Robust optimization in electromagnetic scattering problems*, Journal of Applied Physics, 101 (2007), p. 074507. (Cited on p. 333)

[255] ——, *Nonconvex robust optimization for problems with constraints*, INFORMS Journal on Computing, 22 (2010), pp. 44–58. (Cited on p. 334)

[256] D. BERTSIMAS AND R. SHIODA, *Classification and regression via integer optimization*, Operations Research, 55 (2007), pp. 252–271. (Cited on p. 94)

[257] D. BERTSIMAS AND M. SIM, *The price of robustness*, Operations Research, 52 (2004), pp. 35–53. (Cited on pp. 333, 336, 337, 349, 394)

[258] M. BESTER, I. NIEUWOUDT, AND J. H. VAN VUUREN, *Finding good nurse duty schedules: A case study*, Journal of Scheduling, 10 (2007), pp. 387–405. (Cited on p. 96)

[259] J. T. BETTS, *Practical Methods for Optimal Control and Estimation Using Nonlinear Programming*, SIAM Series on Advances in Design and Control 19, Philadelphia, 2010. (Cited on p. 268)

[260] L. T. BIEGLER, *An overview of simultaneous strategies for dynamic optimization*, Chemical Engineering and Processing: Process Intensification, 46 (2007), pp. 1043–1053. (Cited on p. 328)

[261] ——, *Nonlinear Programming: Concepts, Algorithms and Applications to Chemical Processes*, SIAM, Philadelphia, 2010. (Cited on pp. xviii, 222, 238, 241, 268)

[262] L. T. BIEGLER AND I. E. GROSSMANN, *Retrospective on optimization*, Computers and Chemical Engineering, 28 (2004), pp. 1169–1192. (Cited on p. 316)

[263] L. T. BIEGLER AND V. M. ZAVALA, *Large-scale nonlinear programming using IPOPT: An integrating framework for enterprise-wide dynamic optimization*, Computers and Chemical Engineering, 33 (2009), pp. 575–582. (Cited on p. 259)

[264] D. BIENSTOCK, *Computational study of a family of mixed-integer quadratic programming problems*, Mathematical Programming, 74 (1996), pp. 121–140. (Cited on p. 292)

[265] ——, *Progress on solving power flow problems*, Optima: Mathematical Optimization Society Newsletter, 93 (2013), pp. 1–8. (Cited on pp. 187, 189)

[266] D. BIENSTOCK AND S. MATTIA, *Using mixed-integer programming to solve power grid blackout problems*, Discrete Optimization, 4, pp. 115–141. (Cited on p. 292)

[267] M. BILBAO AND E. ALBA, *GA and PSO applied to wind energy optimization*, in CACIC Conference Proceedings, Jujuy, Argentina, 2009. (Cited on p. 367)

[268] ——, *Simulated annealing for optimization of wind farm annual profit*, in 2009 2nd International Symposium on Logistics and Industrial Informatics, IEEE, September 2009, pp. 1–5. (Cited on p. 367)

[269] J. BILLINGS, J. DIXON, T. MIJANOVICH, AND D. WENNBERG, *Case finding for patients at risk of readmission to hospital: Development of algorithm to identify high risk patients*, BMJ, 333 (2006), p. 327. (Cited on p. 96)

[270] C. BILLINGTON AND B. JOHNSON, *Creating and leveraging options in the high technology supply chain*, in The Practice of Supply Chain Management: Where Theory and Application Converge, T. P. Harrison, H. L. Lee, and J. J. Neale, eds., Kluwer Academic, 2003, pp. 157–174. (Cited on p. 458)

[271] R. BILLINTON AND M. FOTUHI-FIRUZABAD, *A reliability framework for generating unit commitment*, Electric Power Systems Research, 56 (2000), pp. 81–88. (Cited on p. 360)

[272] R. BILLINTON AND R. KARKI, *Capacity reserve assessment using system well-being analysis*, IEEE Transactions on Power Systems, 14 (1999), pp. 433–438. (Cited on p. 360)

[273] S. C. BILLUPS, J. LARSON, AND P. GRAF, *Derivative-free optimization of expensive functions with computational error using weighted regression*, SIAM Journal on Optimization, 23 (2013), pp. 27–53. (Cited on p. 501)

[274] T. A. BINKOWSKI, P. FREEMAN, AND J. LIANG, *pvSOAR: Detecting similar surface patterns of pocket and void surfaces of amino acid residues on proteins*, Nucleic Acids Research, 32 (2004), pp. W555–8. (Cited on p. 520)

[275] S. I. BIRBIL, J. B. G. FRENK, J. A. S. GROMICHO, AND S. ZHANG, *The role of robust optimization in single-leg airline revenue management*, Management Science, 55 (2009), pp. 148–163. (Cited on p. 334)

[276] S. I. BIRBIL, J. FRENK, B. KAYNAR, AND N. NOYAN, *Risk measures and their applications in asset management*, in The VaR Implementation Handbook, G. N. Gregoriou, ed., The McGraw-Hill Companies, New York, 2009, pp. 311–337. (Cited on p. 431)

[277] J. R. BIRGE, *Decomposition and partitioning methods for multistage stochastic linear programs*, Operations Research, 33 (1985), pp. 989–1007. (Cited on p. 391)

[278] J. R. BIRGE, F. X. KÄRTNER, AND O. NOHADANI, *Improving thin-film manufacturing yield with robust optimization*, Applied Optics, 50 (2011), pp. C36–C40. (Cited on p. 333)

[279] J. R. BIRGE AND F. LOUVEAUX, *Introduction to Stochastic Programming*, Springer-Verlag, New York, 2nd ed., 2011. (Cited on pp. 391, 427)

[280] A. BISCHI, L. TACCARI, E. MARTELLI, E. AMALDI, G. MANZOLINI, P. SILVA, S. CAMPANARI, AND E. MACCHI, *A detailed MILP optimization model for combined cooling, heat and power system operation planning*, Energy, 74 (2014), pp. 12–26. (Cited on pp. 304, 314)

[281] ———, *A rolling-horizon MILP optimization method for the operational scheduling of cogeneration systems with incentives*, in Proceedings of the 28th International Conference on Efficiency, Cost, Optimization, Simulation and Environmental Impact of Energy Systems, ECOS, 2015. (Cited on p. 313)

[282] D. BISH, E. CHAMBERLAYNE, AND H. RAKHA, *Optimizing network flows with congestion-based flow reductions*, Networks and Spatial Economics, 13 (2013), pp. 283–306. (Cited on p. 488)

[283] J. BISSCHOP AND R. ENTRIKEN, *AIMMS The Modeling System*, Paragon Decision Technology, 1993. (Cited on p. 287)

[284] R. BIXBY, *Solving real-world linear programs: A decade and more of progress*, Operations Research, 50 (2002), pp. 3–15. (Cited on p. 11)

[285] ———, *Computational Mixed Integer Programming*, Seminar on Algorithm Engineering, Dagstuhl, Germany, September, 2013, pp. 22–27. (Cited on p. 56)

[286] R. BIXBY, T. ACHTERBERG, E. ROTHBERG, AND Z. GU, *Recent Advances in Computational Linear and Mixed Integer Programming*, INFORMS Optimization Society Meeting, Atlanta, March 2008. (Cited on p. 56)

[287] R. BIXBY AND E. ROTHBERG, *Progress in computational mixed integer programming—A look back from the other side of the tipping point*, Annals of Operations Research, 149 (2007), pp. 37–41. (Cited on p. 91)

[288] R. BIXBY AND M. J. SALTZMAN, *Recovering an Optimal LP Basis from an Interior Point Solution*, technical report 607, Department of Mathematical Sciences, Clemson University, Clemson, SC, 1992. (Cited on p. 10)

[289] J. T. BLAKE AND J. DONALD, *Mount Sinai hospital uses integer programming to allocate operating room time*, Interfaces, 32 (2002), pp. 63–73. (Cited on p. 95)

[290] M. BOCCIA, C. MANNINO, AND I. VASILIEV, *The dispatching problem on multitrack territories: Heuristic approaches based on mixed integer linear programming*, Networks, 62 (2013), pp. 315–326. (Cited on p. 69)

[291] C. E. BODINGTON AND T. E. BAKER, *A history of mathematical programming in the petroleum industry*, Interfaces, 20 (1990), pp. 117–127. (Cited on pp. 37, 39, 40, 41)

[292] H. L. BODLAENDER, P. GRITZMANN, V. KLEE, AND J. VAN LEEUWEN, *Computational complexity of norm-maximization*, Combinatorica, 10, pp. 203–225. (Cited on p. 167)

[293] P. T. BOGGS AND J. W. TOLLE, *Sequential quadratic programming*, Acta Numerica, 4 (1995), pp. 1–51. (Cited on p. 229)

[294] S. BOGNER, A. BULGAC, J. CARLSON, J. ENGE, G. FANN, R. J. FURNSTAHL, S. GANDOLFI, G. HAGEN, M. HOROI, C. JOHNSON, M. KORTELAINEN, E. LUSK, P. MARIS, H. NAM, P. NAVRATIL, W. NAZAREWICZ, E. NG, G. P. A. NOBRE, E. ORMAND, T. PAPENBROCK, J. PEI, S. C. PIEPER, S. QUAGLIONI, K. ROCHE, J. SARICH, N. SCHUNCK, M. SOSONKINA, J. TERASAKI, I. THOMPSON, J. P. VARY, AND S. M. WILD, *Computational nuclear quantum many-body problem: The UNEDF project*, Computer Physics Communications, 184 (2013), pp. 2235–2250. (Cited on p. 536)

[295] N. BOLAND, A. C. EBERHARD, F. G. ENGINEER, M. FISCHETTI, M. W. P. SAVELSBERGH, AND A. TSOUKALAS, *Boosting the feasibility pump*, Mathematical Programming Computation, 6 (2014), pp. 255–279. (Cited on p. 57)

[296] N. BOLAND, T. KALINOWSKI, AND F. RIGTERINK, *Discrete flow pooling problems in coal supply chains*, in MODSIM2015, 21st International Congress on Modelling and Simulation, December 2015, T. Weber, M. J. McPhee, and R. S. Anderssen, eds., Modelling and Simulation Society of Australia and New Zealand, 2015, pp. 1710–1716. (Cited on p. 212)

[297] N. BOLAND, T. KALINOWSKI, F. RIGTERINK, AND M. SAVELSBERGH, *A special case of the generalized pooling problem arising in the mining industry*, Optimization Online, 2015. (Cited on p. 212)

[298] I. M. BOMZE, *Evolution towards the maximum clique*, Journal of Global Optimization, 10 (1997), pp. 143–164. (Cited on p. 165)

[299] ——, *Copositive optimization—Recent developments and applications*, European Journal of Operational Research, 216 (2012), pp. 509–520. (Cited on p. 113)

[300] I. M. BOMZE AND E. DE KLERK, *Solving standard quadratic optimization problems via linear, semidefinite and copositive programming*, Journal of Global Optimization, 24 (2002), pp. 163–185. (Cited on p. 169)

[301] P. BONAMI, *Lift-and-project cuts for mixed integer convex programs*, in Integer Programming and Combinatorial Optimization, O. Günlük and G. J. Woeginger, eds., Lecture Notes in Computer Science, volume 6255, Springer, 2011, pp. 52–64. (Cited on p. 287)

[302] P. BONAMI, L. T. BIEGLER, A. R. CONN, G. CORNUÉJOLS, I. E. GROSSMANN, C. D. LAIRD, J. LEE, A. LODI, F. MARGOT, N. SAWAYA, AND A. WÄCHTER, *An algorithmic framework for convex mixed integer nonlinear programs*, Discrete Optimization, 5 (2008), pp. 186–204. (Cited on pp. 279, 288)

[303] P. BONAMI, G. CORNUÉJOLS, A. LODI, AND F. MARGOT, *A feasibility pump for mixed integer nonlinear programs*, Mathematical Programming, 119 (2009), pp. 331–352. (Cited on p. 286)

[304] P. BONAMI AND J. P. GONÇALVES, *Heuristics for convex mixed integer nonlinear programs*, Computational Optimization and Applications, 51 (2012), pp. 729–747. (Cited on p. 286)

[305] P. BONAMI, M. R. KILINÇ, AND J. LINDEROTH, *Algorithms and software for convex mixed integer nonlinear programs*, in [1160], pp. 1–39. (Cited on pp. 274, 315)

[306] P. BONAMI, J. LEE, S. LEYFFER, AND A. WÄCHTER, *More Branch-and-Bound Experiments in Convex Nonlinear Integer Programming*. Preprint ANL/MCS-P1949-0911, Argonne National Laboratory, Mathematics and Computer Science Division, 2011. (Cited on p. 285)

[307] P. BONAMI AND M. A. LEJEUNE, *An exact solution approach for portfolio optimization problems under stochastic and integer constraints*, Operations Research, 57 (2009), pp. 656–670. (Cited on pp. 427, 429, 430)

[308] P. BONAMI, J. LINDEROTH, AND A. LODI, *Disjunctive cuts for mixed integer nonlinear programming problems*, in Progress in Combinatorial Optimization: Recent Progress, R. Mahjoub, ed., Wiley-ISTE, 2011, pp. 521–544. (Cited on p. 287)

[309] P. BONAMI, A. OLIVARES, M. SOLER, AND E. STAFFETTI, *Multiphase mixed-integer optimal control approach to aircraft trajectory optimization*, Journal of Guidance, Control, and Dynamics, 36 (2013), pp. 1267–1277. (Cited on pp. 295, 300)

[310] O. BONI, A. BEN-TAL, AND A. NEMIROVSKI, *Robust solutions to conic quadratic problems and their applications*, Optimization and Engineering, 9 (2008), pp. 1–18. (Cited on pp. 333, 338)

[311] D. BONINI, C. DUPRÉ, AND G. GRANGER, *How ERASMUS can support an increase in capacity in 2020*, in Proceedings of the 7th International Conference on Computing, Communications and Control Technologies: CCCT, 2009. (Cited on p. 294)

[312] A. J. BOOKER, J. E. DENNIS, JR., P. D. FRANK, D. B. SERAFINI, V. TORCZON, AND M. W. TROSSET, *A rigorous framework for optimization of expensive functions by surrogates*, Structural and Multidisciplinary Optimization, 17 (1998), pp. 1–13. (Cited on p. 506)

[313] R. BOORSTYN AND H. FRANK, *Large-scale network topological optimization*, IEEE Transactions on Communications, 25 (1977), pp. 29–47. (Cited on p. 292)

[314] B. BORCHERS, *CSDP, a C library for semidefinite programming*, Optimization Methods and Software, 11 (1999), pp. 613–623. (Cited on p. 117)

[315] B. BORCHERS AND J. E. MITCHELL, *An improved branch and bound algorithm for mixed integer nonlinear programs*, Computers and Operations Research, 21 (1994), pp. 359–368. (Cited on pp. 279, 292)

[316] B. BORCHERS AND J. G. YOUNG, *Implementation of a primal-dual method for SDP on a shared memory parallel architecture*, Computational Optimization and Applications, 37 (2007), pp. 355–369. (Cited on p. 10)

[317] A. BORGHETTI, C. D'AMBROSIO, A. LODI, AND S. MARTELLO, *An MILP approach for short-term hydro scheduling and unit commitment with head-dependent reservoir*, IEEE Transactions on Power Systems, 23 (2008), pp. 1115–1124. (Cited on p. 304)

[318] K. H. BORGWARDT, *The Simplex Method—A Probabilistic Analysis*, Springer Verlag, Berlin, Heidelberg, New York, 1987. (Cited on p. 8)

[319] R. BORNDÖRFER, B. EROL, T. GRAFFAGNINO, T. SCHLECHTE, AND E. SWARAT, *Aggregation Methods for Railway Networks*, technical report, November 2010. ZIB-Report 10-23. (Cited on p. 70)

[320] T. BORTFELD, *IMRT: a review and preview*, Physics in Medicine and Biology, 51 (2006), pp. R363–R379. (Cited on p. 345)

[321] T. BORTFELD, T. C. Y. CHAN, A. TROFIMOV, AND J. N. TSITSIKLIS, *Robust management of motion uncertainty in intensity-modulated radiation therapy*, Operations Research, 56 (2008), pp. 1461–1473. (Cited on pp. 334, 349)

[322] D. M. BORTZ AND C. T. KELLEY, *The simplex gradient and noisy optimization problems*, in Computational Methods in Optimal Design and Control, vol. 24 of Progress in Systems and Control Theory, J. T. Borggaard, J. Burns, E. Cliff, and S. Schreck, eds., Birkhäuser, Boston, 1998, pp. 77–90. (Cited on pp. 500, 512)

[323] J. BORWEIN AND H. WOLKOWICZ, *Characterization of optimality for the abstract convex program with finite dimensional range*, Journal of the Australian Mathematical Society (Series A), 30 (1981), pp. 390–411. (Cited on p. 115)

[324] ——, *Facial reduction for a cone-convex programming problem*, Journal of the Australian Mathematical Society (Series A), 30 (1981), pp. 369–380. (Cited on p. 115)

[325] ——, *Regularizing the abstract convex program*, Journal of Mathematical Analysis and Applications, 83 (1981), pp. 495–530. (Cited on p. 115)

[326] B. E. BOSER, I. M. GUYON, AND V. N. VAPNIK, *A training algorithm for optimal margin classifiers*, in Proceedings of the Fifth Annual Workshop on Computational Learning Theory, COLT 1992, pp. 144–152. (Cited on p. 31)

[327] J. F. BOSHIER, G. B. MANNING, AND E. G. READ, *Scheduling releases from New Zealand's hydro reservoirs*, Transactions of IPENZ, 10 (1983), pp. 33–41. (Cited on p. 413)

[328] K. BOUDT, P. CARL, AND B. G. PETERSON, *Asset allocation with conditional value-at-risk budgets*, Journal of Risk, 15 (2013), pp. 39–68. (Cited on p. 435)

[329] M. P. BOYCE, *Handbook for Cogeneration and Combined Cycle Power Plants*, ASME International, 2010. (Cited on p. 303)

[330] S. BOYD AND C. BARRATT, *Linear Controller Design: Limits of Performance*, Prentice-Hall, 1991. (Cited on p. 27)

[331] S. BOYD, L. EL GHAOUI, E. FERON, AND V. BALAKRISHNAN, *Linear Matrix Inequalities in System and Control Theory*, vol. 15 of Studies in Applied Mathematics, SIAM, Philadelphia, PA, 1994. (Cited on p. 110)

[332] S. BOYD AND L. VANDENBERGHE, *Convex Optimization*, Cambridge University Press, Cambridge, 2004. (Cited on pp. 11, 31, 107, 335, 336, 337, 354)

[333] P. G. BOYVALENKOV, D. P. DANEV, AND S. P. BUMOVA, *Upper bounds on the minimum distance of spherical codes*, IEEE Transactions on Information Theory, 42 (1996), pp. 1576–1581. (Cited on p. 34)

[334] C. BRAGALLI, C. D'AMBROSIO, J. LEE, A. LODI, AND P. TOTH, *An MINLP solution method for a water network problem*, in Algorithms—ESA 2006: Proceedings of 14th Annual European Symposium, Zurich, Switzerland, Springer, 2006, pp. 696–707. (Cited on p. 292)

[335] C. C. BRANAS, E. J. MACKENZIE, AND C. S. REVELLE, *A trauma resource allocation model for ambulances and hospitals*, Health Services Research, 35 (2000), pp. 489–507. (Cited on p. 95)

[336] *Branch and Reduce Optimization Navigator (BARON)*. http://archimedes.cheme.cmu.edu/?q=baron. (Cited on p. 196)

[337] M. L. BRANDEAU, *Allocating resources to control infectious diseases*, in Operations Research and Health Care: A Handbook of Methods and Applications, M. L. Brandeau, F. Sainfort, and W. P. Pierskalla, eds., Springer, Boston, 2004, pp. 443–464. (Cited on p. 96)

[338] M. L. BRANDEAU AND S. S. CHIU, *An overview of representative problems in location research*, Management Science, 35 (1989), pp. 645–674. (Cited on p. 450)

[339] A. BREARLEY, G. MITRA, AND H. P. WILLIAMS, *Analysis of mathematical programming problems prior to applying the simplex algorithm*, Mathematical Programming, 8 (1975), pp. 54–83. (Cited on p. 281)

[340] R. BREKELMANS, L. DRIESSEN, H. HAMERS, AND D. DEN HERTOG, *Constrained optimization involving expensive function evaluations: A sequential approach*, European Journal of Operational Research, 160 (2005), pp. 121–138. (Cited on p. 505)

[341] K. BRITTON, Y. TAKAI, M. MITSUYA, K. NEMOTO, Y. OGAWA, AND S. YAMADA, *Evaluation of inter- and intrafraction organ motion during intensity modulated radiation therapy (IMRT) for localized prostate cancer measured by a newly developed on-board image-guided system*, Radiation Medicine, 23 (2005), pp. 14–24. (Cited on p. 347)

[342] A. BROOKE, D. KENDRICK, AND A. MEERAUS, *GAMS: A Users Guide, Release 23.3*, The Scientific Press, South San Francisco, 2010. (Cited on p. 82)

[343] A. BROOKE, D. KENDRICK, A. MEERAUS, AND R. RAMAN, *GAMS, A User's Guide*, GAMS Development Corporation, 1992. (Cited on p. 287)

[344] J. P. BROOKS AND E. K. LEE, *Analysis of the consistency of a mixed integer programming-based multi-category constrained discriminant model*, Annals of Operations Research, 174 (2010), pp. 147–168. (Cited on p. 99)

[345] ———, *Solving a multigroup mixed-integer programming-based constrained discrimination model*, INFORMS Journal on Computing, 26 (2014), pp. 567–585. (Cited on pp. 99, 102)

[346] G. BROWN, M. CARLYLE, J. SALMERÓN, AND K. WOOD, *Defending critical infrastructure*, Interfaces, 36 (2006), pp. 530–544. (Cited on p. 341)

[347] G. G. BROWN, W. M. CARLYLE, R. C. HARNEY, E. M. SKROCH, AND R. K. WOOD, *Interdicting a nuclear-weapons project*, Operations Research, 57 (2009), pp. 866–877. (Cited on p. 341)

[348] A. M. BRUCKSTEIN, D. L. DONOHO, AND M. ELAD, *From sparse solutions of systems of equations to sparse modeling of signals and images*, SIAM Review, 51 (2009), pp. 34–81. (Cited on p. 33)

[349] M. E. BRUNI, D. CONFORTI, N. SICILIA, AND S. TROTTA, *A new organ transplantation location–allocation policy: A case study of Italy*, Health Care Management Science, 9 (2006), pp. 125–142. (Cited on p. 95)

[350] J. C. BRUNO, F. FERNANDEZ, F. CASTELLS, AND I. E. GROSSMANN, *MINLP model for optimal synthesis and operation of utility plants*, Transactions of the Institution of Chemical Engineers, 76 (1998), pp. 246–258. (Cited on p. 77)

[351] ———, *A rigorous MINLP model for the optimal synthesis and operation of utility plants*, Chemical Engineering Research and Design, 76 (1998), pp. 246–258. (Cited on p. 325)

[352] L. F. BUENO, A. FRIEDLANDER, J. M. MARTÍNEZ, AND F. N. C. SOBRAL, *Inexact restoration method for derivative-free optimization with smooth constraints*, SIAM Journal on Optimization, 23 (2013), pp. 1189–1213. (Cited on p. 505)

[353] M. D. BUHMANN, *Radial Basis Functions: Theory and Implementations*, Cambridge University Press, Cambridge, 2003. (Cited on pp. 192, 497)

[354] W. A. BUKHSH, A. GROTHEY, K. I. M. MCKINNON, AND P. A. TRODDEN, *Local solutions of the optimal power flow problem*, IEEE Transactions on Power Systems, 28 (2013), pp. 4780–4788. (Cited on pp. 192, 195)

[355] R. S. BURACHIK AND A. RUBINOV, *On the absence of duality gap for Lagrange-type functions*, Journal of Industrial and Management Optimization, 1 (2005), pp. 33–38. (Cited on p. 169)

[356] S. BURER AND A. N. LETCHFORD, *Non-convex mixed-integer nonlinear programming: A survey*, Surveys in Operations Research and Management Science, 17 (2012), pp. 97–106. (Cited on pp. 273, 274, 315)

[357] S. BURER AND R. D. C. MONTEIRO, *A nonlinear programming algorithm for solving semidefinite programs via low-rank factorization*, Mathematical Programming, 95 (2003), pp. 329–357. (Cited on p. 118)

[358] S. BURER AND A. SAXENA, *The MILP road to MIQCP*, in [1160], pp. 373–405. (Cited on p. 273)

[359] C. J. C. BURGES, *A tutorial on support vector machines for pattern recognition*, Data Mining and Knowledge Discovery, 2 (1998), pp. 121–167. (Cited on p. 30)

[360] E. K. BURKE, T. CURTOIS, G. POST, R. QU, AND B. VELTMAN, *A hybrid heuristic ordering and variable neighbourhood search for the nurse rostering problem*, European Journal of Operational Research, 188 (2008), pp. 330–341. (Cited on p. 95)

[361] J. V. BURKE, A. S. LEWIS, AND M. L. OVERTON, *A robust gradient sampling algorithm for nonsmooth, nonconvex optimization*, SIAM Journal on Optimization, 15 (2005), pp. 751–779. (Cited on p. 223)

[362] F. BURKOWSKI, Y.-L. CHEUNG, AND H. WOLKOWICZ, *Efficient use of semidefinite programming for selection of rotamers in protein conformations*, INFORMS Journal on Computing, 26 (2014), pp. 748–766. (Cited on p. 115)

[363] J. F. BURRI, S. D. WILSON, AND V. I. MANOUSIOUTHAKIS, *Infinite dimensional state-space approach to reactor network synthesis: Application to attainable region construction*, Computers and Chemical Engineering, 26 (2002), pp. 849–862. (Cited on p. 320)

[364] D. BURSHTEIN AND I. GOLDENBERG, *Improved linear programming decoding of LDPC codes and bounds on the minimum and fractional distance*, IEEE Transactions on Information Theory, 57 (2011), pp. 7386–7402. (Cited on p. 35)

[365] F. BUSHMAN, *Retroviral integration and human gene therapy*, Journal of Clinical Investigation, 117 (2007), pp. 2083–2086. (Cited on p. 182)

[366] F. BUSHMAN, M. LEWINSKI, A. CIUFFI, S. BARR, J. LEIPZIG, S. HANNENHALLI, AND C. HOFFMANN, *Genome-wide analysis of retroviral DNA integration*, Biophysical Journal, 3 (2005), pp. 848–858. (Cited on p. 182)

[367] C. BÜSING AND F. D'ANDREAGIOVANNI, *New results about multi-band uncertainty in robust optimization*, in Experimental Algorithms, Springer, 2012, pp. 63–74. (Cited on p. 333)

[368] M. BUSS, O. STRYK, R. BULIRSCH, AND G. SCHMIDT, *Towards hybrid optimal control*, at—Automatisierungstechnik, 48 (2000), pp. 448–459. (Cited on p. 328)

[369] M. R. BUSSIECK AND A. DRUD, *SBB: A New Solver for Mixed Integer Nonlinear Programming*. Invited talk, Recent Advances in Nonlinear Mixed Integer Optimization, INFORMS, 2000. (Cited on p. 289)

[370] M. R. BUSSIECK, A. S. DRUD, AND A. MEERAUS, *MINLPLib—A collection of test models for mixed-integer nonlinear programming*, INFORMS Journal on Computing, 15 (2003), pp. 114–119. (Cited on pp. 290, 291)

[371] M. R. BUSSIECK, A. S. DRUD, A. MEERAUS, AND A. PRUESSNER, *Quality assurance and global optimization*, in Global Optimization and Constraint Satisfaction, volume 2861 of Lecture Notes in Computer Science, C. Bliek, C. Jermann, and A. Neumaier, eds., Springer, Berlin, Heidelberg, 2003, pp. 223–238. (Cited on p. 291)

[372] M. R. BUSSIECK AND S. VIGERSKE, *MINLP solver software*, in [485], 2011. (Cited on pp. 274, 288, 315)

[373] S. BUSYGIN, S. BUTENKO, AND P. M. PARDALOS, *A heuristic for the maximum independent set problem based on optimization of a quadratic over a sphere*, Journal of Combinatorial Optimization, 6 (2002), pp. 287–297. (Cited on p. 165)

[374] R. H. BYRD, F. E. CURTIS, AND J. NOCEDAL, *Infeasibility detection and SQP methods for nonlinear optimization*, SIAM Journal on Optimization, 20 (2010), pp. 2281–2299. (Cited on p. 229)

[375] R. H. BYRD, N. I. M. GOULD, J. NOCEDAL, AND R. A. WALTZ, *An algorithm for nonlinear optimization using linear programming and equality constrained subproblems*, Mathematical Programming, 100 (2003), pp. 27–48. (Cited on p. 233)

[376] R. H. BYRD, J. NOCEDAL, AND R. A. WALTZ, *KNITRO: An integrated package for nonlinear optimization*, in Large Scale Nonlinear Optimization, G. Di Pillo and M. Roma, eds., vol. 83 of Nonconvex Optimization and Its Applications, Springer, 2006, pp. 35–59. (Cited on pp. 10, 289)

[377] ——, *Steering exact penalty methods for nonlinear programming*, Optimization Methods and Software, 23 (2008), pp. 197–213. (Cited on p. 229)

[378] J. A. CABALLERO AND I. E. GROSSMANN, *Aggregated models for integrated distillation systems*, Industrial and Engineering Chemistry Research, 38 (1999), pp. 2330–2344. (Cited on p. 323)

[379] ——, *Generalized disjunctive programming model for the optimal synthesis of thermally linked distillation columns*, Industrial and Engineering Chemistry Research, 40 (2001), pp. 2260–2274. (Cited on pp. 77, 324)

[380] ——, *Thermodynamically equivalent configurations for thermally coupled distillation*, AIChE Journal, 49 (2003), pp. 2864–2884. (Cited on p. 323)

[381] ———, *Design of distillation sequences: From conventional to fully thermally coupled distillation systems*, Computers and Chemical Engineering, 28 (2004), pp. 2307–2329. (Cited on p. 324)

[382] ———, *Structural considerations and modeling in the synthesis of heat-integrated-thermally coupled distillation sequences*, Industrial and Engineering Chemistry Research, 45 (2006), pp. 8454–8474. (Cited on p. 324)

[383] V. CACCHIANI, D. HUISMAN, M. KIDD, L. KROON, P. TOTH, L. VEELENTURF, AND J. WAGENAAR, *An overview of recovery models and algorithms for real-time railway rescheduling*, Transportation Research Part B, 63 (2014), pp. 15–37. (Cited on pp. 66, 68, 70)

[384] V. CACCHIANI AND P. TOTH, *Nominal and robust train timetabling problems*, European Journal of Operational Research, 219 (2012), pp. 727–737. (Cited on p. 67)

[385] G. P. CACHON, *Supply chain coordination with contracts*, in Supply Chain Management: Design, Coordination and Operation, A. G. de Kok and S. C. Graves, eds., vol. 11 of Handbooks in Operations Research and Management Science, North-Holland, 2003, ch. 6. (Cited on pp. 453, 454, 455)

[386] S. CAFIERI, *Maximizing the number of solved aircraft conflicts through velocity regulation*, in MAGO 2014, 12th Global Optimization Workshop, Malaga, Spain, September 2014, pp. 1–4. (Cited on p. 297)

[387] S. CAFIERI, A. ALONSO-AYUSO, L. ESCUDERO, AND F. J. MARTIN-CAMPO, *Aircraft conflict avoidance by mixed-integer nonlinear optimization models combining turn and velocity change maneuvers*, in ICCOPT 2013, 4th International Conference on Continuous Optimization, Lisbon, Portugal, July 2013. (Cited on p. 299)

[388] S. CAFIERI AND N. DURAND, *Aircraft deconfliction with speed regulation: New models from mixed-integer optimization*, Journal of Global Optimization, 58 (2014), pp. 613–629. (Cited on pp. 295, 296, 297, 299, 300, 301)

[389] G. CAIMI, M. FUCHSBERGER, M. LAUMANNS, AND M. LÜTHI, *A model predictive control approach for discrete-time rescheduling in complex central railway station areas*, Computers and Operations Research, 39 (2012), pp. 2578–2593. (Cited on p. 69)

[390] G. C. CALAFIORE AND L. EL GHAOUI, *On distributionally robust chance-constrained linear programs*, Journal of Optimization Theory and Applications, 130 (2006), pp. 1–22. (Cited on pp. 344, 391)

[391] B. CALVERT AND M. K. VAMANAMURTHY, *Local and global extrema for functions of several variables*, Journal of Australian Mathematical Society (Series A), 29 (1980), pp. 362–368. (Cited on p. 168)

[392] R. CALVIN, C. RAY, AND V. RHYNE, *The design of optimal convolutional filters via linear programming*, IEEE Transactions on Geoscience Electronics, 7 (1969), pp. 142–145. (Cited on p. 28)

[393] K. V. CAMARDA AND C. D. MARANAS, *Optimization in polymer design using connectivity indices*, Industrial and Engineering Chemistry Research, 38 (1999), pp. 1884–1892. (Cited on p. 329)

[394] E. J. CANDÈS, J. ROMBERG, AND T. TAO, *Robust uncertainty principles: Exact signal reconstruction from highly incomplete frequency information*, IEEE Transactions on Information Theory, 52 (2006), pp. 489–509. (Cited on pp. 32, 33)

[395] ———, *Stable signal recovery from incomplete and inaccurate measurements*, Communications on Pure and Applied Mathematics, 59 (2006), pp. 1207–1223. (Cited on p. 33)

[396] E. J. CANDÈS AND T. TAO, *Decoding by linear programming*, IEEE Transactions on Information Theory, 51 (2005), pp. 4203–4215. (Cited on pp. 31, 33)

[397] ———, *Near-optimal signal recovery from random projections and universal encoding strategies*, IEEE Transactions on Information Theory, 52 (2006). (Cited on pp. 32, 33)

[398] E. J. CANDÈS AND M. B. WAKIN, *An introduction to compressive sampling*, IEEE Signal Processing Magazine, 25 (2008), pp. 21–30. (Cited on p. 33)

[399] W. CAO, M. ÇELIK, Ö. ERGUN, J. SWANN, AND N. VILJOEN, *Challenges in service network expansion: An application in donated breastmilk banking in South Africa*, Socio-Economic Planning Sciences, 53 (2016), pp. 33–48. (Cited on pp. 480, 481, 482, 483)

[400] W. CAO, G. LIM, X. LI, Y. LI, X. R. ZHU, AND X. ZHANG, *Incorporating deliverable monitor unit constraints into spot intensity optimization in intensity-modulated proton therapy treatment planning*, Physics in Medicine and Biology, 58 (2013), pp. 5113–5125. (Cited on p. 94)

[401] F. CAPITANESCU, J. L. M. RAMOS, P. PANCIATICI, D. KIRSCHEN, A. M. MARCOLINI, L. PLATBROOD, AND L. WEHENKEL, *State-of-the-art, challenges, and future trends in security constrained optimal power flow*, Electric Power Systems Research, 81 (2011), pp. 1731–1741. (Cited on p. 191)

[402] P. CAPPANERA AND M. P. SCAPARRA, *Optimal allocation of protective resources in shortest-path networks*, Transportation Science, 45 (2011), pp. 64–80. (Cited on p. 480)

[403] A. CAPRARA, M. FISCHETTI, AND P. TOTH, *A heuristic algorithm for the set covering problem*, in Integer Programming and Combinatorial Optimization, vol. 1084 of Lecture Notes in Computer Science, W. H. Cunningham, S. T. McCormick, and M. Queyranne, eds., Springer, Berlin, Heidelberg, 1996, pp. 72–84. (Cited on p. 57)

[404] C. CARAMANIS, *Adaptable Optimization: Theory and Algorithms*, PhD thesis, Massachusetts Institute of Technology, 2006. (Cited on p. 342)

[405] L. E. CÁRDENAS-BARRÓN, K.-J. CHUNG, AND G. TREVIÑO-GARZA, *100th anniversary of EOQ (special issue)*, International Journal of Production Economics, 155 (2014). (Cited on p. 440)

[406] M. CAREY, *Optimal time-varying flows on congested networks*, Operations Research, 35 (1987), pp. 58–69. (Cited on p. 488)

[407] J. CARPENTIER, *Contribution à l'étude du dispatching économique*, Bulletin de la Société française des électriciens, 3 (1962), pp. 431–447. (Cited on pp. 187, 188)

[408] ———, *Optimal power flows*, International Journal of Electrical Power and Energy Systems, 1 (1979), pp. 3–15. (Cited on pp. 187, 188)

[409] R. CARTER, *Nonsequential Dynamic Programming for Optimizing Pipelines*, 1999. Presentation at the SIAM Conference on Optimization, Atlanta, GA. (Cited on p. 507)

[410] R. CARTER, J. M. GABLONSKY, A. PATRICK, C. T. KELLEY, AND O. J. ESLINGER, *Algorithms for noisy problems in gas transmission pipeline optimization*, Optimization and Engineering, 2 (2001), pp. 139–157. (Cited on p. 508)

[411] C. CARTIS, N. I. M. GOULD, AND PH. L. TOINT, *On the complexity of steepest descent, Newton's and regularized Newton's methods for nonconvex unconstrained optimization*, SIAM Journal on Optimization, 20 (2010), pp. 2833–2852. (Cited on p. 501)

[412] ——, *On the oracle complexity of first-order and derivative-free algorithms for smooth nonconvex minimization*, SIAM Journal on Optimization, 22 (2012), pp. 66–86. (Cited on p. 501)

[413] M. CASIRAGHI, F. ALBERTINI, AND A. J. LOMAX, *Advantages and limitations of the "worst case scenario" approach in IMPT treatment planning*, Physics in Medicine and Biology, 58 (2013), pp. 1323–1339. (Cited on pp. 353, 355)

[414] I. CASTILLO, J. WESTERLUND, S. EMET, AND T. WESTERLUND, *Optimization of block layout design problems with unequal areas: A comparison of MILP and MINLP optimization methods*, Computers and Chemical Engineering, 30 (2005), pp. 54–69. (Cited on p. 292)

[415] P. M. CASTRO, *Tightening piecewise McCormick relaxations for bilinear problems*, Computers and Chemical Engineering, 72 (2014), pp. 300–311. (Cited on p. 284)

[416] P. M. CASTRO, A. P. BARBOSA-PÓVOA, AND H. A. MATOS, *Optimal periodic scheduling of batch plants using RTN-based discrete and continuous-time formulations: A case study approach*, Industrial and Engineering Chemistry Research, 42 (2003), pp. 3346–3360. (Cited on p. 328)

[417] P. M. CASTRO AND I. E. GROSSMANN, *Generalized disjunctive programming as a systematic modeling framework to derive scheduling formulations*, Industrial and Engineering Chemistry Research, 51 (2012), pp. 5781–5792. (Cited on p. 328)

[418] M. ÇELIK, Ö. ERGUN, B. JOHNSON, P. KESKINOCAK, Á. LORCA, P. PEKGÜN, AND J. SWANN, *Humanitarian logistics*, in INFORMS TutORials in Operations Research, P. Mirchandani and C. Smith, eds., 2012, pp. 18–49. (Cited on pp. 479, 480)

[419] M. ÇELIK, Ö. ERGUN, AND J. SWANN, *Perishable Commodity Allocation in a Humanitarian Supply Chain with Multiple Demand Types: Effects of Usage Policy and Centralization*, technical report, School of Industrial and Systems Engineering, Georgia Institute of Technology, Atlanta, GA, 2014. (Cited on pp. 484, 485, 486)

[420] J. CERDA, A. W. WESTERBERG, D. MASON, AND B. LINNHOFF, *Minimum utility usage in heat exchanger network synthesis: A transportation problem*, Chemical Engineering Science, 38 (1983), pp. 373–387. (Cited on p. 324)

[421] S. CERIA AND K. SIVARAMAKRISHNAN, *Portfolio optimization*, in Investment Risk and Uncertainty: Advanced Risk Awareness Techniques for the Intelligent Investor, S. P. Greiner, ed., John Wiley & Sons, Hoboken, NJ, USA, 2013, pp. 429–464. (Cited on p. 153)

[422] S. CERIA AND J. SOARES, *Convex programming for disjunctive optimization*, Mathematical Programming, 86 (1999), pp. 595–614. (Cited on p. 287)

[423] S. CERIA AND R. A. STUBBS, *Incorporating estimation errors into portfolio selection: Robust portfolio construction*, Journal of Asset Management, 7 (2006), pp. 109–127. (Cited on pp. 151, 156, 157)

[424] M. T. CEZIK AND G. IYENGAR, *Cuts for mixed 0-1 conic programming*, Mathematical Programming, 104 (2005), pp. 179–202. (Cited on p. 286)

[425] S. CHAIMATANAN, D. DELAHAYE, AND M. MONGEAU, *A hybrid metaheuristic optimization algorithm for strategic planning of 4D aircraft trajectories at the continental scale*, IEEE Computational Intelligence Magazine, 9 (2014), pp. 46–61. (Cited on pp. 295, 300, 301)

[426] T. C. CHAN, T. BORTFELD, AND J. N. TSITSIKLIS, *A robust approach to IMRT optimization*, Physics in Medicine and Biology, 51 (2006), pp. 2567–2583. (Cited on pp. 348, 349)

[427] T. C. Y. CHAN, H. MAHMOUDZADEH, AND T. G. PURDIE, *A robust-CVaR optimization approach with application to breast cancer therapy*, European Journal of Operational Research, 238 (2014), pp. 876–885. (Cited on pp. 334, 350)

[428] T. C. Y. CHAN AND P. A. MAR, *Stability and Continuity in Robust Linear and Linear Semi-infinite Optimization*, arXiv preprint arXiv:1509.06640 (2015). (Cited on p. 344)

[429] T. C. CHAN AND V. V. MIŠIĆ, *Adaptive and robust radiation therapy optimization for lung cancer*, European Journal of Operational Research, 231 (2013), pp. 745–756. (Cited on pp. 334, 343, 344, 350)

[430] T. C. Y. CHAN, Z.-J. M. SHEN, AND A. SIDDIQ, *Robust Facility Location under Demand Location Uncertainty*, arXiv preprint arXiv:1507.04397 (2015). (Cited on pp. 334, 342)

[431] T. C. CHAN, J. N. TSITSIKLIS, AND T. BORTFELD, *Optimal margin and edge-enhanced intensity maps in the presence of motion and uncertainty*, Physics in Medicine and Biology, 55 (2010), pp. 515–533. (Cited on p. 353)

[432] W. A. CHAOVALITWONGSE, Y.-J. FAN, AND R. C. SACHDEO, *Novel optimization models for abnormal brain activity classification*, Operations Research, 56 (2008), pp. 1450–1460. (Cited on p. 94)

[433] G. W. CHARACKLIS, B. R. KIRSCH, J. RAMSEY, K. E. M. DILLARD, AND C. T. KELLEY, *Developing portfolios of water supply transfers*, Water Resources Research, 42 (2006), pp. W05403-1–W05403-14. (Cited on pp. 508, 515)

[434] A. CHARNES, W. W. COOPER, AND G. H. SYMONDS, *Cost horizons and certainty equivalents: An approach to stochastic programming of heating oil*, Management Science, 4 (1958), pp. 235–263. (Cited on p. 391)

[435] B. Y. CHEN, R. CHEN, AND K. SCHEINBERG, *Aligning ligand binding cavities by optimizing superposed volume*, in Proceedings of the IEEE International Conference on Bioinformatics and Biomedicine (BIBM 2012), 2012, pp. 1–5. (Cited on pp. xxi, 520, 526)

[436] B. Y. CHEN, V. Y. FOFANOV, D. H. BRYANT, B. D. DODSON, D. M. KRISTENSEN, A. M. LISEWSKI, M. KIMMEL, O. LICHTARGE, AND L. E. KAVRAKI, *The MASH pipeline for protein function prediction and an algorithm for the geometric refinement of 3D motifs*, Journal of Computational Biology, 14 (2007), pp. 791–816. (Cited on p. 520)

[437] B. Y. CHEN AND B. HONIG, *VASP: A volumetric analysis of surface properties yields insights into protein-ligand binding specificity*, PLOS Computational Biology, 6 (2010), p. e1000881. (Cited on p. 520)

[438] C. CHEN, *Simulated annealing-based optimal wind-thermal coordination scheduling*, IET Generation, Transmission & Distribution, 1 (2007), pp. 447–455. (Cited on p. 367)

[439] C. CHEN AND R. H. KWON, *Robust portfolio selection for index tracking*, Computers and Operations Research, 39 (2012), pp. 829–837. (Cited on p. 333)

[440] C.-L. CHEN AND P.-S. HUNG, *Simultaneous synthesis of flexible heat-exchange networks with uncertain source-stream temperatures and flow rates*, Industrial and Engineering Chemistry Research, 43 (2004), pp. 5916–5928. (Cited on p. 324)

[441] F. CHEN AND Y. S. ZHENG, *Lower bounds for multi-echelon stochastic inventory systems*, Management Science, 40 (1994), pp. 1426–1443. (Cited on p. 449)

[442] L. CHEN, M. GENDREAU, M. H. HÀ, AND A. LANGEVIN, *A robust optimization approach for the road network daily maintenance routing problem with uncertain service time*, Transportation Research Part E: Logistics and Transportation Review, 85 (2016), pp. 40–51. (Cited on p. 333)

[443] L. CHEN AND E. MACDONALD, *A new model for wind farm layout optimization with landowner decisions*, in Proceedings of the ASME 2011 International Design Engineering Technical Conferences and Computers and Information in Engineering Conference, Washington, DC, 2011. (Cited on p. 367)

[444] R. CHEN, *Stochastic Derivative-Free Optimization of Noisy Functions*, PhD thesis, Lehigh University, 2015. (Cited on pp. xxi, 523)

[445] R. CHEN, M. MENICKELLY, AND K. SCHEINBERG, *Stochastic Optimization Using a Trust-Region Method and Random Models*, Technical Report ISE 15T-002, Department of Industrial and Systems Engineering, Lehigh University, 2015. (Cited on p. 502)

[446] S. S. CHEN, D. L. DONOHO, AND M. A. SAUNDERS, *Atomic decomposition by basis pursuit*, SIAM Journal on Scientific Computing, 20 (1998), pp. 33–61. (Cited on pp. 31, 32)

[447] W. CHEN, J. UNKELBACH, A. TROFIMOV, T. MADDEN, H. KOOY, T. BORTFELD, AND D. CRAFT, *Including robustness in multi-criteria optimization for intensity-modulated proton therapy*, Physics in Medicine and Biology, 57 (2012), pp. 591–608. (Cited on p. 353)

[448] X. CHEN AND C. T. KELLEY, *Optimization with hidden constraints and embedded Monte Carlo computations*, Optimization and Engineering, 17(1) (2016), pp. 157–175. (Cited on pp. 501, 510, 513, 514, 515)

[449] X. CHEN, Z. PANG, AND L. PAN, *Coordinating inventory control and pricing strategies for perishable products*, Operations Research, 62 (2014), pp. 284–300. (Cited on p. 473)

[450] X. CHEN, M. SIM, P. SUN, AND J. ZHANG, *A linear decision-based approximation approach to stochastic programming*, Operations Research, 56 (2008), pp. 344–357. (Cited on p. 342)

[451] X. CHEN AND Y. ZHANG, *Uncertain linear programs: Extended affinely adjustable robust counterparts*, Operations Research, 57 (2009), pp. 1469–1482. (Cited on pp. 340, 342)

[452] Y. CHEN, T. A. ADAMS, AND P. I. BARTON, *Optimal design and operation of static energy polygeneration systems*, Industrial and Engineering Chemistry Research, 50 (2010), pp. 5099–5113. (Cited on p. 318)

[453] Y. CHEN, I. E. GROSSMANN, AND D. C. MILLER, *Computational strategies for large-scale MILP transshipment models for heat exchanger network synthesis*, Computers and Chemical Engineering, 82 (2015), pp. 68–83. (Cited on p. 77)

[454] Y.-L. CHEUNG, S. SCHURR, AND H. WOLKOWICZ, *Preprocessing and regularization for degenerate semidefinite programs*, in Computational and Analytical Mathematics, Springer, 2013, pp. 251–303. (Cited on p. 115)

[455] M. CHIANG, S. H. LOW, R. CALDERBANK, AND J. C. DOYLE, *Layering as optimization decomposition: A mathematical theory of network architectures*, Proceedings of the IEEE, 95 (2007), pp. 255–312. (Cited on p. 36)

[456] S. CHICK, H. MAMANI, AND D. SIMCHI-LEVI, *Supply chain coordination and influenza vaccination*, Operations Research, 56 (2008), pp. 1493–1506. (Cited on pp. 471, 472)

[457] C. M. CHIN AND R. FLETCHER, *On the global convergence of an SLP-filter algorithm that takes EQP steps*, Mathematical Programming, 96 (2003), pp. 161–177. (Cited on p. 233)

[458] S. CHO, *The optimal composition of influenza vaccines subject to random production yields*, Manufacturing and Service Operations Management, 12 (2010), pp. 256–277. (Cited on p. 472)

[459] A. K. Y. CHOI, A. K. C. BERESFORD, S. J. PETTIT, AND F. BAYUSUF, *Humanitarian aid distribution in East Africa: A study in supply chain volatility and fragility*, Supply Chain Forum: An International Journal, 11 (2010), pp. 20–31. (Cited on p. 488)

[460] T. D. CHOI, O. J. ESLINGER, C. T. KELLEY, J. W. DAVID, AND M. ETHERIDGE, *Optimization of automotive valve train components with implicit filtering*, Optimization and Engineering, 1 (2000), pp. 9–27. (Cited on p. 508)

[461] T. D. CHOI AND C. T. KELLEY, *Superlinear convergence and implicit filtering*, SIAM Journal on Optimization, 10 (2000), pp. 1149–1162. (Cited on p. 513)

[462] T.-M. CHOI, ed., *Handbook of EOQ Inventory Problems*, Springer, New York, 2013. (Cited on p. 440)

[463] S. CHOPRA AND P. MEINDL, *Supply Chain Management: Strategy, Planning and Operation*, 5th edition, Prentice Hall, 2012. (Cited on p. 439)

[464] V. K. CHOPRA AND W. T. ZIEMBA, *The effects of errors in means, variances, and covariances on optimal portfolio choice*, Journal of Portfolio Management, 19 (1993), pp. 6–11. (Cited on p. 155)

[465] S. CHOWDHURY, J. ZHANG, A. MESSAC, AND L. CASTILLO, *Unrestricted wind farm layout optimization (UWFLO): Investigating key factors influencing the maximum power generation*, Renewable Energy, 38 (2012), pp. 16–30. (Cited on p. 367)

[466] A. CHRISTIDIS, C. KOCH, L. POTTEL, AND G. TSATSARONIS, *The contribution of heat storage to the profitable operation of combined heat and power plants in liberalized electricity markets*, Energy, 41 (2012), pp. 75–82. (Cited on p. 304)

[467] M. CHRISTODOULOU AND C. COSTOULAKIS, *Nonlinear mixed integer programming for aircraft collision avoidance in free flight*, in MELECON 2004 Proceedings—The 12th IEEE Mediterranean Electrotechnical Conference, vol. 1, IEEE, May 2004, pp. 327–330. (Cited on pp. 295, 298, 300)

[468] T. CHRISTOF AND A. LOBEL, *The PORTA Manual Page, v. 1.4.1*. Technical report, ZIB, Berlin, 1997. (Cited on p. 79)

[469] M. CHU, Y. ZINCHENKO, S. G. HENDERSON, AND M. B. SHARPE, *Robust optimization for intensity modulated radiation therapy treatment planning under uncertainty*, Physics in Medicine and Biology, 50 (2005), pp. 5463–5477. (Cited on pp. 334, 351, 352)

[470] R. CHURCH AND C. REVELLE, *The maximal covering location problem*, Papers of the Regional Science Association, 32 (1974), pp. 101–118. (Cited on p. 453)

[471] N. CHURI AND L. E. K. ACHENIE, *Novel mathematical programming model for computer aided molecular design*, Industrial and Engineering Chemistry Research, 35 (1996), pp. 3788–3794. (Cited on p. 329)

[472] ———, *On the use of a mixed integer non-linear programming model for refrigerant design*, International Transactions in Operational Research, 4 (1997), pp. 45–54. (Cited on p. 329)

[473] ——, *The optimal design of refrigerant mixtures for a two-evaporator refrigeration system*, Computers and Chemical Engineering, 21 (1997), pp. S349–S354. (Cited on p. 329)

[474] V. CHVÁTAL, *Edmonds polytopes and a hierarchy of combinatorial problems*, Discrete Mathematics, 4 (1973), pp. 305–337. (Cited on pp. 53, 286)

[475] S. CICERONE, G. DI. STEFANO, M. SCHACHTEBECK, AND A. SCHÖBEL, *Multi-stage recovery robustness for optimization problems: A new concept for planning under disturbances*, Information Sciences, 190 (2012), pp. 107–126. (Cited on pp. 333, 340)

[476] A. R. CIRIC AND C. A. FLOUDAS, *Application of the simultaneous match-network optimization approach to the pseudo-pinch problem*, Computers and Chemical Engineering, 14 (1990), pp. 241–250. (Cited on p. 324)

[477] ——, *Heat exchanger network synthesis without decomposition*, Computers and Chemical Engineering, 15 (1991), pp. 385–396. (Cited on p. 324)

[478] I. CIVELEK, I. KARAESMEN, AND A. SCHELLER-WOLF, *Blood Platelet Inventory Management with Protection Levels*, working paper, Tepper School of Business, 2008. (Cited on p. 473)

[479] J. F. CLAERBOUT AND F. MUIR, *Robust modeling with erratic data*, Geophysics, 38 (1973), pp. 826–844. (Cited on p. 33)

[480] A. J. CLARK AND H. SCARF, *Optimal policies for a multi-echelon inventory problem*, Management Science, 6 (1960), pp. 475–490. (Cited on p. 449)

[481] F. CLARKE, *Functional Analysis, Calculus of Variations and Optimal Control*, Springer-Verlag, London, 2013. (Cited on pp. 122, 124, 130)

[482] F. H. CLARKE, *Optimization and Nonsmooth Analysis*, John Wiley & Sons, New York, 1983. Reprinted as Classics in Applied Mathematics 5, SIAM, Philadelphia., 1990. (Cited on pp. 500, 510, 513)

[483] *CMU–IBM Cyber-Infrastructure for MINLP Collaborative Site*. http://www.minlp.org. (Cited on p. 290)

[484] *CMU–IBM Open Source MINLP Project Test Set*, http://egon.cheme.cmu.edu/ibm/page.htm. (Cited on pp. 290, 291)

[485] J. J. COCHRAN, L. A. COX, JR., P. KESKINOCAK, J. P. KHAROUFEH, AND J. COLE SMITH, eds., *Wiley Encyclopedia of Operations Research and Management Science*, John Wiley & Sons, 2011. (Cited on pp. 543, 564, 644, 647, 649)

[486] G. CODATO AND M. FISCHETTI, *Combinatorial Benders' cuts for mixed-integer linear programming*, Operations Research, 54 (2006), pp. 756–766. (Cited on p. 71)

[487] H. COHN AND N. ELKIES, *New upper bounds on sphere packing I*, Annals of Mathematics, 157 (2003), pp. 689–714. (Cited on p. 34)

[488] Y. COLOMBANI AND S. HEIPCKE, *Mosel: An extensible environment for modeling and programming solutions*, in Proceedings of CP-AI-OR, volume 2, 2002, pp. 277–290. (Cited on p. 289)

[489] B. COLSON, *Trust-Region Algorithms for Derivative-Free Optimization and Nonlinear Bilevel Programming*, PhD thesis, Département de Mathématique, FUNDP, Namur, 2003. (Cited on p. 505)

[490] B. COLSON, P. MARCOTTE, AND G. SAVARD, *Bilevel programming: A survey*, 4OR, 3 (2005), pp. 87–107. (Cited on p. 173)

[491] B. COLSON AND PH. L. TOINT, *Optimizing partially separable functions without derivatives*, Optimization Methods and Software, 20 (2005), pp. 493–508. (Cited on p. 530)

[492] M. COLVIN AND C. T. MARAVELIAS, *A stochastic programming approach for clinical trial planning in new drug development*, Computers and Chemical Engineering, 32 (2008), pp. 2626–2642. (Cited on p. 394)

[493] M. CONFORTI, G. CORNUÉJOLS, AND G. ZAMBELLI, *Integer Programming*, Springer, 2014. (Cited on p. 49)

[494] M. CONFORTI, G. CORNUÉJOLS, AND G. ZAMBELLI, *Polyhedral approaches to mixed integer linear programming*, in [1017], pp. 342–385. (Cited on p. 286)

[495] J. CONG, J. R. SHINNERL, M. XIE, T. KONG, AND X. YUAN, *Large-scale circuit placement*, ACM Transactions on Design Automation of Electronic Systems, 10 (2005), pp. 389–430. (Cited on p. 34)

[496] A. R. CONN, N. I. M. GOULD, AND PH. L. TOINT, *Trust-Region Methods*, MOS-SIAM Series on Optimization 1, SIAM, Philadelphia, 2000. (Cited on pp. 228, 496, 505, 506, 522)

[497] A. R. CONN, K. SCHEINBERG, AND PH. L. TOINT, *On the convergence of derivative-free methods for unconstrained optimization*, in Approximation Theory and Optimization, Tributes to M. J. D. Powell, M. D. Buhmann and A. Iserles, eds., Cambridge University Press, Cambridge, 1997, pp. 83–108. (Cited on p. 497)

[498] ——, *A derivative free optimization algorithm in practice*, in Proceedings of 7th AIAA/USAF/NASA/ISSMO Symposium on Multidisciplinary Analysis and Optimization, St Louis, MO, 1998. (Cited on pp. 497, 505, 507)

[499] A. R. CONN, K. SCHEINBERG, AND L. N. VICENTE, *Geometry of interpolation sets in derivative free optimization*, Mathematical Programming, 111 (2008), pp. 141–172. (Cited on pp. 496, 497)

[500] ——, *Geometry of sample sets in derivative free optimization: Polynomial regression and underdetermined interpolation*, IMA Journal of Numerical Analysis, 28 (2008), pp. 721–748. (Cited on p. 497)

[501] A. R. CONN, K. SCHEINBERG, AND L. N. VICENTE, *Global convergence of general derivative-free trust-region algorithms to first- and second-order critical points*, SIAM Journal on Optimization, 20 (2009), pp. 387–415. (Cited on p. 497)

[502] ——, *Introduction to Derivative-Free Optimization* 8, MOS-SIAM Series on Optimization, SIAM, Philadelphia, 2009. (Cited on pp. 223, 495, 496, 497, 498, 500, 506, 521, 522, 524, 530, 531)

[503] A. R. CONN AND PH. L. TOINT, *An algorithm using quadratic interpolation for unconstrained derivative free optimization*, in Nonlinear Optimization and Applications, G. Di Pillo and F. Gianessi, eds., Plenum Publishing, New York, 1996, pp. 27–47. (Cited on p. 496)

[504] A. R. CONN AND L. N. VICENTE, *Bilevel derivative-free optimization and its application to robust optimization*, Optimization Methods and Software, 27 (2012), pp. 561–577. (Cited on p. 530)

[505] G. CONSTABLE AND B. SOMERVILLE, *A Century of Innovation: Twenty Engineering Achievements that Transformed our Lives*, The National Academies Press, 2003. (Cited on p. 357)

[506] G. C. CONTAXIS, C. DELKIS, AND G. KORRES, *Decoupled optimal load flow using linear or quadratic programming*, IEEE Transactions on Power Systems, 1 (1986), pp. 1–7. (Cited on p. 191)

[507] R. D. COOK, D. S. MALKUS, AND M. E. PLESHA, *Concepts and Applications of Finite Element Analysis*, John Wiley & Sons, 1989. (Cited on p. 15)

[508] S. A. COOK, *The complexity of theorem-proving procedures*, in Proceedings of the Third Annual ACM Symposium on the Theory of Computing, New York, 1971, ACM Press, pp. 151–158. (Cited on p. 52)

[509] W. J. COOK, W. H. CUNNINGHAM, W. R. PULLEYBLANK, AND A. SCHRIJVER, *Combinatorial Optimization*, Wiley, New York, 1998. (Cited on p. 52)

[510] I. D. COOPE AND C. J. PRICE, *On the convergence of grid-based methods for unconstrained optimization*, SIAM Journal on Optimization, 11 (2001), pp. 859–869. (Cited on p. 499)

[511] ——, *Positive bases in numerical optimization*, Computational Optimization and Applications, 21 (2002), pp. 169–175. (Cited on p. 499)

[512] F. CORMAN, A. D'ARIANO, D. PACCIARELLI, AND M. PRANZO, *Optimal inter-area coordination of train rescheduling decisions*, Transportation Research E, Logistics and Transportation Review, 48, pp. 71–88. (Cited on p. 71)

[513] W. D. CORNELL, P. CIEPLAK, C. I. BAYLY, I. R. GOULD, K. M. MERZ, D. M. FERGUSON, D. C. SPELLMEYER, T. FOX, J. W. CALDWELL, AND P. A. KOLLMAN, *A second generation force field for the simulation of proteins, nucleic acids and organic molecules*, Journal of the American Chemical Society, 117 (1995), pp. 5179–5197. (Cited on p. 180)

[514] G. CORNUEJOLS AND R. TÜTÜNCÜ, *Optimization Methods in Finance*, Cambridge University Press, Cambridge, 2006. (Cited on pp. 149, 156, 427)

[515] C. CORTES AND V. VAPNIK, *Support-vector networks*, Machine Learning, 20 (1995), pp. 273–297. (Cited on p. 30)

[516] M. J. CÔTÉ, S. S. SYAM, W. B. VOGEL, AND D. C. COWPER, *A mixed integer programming model to locate traumatic brain injury treatment units in the Department of Veterans Affairs: A case study*, Health Care Management Science, 10 (2007), pp. 253–267. (Cited on pp. 94, 95)

[517] R. W. COTTLE AND G. B. DANTZIG, *Complementary pivot theory of mathematical programming*, in Mathematics of the Decision Sciences, Part 1, G. B. Dantzig and A. F. Veinott, eds., American Mathematical Society, Providence, RI, 1968, pp. 115–136. (Cited on p. 7)

[518] R. W. COTTLE, J. S. PANG, AND R. E. STONE, *The Linear Complementarity Problem*, Academic Press, San Diego, 1992. (Cited on p. 7)

[519] A. COUSIN AND E. DI BERNARDINO, *On multivariate extensions of conditional-tail-expectation*, Insurance: Mathematics and Economics, 55 (2014), pp. 272–282. (Cited on p. 432)

[520] T. M. COVER, *Geometrical and statistical properties of systems of linear inequalities with applications in pattern recognition*, IEEE Transactions on Electronic Computers, EC-14 (1965), pp. 326–334. (Cited on p. 30)

[521] A. COZAD, N. V. SAHINIDIS, AND D. C. MILLER, *Learning surrogate models for simulation-based optimization*, AIChE Journal, 60 (2014), pp. 2211–2227. (Cited on p. 318)

[522] D. L. CRAFT, T. F. HALABI, H. A. SHIH, AND T. R. BORTFELD, *Approximating convex Pareto surfaces in multiobjective radiotherapy planning*, Medical Physics, 33 (2006), pp. 3399–3407. (Cited on p. 94)

[523] N. L. CRAIG, R. CRAIGIE, M. GELLERT, AND A. M. LAMBOWITZ, *Mobile DNA* II, ASM Press, Washington, D.C., 2007. (Cited on p. 182)

[524] P. CRAMA, B. REYCK, AND Z. DEGRAEVE., *Milestone payments or royalties? Contract design for R&D licensing*, Operations Research, 56 (2008), pp. 1539–1552. (Cited on pp. 474, 475)

[525] G. P. CRESPI, D. KUROIWA, AND M. ROCCA, *Quasiconvexity of set-valued maps assures well-posedness of robust vector optimization*, Annals of Operations Research, published online February 21, 2015. (Cited on p. 344)

[526] J. CURRENT, M. S. DASKIN, AND D. SCHILLING, *Discrete network location models*, in [628], ch. 3. (Cited on p. 450)

[527] J. CURRIE AND D. I. WILSON, *OPTI: Lowering the barrier between open source optimizers and the industrial MATLAB user*, in Proceedings of Foundations of Computer-Aided Process Operations 2012, N. Sahinidis and J. Pinto, eds., Savannah, GA, 2012, pp. 8–11. (Cited on p. 288)

[528] F. E. CURTIS, O. SCHENK, AND A. WÄCHTER, *An interior-point algorithm for large-scale nonlinear optimization with inexact step computations*, SIAM Journal on Scientific Computing, 32 (2010), pp. 3447–3475. (Cited on p. 231)

[529] A. L. CUSTÓDIO, J. E. DENNIS, JR., AND L. N. VICENTE, *Using simplex gradients of nonsmooth functions in direct search methods*, IMA Journal of Numerical Analysis, 28 (2008), pp. 770–784. (Cited on p. 500)

[530] A. L. CUSTÓDIO AND J. F. A. MADEIRA, *GLODS: Global and local optimization using direct search*, Journal of Global Optimization, 62 (2015), pp. 1–28. (Cited on p. 506)

[531] A. L. CUSTÓDIO, J. F. A. MADEIRA, A. I. F. VAZ, AND L. N. VICENTE, *Direct multisearch for multiobjective optimization*, SIAM Journal on Optimization, 21 (2011), pp. 1109–1140. (Cited on p. 506)

[532] A. L. CUSTÓDIO, H. ROCHA, AND L. N. VICENTE, *Incorporating minimum Frobenius norm models in direct search*, Computational Optimization and Applications, 46 (2010), pp. 265–278. (Cited on p. 506)

[533] A. L. CUSTÓDIO AND L. N. VICENTE, *Using sampling and simplex derivatives in pattern search methods*, SIAM Journal on Optimization, 18 (2007), pp. 537–555. (Cited on p. 499)

[534] C. R. CUTLER AND B. L. RAMAKER, *Dynamic Matrix Control—A Computer Control Algorithm*, AIChE National Meeting, Houston, TX, 1979. (Cited on pp. 41, 43)

[535] J. CZYZYK, M. P. MESNIER, AND J. J. MORÉ, *The NEOS server*, Computing in Science and Engineering, 5 (1998), pp. 68–75. (Cited on p. 288)

[536] C. F. DAGANZO, *Logistics Systems Analysis*. 4th edition, Springer, 2005. (Cited on p. 465)

[537] ———, *The cell transmission model, part II: Network traffic*, Transportation Research Part B: Methodological, 29 (1995), pp. 79–93. (Cited on p. 488)

[538] O. DAGLIYAN, F. UNEY-YUKSEKTEPE, I. H. KAVAKLI, AND M. TURKAY, *Optimization based tumor classification from microarray gene expression data*, PLoS One, 6 (2011), p. e14579. (Cited on p. 94)

[539] M. A. DAHLEH AND I. J. DIAZ-BOBILLO, *Control of Uncertain Systems. A Linear Programming Approach*, Prentice Hall, 1995. (Cited on p. 27)

[540] M. M. DAICHENDT AND I. E. GROSSMANN, *Integration of hierarchical decomposition and mathematical programming for the synthesis of process flowsheets*, Computers and Chemical Engineering, 22 (1998), pp. 147–175. (Cited on p. 318)

[541] M. D. DAILY, D. MASICA, A. SIVASUBRAMANIAN, S. SOMAROUTHU, AND J. J. GRAY, *CAPRI rounds 3-5 reveal promising successes and future challenges for RosettaDock*, Proteins: Structure, Function, and Bioinformatics, 60 (2005), pp. 181–186. (Cited on p. 181)

[542] R. J. DAKIN, *A tree search algorithm for mixed integer programming problems*, Computer Journal, 8 (1965), pp. 250–255. (Cited on p. 275)

[543] C. D'AMBROSIO, A. FRANGIONI, L. LIBERTI, AND A. LODI, *Experiments with a feasibility pump approach for nonconvex MINLPs*, in Experimental Algorithms, P. Festa, ed., Springer, 2010, pp. 350–360. (Cited on p. 286)

[544] C. D'AMBROSIO AND A. LODI, *Mixed integer nonlinear programming tools: A practical overview*, 4OR, 9 (2011), pp. 329–349. (Cited on p. 274)

[545] ———, *Mixed integer nonlinear programming tools: An updated practical overview*, Annals of Operations Research, 204 (2013), pp. 301–320. (Cited on p. 315)

[546] E. DANNA, E. ROTHBERG, AND C. LE PAPE, *Exploring relaxation induced neighborhoods to improve MIP solutions*, Mathematical Programming, 102 (2005), pp. 71–90. (Cited on pp. 58, 286)

[547] G. DANNINGER AND I. M. BOMZE, *Using copositivity for global optimality criteria in concave quadratic programming problems*, Mathematical Programming, 62 (1993), pp. 575–580. (Cited on p. 168)

[548] G. B. DANTZIG, *Linear programming under uncertainty*, Management Science, 1 (1955), pp. 197–206. (Cited on p. 390)

[549] ———, *Linear Programming and Extensions*, Princeton University Press, Princeton, NJ, 1963. (Cited on pp. 3, 5)

[550] M. L. DARBY, M. NIKOLAOU, J. JONES, AND D. NICHOLSON, *RTO: An overview and assessment of current practice*, Journal of Process Control, 21 (2011), pp. 874–884. (Cited on p. 44)

[551] A. D'ARIANO, *Improving Real-Time Train Dispatching: Models, Algorithms and Applications*, TRAIL Thesis Series no. T2008/6, The Netherlands TRAIL Research School, 2008. (Cited on p. 66)

[552] A. D'ARIANO, D. PACCIARELLI, AND M. PRANZO, *A branch and bound algorithm for scheduling trains in a railway network*, European Journal of Operational Research, 183 (2007), pp. 643–657. (Cited on p. 69)

[553] S. K. DAS AND L. B. MARKS, *Selection of coplanar or noncoplanar beams using three-dimensional optimization based on maximum beam separation and minimized nontarget irradiation*, International Journal of Radiation Oncology • Biology • Physics, 38 (1997), pp. 643–655. (Cited on p. 95)

[554] M. S. DASKIN, *What you should know about location modeling*, Naval Research Logistics, 55 (2008), pp. 283–294. (Cited on p. 450)

[555] ———, *Network and Discrete Location: Models, Algorithms, and Applications*, 2nd edition, Wiley, New York, 2013. (Cited on p. 450)

[556] M. S. DASKIN, C. R. COULLARD, AND Z.-J. M. SHEN, *An inventory-location model: Formulation, solution algorithm and computational results*, Annals of Operations Research, 110 (2002), pp. 83–106. (Cited on p. 441)

[557] J. W. DAVID, C. Y. CHENG, T. D. CHOI, C. T. KELLEY, AND J. GABLONSKY, *Optimal Design of High Speed Mechanical Systems*, technical report, North Carolina State University, Center for Research in Scientific Computation, 1997. CRSC-TR97-18. (Cited on pp. 507, 508)

[558] J. W. DAVID, C. T. KELLEY, AND C. Y. CHENG, *Use of an implicit filtering algorithm for mechanical system parameter identification*, Modeling of CI and SI Engines, pp. 189–194, Society of Automotive Engineers, Washington, DC, 1996. SAE International Congress and Exposition Conference Proceedings. SAE Paper 960358. (Cited on pp. 507, 508)

[559] G. DAVIDSON, I. MOSCOVICE, AND D. REMUS, *Hospital size, uncertainty, and pay-for-performance*, Health Care Financing Review, 29 (2006), pp. 45–57. (Cited on p. 96)

[560] T. N. DAVIDSON, *Enriching the art of FIR filter design via convex optimization*, IEEE Signal Processing Magazine, 27 (2010), pp. 89–101. (Cited on p. 30)

[561] T. N. DAVIDSON, Z.-Q. LUO, AND J. F. STURM, *Linear matrix inequality formulation of spectral mask constraints*, IEEE Transactions on Signal Processing, 50 (2002), pp. 2702–2715. (Cited on pp. 29, 30)

[562] B. J. DAVIS, L. A. TAYLOR, AND V. I. MANOUSIOUTHAKIS, *Identification of the attainable region for batch reactor networks*, Industrial and Engineering Chemistry Research, 47 (2008), pp. 3388–3400. (Cited on p. 320)

[563] C. DAVIS, *Theory of positive linear dependence*, American Journal of Mathematics, 76 (1954), pp. 733–746. (Cited on p. 498)

[564] E. DAVIS AND M. IERAPETRITOU, *A kriging method for the solution of nonlinear programs with black-box functions*, AIChE Journal, 53 (2007), pp. 2001–2012. (Cited on p. 318)

[565] L. B. DAVIS, F. SAMANLIOGLU, X. QU, AND S. ROOT, *Inventory planning and coordination in disaster relief efforts*, International Journal of Production Economics, 141 (2013), pp. 561–573. (Cited on p. 484)

[566] M. L. DE GRANO, D. MEDEIROS, AND D. EITEL, *Accommodating individual preferences in nurse scheduling via auctions and optimization*, Health Care Management Science, 12 (2009), pp. 228–242. (Cited on p. 95)

[567] E. DE KLERK, *Aspects of Semidefinite Programming: Interior Point Algorithms and Selected Applications*, Kluwer Academic, Dordrecht, Netherlands, 2002. (Cited on p. 11)

[568] E. DE KLERK, C. ROOS, AND T. TERLAKY, *Semidefinite Problems in Truss Topology Optimization*, Technical Report 95-128, Faculty of Technical Mathematics and Informatics, T. U. Delft, The Netherlands, 1995. (Cited on p. 11)

[569] ———, *Initialization in semidefinite programming via a self-dual, skew-symmetric embedding*, Operations Research Letters, 20 (1997), pp. 213–221. (Cited on p. 11)

[570] ———, *Infeasible-start semidefinite programming algorithms via self-dual embeddings*, in Topics in Semidefinite and Interior Point Methods, P. M. Pardalos and H. Wolkowicz, eds., vol. 18 of Fields Institute Communications, AMS, Providence, RI, 1998, pp. 215–236. (Cited on p. 11)

[571] ———, *On primal-dual path-following algorithms for semidefinite programming*, in New Trends in Mathematical Programming, F. Gianessi, S. Komlósi, and T. Rapcsák, eds., Kluwer Academic, Dordrecht, Netherlands, 1998, pp. 137–157. (Cited on p. 11)

[572] R. DE SÁ FERREIRA, *A Mixed-Integer Linear Programming Approach to the AC Optimal Power Flow in Distribution Systems*, PhD thesis, Universidade Federal do Rio de Janeiro, 2013. (Cited on p. 198)

[573] S. DEHGHAN, N. AMJADY, AND A. KAZEMI, *Two-stage robust generation expansion planning: A mixed integer linear programming model*, IEEE Transactions on Power Systems, 29 (2013), pp. 584–597. (Cited on p. 365)

[574] O. DEKEL, O. SHAMIR, AND L. XIAO, *Learning to classify with missing and corrupted features*, Machine Learning, 81 (2010), pp. 149–178. (Cited on p. 94)

[575] E. DELAGE AND Y. YE, *Distributionally robust optimization under moment uncertainty with application to data-driven problems*, Operations Research, 58 (2010), pp. 595–612. (Cited on pp. 333, 344, 368, 391)

[576] D. DELAHAYE AND S. PUECHMOREL, *Modeling and Optimization of Air Traffic*, John Wiley & Sons, 2013. (Cited on p. 293)

[577] P. DELSARTE, J. M. GOETHALS, AND J. J. SEIDEL, *Spherical codes and designs*, Geometriae Dedicata, 6 (1977), pp. 363–388. (Cited on p. 34)

[578] S. DEMPE, *Foundations of Bilevel Programming*, Kluwer Academic, 2002. (Cited on p. 173)

[579] D. DEN HERTOG, *Interior Point Approach to Linear, Quadratic and Convex Programming*, vol. 277 of Mathematics and Its Applications, Kluwer Academic, Dordrecht, Netherlands, 1994. (Cited on pp. 4, 7, 8, 11)

[580] G. DENG AND M. C. FERRIS, *Adaptation of the UOBYQA algorithm for noisy functions*, in Proceedings of the 2006 Winter Simulation Conference, L. F. Perrone, F. P. Weiland, J. Liu, B. G. Lawson, D. M. Nicol, and R. M. Fujimoto, eds., 2006, pp. 312–319. (Cited on p. 501)

[581] P. DENHOLM AND R. SIOSHANSI, *The value of compressed air energy storage with wind in transmission-constrained electric power systems*, Energy Policy, 37 (2009), pp. 3149–3158. (Cited on p. 463)

[582] B. DENIZ, I. KARAESMEN, AND A. SCHELLER-WOLF, *Managing perishables with substitution: Issuance and replenishment heuristics*, Manufacturing and Service Operations Management, 12 (2010), pp. 319–329. (Cited on p. 473)

[583] J. E. DENNIS, JR., C. J. PRICE, AND I. D. COOPE, *Direct search methods for nonlinearly constrained optimization using filters and frames*, Optimization and Engineering, 5 (2004), pp. 123–144. (Cited on p. 505)

[584] J. E. DENNIS, JR. AND V. TORCZON, *Direct search methods on parallel machines*, SIAM Journal on Optimization, 1 (1991), pp. 448–474. (Cited on p. 499)

[585] D. DENTCHEVA AND A. RUSZCZYŃSKI, *Optimization with stochastic dominance constraints*, SIAM Journal on Optimization, 14 (2003), pp. 548–566. (Cited on p. 427)

[586] ———, *Portfolio optimization with stochastic dominance constraints*, Journal of Banking and Finance, 30 (2006), pp. 433–451. (Cited on pp. 427, 428)

[587] S. DEO AND C. J. CORBETT, *Cournot competition under yield uncertainty: The case of the US influenza vaccine market*, Manufacturing and Service Operations Management, 11 (2009), pp. 563–576. (Cited on p. 472)

[588] ——, *Dynamic Allocation of Scarce Resources Under Supply Uncertainty*, technical report, 2014. Available at SSRN 1619408. (Cited on p. 491)

[589] C. W. DeWITT, L. S. LASDON, A. D. WAREN, D. A. BRENNER, AND S. A. MEL-HEM, *OMEGA: An improved gasoline blending system for Texaco*, Interfaces, 19 (1989), pp. 85–101. (Cited on p. 208)

[590] S. DEY AND A. GUPTE, *Analysis of MILP techniques for the pooling problem*, Operations Research, 63 (2015), pp. 412–427. (Cited on p. 214)

[591] A. DEZA, E. NEMATOLLAHI, R. PEYGHAMI, AND T. TERLAKY, *The central path visits all the vertices of the Klee-Minty cube*, Optimization Methods and Software, 21 (2006), pp. 851–865. (Cited on p. 9)

[592] A. DEZA, E. NEMATOLLAHI, AND T. TERLAKY, *How good are interior point methods? Klee-Minty cubes tighten iteration-complexity bounds*, Mathematical Programming, 113 (2008), pp. 1–14. (Cited on p. 9)

[593] A. DEZA, T. TERLAKY, AND Y. ZINCHENKO, *Central path curvature and iteration-complexity for redundant Klee-Minty cubes*, in Advances in Mechanics and Mathematics III., D. Gao and H. Sherali, eds., Springer Science, 2000, pp. 195–233. (Cited on p. 9)

[594] *DFL: Derivative-Free Library*, http://www.dis.uniroma1.it/~lucidi/DFL. (Cited on pp. 499, 505, 506)

[595] *DFO*, https://projects.coin-or.org/Dfo. (Cited on pp. 498, 503)

[596] P. J. C. DICKINSON AND L. GIJBEN, *On the computational complexity of membership problems for the completely positive cone and its dual*, Computational Optimization and Applications, 57 (2014), pp. 403–415. (Cited on p. 113)

[597] P. A. DiMAGGIO, S. R. McALLISTER, C. A. FLOUDAS, X.-J. FENG, J. D. RABI-NOWITZ, AND H. A. RABITZ, *Biclustering via optimal re-ordering of data matrices in systems biology: Rigorous methods and comparative studies*, BMC Bioinformatics, 9 (2008), pp. 1–16. (Cited on p. 181)

[598] ——, *Network flow model for biclustering via optimal re-ordering of data matrices*, Journal of Global Optimization, 47 (2010), pp. 343–354. (Cited on p. 181)

[599] M. A. DINIZ-EHRHARDT, J. M. MARTÍNEZ, AND L. G. PEDROSO, *Derivative-free methods for nonlinear programming with general lower-level constraints*, Computational and Applied Mathematics, 30 (2011), pp. 19–52. (Cited on p. 504)

[600] W. DINKLEBACH, *On nonlinear fractional programming*, Management Science, 13 (1967), pp. 492–498. (Cited on p. 167)

[601] Y. DIOUANE, S. GRATTON, AND L. N. VICENTE, *Globally convergent evolution strategies*, Mathematical Programming, 152 (2015), pp. 467–490. (Cited on p. 506)

[602] *Direct Multisearch (DMS)*, http://www.mat.uc.pt/dms. (Cited on p. 506)

[603] S. P. DIRKSE AND M. C. FERRIS, *The path solver: A nonmonotone stabilization scheme for mixed complementarity problems*, Optimization Methods and Software, 5 (1995), pp. 123–156. (Cited on p. 196)

[604] M. DODANGEH AND L. N. VICENTE, *Worst case complexity of direct search under convexity*, Mathematical Programming, 155 (2016), pp. 307–332. (Cited on p. 501)

[605] M. DODANGEH, L. N. VICENTE, AND Z. ZHANG, *On the optimal order of worst case complexity of direct search*, Optimization Letters, 10 (2016), pp. 699–708. (Cited on p. 502)

[606] M. F. DOHERTY AND M. F. MALONE, *Conceptual Design of Distillation Systems*, McGraw-Hill, New York, 2001. (Cited on p. 238)

[607] E. D. DOLAN, J. J. MORÉ, AND T. S. MUNSON, *Optimality measures for performance profiles*, SIAM Journal on Optimization, 16 (2006), pp. 891–909. (Cited on p. 291)

[608] T. DOLLEVOET, F. CORMAN, A. D'ARIANO, AND D. HUISMAN, *An Iterative Optimization Framework for Delay Management and Train Scheduling*. No. EI 2012-10, Econometric Institute Report from Erasmus University, Rotterdam, Econometric Institute, 2012. (Cited on p. 71)

[609] T. DOLLEVOET, D. HUISMAN, L. KROON, M. SCHMIDT, AND A. SCHÖBEL, *Delay management including capacities of stations*, Transportation Science, 49 (2015), pp. 185–203. (Cited on pp. 69, 71)

[610] H. W. DOMMEL AND W. F. TINNEY, *Optimal power flow solutions*, IEEE Transactions on Power Apparatus and Systems, 87 (1962), pp. 1866–1876. (Cited on p. 188)

[611] H.-G. DONG, C.-Y. LIN, AND C.-T. CHANG, *Simultaneous optimization approach for integrated water-allocation and heat-exchange networks*, Chemical Engineering Science, 63 (2008), pp. 3664–3678. (Cited on p. 327)

[612] ———, *Simultaneous optimization strategy for synthesizing heat exchanger networks with multi-stream mixers*, Chemical Engineering Research and Design, 86 (2008), pp. 299–309. (Cited on p. 327)

[613] D. L. DONOHO, *Compressed sensing*, IEEE Transactions on Information Theory, 52 (2006), pp. 1289–1306. (Cited on pp. 32, 33)

[614] D. L. DONOHO AND X. HUO, *Uncertainty principles and ideal atomic decomposition*, IEEE Transactions on Information Theory, 47 (2001), pp. 2845–2862. (Cited on p. 33)

[615] D. L. DONOHO AND J. TANNER, *Sparse nonnegative solutions of underdetermined systems by linear programming*, Proceedings of the National Academy of Sciences of the United States of America, 102 (2005), pp. 9446–9451. (Cited on p. 33)

[616] S. DONOVAN, *Wind Farm Optimization*, in 40th Annual Conference, Operational Research Society of New Zealand, Wellington, New Zealand, 2005. (Cited on p. 370)

[617] ———, *An improved mixed integer programming model for wind farm layout optimisation*, in 41st Annual Conference, Operational Research Society of New Zealand, 2006, pp. 143–151. (Cited on pp. 367, 369, 370)

[618] S. DONOVAN, G. NATES, H. WATERER, AND R. ARCHER, *Mixed Integer Programming Models for Wind Farm Design*. http://coral.ie.lehigh.edu/~jeff/mip-2008/talks/waterer.pdf, 2008. (Cited on p. 370)

[619] W. S. DORN, R. E. GOMORY, AND H. J. GREENBERG, *Automatic design of optimal structures*, Journal de Mécanique, 3 (1964), pp. 25–52. (Cited on pp. 13, 14, 16)

[620] R. DOUGHERTY, C. FREILING, AND K. ZEGER, *Six new non-Shannon information inequalities*, in 2006 IEEE International Symposium on Information Theory, July 2006, pp. 233–236. (Cited on p. 35)

[621] A. W. DOWLING AND L. T. BIEGLER, *Optimization-based process synthesis for sustainable power generation*, Chemical Engineering Transactions, 35 (2013), pp. 1–12. (Cited on p. 246)

[622] ———, *A framework for efficient large scale equation-oriented flowsheet optimization*, Computers and Chemical Engineering, 72 (2015), pp. 3–20. (Cited on p. 246)

[623] J. DOWNWARD, *Targeting RAS signaling pathways in cancer therapy*, Nature Reviews Cancer, 3 (2003), pp. 11–22. (Cited on p. 519)

[624] A. DÖYEN, N. ARAS, AND G. BARBAROSOĞLU, *A two-echelon stochastic facility location model for humanitarian relief logistics*, Optimization Letters, 6 (2012), pp. 1123–1145. (Cited on p. 480)

[625] S. J. DOYLE AND R. SMITH, *Targeting water reuse with multiple contaminants*, Process Safety and Environmental Protection, 75 (1997), pp. 181–189. (Cited on p. 327)

[626] T. DREZNER, Z. DREZNER, AND J. GUYSE, *Equitable service by a facility: Minimizing the Gini coefficient*, Computers and Operations Research, 36 (2009), pp. 3240–3246. (Cited on p. 480)

[627] Z. DREZNER, ed., *Facility Location: A Survey of Applications and Methods*, Springer-Verlag, New York, 1995. (Cited on pp. 450, 582)

[628] Z. DREZNER AND H. W. HAMACHER, eds., *Facility Location: Applications and Theory*, Springer-Verlag, New York, 2002. (Cited on pp. 450, 574, 580)

[629] Z. DREZNER, K. KLAMROTH, A. SCHÖBEL, AND G. O. WESOLOWSKY, *The Weber problem*, in [628], ch. 1. (Cited on p. 450)

[630] A. DRUD, *CONOPT: A System for Large Scale Nonlinear Optimization, Reference Manual for CONOPT*, ARKI Consulting and Development A/S. (Cited on p. 242)

[631] ———, *CONOPT: A GRG code for large sparse dynamic nonlinear optimization problems*, Mathematical Programming, 31 (1985), pp. 153–191. (Cited on p. 290)

[632] ———, *CONOPT—A large scale GRG Code*, ORSA Journal on Computing, 6 (1994), pp. 207–216. (Cited on pp. 196, 242)

[633] E. DRURY, P. DENHOLM, AND R. SIOSHANSI, *The value of compressed air energy storage in energy and reserve markets*, Energy, 36 (2011), pp. 4959–4973. (Cited on pp. 462, 463)

[634] W. D. D'SOUZA, R. R. MEYER, AND L. SHI, *Selection of beam orientations in intensity-modulated radiation therapy using single-beam indices and integer programming*, Physics in Medicine and Biology, 49 (2004), pp. 3465–3481. (Cited on p. 95)

[635] B. L. DU PONT, J. CAGAN, AND P. MORIARTY, *Optimization of wind farm layout and wind turbine geometry using a multi-level extended pattern search algorithm that accounts for variation in wind shear profile shape*, in ASME International Design Engineering Technical Conferences & Computers and Information in Engineering Conference IDETC/CIE 2012, Chicago, IL, 2012. (Cited on p. 369)

[636] B. DUMITRESCU, *Positive Trigonometric Polynomials and Signal Processing Applications*, Springer, 2007. (Cited on pp. 29, 30)

[637] N. DUNLAP, A. MCINTOSH, K. SHENG, W. YANG, B. TURNER, A. SHOUSHTARI, J. SHEEHAN, D. R. JONES, W. LU, AND K. RUCHALA, *Helical tomotherapy-based STAT stereotactic body radiation therapy: Dosimetric evaluation for a real-time SBRT treatment planning and delivery program*, Medical Dosimetry, 35 (2011), pp. 312–319. (Cited on p. 94)

[638] G. DÜNNEBIER AND C. C. PANTELIDES, *Optimal design of thermally coupled distillation columns*, Industrial and Engineering Chemistry Research, 38 (1999), pp. 162–176. (Cited on p. 321)

[639] J. DUPAČOVÁ, *On minimax solution of stochastic linear programming problems*, Časopis pro pěstování matematiky, 91 (1966), pp. 423–430. (Cited on p. 391)

[640] ——, *Uncertainties in minimax stochastic programs*, Optimization, 60 (2011), pp. 1235–1250. (Cited on p. 344)

[641] M. DÜR, *Copositive programming—A survey*, in Recent Advances in Optimization and its Applications in Engineering, Springer, 2010, pp. 3–20. (Cited on p. 113)

[642] M. DÜR, R. HORST, AND M. LOCATELLI, *Necessary and sufficient global optimality conditions for convex maximization revisited*, Journal of Mathematical Analysis and Applications, 217 (1998), pp. 637–649. (Cited on p. 168)

[643] M. A. DURAN AND I. E. GROSSMANN, *An outer-approximation algorithm for a class of mixed-integer nonlinear programs*, Mathematical Programming, 36 (1986), pp. 307–339. (Cited on pp. 277, 292)

[644] ——, *Simultaneous optimization and heat integration of chemical processes*, AIChE Journal, 32 (1986), pp. 123–138. (Cited on p. 324)

[645] S. DURAN, M. A. GUTIERREZ, AND P. KESKINOCAK, *Pre-positioning of emergency items for CARE International*, Interfaces, 41 (2011), pp. 223–237. (Cited on p. 480)

[646] A. DUTTA, A. K. DIKSHIT, S. RAY, AND M. BANDYOPADHYAY, *Impact of SO_2 emission limits on petroleum refinery operations I: A linear programming model*, Journal of Environmental Systems, 29 (2002), pp. 15–38. (Cited on p. 40)

[647] A. P. DUVEDI AND L. E. K. ACHENIE, *Designing environmentally safe refrigerants using mathematical programming*, Chemical Engineering Science, 51 (1996), pp. 3727–3739. (Cited on p. 329)

[648] M. DVOŘÁK AND P. HAVEL, *Combined heat and power production planning under liberalized market conditions*, Applied Thermal Engineering, 43 (2012), pp. 163–173. (Cited on p. 304)

[649] M. DYER AND L. WOLSEY, *Formulating the single machine sequencing problem with release dates as a mixed integer program*, Discrete Applied Mathematics, 26 (1990), pp. 255–270. (Cited on p. 69)

[650] B. VAN DYKE AND T. J. ASAKI, *Using QR decomposition to obtain a new instance of mesh adaptive direct search with uniformly distributed polling directions*, Journal of Optimization Theory and Applications, 159 (2013), pp. 805–821. (Cited on p. 512)

[651] T. EASTON, K. HOOKER, AND E. K. LEE, *Facets of the independent set polytope*, Mathematical Programming, 98 (2003), pp. 177–199. (Cited on pp. 94, 102)

[652] F. EDGEWORTH, *The mathematical theory of banking*, Journal of the Royal Statistical Society, 51 (1888), pp. 113–127. (Cited on p. 443)

[653] J. EDMONDS, *Maximum matching and a polyhedron with 0-1 vertices*, Journal of Research of the National Bureau of Standards, 69B (1965), pp. 125–130. (Cited on p. 52)

[654] M. EHRGOTT, *Multicriteria Optimization*, Lecture Notes in Economics and Mathematical Systems, Springer, 2005. (Cited on p. 173)

[655] M. EHRGOTT, Ç. GÜLER, H. W. HAMACHER, AND L. SHAO, *Mathematical optimization in intensity modulated radiation therapy*, Annals of Operations Research, 175 (2010), pp. 309–365. (Cited on p. 95)

[656] R. EHRHARDT AND C. MOSIER, *A revision of the power approximation for computing* (s, S) *policies*, Management Science, 30 (1984), pp. 618–622. (Cited on p. 446)

[657] H. A. EISELT AND G. LAPORTE, *Objectives in location problems*, in [627], pp. 151–180. (Cited on p. 480)

[658] H. A. EISELT AND VLADIMIR MARIANOV, eds., *Foundations of Location Analysis*, Springer-Verlag, 2011. (Cited on pp. 450, 647)

[659] H. A. EISELT AND C.-L. SANDBLOM, *Decision Analysis, Location Models, and Scheduling Problems*, Springer, New York, 2004. (Cited on p. 450)

[660] A. EKSTRÖM, G. BAARDSEN, C. FORSSÉN, G. HAGEN, M. HJORTH-JENSEN, G. R. JANSEN, R. MACHLEIDT, W. NAZAREWICZ, T. PAPENBROCK, J. SARICH, AND S. M. WILD, *Optimized chiral nucleon-nucleon interaction at next-to-next-to-leading order*, Physical Review Letters, 110 (2013), p. 192502. (Cited on p. 537)

[661] L. H. EL GHAOUI AND H. LIBERT, *Robust solutions to least-square problems with uncertain data*, SIAM Journal on Matrix Analysis and Applications, 18 (1997), pp. 1035–1064. (Cited on p. 394)

[662] L. H. EL GHAOUI, F. OUSTRY, AND H. LEBRET, *Robust solutions to uncertain semidefinite programs*, SIAM Journal on Optimization, 9 (1998), pp. 33–52. (Cited on pp. 344, 394)

[663] M. ELAD, *Sparse and Redundant Representations: From Theory to Applications in Signal and Image Processing*, Springer, 2010. (Cited on pp. 31, 33)

[664] Y. C. ELDAR AND G. KUTYNIOK, eds., *Compressed Sensing: Theory and Applications*, Cambridge University Press, 2012. (Cited on p. 33)

[665] S. ELHEDHLI, *Service system design with immobile servers, stochastic demand, and congestion*, Manufacturing and Service Operations Management, 8 (2006), pp. 92–97. (Cited on p. 292)

[666] A. M. ELICECHE, S. M. CORVALÁN, AND P. MARTÍNEZ, *Environmental life cycle impact as a tool for process optimisation of a utility plant*, Computers and Chemical Engineering, 31 (2007), pp. 648–656. (Cited on p. 292)

[667] A. ELKAMEL, M. BA-SHAMMAKH, P. DOUGLAS, AND E. CROISET, *An optimization approach for integrating planning and* CO_2 *emission reduction in the petroleum refining industry*, Industrial and Engineering Chemistry Research, 47 (2008), pp. 760–776. (Cited on p. 327)

[668] C. N. ELKINTON, J. F. MANWELL, AND J. G. MCGOWAN, *Algorithms for offshore wind farm layout optimization*, Wind Engineering, 32 (2008), pp. 67–84. (Cited on p. 367)

[669] J. ELLEPOLA, N. THIJSSEN, J. GRIEVINK, G. BAAK, A. AVHALE, AND J. SCHIJNDEL, *Development of a synthesis tool for gas-to-liquid complexes*, Computers and Chemical Engineering, 42 (2012), pp. 2–14. (Cited on p. 318)

[670] ENVIRONMENT CANADA, *Engineering Climate Datasets*, http://climate.weather.gc.ca/prods_servs/engineering_e.html, March 2014. (Cited on p. 371)

[671] G. D. EPPEN, *Effects of centralization on expected costs in a multi-location newsboy problem*, Management Science, 25 (1979), pp. 498–501. (Cited on p. 486)

[672] T. ERICSON AND V. ZINOVIEV, *Codes on Euclidean Spheres*, vol. 63 of North Holland Mathematical Library, Elsevier Science B. V., Amsterdam, 2001. (Cited on p. 34)

[673] D. ERLENKOTTER, *A dual-based procedure for uncapacitated facility location*, Operations Research, 26 (1978), pp. 992–1009. (Cited on p. 452)

[674] ———, *Note—An early classic misplaced: Ford W. Harris's economic order quantity model of 1915*, Management Science, 35 (1989), pp. 898–900. (Cited on p. 440)

[675] ———, *Ford Whitman Harris and the economic order quantity model*, Operations Research, 38 (1990), pp. 937–946. (Cited on p. 440)

[676] ———, *Ford Whitman Harris's economical lot size model*, International Journal of Production Economics, 155 (2014), pp. 12–15. (Cited on p. 440)

[677] P. M. ESFAHANI AND D. KUHN, *Data-Driven Distributionally Robust Optimization Using the Wasserstein Metric: Performance Guarantees and Tractable Reformulations*, arXiv preprint arXiv:1505.05116 (2015). (Cited on p. 344)

[678] W. R. ESPOSITO AND C. A. FLOUDAS, *Deterministic global optimization in isothermal reactor network synthesis*, Journal of Global Optimization, 22 (2002), pp. 59–95. (Cited on p. 320)

[679] EUROPEAN COMMISSION, *EU Transport in Figures*, Statistical Pocketbook, 2013. (Cited on p. 65)

[680] L. EVERS, T. DOLLEVOET, A. I. BARROS, AND H. MONSUUR, *Robust UAV mission planning*, Annals of Operations Research, 222 (2014), pp. 293–315. (Cited on p. 333)

[681] Y. G. EVTUSHENKO AND M. A. POSYPKIN, *A deterministic approach to global box-constrained optimization*, Optimization Letters, 7 (2013), pp. 819–829. (Cited on p. 172)

[682] G. A. EZZELL, *Genetic and geometric optimization of three-dimensional radiation therapy treatment planning*, Medical Physics, 23 (1996), pp. 293–305. (Cited on p. 95)

[683] F. J. FABOZZI, D. HUANG, AND G. ZHOU, *Robust portfolios: Contributions from operations research and finance*, Annals of Operations Research, 176 (2010), pp. 191–220. (Cited on p. 431)

[684] F. J. FABOZZI, P. N. KOLM, D. PACHAMANOVA, AND S. M. FOCARDI, *Robust Portfolio Optimization and Management*, John Wiley & Sons, Hoboken, NJ, 2007. (Cited on pp. 149, 156)

[685] J. E. FALK AND K. R. HOFFMAN, *A successive underestimation method for concave minimization problems*, Mathematics of Operations Research, 1 (1976), pp. 251–259. (Cited on p. 171)

[686] ———, *Concave minimization via collapsing polytopes*, Operations Research, 34 (1986), pp. 919–929. (Cited on p. 171)

[687] D. E. FARRAR, *The Investment Decision under Uncertainty*, Prentice-Hall, Englewood Cliffs, NJ, USA, 1962. (Cited on p. 152)

[688] G. FASANO, G. LIUZZI, S. LUCIDI, AND F. RINALDI, *A linesearch-based derivative-free approach for nonsmooth constrained optimization*, SIAM Journal on Optimization, 24 (2014), pp. 959–992. (Cited on pp. 503, 505)

[689] G. FASANO, J. L. MORALES, AND J. NOCEDAL, *On the geometry phase in model-based algorithms for derivative-free optimization*, Optimization Methods and Software, 24 (2009), pp. 145–154. (Cited on p. 497)

[690] H. O. FATTORINI, *Infinite Dimensional Optimization and Control Theory*, Cambridge University Press, Cambridge, 1999. (Cited on p. 122)

[691] FDA, *A Review of FDA's Approach to Medical Product Shortages*. U.S. FDA, U.S. Department of Health and Human Services, 2011. (Cited on p. 477)

[692] A. FEDERGRUEN, G. PRASTACOS, AND P. H. ZIPKIN, *An allocation and distribution model for perishable products*, Operations Research, 34 (1986), pp. 75–82. (Cited on p. 484)

[693] A. FEDERGRUEN AND M. TZUR, *A simple forward algorithm to solve general dynamic lot sizing models with n periods in $O(n \log n)$ or $O(n)$ time*, Management Science, 37 (1991), pp. 909–925. (Cited on p. 442)

[694] A. FEDERGRUEN AND N. YANG, *Selecting a portfolio of suppliers under demand and supply risks*, Operations Research, 56 (2008), pp. 916–936. (Cited on p. 459)

[695] A. FEDERGRUEN AND Y.-S. ZHENG, *An efficient algorithm for computing an optimal (r, Q) policy in continuous review stochastic inventory systems*, Operations Research, 40 (1992), pp. 808–813. (Cited on p. 447)

[696] A. FEDERGRUEN AND P. ZIPKIN, *Computational issues in an infinite-horizon, multiechelon inventory model*, Operations Research, 32 (1984), pp. 818–836. (Cited on p. 449)

[697] M. FEINBERG AND D. HILDEBRANDT, *Optimal reactor design from a geometric viewpoint—I. Universal properties of the attainable region*, Chemical Engineering Science, 52 (1997), pp. 1637–1665. (Cited on p. 320)

[698] J. FELDMAN, T. MALKIN, R. A. SERVEDIO, C. STEIN, AND M. J. WAINWRIGHT, *LP decoding corrects a constant fraction of errors*, IEEE Transactions on Information Theory, 53 (2007), pp. 82–89. (Cited on p. 35)

[699] J. FELDMAN AND C. STEIN, *LP decoding achieves capacity*, in Proceedings of the Sixteenth Annual ACM-SIAM Symposium on Discrete Algorithms (SODA '05), SIAM Philadelphia, 2005, pp. 460–469. (Cited on p. 35)

[700] J. FELDMAN, M. J. WAINWRIGHT, AND D. R. KARGER, *Using linear programming to decode binary linear codes*, IEEE Transactions on Information Theory, 51 (2005), pp. 954–972. (Cited on p. 35)

[701] G. FELICI AND C. GENTILE, *A polyhedral approach for the staff rostering problem*, Management Science, 50 (2004), pp. 381–393. (Cited on p. 96)

[702] F. FELTUS, E. LEE, J. COSTELLO, C. PLASS, AND P. VERTINO, *Predicting aberrant CpG island methylation*, Proceedings of the National Academy of Sciences, 100 (2003), pp. 12253–12258. (Cited on pp. 94, 99)

[703] M. FERDINAND AND D. LINK, *Get seaports ready for disaster—strengthening preparedness at African seaports by improving performance*, in Managing Humanitarian Supply Chains—Strategies, Practices and Research, B. Hellingrath, D. Link, and A. Widera, eds., DVV Media Group GmbH, 2013, pp. 33–45. (Cited on p. 488)

[704] J. FERRIO AND J. M. WASSICK, *Chemical supply chain network optimization*, Computers and Chemical Engineering, 32 (2008), pp. 2481–2504. (Cited on pp. 86, 88)

[705] M. C. FERRIS, J. LIM, AND D. M. SHEPARD, *Radiosurgery treatment planning via nonlinear programming*, Annals of Operations Research, 119 (2003), pp. 247–260. (Cited on p. 94)

[706] FICO, *Xpress-SLP Program Reference Manual*, 2008. Release 1.41 edition. (Cited on p. 289)

[707] ——, *Xpress-Optimizer Reference Manual*, 2009. 20.0 edition. (Cited on p. 289)

[708] T. P. FILOMENA AND M. A. LEJEUNE, *Stochastic portfolio optimization with proportional transaction costs: Convex reformulations and computational experiments*, Operations Research Letters, 40 (2012), pp. 207–212. (Cited on pp. 427, 429)

[709] ——, *Warm-start heuristic for stochastic portfolio optimization with fixed and proportional transaction costs*, Journal of Optimization Theory and Applications, 161 (2014), pp. 308–329. (Cited on p. 429)

[710] D. E. FINKEL, *DIRECT Optimization Algorithm User Guide*, http://www4.ncsu.edu/ctk/Finkel_Direct/DirectUserGuide_pdf.pdf 2003. (Cited on p. 506)

[711] D. E. FINKEL AND C. T. KELLEY, *Additive scaling and the* DIRECT *algorithm*, Journal of Global Optimization, 36 (2006), pp. 597–608. (Cited on p. 172)

[712] ——, *Convergence analysis of sampling methods for perturbed Lipschitz functions*, Pacific Journal of Optimization, 5 (2009), pp. 339–350. (Cited on pp. 508, 510, 513)

[713] C. FIORINO, M. RENI, A. BOLOGNESI, G. CATTANEO, AND R. CALANDRINO, *Intra-and inter-observer variability in contouring prostate and seminal vesicles: Implications for conformal treatment planning*, Radiotherapy and Oncology, 47 (1998), pp. 285–292. (Cited on p. 347)

[714] G. A. G. FISCHER AND L. T. BIEGLER, *Fast NMPC Applied to Industrial High Purity Propylene Distillation*. AIChE Annual Meeting, Minneapolis, MN, 2011. (Cited on p. 247)

[715] M. FISCHETTI, F. GLOVER, AND A. LODI, *The feasibility pump*, Mathematical Programming, 104 (2005), pp. 91–104. (Cited on p. 286)

[716] M. FISCHETTI AND A. LODI, *Local branching*, Mathematical Programming, 98 (2002), pp. 23–47. (Cited on pp. 58, 286)

[717] M. FISCHETTI AND M. MONACI, *Light robustness*, in Robust and Online Large-scale Optimization, Springer, 2009, pp. 61–84. (Cited on p. 333)

[718] ——, *Exploiting erraticism in search*, Operations Research, 62 (2014), pp. 114–122. (Cited on p. 62)

[719] M. FISCHETTI AND D. SALVAGNIN, *Feasibility pump 2.0*, Mathematical Programming Computation, 1 (2009), pp. 201–222. (Cited on p. 57)

[720] P. C. FISHBURN, *Mean-risk analysis with risk associated with below-target returns*, The American Economic Review, 67 (1977), pp. 116–126. (Cited on p. 428)

[721] M. FISHER AND A. RAMAN, *Reducing the cost of demand uncertainty through accurate response to early sales*, Operations Research, 44 (1996), pp. 87–99. (Cited on p. 458)

[722] M. L. FISHER, *The Lagrangian relaxation method for solving integer programming problems*, Management Science, 27 (1981), pp. 1–18. (Cited on p. 452)

[723] ——, *An applications oriented guide to Lagrangian relaxation*, Interfaces, 15 (1985), pp. 10–21. (Cited on p. 452)

[724] R. FLETCHER, *Practical Methods of Optimization*, Wiley, Chichester, 1987. (Cited on pp. 222, 510)

[725] R. FLETCHER AND S. LEYFFER, *User Manual for SQP*, technical report, University of Dundee. Numerical Analysis Report NA-181. (Cited on p. 288)

[726] ——, *Solving mixed integer nonlinear programs by outer approximation*, Mathematical Programming, 66 (1994), pp. 327–349. (Cited on p. 277)

[727] ——, *Nonlinear programming without a penalty function*, Mathematical Programming, 91 (2002), pp. 239–269. (Cited on p. 505)

[728] ——, *Solving mathematical programs with complementarity constraints as nonlinear programs*, Optimization Methods and Software, 19 (2004), pp. 15–40. (Cited on p. 235)

[729] R. FLETCHER AND W. MORTON, *Initialising distillation column models*, Computers and Chemical Engineering, 23 (2000), pp. 1811–1824. (Cited on pp. 238, 240)

[730] H. FLEUREN, C. GOOSSENS, M. HENDRIKS, M.-C. LOMBARD, I. MEUFFELS, AND J. POPPELAARS, *Supply chain–wide optimization at TNT Express*, Interfaces, 43 (2013), pp. 5–20. (Cited on p. 60)

[731] A. FLORES-TLACUAHUAC AND L. T. BIEGLER, *Simultaneous mixed-integer dynamic optimization for integrated design and control*, Computers and Chemical Engineering, 31 (2007), pp. 648–656. (Cited on p. 291)

[732] A. FLORES-TLACUAHUAC AND I. E. GROSSMANN, *Simultaneous cyclic scheduling and control of a multiproduct CSTR*, Industrial and Engineering Chemistry Research, 45 (2006), pp. 6698–6712. (Cited on p. 328)

[733] M. FLORIAN, J. K. LENSTRA, AND A. H. G. RINNOOY KAN, *Deterministic production planning: Algorithms and complexity*, Management Science, 26 (1980), pp. 669–679. (Cited on p. 443)

[734] C. A. FLOUDAS, *Nonlinear and Mixed-Integer Optimization: Fundamentals and Applications*, Oxford University Press, 1995. (Cited on p. 274)

[735] ——, *Deterministic Global Optimization: Theory, Method and Applications*, Kluwer Academic, 2000. (Cited on pp. 173, 274)

[736] C. A. FLOUDAS AND A. R. CIRIC, *Strategies for overcoming uncertainties in heat exchanger network synthesis*, Computers and Chemical Engineering, 13 (1989), pp. 1133–1152. (Cited on p. 324)

[737] ——, *Corrigendum: Strategies for overcoming uncertainties in heat exchanger network synthesis*, Computers and Chemical Engineering, 14 (1990). (Cited on p. 324)

[738] C. A. FLOUDAS, A. R. CIRIC, AND I. E. GROSSMANN, *Automatic synthesis of optimum heat exchanger network configurations*, AIChE Journal, 32 (1986), pp. 276–290. (Cited on p. 324)

[739] C. A. FLOUDAS AND C. E. GOUNARIS, *A review of recent advances in global optimization*, Journal of Global Optimization, 45 (2009), pp. 3–38. (Cited on pp. 173, 329)

[740] C. A. FLOUDAS AND X. LIN, *Continuous-time versus discrete-time approaches for scheduling of chemical processes: A review*, Computers and Chemical Engineering, 28 (2004), pp. 2109–2129. (Cited on p. 328)

[741] C. A. FLOUDAS AND P. M. PARDALOS, eds., *Recent Advances in Global Optimization*, Princeton University Press, 1992. (Cited on p. 173)

[742] C. A. FLOUDAS, P. M. PARDALOS, C. S. ADJIMAN, W. R. ESPOSITO, Z. H. GUMUS, S. T. HARDING, J. L. KLEPEIS, C. A. MEYER, AND C. A. SCHWEIGER, *Handbook of Test Problems in Local and Global Optimization*, Kluwer Academic, Dordrecht, Netherlands, 1999. (Cited on p. 173)

[743] C. A. FLOUDAS AND G. E. PAULES, *A mixed-integer nonlinear programming formulation for the synthesis of heat-integrated distillation sequences*, Computers and Chemical Engineering, 12 (1988), pp. 531–546. (Cited on p. 323)

[744] J. FORREST AND R. LOUGEE-HEIMER, *CBC User Guide*, http://www.coinor.org/Cbc. (Cited on p. 288)

[745] A. FORSGREN, P. E. GILL, AND M. H. WRIGHT, *Interior methods for nonlinear optimization*, SIAM Review, 44 (2002), pp. 525–597. (Cited on p. 233)

[746] R. FORTET, *Applications de l'algèbre de Boole en recherche opérationnelle*, Revue française de recherche opérationnelle, 4 (1960), pp. 17–26. (Cited on p. 298)

[747] L. R. FOULDS, D. HAUGLAND, AND K. JÖRNSTEN, *A bilinear approach to the pooling problem*, Optimization, 24 (1992), pp. 165–180. (Cited on pp. 214, 216)

[748] R. FOURER, *Algebraic modeling languages for optimization*, in Encyclopedia of Operations Research and Management Science, S. I. Gass and M. C. Fu, eds., Springer Science, Heidelberg, 2013, pp. 43–51. (Cited on p. 10)

[749] R. FOURER, D. GAY, AND B. KERNIGHAN, *AMPL: A Modeling Language for Mathematical Programming*, Scientific Press series, Thomson/Brooks/Cole, 2003. (Cited on p. 287)

[750] A. FRANGIONI AND C. GENTILE, *Perspective cuts for a class of convex 0-1 mixed integer programs*, Mathematical Programming, 106 (2006), pp. 225–236. (Cited on pp. 287, 292)

[751] A. FRANGIONI, C. GENTILE, AND F. LACALANDRA, *Tighter approximated MILP formulations for unit commitment problems*, IEEE Transactions on Power Systems, 24 (2009), pp. 105–113. (Cited on pp. 304, 313)

[752] ———, *Sequential Lagrangian-MILP approaches for unit commitment problems*, International Journal of Electrical Power and Energy Systems, 33 (2011), pp. 585–593. (Cited on p. 304)

[753] S. FRANK AND S. REBENNACK, *An introduction to optimal power flow: Theory, formulation, and examples*, IIE Transactions, 48 (2016), pp. 1172–1197. (Cited on pp. 187, 189, 190, 191, 197)

[754] ———, *Optimal design of mixed AC-DC distribution systems for commercial buildings: A nonconvex generalized Benders decomposition approach*, European Journal of Operational Research, 242 (2015), pp. 710–729. (Cited on p. 205)

[755] S. FRANK, I. STEPONAVICE, AND S. REBENNACK, *Optimal power flow: A bibliographic survey* I, *formulations and deterministic methods*, Energy Systems, 3 (2012), pp. 221–258. (Cited on pp. 187, 192, 197)

[756] ———, *Optimal power flow: A bibliographic survey* II, *non-deterministic and hybrid methods*, Energy Systems, 3 (2012), pp. 259–289. (Cited on pp. 187, 192)

[757] H. FRANKOWSKA AND F. RAMPAZZO, *Filippov's and Filippov-Ważewski's theorems on closed domains*, Journal of Differential Equations, 161 (2000), pp. 449–478. (Cited on p. 126)

[758] S. FRAUSTO-HERNÁNDEZ, V. RICO-RAMIREZ, A. JIMÉNEZ-GUTIÉRREZ, AND S. HERNÁNDEZ-CASTRO, *MINLP synthesis of heat exchanger networks considering pressure drop effects*, Computers and Chemical Engineering, 27 (2003), pp. 1143–1152. (Cited on p. 324)

[759] A. FREDRIKSSON, *A characterization of robust radiation therapy treatment planning methods—From expected value to worst case optimization*, Medical Physics, 39 (2012), pp. 5169–5181. (Cited on pp. 348, 353)

[760] A. FREDRIKSSON, *Robust Optimization of Radiation Therapy Accounting for Geometric Uncertainty*, PhD thesis, KTH Royal Institute of Technology, Stockholm, Sweden, 2013. (Cited on pp. xx, 346)

[761] A. FREDRIKSSON AND R. BOKRANTZ, *A critical evaluation of worst case optimization methods for robust intensity-modulated proton therapy planning*, Medical Physics, 41 (2014), pp. 081701-1–081701-11. (Cited on p. 355)

[762] A. FREDRIKSSON, A. FORSGREN, AND B. HÅRDEMARK, *Minimax optimization for handling range and setup uncertainties in proton therapy*, Medical Physics, 38 (2011), pp. 1672–1684. (Cited on pp. 334, 353, 354)

[763] ——, *Maximizing the probability of satisfying the clinical goals in radiation therapy treatment planning under setup uncertainty*, Medical Physics, 42 (2015), pp. 3992–3999. (Cited on p. 351)

[764] F. FRIEDLER, K. TARJAN, Y. W. HUANG, AND L. T. FAN, *Graph-theoretic approach to process synthesis: Axioms and theorems*, Chemical Engineering Science, 47 (1992), pp. 1973–1988. (Cited on p. 318)

[765] ——, *Graph-theoretic approach to process synthesis: Polynomial algorithm for maximal structure generation*, Computers and Chemical Engineering, 17 (1993), pp. 929–942. (Cited on p. 318)

[766] B. E. FRIES, *Optimal ordering policy for a perishable commodity with fixed lifetime*, Operations Research, 23 (1975), pp. 46–61. (Cited on p. 473)

[767] Y. FU, M. SHAHIDEHPOUR, AND Z. LI, *Security-constrained unit commitment with AC constraints*, IEEE Transactions on Power Systems, 20 (2005), pp. 1538–1550. (Cited on p. 191)

[768] A. FÜGENSCHUH, B. GEISSLER, R. GOLLMER, C. HAYN, R. HENRION, B. HILLER, J. HUMPOLA, T. KOCH, T. LEHMANN, A. MARIN, R. MIRCOV, A. MOESI, J. RÖVEKAMP, L. SCHEWE, M. SCHMIDT, R. SCHULTZ, R. SCHWARZ, J. SCHWEIGER, C. STANGL, M. C. STEINBACH, AND B. M. WILLERT, *Mathematical optimization for challenging network planning problems in unbundled liberalized gas markets*, Energy Systems, 5 (2014), pp. 449–473. (Cited on p. 205)

[769] A. FÜGENSCHUH, H. HOMFELD, H. SCHÜLLDORF, AND S. VIGERSKE, *Mixed-integer nonlinear problems in transportation applications*, in Proceedings of the 2nd International Conference on Engineering Optimization, volume 50, 2010. (Cited on p. 292)

[770] H. K. FUNG, C. A. FLOUDAS, M. S. TAYLOR, L. ZHANG, AND D. MORIKIS, *Toward full-sequence de novo protein design with flexible templates for human beta-defensin-2*, Biophysical Journal, 94 (2008), pp. 584–599. (Cited on p. 180)

[771] H. K. FUNG, S. RAO, C. A. FLOUDAS, O. PROKOPYEV, P. M. PARDALOS, AND F. RENDL, *Comparison studies of quadratic assignment like formulations for the in silico sequence selection problem in de novo protein design*, Journal of Combinatorial Optimization, 10 (2005), pp. 41-60. (Cited on p. 180)

[772] H. K. FUNG, M. S. TAYLOR, AND C. A. FLOUDAS, *Novel formulations for the sequence selection problem in de novo protein design with flexible templates*, Optimization Methods and Software, 22 (2007), pp. 51-71. (Cited on p. 180)

[773] K. C. FURMAN AND I. P. ANDROULAKIS, *A novel MINLP-based representation of the original complex model for predicting gasoline emissions*, Computers and Chemical Engineering, 32 (2008), pp. 2857-2876. (Cited on pp. 208, 212, 216)

[774] K. C. FURMAN AND N. V. SAHINIDIS, *A critical review and annotated bibliography for heat exchanger network synthesis in the 20th century*, Industrial and Engineering Chemistry Research, 41 (2002), pp. 2335-2370. (Cited on p. 324)

[775] J. M. GABLONSKY AND C. T. KELLEY, *A locally-biased form of the DIRECT algorithm*, Journal of Global Optimization, 21 (2001), pp. 27-37. (Cited on p. 172)

[776] V. GABREL, C. MURAT, AND A. THIELE, *Recent advances in robust optimization: An overview*, European Journal of Operational Research, 235 (2014), pp. 471-483. (Cited on p. 334)

[777] V. GAITSGORY AND M. QUINCAMPOIX, *Linear programming approach to deterministic infinite horizon optimal control problems with discounting*, SIAM Journal on Control and Optimization, 48 (2009), pp. 2480-2512. (Cited on pp. 122, 126)

[778] A. GAIVORONSKI AND G. PFLUG, *Value-at-risk in portfolio optimization: Properties and computational approach*, Journal of Risk, 7 (2005), pp. 1-31. (Cited on p. 427)

[779] R. J. GALLAGHER, E. K. LEE, AND D. A. PATTERSON, *Constrained discriminant analysis via 0/1 mixed integer programming*, Annals of Operations Research, 74 (1997), pp. 65-88. (Cited on pp. 94, 97, 99)

[780] G. GALLEGO AND D. SHAW, *On the complexity of the economic lot scheduling problem with general cyclic schedules*, IIE Transactions, 29 (1997), pp. 109-113. (Cited on p. 442)

[781] M. R. GALLI AND J. CERDÁ, *A designer-controlled framework for the synthesis of heat exchanger networks involving non-isothermal mixers and multiple units over split streams*, Computers and Chemical Engineering, 22 (1998), pp. S813-S816. (Cited on p. 324)

[782] ——, *Synthesis of structural-constrained heat exchanger networks-I Series networks*, Computers and Chemical Engineering, 22 (1998), pp. 819-839. (Cited on p. 324)

[783] ——, *Synthesis of structural-constrained heat exchanger networks-II Split networks*, Computers and Chemical Engineering, 22 (1998), pp. 1017-1035. (Cited on p. 324)

[784] J. GALLIEN, Z. LEUNG, AND P. YADAV, *Rationality and Transparency in the Distribution of Essential Drugs in Sub-Saharan Africa: Analysis and Design of an Inventory Control System for Zambia*, technical report, London Business School, London, United Kingdom, 2014. (Cited on p. 491)

[785] C. E. GARCIA AND A. M. MORSHEDI, *Quadratic programming solution of dynamic matrix control (QDMC)*, Chemical Engineering Communications, 46 (1986), pp. 73-87. (Cited on p. 42)

[786] D. T. GARDNER AND J. S. ROGERS, *Planning electric power systems under demand uncertainty with different technology lead times*, Management Science, 45 (1999), pp. 1289-1306. (Cited on p. 458)

[787] M. R. GAREY AND D. S. JOHNSON, *Computers and Intractability: A Guide to the Theory of NP-completeness*, W. H. Freeman and Company, New York, 1979. (Cited on p. 166)

[788] R. GARMANJANI, D. JÚDICE, AND L. N. VICENTE, *Trust-region methods without using derivatives: Worst case complexity and the nonsmooth case*, SIAM Journal on Optimization, 26 (2016), pp. 1987–2011. (Cited on p. 501)

[789] R. GARMANJANI AND L. N. VICENTE, *Smoothing and worst-case complexity for direct-search methods in nonsmooth optimization*, IMA Journal of Numerical Analysis, 33 (2013), pp. 1008–1028. (Cited on pp. 500, 501)

[790] M. GAVIANO, D. E. KVASOV, D. LERA, AND Y. D. SERGEYEV, *Algorithm 829: Software for generation of classes of test functions with known local and global minima for global optimization*, ACM Transactions on Mathematical Software, 29 (2003), pp. 469–480. (Cited on p. 173)

[791] R. L. GEDDES, *A general index of fractional distillation power for hydrocarbon mixtures*, AIChE Journal, 4 (1958), pp. 389–392. (Cited on p. 327)

[792] B. GEISSLER, A. MARTIN, A. MORSI, AND L. SCHEWE, *Using piecewise linear functions for solving MINLPs*, in [1160], pp. 287–314. (Cited on p. 284)

[793] Q. GEMINE, D. ERNST, Q. LOUVEAUX, AND B. CORNÉLUSSE, *Relaxations for multiperiod optimal power flow problems with discrete decision variables*, in Proceedings of the 18th Power Systems Computation Conference, 2014, p. 7. (Cited on p. 199)

[794] Y. GENIN, Y. HACHEZ, YU. NESTEROV, AND P. VAN DOOREN, *Convex optimization over positive polynomials and filter design*, in Proceedings of the 2000 UKACC International Conference on Control, Cambridge University, 2000. (Cited on p. 30)

[795] A. GEOFFRION, *Generalized Benders decomposition*, Journal of Optimization Theory and Applications, 10 (1972), pp. 237–260. (Cited on p. 279)

[796] A. GEORGHIOU, W. WIESEMANN, AND D. KUHN, *Generalized decision rule approximations for stochastic programming via liftings*, Mathematical Programming, 152 (2015), pp. 301–338. (Cited on p. 342)

[797] A. GESER, C. MUNOZ, G. DOWEK, AND F. KIRCHNER, *Air Traffic Conflict Resolution and Recovery*, technical report, NASA/CR-2000-ICASE Report No. 2002-12, 2002. (Cited on p. 297)

[798] G. GHIANI, G. LAPORTE, AND R. MUSMANNO, *Introduction to Logistics Systems Management*, 2nd edition, Wiley, Hoboken, NJ, 2013. (Cited on p. 439)

[799] I. GIANNIKOS, *A multiobjective programming model for locating treatment sites and routing hazardous wastes*, European Journal of Operational Research, 104 (1998), pp. 333–342. (Cited on p. 480)

[800] M. GILBERT AND A. TYAS, *Layout optimization of large-scale pin-jointed frames*, Engineering Computations, 20 (2003), pp. 1044–1064. (Cited on p. 22)

[801] P. E. GILL, W. MURRAY, AND M. A. SAUNDERS, *SNOPT: An SQP algorithm for large-scale constrained optimization*, SIAM Journal on Optimization, 12 (2002), pp. 979–1006. (Cited on p. 196)

[802] ——, *SNOPT: An SQP algorithm for large-scale constrained optimization*, SIAM Review, 47 (2005), pp. 99–131. (Cited on p. 251)

[803] P. E. GILL, W. MURRAY, AND M. H. WRIGHT, *Practical Optimization*, Academic Press, New York, 1981. (Cited on p. 222)

[804] P. GILMORE AND C. T. KELLEY, *An implicit filtering algorithm for optimization of functions with many local minima*, SIAM Journal on Optimization, 5 (1995), pp. 269–285. (Cited on pp. 508, 510)

[805] P. GIRARDEAU, V. LECLERC, AND A. B. PHILPOTT, *On the convergence of decomposition methods for multi-stage stochastic convex programs*, Mathematics of Operations Research, 40 (2015), pp. 130–145. (Cited on p. 416)

[806] A. GIRIDHAR AND R. AGRAWAL, *Synthesis of distillation configurations: I. Characteristics of a good search space*, Computers and Chemical Engineering, 34 (2010), pp. 73–83. (Cited on p. 324)

[807] ———, *Synthesis of distillation configurations: II. A search formulation for basic configurations*, Computers and Chemical Engineering, 34 (2010), pp. 84–95. (Cited on p. 324)

[808] K. GKRITZA, *Iowa's Renewable Energy and Infrastructure Impacts*, technical report, Center for Transportation Research and Education (CTRE), Iowa Highway Research Board (TR-593), and Iowa Department of Transportation (InTrans Project 08-334), 2010. (Cited on p. 466)

[809] D. GLASSER, C. CROWE, AND D. HILDEBRANDT, *A geometric approach to steady flow reactors: The attainable region and optimization in concentration space*, Industrial and Engineering Chemistry Research, 26 (1987), pp. 1803–1810. (Cited on p. 320)

[810] P. GLASSERMAN, *Measuring marginal risk contributions in credit portfolios*, Journal of Computational Finance, 9 (2006), pp. 1–41. (Cited on pp. 432, 433, 434)

[811] P. GLASSERMAN AND X. XU, *Robust risk measurement and model risk*, Quantitative Finance, 14 (2014), pp. 29–58. (Cited on p. 333)

[812] S. GLIED AND L. BUSH, *Economic Analysis of the Causes of Drug Shortages*, technical report, Office of the Assistant Secretary for Planning and Evaluation (ASPE), U.S. Department of Health & Human Services, 2011. (Cited on p. 477)

[813] F. GLINEUR AND T. TERLAKY, *Conic formulation for l_p-norm optimization*, Journal of Optimization Theory and Applications, 122 (2004), pp. 285–307. (Cited on p. 112)

[814] *GLODS: Global and Local Optimization Using Direct Search*, http://ferrari.dmat.fct.unl.pt/personal/alcustodio/GLODS.htm. (Cited on p. 506)

[815] B. M. GLOVER, Y. ISHIZUKA, V. JEYAKUMAR, AND H. D. TUAN, *Complete characterizations of global optimality for problems involving the pointwise minimum of sublinear functions*, SIAM Journal on Optimization, 6 (1996), pp. 362–372. (Cited on p. 168)

[816] F. GLOVER, *Tabu search: A tutorial*, Interfaces, 20 (1990), pp. 74–94. (Cited on p. 57)

[817] F. GLOVER AND G. A. KOCHENBERGER, eds., *Handbook of Metaheuristics*, Kluwer Academic, 2003. (Cited on p. 173)

[818] V. GOEL, K. FURMAN, J. H. SONG, AND A. S. EL-BAKRY, *Large neighborhood search for LNG inventory routing*, Journal of Heuristics, 18 (2012), pp. 821–848. (Cited on pp. xxiii, 81, 83, 84, 85)

[819] M. GOERIGK AND A. SCHÖBEL, *A scenario-based approach for robust linear optimization.*, in TAPAS, Springer, 2011, pp. 139–150. (Cited on p. 333)

[820] J. GOH AND M. SIM, *Distributionally robust optimization and its tractable approximations*, Operations Research, 58 (2010), pp. 902–917. (Cited on pp. 342, 344)

[821] D. GOLDFARB AND G. IYENGAR, *Robust portfolio selection problems*, Mathematics of Operations Research, 28 (2003), pp. 1–38. (Cited on pp. 156, 333)

[822] A. J. GOLDMAN AND A. W. TUCKER, *Theory of linear programming*, in Linear Inequalities and Related Systems, H. W. Kuhn and A. W. Tucker, eds., no. 38 in Annals of Mathematical Studies, Princeton University Press, Princeton, New Jersey, 1956, pp. 53–97. (Cited on p. 4)

[823] R. GOMORY, *An algorithm for integer solutions to linear programs*, in Recent Advances in Mathematical Programming, R. Graves and P. Wolfe, eds., McGraw-Hill, New York, 1963, pp. 269–302. (Cited on pp. 53, 54)

[824] R. GOMORY AND E. JOHNSON, *Some continuous functions related to corner polyhedra*, Mathematical Programming, 3 (1972), pp. 28–85. (Cited on p. 54)

[825] J. GONDZIO AND T. TERLAKY, *A computational view of interior point methods for linear programming*, in Advances in Linear and Integer Programming, J. E. Beasley, ed., Oxford University Press, Oxford, 1996, pp. 103–185. (Cited on p. 10)

[826] J. S. GONZÁLEZ, Á. G. G. RODRÍGUEZ, J. C. MORA, M. B. PAYÁN, AND J. M. R. SANTOS, *Overall design optimization of wind farms*, Renewable Energy, 36 (2011), pp. 1973–1982. (Cited on p. 367)

[827] J. S. GONZÁLEZ, A. G. G. RODRIGUEZ, J. C. MORA, J. R. SANTOS, AND M. B. PAYAN, *Optimization of wind farm turbines layout using an evolutive algorithm*, Renewable Energy, 35 (2010), pp. 1671–1681. (Cited on p. 367)

[828] H. GOOI, D. MENDES, K. BELL, AND D. KIRSCHEN, *Optimum scheduling of spinning reserve*, IEEE Transactions on Power Systems, 14 (1999), pp. 1485–1490. (Cited on p. 360)

[829] A. GOPALAKRISHNAN, A. U. RAGHUNATHAN, D. NIKOVSKI, AND L. T. BIEGLER, *Global optimization of optimal power flow using a branch and bound algorithm*, in Annual Allerton Conference on Communication, Control, and Computing, IEEE, October 2012, pp. 609–616. (Cited on p. 204)

[830] B. L. GORISSEN, H. BLANC, D. DEN. HERTOG, AND A. BEN-TAL, *Technical note— Deriving robust and globalized robust solutions of uncertain linear programs with general convex uncertainty sets*, Operations Research, 62 (2014), pp. 672–679. (Cited on p. 337)

[831] B. L. GORISSEN AND D. DEN. HERTOG, *Robust counterparts of inequalities containing sums of maxima of linear functions*, European Journal of Operational Research, 227 (2013), pp. 30–43. (Cited on p. 334)

[832] B. L. GORISSEN, İ. YANIKOĞLU, AND D. DEN HERTOG, *A practical guide to robust optimization*, Omega, 53 (2015), pp. 124–137. (Cited on p. 334)

[833] A. P. GORYASHKO AND A. S. NEMIROVSKI, *Robust energy cost optimization of water distribution system with uncertain demand*, Automation and Remote Control, 75 (2014), pp. 1754–1769. (Cited on p. 333)

[834] N. I. M. GOULD AND PH. L. TOINT, *Nonlinear programming without a penalty function or a filter*, Mathematical Programming, 122 (2010), pp. 155–196. (Cited on p. 505)

[835] C. E. GOUNARIS, R. MISENER, AND C. A. FLOUDAS, *Computational comparison of piecewise-linear relaxations for pooling problems*, Industrial and Engineering Chemistry Research, 48 (2009), pp. 5742–5766. (Cited on pp. 199, 216, 217)

[836] C. E. GOUNARIS, W. WIESEMANN, AND C. A. FLOUDAS, *The robust capacitated vehicle routing problem under demand uncertainty*, Operations Research, 61 (2013), pp. 677–693. (Cited on p. 333)

[837] S. GRADY, *Placement of wind turbines using genetic algorithms*, Renewable Energy, 30 (2005), pp. 259–270. (Cited on pp. 367, 370)

[838] M. GRANT AND S. BOYD, *Graph implementations for nonsmooth convex programs*, in Recent Advances in Learning and Control, V. Blondel, S. Boyd, and H. Kimura, eds., Lecture Notes in Control and Information Sciences, Springer-Verlag, 2008, pp. 95–110, http://stanford.edu/~boyd/papers/graph_dcp.html. (Cited on pp. 117, 119)

[839] ———, *CVX: MATLAB Software for Disciplined Convex Programming, Version 2.1*, http://cvxr.com/cvx, Mar. 2014. (Cited on pp. 117, 119)

[840] M. GRANT, S. BOYD, AND Y. YE, *Disciplined Convex Programming*, Springer, 2006. (Cited on p. 119)

[841] S. GRATTON, C. W. ROYER, L. N. VICENTE, AND Z. ZHANG, *Direct search based on probabilistic descent*, SIAM Journal on Optimization, 25 (2015), pp. 1515–1541. (Cited on p. 502)

[842] S. GRATTON, A. SARTENAER, AND PH. L. TOINT, *Recursive trust-region methods for multiscale nonlinear optimization*, SIAM Journal on Optimization, 19 (2008), pp. 414–444. (Cited on p. 501)

[843] S. GRATTON, PH. L. TOINT, AND A. TROELTZCH, *An active-set trust-region method for derivative-free nonlinear bound-constrained optimization*, Optimization Methods and Software, 21 (2011), pp. 873–894. (Cited on pp. 498, 503)

[844] S. GRATTON AND L. N. VICENTE, *A merit function approach for direct search*, SIAM Journal on Optimization, 24 (2014), pp. 1980–1998. (Cited on p. 505)

[845] ———, *A surrogate management framework using rigorous trust-region steps*, Optimization Methods and Software, 29 (2014), pp. 10–23. (Cited on p. 506)

[846] S. GRATTON, L. N. VICENTE, AND Z. ZHANG, *A Subspace Decomposition Framework for Nonlinear Optimization*, technical report, in preparation. (Cited on p. 506)

[847] S. C. GRAVES, *Safety stocks in manufacturing systems*, Journal of Manufacturing and Operations Management, 1 (1988), pp. 67–101. (Cited on p. 450)

[848] S. C. GRAVES AND S. P. WILLEMS, *Optimizing strategic safety stock placement in supply chains*, Manufacturing and Service Operations Management, 2 (2000), pp. 68–83. (Cited on p. 450)

[849] ———, *Erratum: Optimizing strategic safety stock placement in supply chains*, Manufacturing and Service Operations Management, 5 (2003), pp. 176–177. (Cited on p. 450)

[850] J. J. GRAY, S. MOUGHON, C. WANG, O. SCHUELER-FURMAN, B. KUHLMAN, C. A. ROHL, AND D. BAKER, *Protein-protein docking with simultaneous optimization of rigid-body displacement and side-chain conformations*, Journal of Molecular Biology, 331 (2003), pp. 281–299. (Cited on p. 181)

[851] H. J. GREENBERG, *Analyzing the pooling problem*, ORSA Journal on Computing, 7 (1995), pp. 205–217. (Cited on p. 212)

[852] H. J. GREENBERG AND A. G. KONHEIM, *Linear and nonlinear methods in pattern classification*, IBM Journal of Research and Development, 8 (1964), pp. 299–307. (Cited on p. 30)

[853] A. GRIEWANK AND A. WALTHER, *Evaluating Derivatives: Principles and Techniques of Algorithmic Differentiation*, 2nd edition, SIAM, 2008. (Cited on p. 223)

[854] J. GRIFFIN, P. KESKINOCAK, H. SMALLEY, AND J. SWANN, *Designing a Malaria Intervention Supply Chain*, technical report, School of Industrial and Systems Engineering, Georgia Institute of Technology, Atlanta, GA, 2010. (Cited on p. 480)

[855] J. D. GRIFFIN, T. G. KOLDA, AND R. M. LEWIS, *Asynchronous parallel generating set search for linearly constrained optimization*, SIAM Journal on Scientific Computing, 30 (2008), pp. 1892–1924. (Cited on p. 506)

[856] G. R. GRIMMETT AND D. R. STIRZAKER, *Probability and Random Processes*, 2nd edition, Oxford University Press, 1992. (Cited on p. 412)

[857] I. E. GROSSMANN, *Mixed-integer programming approach for the synthesis of integrated process flowsheets*, Computers and Chemical Engineering, 9 (1985), pp. 463–482. (Cited on p. 318)

[858] ———, *MINLP optimization strategies and algorithms for process synthesis*, in Foundations of Computer-Aided Process Design: Proceedings of the Third International Conference on Foundations of Computer-Aided Process Design, J. J. Siirola, I. E. Grossmann, and G. Stephanopoulos, eds., Elsevier, 1990, p. 105. (Cited on p. 318)

[859] ———, *Mixed-integer nonlinear programming techniques for the synthesis of engineering systems*, Research in Engineering Design, 1 (1990), pp. 205–228. (Cited on p. 318)

[860] ———, *Review of nonlinear mixed-integer and disjunctive programming techniques*, Optimization and Engineering, 3 (2002), pp. 227–252. (Cited on pp. 274, 315, 316)

[861] ———, *Enterprise-wide optimization: A new frontier in process systems engineering*, AIChE Journal, 51 (2005), pp. 1846–1857. (Cited on p. 77)

[862] I. E. GROSSMANN, P. A. AGUIRRE, AND M. BARTTFELD, *Optimal synthesis of complex distillation columns using rigorous models*, Computers and Chemical Engineering, 29 (2005), pp. 1203–1215. (Cited on p. 238)

[863] I. E. GROSSMANN, J. A. CABALLERO, AND H. YEOMANS, *Mathematical programming approaches for the synthesis of chemical process systems*, Korean Journal of Chemical Engineering, 16 (1999), pp. 407–426. (Cited on pp. 77, 238, 243)

[864] ———, *Advances in mathematical programming for the synthesis of process systems*, Latin American Applied Research, 30 (2000), pp. 263–284. (Cited on pp. 316, 317)

[865] I. E. GROSSMANN AND Z. KRAVANJA, *Mixed–integer nonlinear programming: A survey of algorithms and applications*, in Large–Scale Optimization with Applications, Part II: Optimal Design and Control, L. T. Biegler, T. F. Coleman, A. R. Conn, and F. N. Santosa, eds., Springer, New York, Berlin, 1997. (Cited on p. 274)

[866] I. E. GROSSMANN AND F. TRESPALACIOS, *Systematic modeling of discrete-continuous optimization models through generalized disjunctive programming*, AIChE Journal, 59 (2013), pp. 3276–3295. (Cited on pp. 315, 316)

[867] I. E. GROSSMANN, H. YEOMANS, AND Z. KRAVANJA, *A rigorous disjunctive optimization model for simultaneous flowsheet optimization and heat integration*, Computers and Chemical Engineering, 22 (1998), pp. S157–S164. (Cited on p. 324)

[868] X. GUAN, S. GUO, AND Q. ZHAI, *The conditions for obtaining feasible solutions to security-constrained unit commitment problems*, IEEE Transactions on Power Systems, 20 (2005), pp. 1746–1756. (Cited on p. 191)

[869] Y. GUAN AND J. WANG, *Uncertainty sets for robust unit commitment*, IEEE Transactions on Power Systems, 29 (2014), pp. 1439–1440. (Cited on p. 363)

[870] O. J. GUERRA AND G. A. C. LE ROUX, *Improvements in petroleum refinery planning: 1. formulation of process models*, Industrial and Engineering Chemistry Research, 50 (2011), pp. 13403–13418. (Cited on pp. 38, 39, 40)

[871] N. GÜLPINAR, E. CANAKOGLU, AND D. PACHAMANOVA, *Robust investment decisions under supply disruption in petroleum markets*, Computers and Operations Research, 44 (2014), pp. 75–91. (Cited on p. 333)

[872] E. A. E. GUMMA, M. H. A. HASHIM, AND M. MONTAZ ALI, *A derivative-free algorithm for linearly constrained optimization problems*, Computational Optimization and Applications, 57 (2014), pp. 599–621. (Cited on p. 503)

[873] M. GUNARATNAM, A. ALVA-ARGAEZ, A. KOKOSSIS, J.-K. KIM, AND R. SMITH, *Automated design of total water systems*, Industrial and Engineering Chemistry Research, 44 (2005), pp. 588–599. (Cited on p. 327)

[874] T. GUNDERSEN, S. DUVOLD, AND A. HASHEMI-AHMADY, *An extended vertical MILP model for heat exchanger network synthesis*, Computers and Chemical Engineering, 20 (1996), pp. S97–S102. (Cited on p. 324)

[875] T. GUNDERSEN AND I. E. GROSSMANN, *Improved optimization strategies for automated heat exchanger network synthesis through physical insights*, Computers and Chemical Engineering, 14 (1990), pp. 925–944. (Cited on p. 324)

[876] O. GÜNLÜK, J. LEE, AND R. WEISMANTEL, *MINLP Strengthening for Separable Convex Quadratic Transportation-Cost UFL*, Technical Report RC24213 (W0703-042), IBM Research Division, March 2007. (Cited on p. 292)

[877] O. GÜNLÜK AND J. LINDEROTH, *Perspective reformulations of mixed integer nonlinear programs with indicator variables*, Mathematical Programming, 124 (2010), pp. 183–205. (Cited on p. 313)

[878] P. GÜNTERT, *Automated NMR structure calculation with CYANA*, in Protein NMR Techniques, A. K. Downing, ed., Volume 278 of Methods in Molecular Biology, Humana Press, 2004, pp. 353–378. (Cited on p. 180)

[879] P. GÜNTERT, C. MUMENTHALER, AND K. WÜTHRICH, *Torsion angle dynamics for NMR structure calculation with the new program DYANA*, Journal of Molecular Biology, 273 (1997), pp. 283–298. (Cited on p. 180)

[880] Z. GUO, K. SCHEINBERG, J. HONG, AND B. Y. CHEN, *Superposition of protein structures using electrostatic isopotentials*, in IEEE International Conference on Bioinformatics and Biomedicine (BIBM), 2015, pp. 75–82. (Cited on pp. xxi, 525, 526)

[881] A. GUPTA AND C. D. MARANAS, *A two-stage modeling and solution framework for multisite midterm planning under demand uncertainty*, Industrial and Engineering Chemistry Research, 39 (2000a), pp. 3799–3813. (Cited on p. 393)

[882] A. GUPTA, C. D. MARANAS, AND C. M. MCDONALD, *Mid-term supply chain planning under demand uncertainty: customer demand satisfaction and inventory management*, Computers and Chemical Engineering, 24 (2000b), pp. 2613–2621. (Cited on p. 393)

[883] O. K. GUPTA AND A. RAVINDRAN, *Branch and bound experiments in convex nonlinear integer programming*, Management Science, 31 (1985), pp. 1533–1546. (Cited on p. 275)

[884] A. GUPTE, S. AHMED, M. S. CHEON, AND S. DEY, *Relaxations and discretizations for the pooling problem*, Journal of Global Optimization, 67 (2017), pp. 631–669. (Cited on p. 328)

[885] A. GUPTE, S. AHMED, S. DEY, AND M. S. CHEON, *Pooling problems: Relaxations and discretizations.* Optimization Online, 2015. (Cited on pp. 208, 212, 216)

[886] E. GUTIN, D. KUHN, AND W. WIESEMANN, *Interdiction games on Markovian PERT networks*, Management Science, 61 (2014), pp. 999–1017. (Cited on p. 341)

[887] S. HACEIN-BEY-ABINA, C. VON KALLE, M. SCHMIDT, F. LE DEIST, N. WULFFRAAT, E. MCINTYRE, I. RADFORD, J. L. VILLEVAL, C. C. FRASER, M. CAVAZZANA-CALVO, AND A. FISCHER, *A serious adverse event after successful gene therapy for X-linked severe combined immunodeficiency*, New England Journal of Medicine, 348 (2003), pp. 255–256. (Cited on p. 182)

[888] S. HACEIN-BEY-ABINA, C. VON KALLE, M. SCHMIDT, M. P. MCCORMACK, N. WULFFRAAT, P. LEBOULCH, A. LIM, C. S. OSBORNE, R. PAWLIUK, E. MORILLON, R. SORENSEN, A. FORSTER, P. FRASER, J. I. COHEN, G. DE SAINT BASILE, I. ALEXANDER, U. WINTERGERST, T. FREBOURG, A. AURIAS, D. STOPPA-LYONNET, S. ROMANA, I. RADFORD-WEISS, F. GROSS, F. VALENSI, E. DELABESSE, E. MACINTYRE, F. SIGAUX, J. SOULIER, L. E. LEIVA, M. WISSLER, C. PRINZ, T. H. RABBITTS, F. LE DEIST, A. FISCHER, AND M. CAVAZZANA-CALVO, *LMO2-associated clonal T cell proliferation in two patients after gene therapy for SCID-X1*, Science, 302 (2003), pp. 415–419. (Cited on p. 182)

[889] A. R. G. HADIGHEH AND T. TERLAKY, *Sensitivity analysis in convex quadratic optimization*, Optimization, 54(1) (2005), pp. 59–79. (Cited on pp. 4, 11)

[890] R. T. HAFTKA, *Optimization of flexible wing structures subject to strength and induced drag constraints*, AIAA Journal, 14 (1977), pp. 1106–1977. (Cited on p. 250)

[891] R. HAIJEMA, J. WAL, AND N. M. VAN DIJK, *Blood platelet production: A multi-type perishable inventory problem*, in Operations Research Proceedings 2004, H. Fleuren, D. Hertog, and P. Kort, eds., Springer, Berlin, Heidelberg, 2005, pp. 84–92. (Cited on p. 473)

[892] ———, *Blood platelet production: Optimization by dynamic programming and simulation*, Computers & Operations Research, 34 (2007), pp. 760–779. (Cited on p. 473)

[893] L. HAJIBABAI, Y. BAI, AND Y. OUYANG, *Joint optimization of freight facility location and pavement infrastructure rehabilitation under network traffic equilibrium*, Transportation Research Part B, 63 (2014), pp. 38–52. (Cited on p. 466)

[894] L. HAJIBABAI AND Y. OUYANG, *Integrated planning of supply chain networks and multimodal transportation infrastructure expansion: Model development and application to the biofuel industry*, Computer-aided Civil and Infrastructure Engineering, 28 (2013), pp. 247–259. (Cited on p. 466)

[895] A. H. HAJIMIRAGHA, C. A. CANIZARES, M. W. FOWLER, S. M. MOAZENI, AND A. ELKAMEL, *A robust optimization approach for planning the transition to plug-in hybrid electric vehicles*, IEEE Transactions on Power Systems, 26 (2011), pp. 2264–2274. (Cited on p. 333)

[896] S. L. HAKIMI, *Optimum distribution of switching centers in a communication network and some related graph theoretic problems*, Operations Research, 13 (1965), pp. 462–475. (Cited on p. 453)

[897] T. S. HALE AND C. R. MOBERG, *Location science research: A review*, Annals of Operations Research, 123 (2003), pp. 21–35. (Cited on p. 450)

[898] S. N. HALL, S. H. JACOBSON, AND E. C. SEWELL, *An analysis of pediatric vaccine formulary selection problems*, Operations Research, 56 (2008), pp. 1348–1365. (Cited on p. 96)

[899] G. A. HANASUSANTO, D. KUHN, S. W. WALLACE, AND S. ZYMLER, *Distributionally robust multi-item newsvendor problems with multimodal demand distributions*, Mathematical Programming, 152 (2015), pp. 1–32. (Cited on p. 344)

[900] G. A. HANASUSANTO, D. KUHN, AND W. WIESEMANN, *K-adaptability in two-stage robust binary programming*, Operations Research, 63 (2015), pp. 877–891. (Cited on p. 343)

[901] G. Y. HANDLER AND P. B. MIRCHANDANI, *Location on Networks*, MIT Press, Cambridge, MA, 1979. (Cited on p. 450)

[902] E. HANSEN AND G. W. WALSTER, *Global Optimization Using Interval Analysis*, 2nd edition, Marcel Dekker, Inc., 2004. (Cited on pp. 169, 173)

[903] N. HANSEN, A. OSTERMEIER, AND A. GAWELCZYK, *On the adaptation of arbitrary normal mutation distributions in evolution strategies: The generating set adaptation*, in Proceedings of the Sixth International Conference on Genetic Algorithms, Pittsburgh, L. Eshelman, ed., 1995, pp. 57–64. (Cited on p. 506)

[904] P. HANSEN AND B. JAUMARD, *Lipschitz optimization*, in [957], pp. 407–493. (Cited on pp. 169, 172)

[905] P. C. HANSEN, J. G. NAGY, AND D. P. O'LEARY, *Deblurring Images. Matrices, Spectra, and Filtering*, Fundamentals of Algorithms 3, SIAM, Philadelphia, 2006. (Cited on p. 32)

[906] J. HANSMAN, *Impact of NextGen integration on improving efficiency and safety of operations*, in Proceedings of TRB: the 91st Annual Meeting of the Transportation Research Board, Washington, D.C., 2012. (Cited on p. 293)

[907] J. HARANT, *Some news about the independence number of a graph*, Discussiones Mathematicae Graph Theory, 20 (2000), pp. 71–79. (Cited on p. 165)

[908] J. HARANT, A. PRUCHNEWSKI, AND M. VOIGT, *On dominating sets and independent sets of graphs*, Combinatorics, Probability and Computing, 8 (1999), pp. 547–553. (Cited on p. 165)

[909] W. HARE AND J. NUTINI, *A derivative-free approximate gradient sampling algorithm for finite minimax problems*, Computational Optimization and Applications, 56 (2013), pp. 1–38. (Cited on p. 500)

[910] I. HARJUNKOSKI, C. MARAVELIAS, P. BONGERS, P. CASTRO, S. ENGELL, I. E. GROSSMANN, J. HOOKER, C. MÉNDEZ, G. SAND, AND J. WASSICK, *Scope for industrial applications of production scheduling models and solution methods*, Computers and Chemical Engineering, 62 (2014), pp. 161–193. (Cited on pp. 77, 327, 328)

[911] I. HARJUNKOSKI, R. PÖRN, AND T. WESTERLUND, *MINLP: Trim-loss problem*, in Encyclopedia of Optimization, C. A. Floudas and P. M. Pardalos, eds., Springer, 2009, pp. 2190–2198. (Cited on p. 292)

[912] P. HARPER, A. SHAHANI, J. GALLAGHER, AND C. BOWIE, *Planning health services with explicit geographical considerations: A stochastic location–allocation approach*, Omega, 33 (2005), pp. 141–152. (Cited on p. 95)

[913] F. W. HARRIS, *How many parts to make at once*, Factory: The Magazine of Management, 10 (1913), pp. 135–136. Reprinted in Operations Research 38 (1990), pp. 947–950. (Cited on p. 440)

[914] P. M. J. HARRIS, *Pivot selection methods of the Devex LP code*, Mathematical Programming Study, 4 (1975), pp. 30–57. (Cited on p. 9)

[915] W. E. HART, J.-P. WATSON, AND D. L. WOODRUFF, *Pyomo: Modeling and solving mathematical programs in Python*, Mathematical Programming Computation, 3 (2011), pp. 219–260. (Cited on p. 288)

[916] P. HARTMAN, *On functions representable as a difference of convex functions*, Pacific Journal of Mathematics, 9 (1959), pp. 707–713. (Cited on p. 283)

[917] M. HASAN AND I. KARIMI, *Piecewise linear relaxation of bilinear programs using bivariate partitioning*, AIChE Journal, 56 (2010), pp. 1880–1893. (Cited on p. 284)

[918] O. HASAN, D. O. MELTZER, S. A. SHAYKEVICH, C. M. BELL, P. J. KABOLI, A. D. AUERBACH, T. B. WETTERNECK, V. M. ARORA, J. ZHANG, AND J. L. SCHNIPPER, *Hospital readmission in general medicine patients: A prediction model*, Journal of General Internal Medicine, 25 (2010), pp. 211–219. (Cited on p. 96)

[919] T. HASTIE, R. TIBSHIRANI, AND J. FRIEDMAN, *The Elements of Statistical Learning. Data Mining, Inference, and Prediction*, Springer-Verlag, New York, 2001. (Cited on p. 30)

[920] T. HASTIE, R. TIBSHIRANI, AND M. WAINWRIGHT, *Statistical Learning with Sparsity. The Lasso and Generalizations*, CRC Press, 2015. (Cited on pp. 31, 33)

[921] C. A. HAVERLY, *Studies of the behaviour of recursion for the pooling problem*, ACM SIGMAP Bulletin, 25 (1978), pp. 19–28. (Cited on pp. 209, 212, 216)

[922] ———, *Behaviour of recursion model—More studies*, ACM SIGMAP Bulletin, 26 (1979), pp. 22–28. (Cited on pp. 209, 212, 216)

[923] S. HE, Z. LI, AND S. ZHANG, *Approximation algorithms for homogeneous polynomial optimization with quadratic constraints*, Mathematical Programming, 125 (2010), pp. 353–383. (Cited on p. 173)

[924] M. HEINKENSCHLOSS AND L. N. VICENTE, *Analysis of inexact trust-region SQP algorithms*, SIAM Journal on Optimization, 12 (2001), pp. 283–302. (Cited on p. 231)

[925] C. HELMBERG AND K. C. KIWIEL, *A spectral bundle method with bounds*, Mathematical Programming, Series A, 93 (2002), pp. 173–194. (Cited on p. 118)

[926] C. HELMBERG AND F. RENDL, *A spectral bundle method for semidefinite programming*, SIAM Journal on Optimization, 10 (2000), pp. 673–696. (Cited on p. 118)

[927] H. D. HELMS, *Digital filters with equiripple or minimax responses*, IEEE Transactions on Audio and Electroacoustics, AU-19 (1971), pp. 87–93. (Cited on p. 28)

[928] K. HELSGAUN, *An effective implementation of the Lin-Kernighan traveling salesman heuristic*, EJOR, 126 (2000), pp. 106–130. (Cited on p. 57)

[929] R. HEMMECKE, M. KÖPPE, J. LEE, AND R. WEISMANTEL, *Nonlinear integer programming*, in [1017], pp. 561–618. (Cited on p. 274)

[930] W. S. HEMP, *Optimum Structures*, Claredon Press, Oxford, 1973. (Cited on pp. 13, 16)

[931] C. A. HENAO AND C. T. MARAVELIAS, *Surrogate-based superstructure optimization framework*, AIChE Journal, 57 (2011), pp. 1216–1232. (Cited on p. 318)

[932] E. M. T. HENDRIX AND B. G.-TÓTH, *Introduction to Nonlinear and Global Optimization*, Springer, 2010. (Cited on p. 173)

[933] D. HENRION AND M. KORDA, *Convex computation of the region of attraction of polynomial control systems*, IEEE Transactions on Automatic Control, 59 (2014), pp. 297–312. (Cited on p. 126)

[934] D. HENRION, J. B. LASSERRE, AND J. LÖFBERG, *GloptiPoly 3: Moments, optimization and semidefinite programming*, Optimization Methods and Software, 24 (2009), pp. 761–779. (Cited on p. 120)

[935] P. V. HENTENRYCK, R. BENT, AND C. COFFRIN, *Strategic planning for disaster recovery with stochastic last mile distribution*, in Lecture Notes in Computer Science: Integration of AI and OR Techniques in Constraint Programming for Combinatorial Optimization Problems, A. Lodi, M. Milano, and P. Toth, eds., Springer, Berlin, 2010, pp. 318–333. (Cited on p. 484)

[936] G. P. HENZE, C. FELSMANN, AND G. KNABE, *Evaluation of optimal control for active and passive building thermal storage*, International Journal of Thermal Sciences, 43 (2004), pp. 173–183. (Cited on p. 259)

[937] J. HERBERT-ACERO, J. FRANCO-ACEVEDO, M. VALENZUELA-RENDÓN, AND O. PROBST-OLESZEWSKI, *Linear wind farm layout optimization through computational intelligence*, Lecture Notes in Computer Science, 5845 (2009), pp. 692–703. (Cited on p. 367)

[938] R. HERNANDEZ-SUAREZ, J. CASTELLANOS-FERNANDEZ, AND J. M. ZAMORA, *Superstructure decomposition and parametric optimization approach for the synthesis of distributed wastewater treatment networks*, Industrial and Engineering Chemistry Research, 43 (2004), pp. 2175–2191. (Cited on p. 327)

[939] J. HIGLE AND S. SEN, *Stochastic Decomposition: A Statistical Method for Large Scale Stochastic Linear Programming*, Springer, 1996. (Cited on p. 391)

[940] H. HIJAZI, P. BONAMI, AND A. OUOROU, *An outer-inner approximation for separable mixed-integer nonlinear programs*, INFORMS Journal on Computing, 26 (2013), pp. 31–44. (Cited on p. 285)

[941] D. HILDEBRANDT AND L. T. BIEGLER, *Synthesis of Chemical Reactor Networks*, technical report, Carnegie Mellon University, 1995. (Cited on p. 319)

[942] D. HILDEBRANDT AND D. GLASSER, *The attainable region and optimal reactor structures*, Chemical Engineering Science, 45 (1990), pp. 2161–2168. (Cited on p. 320)

[943] D. HILDEBRANDT, D. GLASSER, AND C. M. CROWE, *Geometry of the attainable region generated by reaction and mixing: With and without constraints*, Industrial and Engineering Chemistry Research, 29 (1990), pp. 49–58. (Cited on p. 320)

[944] J.-B. HIRIART-URRUTY, *When is a point satisfying $\nabla f(x) = 0$ a global minimum of f?* American Mathematical Monthly, 93 (1986), pp. 556–558. (Cited on p. 167)

[945] ——, *From convex optimization to nonconvex optimization. Part I: Necessary and sufficient conditions for global optimality*, in Nonsmooth Optimization and Related Topics, F. H. Clarke et al., ed., Plenum Press, 1989, pp. 219–239. (Cited on pp. 167, 168)

[946] M. J. HODGSON, G. LAPORTE, AND F. SEMET, *A covering tour model for planning mobile health care facilities in Suhumdistrict, Ghama*, Journal of Regional Science, 38 (1998), pp. 621–638. (Cited on p. 95)

[947] A. J. HOFFMAN, *Cycling in the Simplex Algorithm*, Technical Report 2974, National Bureau of Standards, Gaithersburg, MD, 1953. (Cited on p. 6)

[948] A. L. HOFFMANN, D. DEN HERTOG, A. Y. SIEM, J. KAANDERS, AND H. HUIZENGA, *Convex reformulation of biologically-based multi-criteria IMRT optimization including fractionation effects*, Physics in Medicine and Biology, 53 (2008), pp. 6345–6362. (Cited on p. 95)

[949] J. HOLGUIN-VERAS, N. PEREZ, M. JALLER, L. VAN WASSENHOVE, AND F. AROS-VERA, *On the appropriate objective function for post-disaster humanitarian logistics response*, Journal of Operations Management, 31 (2013), pp. 262–280. (Cited on p. 491)

[950] L. HOLM AND C. SANDER, *Mapping the protein universe*, Science, 273 (1996), pp. 595–603. (Cited on p. 520)

[951] K. HOLMSTRÖM AND M. M. EDVALL, *The TomLab optimization environment*, in Modeling Languages in Mathematical Optimization, Springer, 2004, pp. 369–376. (Cited on p. 288)

[952] C. HOLOHAN, S. VAN SCHAEYBROECK, D. B. LONGLEY, AND P. G. JOHNSTON, *Cancer drug resistance: An evolving paradigm*, Nature Reviews Cancer, 13 (2013), pp. 714–726. (Cited on p. 519)

[953] W. J. HOPP AND M. L. SPEARMAN, *Factory Physics*, Waveland Press, 2011. (Cited on p. 489)

[954] *HOPSPACK: Hybrid Optimization Parallel Search Package*, https://software.sandia.gov/trac/hopspack/wiki. (Cited on pp. 499, 503, 506)

[955] F. HORN, *Attainable and non-attainable regions in chemical reaction technique*, in Third European Symposium on Chemical Reaction Engineering, J. Hoogschagen, ed., Pergamon Press, London, 1965, pp. 1–10. (Cited on p. 320)

[956] R. A. HORN AND C. R. JOHNSON, *Matrix Analysis*, Cambridge University Press, Cambridge, 1990. (Cited on pp. 108, 112, 115)

[957] R. HORST AND P. M. PARDALOS, eds., *Handbook of Global Optimization*, Kluwer Academic, 1995. (Cited on pp. 169, 173, 554, 597, 631, 637, 652, 654)

[958] R. HORST, P. M. PARDALOS, AND N. V. THOAI, *Introduction to Global Optimization*, 2nd edition, Kluwer Academic, 2000. (Cited on pp. 166, 167, 173)

[959] R. HORST AND H. TUY, *Global Optimization: Deterministic Approaches*, 3rd edition, Springer, 1996. (Cited on pp. 170, 171, 173, 280, 328)

[960] P. HORTON AND K. NAKAI, *A probabilistic classification system for predicting the cellular localization sites of proteins*, in Proceedings of the Fourth International Conference on Intelligent Systems for Molecular Biology, 1996, pp. 109–115. (Cited on p. 94)

[961] P. D. HOUGH, T. G. KOLDA, AND V. J. TORCZON, *Asynchronous parallel pattern search for nonlinear optimization*, SIAM Journal on Scientific Computing, 23 (2001), pp. 134–156. (Cited on p. 506)

[962] J. HU, M. PRANDINI, AND S. SASTRY, *Aircraft conflict prediction in the presence of a spatially correlated wind field*, IEEE Transactions on Intelligent Transportation Systems, 6 (2005), pp. 326–340. (Cited on p. 301)

[963] W. HU AND A. KERCHEVAL, *Portfolio optimization for t and skewed t returns*, Quantitative Finance, 10 (2010), pp. 91–105. (Cited on p. 429)

[964] M. HUANG, K. SMILOWITZ, AND B. BALCIK, *Models for relief routing: Equity, efficiency and efficacy*, Transportation Research Part E: Logistics and Transportation Review, 48 (2011), pp. 2–18. (Cited on pp. 483, 491)

[965] R. HUANG, V. ZAVALA, AND L. BIEGLER, *Advanced step nonlinear model predictive control for air separation units*, Journal of Process Control, 19 (2009), pp. 678–685. (Cited on p. 247)

[966] X. X. HUANG AND X. Q. YANG, *A unified augmented Lagrangian approach to duality and exact penalization*, Mathematics of Operations Research, 28 (2003), pp. 533–552. (Cited on p. 169)

[967] P. J. HUBER, *Robust estimation of a location parameter*, The Annals of Mathematical Statistics, 35 (1964), pp. 73–101. (Cited on p. 31)

[968] M. HUNEAULT AND F. D. GALIANA, *A survey of the optimal power flow literature*, IEEE Transactions on Power Systems, 6 (1991), pp. 762–770. (Cited on p. 187)

[969] C. HURKMANS, P. REMEIJER, J. LEBESQUE, AND B. MIJNHEER, *Set-up verification using portal imaging: Review of current clinical practice*, Radiotherapy and Oncology, 58 (2001), pp. 105–120. (Cited on p. 347)

[970] A. P. HURTER AND J. S. MARTINICH, *Facility Location and the Theory of Production*, Kluwer Academic, Boston, 1989. (Cited on p. 450)

[971] W. HUYER AND A. NEUMAIER, *Global optimization by multilevel coordinate search*, Journal of Global Optimization, 14 (1999), pp. 331–355. (Cited on p. 506)

[972] J. T. HWANG, D. Y. LEE, J. W. CUTLER, AND J. R. R. A. MARTINS, *Large-scale multidisciplinary optimization of a small satellite's design and operation*, Journal of Spacecraft and Rockets, 51 (2014), pp. 1648–1663. (Cited on p. 257)

[973] D. A. IANCU AND N. TRICHAKIS, *Pareto efficiency in robust optimization*, Management Science, 60 (2014), pp. 130–147. (Cited on pp. 333, 340)

[974] IBM CORPORATION, *IBM ILOG CPLEX Optimization Studio Documentation*, 2014. (Cited on p. 22)

[975] T. ILLÉS AND T. TERLAKY, *Pivot versus interior point methods: Pros and cons*, European Journal of Operational Research, 140 (2002), pp. 170–190. (Cited on pp. xvii, 5, 8)

[976] *Implicit Filtering*, http://www4.ncsu.edu/~ctk/iffco.html. (Cited on p. 500)

[977] INFORMS, *MISO Wins INFORMS Edelman Award*, www.informs.org/About-INFORMS/News-Room/Press-Releases/Edelman-Winner-2011, 2011. (Cited on p. 459)

[978] *ITIP—Information Theoretic Inequality Prover*, http://user-www.ie.cuhk.edu.hk/~ITIP/, 1996. (Cited on p. 35)

[979] J. IVAKPOUR AND N. KASIRI, *Synthesis of distillation column sequences for nonsharp separations*, Industrial and Engineering Chemistry Research, 48 (2009), pp. 8635–8649. (Cited on p. 324)

[980] G. N. IYENGAR, *Robust dynamic programming*, Mathematics of Operations Research, 30 (2005), pp. 257–280. (Cited on p. 344)

[981] R. A. JABR, *Adjustable robust OPF with renewable energy sources*, IEEE Transactions on Power Systems, 28 (2013), pp. 4742–4751. (Cited on p. 365)

[982] ———, *Robust transmission network expansion planning with uncertain renewable generation and loads*, IEEE Transactions on Power Systems, 28 (2013), pp. 4558–4567. (Cited on p. 365)

[983] M. JACH, D. MICHAELS, AND R. WEISMANTEL, *The convex envelope of $(n-1)$-convex functions*, SIAM Journal on Optimization, 19 (2008), pp. 1451–1466. (Cited on p. 283)

[984] S. H. JACOBSON, E. C. SEWELL, R. DEUSON, AND B. G. WENIGER, *An integer programming model for vaccine procurement and delivery for childhood immunization: A pilot study*, Health Care Management Science, 2 (1999), pp. 1–9. (Cited on p. 96)

[985] V. JACOBSON, *Congestion avoidance and control*, ACM SIGCOMM Computer Communication Review, 18 (1988), pp. 314–329. (Cited on p. 36)

[986] S. L. JANAK, C. A. FLOUDAS, J. KALLRATH, AND N. VORMBROCK, *Production scheduling of a large-scale industrial batch plant. I. Short-term and medium-term scheduling*, Industrial and Engineering Chemistry Research, 45 (2006a), pp. 8234–8252. (Cited on pp. 394, 395)

[987] ———, *Production scheduling of a large-scale industrial batch plant. II. Reactive scheduling*, Industrial and Engineering Chemistry Research, 45 (2006b), pp. 8253–8269. (Cited on pp. 394, 395)

[988] S. L. JANAK, X. LIN, AND C. A. FLOUDAS, *A new robust optimization approach for scheduling under uncertainty: II. Uncertainty with known probability distribution*, Computers and Chemical Engineering, 31 (2007), pp. 171–195. (Cited on p. 394)

[989] B. JANSEN, *Interior Point Techniques in Optimization. Complexity, Sensitivity and Algorithms*, vol. 6 of Applied Optimization, Kluwer Academic, Dordrecht, Netherlands, 1996. (Cited on p. 11)

[990] F. JARRE, M. KOČVARA, AND J. ZOWE, *Optimal truss design by interior-point methods*, SIAM Journal on Optimization, 8 (1998), pp. 1084–1107. (Cited on pp. 22, 144)

[991] I. JEHAN, H. HARRIS, S. SALAT, A. ZEB, N. MOBEEN, O. PASHA, E. MCCLURE, J. MOORE, L. L. WRIGHT, AND R. L. GOLDBERG, *Neonatal mortality, risk factors and causes: A prospective population-based cohort study in urban Pakistan*, Bulletin of the World Health Organization, 87 (2009), pp. 130–138. (Cited on p. 481)

[992] N. O. JENSEN, *A Note on Wind Generator Interaction*, technical report, Risø National Laboratory, 1983. (Cited on p. 368)

[993] K. Y. JEONG, J. D. HONG, AND Y. XIE, *Design of emergency logistics networks, taking efficiency, risk and robustness into consideration*, International Journal of Logistics Research and Applications, 17 (2014), pp. 1–22. (Cited on p. 487)

[994] R. C. JEROSLOW, *There cannot be any algorithm for integer programming with quadratic constraints*, Operations Research, 21 (1973), pp. 221–224. (Cited on p. 274)

[995] V. JEYAKUMAR, A. M. RUBINOV, AND Z. Y. WU, *Sufficient global optimality conditions for non-convex quadratic minimization problems with box constraints*, Journal of Global Optimization, 36 (2006), pp. 471–481. (Cited on p. 168)

[996] ———, *Non-convex quadratic minimization problems with quadratic constraints: Global optimality conditions*, Mathematical Programming, 110 (2007), pp. 521–541. (Cited on p. 168)

[997] J. JEŻOWSKI, *Review of water network design methods with literature annotations*, Industrial and Engineering Chemistry Research, 49 (2010), pp. 4475–4516. (Cited on p. 326)

[998] J. JEŻOWSKI, R. BOCHENEK, AND G. POPLEWSKI, *On application of stochastic optimization techniques to designing heat exchanger- and water networks*, Chemical Engineering and Processing: Process Intensification, 46 (2007), pp. 1160–1174. (Cited on p. 327)

[999] J. JEŻOWSKI AND F. FRIEDLER, *A simple approach for maximum heat recovery calculations*, Chemical Engineering Science, 47 (1992), pp. 1481–1494. (Cited on p. 324)

[1000] Z. JIA AND H. ZHAO, *Mitigating the U.S. drug shortages through Pareto-improving contracts*, Production and Operations Management, Forthcoming, 2017. (Cited on p. 478)

[1001] R. JIANG, J. WANG, AND Y. GUAN, *Robust unit commitment with wind power and pumped storage hydro*, IEEE Transactions on Power Systems, 27 (2012), pp. 800–810. (Cited on pp. 333, 358, 360, 362)

[1002] R. JIANG, M. ZHANG, AND Y. GUAN, *Two-stage minimax regret robust unit commitment*, IEEE Transactions on Power Systems, 28 (2013), pp. 2271–2282. (Cited on p. 363)

[1003] R. JIANG, M. ZHANG, G. LI, AND Y. GUAN, *Two-stage network constrained robust unit commitment problem*, European Journal of Operational Research, 234 (2014), pp. 751–762. (Cited on pp. 358, 360, 362)

[1004] D. JIBETEAN AND E. DE KLERK, *Global optimization of rational functions: A semidefinite programming approach*, Mathematical Programming, 106 (2006), pp. 93–109. (Cited on p. 169)

[1005] K. G. JOBACK, *Designing Molecules Possessing Desired Physical Property Values*, PhD thesis, Massachusetts Institute of Technology, 1989. (Cited on p. 329)

[1006] K. G. JOBACK AND G. STEPHANOPOULOS, *Searching spaces of discrete solutions: The design of molecules possessing desired physical properties*, Advances in Chemical Engineering, 21 (1995), pp. 257–311. (Cited on p. 329)

[1007] N. J. JOBST, M. D. HORNIMAN, C. A. LUCAS, AND G. MITRA, *Computational aspects of alternative portfolio selection models in the presence of discrete asset choice constraints*, Quantitative Finance, 1 (2001), pp. 489–501. (Cited on p. 292)

[1008] M. P. JOHNSON, *Single-period location models for subsidized housing: Project-based subsidies*, Socio-Economic Planning Sciences, 40 (2006), pp. 249–274. (Cited on p. 480)

[1009] M. P. JOHNSON, W. L. GORR, AND S. ROEHRIG, *Location of service facilities for the elderly*, Annals of Operations Research, 136 (2005), pp. 324–349. (Cited on p. 480)

[1010] M. JOLY, *Refinery production planning and scheduling: The refining core business*, Brazilian Journal of Chemical Engineering, 29 (2012), pp. 371–384. (Cited on p. 41)

[1011] D. R. JONES, *The direct global optimization algorithm*, in The Encyclopedia of Optimization, C. A. Floudas and P. M. Pardalos, eds., Kluwer Academic, 2001, pp. 431–440. (Cited on p. 172)

[1012] D. R. JONES, C. D. PERTTUNEN, AND B. E. STUCKMAN, *Lipschitzian optimization without the Lipschitz constant*, Journal of Optimization Theory and Applications, 79 (1993), pp. 157–181. (Cited on pp. 172, 506, 508, 513)

[1013] P. JORION, *Value at Risk: The New Benchmark for Controlling Market Risk*, McGraw-Hill, New York, 1997. (Cited on p. 432)

[1014] C. JOSZ, J. MAEGHT, P. PANCIATICI, AND J. C. GILBERT, *Application of the moment-SOS approach to global optimization of the OPF problem*, IEEE Transactions on Power Systems, 30 (2015), pp. 463–470. (Cited on p. 204)

[1015] K. E. JOYNT, E. J. ORAV, AND A. K. JHA, *Thirty-day readmission rates for Medicare beneficiaries by race and site of care*, Journal of the American Medical Association, 305 (2011), pp. 675–681. (Cited on p. 96)

[1016] J. J. JÚDICE, H. D. SHERALI, I. M. RIBEIRO, AND A. M. FAUSTINO, *A complementarity-based partitioning and disjunctive cut algorithm for mathematical programming problems with equilibrium constraints*, Journal of Global Optimization, 36 (2006), pp. 89–114. (Cited on p. 287)

[1017] M. JÜNGER, T. M. LIEBLING, D. NADDEF, G. L. NEMHAUSER, W. R. PULLEYBLANK, G. REINELT, G. RINALDI, AND L. A. WOLSEY, eds., *50 Years of Integer Programming* 1958–2008, Springer, Berlin, Heidelberg, 2010. (Cited on pp. 572, 598, 654)

[1018] A. B. KAHNG AND Q. WANG, *Implementation and extensibility of an analytic placer*, IEEE Transactions on Computer-Aided Design of Integrated Circuits and Systems, 24 (2005), pp. 734–747. (Cited on p. 34)

[1019] G. KAIBEL, *Distillation columns with vertical partitions*, Chemical Engineering and Technology, 10 (1987), pp. 92–98. (Cited on p. 323)

[1020] M. KALKBRENER, *An axiomatic approach to capital allocation*, Mathematical Finance, 15 (2005), pp. 425–437. (Cited on p. 432)

[1021] M. KALKBRENER, H. LOTTER, AND L. OVERBECK, *Sensible and efficient asset allocation for credit portfolios*, Risk, 17 (2004), pp. 19–24. (Cited on p. 433)

[1022] P. KALL AND S. W. WALLACE, *Stochastic Programming*, Springer, 1994. (Cited on p. 391)

[1023] J. KALLRATH, *Mixed integer optimization in the chemical process industry: Experience, potential, and future perspectives*, Chemical Engineering Research and Design, 78 (2000), pp. 809–822. (Cited on p. 208)

[1024] ——, *Planning and scheduling in the process industry*, OR Spectrum, 24 (2002), pp. 219–250. (Cited on p. 327)

[1025] ——, *Modeling Languages in Mathematical Optimization*, Kluwer Academic, 2004. (Cited on p. 91)

[1026] ——, *Solving planning and design problems in the process industry using mixed integer and global optimization*, Annals of Operations Research, 140 (2005), pp. 339–373. (Cited on p. 317)

[1027] J. KALLRATH AND S. REBENNACK, *Computing area-tight piecewise linear overestimators, underestimators and tubes for univariate functions*, in Optimization in Science and Engineering, Springer, New York, 2014, pp. 273–292. (Cited on p. 205)

[1028] R. S. KAMATH, L. T. BIEGLER, AND I. E. GROSSMANN, *An equation-oriented approach for handling thermodynamics based on cubic equation of state in process optimization*, Computers and Chemical Engineering, 34 (2010), pp. 2085–2096. (Cited on pp. xviii, 242, 243, 246)

[1029] B. KAMM AND M. KAMM, *Biorefineries—Multi product processes*, in White Biotechnology, Springer, 2007, pp. 175–204. (Cited on p. 318)

[1030] Y. KAMP AND C. J. WELLEKENS, *Optimal design of minimum-phase FIR filters*, IEEE Transactions on Acoustics, Speech, and Signal Processing, 31 (1983), pp. 922–926. (Cited on p. 29)

[1031] S. KANG, H. ÖNAL, Y. OUYANG, J. SCHEFFRAN, AND D. TURSUN, *Optimizing the biofuels infrastructure: Transportation networks and biorefinery locations in Illinois*, in Handbook of Bioenergy Economics and Policy. Series: Natural Resource Management and Policy, vol. 33, M. Khanna, J. Scheffran, and D. Zilberman, eds., Springer, 2010. (Cited on p. 465)

[1032] A. KANNAN AND S. M. WILD, *Obtaining Quadratic Models of Noisy Functions*, Technical Report ANL/MCS-P1975-1111, Argonne National Laboratory, 2011. (Cited on p. 501)

[1033] ———, *Benefits of deeper analysis in simulation-based groundwater optimization problems*, in Proceedings of the XIX International Conference on Computational Methods in Water Resources (CMWR 2012), June 2012. (Cited on p. 530)

[1034] R. KANNAN AND C. MONMA, *On the computational complexity of integer programming problems*, in Optimization and Operations Research, volume 157 of Lecture Notes in Economics and Mathematical Systems, R. Henn, B. Korte, and W. Oettli, eds., Springer, 1978, pp. 161–172. (Cited on p. 274)

[1035] Y. KANNO, M. OHSAKI, AND J. ITO, *Large-deformation and friction analysis of nonlinear elastic cable networks by second-order cone programming*, International Journal for Numerical Methods in Engineering, 55 (2002), pp. 1079–1114. (Cited on p. 144)

[1036] D. KANSAGARA, H. ENGLANDER, A. SALANITRO, D. KAGEN, C. THEOBALD, M. FREEMAN, AND S. KRIPALANI, *Risk prediction models for hospital readmission: A systematic review*, JAMA, 306 (2011), pp. 1688–1698. (Cited on p. 97)

[1037] I. Z. KARAESMEN, A. SCHELLER-WOLF, AND B. DENIZ, *Managing perishable and aging inventories: Review and future research directions*, in Planning Production and Inventories in the Extended Enterprise, K. G. Kempf, P. Keskinocak, and R. Uzsoy, eds., Springer, Berlin, 2011, pp. 393–436. (Cited on pp. 473, 485)

[1038] N. KARMARKAR, *A new polynomial-time algorithm for linear programming*, Combinatorica, 4 (1984), pp. 373–395. (Cited on p. 11)

[1039] N. KARMITSA, A. BAGIROV, AND M. M. MÄKELÄ, *Comparing different nonsmooth minimization methods and software*, Optimization Methods and Software, 27 (2012), pp. 131–153. (Cited on p. 223)

[1040] R. KARP, *Reducibility among combinatorial problems*, in Complexity of Computer Computations, R. Miller and J. Thatcher, eds., Plenum Press, New York, 1972, pp. 85–103. (Cited on p. 52)

[1041] R. KARUPPIAH AND I. E. GROSSMANN, *Global optimization for the synthesis of integrated water systems in chemical processes*, Computers and Chemical Engineering, 30 (2006), pp. 650–673. (Cited on pp. 212, 284, 292)

[1042] ———, *Global optimization of multiscenario mixed integer nonlinear programming models arising in the synthesis of integrated water networks under uncertainty*, Computers and Chemical Engineering, 32 (2008), pp. 145–160. (Cited on pp. 326, 327)

[1043] S. KATAOKA, *A stochastic programming model*, Econometrica, 31 (1963), pp. 181–196. (Cited on pp. 427, 429)

[1044] M. A. KAY, *State-of-the-art gene-based therapies: The road ahead*, Nature Reviews Genetics, 12 (2011), pp. 316–328. (Cited on p. 182)

[1045] R. B. KEARFOTT, *Rigorous Global Search: Continuous Problems*, Kluwer Academic, 1996. (Cited on p. 173)

[1046] S. S. KEERTHI AND E. G. GILBERT, *Optimal infinite-horizon feedback laws for a general class of constrained discrete-time systems: Stability and moving-horizon approximations*, Journal of Optimization Theory and Applications, 57 (1988), pp. 265–293. (Cited on p. 44)

[1047] A. B. KEHA, I. R. DE FARIAS, JR., AND G. L. NEMHAUSER, *A branch-and-cut algorithm without binary variables for nonconvex piecewise linear optimization*, Operations Research, 54 (2006), pp. 847–858. (Cited on p. 284)

[1048] C. T. KELLEY, *Iterative Methods for Optimization*, Frontiers in Applied Mathematics 18, SIAM, Philadelphia, 1999. (Cited on p. 500)

[1049] ———, *Implicit filtering and nonlinear least squares problems*, in System Modeling and Optimization XX, vol. 130 of IFIP—The International Federation for Information Processing, E. W. Sachs and R. Tichatschke, eds., Springer, 2003, pp. 71–90. (Cited on pp. 530, 532)

[1050] ———, *Implicit Filtering*. Software Environments and Tools, 23, SIAM, Philadelphia, 2011. (Cited on pp. xxi, 500, 507, 508, 509, 510, 511, 512, 513, 514, 515, 530, 532)

[1051] F. KELLY, *Charging and rate control for elastic traffic*, European Transactions on Telecommunications, 8 (1997), pp. 33–37. (Cited on p. 36)

[1052] F. P. KELLY, A. K. MAULLOO, AND D. K. H. TAN, *Rate control for communication networks: Shadow prices, proportional fairness and stability*, Journal of the Operational Research Society, 49 (1997), pp. 237–252. (Cited on p. 36)

[1053] G. J. KENNEDY, G. K. W. KENWAY, AND J. R. R. A. MARTINS, *High aspect ratio wing design: Optimal aerostructural tradeoffs for the next generation of materials*, in Proceedings of the AIAA Science and Technology Forum and Exposition (SciTech), National Harbor, MD, January 2014. AIAA-2014-0596. (Cited on pp. xix, 254)

[1054] G. J. KENNEDY AND J. R. R. A. MARTINS, *A parallel finite-element framework for large-scale gradient-based design optimization of high-performance structures*, Finite Elements in Analysis and Design, 87 (2014), pp. 56–73. (Cited on p. 255)

[1055] G. K. W. KENWAY, G. J. KENNEDY, AND J. R. R. A. MARTINS, *Scalable parallel approach for high-fidelity steady-state aeroelastic analysis and derivative computations*, AIAA Journal, 52 (2014), pp. 935–951. (Cited on p. 255)

[1056] G. K. W. KENWAY AND J. R. R. A. MARTINS, *Multi-point high-fidelity aerostructural optimization of a transport aircraft configuration*, Journal of Aircraft, 51 (2014), pp. 144–160. (Cited on p. 256)

[1057] ———, *High-fidelity aerostructural optimization considering buffet onset*, in Proceedings of the 16th AIAA/ISSMO Multidisciplinary Analysis and Optimization Conference, Dallas, TX, June 2015. AIAA 2015-2790. (Cited on p. 256)

[1058] A. KHAJAVIRAD AND N. V. SAHINIDIS, *Convex envelopes of products of convex and component-wise concave functions*, Journal of Global Optimization, 52 (2012), pp. 391–409. (Cited on p. 283)

[1059] ——, *Convex envelopes generated from finitely many compact convex sets*, Mathematical Programming, 137 (2013), pp. 371–408. (Cited on p. 283)

[1060] J. KIEFER AND J. WOLFOWITZ, *Stochastic estimation of the maximum of a regression function*, The Annals of Mathematical Statistics, 23 (1952), pp. 462–466. (Cited on p. 390)

[1061] M. R. KILINÇ, *Disjunctive Cutting Planes and Algorithms for Convex Mixed Integer Nonlinear Programming*, PhD thesis, University of Wisconsin-Madison, 2011. (Cited on pp. 285, 287)

[1062] M. R. KILINÇ, J. LINDEROTH, AND J. LUEDTKE, *Effective Separation of Disjunctive Cuts for Convex Mixed Integer Nonlinear Programs*, technical report, University of Wisconsin-Madison, 2010. (Cited on p. 287)

[1063] M. R. KILINÇ, J. LINDEROTH, J. LUEDTKE, AND A. MILLER, *Strong-branching inequalities for convex mixed integer nonlinear programs*, Computational Optimization and Applications, 59 (2014), pp. 639–665. (Cited on p. 287)

[1064] J. KIM, S. M. SEN, AND C. T. MARAVELIAS, *An optimization-based assessment framework for biomass-to-fuel conversion strategies*, Energy and Environmental Science, 6 (2013), pp. 1093–1104. (Cited on p. 318)

[1065] S. KIM AND M. KOJIMA, *Exact solutions of some nonconvex quadratic optimization problems via SDP and SOCP relaxations*, Computational Optimization and Applications, 26 (2003), pp. 143–154. (Cited on p. 111)

[1066] S.-J. KIM, K. KOH, S. BOYD, AND D. GORINEVSKY, ℓ_1 *trend filtering*, SIAM Review, 51 (2009), pp. 339–360. (Cited on p. 33)

[1067] K. KINOSHITA AND H. NAKAMURA, *Identification of the ligand binding sites on the molecular surface of proteins*, Protein Science, 14 (2005), pp. 711–718. (Cited on p. 520)

[1068] S. E. KIRBY, S. M. DENNIS, U. W. JAYASINGHE, AND M. F. HARRIS, *Patient related factors in frequent readmissions: The influence of condition, access to services and patient choice*, BMC Health Services Research, 10 (2010), p. 216. (Cited on p. 96)

[1069] B. R. KIRSCH, G. W. CHARACKLIS, K. E. M. DILLARD, AND C. T. KELLEY, *More efficient optimization of long-term water supply portfolios*, Water Resources Research, 45 (2009), pp. W03414-1–W03414-12. (Cited on p. 508)

[1070] K. C. KIWIEL, *A nonderivative version of the gradient sampling algorithm for nonsmooth nonconvex optimization*, SIAM Journal on Optimization, 20 (2010), pp. 1983–1994. (Cited on p. 500)

[1071] V. KLEE AND G. J. MINTY, *How good is the simplex algorithm*, in Inequalities III, O. Shisha, ed., Academic Press, 1972, pp. 159–175. (Cited on pp. 8, 9)

[1072] J. L. KLEPEIS, C. A. FLOUDAS, D. MORIKIS, C. G. TSOKOS, E. ARGYROPOULOS, L. SPRUCE, AND J. D. LAMBRIS, *Integrated computational and experimental approach for lead optimization and design of compstatin variants with improved activity*, Journal of the American Chemical Society, 125 (2003), pp. 8422–8423. (Cited on pp. 176, 179, 180)

[1073] J. L. KLEPEIS, C. A. FLOUDAS, D. MORIKIS, C. G. TSOKOS, AND J. D. LAMBRIS, *Design of peptide analogues with improved activity using a novel de novo protein design approach*, Industrial and Engineering Chemistry Research, 43 (2004), pp. 3817–3826. (Cited on pp. 176, 179, 180)

[1074] A. J. KLEYWEGT, A. SHAPIRO, AND T. HOMEM-DE-MELLO, *The sample average approximation method for stochastic discrete optimization*, SIAM Journal on Optimization, 12 (2001), pp. 479–502. (Cited on p. 390)

[1075] A. KLOSE AND A. DREXL, *Facility location models for distribution system design*, European Journal of Operational Research, 162 (2005), pp. 4–29. (Cited on p. 450)

[1076] T. KOCH, T. ACHTERBERG, E. ANDERSEN, O. BASTERT, T. BERTHOLD, R. E. BIXBY, E. DANNA, G. GAMRATH, A. M. GLEIXNER, M. AMBROS, S. HEINZ, A. LODI, H. MITTELMANN, T. RALPHS, D. SALVAGNIN, D. E. STEFFY, AND K. WOLTER, *MIPLIB 2010*, Mathematical Programming Computation, 3 (2011), pp. 103–163. (Cited on p. 55)

[1077] G. R. KOCIS AND I. E. GROSSMANN, *Computational experience with DICOPT solving MINLP problems in process systems engineering*, Computers and Chemical Engineering, 13 (1989), pp. 307–315. (Cited on p. 289)

[1078] ———, *A modelling and decomposition strategy for the MINLP optimization of process flowsheets*, Computers and Chemical Engineering, 13 (1989), pp. 797–819. (Cited on p. 318)

[1079] M. KOČVARA, *On the modelling and solving of the truss design problem with global stability constraints*, Structural and Multidisciplinary Optimization, 23 (2000), pp. 189–203. (Cited on p. 144)

[1080] M. KOČVARA AND M. STINGL, *PENNON: A code for convex nonlinear and semidefinite programming*, Optimization Methods and Software, 18 (2003), pp. 317–333. (Cited on p. 118)

[1081] ———, *PENNON: Software for linear and nonlinear matrix inequalities*, in [87], pp. 755–791. (Cited on pp. 118, 147)

[1082] M. KOČVARA, M. STINGL, AND J. ZOWE, *Free material optimization: Recent progress*, Optimization, 57 (2008), pp. 79–100. (Cited on p. 144)

[1083] C. A. KOCZOR, E. K. LEE, R. A. TORRES, A. BOYD, J. D. VEGA, K. UPPAL, F. YUAN, E. J. FIELDS, A. M. SAMAREL, AND W. LEWIS, *Detection of differentially methylated gene promoters in failing and nonfailing human left ventricle myocardium using computation analysis*, Physiological Genomics, 45 (2013), pp. 597–605. (Cited on p. 99)

[1084] A. C. KOKOSSIS AND C. A. FLOUDAS, *Optimization of complex reactor networks—I. Isothermal operation*, Chemical Engineering Science, 45 (1990), pp. 595–614. (Cited on p. 319)

[1085] ———, *Synthesis of isothermal reactor–separator–recycle systems*, Chemical Engineering Science, 46 (1991), pp. 1361–1383. (Cited on p. 319)

[1086] ———, *Optimization of complex reactor networks · II. Nonisothermal operation*, Chemical Engineering Science, 49 (1994), pp. 1037–1051. (Cited on p. 319)

[1087] B. F. KOLANOWSKI, *Small-Scale Cogeneration Handbook*, 4th edition, Fairmont Press, 2011. (Cited on p. 303)

[1088] F. KOLBERT AND L. WORMALD, *Robust portfolio optimization using second-order cone programming*, in Optimizing Optimization: The Next Generation of Optimization Applications and Theory, Stephen Satchell, ed., Elsevier Academic Press, Burlington, MA, 2009, pp. 3–22. (Cited on p. 153)

[1089] T. G. KOLDA, R. M. LEWIS, AND V. TORCZON, *Optimization by direct search: New perspectives on some classical and modern methods*, SIAM Review, 45 (2003), pp. 385–482. (Cited on pp. 499, 503, 509, 510)

[1090] S. P. KOLODZIEJ, I. E. GROSSMANN, K. C. FURMAN, AND N. W. SAWAYA, *A discretization-based approach for the optimization of the multiperiod blend scheduling problem*, Computers and Chemical Engineering, 53 (2013), pp. 122–142. (Cited on pp. 208, 212, 328)

[1091] T. KOLTAI AND T. TERLAKY, *The difference between managerial and mathematical interpretation of sensitivity analysis results in linear programming*, International Journal of Production Economics, 65 (200), pp. 257–274. (Cited on p. 11)

[1092] E. KONDILI, C. C. PANTELIDES, AND R. W. H. SARGENT, *A general algorithm for short-term scheduling of batch operations—I. MILP formulation*, Computers and Chemical Engineering, 17 (1993), pp. 211–227. (Cited on pp. 77, 328)

[1093] J. KONEČNÝ AND P. RICHTÁRIK, *Simple Complexity Analysis of Simplified Direct Search*, Technical Report arXiv:1410.0390v2, School of Mathematics, University of Edinburgh, 2014. (Cited on p. 501)

[1094] H. KONNO, *A cutting plane algorithm for solving bilinear programs*, Mathematical Programming, 11 (1976), pp. 14–27. (Cited on p. 362)

[1095] A. E. Ş. KONUKMAN, M. C. ÇAMURDAN, AND U. AKMAN, *Simultaneous flexibility targeting and synthesis of minimum-utility heat-exchanger networks with superstructure-based MILP formulation*, Chemical Engineering and Processing: Process Intensification, 41 (2002), pp. 501–518. (Cited on p. 324)

[1096] M. KÖPPE, *On the complexity of nonlinear mixed-integer optimization*, in [1160], pp. 533–557. (Cited on p. 274)

[1097] A. P. R. KOPPOL, M. BAGAJEWICZ, B. J. DERICKS, AND M. J. SAVELSKI, *On zero water discharge solutions in the process industry*, Advances in Environmental Research, 8 (2004), pp. 151–171. (Cited on p. 327)

[1098] L. KORNISH AND R. KEENEY, *Repeated commit-or-defer decisions with a deadline: The influenza vaccine composition*, Operations Research, 56 (2008), pp. 527–541. (Cited on p. 472)

[1099] K. O. KORTANEK AND P. MOULIN, *Semi-infinite programming in orthogonal wavelet filter design*, in Semi-Infinite Programming, R. Reemtsen and J.-J. Rückmann, eds., vol. 25 of Nonconvex Optimization and Its Applications, Kluwer Academic, 1998, ch. 10, pp. 323–360. (Cited on p. 29)

[1100] M. KORTELAINEN, T. LESINSKI, J. MORÉ, W. NAZAREWICZ, J. SARICH, N. SCHUNCK, M. V. STOITSOV, AND S. M. WILD, *Nuclear energy density optimization*, Physical Review C, 82(2):024313, 2010. (Cited on pp. 537, 538, 539)

[1101] M. KORTELAINEN, J. MCDONNELL, W. NAZAREWICZ, E. OLSEN, P.-G. REINHARD, J. SARICH, N. SCHUNCK, S. M. WILD, D. DAVESNE, J. ERLER, AND A. PASTORE, *Nuclear energy density optimization: Shell structure*, Physical Review C, 89 (2014), p. 054314. (Cited on p. 537)

[1102] M. KORTELAINEN, J. MCDONNELL, W. NAZAREWICZ, P.-G. REINHARD, J. SARICH, N. SCHUNCK, M. V. STOITSOV, AND S. M. WILD, *Nuclear energy density optimization: Large deformations*, Physical Review C, 85 (2012), p. 024304. (Cited on pp. xxii, 537)

[1103] J. T. KRASEMANN, *Design of an effective algorithm for fast response to the re-scheduling of railway traffic during disturbances*, Transportation Research Part C, 20 (2012), pp. 62–78. (Cited on p. 69)

[1104] Z. KRAVANJA AND I. E. GROSSMANN, *Multilevel-hierarchical MINLP synthesis of process flowsheets*, Computers and Chemical Engineering, 21 (1997), pp. S421–S426. (Cited on p. 318)

[1105] N. KRISLOCK AND H. WOLKOWICZ, *Explicit sensor network localization using semidefinite representations and facial reductions*, SIAM Journal on Optimization, 20 (2010), pp. 2679–2708. (Cited on p. 115)

[1106] P. KROKHMAL, J. PALMQUIST, AND S. URYASEV, *Portfolio optimization with conditional value-at-risk objective and constraints*, Journal of Risk, 4 (2002), pp. 43–68. (Cited on pp. 152, 427, 431)

[1107] P. KROKHMAL, S. URYASEV, AND G. ZRAZHEVSKY, *Risk management for hedge fund portfolios: A comparative analysis of linear rebalancing strategies*, Journal of Alternative Investments, 5 (2002), pp. 10–29. (Cited on p. 394)

[1108] P. KROKHMAL, M. ZABARANKIN, AND S. URYASEV, *Modeling and optimization of risk*, Surveys in Operations Research and Management Science, 16 (2011), pp. 49–66. (Cited on p. 427)

[1109] L. KROON, D. HUISMAN, E. ABBINK, P.-J. FIOOLE, M. FISCHETTI, G. MARÓTI, A. SCHRIJVER, A. STEENBEEK, AND R. YBEMA, *The new Dutch timetable: The OR revolution*, Interfaces, 39 (2009), pp. 6–17. (Cited on p. 58)

[1110] J. KUCHAR AND L. YANG, *A review of conflict detection and resolution modeling methods*, IEEE Transactions on Intelligent Transportation Systems, 1 (2000), pp. 179–189. (Cited on p. 294)

[1111] B. KUHLMAN AND D. BAKER, *Native protein sequences are close to optimal for their structures*, Proceedings of the National Academy of Sciences of the United States of America, 97 (2000), pp. 10383–10388. (Cited on p. 181)

[1112] A. KUSIAK AND W. LI, *Estimation of wind speed: A data-driven approach*, Journal of Wind Engineering and Industrial Aerodynamics, 98 (2010), pp. 559–567. (Cited on p. 367)

[1113] A. KUSIAK AND Z. SONG, *Design of wind farm layout for maximum wind energy capture*, Renewable Energy, 35 (2010), pp. 685–694. (Cited on p. 370)

[1114] W. Y. KWONG, P. Y. ZHANG, D. ROMERO, J. MORAN, M. MORGENROTH, AND C. AMON, *Multi-objective optimization of wind farm layouts under energy generation and noise propagation*, in Proceedings of the ASME 2012 International Design Engineering Technical Conferences and Computers and Information in Engineering Conference IDETC/CIE 2012, Chicago, 2012. (Cited on pp. 367, 370)

[1115] J. C. LAGARIAS, B. POONEN, AND M. H. WRIGHT, *Convergence of the restricted Nelder–Mead algorithm in two dimensions*, SIAM Journal on Optimization, 22 (2012), pp. 501–532. (Cited on p. 498)

[1116] J. C. LAGARIAS, J. A. REEDS, M. H. WRIGHT, AND P. E. WRIGHT, *Convergence properties of the Nelder–Mead simplex method in low dimensions*, SIAM Journal on Optimization, 9 (1998), pp. 112–147. (Cited on p. 498)

[1117] R. LAHDELMA AND H. HAKONEN, *An efficient linear programming algorithm for combined heat and power production*, European Journal of Operational Research, 148 (2003), pp. 141–151. (Cited on p. 304)

[1118] C. D. LAIRD, L. T. BIEGLER, AND B. VAN BLOEMEN WAANDERS, *A mixed integer approach for obtaining unique solutions in source inversion of drinking water networks*, Journal of Water Resources Planning and Management, 132 (2006), pp. 242–251. (Cited on p. 292)

[1119] A. LAKSHMANAN AND L. T. BIEGLER, *Synthesis of optimal chemical reactor networks*, Industrial and Engineering Chemistry Research, 35 (1996), pp. 1344–1353. (Cited on p. 320)

[1120] L. LAMORGESE AND C. MANNINO, *An exact decomposition approach for the real-time train dispatching problem*, Operations Research, 63 (2015), pp. 48–64. (Cited on p. 72)

[1121] ———, *The track formulation for the train dispatching problem*, Electronic Notes in Discrete Mathematics, 41 (2013), pp. 559–566. (Cited on p. 72)

[1122] L. LAMORGESE, C. MANNINO, AND M. PIACENTINI, *Optimal train dispatching by Benders'-like reformulation*, Transportation Science, 50.3 (2016), pp. 910–925. (Cited on p. 72)

[1123] G. LAN, A. NEMIROVSKI, AND A. SHAPIRO, *Validation analysis of mirror descent stochastic approximation method*, Mathematical Programming, 134 (2012), pp. 425–458. (Cited on p. 390)

[1124] A. H. LAND AND A. G. DOIG, *An automatic method for solving discrete programming problems*, Econometrica, 28 (1960), pp. 497–520. (Cited on pp. 53, 274)

[1125] Y.-D. LANG AND L. T. BIEGLER, *A distributed stream method for tray optimization*, AIChE Journal, 48 (2002), pp. 582–595. (Cited on p. 243)

[1126] K. M. LANGEN AND D. T. L. JONES, *Organ motion and its management*, International Journal of Radiation Oncology • Biology • Physics, 50 (2001), pp. 265–278. (Cited on p. 348)

[1127] M. LANGER AND J. LEONG, *Optimization of beam weights under dose-volume restrictions*, International Journal of Radiation Oncology • Biology • Physics, 13 (1987), pp. 1255–1260. (Cited on pp. 94, 95)

[1128] L. S. LASDON, A. D. WAREN, S. SARKAR, AND F. PALACIOS, *Solving the pooling problem using generalized reduced gradient and successive linear programming algorithms*, ACM SIGMAP Bulletin, 27 (1979), pp. 9–15. (Cited on p. 212)

[1129] A. LASRY, G. S. ZARIC, AND M. W. CARTER, *Multi-level resource allocation for HIV prevention: A model for developing countries*, European Journal of Operational Research, 180 (2007), pp. 786–799. (Cited on p. 96)

[1130] W. LASSCHUIT AND N. THIJSSEN, *Supporting supply chain planning and scheduling decisions in the oil and chemical industry*, Computers and Chemical Engineering, 28 (2004), pp. 863–870. (Cited on p. 91)

[1131] J. B. LASSERRE, *Semidefinite programming vs. LP relaxations for polynomial programming*, Mathematics of Operations Research, 27, pp. 347–360. (Cited on p. 173)

[1132] ———, *Optimisation globale et théorie des moments*, Comptes Rendus de l'Académie des Sciences Paris, Série I, Mathématique, 331 (2000), pp. 929–934. (Cited on p. 121)

[1133] ———, *Global optimization with polynomials and the problem of moments*, SIAM Journal on Optimization, 11 (2001), pp. 796–817. (Cited on pp. 119, 121, 173)

[1134] ——, *Moments, Positive Polynomials and Their Applications*, Imperial College Press, London, UK, 2010. (Cited on p. 121)

[1135] J. B. LASSERRE, D. HENRION, C. PRIEUR, AND E. TRÉLAT, *Nonlinear optimal control via occupation measures and LMI-relaxations*, SIAM Journal on Control and Optimization, 47 (2008), pp. 1643–1666. (Cited on pp. 121, 122)

[1136] J. LAVAEI AND S. H. LOW, *Zero duality gap in optimal power flow problem*, IEEE Transactions on Power Systems, 27 (2012), pp. 92–107. (Cited on pp. 203, 204)

[1137] V. LAVRIC, P. IANCU, AND V. PLEŞU, *Genetic algorithm optimisation of water consumption and wastewater network topology*, Journal of Cleaner Production, 13 (2005), pp. 1405–1415. (Cited on p. 327)

[1138] S. LE DIGABEL, *Algorithm 909: NOMAD: Nonlinear optimization with the MADS algorithm*, ACM Transactions on Mathematical Software, 37 (2011), pp. 1–15. (Cited on p. 500)

[1139] S. LE DIGABEL AND S. M. WILD, *A Taxonomy of Constraints in Simulation-Based Optimization*, Preprint ANL/MCS-P5350-0515, Argonne National Laboratory, Mathematics and Computer Science Division, 2015. (Cited on p. 535)

[1140] A. LEAVER-FAY, M. TYKA, S. M. LEWIS, O. F. LANGE, J. THOMPSON, R. JACAK, K. KAUFMAN, P. D. RENFREW, C. A. SMITH, W. SHEFFLER, I. W. DAVIS, S. COOPER, A. TREUILLE, D. J. MANDELL, F. RICHTER, Y. E. BAN, S. J. FLEISHMAN, J. E. CORN, D. E. KIM, S. LYSKOV, M. BERRONDO, S. MENTZER, Z. POPOVIĆ, J. J. HAVRANEK, J. KARANICOLAS, R. DAS, J. MEILER, T. KORTEMME, J. J. GRAY, B. KUHLMAN, D. BAKER, AND P. BRADLEY, *Rosette3: An object-oriented software suite for the simulation and design of macromolecules*, Methods in Enzymology, 487 (2011), pp. 545–574. (Cited on p. 181)

[1141] H. LEBRET AND S. BOYD, *Antenna array pattern synthesis via convex optimization*, IEEE Transactions on Signal Processing, 45 (1997), pp. 526–532. (Cited on p. 30)

[1142] C. LEE, C. LIU, S. MEHROTRA, AND M. SHAHIDEHPOUR, *Modeling transmission line constraints in two-stage robust unit commitment problem*, IEEE Transactions on Power Systems, 29 (2014), pp. 1221–1231. (Cited on p. 362)

[1143] E. K. LEE, *Large-scale optimization-based classification models in medicine and biology*, Annals of Biomedical Engineering, 35 (2007), pp. 1095–1109. (Cited on pp. 94, 99)

[1144] E. K. LEE, H. Y. ATALLAH, M. D. WRIGHT, E. T. POST, C. THOMAS IV, D. T. WU, AND L. L. HALEY, *Transforming emergency department workflow and patient care*, Interfaces, 45 (2015), pp. 58–82. (Cited on pp. 95, 97)

[1145] E. K. LEE, T. FOX, AND I. CROCKER, *Optimization of radiosurgery treatment planning via mixed integer programming*, Medical Physics, 27 (2000), pp. 995–1004. (Cited on pp. 94, 95)

[1146] ——, *Integer programming applied to intensity-modulated radiation therapy treatment planning*, Annals of Operations Research, 119 (2003), pp. 165–181. (Cited on pp. 94, 95, 97)

[1147] ——, *Simultaneous beam geometry and intensity map optimization in intensity-modulated radiation therapy*, International Journal of Radiation Oncology • Biology • Physics, 64 (2006), pp. 301–320. (Cited on pp. 94, 95)

[1148] E. K. LEE, A. Y. FUNG, J. P. BROOKS, AND M. ZAIDER, *Automated planning volume definition in soft-tissue sarcoma adjuvant brachytherapy*, Physics in Medicine and Biology, 47 (2002), pp. 1891–1910. (Cited on p. 94)

[1149] E. K. LEE, R. J. GALLAGHER, AND D. PATTERSON, *A linear programming approach to discriminant analysis with a reserved judgment region*, INFORMS Journal on Computing, 15 (2003), pp. 23–41. (Cited on pp. 94, 97)

[1150] E. K. LEE, R. J. GALLAGHER, A. M. CAMPBELL, AND M. R. PRAUSNITZ, *Prediction of ultrasound-mediated disruption of cell membranes using machine learning techniques and statistical analysis of acoustic spectra*, IEEE Transactions on Biomedical Engineering, 51 (2004), pp. 82–89. (Cited on p. 94)

[1151] E. K LEE, S. MAHESHWARY, AND X. WEI, *Facets of Conflict Hypergraphs*, technical report, Georgia Tech, submitted for publication, 2016. (Cited on p. 102)

[1152] E. K. LEE, S. MAHESHWARY, J. MASON, AND W. GLISSON, *Large-scale dispensing for emergency response to bioterrorism and infectious-disease outbreak*, Interfaces, 36 (2006), pp. 591–607. (Cited on p. 96)

[1153] E. K. LEE, F. PIETZ, B. BENECKE, J. MASON, AND G. BUREL, *Advancing public health and medical preparedness with operations research*, Interfaces, 43 (2013), pp. 79–98. (Cited on pp. 95, 96)

[1154] E. K. LEE AND T.-L. WU, *Classification and disease prediction via mathematical programming*, in Handbook of Optimization in Medicine, P. M. Pardalos and H. E. Romeijn, eds., Springer, New York, 2009, pp. 381–430. (Cited on p. 99)

[1155] E. K. LEE, T.-L. WU, F. GOLDSTEIN, AND A. LEVEY, *Predictive model for early detection of mild cognitive impairment and Alzheimer's disease*, in Optimization and Data Analysis in Biomedical Informatics, P. M. Pardalos, T. F. Coleman, and P. Xanthopoulos, eds., Springer, New York, 2012, pp. 83–97. (Cited on p. 94)

[1156] E. K. LEE, F. YUAN, D. A. HIRSH, M. D. MALLORY, AND H. K. SIMON, *A clinical decision tool for predicting patient care characteristics: Patients returning within 72 hours in the emergency department*, AMIA Annual Symposium Proceedings, American Medical Informatics Association, 2012 (2012), pp. 495–504. (Cited on pp. xxiii, 97, 98, 99)

[1157] E. K. LEE, F. YUAN, A. TEMPLETON, R. YAO, K. KIEL, AND J. C. CHU, *Biological planning for high-dose-rate brachytherapy: Application to cervical cancer treatment*, Interfaces, 43 (2013), pp. 462–476. (Cited on pp. 99, 101, 102)

[1158] E. K. LEE AND M. ZAIDER, *Mixed integer programming approaches to treatment planning for brachytherapy—Application to permanent prostate implants*, Annals of Operations Research, 119 (2003), pp. 147–163. (Cited on p. 94)

[1159] ———, *Operations research advances cancer therapeutics*, Interfaces, 38 (2008), pp. 5–25. (Cited on p. 94)

[1160] J. LEE AND S. LEYFFER, *Mixed Integer Nonlinear Programming*, volume 154 of The IMA Volumes in Mathematics and Its Applications, Springer, 2012. (Cited on pp. 274, 553, 560, 563, 590, 609, 636)

[1161] S. LEE AND I. E. GROSSMANN, *New algorithms for nonlinear generalized disjunctive programming*, Computers and Chemical Engineering, 24 (2000), pp. 2125–2141. (Cited on p. 316)

[1162] A. LEIRAS, S. HAMACHER, AND A. ELKAMEL, *Petroleum refinery operational planning using robust optimization*, Engineering Optimization, 42 (2010), pp. 1119–1131. (Cited on p. 333)

[1163] M. LEJEUNE, *A Unified Approach for Cycle Service Levels*, technical report, George Washington University, 2009. (Cited on p. 292)

[1164] ———, *A VaR Black-Litterman model for the construction of absolute return fund-of-funds*, Quantitative Finance, 11 (2011), pp. 1489–1501. (Cited on p. 430)

[1165] ———, *Game theoretical approach for reliable enhanced indexation*, Decision Analysis, 9 (2012), pp. 146–155. (Cited on p. 430)

[1166] C. LEMARÉCHAL, A. OUOROU, AND G. PETROU, *Robust network design in telecommunications under polytope demand uncertainty*, European Journal of Operational Research, 206 (2010), pp. 634–641. (Cited on p. 333)

[1167] B. C. LESIEUTRE, D. K. MOLZAHN, A. R. BORDEN, AND C. L. DEMARCO, *Examining the limits of the application of semidefinite programming to power flow problems*, in 49th Annual Allerton Conference on Communication, Control, and Computing, 2011, pp. 1492–1499. (Cited on p. 204)

[1168] Z. LEUNG, A. CHEN, P. YADAV, AND J. GALLIEN, *The Impact of Inventory Management on Stock-Outs of Essential Drugs in Sub-Saharan Africa: Secondary Analysis of a Field Experiment in Zambia*, working paper, London Business School, London, United Kingdom, 2014. (Cited on p. 491)

[1169] M. K. LEWINSKI AND F. D. BUSHMAN, *Retroviral DNA integration—Mechanism and consequences*, Advances in Genetics, 55 (2005), pp. 147–181. (Cited on p. 182)

[1170] B. M. LEWIS, A. L. ERERA, M. A. NOWAK, AND CHELSEA C. WHITE III, *Managing inventory in global supply chains facing port-of-entry disruption risks*, Transportation Science, 47 (2013), pp. 162–180. (Cited on p. 488)

[1171] R. M. LEWIS AND V. TORCZON, *Pattern search algorithms for bound constrained minimization*, SIAM Journal on Optimization, 9 (1999), pp. 1082–1099. (Cited on pp. 503, 510)

[1172] ———, *Pattern search methods for linearly constrained minimization*, SIAM Journal on Optimization, 10 (2000), pp. 917–941. (Cited on pp. 503, 511)

[1173] ———, *A globally convergent augmented Lagrangian pattern search algorithm for optimization with general constraints and simple bounds*, SIAM Journal on Optimization, 12 (2002), pp. 1075–1089. (Cited on p. 504)

[1174] S. LEYFFER, *User Manual for MINLP_BB*, http://www.mcs.anl.gov/~leyffer/MINLP_manual.pdf. (Cited on p. 289)

[1175] ———, *Integrating SQP and branch-and-bound for mixed integer nonlinear programming*, Computational Optimization and Applications, 18 (2001), pp. 295–309. (Cited on p. 279)

[1176] S. LEYFFER, A. SARTENAER, AND E. WANUFELLE, *Branch-and-Refine for Mixed-Integer Nonconvex Global Optimization*, Preprint ANL/MCS-P1547-0908, Mathematics and Computer Science Division, Argonne National Laboratory, 2008. (Cited on p. 284)

[1177] A. LI AND L. TUNÇEL, *Some applications of symmetric cone programming in financial mathematics*, Transactions on Operational Research, 17 (2006), pp. 1–19. (Cited on p. 149)

[1178] B.-H. LI AND C.-T. CHANG, *A simple and efficient initialization strategy for optimizing water-using network designs*, Industrial and Engineering Chemistry Research, 46 (2007), pp. 8781–8786. (Cited on p. 327)

[1179] C. LI AND C.-K. KOH, *Recursive function smoothing of half-perimeter wirelength for analytical placement*, in Proceedings of the 8th International Symposium on Quality Electronic Design (ISQED'07), 2007. (Cited on p. 34)

[1180] C. LI, H. XIONG, J. ZOU, AND C. W. CHEN, *Distributed robust optimization for scalable video multirate multicast over wireless networks*, IEEE Transactions on Circuits and Systems for Video Technology, 22 (2012), pp. 943–957. (Cited on p. 333)

[1181] D. LI, X. L. SUN, AND C. L. LIU, *An exact solution method for unconstrained quadratic 0–1 programming: A geometric approach*, Journal of Global Optimization, 52 (2012), pp. 797–829. (Cited on p. 168)

[1182] F. LI, X. ZHOU, J. MA, AND S. T. WONG, *Multiple nuclei tracking using integer programming for quantitative cancer cell cycle analysis*, IEEE Transactions on Medical Imaging, 29 (2010), pp. 96–105. (Cited on p. 94)

[1183] H. LI, J. P. BISSONNETTE, T. PURDIE, AND T. C. Y. CHAN, *Robust PET-guided intensity-modulated radiation therapy*, Medical Physics, 42 (2015), pp. 4863–4871. (Cited on p. 334)

[1184] L.-J. LI, R.-J. ZHOU, AND H.-G. DONG, *State-time-space superstructure-based MINLP formulation for batch water-allocation network design*, Industrial and Engineering Chemistry Research, 49 (2009), pp. 236–251. (Cited on p. 327)

[1185] L.-J. LI, R.-J. ZHOU, H.-G. DONG, AND I. E. GROSSMANN, *Separation network design with mass and energy separating agents*, Computers and Chemical Engineering, 35 (2011), pp. 2005–2016. (Cited on p. 327)

[1186] W. LI, C.-W. HUI, AND A. LI, *Integrating CDU, FCC and product blending models into refinery planning*, Computers and Chemical Engineering, 29 (2005), pp. 2010–2028. (Cited on p. 40)

[1187] X. LI, E. ARMAGAN, A. TOMASGARD, AND P. BARTON, *Stochastic pooling problem for natural gas production network design and operation under uncertainty*, AIChE Journal, 57 (2011), pp. 2120–2135. (Cited on p. 212)

[1188] X. LI, A. SUNDARAMOORTHY, AND P. I. BARTON, *Nonconvex generalized Benders decomposition*, in Optimization in Science and Engineering, Springer, 2014, pp. 307–331. (Cited on p. 283)

[1189] Z. LI, R. DING, AND C. A. FLOUDAS, *A comparative theoretical and computational study on robust counterpart optimization*: I. Robust linear optimization and robust mixed integer linear optimization, Industrial and Engineering Chemistry Research, 50 (2011), pp. 10567–10603. (Cited on pp. 333, 334, 394)

[1190] Z. LI AND M. IERAPETRITOU, *Process scheduling under uncertainty: Review and challenges*, Computers and Chemical Engineering, 32 (2008), pp. 715–727. (Cited on p. 394)

[1191] Z. LI, Q. TANG, AND C. A. FLOUDAS, *A comparative theoretical and computational study on robust counterpart optimization*: II. Probabilistic guarantees on constraint satisfaction, Industrial and Engineering Chemistry Research, 51 (2012), pp. 6769–6788. (Cited on p. 394)

[1192] L. LIBERTI, *Reformulations in mathematical programming: Automatic symmetry detection and exploitation*, Mathematical Programming, 131 (2012), pp. 273–304. (Cited on p. 285)

[1193] L. LIBERTI, S. CAFIERI, AND F. TARISSAN, *Reformulations in mathematical programming: A computational approach*, in Foundations of Computational Intelligence, vol. 3, A. Abraham, A.-E. Hassanien, P. Siarry, and A. Engelbrecht, eds., vol. 203 of Studies in Computational Intelligence, Springer, Berlin, Heidelberg, 2009, pp. 153–234. (Cited on p. 299)

[1194] L. LIBERTI, N. MLADENOVIĆ, AND G. NANNICINI, *A recipe for finding good solutions to MINLPs*, Mathematical Programming Computation, 3 (2011), pp. 349–390. (Cited on p. 286)

[1195] C. LIEBCHEN, M. LÜBBECKE, R. H. MÖHRING, AND S. STILLER, *Recoverable Robustness*, technical report, ARRIVAL-Project, August 2007. (Cited on pp. 333, 340)

[1196] R. LIKOVER, *Energy Market Accounting: Day-Ahead and Real-Time*, https://www.iso-ne.com/static-assets/documents/2016/10/20160926-14-wem/0/-energy-market_accounting.pdf, 2013. (Cited on p. 373)

[1197] C.-J. LIN AND J. J. MORÉ, *Newton's method for large bound-constrained optimization problems*, SIAM Journal on Optimization, 9 (1999), pp. 1100–1127. (Cited on p. 535)

[1198] S. LIN AND B. W. KERNIGHAN, *An effective heuristic algorithm for the traveling salesman problem*, Operations Research, 21 (1973), pp. 498–516. (Cited on p. 57)

[1199] X. LIN, S. L. JANAK, AND C. A. FLOUDAS, *A new robust optimization approach for scheduling under uncertainty*: I. Bounded uncertainty, Computers and Chemical Engineering, 28 (2004), pp. 1069–1085. (Cited on p. 394)

[1200] Y. LIN AND L. SCHRAGE, *The global solver in the LINDO API*, Optimization Methods and Software, 24 (2009), pp. 657–668. (Cited on pp. 196, 290)

[1201] J. LINDEROTH, A. SHAPIRO, AND S. WRIGHT, *The empirical behavior of sampling methods for stochastic programming*, Annals of Operations Research, 142 (2006), pp. 215–241. (Cited on p. 390)

[1202] P. LINO, L. A. N. BARROSO, M. V. F. PEREIRA, R. KELMAN, AND M. H. C. FAMPA, *Bid-based dispatch of hydrothermal systems in competitive markets*, Annals of Operations Research, 120 (2003), pp. 81–97. (Cited on p. 410)

[1203] J. LIU, A. B. GERSHMAN, Z.-Q. LUO, AND K. M. WONG, *Adaptive beamforming with sidelobe control: A second-order cone programming approach*, IEEE Signal Processing Letters, 10 (2003), pp. 331–334. (Cited on p. 30)

[1204] P. LIU, M. C. GEORGIADIS, AND E. N. PISTIKOPOULOS, *Advances in energy systems engineering*, Industrial and Engineering Chemistry Research, 50 (2010), pp. 4915–4926. (Cited on p. 318)

[1205] P. LIU, E. N. PISTIKOPOULOS, AND Z. LI, *A multi-objective optimization approach to polygeneration energy systems design*, AIChE Journal, 56 (2010), pp. 1218–1234. (Cited on p. 318)

[1206] T. LIU AND J. S. L. LAM, *Impact of Port Disruption on Transportation Network*, working paper, 2012. (Cited on p. 488)

[1207] W. LIU, X. ZHANG, Y. LI, AND R. MOHAN, *Robust optimization of intensity modulated proton therapy*, Medical Physics, 39 (2012), pp. 1079–1091. (Cited on p. 355)

[1208] Z. LIU, F. MAO, J.-T. GUO, B. YAN, P. WANG, Y. QU, AND Y. XU, *Quantitative evaluation of protein-DNA interactions using an optimized knowledge-based potential*, Nucleic Acids Research, 33 (2005), pp. 546–558. (Cited on p. 178)

[1209] G. LIUZZI AND S. LUCIDI, *A derivative-free algorithm for inequality constrained nonlinear programming via smoothing of an ℓ_∞ penalty function*, SIAM Journal on Optimization, 20 (2009), pp. 1–29. (Cited on p. 505)

[1210] G. LIUZZI, S. LUCIDI, AND V. PICCIALLI, *A direct-based approach exploiting local minimizations for the solution of large-scale global optimization problems*, Computational Optimization and Applications, 45 (2010), pp. 353–375. (Cited on p. 172)

[1211] G. LIUZZI, S. LUCIDI, AND F. RINALDI, *Derivative-free methods for bound constrained mixed-integer optimization*, Computational Optimization and Applications, 53 (2012), pp. 505–526. (Cited on p. 506)

[1212] G. LIUZZI, S. LUCIDI, AND M. SCIANDRONE, *Sequential penalty derivative-free methods for nonlinear constrained optimization*, SIAM Journal on Optimization, 20 (2010), pp. 2614–2635. (Cited on p. 505)

[1213] M. S. LOBO AND S. BOYD, *The Worst-Case Risk of a Portfolio*, technical report, Stanford University, 2000. (Cited on p. 156)

[1214] M. S. LOBO, L. VANDENBERGHE, S. BOYD, AND H. LEBRET, *Applications of second-order cone programming*, Linear Algebra and Its Applications, 284 (1998), pp. 193–228. (Cited on p. 159)

[1215] M. LOCATELLI AND F. SCHOEN, *Global Optimization: Theory, Algorithms, and Applications*, MOS-SIAM Series on Optimizations, 15, SIAM, Philadelphia, 2013. (Cited on pp. 173, 506)

[1216] J. LODREE AND S. TASKIN, *Supply chain planning for hurricane response with wind speed information updates*, Computers and Operations Research, 36 (2009), pp. 2–15. (Cited on p. 484)

[1217] J. LOFBERG, *YALMIP: A toolbox for modeling and optimization in MATLAB*, in International Symposium on Computer Aided Control Systems Design, IEEE, 2004, pp. 284–289. (Cited on pp. 119, 147, 288)

[1218] A LOMAX, *Intensity modulated proton therapy and its sensitivity to treatment uncertainties 2: The potential effects of inter-fraction and inter-field motions*, Physics in Medicine and Biology, 53 (2008), pp. 1043–1056. (Cited on p. 347)

[1219] A. LOMAX, E. PEDRONI, H. RUTZ, AND G. GOITEIN, *The clinical potential of intensity modulated proton therapy*, Zeitschrift für Medizinische Physik, 14 (2004), pp. 147–152. (Cited on p. 354)

[1220] Á. LORCA, M. ÇELIK, Ö. ERGUN, AND P. KESKINOCAK, *An optimization-based decision-support tool for post-disaster debris operations*, Production and Operations Management, published online October 1, 2016. (Cited on p. 491)

[1221] A. LORCA AND X. A. SUN, *Adaptive robust optimization and dynamic uncertainty sets for multi-period economic dispatch with significant wind*, IEEE Transactions on Power Systems, 30 (2014), pp. 1702–1713. (Cited on p. 364)

[1222] A. LORCA, X. A. SUN, E. LITVINOV, AND T. ZHENG, *Multistage adaptive robust optimization for the unit commitment problem*, Operations Research, 64 (2016), pp. 32–51. (Cited on p. 364)

[1223] W. E. LORENSEN AND H. E. CLINE, *Marching cubes: A high resolution 3D surface construction algorithm*, in Proceedings of the 14th Annual Conference on Computer Graphics and Interactive Techniques (SIGGRAPH '87), 1987, pp. 163–170. vol. 21. (Cited on p. 520)

[1224] I. LOTERO, F. TRESPALACIOS, I. E. GROSSMANN, D. J. PAPAGEORGIU, AND M.-S. CHEON, *An MILP–MINLP decomposition method for the global optimization of a source based model of the multiperiod blending problem*, Computers and Chemical Engineering, 87 (2016), pp. 13–35. (Cited on p. 212)

[1225] R. LOUGEE-HEIMER, *The Common Optimization INterface for Operations Research*, IBM Journal of Research and Development, 47 (2003), pp. 57–66. (Cited on p. 288)

[1226] L. LOVÁSZ AND A. SCHRIJVER, *Cones of matrices and set-functions and 0-1 optimization*, SIAM Journal on Optimization, 1 (1991), pp. 166–190. (Cited on pp. 55, 286)

[1227] E. M. LOVELADY, M. EL-HALWAGI, AND G. A. KRISHNAGOPALAN, *An integrated approach to the optimisation of water usage and discharge in pulp and paper plants*, International Journal of Environment and Pollution, 29 (2007), pp. 274–307. (Cited on p. 327)

[1228] S. H. LOW, *A duality model of TC and queue management algorithms*, IEEE/ACM Transactions on Networking, 11 (2003), pp. 525–536. (Cited on p. 36)

[1229] S. H. LOW AND D. E. LAPSEY, *Optimization flow control—I: Basic algorithm and convergence*, IEEE/ACM Transactions on Networking, 7 (1999), pp. 861–874. (Cited on p. 36)

[1230] J. LU, L.-Z. LIAO, A. NERODE, AND J. H. TAYLOR, *Optimal control of systems with continuous and discrete states*, in Proceedings of the 32nd IEEE Conference on Decision and Control, IEEE, 1993, pp. 2292–2297. (Cited on p. 328)

[1231] J. Z. LU, *Challenging control problems and emerging technologies in enterprise optimization*, Control Engineering Practice, 11 (2003), pp. 847–858. (Cited on p. 45)

[1232] S. LUAN, N. SWANSON, Z. CHEN, AND L. MA, *Dynamic gamma knife radiosurgery*, Physics in Medicine and Biology, 54 (2009), pp. 1579–91. (Cited on p. 94)

[1233] A. LUCIA AND A. KUMAR, *Distillation optimization*, Computers and Chemical Engineering, 12 (1988), pp. 1263–1266. (Cited on pp. 238, 240)

[1234] S. LUCIDI AND M. SCIANDRONE, *On the global convergence of derivative-free methods for unconstrained optimization*, SIAM Journal on Optimization, 13 (2002), pp. 97–116. (Cited on p. 499)

[1235] S. LUCIDI, M. SCIANDRONE, AND P. TSENG, *Objective-derivative-free methods for constrained optimization*, Mathematical Programming, 92 (2002), pp. 37–59. (Cited on p. 504)

[1236] J. LUEDTKE AND S. AHMED, *A sample approximation approach for optimization with probabilistic constraints*, SIAM Journal on Optimization, 19 (2008), pp. 674–699. (Cited on p. 391)

[1237] J. LUEDTKE, M. NAMAZIFAR, AND J. LINDEROTH, *Some results on the strength of relaxations of multilinear functions*, Mathematical Programming, 136 (2012), pp. 325–351. (Cited on p. 282)

[1238] D. G. LUENBERGER, *Optimization by Vector Space Methods*, Wiley, New York, 1969. (Cited on p. 125)

[1239] ———, *Linear and Nonlinear Programming*, Addison Wesley, 1989. (Cited on pp. 17, 18, 19)

[1240] Z.-Q. LUO, *Applications of convex optimization in signal processing and digital communication*, Mathematical Programming, Series B, 97 (2003), pp. 177–207. (Cited on p. 30)

[1241] Z.-Q. LUO, W.-K. MA, A. M.-C. SO, Y. YE, AND S. ZHANG, *Semidefinite relaxation of quadratic optimization problems*, IEEE Signal Processing Magazine, 27 (2010), pp. 20–34. (Cited on p. 169)

[1242] R. LUSBY, J. LARSEN, M. EHRGOTT, AND D. RYAN, *A set packing inspired method for real-time junction train routing*, Computers and Operations Research, 40 (2013), pp. 713–724. (Cited on p. 69)

[1243] H. LUSS, *On equitable resource allocation problems: A lexicographic minimax approach*, Operations Research, 47 (1999), pp. 361–378. (Cited on p. 480)

[1244] M. LUYBEN AND C. FLOUDAS, *Analyzing the interaction of design and control—1. A multiobjective framework and application to binary distillation synthesis*, Computers and Chemical Engineering, 18 (1994), pp. 933–969. (Cited on p. 292)

[1245] Z. LYU, G. K. KENWAY, AND J. R. R. A. MARTINS, *Aerodynamic shape optimization investigations of the Common Research Model wing benchmark*, AIAA Journal, 53 (2015), pp. 968–985. (Cited on pp. xix, 254)

[1246] Z. LYU, G. K. KENWAY, C. PAIGE, AND J. R. R. A. MARTINS, *Automatic differentiation adjoint of the Reynolds-averaged Navier–Stokes equations with a turbulence model*, in Proceedings of the 21st AIAA Computational Fluid Dynamics Conference, San Diego, CA, Jul 2013. (Cited on p. 253)

[1247] Y. MA, F. BORRELLI, B. HENCEY, A. PACKARD, AND S. BORTOFF, *Model predictive control of thermal energy storage in building cooling systems*, in Proceedings of the 48th IEEE Conference on Decision and Control 2009, held jointly with the 2009 28th Chinese Control Conference. CDC/CCC 2009, December 2009, pp. 392–397. (Cited on p. 259)

[1248] S. MACCHIETTO, O. ODELE, AND O. OMATSONE, *Design on optimal solvents for liquid-liquid extraction and gas absorption processes*, Chemical Engineering Research and Design, 68 (1990), pp. 429–433. (Cited on p. 329)

[1249] J. M. MACIEJOWSKI, *Predictive Control with Constraints*, Pearson Education, Harlow, U.K., 2002. (Cited on p. 43)

[1250] G. N. MAERTENS, S. HARE, AND P. CHEREPANOV, *The mechanism of retroviral integration from X-ray structures of its key intermediates*, Nature, 468 (2010), pp. 326–329. (Cited on p. 182)

[1251] T. L. MAGNANTI, P. MIREAULT, AND R. T. WONG, *Tailoring Benders decomposition for uncapacitated network design*, Mathematical Programming Study, 26 (1986), pp. 112–154. (Cited on p. 450)

[1252] T. L. MAGNANTI, Z.-J. M. SHEN, J. SHU, D. SIMCHI-LEVI, AND C.-P. TEO, *Inventory placement in acyclic supply chain networks*, Operations Research Letters, 34 (2006), pp. 228–238. (Cited on p. 450)

[1253] A. MAHAJAN, S. LEYFFER, AND C. KIRCHES, *Solving Mixed-Integer Nonlinear Programs by QP-Diving*, Preprint ANL/MCS-2071-0312, Argonne National Laboratory, Mathematics and Computer Science Division, 2012. (Cited on pp. 279, 286)

[1254] S. MAHESHWARY, *Facets of Conflict Hypergraphs*. PhD thesis, Georgia Institute of Technology, 2008. (Cited on p. 102)

[1255] S. MAILLARD, T. RONCALLI, AND J. TEÏLETCHE, *The properties of equally weighted risk contribution portfolios*, Journal of Portfolio Management, 36 (2010), pp. 60–70. 10.3905/jpm.2010.36.4.060. (Cited on p. 158)

[1256] K. MAKARYCHEV, Y. MAKARYCHEV, A. ROMASHCHENKO, AND N. VERESHCHAGIN, *A new class of non-Shannon-type inequalities for entropies*, Communications in Information and Systems, 2 (2002), pp. 147–166. (Cited on p. 35)

[1257] S. MAKKONEN AND R. LAHDELMA, *Non-convex power plant modelling in energy optimisation*, European Journal of Operational Research, 171 (2006), pp. 1113–1126. (Cited on p. 304)

[1258] H. MAMANI, E. ADIDA, AND D. DEY, *A general framework for the study of decentralized distribution systems*, IIE Transactions on Healthcare Systems Engineering, 2 (2012), pp. 78–96. (Cited on p. 472)

[1259] H. MAMANI, S. E. CHICK, AND D. SIMCHI-LEVI, *A game-theoretic model of international influenza vaccination coordination*, Management Science, 59 (2013), pp. 1650–1670. (Cited on p. 472)

[1260] D. J. MANDELL, E. A. COUTSIAS, AND T. KORTEMME, *Sub-angstrom accuracy in protein loop reconstruction by robotics-inspired conformational sampling*, Nature Methods, 6 (2009), pp. 551–552. (Cited on p. 181)

[1261] D. J. MANDELL AND T. KORTEMME, *Backbone flexibility in computational protein design*, Current Opinion in Biotechnology, 20 (2009), pp. 420–428. (Cited on p. 181)

[1262] O. L. MANGASARIAN, *Linear and nonlinear separation of patterns by linear programming*, Operations Research, 13 (1965), pp. 444–452. (Cited on p. 30)

[1263] ——, *Multisurface method of pattern separation*, IEEE Transactions on Information Theory, 14 (1968), pp. 801–807. (Cited on p. 30)

[1264] V. MANI AND H. ZHAO, *Impact of Strategic Investments in Specialty Drugs on Financial Performance of Big Pharma Companies*, working paper, Smeal College of Business, The Pennsylvania State University, 2017. (Cited on p. 477)

[1265] C. MANNINO, *Real-time traffic control in railway systems*, in Proceedings of Atmos'11, A. Caprara and S. Kontogiannis, eds., OASICS vol. 20, 2011. (Cited on p. 72)

[1266] C. MANNINO AND A. MASCIS, *Optimal real-time traffic control in metro stations*, Operations Research, 57 (2009), pp. 1026–1039. (Cited on pp. xvii, 66, 75)

[1267] J. F. MANWELL, J. G. MCGOWAN, AND A. L. ROGERS, *Wind Energy Explained: Theory, Design and Application*, 2nd edition, Wiley, 2009. (Cited on pp. 367, 368)

[1268] P. A. MAR AND T. C. Y. CHAN, *Adaptive and robust radiation therapy in the presence of drift*, Physics in Medicine and Biology, 60 (2015), pp. 3599–3615. (Cited on pp. 334, 344, 350)

[1269] C. D. MARANAS, *Optimal computer-aided molecular design: A polymer design case study*, Industrial and Engineering Chemistry Research, 35 (1996), pp. 3403–3414. (Cited on p. 329)

[1270] ——, *Optimal molecular design under property prediction uncertainty*, AIChE Journal, 43 (1997), pp. 1250–1264. (Cited on p. 329)

[1271] ——, *Optimization accounting for property prediction uncertainty in polymer design*, Computers and Chemical Engineering, 21 (1997), pp. S1019–S1024. (Cited on p. 329)

[1272] C. T. MARAVELIAS AND C. SUNG, *Integration of production planning and scheduling: Overview, challenges and opportunities*, Computers and Chemical Engineering, 33 (2009), pp. 1919–1930. (Cited on p. 327)

[1273] H. MARCHAND AND L. WOLSEY, *Aggregation and mixed integer rounding to solve MIPs*, Operations Research, 49 (2001), pp. 363–371. (Cited on p. 286)

[1274] F. MARGOT, *Pruning by isomorphism in branch-and-cut*, Mathematical Programming, 94 (2002), pp. 71–90. (Cited on p. 285)

[1275] V. MARIANOV AND P. TABORGA, *Optimal location of public health centres which provide free and paid services*, Journal of the Operational Research Society, 52 (2001), pp. 391–400. (Cited on p. 95)

[1276] A. MARÍN, *The discrete facility location problem with balanced allocation of customers*, European Journal of Operational Research, 210 (2011), pp. 27–38. (Cited on p. 481)

[1277] H. M. MARKOWITZ, *Portfolio selection*, The Journal of Finance, 7 (1952), pp. 77–91. (Cited on pp. 149, 151)

[1278] ——, *The optimization of a quadratic function subject to linear constraints*, Naval Research Logistics Quarterly, 3 (1956), pp. 111–133. (Cited on pp. 149, 151)

[1279] ——, *Portfolio Selection: Efficient Diversification of Investments*, John Wiley & Sons, New York, 1959. (Cited on p. 428)

[1280] H. M. MARKOWITZ AND A. S. MANNE, *On the solution of discrete programming problems*, Econometrica, 25 (1957), pp. 84–110. (Cited on p. 284)

[1281] H. M. MARKOWITZ AND N. USMEN, *The likelihood of various stock market return distributions, part 1: Principles of inference*, Journal of Risk Uncertainty, 13 (1996), pp. 207–219. (Cited on p. 429)

[1282] T. E. MARLIN AND A. N. HRYMAK, *Real-time operations optimization of continuous processes*, in Chemical Process Control—V, J. C. Kantor, C. E. Garcia, and B. Carnahan, eds., CACHE and AIChE 1997, pp. 156–164. (Cited on p. 44)

[1283] I. MAROS, *Computational Techniques of the Simplex Method*, Springer Science, Heidelberg, 2003. (Cited on p. 9)

[1284] M. T. MARSH AND D. A. SCHILLING, *Equity measurement in facility location analysis: A review and framework*, European Journal of Operational Research, 74 (1994), pp. 1–17. (Cited on p. 480)

[1285] A. MARTIN, M. MÖLLER, AND S. MORITZ, *Mixed integer models for the stationary case of gas network optimization*, Mathematical Programming, 105 (2006), pp. 563–582. (Cited on p. 292)

[1286] M. MARTÍN AND I. E. GROSSMANN, *Energy optimization of bioethanol production via gasification of switchgrass*, AIChE Journal, 57 (2011), pp. 3408–3428. (Cited on p. 318)

[1287] ——, *Process optimization of FT-diesel production from lignocellulosic switchgrass*, Industrial and Engineering Chemistry Research, 50 (2011), pp. 13485–13499. (Cited on p. 318)

[1288] J. M. MARTÍNEZ AND E. A. PILOTTA, *Inexact restoration algorithms for constrained optimization*, Journal of Optimization Theory and Applications, 104 (2000), pp. 135–163. (Cited on p. 505)

[1289] J. M. MARTÍNEZ AND F. N. C. SOBRAL, *Constrained derivative-free optimization on thin domains*, Journal of Global Optimization, 56 (2013), pp. 1217–1232. (Cited on p. 505)

[1290] A. MARTINEZ-MARES AND C. R. FUERTE-ESQUIVEL, *A robust optimization approach for the interdependency analysis of integrated energy systems considering wind power uncertainty*, IEEE Transactions on Power Systems, 28 (2013), pp. 3964–3976. (Cited on p. 333)

[1291] J. R. R. A. MARTINS AND J. T. HWANG, *Review and unification of methods for computing derivatives of multidisciplinary computational models*, AIAA Journal, 51 (2013), pp. 2582–2599. (Cited on p. 252)

[1292] J. R. R. A. MARTINS AND A. B. LAMBE, *Multidisciplinary design optimization: A survey of architectures*, AIAA Journal, 51 (2013), pp. 2049–2075. (Cited on p. 250)

[1293] A. MASCIS AND D. PACCIARELLI, *Job shop scheduling with blocking and no-wait constraints*, European Journal of Operational Research, 143 (2002), pp. 498–517. (Cited on p. 69)

[1294] Y. MASMOUDI, H. CHABCHOUB, S. HANAFI, AND A. REBAÏ, *A mathematical programming based procedure for breast cancer classification*, Journal of Mathematical Modelling and Algorithms, 9 (2010), pp. 247–255. (Cited on p. 94)

[1295] J. L. MATHIEU, P. N. PRICE, S. KILICCOTE, AND M. A. PIETTE, *Quantifying changes in building electricity use, with application to demand response*, IEEE Transactions on Smart Grid, 2 (2011), pp. 507–518. (Cited on p. 259)

[1296] MATLAB VERSION 9.0 (R2016A). The MathWorks Inc., Natick, MA, 2016. (Cited on p. 79)

[1297] *The Matrix Computation Toolbox*, http://www.maths.manchester.ac.uk/~higham/mctoolbox. (Cited on p. 498)

[1298] J. MATTINGLEY AND S. BOYD, *Real-time convex optimization in signal processing*, IEEE Signal Processing Magazine, 27 (2010), pp. 50–61. (Cited on p. 27)

[1299] F. MATUS, *Conditional independences among four random variables III: Final conclusion*, Combinatorics, Probability, and Computing, 8 (1999), pp. 269–276. (Cited on p. 35)

[1300] ——, *Infinitely many information inequalities*, in Proceedings of the IEEE International Symposium on Information Theory, June 2007, pp. 41–44. (Cited on p. 35)

[1301] H. MAUSSER AND O. ROMANKO, *Computing equal risk contribution portfolios*, IBM Journal of Research and Development, 58 (2014), pp. 5:1–5:12. (Cited on pp. 151, 158, 159, 160)

[1302] H. MAUSSER AND D. ROSEN, *Economic credit capital allocation and risk contributions*, in Handbooks in Operations Research and Management Science, vol. 15: Financial Engineering, J. R. Birge and V. Linetsky, eds., Elsevier, 2008, pp. 681–726. (Cited on p. 432)

[1303] D. Q. MAYNE, J. B. RAWLINGS, C. V. RAO, AND P. O. M. SCOKAERT, *Constrained model predictive control: Stability and optimality*, Automatica, 36 (2000), pp. 789–814. (Cited on p. 44)

[1304] G. P. MCCORMICK, *Computability of global solutions to factorable nonconvex programs: Part I—Convex underestimating problems*, Mathematical Programming, 10 (1976), pp. 147–175. (Cited on pp. 169, 198, 199, 214, 280)

[1305] C. M. MCDONALD AND C. A. FLOUDAS, *Global optimization and analysis for the Gibbs free energy function using the UNIFAC, Wilson, and ASOG equations*, Industrial and Engineering Chemistry Research, 34 (1995), pp. 1674–1687. (Cited on p. 246)

[1306] M. D. MCKAY, W. J. CONOVER, AND R. J. BECKMAN, *A comparison of three methods for selecting values of input variables in the analysis of output from a computer code*, Technometrics, 21 (1979), pp. 239–245. (Cited on p. 525)

[1307] K. I. M. MCKINNON, *Convergence of the Nelder–Mead simplex method to a nonstationary point*, SIAM Journal on Optimization, 9 (1998), pp. 148–158. (Cited on p. 498)

[1308] *MCS: Global Optimization by Multilevel Coordinate Search*, http://www.mat.univie.ac.at/~neum/software/mcs. (Cited on p. 506)

[1309] K. A. MCSHANE, C. L. MONMA, AND D. F. SHANNO, *An implementation of a primal-dual interior point method for linear programming*, ORSA Journal on Computing, 1 (1989), pp. 70–83. (Cited on p. 10)

[1310] C. C. MEEWELLA AND D. Q. MAYNE, *An algorithm for global optimization of Lipschitz continuous functions*, Journal of Optimization Theory and Applications, 57 (1988), pp. 307–322. (Cited on p. 172)

[1311] ——, *Efficient domain partitioning algorithms for global optimization of rational and Lipschitz continuous functions*, Journal of Optimization Theory and Applications, 61 (1989), pp. 247–270. (Cited on p. 172)

[1312] S. MEHROTRA, *On the implementation of a primal-dual interior point method*, SIAM Journal on Optimization, 2 (1992), pp. 575–601. (Cited on pp. 6, 10)

[1313] D. MEJIA-GIRALDO AND J. MCCALLEY, *Adjustable decisions for reducing the price of robustness of capacity expansion planning*, IEEE Transactions on Power Systems, 29 (2014), pp. 1573–1582. (Cited on p. 365)

[1314] C. A. MÉNDEZ, J. CERDÁ, I. E. GROSSMANN, I. HARJUNKOSKI, AND M. FAHL, *State-of-the-art review of optimization methods for short-term scheduling of batch processes*, Computers and Chemical Engineering, 30 (2006), pp. 913–946. (Cited on p. 328)

[1315] C. A. MÉNDEZ, I. E. GROSSMANN, I. HARJUNKOSKI, AND P. KABORÉ, *A simultaneous optimization approach for off-line blending and scheduling of oil-refinery operations*, Computers and Chemical Engineering, 30 (2006), pp. 614–634. (Cited on p. 328)

[1316] H. O. METE AND Z. B. ZABINSKY, *Stochastic optimization of medical supply location and distribution in disaster management*, International Journal of Production Economics, 126 (2010), pp. 76–84. (Cited on p. 480)

[1317] C. A. MEYER AND C. A. FLOUDAS, *Trilinear monomials with mixed sign domains: Facets of the convex and concave envelopes*, Journal of Global Optimization, 29 (2004), pp. 125–155. (Cited on p. 282)

[1318] ——, *Convex envelopes for edge-concave functions*, Mathematical Programming, 103 (2005), pp. 207–224. (Cited on p. 282)

[1319] ——, *Global optimization of a combinatorially complex generalized pooling problem*, AIChE Journal, 52 (2006), pp. 1027–1037. (Cited on pp. 212, 216, 326)

[1320] A. G. M. MICHELL, *The limit of economy of material in frame structures*, Philosophical Magazine, 8 (1904), pp. 589–597. (Cited on p. 13)

[1321] K. MIETTINEN, *Nonlinear Multiobjective Optimization*, vol. 12 of International Series in Operations Research and Management Science, Springer, 1998. (Cited on pp. 173, 265)

[1322] A. MIGDALAS, P. M. PARDALOS, AND P. VÄRBRAND, eds., *Multilevel Optimization: Algorithms and Applications*, Kluwer Academic, 1997. (Cited on p. 173)

[1323] *MINLP Library* 2, http://www.gamsworld.org/minlp/minlplib2/html/index.html. (Cited on p. 290)

[1324] *MINLP: Test Problems for Mixed Integer Nonlinear Programming*, https://wiki.mcs.anl.gov/leyffer/index.php/MacMINLP. (Cited on p. 290)

[1325] *MINOTAUR: A Toolkit for Solving Mixed-Integer Nonlinear Optimization*, http://wiki.mcs.anl.gov/minotaur, 2011. (Cited on p. 289)

[1326] J. MINOTT, *Reducing hospital readmissions*, Academy Health, 23 (2008), pp. 1–10. (Cited on p. 96)

[1327] P. B. MIRCHANDANI AND R. L. FRANCIS, eds., *Discrete Location Theory*, Wiley-Interscience, New York, 1990. (Cited on p. 450)

[1328] R. MISENER AND C. A. FLOUDAS, *Advances for the pooling problem: Modeling, global optimization, and computational studies*, Applied and Computational Mathematics, 8 (2009), pp. 3–22. (Cited on pp. 292, 328)

[1329] ———, *Global optimization of large-scale generalized pooling problems: Quadratically constrained MINLP models*, Industrial and Engineering Chemistry Research, 49 (2010), pp. 5424–5438. (Cited on pp. 216, 328)

[1330] ———, *Global optimization of mixed-integer quadratically-constrained quadratic programs (MIQCQP) through piecewise-linear and edge-concave relaxations*, Mathematical Programming, 136 (2012), pp. 155–182. (Cited on p. 284)

[1331] ———, *GloMIQO: Global mixed-integer quadratic optimizer*, Journal of Global Optimization, 57 (2013), pp. 3–50. (Cited on pp. 173, 196)

[1332] ———, *ANTIGONE: Algorithms for continuous/integer global optimization of nonlinear equations*, Journal of Global Optimization, 59 (2014), pp. 503–526. (Cited on p. 289)

[1333] R. MISENER, C. E. GOUNARIS, AND C. A. FLOUDAS, *Global optimization of gas lifting operations: A comparative study of piecewise linear formulations*, Industrial and Engineering Chemistry Research, 48 (2009), pp. 6098–6104. (Cited on p. 216)

[1334] R. MISENER, J. P. THOMPSON, AND C. A. FLOUDAS, *APOGEE: Global optimization of standard, generalized, and extended pooling problems via linear and logarithmic partitioning schemes*, Computers and Chemical Engineering, 35 (2011), pp. 876–892. (Cited on pp. 208, 212, 216)

[1335] B. V. MISHRA, E. MAYER, J. RAISCH, AND A. KIENLE, *Short-term scheduling of batch processes. A comparative study of different approaches*, Industrial and Engineering Chemistry Research, 44 (2005), pp. 4022–4034. (Cited on p. 328)

[1336] MIT ENERGY INITIATIVE, *The Future of the Electric Grid: An Interdisciplinary MIT Study*, http://mitei.mit.edu/publications/reports-studies/future-electric-grid, 2011. (Cited on p. 357)

[1337] J. E. MITCHELL AND S. BRAUN, *Rebalancing an investment portfolio in the presence of convex transaction costs, including market impact costs*, Optimization Methods and Software, 28 (2013), pp. 523–542. (Cited on pp. 151, 153, 155)

[1338] S. MITRA, I. E. GROSSMANN, J. M. PINTO, AND N. ARORA, *Optimal production planning under time-sensitive electricity prices for continuous power-intensive processes*, Computers and Chemical Engineering, 38 (2012), pp. 171–184. (Cited on pp. xvii, 78, 79, 82, 304)

[1339] H. D. MITTELMANN, *Decision Tree for Optimization Software*, http://plato.asu.edu/sub/global.html. (Cited on pp. 173, 222)

[1340] R. H. MLADINEO, *An algorithm for finding the global maximum of a multimodal, multivariate function*, Mathematical Programming, 34 (1986), pp. 188–200. (Cited on p. 172)

[1341] S. MOAZENI, T. F. COLEMAN, AND Y. LI, *Regularized robust optimization: The optimal portfolio execution case*, Computational Optimization and Applications, 55 (2013), pp. 341–377. (Cited on p. 344)

[1342] S. MODARESI, M. R. KILINÇ, AND J. P. VIELMA, *Split cuts and extended formulations for mixed integer conic quadratic programming*, Operations Research Letters, 43 (2015), pp. 10–15. (Cited on p. 286)

[1343] F. MODIGLIANI AND F. E. HOHN, *Production planning over time and the nature of the expectation and planning horizon*, Econometrica, 23 (1955), pp. 46–66. (Cited on p. 461)

[1344] J. A. MOMOH, M. E. EL-HAWARY, AND R. ADAPA, *A review of selected optimal power flow literature to 1993 part I: Nonlinear and quadratic programming approaches*, IEEE Transactions on Power Systems, 14 (1999), pp. 96–104. (Cited on p. 187)

[1345] ———, *A review of selected optimal power flow literature to 1993 part II: Newton, linear programming, and interior point methods*, IEEE Transactions on Power Systems, 14 (1999), pp. 105–111. (Cited on p. 187)

[1346] P. MORALES-VALDÉS, A. FLORES-TLACUAHUAC, AND V. M. ZAVALA, *Analyzing the effects of comfort relaxation on energy demand flexibility of buildings: A multiobjective optimization approach*, Energy and Buildings, 85 (2014), pp. 416–426. (Cited on pp. xix, 263, 264, 266, 268)

[1347] M. MORAR AND P. S. AGACHI, *Review: Important contributions in development and improvement of the heat integration techniques*, Computers and Chemical Engineering, 34 (2010), pp. 1171–1179. (Cited on p. 324)

[1348] M. MORARI AND M. BARIĆ, *Recent developments in the control of constrained hybrid systems*, Computers and Chemical Engineering, 30 (2006), pp. 1619–1631. (Cited on p. 328)

[1349] M. MORARI AND J. H. LEE, *Model predictive control: Past, present and future*, Computers and Chemical Engineering, 23 (1999), pp. 667–682. (Cited on p. 41)

[1350] J. J. MORÉ, B. S. GARBOW, AND K. E. HILLSTROM, *User Guide for MINPACK-1*, Technical Report ANL-80-74, Argonne National Laboratory, Argonne, IL, August 1980. (Cited on p. 530)

[1351] J. J. MORÉ AND D. C. SORENSEN, *Computing a trust region step*, SIAM Journal on Scientific and Statistical Computing, 4 (1983), pp. 553–572. (Cited on p. 536)

[1352] J. J. MORÉ AND S. M. WILD, *Estimating Computational Noise*, SIAM Journal on Scientific Computing, 33 (2011), pp. 1292–1314. (Cited on p. 522)

[1353] J. P. MORGAN, *Creditmetrics^{TM}*, technical report, J. P. Morgan, New York, 1997. (Cited on pp. 427, 428, 432)

[1354] L. F. L. MORO AND I. E. GROSSMANN, *A mixed-integer model predictive control formulation for linear systems*, Computers and Chemical Engineering, 55 (2013), pp. 1–18. (Cited on p. 328)

[1355] L. F. L. MORO, A. C. ZANIN, AND J. M. PINTO, *A planning model for refinery diesel production*, Computers and Chemical Engineering, 22 (1998), pp. S1039–S1042. (Cited on p. 327)

[1356] P. MORSE AND G. KIMBALL, *Methods of Operations Research*, Technology Press of MIT, Cambridge, MA, 1951. (Cited on p. 443)

[1357] P. MOSCATO, R. BERRETTA, A. MENDES, AND C. COTTA, *Genes related with Alzheimer's disease: A comparison of evolutionary search statistical and integer programming approaches*, in Applications of Evolutionary Computing, Springer, 2005, pp. 84–94. (Cited on p. 94)

[1358] G. MOSETTI, C. POLONI, AND B. DIVIACCO, *Optimization of wind turbine positioning in large windfarms by means of a genetic algorithm*, Journal of Wind Engineering and Industrial Aerodynamics, 51 (1994), pp. 105–116. (Cited on pp. 367, 370)

[1359] T. S. MOTZKIN AND E. G. STRAUS, *Maxima for graphs and a new proof of a theorem of Turán*, Canadian Journal of Mathematics, 17 (1965), pp. 533–540. (Cited on p. 165)

[1360] P. MOULIN, M. ANITESCU, K. O. KORTANEK, AND F. A. POTRA, *The role of linear semi-infinite programming in signal-adapted QMF bank design*, IEEE Transactions on Signal Processing, 45 (1995), pp. 2160–2174. (Cited on p. 29)

[1361] S. MUDCHANATONGSUK, F. ORDÓÑEZ, AND J. LIU, *Robust solutions for network design under transportation cost and demand uncertainty*, Journal of the Operational Research Society, 59 (2008), pp. 652–662. (Cited on p. 333)

[1362] T. MUNSON, J. SARICH, S. M. WILD, S. BENSON, AND L. CURFMAN MCINNES, *TAO 2.0 Users Manual*. Technical Memorandum ANL/MCS-TM-322, Argonne National Laboratory, Argonne, IL, 2012. (Cited on p. 535)

[1363] S. MURAT SEN, C. A. HENAO, D. J. BRADEN, J. A. DUMESIC, AND C. T. MARAVELIAS, *Catalytic conversion of lignocellulosic biomass to fuels: Process development and technoeconomic evaluation*, Chemical Engineering Science, 67 (2012), pp. 57–67. (Cited on p. 318)

[1364] L. MURAWSKI AND R. L. CHURCH, *Improving accessibility to rural health services: The maximal covering network improvement problem*, Socio-Economic Planning Sciences, 43 (2009), pp. 102–110. (Cited on p. 95)

[1365] K. G. MURTY, *Solving the fixed charge problem by ranking the extreme points*, Operations Research, 16 (1968), pp. 268–279. (Cited on p. 170)

[1366] K. G. MURTY AND S. N. KABADI, *Some NP-complete problems in quadratic and nonlinear programming*, Mathematical Programming, 39 (1987), pp. 117–129. (Cited on p. 113)

[1367] O. R. MUSIN, *The kissing number in four dimensions*, Annals of Mathematics, 168 (2008), pp. 1–32. (Cited on p. 34)

[1368] K. R. MUSKE AND J. B. RAWLINGS, *Model predictive control with linear models*, AIChE Journal, 39 (1993), pp. 262–287. (Cited on p. 44)

[1369] A. NAGURNEY, J. DONG, AND D. ZHANG, *A supply chain network equilibrium model*, Transportation Research Part E, 38 (2002), pp. 120–142. (Cited on p. 464)

[1370] S. NAHMIAS, *Optimal ordering policies for perishable inventory*—II, Operations Research, 23 (1975), pp. 735–749. (Cited on p. 473)

[1371] ———, *Simple approximations for a variety of dynamic leadtime lost-sales inventory models*, Operations Research, 27 (1979), pp. 904–924. (Cited on p. 448)

[1372] ———, *Perishable Inventory Systems*, International Series in Operations Research and Management Science, vol. 160, Springer, 2011. (Cited on p. 473)

[1373] S. NAHMIAS AND W. P. PIERSKALLA, *Optimal ordering policies for a product that perishes in two periods subject to stochastic demand*, Naval Research Logistics Quarterly, 20 (1973), pp. 207–229. (Cited on p. 473)

[1374] H. I. NAKAYA, J. WRAMMERT, E. K. LEE, L. RACIOPPI, S. MARIE-KUNZE, W. N. HAINING, A. R. MEANS, S. P. KASTURI, N. KHAN, AND G.-M. LI, *Systems biology of vaccination for seasonal influenza in humans*, Nature Immunology, 12 (2011), pp. 786–795. (Cited on p. 99)

[1375] G.-J. NAM AND J. CONG, eds., *Modern Circuit Placement. Best Practices and Results*, Springer, 2007. (Cited on p. 34)

[1376] G. NANNICINI AND P. BELOTTI, *Rounding-based heuristics for nonconvex MINLPs*, Mathematical Programming Computation, 4 (2012), pp. 1–31. (Cited on p. 286)

[1377] G. NANNICINI, P. BELOTTI, AND L. LIBERTI, *A Local Branching Heuristic for MINLPs*, arXiv preprint arXiv:0812.2188, 2008. (Cited on p. 286)

[1378] S. F. NASER AND R. L. FOURNIER, *A system for the design of an optimum liquid-liquid extractant molecule*, Computers and Chemical Engineering, 15 (1991), pp. 397–414. (Cited on p. 329)

[1379] K. V. NATARAJAN AND J. M. SWAMINATHAN, *Inventory management in humanitarian operations: Impact of amount, schedule, and uncertainty in funding*, Manufacturing and Service Operations Management, 16 (2014), pp. 595–603. (Cited on p. 491)

[1380] M. A. NAVARRO-AMORÓS, J. A. CABALLERO, R. RUIZ-FEMENIA, AND I. E. GROSSMANN, *An alternative disjunctive optimization model for heat integration with variable temperatures*, Computers and Chemical Engineering, 56 (2013), pp. 12–26. (Cited on p. 324)

[1381] N. NAVID AND G. ROSENWALD, *Market solutions for managing ramp flexibility with high penetration of renewable resource*, IEEE Transactions on Sustainable Energy, 3 (2012), pp. 784–790. (Cited on p. 364)

[1382] D. N. NAWROCKI, *A brief history of downside risk measures*, The Journal of Investing, 8 (1999), pp. 9–25. (Cited on p. 428)

[1383] M. NDIAYE AND H. ALFARES, *Modeling health care facility location for moving population groups*, Computers and Operations Research, 35 (2008), pp. 2154–2161. (Cited on p. 95)

[1384] A. NEDICH AND A. OZDAGLAR, *A geometric framework for nonconvex optimization duality using augmented Lagrangian functions*, Journal of Global Optimization, 40 (2008), pp. 545–573. (Cited on p. 169)

[1385] S. M. S. NEIRO AND J. M. PINTO, *Multiperiod optimization for production planning of petroleum refineries*, Chemical Engineering Communications, 192 (2005), pp. 62–88. (Cited on p. 327)

[1386] J. A. NELDER AND R. MEAD, *A simplex method for function minimization*, Computer Journal, 7 (1965), pp. 308–313. (Cited on p. 498)

[1387] *Nelder-Mead Optimizer*, http://www4.ncsu.edu/~ctk/darts/nelder.m. (Cited on p. 498)

[1388] G. L. NEMHAUSER, *Integer Programming: The Global Impact*, EURO-INFORMS conference, Rome, July, 2013. (Cited on p. 58)

[1389] G. L. NEMHAUSER, M. W. P. SAVELSBERGH, AND G. C. SIGISMONDI, *MINTO, a Mixed INTeger Optimizer*, Operations Research Letters, 15 (1994), pp. 47–58. (Cited on p. 289)

[1390] G. L. NEMHAUSER AND L. A. WOLSEY, *Integer and Combinatorial Optimization*, Wiley, New York, 1988. (Cited on pp. 49, 316)

[1391] A. NEMIROVSKI, A. JUDITSKY, G. LAN, AND A. SHAPIRO, *Robust stochastic approximation approach to stochastic programming*, SIAM Journal on Optimization, 19 (2009), pp. 1574–1609. (Cited on p. 390)

[1392] A. S. NEMIROVSKY AND D. B. YUDIN, *Problem Complexity and Method Efficiency in Optimization*, John Wiley & Sons, 1983. (Cited on p. 166)

[1393] *NEOS Guide: Companion Site to the NEOS Server*, http://www.neos-guide.org/. (Cited on p. 222)

[1394] Y. NESTEROV, *Introductory Lectures on Convex Optimization: A Basic Course*, Kluwer Academic, Dordrecht, Netherlands, 2004. (Cited on pp. 163, 166, 171, 501)

[1395] ———, *Random Gradient-Free Minimization of Convex Functions*, Technical Report 2011/1, CORE, 2011. (Cited on pp. 500, 501)

[1396] Y. E. NESTEROV AND A. NEMIROVSKII, *Interior-Point Polynomial Algorithms in Convex Programming*, Studies in Applied Mathematics 13, SIAM, Philadelphia, 1994. (Cited on p. 11)

[1397] Y. E. NESTEROV AND M. J. TODD, *Self-scaled barriers and interior-point methods for convex programming*, Mathematics of Operations Research, 22 (1997), pp. 1–42. (Cited on pp. 11, 112)

[1398] ———, *Primal-dual interior-point methods for self-scaled cones*, SIAM Journal on Optimization, 8 (1998), pp. 324–364. (Cited on p. 112)

[1399] A. NEUMAIER, *Global Optimization*, http://www.mat.univie.ac.at/~neum/glopt.html. (Cited on p. 173)

[1400] T. L. NG, X. CAI, AND Y. OUYANG, *Implications of biofuel development for engineering infrastructures in the United States*, Biofuels, Bioproducts and Biorefining, 5 (2011), pp. 581–592. (Cited on p. 465)

[1401] A. NIKANDROV AND C. L. E. SWARTZ, *Sensitivity analysis of LP-MPC cascade control systems*, Journal of Process Control, 19 (2009), pp. 16–24. (Cited on pp. xvii, 45, 46)

[1402] J. NOCEDAL AND S. WRIGHT, *Numerical Optimization*, 2nd edition, Springer, New York, 2006. (Cited on pp. 222, 225, 227, 229)

[1403] O. NOHADANI, J. R. BIRGE, F. X. KÄRTNER, AND D. J. BERTSIMAS, *Robust chirped mirrors*, Applied Optics, 47 (2008), pp. 2630–2636. (Cited on p. 333)

[1404] *NOMAD*, http://www.gerad.ca/nomad. (Cited on pp. 499, 500, 506, 510)

[1405] NORTH AMERICAN ELECTRIC RELIABILITY CORPORATION, *Electricity Supply and Demand Database*, http://www.nerc.com/pa/RAPA/ESD/Pages/default.aspx. (Cited on p. 357)

[1406] Z. NOVAK, Z. KRAVANJA, AND I. E. GROSSMANN, *Simultaneous synthesis of distillation sequences in overall process schemes using an improved MINLP approach*, Computers and Chemical Engineering, 20 (1996), pp. 1425–1440. (Cited on p. 323)

[1407] I. NOWAK AND S. VIGERSKE, *LaGO: A (heuristic) branch and cut algorithm for nonconvex MINLPs*, Central European Journal of Operations Research, 16 (2008), pp. 127–138. (Cited on pp. 283, 289)

[1408] N. NOYAN AND G. RUDOLF, *Optimization with multivariate conditional value-at-risk constraints*, Operations Research, 61 (2013), pp. 990–1013. (Cited on p. 432)

[1409] R. H. NYSTRÖM, R. FRANKE, I. HARJUNKOSKI, AND A. KROLL, *Production campaign planning including grade transition sequencing and dynamic optimization*, Computers and Chemical Engineering, 29 (2005), pp. 2163–2179. (Cited on p. 328)

[1410] T. OBERTOPP, A. SPIEKER, AND E. D. GILLES, *Optimierung hybrider prozesse in der verfahrenstechnik*, Oberhausener UMSICHT-Tage: Rechneranwendungen in der Verfahrenstechnik, UMSICHT-Schriftenreihe, 7 (1998), pp. 5.1–5.18. (Cited on p. 328)

[1411] J. ODDOYE, M. YAGHOOBI, M. TAMIZ, D. JONES, AND P. SCHMIDT, *A multiobjective model to determine efficient resource levels in a medical assessment unit*, Journal of the Operational Research Society, 58 (2007), pp. 1563–1573. (Cited on p. 95)

[1412] O. ODELE AND S. MACCHIETTO, *Computer aided molecular design: A novel method for optimal solvent selection*, Fluid Phase Equilibria, 82 (1993), pp. 47–54. (Cited on p. 329)

[1413] U. OELFKE AND T. BORTFELD, *Inverse planning for photon and proton beams*, Medical Dosimetry, 26 (2001), pp. 113–124. (Cited on p. 353)

[1414] W. OGRYCZAK, *On the lexicographic minimax approach to location problems*, European Journal of Operational Research, 100 (1997), pp. 566–585. (Cited on p. 480)

[1415] ———, *Stochastic Dominance Relation and Linear Risk Measures*, Progress and Business Publication, Cracow, Poland, 1999. (Cited on p. 394)

[1416] W. OGRYCZAK AND A. RUSZCZYŃSKI, *From stochastic dominance to mean-risk models: Semi-deviations as risk measures*, European Journal of Operational Research, 116 (1999), pp. 33–50. (Cited on p. 427)

[1417] J. R. O'HANLEY AND R. L. CHURCH, *Designing robust coverage networks to hedge against worst-case facility losses*, European Journal of Operational Research, 209 (2011), pp. 23–36. (Cited on p. 480)

[1418] H. OHLSSON, L. LJUNG, AND S. BOYD, *Segmentation of ARX-models using sum-of-norms regularization*, Automatica, 46 (2010), pp. 1107–1111. (Cited on p. 33)

[1419] H. OKAJIMA, H. XU, AND H. ZHAO, *Dynamic Dual Capacity Sourcing for New Ethical Drugs*, working paper, Smeal College of Business, The Pennsylvania State University, 2013. (Cited on p. 469)

[1420] ———, *Dynamic Capacity Planning for Multiple Drugs*, working paper, Smeal College of Business, The Pennsylvania State University, 2014. (Cited on p. 470)

[1421] A. ÓLAFSSON AND S. WRIGHT, *Efficient schemes for robust IMRT treatment planning*, Physics in Medicine and Biology, 51 (2006), pp. 5621–5642. (Cited on p. 352)

[1422] J. OLDENBURG AND W. MARQUARDT, *Disjunctive modeling for optimal control of hybrid systems*, Computers and Chemical Engineering, 32 (2008), pp. 2346–2364. (Cited on p. 328)

[1423] J. OLDENBURG, W. MARQUARDT, D. HEINZ, AND D. B. LEINEWEBER, *Mixed-logic dynamic optimization applied to batch distillation process design*, AIChE Journal, 49 (2003), pp. 2900–2917. (Cited on p. 328)

[1424] F. OLDEWURTEL, A. PARISIO, C. N. JONES, M. MORARI, D. GYALISTRAS, M. GWERDER, V. STAUCH, B. LEHMANN, AND K. WIRTH, *Energy efficient building climate control using stochastic model predictive control and weather predictions*, in Proceedings of the American Control Conference, 2010, pp. 5100–5105. (Cited on pp. 259, 260)

[1425] W. ORCHARD-HAYES, *Background Development and Extensions of the Revised Simplex Method*, Rand Report HM 1433, The Rand Corporation, Santa Monica, CA, 1954. (Cited on p. 9)

[1426] C. ORSENIGO AND C. VERCELLIS, *Multivariate classification trees based on minimum features discrete support vector machines*, IMA Journal of Management Mathematics, 14 (2003), pp. 221–234. (Cited on p. 94)

[1427] A. OSEI-ANTO, C. FUND, AND A. JOHN, *Health Care Leader Action Guide to Reduce Avoidable Readmissions*, Commonwealth Fund, 2010. (Cited on p. 96)

[1428] G. M. OSTROVSKY, L. E. K. ACHENIE, AND M. SINHA, *On the solution of mixed-integer nonlinear programming models for computer aided molecular design*, Computers and Chemistry, 26 (2002), pp. 645–660. (Cited on p. 329)

[1429] ———, *A reduced dimension branch-and-bound algorithm for molecular design*, Computers and Chemical Engineering, 27 (2003), pp. 551–567. (Cited on p. 329)

[1430] J. OSTROWSKI, J. LINDEROTH, F. ROSSI, AND S. SMRIGLIO, *Orbital branching*, Mathematical Programming, 126 (2011), pp. 147–178. (Cited on p. 285)

[1431] T. J. OVERBYE, X. CHEN, AND Y. SUN, *A comparison of the AC and DC power flow models for LMP calculations*, in Proceedings of the 37th Annual Hawaii International Conference on System Sciences, IEEE, 2004. (Cited on pp. 197, 198)

[1432] O. OZALTIN, O. PROKOPYEV, A. SCHAEFER, AND M. ROBERTS, *Optimizing the societal benefits of the annual influenza vaccine: A stochastic programming approach*, Operations Research, 59 (2011), pp. 1131–1143. (Cited on p. 472)

[1433] I. OZKARAHAN, *Allocation of surgeries to operating rooms by goal programing*, Journal of Medical Systems, 24 (2000), pp. 339–378. (Cited on pp. 95, 96)

[1434] U. A. OZTURK AND B. A. NORMAN, *Heuristic methods for wind energy conversion system positioning*, Electric Power Systems Research, 70 (2004), pp. 179–185. (Cited on p. 367)

[1435] M. PADBERG AND G. RINALDI, *A branch-and-cut algorithm for the resolution of large-scale symmetric travelling salesman problems*, SIAM Review, 33 (1991), pp. 60–100. (Cited on p. 56)

[1436] N. P. PADHY, *Unit commitment—A bibliographical survey*, IEEE Transactions on Power Systems, 19 (2004), pp. 1196–1205. (Cited on p. 304)

[1437] F. PALACIOS-GOMEZ, L. S. LASDON, AND M. ENGQUIST, *Nonlinear optimization by successive linear programming*, Management Science, 28 (1982), pp. 1106–1120. (Cited on p. 212)

[1438] L. PALLOTTINO, E. FERON, AND A. BICCHI, *Conflict resolution problems for air traffic management systems solved with mixed integer programming*, IEEE Transactions on Intelligent Transportation Systems, 3 (2002), pp. 3–11. (Cited on pp. 294, 295, 296, 300)

[1439] P.-Q. PAN, *Linear Programming Computation*, Springer Science, Heidelberg, Germany, 2014. (Cited on p. 9)

[1440] J.-S. PANG, *Complementary problems*, in [957], pp. 271–338. (Cited on p. 173)

[1441] K. P. PAPALEXANDRI AND E. N. PISTIKOPOULOS, *Generalized modular representation framework for process synthesis*, AIChE Journal, 42 (1996), pp. 1010–1032. (Cited on p. 318)

[1442] S. A. PAPOULIAS AND I. E. GROSSMANN, *A structural optimization approach in process synthesis—I: Utility systems*, Computers and Chemical Engineering, 7 (1983), pp. 695–706. (Cited on p. 324)

[1443] ———, *A structural optimization approach in process synthesis—II: Heat recovery networks*, Computers and Chemical Engineering, 7 (1983), pp. 707–721. (Cited on pp. 318, 324)

[1444] P. M. PARDALOS, *On the passage from local to global in optimization*, in Mathematical Programming: State of the Art 1994, J. R. Birge and K. G. Murty, eds., The University of Michigan, 1994, pp. 220–247. (Cited on pp. 167, 173)

[1445] P. M. PARDALOS AND T. COLEMAN, eds., *Lectures on Global Optimization*, vol. 55 of Fields Institute Communications, American Mathematical Society, 2009. (Cited on pp. 173, 551, 634)

[1446] P. M. PARDALOS AND E. H. ROMEIJN, eds., *Handbook of Global Optimization*, vol. 2, Kluwer Academic, 2002. (Cited on p. 173)

[1447] P. M. PARDALOS AND J. B. ROSEN, *Constrained Global Optimization: Algorithms and Applications*, vol. 268 of Lecture Notes in Computer Science, Springer Verlag, 1987. (Cited on p. 173)

[1448] P. M. PARDALOS AND G. SCHNITGER, *Checking local optimality in constrained quadratic programming is NP-hard*, Operations Research Letters, 7 (1988), pp. 33–35. (Cited on p. 167)

[1449] P. M. PARDALOS AND S. A. VAVASIS, *Quadratic programming with one negative eigenvalue is NP-hard*, Journal of Global Optimization, 1 (1991), pp. 15–22. (Cited on pp. 166, 203)

[1450] ———, *Open questions in complexity theory for numerical optimization*, Mathematical Programming, 57 (1992), pp. 337–339. (Cited on p. 166)

[1451] A. PARISIO, C. D. VECCHIO, AND A. VACCARO, *A robust optimization approach to energy hub management*, International Journal of Electrical Power and Energy Systems, 42 (2012), pp. 98–104. (Cited on p. 333)

[1452] B. R. PARKER, *A program selection/resource allocation model for control of malaria and related parasitic diseases*, Computers and Operations Research, 10 (1983), pp. 375–389. (Cited on p. 96)

[1453] S. C. J. PARKER, L. HANSEN, H. O. ABAAN, T. D. TULLIUS, AND E. H. MARGULIES, *Local DNA topography correlates with functional noncoding regions of the human genome*, Science, 324 (2009), pp. 389–392. (Cited on p. 176)

[1454] P. A. PARRILO, *Semidefinite programming relaxations for semialgebraic problems*, Mathematical Programming, Series B, 96 (2003), pp. 293–320. (Cited on p. 119)

[1455] B. PARSONS, M. MILLIGAN, B. ZAVADIL, D. BROOKS, B. KIRBY, K. DRAGOON, AND J. CALDWELL, *Grid impacts of wind power: A summary of recent studies in the United States*, Wind Energy, 7 (2004), pp. 87–108. (Cited on p. 368)

[1456] H. PARVIN, S. AHMAD-BEYGI, J. HELM, M. VAN OYEN, AND P. LARSON, *Malaria Treatment Distribution in Developing World Health Systems and Application to Malawi*, technical report, School of Industrial and Systems Engineering, University of Michigan, Ann Arbor, MI. (Cited on p. 491)

[1457] B. A. PASTERNACK, *Optimal pricing and return policies for perishable commodities*, Marketing Science, 4 (1985), pp. 166–176. (Cited on p. 454)

[1458] G. PATAKI, *Strong duality in conic linear programming: Facial reduction and extended duals*, in Computational and Analytical Mathematics, Springer, 2013, pp. 613–634. (Cited on p. 115)

[1459] M. V. PATO AND M. MOZ, *Solving a bi-objective nurse rerostering problem by using a utopic Pareto genetic heuristic*, Journal of Heuristics, 14 (2008), pp. 359–374. (Cited on p. 96)

[1460] R. PAULAVIČIUS, Y. D. SERGEYEV, D. E. KVASOV, AND J. ŽILINSKAS, *Globally-biased DISIMPL algorithm for expensive global optimization*, Journal of Global Optimization, 59 (2014), pp. 545–567. (Cited on p. 172)

[1461] R. PAULAVIČIUS AND J. ŽILINSKAS, *Simplicial Global Optimization*, Springer, 2014. (Cited on pp. 172, 173)

[1462] G. E. PAULES AND C. A. FLOUDAS, *Stochastic programming in process synthesis: A two-stage model with MINLP recourse for multiperiod heat-integrated distillation sequences*, Computers and Chemical Engineering, 16 (1992), pp. 189–210. (Cited on p. 323)

[1463] P. PELLEGRINI, G. MARLIÈRE, AND J. RODRIGUEZ, *Optimal train routing and scheduling for managing traffic perturbations in complex junctions*, Transportation Research Part B, 59 (2014), pp. 58–80. (Cited on pp. 68, 69)

[1464] M. V. F. PEREIRA AND L. M. V. G. PINTO, *Multi-stage stochastic optimization applied to energy planning*, Mathematical Programming, 52 (1991), pp. 359–375. (Cited on pp. 406, 411, 416, 417)

[1465] R. E. PEREZ, P. W. JANSEN, AND J. R. R. A. MARTINS, *pyOpt: A Python-based object-oriented framework for nonlinear constrained optimization*, Structural and Multidisciplinary Optimization, 45 (2012), pp. 101–118. (Cited on p. 251)

[1466] I. J. PÉREZ-ARRIAGA, H. RUDNICK, AND M. R. ABBAD, *Electric Energy Systems: Analysis and Operation*, CRC Press, 2009, ch. 1. (Cited on p. 357)

[1467] M. PERSSON AND J. A. PERSSON, *Health economic modeling to support surgery management at a Swedish hospital*, Omega, 37 (2009), pp. 853–863. (Cited on p. 95)

[1468] S. B. PETKOV AND C. D. MARANAS, *Multiperiod planning and scheduling of multiproduct batch plants under demand uncertainty*, Industrial and Engineering Chemistry Research, 36 (1997), pp. 4864–4881. (Cited on p. 393)

[1469] T. PETROULAS AND G. V. REKLAITIS, *Computer-aided synthesis and design of plant utility systems*, AIChE Journal, 30 (1984), pp. 69–78. (Cited on p. 324)

[1470] G. CH. PFLUG AND W. ROMISCH, *Modeling, Measuring and Managing Risk*, World Scientific, 2007. (Cited on p. 391)

[1471] D. PFLUGFELDER, J. WILKENS, AND U. OELFKE, *Worst case optimization: a method to account for uncertainties in the optimization of intensity modulated proton therapy*, Physics in Medicine and Biology, 53 (2008), pp. 1689–1700. (Cited on pp. 354, 355)

[1472] V. PHAM, C. LAIRD, AND M. EL-HALWAGI, *Convex hull discretization approach to the global optimization of pooling problems*, Industrial and Engineering Chemistry Research, 48 (2009), pp. 1973–1979. (Cited on pp. 213, 217)

[1473] D. T. PHAN, *Lagrangian duality and branch-and-bound algorithms for optimal power flow*, Operations Research, 60 (2012), pp. 275–285. (Cited on pp. 203, 204, 205)

[1474] A. B. PHILPOTT, A. DALLAGI, AND E. GALLET, *On cutting-plane algorithms and dynamic programming for hydro-electricity generation*, in Risk Management and Energy Trading, R. M. Kovacevic, G. Ch. Pflug, and M. T. Vespucci, eds., Springer, 2013, pp. 105–127. (Cited on p. 419)

[1475] A. B. PHILPOTT, V. L. DE MATOS, AND E. C. FINARDI, *On solving multistage stochastic programs with coherent risk measures*, Operations Research, 61 (2013), pp. 957–970. (Cited on p. 425)

[1476] A. B. PHILPOTT, M. C. FERRIS, AND R. J.-B. WETS, *Equilibrium, uncertainty and risk in hydrothermal electricity systems*, Mathematical Programming B, 157 (2016), pp. 483–513. (Cited on p. 425)

[1477] A. B. PHILPOTT AND Z. GUAN, *On the convergence of stochastic dual dynamic programming and other methods*, Operations Research Letters, 36 (2008), pp. 450–455. (Cited on p. 416)

[1478] ———, *Models for Estimating the Performance of Electricity Markets with Hydro-Electric Reservoir Storage*, technical report, Electric Power Optimization Centre, University of Auckland, 2013. (Cited on p. 419)

[1479] A. B. PHILPOTT, Z. GUAN, J. KHAZAEI, AND G. ZAKERI, *Production inefficiency of electricity markets with hydro generation*, Utilities Policy, 18 (2010), pp. 174–185. (Cited on pp. xxi, 419, 420)

[1480] P. PHOUNGPHOL, Y. ZHANG, Y. ZHAO, AND B. SRICHANDAN, *Multiclass SVM with ramp loss for imbalanced data classification*, in Proceedings of the IEEE International Conference on Granular Computing (GrC), IEEE, 2012. (Cited on p. 94)

[1481] N. A. PIERCE AND E. WINFREE, *Protein design is NP-hard*, Protein Engineering, 15 (2002), pp. 779–782. (Cited on p. 175)

[1482] W. P. PIERSKALLA, *Supply chain management of blood banks*, in Operations Research and Health Care, A Handbook of Methods and Applications, M. Brandeau, F. Sainfort, and W. P. Pierskella, eds., Kluwer Academic, New York, 2004, pp. 104–145. (Cited on pp. 95, 473)

[1483] R. H. PIKE, *Optimization for Engineering Systems*, Van Nostrand Reinhold, New York, 1986. (Cited on pp. 38, 40)

[1484] M. PINEDO, *Scheduling: Theory, Algorithms, and Systems*, 4th edition, Springer, New York, 2012. (Cited on pp. 57, 327)

[1485] M. PINEDO AND X. CHAO, *Operations Scheduling with Applications in Manufacturing and Services*, McGraw-Hill, 1999. (Cited on p. 57)

[1486] J. PINTÉR, *Software development for global optimization*, in [1445], pp. 183–204. (Cited on p. 173)

[1487] ———, *Branch-and-bound algorithms for solving global optimization problems with Lipschitzian structure*, Optimization, 19 (1988), pp. 101–110. (Cited on p. 172)

[1488] ———, *Global Optimization in Action: Continuous and Lipschitz Optimization. Algorithms, Implementations and Applications*, Kluwer, Dordrecht, 1996. (Cited on p. 173)

[1489] J. M. PINTO AND I. E. GROSSMANN, *Optimal cyclic scheduling of multistage continuous multiproduct plants*, Computers and Chemical Engineering, 18 (1994), pp. 797–816. (Cited on p. 328)

[1490] J. M. PINTO, M. JOLY, AND L. F. L. MORO, *Planning and scheduling models for refinery operations*, Computers and Chemical Engineering, 24 (2000), pp. 2259–2276. (Cited on p. 41)

[1491] J. M. PINTO AND L. F. L. MORO, *A planning model for petroleum refineries*, Brazilian Journal of Chemical Engineering, 17 (2000), pp. 575–586. (Cited on p. 327)

[1492] H. PIRNAY, R. LÓPEZ-NEGRETE, AND L. T. BIEGLER, *Optimal sensitivity based on IPOPT*, Mathematical Programming Computation, 4 (2012), pp. 307–331. (Cited on p. 234)

[1493] E. N. PISTIKOPOULOS AND S. K. STEFANIS, *Optimal solvent design for environmental impact minimization*, Computers and Chemical Engineering, 22 (1998), pp. 717–733. (Cited on p. 329)

[1494] S. A. PIYAVSKII, *An algorithm for finding the absolute minimum of a function*, Theory of Optimal Solutions, 2 (1967), pp. 13–24. In Russian. (Cited on pp. 169, 172)

[1495] ———, *An algorithm for finding the absolute extremum of a function*, USSR Computational Mathematics and Mathematical Physics, 12 (1972), pp. 57–67. (Cited on p. 172)

[1496] Y. POCHET AND F. WARICHET, *A tighter continuous time formulation for the cyclic scheduling of a mixed plant*, Computers and Chemical Engineering, 32 (2008), pp. 2723–2744. (Cited on p. 328)

[1497] Y. POCHET AND L. A. WOLSEY, *Production Planning by Mixed Integer Programming*, Springer, 2006. (Cited on pp. 305, 327)

[1498] I. PÓLIK AND T. TERLAKY, *A survey of the S-lemma*, SIAM Review, 49 (2007), pp. 371–418. (Cited on p. 11)

[1499] ———, *Interior point methods for nonlinear optimization*, in Nonlinear Optimization, G. DiPillo and F. Schoen, eds., C.I.M.E. Lecture Series, Springer-Science, Heidelberg, 2009, pp. 209–271. (Cited on p. 11)

[1500] J. M. PONCE-ORTEGA, A. JIMÉNEZ GUTIÉRREZ, AND I. E. GROSSMANN, *Optimal synthesis of heat exchanger networks involving isothermal process streams*, Computers and Chemical Engineering, 32 (2008), pp. 1918–1942. (Cited on p. 324)

[1501] J. PONDER AND F. RICHARDS, *TINKER molecular modeling package*, Journal of Computational Chemistry, 8 (1987), pp. 1016–1024. (Cited on p. 180)

[1502] A. PONGSAKDI, P. RANGSUNVIGIT, K. SIEMANOND, AND M. J. BAGAJEWICZ, *Financial risk management in the planning of refinery operations*, International Journal of Production Economics, 103 (2006), pp. 64–86. (Cited on pp. 40, 41)

[1503] N. M. K. POON AND J. R. R. A. MARTINS, *An adaptive approach to constraint aggregation using adjoint sensitivity analysis*, Structural and Multidisciplinary Optimization, 34 (2007), pp. 61–73. (Cited on p. 255)

[1504] F. PORTÉ-AGEL, Y.-T. WU, AND C.-H. CHEN, *A numerical study of the effects of wind direction on turbine wakes and power losses in a large wind farm*, Energies, 6 (2013), pp. 5297–5313. (Cited on p. 368)

[1505] E. L. PORTEUS, *Foundations of Stochastic Inventory Theory*, Stanford University Press, Stanford, CA, 2002. (Cited on pp. 439, 446)

[1506] ——, *The newsvendor problem*, in Building Intuition: Insights From Basic Operations Management Models and Principles, D. Chhajed and T. J. Lowe, eds., Springer, 2008, ch. 7, pp. 115–134. (Cited on p. 443)

[1507] K. POSTEK AND D. DEN HERTOG, *Multi-stage adjustable robust mixed-integer optimization via iterative splitting of the uncertainty set*, INFORMS Journal on Computing, 28 (2016), pp. 553–574. (Cited on p. 343)

[1508] F. POTRA AND R. SHENG, *On homogeneous interior-point algorithms for semidefinite programming*, Optimization Methods and Software, 9 (1998), pp. 161–184. (Cited on p. 11)

[1509] M. J. D. POWELL, *A new algorithm for unconstrained optimization*, in Nonlinear Programming, J. B. Rosen, O. L. Mangasarian, and K. Ritter, eds., Academic Press, New York, 1970. (Cited on p. 496)

[1510] ——, *The theory of radial basis function approximation in 1990*, in Advances in Numerical Analysis, vol. II: Wavelets, Subdivision Algorithms and Radial Basis Functions, W. A. Light, ed., Oxford University Press, Cambridge, 1992, pp. 105–210. (Cited on p. 497)

[1511] ——, *A direct search optimization method that models the objective and constraint functions by linear interpolation*, in Advances in Optimization and Numerical Analysis, Proceedings of the Sixth Workshop on Optimization and Numerical Analysis, Oaxaca, Mexico, S. Gomez and J.-P. Hennart, eds., Kluwer Academic, Dordrecht, 1994, pp. 51–67. (Cited on p. 505)

[1512] ——, *Least Frobenius norm updating of quadratic models that satisfy interpolation conditions*, Mathematical Programming, 100 (2004), pp. 183–215. (Cited on pp. 497, 531)

[1513] ——, *Developments of NEWUOA for minimization without derivatives*, IMA Journal of Numerical Analysis, 28 (2008), pp. 649–664. (Cited on pp. 498, 503)

[1514] ——, *The BOBYQA Algorithm for Bound Constrained Optimization without Derivatives*, Technical Report DAMTP 2009/NA06, University of Cambridge, 2009. (Cited on p. 503)

[1515] ——, *On Fast Trust Region Methods for Quadratic Models with Linear Constraints*, Technical Report DAMTP 2014/NA02, University of Cambridge, 2014. (Cited on p. 503)

[1516] W. PRAGER AND G. I. N. ROZVANY, *Optimal layout of grillages*, Journal of Structural Mechanics, 5 (1977), pp. 1–18. (Cited on p. 13)

[1517] S. PRAJNA, A. PAPACHRISTODOULOU, AND P. A. PARRILO, *Introducing SOSTOOLS: A general purpose sum of squares programming solver*, in Proceedings of the 41st IEEE Conference on Decision and Control, IEEE, 2002, pp. 741–746, vol. 1. (Cited on p. 120)

[1518] G. P. PRASTACOS, *Allocation of a perishable product inventory*, Operations Research, 29 (1981), pp. 95–107. (Cited on p. 484)

[1519] A. Prata, J. Oldenburg, A. Kroll, and W. Marquardt, *Integrated scheduling and dynamic optimization of grade transitions for a continuous polymerization reactor*, Computers and Chemical Engineering, 32 (2008), pp. 463–476. (Cited on p. 328)

[1520] F. Preciado-Walters, R. Rardin, M. Langer, and V. Thai, *A coupled column generation, mixed integer approach to optimal planning of intensity modulated radiation therapy for cancer*, Mathematical Programming, 101 (2004), pp. 319–338. (Cited on p. 94)

[1521] A. Prékopa, *Contributions to the theory of stochastic programming*, Mathematical Programming, 4 (1973), pp. 202–221. (Cited on p. 431)

[1522] ——, *Stochastic Programming*, Kluwer, Boston, MA, 1995. (Cited on pp. 391, 427)

[1523] ——, *Multivariate value-at-risk and related topics*, Annals of Operations Research, 193 (2012), pp. 49–69. (Cited on pp. 430, 431)

[1524] G. Pritchard, *Stochastic inflow modeling for hydropower scheduling problems*, European Journal of Operational Research, 246 (2015), pp. 496–504. (Cited on p. 424)

[1525] *PSwarm Solver Home Page*. http://www.norg.uminho.pt/aivaz/pswarm. (Cited on pp. 503, 506)

[1526] J. Puerto, F. Ricca, and A. Scozzari, *Extensive facility location problems on networks with equity objectives*, Discrete Applied Mathematics, 157 (2009), pp. 1069–1085. (Cited on p. 481)

[1527] A. Pugachev and L. Xing, *Pseudo beam's-eye–view as applied to beam orientation selection in intensity-modulated radiation therapy*, International Journal of Radiation Oncology • Biology • Physics, 51 (2001), pp. 1361–1370. (Cited on p. 95)

[1528] S. J. Qin and T. A. Badgwell, *A survey of industrial model predictive control technology*, Control Engineering Practice, 11 (2003), pp. 733–764. (Cited on pp. 41, 44, 45)

[1529] Y. Qin, A. Paul, and A. Alptekinoglu, *Optimal Distribution of Blood Products*, technical report, School of Business Administration, University of Houston, Victoria, TX, 2012. (Cited on p. 484)

[1530] Y. Qin, R. Wang, A. J. Vakharia, Y. Chen, and M. M. H. Seref, *The newsvendor problem: Review and directions for future research*, European Journal of Operational Research, 213 (2011), pp. 361–374. (Cited on p. 484)

[1531] A. Qualizza, P. Belotti, and F. Margot, *Linear programming relaxations of quadratically constrained quadratic programs*, in [1160], pp. 407–426. (Cited on p. 282)

[1532] A. G. Quaranta and A. Zaffaroni, *Robust optimization of conditional value at risk and portfolio selection*, Journal of Banking and Finance, 32 (2008), pp. 2046–2056. (Cited on p. 333)

[1533] T. D. Querec, R. S. Akondy, E. K. Lee, W. Cao, H. I. Nakaya, D. Teuwen, A. Pirani, K. Gernert, J. Deng, and B. Marzolf, *Systems biology approach predicts immunogenicity of the yellow fever vaccine in humans*, Nature Immunology, 10 (2008), pp. 116–125. (Cited on p. 99)

[1534] I. Quesada and I. E. Grossmann, *An LP/NLP based branch-and-bound algorithm for convex MINLP optimization problems*, Computers and Chemical Engineering, 16 (1992), pp. 937–947. (Cited on p. 278)

[1535] ——, *Global optimization of bilinear process networks with multicomponent flows*, Computers and Chemical Engineering, 19 (1995), pp. 1219–1242. (Cited on p. 211)

[1536] A. J. QUIST, R. VAN GEMEERT, J. E. HOOGENBOOM, T. ÍLLES, C. ROOS, AND T. TERLAKY, *Application of nonlinear optimization to reactor core fuel reloading*, Annals of Nuclear Energy, 26 (1998), pp. 423–448. (Cited on p. 292)

[1537] L. R. RABINER, *The design of finite impulse response digital filters using linear programming techniques*, The Bell System Technical Journal, 51 (1972), pp. 1177–1198. (Cited on pp. 28, 29)

[1538] ——, *Linear program design of finite impulse response (FIR) digital filters*, IEEE Transactions on Audio and Electroacoustics, 20 (1972), pp. 280–288. (Cited on pp. 28, 29)

[1539] L. R. RABINER, N. Y. GRAHAM, AND H. D. HELMS, *Linear programming design of IIR digital filters with arbitrary magnitude function*, IEEE Transactions on Acoustics, Speech and Signal Processing, 22 (1974), pp. 117–123. (Cited on p. 29)

[1540] A. U. RAGHUNATHAN AND L. T. BIEGLER, *MPEC formulations and algorithms in process engineering*, Computers and Chemical Engineering, 27 (2003), pp. 1381–1392. (Cited on pp. 242, 243)

[1541] A. U. RAGHUNATHAN, M. S. DIAZ, AND L. T. BIEGLER, *An MPEC formulation for dynamic optimization of distillation operation*, Computers and Chemical Engineering, 28 (2004), pp. 2037–2052. (Cited on pp. 247, 328)

[1542] S.-U. RAHMAN AND D. SMITH, *Deployment of rural health facilities in a developing country*, Journal of the Operational Research Society, 50 (1999), pp. 892–902. (Cited on p. 95)

[1543] D. RAJAGOPAL, S. SEXTON, G. HOCHMAN, D. ROLAND-HOLST, AND D. ZILBERMAN, *Model estimates food-versus-biofuel trade-off*, California Agriculture, 63 (2009), pp. 199–201. (Cited on p. 463)

[1544] R. RAJGARIA, S. R. MCALLISTER, AND C. A. FLOUDAS, *Distance dependent centroid to centroid force fields using high resolution decoys*, Proteins: Structure, Function, and Bioinformatics, 70 (2008), pp. 950–970. (Cited on p. 178)

[1545] R. RAMAN AND I. E. GROSSMANN, *Symbolic integration of logic in mixed-integer linear programming techniques for process synthesis*, Computers and Chemical Engineering, 17 (1993), pp. 909–927. (Cited on pp. 318, 323)

[1546] ——, *Modeling and computational techniques for logic-based integer programming*, Computers & Chemical Engineering, 18 (1994), pp. 563–578. (Cited on p. 316)

[1547] V. S. RAMAN AND C. D. MARANAS, *Optimization in product design with properties correlated with topological indices*, Computers and Chemical Engineering, 22 (1998), pp. 747–763. (Cited on p. 329)

[1548] I. RASHKOVA, P. YADAV, R. ATUN, AND J. GALLIEN, *National Drug Stockout Risks in Africa: The Global Fund Disbursement Process*, 2002–2013, technical report, London Business School, London, United Kingdom, 2014. (Cited on p. 491)

[1549] H. RATSCHEK AND J. ROKNE, *Interval methods*, in [957], pp. 752–828. (Cited on p. 169)

[1550] H. RATSCHEK AND R. L. VOLLER, *What can interval analysis do for global optimization?* Journal of Global Optimization, 1 (1991), pp. 111–130. (Cited on p. 173)

[1551] N. S. RAU, *Optimization Principles: Practical Applications to the Operation and Markets of the Electric Power Industry*, Wiley-IEEE Press, Hoboken, NJ, 2003. (Cited on p. 187)

[1552] M. S. RAUNER AND N. BAJMOCZY, *How many AEDs in which region? An economic decision model for the Austrian Red Cross*, European Journal of Operational Research, 150 (2003), pp. 3–18. (Cited on p. 95)

[1553] D. E. RAVEMARK AND D. W. T. RIPPIN, *Optimal design of a multi-product batch plant*, Computers and Chemical Engineering, 22 (1998), pp. 177–183. (Cited on p. 291)

[1554] J. B. RAWLINGS AND K. R. MUSKE, *The stability of constrained receding horizon control*, IEEE Transactions on Automatic Control, 38 (1993), pp. 1512–1516. (Cited on p. 44)

[1555] E. G. READ AND M. HINDSBERGER, *Constructive dual DP for reservoir optimization*, in Handbook of Power Systems I, Springer, 2010, pp. 3–32. (Cited on p. 420)

[1556] S. REBENNACK AND J. KALLRATH, *Continuous piecewise linear delta-approximations for bivariate and multivariate functions*, Journal of Optimization Theory and Applications, 167 (2015), pp. 102–117. (Cited on p. 205)

[1557] ———, *Continuous piecewise linear delta-approximations for univariate functions: Computing minimal breakpoint systems*, Journal of Optimization Theory and Applications, 167 (2015), pp. 617–643. (Cited on p. 205)

[1558] J. REESE, *Methods for Solving the p-Median Problem: An Annotated Bibliography*, technical report, Department of Mathematics, Trinity University, 2005. (Cited on p. 453)

[1559] P.-E. RÉTHORÉ, P. FUGLSANG, G. C. LARSEN, T. BUHL, T. J. LARSEN, AND H. A. MADSEN, *TopFarm: Multi-fidelity optimization of offshore wind farm*, in Proceedings of the Twenty-first (2011) International Offshore and Polar Engineering Conference, no. 2007, Maui, Hawaii, 2011. (Cited on pp. 367, 370)

[1560] C. S. REVELLE AND H. A. EISELT, *Location analysis: A synthesis and survey*, European Journal of Operational Research, 165 (2005), pp. 1–19. (Cited on p. 450)

[1561] C. S. REVELLE, H. A. EISELT, AND M. S. DASKIN, *A bibliography for some fundamental problem categories in discrete location science*, European Journal of Operational Research, 184 (2008), pp. 817–848. (Cited on p. 450)

[1562] D. REY, S. CONSTANS, R. FONDACCI, AND C. RAPINE, *A mixed integer linear model for potential conflict minimization by speed modulations*, in Proceedings of the International Conference on Research in Air Transportation, Budapest, 2010. (Cited on p. 295)

[1563] D. REY, C. RAPINE, R. FONDACCI, AND N.-E. EL FAOUZI, *Minimization of potential air conflicts through speed regulation*, Transportation Research Record: Journal of the Transportation Research Board, 2300 (2012), pp. 59–67. (Cited on pp. 295, 301)

[1564] B. REZNICK, *Some concrete aspects of Hilbert's 17th problem*, Contemporary Mathematics, 253 (2000), pp. 251–272. (Cited on p. 120)

[1565] J. RICHALET, A. RAULT, J. L. TESTUD, AND J. PAPON, *Model predictive heuristic control: Applications to industrial processes*, Automatica, 14 (1978), pp. 413–428. (Cited on p. 41)

[1566] A. RICHARDS AND J. P. HOW, *Aircraft trajectory planning with collision avoidance using mixed integer linear programming*, in Proceedings of the American Control Conference, Anchorage, Alaska, 2002. (Cited on p. 294)

[1567] M. RIIS AND K. A. ANDERSEN, *Applying the minimax criterion in stochastic recourse programs*, European Journal of Operational Research, 165 (2005), pp. 569–584. (Cited on p. 344)

[1568] A. D. RIKUN, *A convex envelope formula for multilinear functions*, Journal of Global Optimization, 10 (1997), pp. 425–437. (Cited on p. 282)

[1569] L. M. RIOS AND N. V. SAHINIDIS, *Derivative-free optimization: A review of algorithms and comparison of software implementations*, Journal of Global Optimization, 56 (2013), pp. 1247–1293. (Cited on p. 173)

[1570] R. A. RIVAS, *Optimization of Offshore Wind Farm Layouts*, Master's thesis, Technical University of Denmark, 2007. (Cited on p. 367)

[1571] H. ROBBINS AND S. MONRO, *A stochastic approximation method*, The Annals of Mathematical Statistics, 22 (1951), pp. 400–407. (Cited on p. 390)

[1572] R. T. ROCKAFELLAR, *Convex Analysis*, Princeton University Press, Princeton, NJ, 1970. (Cited on p. 138)

[1573] ———, *Generalized directional derivatives and subgradients of nonconvex functions*, Canadian Journal of Mathematics, 32 (1980), pp. 257–280. (Cited on p. 500)

[1574] R. T. ROCKAFELLAR AND S. URYASEV, *Optimization of conditional value-at-risk*, Journal of Risk, 2 (2000), pp. 21–41. (Cited on pp. 150, 350, 394, 427, 430)

[1575] ———, *Conditional value-at-risk for general loss distributions*, Journal of Banking and Finance, 26 (2002), p. 1443. (Cited on p. 394)

[1576] R. T. ROCKAFELLAR AND R. J.-B. WETS, *Scenarios and policy aggregation in optimization under uncertainty*, Mathematics of Operations Research, 16 (1991), pp. 119–147. (Cited on p. 391)

[1577] ———, *Variational Analysis*, Springer, Berlin, 1998. (Cited on p. 500)

[1578] O. RODIONOVA, M. SBIHI, D. DELAHAYE, AND M. MONGEAU, *North Atlantic aircraft trajectory optimization*, IEEE Transactions on Intelligent Transportation Systems, 15 (2014), pp. 2202–2212. (Cited on pp. 295, 300, 301)

[1579] H. M. RODRIGUEZ, A. CANO, AND M. MATZOPOULOS, *Improve engineering via whole-plant design optimization*, Hydrocarbon Processing, December 2010, pp. 43–49. (Cited on p. 91)

[1580] W. ROEMISCH AND S. VIGERSKE, *Recent progress in two-stage mixed-integer stochastic programming with applications to power production planning*, in Handbook of Power Systems I, Energy Systems, S. Rebennack, P. M. Pardalos, M. V. F. Pereira, and N. A. Iliadis, eds., Springer, 2010, pp. 177–208. (Cited on p. 91)

[1581] J. ROGERS AND K. PORTER, *Wind Power and Electricity Markets*, technical report, Utility Wind Integration Group, 2011. (Cited on p. 373)

[1582] T. ROH AND L. VANDENBERGHE, *Discrete transforms, semidefinite programming, and sum-of-squares representations of nonnegative polynomials*, SIAM Journal on Optimization, 16 (2006), pp. 939–964. (Cited on p. 30)

[1583] R. ROHS, X. JIN, S. M. WEST, R. JOSHI, B. HONIG, AND R. S. MANN., *Origins of specificity in protein-DNA recognition*, Annual Review of Biochemistry, 79 (2010), pp. 233–269. (Cited on p. 175)

[1584] D. ROMAN, K. DARBY-DOWMAN, AND G. MITRA, *Portfolio construction based on stochastic dominance and target return distributions*, Mathematical Programming, 108 (2006), pp. 541–569. (Cited on p. 427)

[1585] O. ROMANKO AND H. MAUSSER, *Robust scenario-based value-at-risk optimization*, Annals of Operations Research, 237 (2016), pp. 203–218. (Cited on pp. 150, 153)

[1586] J. ROMBERG, *Imaging via compressive sampling*, IEEE Signal Processing Magazine, 25 (2008), pp. 14–20. (Cited on p. 33)

[1587] H. E. ROMEIJN, R. K. AHUJA, J. F. DEMPSEY, A. KUMAR, AND J. G. LI, *A novel linear programming approach to fluence map optimization for intensity modulated radiation therapy treatment planning*, Physics in Medicine and Biology, 48 (2003), p. 3521. (Cited on p. 94)

[1588] T. RONCALLI, *Introduction to Risk Parity and Budgeting*, Chapman & Hall/CRC. Financial Mathematics Series, 2013. (Cited on pp. 427, 432, 435)

[1589] A. RONG, R. LAHDELMA, AND P. B. LUH, *Lagrangian relaxation based algorithm for tri-generation planning with storages*, European Journal of Operational Research, 188 (2008), pp. 240–257. (Cited on p. 304)

[1590] W. C. ROONEY, B. P. HAUSBERGER, L. T. BIEGLER, AND D. GLASSER, *Convex attainable region projections for reactor network synthesis*, Computers and Chemical Engineering, 24 (2000), pp. 225–229. (Cited on p. 320)

[1591] C. ROOS, T. TERLAKY, AND J.-PH. VIAL, *Interior Point Methods for Linear Optimization*, 2nd edition, Springer, New York, 2006. First edition: C. Roos, T. Terlaky, and J.-Ph. Vial, *Theory and Algorithms for Linear Optimization: An Interior Point Approach*, John Wiley & Sons, Chichester, U.K., 1997. (Cited on pp. 3, 4, 6, 8, 9, 10, 11)

[1592] J. B. ROSEN, *Pattern separation by convex programming*, Journal of Mathematical Analysis and Applications, 10 (1965), pp. 123–134. (Cited on p. 30)

[1593] B. ROTTKEMPER, K. FISCHER, A. BLECKEN, AND C. DANNE, *Inventory relocation for overlapping disaster settings in humanitarian operations*, OR Spectrum, 33 (2011), pp. 721–749. (Cited on p. 484)

[1594] A. D. ROY, *Safety first and the holding assets*, Econometrica, 20 (1952), pp. 431–449. (Cited on pp. 427, 428, 429)

[1595] B. ROY, *Robustness in operational research and decision aiding: A multi-faceted issue*, European Journal of Operational Research, 200 (2010), pp. 629–638. (Cited on p. 333)

[1596] G. I. N. ROZVANY, *On design-dependent constraints and singular topologies*, Structural and Multidisciplinary Optimization, 21 (2001), pp. 164–173. (Cited on p. 25)

[1597] G. I. N. ROZVANY AND T. BIRKER, *On singular topologies in exact layout optimization*, Structural Optimization, 8 (1994), pp. 228–235. (Cited on p. 25)

[1598] D. F. RUDD AND C. C. WATSON, *Strategy of Process Engineering*, Wiley, New York, 1968. (Cited on p. 323)

[1599] L. RUDIN, S. J. OSHER, AND E. FATEMI, *Nonlinear total variation based noise removal algorithms*, Physica D, 60 (1992), pp. 259–268. (Cited on pp. 31, 32)

[1600] J. P. RUIZ AND I. E. GROSSMANN, *Strengthening of lower bounds in the global optimization of bilinear and concave generalized disjunctive programs*, Computers and Chemical Engineering, 34 (2010), pp. 914–930. (Cited on p. 327)

[1601] ———, *A hierarchy of relaxations for nonlinear convex generalized disjunctive programming*, European Journal of Operational Research, 218 (2012), pp. 38–47. (Cited on p. 318)

[1602] ———, *Using convex nonlinear relaxations in the global optimization of nonconvex generalized disjunctive programs*, Computers and Chemical Engineering, 49 (2013), pp. 70–84. (Cited on p. 318)

[1603] R. RUIZ-FEMENIA, A. F. TLACUAHUAC, AND I. E. GROSSMANN, *Logic-based outer-approximation algorithm for solving discrete-continuous dynamic optimization problems*, Industrial and Engineering Chemistry Research, 53 (2013), pp. 5067–5080. (Cited on p. 329)

[1604] A. RUSZCZYŃSKI, *Nonlinear Optimization*, Princeton University Press, Princeton, NJ, 2006. (Cited on p. 222)

[1605] A. RUSZCZYŃSKI AND A. SHAPIRO, *Conditional risk mappings*, Mathematics of Operations Research, 31 (2006), pp. 544–561. (Cited on p. 391)

[1606] ———, *Optimization of convex risk functions*, Mathematics of Operations Research, 31 (2006), pp. 433–452. (Cited on p. 391)

[1607] ———, *Optimization of risk measures*, in Probabilistic and Randomized Methods for Design under Uncertainty, Springer, London, U.K., 2006, pp. 119–157. (Cited on p. 355)

[1608] H. S. RYOO AND N. V. SAHINIDIS, *Global optimization of nonconvex NLPs and MINLPs with applications in process design*, Computers and Chemical Engineering, 19 (1995), pp. 552–566. (Cited on p. 281)

[1609] ———, *A branch-and-reduce approach to global optimization*, Journal of Global Optimization, 8 (1996), pp. 107–138. (Cited on pp. 281, 282)

[1610] G. SAHIN, R. K. AHUJA, AND C. B. CUNHA, *Integer Programming Based Approach for the Train Dispatching Problem*, technical report, Sabanci University, 2010. (Cited on p. 69)

[1611] N. V. SAHINIDIS, *BARON: A general purpose global optimization software package*, Journal of Global Optimization, 8 (1996), pp. 201–205. (Cited on pp. 173, 216, 289)

[1612] ———, *Optimization under uncertainty: State-of-the-art and opportunities*, Computers and Chemical Engineering, 28 (2004), pp. 971–983. (Cited on p. 91)

[1613] N. V. SAHINIDIS AND I. E. GROSSMANN, *Convergence properties of generalized Benders decomposition*, Computers and Chemical Engineering, 15 (1991), pp. 481–491. (Cited on p. 212)

[1614] ———, *MINLP model for cyclic multiproduct scheduling on continuous parallel lines*, Computers and Chemical Engineering, 15 (1991), pp. 85–103. (Cited on p. 328)

[1615] ———, *Reformulation of multiperiod MILP models for planning and scheduling of chemical processes*, Computers and Chemical Engineering, 15 (1991), pp. 255–272. (Cited on p. 328)

[1616] ———, *Reformulation of the multiperiod MILP model for capacity expansion of chemical processes*, Operations Research, 40 (1992), pp. S127–S144. (Cited on p. 77)

[1617] N. V. SAHINIDIS AND M. TAWARMALANI, *Applications of global optimization to process and molecular design*, Computers and Chemical Engineering, 24 (2000), pp. 2157–2169. (Cited on p. 329)

[1618] N. V. SAHINIDIS, M. TAWARMALANI, AND M. YU, *Design of alternative refrigerants via global optimization*, AIChE Journal, 49 (2003), pp. 1761–1775. (Cited on p. 329)

[1619] P. SAIKUMAR, R. MURALI, AND E. P. REDDY, *Role of tryptophan repeats and flanking amino acids in Myb-DNA interactions*, Proceedings of the National Academy of Sciences of the United States of America, 87 (1990), pp. 8452–8456. (Cited on p. 184)

[1620] L. C. SALAS, M. R. CÁRDENAS, AND M. ZHANG, *Inventory policies for humanitarian aid during hurricanes*, Socio-Economic Planning Sciences, 46 (2012), pp. 272–280. (Cited on p. 484)

[1621] PH. R. SAMPAIO AND PH. L. TOINT, *A derivative-free trust-funnel method for equality-constrained nonlinear optimization*, Computational Optimization and Applications, 61 (2015), pp. 25–49. (Cited on p. 506)

[1622] H. SAMUELI, *Linear programming design of digital data transmission filters with arbitrary magnitude specifications*, in Conference Record, International Conference on Communications, IEEE, June 1988, pp. 30.6.1–30.6.5. (Cited on p. 29)

[1623] ———, *On the design of FIR digital data transmission filters with arbitrary magnitude specifications*, IEEE Transactions on Circuits and Systems, 38 (1991), pp. 1563–1567. (Cited on p. 29)

[1624] J. E. SANTIBAÑEZ-AGUILAR, J. B. GONZÁLEZ-CAMPOS, J. M. PONCE-ORTEGA, M. SERNA-GONZÁLEZ, AND M. M. EL-HALWAGI, *Optimal planning of a biomass conversion system considering economic and environmental aspects*, Industrial and Engineering Chemistry Research, 50 (2011), pp. 8558–8570. (Cited on p. 318)

[1625] F. SANTOSA AND W. W. SYMES, *Linear inversion of band-limited reflection seismograms*, SIAM Journal on Scientific and Statistical Computing, 7 (1986), pp. 1307–1330. (Cited on p. 33)

[1626] T. SANTOSO, S. AHMED, M. GOETSCHALCKX, AND A. SHAPIRO, *A stochastic programming approach for supply chain network design under uncertainty*, European Journal of Operational Research, 167 (2005), pp. 96–115. (Cited on p. 390)

[1627] R. W. H. SARGENT, *A functional approach to process synthesis and its application to distillation systems*, Computers and Chemical Engineering, 22 (1998), pp. 31–45. (Cited on pp. 320, 323)

[1628] R. W. H. SARGENT AND K. GAMINIBANDARA, *Optimum design of plate distillation columns*, in Optimization in Action, L. C. W. Dixon, ed., Academic Press, London, 1976, pp. 267–314. (Cited on pp. 320, 323, 324)

[1629] A. T. SARIĆ AND A. M. STANKOVIĆ, *Finitely adaptive linear programming in robust power system optimization*, in Proceedings of Power Tech, 2007 IEEE Lausanne, 2007, pp. 1302–1307. (Cited on p. 333)

[1630] K. SATO, K. TAMURA, AND N. TOMII, *A MIP-based timetable rescheduling formulation and algorithm minimizing further inconvenience to passengers*, Journal of Rail Transport Planning and Management, 3 (2013), pp. 38–53. (Cited on p. 69)

[1631] M. W. P. SAVELSBERGH, *Preprocessing and probing techniques for mixed integer programming problems*, ORSA Journal on Computing, 6 (1994), pp. 445–454. (Cited on pp. 282, 285)

[1632] M. SAVELSKI AND M. BAGAJEWICZ, *On the optimality conditions of water utilization systems in process plants with single contaminants*, Chemical Engineering Science, 55 (2000), pp. 5035–5048. (Cited on p. 326)

[1633] ———, *On the necessary conditions of optimality of water utilization systems in process plants with multiple contaminants*, Chemical Engineering Science, 58 (2003), pp. 5349–5362. (Cited on p. 327)

[1634] T. SAVOLA, T.-M. TVEIT, AND C.-J. FOGELHOLM, *A MINLP model including the pressure levels and multiperiods for CHP process optimisation*, Applied Thermal Engineering, 27 (2007), pp. 1857–1867. (Cited on p. 325)

[1635] N. SAWAYA, *Reformulations, Relaxations and Cutting Planes for Generalized Disjunctive Programming*, PhD thesis, Chemical Engineering Department, Carnegie Mellon University, 2006. (Cited on p. 292)

[1636] N. SAWAYA AND I. E. GROSSMANN, *A hierarchy of relaxations for linear generalized disjunctive programming*, European Journal of Operational Research, 216 (2012), pp. 70–82. (Cited on p. 318)

[1637] A. SAXENA, P. BONAMI, AND J. LEE, *Convex relaxations of non-convex mixed integer quadratically constrained programs: Extended formulations*, Mathematical Programming, 124 (2010), pp. 383–411. (Cited on p. 287)

[1638] ———, *Convex relaxations of non-convex mixed integer quadratically constrained programs: Projected formulations*, Mathematical Programming, 130 (2011), pp. 359–413. (Cited on p. 287)

[1639] M. P. SCAPARRA AND R. L. CHURCH, *A bilevel mixed-integer program for critical infrastructure protection planning*, Computers and Operations Research, 35 (2008), pp. 1905–1923. (Cited on p. 480)

[1640] H. SCARF, *A min-max solution of an inventory problem*, in Studies in the Mathematical Theory of Inventory and Production, K. J. Arrow, S. Karlin, and H. Scarf, eds., Stanford Unversity Press, 1958, pp. 201–209. (Cited on p. 391)

[1641] M. SCHACHTEBECK AND A. SCHÖBEL, *To wait or not to wait—and who goes first? Delay management with priority decisions*, Transportation Science, 44 (2010), pp. 307–321. (Cited on p. 69)

[1642] J. SCHAER AND M. G. STONE, *Face traverses and a volume algorithm for polyhedra*, Lecture Notes in Computer Science, 555 (1991), pp. 290–297. (Cited on p. 520)

[1643] S. SCHAIBLE AND T. IBARAKI, *Fractional programming*, European Journal of Operational Research, 12 (1983), pp. 325–338. (Cited on p. 167)

[1644] K. SCHEINBERG AND PH. L. TOINT, *Self-correcting geometry in model-based algorithms for derivative-free unconstrained optimization*, SIAM Journal on Optimization, 20 (2010), pp. 3512–3532. (Cited on p. 497)

[1645] I. J. SCHOENBERG, *Positive definite functions on spheres*, Duke Mathematical Journal, 9 (1942), pp. 96–108. (Cited on p. 34)

[1646] D. SCHOLZ, *Deterministic Global Optimization: Geometric Branch-and-Bound Methods and Their Applications*, Springer, 2012. (Cited on p. 173)

[1647] L. SCHRAGE, *Optimization Modeling with LINGO*, LINDO Systems, Chicago, IL, 1998. (Cited on p. 290)

[1648] ———, *LINGO User's Guide*, LINDO Systems, Chicago, IL, 2006. (Cited on p. 217)

[1649] A. SCHRIJVER, *Theory of Linear and Integer Programming*, Wiley, New York, 1986. (Cited on p. 49)

[1650] R. SCHULTZ, L. STOUGIE, AND M. H. VAN DER VLERK, *Two-stage stochastic integer programming: A survey*, Statistica Neerlandica, 50 (1993), pp. 404–416. (Cited on p. 391)

[1651] R. SCHULTZ AND S. TIEDEMANN, *Risk aversion via excess probabilities in stochastic programs with mixed-integer recourse*, SIAM Journal on Optimization, 14 (2003), pp. 115–138. (Cited on p. 427)

[1652] L. SCHWARZ AND H. ZHAO, *The unexpected impact of information-sharing on US pharmaceutical supply-chains*, Interfaces, 41 (2011), pp. 354–364. (Cited on p. 475)

[1653] M. SCHWARZ, *Treatment planning in proton therapy*, The European Physical Journal Plus, 126 (2011), pp. 1–10. (Cited on p. 345)

[1654] J. K. SCOTT, M. D. STUBER, AND P. I. BARTON, *Generalized McCormick relaxations*, Journal of Global Optimization, 51 (2011), pp. 569–606. (Cited on p. 169)

[1655] J. SCUTERI, L. FODERO, AND J. PEARSE, *Determining a threshold hospital size for the application of activity-based funding*, BMC Health Services Research, 11 (2011), p. A10. (Cited on p. 96)

[1656] N. SECOMANDI, *Optimal commodity trading with a capacitated storage asset*, Management Science, 56 (2010), pp. 449–467. (Cited on p. 462)

[1657] A. ŞEN AND A. ZHANG, *The newsboy problem with multiple demand classes*, IIE Transactions, 31 (1999), pp. 431–444. (Cited on p. 484)

[1658] S. SEN, *Algorithms for stochastic mixed-integer programming models*, in Discrete Optimization, G. L. Nemhauser, K. Aardal, and R. Weismantel, eds., vol. 12 of Handbooks in Operations Research and Management Science, Elsevier, 2005, pp. 515–558. (Cited on p. 391)

[1659] ———, *Stochastic mixed-integer programming: Beyond Benders' decomposition*, in [485], 2011. (Cited on p. 391)

[1660] O. ŞEREF, W. A. CHAOVALITWONGSE, AND J. P. BROOKS, *Relaxing support vectors for classification*, Annals of Operations Research, 216 (2014), pp. 229–255. (Cited on p. 94)

[1661] Y. D. SERGEYEV, R. G. STRONGIN, AND D. LERA, *Introduction to Global Optimization Exploiting Space-Filling Curves*, Springer, New York, 2013. (Cited on p. 172)

[1662] M. SERNA-GONZÁLEZ, J. M. PONCE-ORTEGA, AND A. JIMÉNEZ-GUTIÉRREZ, *Two-level optimization algorithm for heat exchanger networks including pressure drop considerations*, Industrial and Engineering Chemistry Research, 43 (2004), pp. 6766–6773. (Cited on p. 324)

[1663] M. SERRA, A. ESPUNA, AND L. PUIGJANER, *Control and optimization of the divided wall column*, Chemical Engineering and Processing: Process Intensification, 38 (1999), pp. 549–562. (Cited on p. 323)

[1664] ———, *Controllability of different multicomponent distillation arrangements*, Industrial and Engineering Chemistry Research, 42 (2003), pp. 1773–1782. (Cited on p. 323)

[1665] M. SERRA, M. PERRIER, A. ESPUNA, AND L. PUIGJANER, *Study of the divided wall column controllability: Influence of design and operation*, Computers and Chemical Engineering, 24 (2000), pp. 901–907. (Cited on p. 323)

[1666] ———, *Analysis of different control possibilities for the divided wall column: Feedback diagonal and dynamic matrix control*, Computers and Chemical Engineering, 25 (2001), pp. 859–866. (Cited on p. 323)

[1667] SESAR JOINT UNDERTAKING, *The European ATM Master Plan*, 1st edition, http://www.sesarju.eu/sites/default/files/documents/reports/European_ATM_Master_Plan.pdf, 2009. (Cited on p. 293)

[1668] N. SHAH, Z. LI, AND M. G. IERAPETRITOU, *Petroleum refining operations: Key issues, advances, and opportunities*, Industrial and Engineering Chemistry Research, 50 (2011), pp. 1161–1170. (Cited on pp. 41, 327)

[1669] N. SHAH, C. C. PANTELIDES, AND R. W. H. SARGENT, *Optimal periodic scheduling of multipurpose batch plants*, Annals of Operations Research, 42 (1993), pp. 193–228. (Cited on p. 328)

[1670] V. H. SHAH AND R. AGRAWAL, *A matrix method for multicomponent distillation sequences*, AIChE Journal, 56 (2010), pp. 1759–1775. (Cited on p. 324)

[1671] M. SHAHIDEHPOUR, H. YAMIN, AND Z. LI, *Market Operations in Electric Power Systems: Forecasting, Scheduling, and Risk Management*, John Wiley & Sons, Inc., 2002. (Cited on p. 358)

[1672] S. SHAKKOTAI AND R. SRIKANT, *Network optimization and control*, Foundations and Trends in Networking, 2 (2007), pp. 271–379. (Cited on p. 36)

[1673] L. F. SHAMPINE, *Numerical Solution of Ordinary Differential Equations*, Chapman and Hall, New York, 1994. (Cited on p. 508)

[1674] K. H. SHANG AND J.-S. SONG, *Newsvendor bounds and heuristic for optimal policies in serial supply chains*, Management Science, 49 (2003), pp. 618–638. (Cited on p. 449)

[1675] T. SHAO, S. KRISHNAMURTY, AND G. WILMES, *Preference-based surrogate modeling in engineering design*, AIAA Journal, 45 (2007), pp. 2688–2701. (Cited on p. 318)

[1676] A. SHAPIRO, *Analysis of stochastic dual dynamic programming method*, European Journal of Operational Research, 209 (2011), pp. 63–72. (Cited on p. 411)

[1677] A. SHAPIRO AND S. AHMED, *On a class of minimax stochastic programs*, SIAM Journal on Optimization, 14 (2004), pp. 1237–1249. (Cited on p. 391)

[1678] A. SHAPIRO, D. DENTCHEVA, AND A. RUSZCZYŃSKI, *Lectures on Stochastic Programming: Modeling and Theory*, MPS-SIAM Series on Optimization 9, SIAM, Philadelphia, 2009. (Cited on pp. 391, 427)

[1679] A. SHAPIRO AND A. KLEYWEGT, *Minimax analysis of stochastic programs*, Optimization Methods and Software, 17 (2002), pp. 523–542. (Cited on p. 391)

[1680] P. SHAW, *Using programming and local search methods to solve vehicle routing problems*, in Principles and Practice of Constraint Programming, CP98, M. Maher and J. F. Puget, eds., Springer-Verlag, 1998, pp. 417–431. LNCS 1520. (Cited on p. 58)

[1681] J. SHAWE-TAYLER AND N. CRISTIANINI, *An Introduction to Support Vector Machines and Other Kernel-Based Learning Methods*, Cambridge University Press, 2000. (Cited on p. 30)

[1682] J. P. SHECTMAN AND N. V. SAHINIDIS, *A finite algorithm for global minimization of separable concave programs*, Journal of Global Optimization, 12 (1998), pp. 1–36. (Cited on p. 281)

[1683] Z.-J. M. SHEN, C. R. COULLARD, AND M. S. DASKIN, *A joint location-inventory model*, Transportation Science, 37 (2003), pp. 40–55. (Cited on p. 441)

[1684] C. J. R. SHEPPARD, *Analysis of the Measure-Correlate-Predict Methodology for Wind Resource Assessment*, Master's thesis, Humboldt State University, 2009. (Cited on p. 368)

[1685] H. D. SHERALI, *Convex envelopes of multilinear functions over a unit hypercube and over special discrete sets*, Acta Mathematica Vietnamica, 22 (1997), pp. 245–270. (Cited on p. 282)

[1686] H. D. SHERALI AND W. P. ADAMS, *A hierarchy of relaxations between the continuous and convex hull representations for zero-one programming problems*, SIAM Journal on Discrete Mathematics, 3 (1990), pp. 411–430. (Cited on pp. 55, 211, 286)

[1687] ———, *A Reformulation-Linearization Technique for Solving Discrete and Continuous Nonconvex Problems*, Kluwer Academic, 1999. (Cited on pp. 169, 173, 282)

[1688] H. D. SHERALI AND B. M. FRATICELLI, *Enhancing RLT relaxations via a new class of semidefinite cuts*, Journal of Global Optimization, 22 (2002), pp. 233–261. (Cited on p. 282)

[1689] J. B. SHEU, *An emergency logistics distribution approach for quick response to urgent relief demand in disasters*, Transportation Research Part E: Logistics and Transportation Review, 43 (2007), pp. 687–709. (Cited on p. 487)

[1690] I. N. SHINDYALOV AND P. E. BOURNE, *Protein structure alignment by incremental combinatorial extension (CE) of the optimal path*, Protein Engineering, 11 (1998), pp. 739–47. (Cited on p. 520)

[1691] N. Z. SHOR, *Nondifferentiable Optimization and Polynomial Problems*, Kluwer Academic, 1998. (Cited on p. 173)

[1692] S. SIDDHAYE, K. V. CAMARDA, E. TOPP, AND M. SOUTHARD, *Design of novel pharmaceutical products via combinatorial optimization*, Computers and Chemical Engineering, 24 (2000), pp. 701–704. (Cited on p. 329)

[1693] *SID-PSM*. http://www.mat.uc.pt/sid-psm. (Cited on pp. 499, 506)

[1694] G. SIGL, K. DOLL, AND F. M. JOHANNES, *Analytical placement: A linear or quadratic objective function?*, in Proceedings of the 28th ACM/IEEE Design Automation Conference, 1991, pp. 427–432. (Cited on p. 34)

[1695] E. A. SILVER, D. F. PYKE, AND R. PETERSON, *Inventory Management and Production Planning and Scheduling*, 3rd edition, Wiley, 1998. (Cited on pp. 439, 442)

[1696] D. SIMCHI-LEVI, X. CHEN, AND J. BRAMEL, *The Logic of Logistics*, 3rd edition, Springer, 2013. (Cited on p. 439)

[1697] D. SIMCHI-LEVI, P. KAMINSKI, AND E. SIMCHI-LEVI, *Designing and Managing the Supply Chain: Concepts, Strategies, and Case Studies*, 2nd edition, McGraw-Hill Irwin, New York, 2007. (Cited on p. 439)

[1698] M. SINHA, L. E. K. ACHENIE, AND R. GANI, *Blanket wash solvent blend design using interval analysis*, Industrial and Engineering Chemistry Research, 42 (2003), pp. 516–527. (Cited on p. 329)

[1699] M. SINHA, L. E. K. ACHENIE, AND G. M. OSTROVSKY, *Environmentally benign solvent design by global optimization*, Computers and Chemical Engineering, 23 (1999), pp. 1381–1394. (Cited on p. 329)

[1700] J. SMADBECK, *Advances in Protein Design: Conformational Switch, Multimeric, and Protein-DNA Design*, PhD thesis, Princeton University, 2015. (Cited on pp. xviii, xxiii, 177, 179, 183, 184)

[1701] E. M. B. SMITH, *On the Optimal Design of Continuous Processes*, PhD thesis, Imperial College London, 1996. (Cited on p. 323)

[1702] E. M. B. SMITH AND C. C. PANTELIDES, *Design of reaction/separation networks using detailed models*, Computers and Chemical Engineering, 19 (1995), pp. 83–88. (Cited on pp. 318, 320, 321)

[1703] ———, *A symbolic reformulation/spatial branch-and-bound algorithm for the global optimisation of nonconvex MINLPs*, Computers and Chemical Engineering, 23 (1999), pp. 457–478. (Cited on p. 280)

[1704] H. K. SMITH, G. LAPORTE, AND P. R. HARPER, *Locational analysis: Highlights of growth to maturity*, Journal of the Operational Research Society, 60 (2009), pp. S140–S148. (Cited on p. 450)

[1705] J. SMITH, F. MÖHRING, AND D. LINK, *Making ports more resilient*, InterAction Monthly Developments Magazine, 31 (2013). (Cited on p. 488)

[1706] A. J. SMOLA AND B. SCHÖLKOPF, *A tutorial on support vector regression*, Statistics and Computing, 14 (2004), pp. 199–222. (Cited on p. 30)

[1707] L. V. SNYDER, *Covering problems*, in [658], ch. 6, pp. 109–135. (Cited on p. 453)

[1708] ———, *Introduction to facility location*, in [485], 2011. (Cited on p. 450)

[1709] L. V. SNYDER, Z. ATAN, P. PENG, Y. RONG, A. J. SCHMITT, AND B. SINSOYSAL, *OR/MS models for supply chain disruptions: A review*, IIE Transactions, 48 (2016), pp. 89–109. (Cited on pp. 448, 459)

[1710] L. V. SNYDER AND M. S. DASKIN, *Reliability models for facility location: The expected failure cost case*, Transportation Science, 39 (2005), pp. 400–416. (Cited on p. 480)

[1711] L. V. SNYDER AND Z.-J. M. SHEN, *Fundamentals of Supply Chain Theory*, Wiley, Hoboken, NJ, 2011. (Cited on pp. 439, 446, 447, 448, 450, 454, 455)

[1712] S. SODERSTROM AND A. BRAHME, *Selection of suitable beam orientations in radiation therapy using entropy and Fourier transform measures*, Physics in Medicine and Biology, 37 (1992), p. 911. (Cited on p. 95)

[1713] S. SOLAK, C. SCHERRER, AND A. GHONIEM, *The stop-and-drop problem in nonprofit food distribution networks*, Annals of Operations Research, 221 (2014), pp. 407–426. (Cited on p. 480)

[1714] M. SOLDNER, *Optimization and Measurement in Humanitarian Networks: Bridging the Gap between Research and Practice*, PhD thesis, School of Industrial and Systems Engineering, Georgia Institute of Technology, Atlanta, GA, 2014. (Cited on pp. 489, 490)

[1715] J.-H. SONG AND K. C. FURMAN, *A maritime inventory routing problem: Practical approach*, Computers and Operations Research, 40 (2013), pp. 657–665. (Cited on pp. xvii, 83)

[1716] D. A. SOUTHERN, H. QUAN, AND W. A. GHALI, *Comparison of the Elixhauser and Charlson/Deyo methods of comorbidity measurement in administrative data*, Medical Care, 42 (2004), pp. 355–360. (Cited on p. 96)

[1717] A. L. SOYSTER, *Convex programming with set-inclusive constraints and applications to inexact linear programming*, Operations Research, 21 (1973), pp. 1154–1157. (Cited on p. 394)

[1718] T. A. SRIVER, J. W. CHRISSIS, AND M. A. ABRAMSON, *Pattern search ranking and selection algorithms for mixed variable simulation-based optimization*, European Journal of Operational Research, 198 (2009), pp. 878–890. (Cited on p. 501)

[1719] S. STAGE AND Y. LARSSON, *Incremental cost of water power*, AIEE Transactions (Power Apparatus and Systems), 80 (1961), pp. 361–364. (Cited on pp. 406, 411)

[1720] J. STEEG AND M. SCHRÖDER, *A hybrid approach to solve the periodic home health care problem*, in Operations Research Proceedings 2007, Springer, 2008, pp. 297–302. (Cited on p. 96)

[1721] K. STEIGLITZ, *Optimal design of FIR digital filters with monotone passband response*, IEEE Transactions on Acoustics, Speech, and Signal Processing, ASSP-27 (1979), pp. 643–649. (Cited on p. 29)

[1722] A. A. STINNETT AND A. D. PALTIEL, *Mathematical programming for the efficient allocation of health care resources*, Journal of Health Economics, 15 (1996), pp. 641–653. (Cited on p. 96)

[1723] M. STOLPE, *Truss topology optimization with discrete design variables by outer approximation*, Journal of Global Optimization, 61 (2015), pp. 139–163. (Cited on pp. 145, 146)

[1724] M. STOLPE AND K. SVANBERG, *A stress-constrained truss-topology and material-selection problem that can be solved by linear programming*, Structural and Multidisciplinary Optimization, 27 (2004), pp. 126–129. (Cited on pp. 16, 17, 18)

[1725] D. E. STONEKING, G. L. BILBRO, R. J. TREW, P. GILMORE, AND C. T. KELLEY, *Yield optimization using a GaAs process simulator coupled to a physical device model*, IEEE Transactions on Microwave Theory and Techniques, 40 (1992), pp. 1353–1363. (Cited on p. 508)

[1726] B. STOTT, J. JARDIM, AND O. ALSAÇ, *DC power flow revisited*, IEEE Transactions on Power Systems, 24 (2009), pp. 1290–1300. (Cited on p. 197)

[1727] A. STREET, A. MOREIRA, AND J. M. ARROYO, *Energy and reserve scheduling under a joint generation and transmission security criterion: An adjustable robust optimization approach*, IEEE Transactions on Power Systems, 29 (2014), pp. 3–14. (Cited on p. 363)

[1728] A. STREET, F. OLIVEIRA, AND J. M. ARROYO, *Contingency-constrained unit commitment with $n - k$ security criterion: A robust optimization approach*, IEEE Transactions on Power Systems, 26 (2011), pp. 1581–1590. (Cited on p. 363)

[1729] R. G. STRONGIN, *Numerical Methods in Multi-Extremal Problems: Information-Statistical Algorithms*, Nauka, Moscow. In Russian. (Cited on p. 172)

[1730] R. G. STRONGIN AND YA. D. SERGEYEV, *Global Optimization with Non-convex Constraints: Sequential and Parallel Algorithms*, Kluwer Academic, Dordrecht, Netherlands, 2000. (Cited on pp. 169, 172, 173)

[1731] R. STUBBS AND S. MEHROTRA, *A branch-and-cut method for 0-1 mixed convex programming*, Mathematical Programming, 86 (1999), pp. 515–532. (Cited on p. 287)

[1732] A. STULMAN, *Benefits of centralized stocking for the multi-centre newsboy problem with first come, first served allocation*, Journal of the Operational Research Society, 38 (1987), pp. 827–832. (Cited on p. 486)

[1733] C. STUMMER, K. DOERNER, A. FOCKE, AND K. HEIDENBERGER, *Determining location and size of medical departments in a hospital network: A multiobjective decision support approach*, Health Care Management Science, 7 (2004), pp. 63–71. (Cited on p. 95)

[1734] J. F. STURM, *Primal-dual interior point approach to semidefinite programming*, in High Performance Optimization, J. B. G. Frenk, C. Roos, T. Terlaky, and S. Zhang, eds., Kluwer Academic, Dordrecht, Netherlands, 1999. (Cited on p. 11)

[1735] ———, *Using SeDuMi 1.02, a MATLAB toolbox for optimization over symmetric cones*, Optimization Methods and Software, 11-12 (1999), pp. 625–653. (Cited on pp. 10, 117)

[1736] J. J. STURM, D. A. HIRSH, E. K. LEE, R. MASSEY, B. WESELMAN, AND H. K. SIMON, *Practice characteristics that influence nonurgent pediatric emergency department utilization*, Academic Pediatrics, 10 (2010), pp. 70–74. (Cited on p. 99)

[1737] O. STURSBERG AND S. PANEK, *Control of switched hybrid systems based on disjunctive formulations*, in Hybrid Systems: Computation and Control, Springer, 2002, pp. 421–435. (Cited on p. 328)

[1738] A. SUBRAMANI, P. A. DiMAGGIO, AND C. A. FLOUDAS, *Selecting high quality protein structures from diverse conformational ensembles*, Biophysical Journal, 97 (2009), pp. 1728–1736. (Cited on p. 181)

[1739] A. X. SUN AND D. T. PHAN, *Some optimization models and techniques for electric power system short-term operations*, in [485], pp. 1–17. (Cited on p. 359)

[1740] H. SUN AND H. XU, *Convergence analysis for distributionally robust optimization and equilibrium problems*, Mathematics of Operations Research, 41 (2016), pp. 377–401. (Cited on p. 344)

[1741] P. SUN, L. YANG, AND F. DE VERICOURT, *Selfish drug allocation for containing an international influenza pandemic at the onset*, Operations Research, 57 (2009), pp. 1320–1332. (Cited on p. 472)

[1742] I. SUNGUR, F. ORDÓNEZ, AND M. DESSOUKY, *A robust optimization approach for the capacitated vehicle routing problem with demand uncertainty*, IIE Transactions, 40 (2008), pp. 509–523. (Cited on p. 333)

[1743] G. SVED AND Z. GINOS, *Structural optimization under multiple loading*, International Journal of Mechanical Sciences, 10 (1968), pp. 803–805. (Cited on p. 25)

[1744] G. F. SYMONDS, *Linear programming solves gasoline refining and blending problems*, Industrial and Engineering Chemistry, 48 (1956), pp. 394–401. (Cited on p. 37)

[1745] L. TACCARI, E. AMALDI, A. BISCHI, AND E. MARTELLI, *MINLP Instances for Short-Term Planning of Combined Heat and Power (CHP) Systems*. http://chpminlp.deib.polimi.it, 2017. (Cited on pp. 310, 313)

[1746] N. TAKAMA, T. KURIYAMA, K. SHIROKO, AND T. UMEDA, *Optimal water allocation in a petroleum refinery*, Computers and Chemical Engineering, 4 (1980), pp. 251–258. (Cited on p. 327)

[1747] F. TARDELLA, *Existence and sum decomposition of vertex polyhedral convex envelopes*, Optimization Letters, 2 (2008), pp. 363–375. (Cited on p. 282)

[1748] D. TASCHE, *Conditional Expectation as Quantile Derivative*, working paper, 2000. (Cited on p. 433)

[1749] ———, *Capital allocation for credit portfolios with kernel estimators*, Quantitative Finance, 9 (2009), pp. 581–595. (Cited on pp. 433, 434)

[1750] M. TAWARMALANI, J.-P. P. RICHARD, AND K. CHUNG, *Strong valid inequalities for orthogonal disjunctions and bilinear covering sets*, Mathematical Programming, 124 (2010), pp. 481–512. (Cited on p. 287)

[1751] M. TAWARMALANI, J.-P. P. RICHARD, AND C. XIONG, *Explicit convex and concave envelopes through polyhedral subdivisions*, Mathematical Programming, 138 (2013), pp. 531–577. (Cited on p. 282)

[1752] M. TAWARMALANI AND N. V. SAHINIDIS, *Semidefinite relaxations of fractional programs via novel convexification techniques*, Journal of Global Optimization, 20 (2001), pp. 133–154. (Cited on p. 283)

[1753] ——, *Convex extensions and envelopes of lower semi-continuous functions*, Mathematical Programming, 93 (2002), pp. 247–263. (Cited on p. 283)

[1754] ——, *Convexification and Global Optimization in Continuous and Mixed-Integer Nonlinear Programming: Theory, Algorithms, Software, and Applications*, Kluwer Academic, Dordrecht, Netherlands, 2002. (Cited on pp. 173, 211, 215, 216, 274, 280, 281, 315)

[1755] ——, *Global optimization of mixed-integer nonlinear programs: A theoretical and computational study*, Mathematical Programming, 99 (2004), pp. 563–591. (Cited on pp. 173, 280, 285)

[1756] ——, *A polyhedral branch-and-cut approach to global optimization*, Mathematical Programming, 103 (2005), pp. 225–249. (Cited on pp. 196, 280, 285, 289, 304, 313)

[1757] D. H. S. TAY, H. KHEIREDDINE, D. K. S. NG, M. M. EL-HALWAGI, AND R. R. TAN, *Conceptual synthesis of gasification-based biorefineries using thermodynamic equilibrium optimization models*, Industrial and Engineering Chemistry Research, 50 (2011), pp. 10681–10695. (Cited on p. 318)

[1758] D. H. S. TAY, D. K. S. NG, N. E. SAMMONS JR, AND M. R. EDEN, *Fuzzy optimization approach for the synthesis of a sustainable integrated biorefinery*, Industrial and Engineering Chemistry Research, 50 (2011), pp. 1652–1665. (Cited on p. 318)

[1759] J. TELES, P. M. CASTRO, AND A. Q. NOVAIS, *LP-based solution strategies for the optimal design of industrial water networks with multiple contaminants*, Chemical Engineering Science, 63 (2008), pp. 376–394. (Cited on p. 327)

[1760] T. TERLAKY, *A convergent criss–cross method*, Optimization, 16 (1985), pp. 683–690. (Cited on p. 5)

[1761] ——, ed., *Interior Point Methods of Mathematical Programming*, vol. 5 of Applied Optimization, Kluwer Academic, Dordrecht, Netherlands, 1996. (Cited on p. 11)

[1762] T. TERLAKY AND S. ZHANG, *Pivot rules for linear programming—A survey*, Annals of Operations Research, 46 (1993), pp. 203–233. (Cited on pp. 5, 8)

[1763] S. TERRAZAS-MORENO, A. FLORES-TLACUAHUAC, AND I. E. GROSSMANN, *Simultaneous cyclic scheduling and optimal control of polymerization reactors*, AIChE Journal, 53 (2007), pp. 2301–2315. (Cited on p. 328)

[1764] *Test Case Archive of Optimal Power Flow (OPF) Problems with Local Optima*, http://www.maths.ed.ac.uk/optenergy/LocalOpt/. (Cited on p. 195)

[1765] A. THATTE, A. X. SUN, AND L. XIE, *Robust optimization based economic dispatch for managing system ramp capability*, in HICSS 2014. (Cited on p. 364)

[1766] THERMOFLOW, *THERMOFLEX 24—Design and Simulation of Power Plants*, https://www.thermoflow.com/combinedcycle_TFX.html, 2014. (Cited on pp. xix, 311)

[1767] A. THIELE, *A note on issues of over-conservatism in robust optimization with cost uncertainty*, Optimization, 59 (2010), pp. 1033–1040. (Cited on p. 333)

[1768] E. THOMAS, O. CHAPET, M. L. KESSLER, T. S. LAWRENCE, AND R. K. TEN HAKEN, *Benefit of using biologic parameters (EUD and NTCP) in IMRT optimization for treatment of intrahepatic tumors*, International Journal of Radiation Oncology • Biology • Physics, 62 (2005), pp. 571–578. (Cited on p. 95)

[1769] R. TIBSHIRANI, *Regression shrinkage and selection via the Lasso*, Journal of the Royal Statistical Society, Series B (Methodological), 58 (1996), pp. 267–288. (Cited on pp. 31, 33)

[1770] D. TILMAN, R. SOCOLOW, J. A. FOLEY, J. HILL, E. LARSON, L. LYND, S. PACALA, J. REILLY, T. SEARCHINGER, C. SOMERVILLE, AND R. WILLIAMS, *Beneficial biofuels—The food, energy, and environment trilemma*, Science, 325 (2009), pp. 270–271. (Cited on p. 463)

[1771] R. L. TOBIN AND T. L. FRIESZ, *Spatial competition facility location models: Definition, formulation and solution approach*, Annals of Operations Research, 6 (1986), pp. 47–74. (Cited on p. 464)

[1772] M. J. TODD, *A study of search directions in primal-dual interior-point methods for semidefinite programming*, Optimization Methods and Software, 11/12 (1999), pp. 1–46. (Cited on p. 116)

[1773] K.-C. TOH, M. J. TODD, AND R. H. TÜTÜNCÜ, *On the implementation and usage of SDPT3—A MATLAB software package for semidefinite-quadratic-linear programming, version 4.0*, in [87], pp. 715–754. (Cited on p. 117)

[1774] K. C. TOH, R. H. TÜTÜNCÜ, AND M. J. TODD, *On the implementation of SDPT3 (version 3.1)—A MATLAB software package for semidefinite-quadratic-linear programming*, in Proceedings of the IEEE Conference on Computer-Aided Control System Design, 2004. (Cited on p. 10)

[1775] B. TOMLIN, *On the value of mitigation and contingency strategies for managing supply chain disruption risks*, Management Science, 52 (2006), pp. 639–657. (Cited on p. 459)

[1776] W. TONG, S. CHOWDHURY, J. ZHANG, AND A. MESSAC, *Impact of different wake models on the estimation of wind farm power generation*, in Proceedings of the AIAA/ISSMO Multidisciplinary Analysis and Optimization Conference 2012, Indianapolis, IN, September 2012. (Cited on p. 368)

[1777] V. TORCZON, *Multidirectional Search*, PhD thesis, Rice University, Houston, TX, 1989. (Cited on p. 510)

[1778] ——, *On the convergence of pattern search algorithms*, SIAM Journal on Optimization, 7 (1997), pp. 1–25. (Cited on p. 499)

[1779] A. TÖRN AND A. ŽILINSKAS, eds., *Global Optimization*, Springer, 1989. (Cited on p. 173)

[1780] G. L. TORRES AND V. H. QUINTANA, *An interior-point method for nonlinear optimal power flow using voltage rectangular coordinates*, IEEE Transactions on Power Systems, 13 (1998), pp. 1211–1218. (Cited on p. 190)

[1781] G. L. TORRES, V. H. QUINTANA, AND G. LAMBERT-TORRES, *Optimal power flow in rectangular form via an interior point method*, in Proceedings of the North American Power Symposium, IEEE, 1996. (Cited on p. 190)

[1782] E. TRÉLAT, *Contrôle optimal : théorie et applications*, Vuibert, Paris, 2005. (Cited on p. 122)

[1783] F. TRESPALACIOS AND I. E. GROSSMANN, *Review of mixed-integer nonlinear and generalized disjunctive programming methods*, Chemie Ingenieur Technik, 86 (2014), pp. 991–1012. (Cited on p. 316)

[1784] ——, *Improved big-M reformulation for generalized disjunctive programs*, Computers and Chemical Engineering, 76 (2015), pp. 98–103. (Cited on p. 316)

[1785] J. A. TROPP, *Just relax: Convex programming methods for identifying sparse signals in noise*, IEEE Transactions on Information Theory, 52 (2006), pp. 1030–1051. (Cited on p. 33)

[1786] M. W. TROSSET, *On the Use of Direct Search Methods for Stochastic Optimization*, Technical Report TR00-20, CAAM Technical Report, Rice University, 2000. (Cited on p. 501)

[1787] A. W. S. TRUPOWER, *Open Wind Theoretical Basis and Validation*, technical report, Albany, NY, 2010. (Cited on p. 368)

[1788] M.-J. TSAI AND C.-T. CHANG, *Water usage and treatment network design using genetic algorithms*, Industrial and Engineering Chemistry Research, 40 (2001), pp. 4874–4888. (Cited on p. 327)

[1789] P. TSENG, *Fortified-descent simplicial search method: A general approach*, SIAM Journal on Optimization, 10 (1999), pp. 269–288. (Cited on p. 498)

[1790] D. W. TUFTS, D. RORABACHER, AND W. MOSIER, *Designing simple, effective digital filters*, IEEE Transactions on Audio and Electroacoustics, 18 (1970), pp. 142–158. (Cited on p. 28)

[1791] M. TÜRKAY AND I. E. GROSSMANN, *Disjunctive programming techniques for the optimization of process systems with discontinuous investment costs-multiple size regions*, Industrial and Engineering Chemistry Research, 35 (1996), pp. 2611–2623. (Cited on p. 318)

[1792] ——, *Logic-based MINLP algorithms for the optimal synthesis of process networks*, Computers and Chemical Engineering, 20 (1996), pp. 959–978. (Cited on pp. 292, 318)

[1793] ——, *Logic-based outer-approximation and Benders decomposition algorithms for the synthesis of process networks*, in State of the Art in Global Optimization: Computational Methods and Applications, C. A. Floudas and P. M. Pardalos, eds., Springer, 1996, pp. 585–607. (Cited on p. 316)

[1794] S. D. O. TURNER, D. A. ROMERO, P. Y. ZHANG, C. H. AMON, AND T. C. Y. CHAN, *A new mathematical programming approach to optimize wind farm layouts*, Renewable Energy, 63 (2014), pp. 674–680. (Cited on p. 367)

[1795] R. H. TÜTÜNCÜ AND M. KOENIG, *Robust asset allocation*, Annals of Operations Research, 132 (2004), pp. 157–187. (Cited on p. 156)

[1796] R. H. TÜTÜNCÜ, K.-C. TOH, AND M. J. TODD, *Solving semidefinite-quadratic-linear programs using SDPT3*, Mathematical Programming, 95 (2003), pp. 189–217. (Cited on p. 117)

[1797] H. TUY, *D.C. optimization: Theory, methods and algorithms*, in [957], pp. 149–216. (Cited on p. 173)

[1798] ——, *Concave programming under linear constraints*, Doklady Akademii Nauk, 159 (1964), pp. 32–35. (Cited on pp. 163, 170)

[1799] ——, *On polyhedral annexation method for concave minimization*, in Functional Analysis, Optimization, and Mathematical Economics, L. J. Leifman, ed., Oxford University Press, 1990, pp. 248–260. (Cited on p. 171)

[1800] ——, *Polyhedral annexation, dualization, and dimension reduction technique in global optimization*, Journal of Global Optimization, 1 (1991), pp. 229–244. (Cited on p. 171)

[1801] ——, *Monotonic optimization: Problems and solution approaches*, SIAM Journal on Optimization, 11 (2000), pp. 464–494. (Cited on p. 173)

[1802] ——, *On solving nonconvex optimization problems by reducing the duality gap*, Journal of Global Optimization, 32 (2005), pp. 349–365. (Cited on p. 169)

[1803] Z. UGRAY, L. LASDON, J. PLUMMER, F. GLOVER, J. KELLY, AND R. MARTÍ, *Scatter search and local NLP solvers: A multistart framework for global optimization*, INFORMS Journal on Computing, 19 (2007), pp. 328–340. (Cited on p. 290)

[1804] C. ULLMER, N. KUNDE, A. LASSAHN, G. GRUHN, AND K. SCHULZ, *Wado^{TM}: Water design optimization—Methodology and software for the synthesis of process water systems*, Journal of Cleaner Production, 13 (2005), pp. 485–494. (Cited on p. 327)

[1805] U. N. WORLD FOOD PROGRAMME, *Emergency Field Operations Pocketbook*, 2002. (Cited on p. 488)

[1806] ——, *Construction and Management of the WFP Humanitarian Logistics Base at Djibouti Port*, Operational Document, Special Operation 200358, 2012. (Cited on p. 488)

[1807] ——, *2012 Food Aid Flows*, 2013. (Cited on p. 488)

[1808] F. ÜNEY AND M. TÜRKAY, *A mixed-integer programming approach to multi-class data classification problem*, European Journal of Operational Research, 173 (2006), pp. 910–920. (Cited on p. 94)

[1809] UNICEF, *Childinfo: Monitoring the Situation of Children and Women*, http://www.childinfo.org/malnutrition_status.html. (Cited on p. 479)

[1810] ——, *Infant and Young Child Feeding*, http://www.unicef.org/nutrition/index_breastfeeding.html, 2014. (Cited on pp. 479, 481)

[1811] J. UNKELBACH, T. CHAN, AND T. BORTFELD, *Accounting for range uncertainties in the optimization of intensity modulated proton therapy*, Physics in Medicine and Biology, 52 (2007), pp. 2755–2773. (Cited on pp. 354, 355)

[1812] J. UNKELBACH AND U. OELFKE, *Inclusion of organ movements in IMRT treatment planning via inverse planning based on probability distributions*, Physics in Medicine and Biology, 49 (2004), pp. 4005–4029. (Cited on p. 348)

[1813] R. VAIDYANATHAN, Y. GOWAYED, AND M. EL-HALWAGI, *Computer-aided design of fiber reinforced polymer composite products*, Computers and Chemical Engineering, 22 (1998), pp. 801–808. (Cited on p. 329)

[1814] J. VAN DE KLUNDERT, P. MULS, AND M. SCHADD, *Optimizing sterilization logistics in hospitals*, Health Care Management Science, 11 (2008), pp. 23–33. (Cited on p. 95)

[1815] C. VAN DE PANNE, *Methods for Linear and Quadratic Programming*, Studies in Mathematics and Managerial Economics, North Holland, Amsterdam, Netherlands, 1975. (Cited on p. 4)

[1816] E. VAN DER WEIDE, G. KALITZIN, J. SCHLUTER, AND J. J. ALONSO, *Unsteady turbomachinery computations using massively parallel platforms*, in Proceedings of the 44th AIAA Aerospace Sciences Meeting and Exhibit, Reno, NV, 2006. AIAA 2006-0421. (Cited on p. 253)

[1817] N. M. VAN DIJK AND E. VAN DER SLUIS, *Pooling is not the answer*, European Journal of Operational Research, 17 (2008), pp. 415–421. (Cited on p. 487)

[1818] ——, *To pool or not to pool in call centers*, Production and Operations Management, 17 (2008), pp. 296–305. (Cited on p. 487)

[1819] M. VAN HERK, *Errors and margins in radiotherapy*, Seminars in Radiation Oncology, 14 (2004), pp. 56–64. (Cited on p. 347)

[1820] R. M. VAN SLYKE AND R. WETS, *L-shaped linear programs with applications to optimal control and stochastic programming*, SIAM Journal on Applied Mathematics, 17 (1969), pp. 638–663. (Cited on p. 391)

[1821] L. VANDENBERGHE AND S. BOYD, *Semidefinite programming*, SIAM Review, 38 (1996), pp. 49–95. (Cited on p. 11)

[1822] F. VANDERBECK AND L. A. WOLSEY, *Reformulation and decomposition of integer programs*, in [1017], pp. 431–502. (Cited on p. 71)

[1823] R. J. VANDERBEI, *Linear Programming: Foundations and Extensions*, Kluwer, Boston, 1996. (Cited on p. 352)

[1824] ——, *LOQO: An interior point code for quadratic programming*, Optimization Methods and Software, 11/12 (1999), pp. 451–484. (Cited on p. 117)

[1825] ——, *LOQO user's manual—Version 3.10*, Optimization Methods and Software, 11/12 (1999), pp. 485–514. (Cited on p. 117)

[1826] ——, *LOQO User's Guide - Version 4.05*, Princeton University, School of Engineering and Applied Science, Department of Operations Research and Financial Engineering, Princeton, NJ, 2006. (Cited on p. 10)

[1827] R. J. VANDERBEI AND D. F. SHANNO, *An interior point algorithm for nonconvex nonlinear programming*, Computational Optimization and Applications, 13 (1999), pp. 231–252. (Cited on p. 289)

[1828] M. VANSUCH, J. M. NAESSENS, R. J. STROEBEL, J. M. HUDDLESTON, AND A. R. WILLIAMS, *Effect of discharge instructions on readmission of hospitalised patients with heart failure: Do all of the Joint Commission on Accreditation of Healthcare Organizations heart failure core measures reflect better care?* Quality and Safety in Health Care, 15 (2006), pp. 414–417. (Cited on p. 96)

[1829] V. N. VAPNIK, *The Nature of Statistical Learning Theory*, 2nd edition, Springer, 2000. (Cited on p. 30)

[1830] J. C. VASSBERG, M. A. DEHAAN, S. M. RIVERS, AND R. A. WAHLS, *Development of a common research model for applied CFD validation studies*, in Proceedings of the 26th AIAA Applied Aerodynamics Conference, Honolulu, HI, 2008. AIAA 2008-6919. (Cited on p. 253)

[1831] S. VAVASIS, *Complexity issues in global optimization: A survey*, in [957], pp. 27–41. (Cited on p. 167)

[1832] ——, *Nonlinear Optimization: Complexity Issues*, Oxford University Press, 1991. (Cited on p. 167)

[1833] A. I. F. VAZ AND L. N. VICENTE, *A particle swarm pattern search method for bound constrained global optimization*, Journal of Global Optimization, 39 (2007), pp. 197–219. (Cited on p. 506)

[1834] A. VECCHIETTI AND I. E. GROSSMANN, *LOGMIP: A disjunctive 0-1 non-linear optimizer for process system models*, Computers and Chemical Engineering, 23 (1999), pp. 555–565. (Cited on pp. 291, 318)

[1835] A. E. VELA, S. SOLAK, J. B. CLARKE, W. E. SINGHOSE, E. R. BARNES, AND E. L. JOHNSON, *Near real-time fuel-optimal en route conflict resolution*, IEEE Transactions on Intelligent Transportation Systems, 11 (2010), pp. 826–837. (Cited on p. 301)

[1836] A. VELA, S. SOLAK, W. SINGHOSE, AND J.-P. CLARKE, *A mixed-integer program for flight-level assignment and speed control for conflict resolution*, in Proceedings of the Joint 48th IEEE Conference on Decision and Control and 28th Chinese Control Conference, Shanghai, IEEE, 2009. (Cited on p. 295)

[1837] P. M. VERDERAME, J. A. ELIA, J. LI, AND C. A. FLOUDAS, *Planning and scheduling under uncertainty: A review across multiple sectors*, Industrial and Engineering Chemistry Research, 49 (2010), pp. 3993–4017. (Cited on p. 394)

[1838] P. M. VERDERAME AND C. A. FLOUDAS, *Integrated operational planning and medium-term scheduling for large-scale industrial batch plants*, Industrial and Engineering Chemistry Research, 47 (2008), pp. 4845–4860. (Cited on pp. 394, 395)

[1839] ——, *Operational planning framework for multisite production and distribution networks*, Computers and Chemical Engineering, 33 (2009), pp. 1036–1050. (Cited on p. 395)

[1840] ——, *Operational planning of large-scale industrial batch plants under demand due date and amount uncertainty: I. Robust optimization framework*, Industrial and Engineering Chemistry Research, 48 (2009), pp. 7214–7231. (Cited on pp. xx, 394, 395, 397, 399)

[1841] ——, *Operational planning of large-scale industrial batch plants under demand due date and amount uncertainty: II. Conditional value-at-risk framework*, Industrial and Engineering Chemistry Research, 49 (2010), pp. 260–275. (Cited on pp. xxiv, 394, 401, 403)

[1842] W. VERHEYEN AND N. ZHANG, *Design of flexible heat exchanger network for multiperiod operation*, Chemical Engineering Science, 61 (2006), pp. 7730–7753. (Cited on p. 324)

[1843] V. VERTER AND S. D. LAPIERRE, *Location of preventive health care facilities*, Annals of Operations Research, 110 (2002), pp. 123–132. (Cited on p. 95)

[1844] B. VERWEIJ, S. AHMED, A. J. KLEYWEGT, G. NEMHAUSER, AND A. SHAPIRO, *The sample average approximation method applied to stochastic routing problems: A computational study*, Computational and Applied Optimization, 24 (2003), pp. 289–333. (Cited on p. 390)

[1845] L. N. VICENTE, *Worst case complexity of direct search*, EURO Journal on Computational Optimization, 1 (2013), pp. 143–153. (Cited on p. 501)

[1846] L. N. VICENTE AND A. L. CUSTÓDIO, *Analysis of direct searches for discontinuous functions*, Mathematical Programming, 133 (2012), pp. 299–325. (Cited on pp. 500, 503, 504)

[1847] J. P. VIELMA, S. AHMED, AND G. NEMHAUSER, *Mixed-integer models for nonseparable piecewise-linear optimization: Unifying framework and extensions*, Operations Research, 58 (2010), pp. 303–315. (Cited on p. 284)

[1848] J. P. VIELMA AND G. L. NEMHAUSER, *Modeling disjunctive constraints with a logarithmic number of binary variables and constraints*, Mathematical Programming, 128 (2011), pp. 49–72. (Cited on pp. 205, 216, 284)

[1849] S. VIGERSKE, *Decomposition in Multistage Stochastic Programming and a Constraint Integer Programming Approach to Mixed-Integer Nonlinear Programming*, PhD thesis, Humboldt-Universität zu Berlin, 2013. (Cited on p. 290)

[1850] ———, *(MI)NLPLib 2*, http://www.gams.com/presentations/2015_ismp.pdf, 2015. 22nd International Symposium on Mathematical Programming. (Cited on p. 290)

[1851] C. VILLANI, *Topics in Optimal Transportation*, AMS, Providence, RI, 2003. (Cited on p. 127)

[1852] R. B. VINTER, *Convex duality and nonlinear optimal control*, SIAM Journal on Control and Optimization, 31 (1993), pp. 518–538. (Cited on p. 129)

[1853] J. VISWANATHAN AND I. E. GROSSMANN, *A combined penalty function and outer-approximation method for MINLP optimization*, Computers and Chemical Engineering, 14 (1990), pp. 769–782. (Cited on p. 289)

[1854] ———, *Optimal feed locations and number of trays for distillation columns with multiple feeds*, Industrial and Engineering Chemistry Research, 32 (1993), pp. 2942–2949. (Cited on p. 320)

[1855] M. VLEDDER, J. FRIEDMAN, M. SJÖBLOM, T. BROWN, AND P. YADAV, *Optimal Supply Chain Structure for Distributing Essential Drugs in Low Income Countries: Results From a Randomized Experiment*, technical report, Ross School of Business, University of Michigan, 2015. Available at SSRN 2585671. (Cited on p. 491)

[1856] C. R. VOGEL, *Computational Methods for Inverse Problems*, Frontiers in Applied Mathematics 23, SIAM, Philadelphia, 2002. (Cited on p. 32)

[1857] S. A. VOROBYOV, A. B. GERSHMAN, AND Z.-Q. LUO, *Robust adaptive beamforming using worst-case performance optimization: A solution to the signal mismatch problem*, IEEE Transactions on Signal Processing, 51 (2003), pp. 313–324. (Cited on p. 30)

[1858] A. WÄCHTER, *An Interior Point Algorithm for Large-Scale Nonlinear Optimization with Applications in Process Engineering*, PhD thesis, Carnegie Mellon University, Pittsburgh, PA, 2002. (Cited on pp. 10, 196)

[1859] A. WÄCHTER AND L. T. BIEGLER, *On the implementation of a primal-dual interior point filter line search algorithm for large-scale nonlinear programming*, Mathematical Programming, 106 (2006), pp. 25–57. (Cited on pp. 10, 268, 288)

[1860] A. WAGELMANS, S. VAN HOESEL, AND A. KOLEN, *Economic lot sizing: An $O(n \log n)$ algorithm that runs in linear time in the Wagner-Whitin case*, Operations Research, 40 (1992), pp. S145–S156. (Cited on p. 442)

[1861] H. M. WAGNER AND T. M. WHITIN, *Dynamic version of the economic lot size model*, Management Science, 5 (1958), pp. 89–96. (Cited on p. 442)

[1862] G. M. WAKE, N. BOLAND, AND L. S. JENNINGS, *Mixed integer programming approaches to exact minimization of total treatment time in cancer radiotherapy using multileaf collimators*, Computers and Operations Research, 36 (2009), pp. 795–810. (Cited on p. 95)

[1863] H. WAKI AND M. MURAMATSU, *Facial reduction algorithms for conic optimization problems*, Journal of Optimization Theory and Applications, 158 (2013), pp. 188–215. (Cited on p. 115)

[1864] K. WAŁCZYK AND J. JEŻOWSKI, *A single stage approach for designing water networks with multiple contaminants*, Computer Aided Chemical Engineering, 25 (2008), pp. 719–724. (Cited on p. 327)

[1865] S. W. WALLACE AND W. T. ZIEMBA, eds., *Applications of Stochastic Programming*, MOS-SIAM Series on Optimization 5, SIAM, Philadelphia, 2005. (Cited on p. 391)

[1866] C. WAN, J. WANG, G. YANG, AND X. ZHANG, *Optimal micro-siting of wind farms by particle swarm optimization*, in Advances in Swarm Intelligence: Lecture Notes in Computer Science, vol. 6145, Springer, 2010, pp. 198–205. (Cited on p. 367)

[1867] B. WANG, B. H. GEBRESLASSIE, AND F. YOU, *Sustainable design and synthesis of hydrocarbon biorefinery via gasification pathway: Integrated life cycle assessment and technoeconomic analysis with multiobjective superstructure optimization*, Computers and Chemical Engineering, 52 (2013), pp. 55–76. (Cited on p. 318)

[1868] G. G. WANG AND S. SHAN, *Review of metamodeling techniques in support of engineering design optimization*, Journal of Mechanical Design, 129 (2007), pp. 370–380. (Cited on p. 318)

[1869] Q. WANG, J. P. WATSON, AND Y. GUAN, *Two-stage robust optimization for $n - k$ contingency-constrained unit commitment*, IEEE Transactions on Power Systems, 28 (2013), pp. 2366–2375. (Cited on pp. 362, 363)

[1870] S. WANG, *Dynamic simulation of building VAV air-conditioning system and evaluation of EMCS on-line control strategies*, Building and Environment, 34 (1999), pp. 681–705. (Cited on p. 260)

[1871] S. WANG, F. DE VERICOURT, AND P. SUN, *Decentralized resource allocation to control an epidemic: A game theoretic approach*, Mathematical Biosciences, 222 (2009), pp. 1–22. (Cited on p. 472)

[1872] X. L. WANG, Y. OUYANG, H. YANG, AND Y. BAI, *Optimal biofuel supply chain design under consumption mandates with renewable identification numbers*, Transportation Research Part B, 57 (2013), pp. 158–171. (Cited on p. 465)

[1873] Y. WANG AND L. E. K. ACHENIE, *Computer aided solvent design for extractive fermentation*, Fluid Phase Equilibria, 201 (2002), pp. 1–18. (Cited on p. 329)

[1874] ———, *A hybrid global optimization approach for solvent design*, Computers and Chemical Engineering, 26 (2002), pp. 1415–1425. (Cited on p. 329)

[1875] Y. P. WANG AND R. SMITH, *Design of distributed effluent treatment systems*, Chemical Engineering Science, 49 (1994), pp. 3127–3145. (Cited on p. 326)

[1876] ———, *Wastewater minimisation*, Chemical Engineering Science, 49 (1994), pp. 981–1006. (Cited on p. 326)

[1877] P. C. WANKAT, *Separation Process Engineering: Includes Mass Transfer Analysis*, Prentice Hall, Upper Saddle River, NJ, 2011. (Cited on p. 238)

[1878] J. WARRINGTON, P. GOULART, S. MARIÉTHOZ, AND M. MORARI, *Policy-based reserves for power systems*, IEEE Transactions on Power Systems, 28 (2013), pp. 4427–4437. (Cited on p. 365)

[1879] S. WEBB, *Motion effects in (intensity modulated) radiation therapy: a review*, Physics in Medicine and Biology, 51 (2006), pp. R403–R425. (Cited on p. 347)

[1880] M. WEDENBERG, *Assessing the uncertainty in QUANTEC's dose–response relation of lung and spinal cord with a bootstrap analysis*, International Journal of Radiation Oncology • Biology • Physics, 87 (2013), pp. 795–801. (Cited on p. 347)

[1881] E. WEISZFELD, *Sur le point pour lequel la somme des distances de n points donnés est minimum*, Tôhoku Mathematical Journal—First Series, 43 (1937), pp. 355–386. (Cited on p. 451)

[1882] R. WERNER, *Consistency of Robust Optimization with Applications to Portfolio Optimization*, Optimization Online, 2010. (Cited on p. 344)

[1883] T. WESTERLUND AND F. PETTERSSON, *An extended cutting plane method for solving convex MINLP problems*, Computers and Chemical Engineering, 19 (1995), pp. 131–136. (Cited on p. 279)

[1884] T. WESTERLUND AND R. PÖRN, *Solving pseudo-convex mixed integer optimization problems by cutting plane techniques*, Optimization and Engineering, 3 (2002), pp. 253–280. (Cited on p. 288)

[1885] G. P. WESTERT, R. J. LAGOE, I. KESKIMÄKI, A. LEYLAND, AND M. MURPHY, *An international study of hospital readmissions and related utilization in Europe and the USA*, Health Policy, 61 (2002), pp. 269–278. (Cited on p. 96)

[1886] T. M. WHITIN, *The Theory of Inventory Management*, Princeton University Press, Princeton, NJ, 1953. (Cited on p. 443)

[1887] D. S. WICAKSONO AND I. A. KARIMI, *Piecewise MILP under- and overestimators for global optimization of bilinear programs*, AIChE Journal, 54 (2008), pp. 991–1008. (Cited on pp. 199, 216)

[1888] W. WIESEMANN, D. KUHN, AND B. RUSTEM, *Robust Markov decision processes*, Mathematics of Operations Research, 38 (2013), pp. 153–183. (Cited on p. 344)

[1889] W. WIESEMANN, D. KUHN, AND M. SIM, *Distributionally robust convex optimization*, Operations Research, 62 (2014), pp. 1358–1376. (Cited on p. 344)

[1890] S. M. WILD, *MNH: A derivative-free optimization algorithm using minimal norm Hessians*, in Proceedings of the Tenth Copper Mountain Conference on Iterative Methods, http://grandmaster.colorado.edu/~copper/2008/SCWinners/Wild.pdf, April 2008. (Cited on p. 531)

[1891] S. M. WILD, R. G. REGIS, AND C. A. SHOEMAKER, *ORBIT: Optimization by radial basis function interpolation in trust-regions*, SIAM Journal on Scientific Computing, 30 (2008), pp. 3197–3219. (Cited on p. 498)

[1892] S. M. WILD AND C. SHOEMAKER, *Global convergence of radial basis function trust region derivative-free algorithms*, SIAM Journal on Optimization, 21 (2011), pp. 761–781. (Cited on p. 498)

[1893] S. M. WILD AND C. A. SHOEMAKER, *Global convergence of radial basis function trust-region algorithms for derivative-free optimization*, SIAM Review, 55 (2013), pp. 349–371. (Cited on pp. 530, 534, 535)

[1894] H. P. WILLIAMS, *Model Building in Mathematical Programming*, Wiley, 1999. (Cited on p. 274)

[1895] D. WINFIELD, *Function and Functional Optimization by Interpolation in Data Tables*, PhD thesis, Harvard University, USA, 1969. (Cited on p. 496)

[1896] E. A. WOLFF AND S. SKOGESTAD, *Operation of integrated three-product (Petlyuk) distillation columns*, Industrial and Engineering Chemistry Research, 34 (1995), pp. 2094–2103. (Cited on p. 323)

[1897] H. WOLKOWICZ AND M. F. ANJOS, *Semidefinite programming for discrete optimization and matrix completion problems*, Discrete Applied Mathematics, 123 (2002), pp. 513–577. (Cited on p. 109)

[1898] H. WOLKOWICZ, R. SAIGAL, AND L. VANDENBERGHE, eds., *Handbook of Semidefinite Programming: Theory, Algorithms, and Applications*, Kluwer Academic, 2000. (Cited on pp. 11, 107, 113)

[1899] L. WOLSEY, *Integer Programming*, Wiley, New York, 1998. (Cited on p. 49)

[1900] K. S. WON AND T. RAY, *A framework for design optimization using surrogates*, Engineering Optimization, 37 (2005), pp. 685–703. (Cited on p. 318)

[1901] A. J. WOOD AND B. F. WOLLENBERG, *Power Generation, Operation, and Control*, John Wiley & Sons, New York, 1996. (Cited on pp. 187, 189)

[1902] A. J. WOOD, B. F. WOLLENBERG, AND G. B. SHEBLE, *Power Generation, Operation and Control*, 3rd edition, John Wiley & Sons, 2013. (Cited on p. 358)

[1903] WORKING GROUP ON A COMMON FORMAT, *Common format for exchange of solved load flow data*, IEEE Transactions on Power Apparatus and Systems, 6 (1973), pp. 1916–1925. (Cited on p. 188)

[1904] M. H. WRIGHT, *The interior-point revolution in optimization: History, recent developments, and lasting consequences*, Bulletin (New Series) of the American Mathematical Society, 42 (2004), pp. 39–56. (Cited on p. 11)

[1905] S. J. WRIGHT, *Primal-Dual Interior-Point Methods*, SIAM, Philadelphia, 1997. (Cited on pp. 6, 7, 10)

[1906] D. WU AND M. IERAPETRITOU, *Hierarchical approach for production planning and scheduling under uncertainty*, Chemical Engineering and Processing, 46 (2007), pp. 1129–1140. (Cited on p. 394)

[1907] J. WU, A. WEI, AND S. PERELSON, *Optimization of influenza vaccine selection*, Operations Research, 53 (2005), pp. 456–476. (Cited on p. 472)

[1908] O. Q. WU AND R. KAPUSCINSKI, *Curtailing intermittent generation in electrical systems*, Manufacturing and Service Operations Management, 15 (2013), pp. 578–595. (Cited on pp. 460, 461, 462)

[1909] O. Q. WU, D. D. WANG, AND Z. QIN, *Seasonal energy storage operations with limited flexibility: The price-adjusted rolling intrinsic policy*, Manufacturing and Service Operations Management, 14 (2012), pp. 455–471. (Cited on p. 462)

[1910] S.-P. WU, S. BOYD, AND L. VANDENBERGHE, *FIR filter design via spectral factorization and convex optimization*, in Applied and Computational Control, Signals, and Circuits, vol. 1, B. Datta, ed., Birkhauser, 1998, pp. 215–245. (Cited on pp. 29, 30)

[1911] L. XIE AND P. E. BOURNE, *Detecting evolutionary relationships across existing fold space, using sequence order-independent profile-profile alignments.*, Proceedings of the National Academy of Sciences of the United States of America, 105 (2008), pp. 5441–5446. (Cited on p. 520)

[1912] L. XIE, Y. GU, X. ZHU, AND M. GENTON, *Short-term spatio-temporal wind power forecast in robust look-ahead power system dispatch*, IEEE Transactions on Smart Grid, 5 (2013), pp. 511–520. (Cited on p. 364)

[1913] P. XIONG AND P. JIRUTITIJAROEN, *An adjustable robust optimization approach for unit commitment under outage contingencies*, in Proceedings of the Power and Energy Society General Meeting, IEEE, 2012, pp. 1–8. (Cited on p. 333)

[1914] Y. XIONG, W. CHEN, D. APLEY, AND X. DING, *A non-stationary covariance-based kriging method for metamodelling in engineering design*, International Journal for Numerical Methods in Engineering, 71 (2007), pp. 733–756. (Cited on p. 318)

[1915] H. XU, C. CARAMANIS, AND S. MANNOR, *A distributional interpretation of robust optimization*, Mathematics of Operations Research, 37 (2012), pp. 95–110. (Cited on p. 344)

[1916] J. XU, M. P. JOHNSON, P. S. FISHBECK, M. J. SMALL, AND J. M. VAN BRIESEN, *Robust placement of sensors in dynamic water distribution systems*, European Journal of Operational Research, 202 (2010), pp. 707–716. (Cited on p. 480)

[1917] L. XU, V. MANI, AND H. ZHAO, *Not a Box of Nuts and Bolts: The Distribution Channel Decision of the Rising Specialty Drugs*, working paper, Smeal College of Business, The Pennsylvania State University, 2016. (Cited on p. 477)

[1918] L. XU AND H. ZHAO, *Gatekeeper or Roadblock: Optimizing Evidence Generation and Access to New Drugs Through Accelerated Approval*, working paper, Smeal College of Business, The Pennsylvania State University, 2016. (Cited on p. 475)

[1919] X. XU, P. HUNG, AND Y. YE, *A simplified homogeneous and self-dual linear programming algorithm and its implementation*, Annals of Operations Research, 62 (1996), pp. 151–172. (Cited on pp. 8, 10)

[1920] Y. XU, K.-L. HSIUNG, X. LI, L. T. PILEGGI, AND S. P. BOYD, *Regular analog/RF integrated circuits design using optimization with recourse including ellipsoidal uncertainty*, IEEE Transactions on Computer-Aided Design of Integrated Circuits and Systems, 28 (2009), pp. 623–637. (Cited on p. 333)

[1921] S. YAMAMOTO AND I. HASHIMOTO, *Present status and future needs: The view from the Japanese industry*, in Chemical Process Control CPC IV, Y. Arkun and W. H. Ray, eds., CACHE and AIChE, 1991, pp. 1–28. (Cited on p. 41)

[1922] M. YAMASHITA, K. FUJISAWA, M. FUKUDA, K. KOBAYASHI, K. NAKATA, AND M. NAKATA, *Latest developments in the SDPA family for solving large-scale SDPs*, in [87], pp. 687–713. (Cited on p. 118)

[1923] M. YAMASHITA, K. FUJISAWA, AND M. KOJIMA, *Implementation and evaluation of SDPA 6.0 (semidefinite programming algorithm 6.0)*, Optimization Methods and Software, 18 (2003), pp. 491–505. (Cited on p. 118)

[1924] A.-S. YANG AND B. HONIG, *An integrated approach to the analysis and modeling of protein sequences and structures. I. Protein structural alignment and a quantitative measure for protein structural distance*, Journal of Molecular Biology, 301 (2000), pp. 665–678. (Cited on pp. 520, 526)

[1925] C. A. YANO AND H. L. LEE, *Lot sizing with random yields: A review*, Operations Research, 43 (1995), pp. 311–334. (Cited on pp. 448, 459)

[1926] Y. YE, *Interior-Point Algorithms: Theory and Analysis*, Wiley-Interscience Series in Discrete Mathematics and Optimization, John Wiley and Sons, New York, 1997. (Cited on pp. 6, 7, 8, 9, 10, 11)

[1927] Y. YE AND A. GODZIK, *Multiple flexible structure alignment using partial order graphs*, Bioinformatics, 21 (2005), pp. 2362–2369. (Cited on p. 520)

[1928] Y. YE, M. J. TODD, AND S. MIZUNO, *An $O(\sqrt{n}L)$-iteration homogeneous and self-dual linear programming algorithm*, Mathematics of Operations Research, 19 (1994), pp. 53–67. (Cited on p. 8)

[1929] Y. YE AND S. ZHANG, *New results on quadratic minimization*, SIAM Journal on Optimization, 14 (2003), pp. 245–267. (Cited on p. 169)

[1930] T. F. YEE AND I. E. GROSSMANN, *Simultaneous optimization models for heat integration—II. Heat exchanger network synthesis*, Computers and Chemical Engineering, 14 (1990), pp. 1165–1184. (Cited on pp. 77, 292, 324)

[1931] T. F. YEE, I. E. GROSSMANN, AND Z. KRAVANJA, *Simultaneous optimization models for heat integration—I. Area and energy targeting and modeling of multi-stream exchangers*, Computers and Chemical Engineering, 14 (1990), pp. 1151–1164. (Cited on p. 324)

[1932] H. YEOMANS AND I. E. GROSSMANN, *A systematic modeling framework of superstructure optimization in process synthesis*, Computers and Chemical Engineering, 23 (1999), pp. 709–731. (Cited on pp. 77, 318, 323)

[1933] ———, *Disjunctive programming models for the optimal design of distillation columns and separation sequences*, Industrial and Engineering Chemistry Research, 39 (2000), pp. 1637–1648. (Cited on pp. 318, 321, 322, 323)

[1934] R. W. YEUNG, *A framework for linear information inequalities*, IEEE Transactions on Information Theory, 43 (1997), pp. 1924–1934. (Cited on p. 35)

[1935] ———, *On entropy, information inequalities, and groups*, in Communications, Information and Network Security, V. K. Bhargava, H. V. Poor, V. Tarokh, and S. S. Yoon, eds., vol. 712 of The Springer International Series in Engineering and Computer Science, Springer, 2003, pp. 333–359. (Cited on p. 35)

[1936] C.-M. YING AND B. JOSEPH, *Performance and stability analysis of LP-MPC and QP-MPC cascade control systems*, AIChE Journal, 45 (1999), pp. 1521–1534. (Cited on pp. 45, 46)

[1937] F. YOU, P. M. CASTRO, AND I. E. GROSSMANN, *Dinkelbach's algorithm as an efficient method to solve a class of MINLP models for large-scale cyclic scheduling problems*, Computers and Chemical Engineering, 33 (2009), pp. 1879–1889. (Cited on p. 328)

[1938] F. YOU AND I. E. GROSSMANN, *Design of responsive supply chains under demand uncertainty*, Computers and Chemical Engineering, 32 (2008), pp. 3090–3111. (Cited on p. 394)

[1939] F. YOU, J. M. WASSICK, AND I. E. GROSSMANN, *Risk management for a global supply chain planning under uncertainty: Models and algorithms*, AIChE Journal, 55 (2009), pp. 931–946. (Cited on p. 394)

[1940] L.-Y. YU, X.-D. JI, AND S.-Y. WANG, *Stochastic programming models in financial optimization: A survey*, Advanced Modeling and Optimization, 5 (2003), pp. 1–26. (Cited on p. 149)

[1941] M. YUAN AND Y. LIN, *Model selection and estimation in regression with grouped variables*, Journal of the Royal Statistical Society, Series B. Statistical Methodology, 68 (2006), pp. 49–67. (Cited on p. 33)

[1942] X. YUAN, L. PIBOULEAU, AND S. DOMENECH, *Experiments in process synthesis via mixed-integer programming*, Chemical Engineering and Processing: Process Intensification, 25 (1989), pp. 99–116. (Cited on p. 324)

[1943] Z. B. ZABINSKY, *Stochastic Adaptive Search for Global Optimization*, Kluwer Academic, 2003. (Cited on p. 173)

[1944] H. ZABIRI AND Y. SAMYUDIA, *A hybrid formulation and design of model predictive control for systems under actuator saturation and backlash*, Journal of Process Control, 16 (2006), pp. 693–709. (Cited on p. 328)

[1945] E. ZAFIRIOU, *Robust model predictive control of processes with hard constraints*, Computers and Chemical Engineering, 14 (1990), pp. 359–371. (Cited on p. 44)

[1946] M. ZAIDER AND G. MINERBO, *Tumour control probability: A formulation applicable to any temporal protocol of dose delivery*, Physics in Medicine and Biology, 45 (2000), pp. 279–293. (Cited on p. 99)

[1947] J. M. ZAMORA AND I. E. GROSSMANN, *Continuous global optimization of structured process systems models*, Computers and Chemical Engineering, 22 (1998), pp. 1749–1770. (Cited on p. 327)

[1948] W. I. ZANGWILL, *A deterministic multi-period production scheduling model with backlogging*, Management Science, 13 (1966), pp. 105–119. (Cited on p. 443)

[1949] V. M. ZAVALA, *Real-time optimization strategies for building systems*, Industrial and Engineering Chemistry Research, 52 (2013), pp. 3137–3150. (Cited on pp. xix, 261, 263, 266, 267)

[1950] V. M. ZAVALA, D. SKOW, T. CELINSKI, AND P. DICKINSON, *Techno-Economic Evaluation of a Next-Generation Building Energy Management System*, technical report, ANL/MCS-TM-313, Argonne National Laboratory, 2011. (Cited on pp. 259, 260)

[1951] B. ZENG AND L. ZHAO, *Solving two-stage robust optimization problems using a column-and-constraint generation method*, Operations Research Letters, 41 (2013), pp. 457–461. (Cited on p. 362)

[1952] S. A. ZENIOS, ed., *Financial Optimization*, Cambridge University Press, Cambridge, U.K., 1996. (Cited on p. 149)

[1953] H. ZHANG AND A. R. CONN, *On the local convergence of a derivative-free algorithm for least-squares minimization*, Computational Optimization and Applications, 51 (2012), pp. 481–507. (Cited on pp. 530, 533, 539)

[1954] H. ZHANG, A. R. CONN, AND K. SCHEINBERG, *A derivative-free algorithm for least-squares minimization*, SIAM Journal on Optimization, 20 (2010), pp. 3555–3576. (Cited on pp. 530, 533)

[1955] J. ZHANG, M. DONG, AND F. F. CHEN, *A bottleneck Steiner tree based multi-objective location model and intelligent optimization of emergency logistics systems*, Robotics and Computer-Integrated Manufacturing, 29 (2013), pp. 48–55. (Cited on p. 488)

[1956] J. ZHANG, X. X. ZHU, AND G. P. TOWLER, *A simultaneous optimization strategy for overall integration in refinery planning*, Industrial and Engineering Chemistry Research, 40 (2001), pp. 2640–2653. (Cited on p. 327)

[1957] J.-K. ZHANG, T. N. DAVIDSON, AND K. M. WONG, *Efficient design of orthonormal wavelet bases for signal representation*, IEEE Transactions on Signal Processing, 52 (2004), pp. 1983–1996. (Cited on p. 29)

[1958] P. Y. ZHANG, *Topics in Wind Farm Layout Optimization*, Master's thesis, University of Toronto, 2013. (Cited on p. 367)

[1959] P. Y. ZHANG, D. A. ROMERO, J. C. BECK, AND C. H. AMON, *Solving wind farm layout optimization with mixed integer programming and constraint programming*, in Integration of AI and OR Techniques in Constraint Programming for Combinatorial Optimization Problems—10th International Conference (CPAIOR), Yorktown Heights, NY, 2013, Springer. (Cited on pp. 367, 370)

[1960] S. ZHANG, *Quadratic maximization and semidefinite relaxation*, Mathematical Programming, 87 (2000), pp. 453–465. (Cited on p. 169)

[1961] W. ZHANG, F. LI, AND L. M. TOLBERT, *Review of reactive power planning: Objectives, constraints, and algorithms*, IEEE Transactions on Power Systems, 22 (2007), pp. 2177–2186. (Cited on p. 191)

[1962] Z. ZHANG AND R. W. YEUNG, *A non-Shannon-type conditional inequality of information quantities*, IEEE Transactions on Information Theory, 43 (1997), pp. 1982–1986. (Cited on p. 35)

[1963] ——, *On characterization of entropy function via information inequalities*, IEEE Transactions on Information Theory, 44 (1998), pp. 1440–1452. (Cited on p. 35)

[1964] C. ZHAO AND Y. GUAN, *Unified stochastic and robust unit commitment*, IEEE Transactions on Power Systems, 28 (2013), pp. 3353–3361. (Cited on p. 363)

[1965] H. ZHAO, C. XIONG, S. GAVIRNENI, AND A. FEIN, *Fee-for-service contracts in pharmaceutical distribution supply chains: Design, analysis, and management*, Manufacturing and Service Operations Management, 14 (2012), pp. 685–699. (Cited on p. 476)

[1966] H. ZHAO, Y. YANG, AND Y. ZHOU, *Structure-based prediction of DNA-binding proteins by structural alignment and a volume-fraction corrected DFIRE-based energy function*, Bioinformatics, 26 (2010), pp. 1857–1863. (Cited on pp. 176, 181)

[1967] L. ZHAO AND B. ZENG, *Robust unit commitment problem with demand response and wind energy*, in Proceedings of the Power and Energy Society General Meeting, IEEE, 2012, pp. 1–8. (Cited on pp. 358, 360)

[1968] M. ZHAO, Z. CHEN, AND F. BLAABJERG, *Optimisation of electrical system for offshore wind farms via genetic algorithm*, IET Renewable Power Generation, 3 (2009), pp. 205–216. (Cited on p. 367)

[1969] X.-Y. ZHAO, D. SUN, AND K.-C. TOH, *A Newton-CG augmented Lagrangian method for semidefinite programming*, SIAM Journal on Optimization, 20 (2010), pp. 1737–1765. (Cited on p. 118)

[1970] Q. P. ZHENG, S. REBENNACK, N. A. ILIADIS, AND P. M. PARDALOS, *Optimization models in the natural gas industry*, in Handbook of Power Systems I, Springer, Berlin, Heidelberg, 2010, pp. 121–148. (Cited on p. 205)

[1971] T. ZHENG, J. ZHAO, E. LITVINOV, AND F. ZHAO, *Robust optimization and its application to power system operation*, in Proceedings of CIGRE, 2012. (Cited on p. 364)

[1972] Y.-S. ZHENG, *On properties of stochastic inventory systems*, Management Science, 38 (1992), pp. 87–103. (Cited on p. 448)

[1973] Y.-S. ZHENG AND A. FEDERGRUEN, *Finding optimal (s, S) policies is about as simple as evaluating a single policy*, Operations Research, 39 (1991), pp. 654–665. (Cited on p. 446)

[1974] A. ZHIGLJAVSKY AND A. ŽILINSKAS, *Stochastic Global Optimization*, Springer, 2008. (Cited on p. 173)

[1975] D. ZHOU, L. LEUNG, AND W. PIERSKALLA, *Inventory management of platelets in hospitals: Optimal inventory policy for perishable products with regular and optional expedited replenishments*, Manufacturing and Service Operations Management, 13 (2011), pp. 420–438. (Cited on p. 473)

[1976] R.-J. ZHOU, L.-J. LI, W. XIAO, AND H.-G. DONG, *Simultaneous optimization of batch process schedules and water-allocation network*, Computers and Chemical Engineering, 33 (2009), pp. 1153–1168. (Cited on p. 327)

[1977] W. ZHOU AND V. I. MANOUSIOUTHAKIS, *Non-ideal reactor network synthesis through ideas: Attainable region construction*, Chemical Engineering Science, 61 (2006), pp. 6936–6945. (Cited on p. 320)

[1978] Z. ZHOU, P. LIU, Z. LI, E. N. PISTIKOPOULOS, AND M. C. GEORGIADIS, *Impacts of equipment off-design characteristics on the optimal design and operation of combined cooling, heating and power systems*, Computers and Chemical Engineering, 48 (2013), pp. 40–47. (Cited on pp. 304, 314)

[1979] J. ZHU, *Optimization of Power System Operation*, Wiley-IEEE Press, Piscataway, NJ, 2009. (Cited on pp. 187, 197)

[1980] S. ZHU, D. LI, AND X. SUN, *Portfolio selection with marginal risk control*, The Journal of Computational Finance, 14 (2010), pp. 3–28. (Cited on pp. 427, 432, 435)

[1981] M. ZIBULEVSKY AND M. ELAD, *L1-L2 optimization in signal and image processing*, IEEE Signal Processing Magazine, 27 (2010), pp. 76–88. (Cited on p. 33)

[1982] J. C. ZIEMS AND S. ULBRICH, *Adaptive multilevel inexact SQP methods for PDE-constrained optimization*, SIAM Journal on Optimization, 21 (2011), pp. 1–40. (Cited on p. 231)

[1983] R. D. ZIMMERMAN AND C. E. MURILLO-SÁNCHEZ, *MATPOWER: Steady-state operations, planning, and analysis tools for power systems research and education*, IEEE Transactions on Power Systems, 26 (2011), pp. 12–19. (Cited on p. 188)

[1984] P. H. ZIPKIN, *Foundations of Inventory Management*, Irwin/McGraw-Hill, New York, 2000. (Cited on pp. 439, 446, 447, 449)

[1985] S. ZYMLER, D. KUHN, AND B. RUSTEM, *Distributionally robust joint chance constraints with second-order moment information*, Mathematical Programming, 137 (2013), pp. 167–198. (Cited on p. 344)

Index